Atomic masses are based on carbon-12.
Numbers in parentheses are the mass
numbers of the most stable isotopes.

					VIIIA or 0
					2 Helium **He** 4.003

IIIA	IVA	VA	VIA	VIIA	
5 Boron **B** 10.81	6 Carbon **C** 12.01	7 Nitrogen **N** 14.01	8 Oxygen **O** 16.00	9 Fluorine **F** 19.00	10 Neon **Ne** 20.18
13 Aluminum **Al** 26.98	14 Silicon **Si** 28.09	15 Phosphorus **P** 30.97	16 Sulfur **S** 32.07	17 Chlorine **Cl** 35.45	18 Argon **Ar** 39.95

IB	IIB						
29 Copper **Cu** 63.55	30 Zinc **Zn** 65.39	31 Gallium **Ga** 69.72	32 Germanium **Ge** 72.59	33 Arsenic **As** 74.92	34 Selenium **Se** 78.96	35 Bromine **Br** 79.90	36 Krypton **Kr** 83.80
47 Silver **Ag** 107.9	48 Cadmium **Cd** 112.4	49 Indium **In** 114.8	50 Tin **Sn** 118.7	51 Antimony **Sb** 121.8	52 Tellurium **Te** 127.6	53 Iodine **I** 126.9	54 Xenon **Xe** 131.3
79 Gold **Au** 197.0	80 Mercury **Hg** 200.6	81 Thallium **Tl** 204.4	82 Lead **Pb** 207.2	83 Bismuth **Bi** 209.0	84 Polonium **Po** (209)	85 Astatine **At** (210)	86 Radon **Rn** (222)
111							

64 Gadolinium **Gd** 157.3	65 Terbium **Tb** 158.9	66 Dysprosium **Dy** 162.5	67 Holmium **Ho** 164.9	68 Erbium **Er** 167.3	69 Thulium **Tm** 168.9	70 Ytterbium **Yb** 173.0	71 Lutetium **Lu** 175.0
96 Curium **Cm** 247	97 Berkelium **Bk** 247	98 Californium **Cf** 249	99 Einsteinium **Es** 254	100 Fermium **Fm** 253	101 Mendelevium **Md** (256)	102 Nobelium **No** (253)	103 Lawrencium **Lr** (257)

EIGHTH EDITION

Chemistry for the Health Sciences

GEORGE I. SACKHEIM

Associate Professor Emeritus
University of Illinois, Chicago

DENNIS D. LEHMAN

Professor of Chemistry
Harold Washington College, Chicago

Prentice
Hall

Prentice Hall
Upper Saddle River, New Jersey 07458

Library of Congress Cataloging-in-Publication Data

Sackheim, George I.
 Chemistry for the health sciences / George I. Sackheim, Dennis D. Lehman. — 8th ed.
      p.   cm.
 Includes index.
 ISBN 0–13–744319–6
 1. Chemistry. 2. Medicine.   I. Lehman, Dennis D.   II. Title.
 [DNLM: 1. Biochemistry. 2. Chemistry, Inorganic. 3. Chemistry,
Organic.   QU 4 S121c   1998]
 QD33.S13   1998
 540'.24'61—dc21
 DNLM/DLC
 for Library of Congress   97–23061
                                                              CIP

Acquisition Editor: Ben Roberts
Editor-in-Chief: Paul F. Corey
Editorial Director: Tim Bozik
Assistant Vice President of Production and Manufacturing: David W. Riccardi
Executive Managing Editor: Kathleen Schiaparelli
Manufacturing Manager: Trudy Pisciotti
Production Project Coordinator: Accu•color, Inc.
Full Service Liaison/Manufacturing Buyer: Benjamin D. Smith
Associate Editor: Mary Hornby
Creative Director: Paula Maylahn
Art Director: Joseph Sengotta
Art Manager: Gus Vibal
Interior Design: Donna Wickes
Cover Design: Joseph Sengotta
Cover Photo: © Lori Adamski Peek/Tony Stone Images
Marketing Manager: Linda Taft MacKinnon
Photo Researcher: Stuart Kenter Associates
Photo Reasearch Administrator: Melinda Reo

Printed in the United States of America

10   9   8   7

ISBN 0-13-744319-6

Prentice-Hall International (UK) Limited, *London*
Prentice-Hall of Australia Pty. Limited, *Sydney*
Prentice-Hall of Canada, Inc., *Toronto*
Prentice-Hall Hispanoamericana, S. A., *Mexico*
Prentice-Hall of India Private Limited, *New Delhi*
Prentice-Hall of Japan, Inc., *Tokyo*
Prentice-Hall Asia Pte. Ltd., *Singapore*
Editora Prentice-Hall do Brasil, Ltda., *Rio de Janeiro*

Contents

24 *Enzymes*   419

25 *Digestion*   438

26 *Metabolism of Carbohydrates*   452

27 *Metabolism of Fats*   478

28 *Metabolism of Proteins*   492

29 *Body Fluids: Urine*   509

Preface

This textbook of chemistry is designed primarily for first-year students in various health-related programs—nursing, dietetics, laboratory technology, inhalation therapy, dental hygiene, dental assisting, medical assisting, dental technology, and so on. Emphasis is placed on *practical* aspects of inorganic chemistry, organic chemistry, and biochemistry. Theoretic topics are dealt with only as an aid to understanding bodily processes in the human.

Organization of the Text

Part I, "Inorganic Chemistry," stresses relationships with the life processes that are the subject of Part III, "Biochemistry." Among these related topics and processes are the following:

1. Acids, bases, salts, and electrolytes/acid–base balance and electrolyte balance in the body
2. Oxidation–reduction/biologic oxidation–reduction reactions in the mitochondria of cells
3. Solutions/the solvent action involved in digestion
4. Colloids/the nature and properties of proteins, amino acids, and nucleic acids
5. Covalent compounds/the bonds that must be broken and rearranged in the formulation of high-energy phosphate bonds
6. Emulsions/the need for emulsification of fats before digestion
7. Nuclear chemistry and radioactivity/biologic effects of radiation on cells and organs

Part II, "Organic Chemistry," introduces the various classes of organic compounds—hydrocarbons, alcohols, ethers, esters, acids, aldehydes, ketones, amines, amides, thiols, and aromatic compounds. In addition, the text discussions relate such compounds to carbohydrates, fats, proteins, vitamins, hormones, and nucleic acids.

Part III, "Biochemistry," deals with the chemical and molecular basis of life itself. The various chemical processes taking place in the body are described in terms of both normal and abnormal metabolism. The role of ATP, the principal direct source of energy for the body, is stressed throughout the chapters on metabolism. The formation and decomposition of this compound, and the energies involved, serve to indicate the complexity of "normal" processes. Also emphasized is the role of coenzymes, such as CoA, in metabolism. Discussions of excesses and deficiencies of vitamins and hormones are designed to demonstrate the involved interrelationships in the body's metabolic processes. In the section on heredity the combination of chemistry and the molecular basis of life is evidenced by advances in our understanding of DNA structure and the replication of DNA and RNA.

Chapter Objectives at the beginning of each chapter indicate the topics to be discussed and, in general, how they are related. The Summary at the end of each chapter identifies the particularly important aspects of the subject matter. The Questions and Problems at the end of the chapter may be used for review or assigned as homework. Finally, the Practice Test will aid students in checking their understanding of the topics of the chapter. Answers to odd-numbered questions and the practice test questions are given at the end of the book.

Changes in This Edition

New features in this edition include many example exercises and practice problems, case histories and questions related to them, thought problems, and medical application problems. New topics introduced include the usage of significant figures; bone density and osteoporosis; density and body fat; interpretation of nutritional data on food packages; specific heat; energy involved in changes of temperature and state; and energy involved in various body activities.

Additional material provided in this edition include calorimetric calculations; freezing, boiling, condensation, and sublimation; electron-dot structures; shapes of molecules; VSEPR theory; logarithmic problems solved with a calculator; fullerene compounds; black and white photography; a table of normal blood values for use with various case histories; and a table of disorders affecting red blood cells.

Topics given expanded treatment include Dalton's atomic theory; groups and periods in the periodic table of the elements; metals, nonmetals, and metalloids; measurement of air pressure; Gay-Lussac's law; ideal gas law; calculation of rate of diffusion; greenhouse effect; use of nitric oxide as a vasodilator; hydrogen bonding; saturated solution and gout; ionization; comparisions of properties of inorganic and organic compounds; reactions of alkanes; comparison of octane ratings and branching; naming, structure, and reactions of alkenes and alkynes; cis-trans isomers; naming alcohols, ethers, esters, carboxylic acids, and amines; halogenated anesthetics; structures of aldehydes, ketones, carboxylic acids, esters, amines, and amides; DNA fingerprinting; D and L enantiomers; diastereomers; chiral drugs; fat in the diet; cholesterol levels and anticholesterol drugs; the Human Genome Project; lactose intolerance; lactacidosis; poisons that inhibit the respiratory chain; and naming of thiols and mercaptans.

The order of some chapters has been rearranged; the chapter on radioactivity is now at the end of Part II, "Inorganic Chemistry"; the chapter on acids and bases and the chapter on salts have been combined; and the chapter on nucleic acids and heredity now follows the chapter on proteins.

The discussions on immunology and heterocyclic compounds have been deleted; the section covering aromatic compounds has been integrated into the text discussion of aliphatic compounds; exponents integrated into text, incorporation of heterocyclic amines into chapter on amines.

Supplements

A comprehensive set of laboratory experiments (Sackheim and Lehman, *Laboratory Chemisty for the Health Sciences*, Prentice Hall Publishing Company) has been designed to supplement various topics in this textbook. Performing these experiments will aid the student in understanding the fundamental concepts involved. An Instructor's Manual, also available from Prentice Hall, contains answers to the text questions and a test bank.

We appreciate the significant contributions made by the following reviewers who commented on the manuscript for the current edition. Their input and suggestions have proved to be most helpful.

Richard F. Jones, *Sinclair Community College*
V.M. Doctor, *Prairie View A&M University*
Harvey Yablonsky, *City University of New York*
Cosmas Okoro, *University of Arkansas at Pine Bluff*
Ricardo E. Rodriguez, *Texas Wesleyan University*
Peggy J. McClure, *Midlands Technical College*
Theodore Richerzhagen, *Walla Walla Community College*
Sharon Kapica, *County College of Morris*

We are also grateful to Levi Hicks, Sharon Oliver, and Kathleen Kuchta for their valuable assistance in preparing the manuscript and for their clerical expertise.

George I. Sackheim
Dennis D. Lehman

1

Units of Measurement

Objectives

- To understand the use of the SI system of measurement
- To understand the use of significant figures
- To convert from one system of measurement to another
- To understand the use of the factor label method
- To perform density calculations
- To interconvert temperatures in the Fahrenheit, Celsius, and Kelvin scales

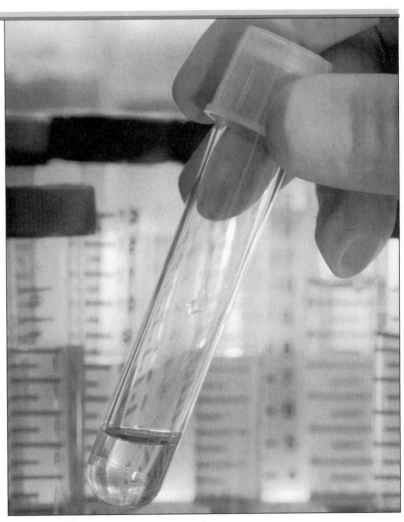

Workers in all fields of science, medicine, and technology rely on fundamental units of measurement on a daily basis.

1-1 The International System (SI)

We use measurements every day of our lives. We measure gasoline by the gallon and milk by the quart; we buy meat by the pound and pay for postage by the ounce. A recipe may call for a cup of flour, a tablespoonful of butter, and $\frac{1}{4}$ teaspoonful of salt. In chemistry, all measurements are made in the metric system or in its expanded modernized version, the SI system, from the French *le Système Internationale d'Unités*. The main advantage of the SI system is that it is a decimal system; that is, the units are all multiples of ten of larger or smaller units. In the English system there is no such common number. Recall that there are 3 feet in a yard, 4 quarts in a gallon, and 16 ounces in a pound.

Mass and Weight

The mass of an object is a measure of the amount of matter it contains. The weight of an object depends on the pull of gravity. Mass remains constant regardless of location, whereas weight can vary slightly from place to place on the surface of the Earth. For our purposes, however, we will use the terms mass and weight interchangeably.

Fundamental Units

The fundamental unit of length in the SI system is the meter. A meter is a little longer than a yard. The word meter is abbreviated as m.

The fundamental SI unit of mass (weight) is the kilogram. A kilogram is slightly more than 2 pounds. The basic laboratory unit of mass (weight) is the gram. A gram is $\frac{1}{454}$ pound.

The fundamental SI unit of time is the second(s). Decimal-based multiples and submultiples can also be used, but the older units such as minutes, hours, and days are still in common use.

The fundamental SI unit of temperature is the Kelvin (K). Note that the word degree and the symbol for degree are not used. In medical applications, however, temperature is usually measured in degrees Celsius (°C). We will discuss the relationship between temperatures in kelvins and degrees Celsius later in this chapter.

Derived Units

Volume is not a fundamental SI unit because it can be derived from length (the volume of a rectangular object can be calculated from the relationship: volume = length × width × height). In the SI system, volume is expressed in the unit cubic meter (m^3). One cubic meter contains approximately 250 gallons. A more common unit of volume is the liter (often spelled out, but abbreviated by chemists as L and by some others as l). A liter is slightly more than 1 quart.

The basic units most commonly used in medicine are the meter, the gram, and the liter. For this reason, these are the units used throughout this text.

In the SI system prefixes are used to designate various multiples or submultiples. The most commonly used prefixes are indicated in color in Table 1-1. For example, kilo- means one thousand and centi- means one-hundredth. Thus, a kilometer (km) is one thousand meters, or 1000 m, and a centigram (cg) is $\frac{1}{100}$ gram, or 0.01 g. Likewise, a milliliter (mL) is 0.001 L (see Table 1-2). Another unit of volume is the cubic centimeter (cm^3); 1 cm^3 is equal to 1 mL, so these two units may be used

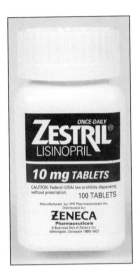

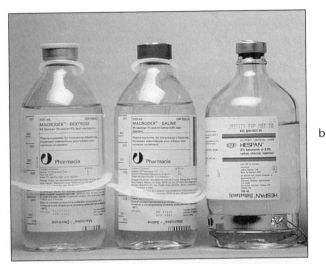

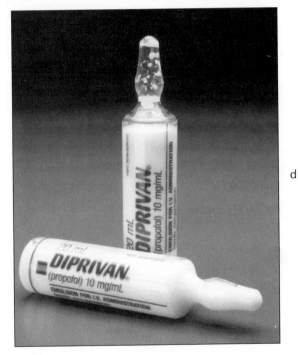

a

b

c

d

FIGURE 1-1

Examples of units of measurement commonly encountered in the medical profession. (a) Prescription pills and capsules are measured by their mass in milligrams. (b) The concentration of solutes within a liquid may also be measured: These intravenous solutions contain (left) 5% dextrose, (middle) 6% dextran 70 and 0.9% sodium chloride, (right) 6% hetastarch and 0.9% sodium chloride. (c) Injectable medication is measured by its volume in milliliters (mL). (d) Ampule of propofol (Diprivan) contains a concentration of 10 mg/mL.

interchangeably. Although the term cubic centimeter is abbreviated as cm^3 by chemists, the older abbreviation cc is still seen at times. Likewise 1 MCi (1 megacurie) means 1 million Ci, and 1 µg or 1 mcg (1 microgram) means 0.000001 g. Note that the letter m indicates both meter and milli-; however, when used by itself, it indicates meter.

Table 1-1   Some Prefixes Used in the SI System

Prefix	Abbreviation	Decimal Expression	Exponential Expression
mega	M	1,000,000	10^6
kilo	k	1,000	10^3
hekto	h	100	10^2
deka	da	10	10^1
—	—	1	10^0
deci	d	0.1	10^{-1}
centi	c	0.01	10^{-2}
milli	m	0.001	10^{-3}
micro	µ (or mc)	0.000001	10^{-6}
nano	n	0.000000001	10^{-9}

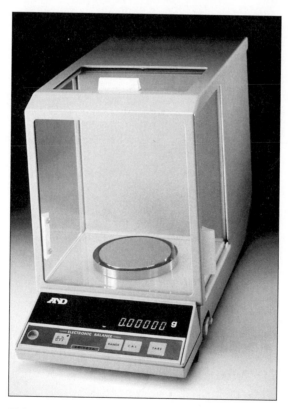

FIGURE 1-2
A modern laboratory balance.

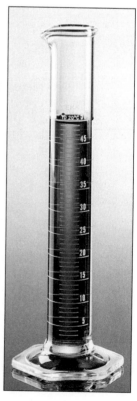

FIGURE 1-3
Graduated cylinder in use.

Table 1-2 Metric Conversions

Mass (weight)

1 g = 1000 mg	one gram = one thousand milligrams
1 g = 100 cg	one gram = one hundred centigrams
1 g = 10 dg	one gram = ten decigrams
1 kg = 1000 g	one kilogram = one thousand grams

Length

1 m = 1000 mm	one meter = one thousand millimeters
1 m = 100 cm	one meter = one hundred centimeters
1 m = 10 dm	one meter = ten decimeters
1 km = 1000 m	one kilometer = one thousand meters

Volume

1 L = 1000 mL	one liter = one thousand milliliters
1 L = 100 cL	one liter = one hundred centiliters
1 L = 10 dL	one liter = ten deciliters
1 kL = 1000 L	one kiloliter = one thousand liters

Exercise 1-1

Complete the equation for each metric relationship.

a. 1 liter = _____ mL

b. 1 cm = _____ meters

Solution

Refer to the metric prefixes in Table 1-1 if necessary.

a. The prefix milli is 0.001 of a basic unit; 1000 milli units are in 1 basic unit, therefore, 1 liter = 1000 mL.

b. The prefix centi is 0.01 of a basic unit; 100 centi units are in 1 basic unit, therefore, 1 cm = 0.01 meters.

Self-test

a. 1 kg = _____ g

b. 1 sec = _____ μsec

c. 1 mm = _____ m

Answers

a. 1 kg = 1000 g

b. 1 sec = 1,000,000 μsec

c. 1 mm = 0.001 m

1-2 Factor Label Method

Consider the following English-English conversion units.

$$12 \text{ in} = 1 \text{ ft}$$
$$3 \text{ ft} = 1 \text{ yd}$$
$$1760 \text{ yd} = 1 \text{ mi}$$

When asked to convert 6 ft to inches, many students merely multiply by 12. However, mathematically, the correct procedure is to multiply by the conversion unit 12 in/1 ft so that the result is correct both numerically and in units.

$$6 \text{ ft} = 6 \text{ ft} \times \frac{12 \text{ in}}{1 \text{ ft}} = 72 \text{ in}$$

Likewise, to convert 657.5 yd to mi,

$$657.5 \text{ yd} \times \frac{1 \text{ mi}}{1760 \text{ yd}} = 0.3736 \text{ mi}$$

Exercise 1-2

Convert 68 grams to kilograms.

Solution

$$68 \text{ g} \times \frac{1 \text{ kg}}{1000 \text{ g}} = 0.068 \text{ kg}$$

Self-test

a. The average amount of zinc in 100 mL of normal blood is 120 μg. Express this amount in grams.
b. A premature infant weighs 1.45 kg. Express this weight in grams; in milligrams.
c. The most common protein in the body is collagen. Collagen is composed of fibrous rods about 150 nm in diameter. Express this diameter in meters; in centimeters; in millimeters.

Answers

a. 0.000120 g   b. 1450 g; 1,450,000 mg
c. 0.000000150 m; 0.0000150 cm; 0.000150 mm

FIGURE 1-4
Metric units.

1-3 Exponential Numbers

Changing Exponential Numbers to Common Numbers

Mathematicians have developed a shorthand method for expressing very large or very small numbers. This system involves the use of a base number, 10, raised to some power. A power, or exponent, indicates how many times the base number, 10, is repeated as a factor (see Table 1-3). Thus,

1×10^2 (10 repeated as a factor 2 times) $= 1 \times 10 \times 10 = 100$
1×10^3 (10 repeated as a factor 3 times) $= 1 \times 10 \times 10 \times 10 = 1000$
1×10^5 (10 repeated as a factor 5 times) $= 1 \times 10 \times 10 \times 10 \times 10 \times 10 = 100{,}000$

Negative exponents are used to indicate numbers less than 1. A negative exponent indicates the reciprocal of the same number with a positive exponent:

$$1 \times 10^{-1} = \frac{1}{10^1} = 0.1$$

$$1 \times 10^{-2} = 1 \times \frac{1}{10^2} = 0.01$$

$$1 \times 10^{-4} = 1 \times \frac{1}{10^4} = 0.0001$$

Note: The positive exponent indicates how many places the decimal point must be moved to the right (from the number 1). Also, note that the negative exponent

Table 1-3 Powers of Ten	
Exponential Number	**Ordinary Number**
$10^6 = 10 \times 10 \times 10 \times 10 \times 10 \times 10$	1,000,000
$10^3 = 10 \times 10 \times 10$	1,000
$10^2 = 10 \times 10$	100
$10^1 = 10$	10
$10^0 = 1$	1
$10^{-1} = \dfrac{1}{10^1} = \dfrac{1}{10}$	0.1
$10^{-2} = \dfrac{1}{10^2} = \dfrac{1}{10} \times \dfrac{1}{10} = \dfrac{1}{100}$	0.01
$10^{-3} = \dfrac{1}{10^3} = \dfrac{1}{10} \times \dfrac{1}{10} \times \dfrac{1}{10} = \dfrac{1}{1000}$	0.001
$10^{-6} = \dfrac{1}{10^6} = \dfrac{1}{10} \times \dfrac{1}{10} \times \dfrac{1}{10} \times \dfrac{1}{10} \times \dfrac{1}{10} \times \dfrac{1}{10} = \dfrac{1}{1{,}000{,}000}$	0.000001

indicates how many places the decimal point must be moved to the left (including the number 1).

Any number that is not an exact power of 10 can be expressed as a product of two numbers, one of which is a power of 10. The other number is always written with just one figure to the left of the decimal point.

Exercise 1-3

Exercise 1-3a

Express 6.2×10^3 as a common number.

A positive three $(+3)$ exponent indicates that the decimal point should be moved three places to the right. Thus,

$$6.2 \times 10^3 = (6.200) = 6200$$

Exercise 1-3b

Change 8.45×10^5 to a common number.

A positive five $(+5)$ exponent indicates that the decimal point should be moved five places to the right.

$$8.45 \times 10^5 = (8.45000) = 845{,}000$$

Exercise 1-3c

Change 1.27×10^{-2} to a common number.

A negative two (-2) exponent indicates that the decimal point should be moved two places to the left.

$$1.27 \times 10^{-2} = (001.27) = 0.0127$$

Exercise 1-3d

Change 3.5×10^{-9} to a common number.

$$3.5 \times 10^{-9} = (0000000003.5) = 0.0000000035$$

Self-test

Write the following as common numbers.
a. 6.02×10^{-5} b. 3.846×10^2

Answers

a. 0.0000602   b. 384.6

Changing Common Numbers to Exponential Numbers

When a common number is changed to an exponential number, the decimal point is moved so that there is just one digit to the left of it. The exponent corresponds to the number of places the decimal point must be moved. If the decimal point is moved to the left, the exponent is positive; if it is moved to the right, the exponent is negative.

Exercise 1-4

Exercise 1-4a

Change 4000 to an exponential number.

The decimal point must be moved to the left three places for just one digit to remain before that decimal point. Three places to the left indicates an exponent of 3 (+3).

$$4000 = (4000) = 4 \times 10^3$$

Exercise 1-4b

Change 604 to an exponential number.

$$604 = (604) = 6.04 \times 10^2$$

Exercise 1-4c

Change 0.00037 to an exponential number.

$$0.00037 = (0.00037) = 3.7 \times 10^{-4}$$

The negative exponent indicates that the decimal point has been moved to the right.

Self-test

Write the following as exponential numbers:
a. 0.0000913   b. 6,780,000

Answers

a. 9.13×10^{-5} b. 6.78×10^6

FIGURE 1-5
Student being weighed.

1-4 Significant Figures

Imagine the following laboratory situation. An instructor asks a group of students to weigh one of their classmates. The following results were obtained:

125.6 lb	125.7 lb	125.5 lb
125.4 lb	125.8 lb	125.7 lb
125.3 lb	126.9 lb	125.3 lb

All of the students had the same approximate weight, but which one is correct?

By convention, the numbers in a measurement, including the final estimated digit, are called **significant figures**. The number of significant figures in a measurement will depend on the type of measuring device being used and the skill of the operator. In the above weights there are four significant figures.

Figure 1-6 illustrates a measurement of 25 mL (two significant figures). In a blood cholesterol level of 185, there are three significant figures. In a serum potassium level of 3.9, there are two significant figures. In a urine pH reading of 5.9, there are two significant figures. In a urine density reading of 1.016 g/mL, there are four significant figures. Therefore, a significant figure is one that is known to be reliable.

FIGURE 1-6
Fluid in syringe.

Rules in Determining the Number of Significant Figures

1. *All digits that are not zero are significant, regardless of the location of a decimal point.*
   1.59 contains three significant figures
   15.9 contains three significant figures
   3.14159 contains six significant figures

2. *Zeros between nonzero digits are significant.*
   20.6 contains three significant figures
   100.25 contains five significant figures

3. *Zeros to the left of a decimal point or to the left of a number are not significant;* they merely indicate the placement of the decimal point.
   0.517 contains three significant figures
   0.0039 contains two significant figures

4. *Zeros to the right of a decimal point are significant if they are at the end of a number.*
   59.0 contains three significant figures
   16.0000 contains six significant figures

5. *When a number ends in a zero or zeros that are not to the right of a decimal point, the end zero or zeros may or may not be significant.* Thus, 760 may or may not contain three significant figures, depending on the accuracy of the measurement. To avoid any ambiguity in the number of significant figures, such a number should be written using exponential notation.
   760 (three significant figures) is written as 7.60×10^2
   760 (two significant figures) is written as 7.6×10^2
   16,000 (three significant figures) is written as 1.60×10^4

6. *Exact numbers.* Exact numbers are numbers that can be defined or that are used in counting. For example, in the conversion factor 12 inches = 1 foot, there are exactly 12 inches in each foot. There is no ambiguity in the number 12, so it can be used with as many significant figures as necessary in a calculation. Likewise, in the conversion factor 1 kilogram = 1000 grams, the number 1000 is an exact number and can have as many significant figures as needed.

   When counting the number of tablets in a medication order, that number is also an exact number. For example, if a patient is to receive 3 tablets every morning, then 3 is an exact number and can be used with as many significant figures as necessary.

Rounding Off

The following rules are used when rounding off a number:

1. *When the number dropped is less than 5, the preceding number remains unchanged.* Thus, 1.874 is rounded off to 1.87 and 0.056452 is rounded off to 0.05645.

2. *When the number dropped is more than 5, 1 is added to the preceding number.* Thus, 4.7438 is rounded off to 4.744 and 0.0526 is rounded off to 0.053.

3. *When the number dropped is exactly 5, if the preceding number is even, it remains unchanged; if the preceding number is odd, 1 is added to it.* Thus, 7.245 is rounded off to 7.24 and 13.7735 is rounded off to 13.774. For a number such as may be obtained from a hand calculator, rounding off successively, we have:

9.1703845
9.170384
9.17038
9.1704
9.170
9.17
9.2
9

Addition and Subtraction

In addition and in subtraction, the answer should be rounded off so as to contain the same number of decimal places as in the number with the least number of decimal places. Thus, to add,

$$\begin{array}{r} 12.376 \\ + \ 9.51 \\ \hline 21.886 \end{array} \quad \text{(to be rounded off)}$$

The least number of decimal places in the above problem is two, so the answer should be rounded off to two decimal places, or

$$21.89 \qquad \text{(correct answer)}$$

To subtract,

$$\begin{array}{r} 127.5093 \\ - \ 61.425 \\ \hline 66.0843 \end{array} \quad \text{(to be rounded off)}$$

The least number of decimal places in this example is three, so the answer should be rounded off to three decimal places, or

$$66.084 \qquad \text{(correct answer)}$$

Multiplication and Division

In multiplication and division, the answer should be rounded off so as to contain only as many significant figures as in the number with the least number of significant figures. Thus, to multiply,

$$\begin{array}{r} 5.29 \\ \times \ 11.276 \\ \hline 59.64475 \end{array} \quad \text{(calculator answer)}$$

The least number of significant figures in this problem is three, so the answer should be rounded off to three significant figures, or

$$59.6 \qquad \text{(correct answer)}$$

Likewise, to divide,

$$127.62 \div 25.67513 = 4.9705687 \qquad \text{(calculator answer)}$$

The answer should be rounded off to five significant figures, or

$$4.9706 \qquad \text{(correct answer)}$$

Exercise 1-5

Find the number of significant figures in the following measurements.
a. 52.0 mL   b. 52 mL   c. 520 mL
d. 5200 mL   e. 0.000520 mL

Solution

To determine the number of significant figures, count all numbers from left to right starting with the first nonzero number. All zeros at the end of a number are placeholders unless a decimal point is present.
a. 3   b. 2
c. 2 (zero is placeholder)   d. 2 (both zeros are placeholders)
e. 3 (zero at the far right is significant, other zeros are placeholders)

Self-test

Write the number of significant figures in the following measurements.
a. 0.014 L   b. 6.073 g   c. 1400 mL

Answers

a. 2   b. 4   c. 2

Exercise 1-6

Round off the following numbers to three significant figures.
a. 10.071   b. 0.008695   c. 51,428

Solution

Count three significant numbers from left to right to locate the first nonsignificant number. If the nonsignificant numbers are less than 5, drop all nonsignificant numbers. If the nonsignificant numbers are greater than 5, add 1 to the last significant number. If the nonsignificant number is exactly 5, add 1 if the preceding number is odd and leave unchanged if the preceding number is even.
a. 10.1   b. 0.00870   c. 51,400

Self-test

Round off the following numbers to two significant figures.
a. 1148   b. 0.0385   c. 20.72

Answers

a. 1100   b. 0.038   c. 21

Exercise 1-7

Perform the following mathematical operation. Give the answer with the correct number of significant figures.
a. 11.73 g + 6.8 g + 120 g = _____
b. 150 mL − 6.8 mL = _____

c. 2.6 cm × 5.2 cm × 11.1 cm = _____
d. 8.238 g ÷ 0.92 mL = _____

Solution

In addition or subtraction, the number of significant figures is determined by the number with the fewest decimal places.

a. 11.73 g measures to the hundredths place, 6.8 g measures to the tenths place, and 120 g measures to the tens place. Therefore, the answer is limited to the tens place: 11.73 g + 6.8 g + 120 g = 138.53 g; this rounds off to 140 g (tens place).

b. The volume of 150 mL measures to the tens place, whereas the volume 6.8 mL measures to the tenths place. The answer is rounded off to the tens place: 140 mL.

c. In multiplication and division the answer is determined by the data with the least number of significant figures. Both 2.6 cm and 5.2 cm have two significant figures, whereas the number 11.1 cm has three significant figures. The answer can only have two significant figures: 2.6 cm × 5.2 cm × 11.1 cm = 150.072 cm³; this rounds off to 150 cm³.

d. The number 8.238 g has four significant figures and 0.92 mL has two significant figures. The answer must contain two significant figures: 8.238 g ÷ 0.92 mL = 8.9543; this rounds off to 9.0 g/mL.

Self-test

Perform the following mathematical operations and round off your answer to the correct number of significant figures.

a. 10.1 mL − 0.20 mL = _____
b. 131 g + 456 g + 1200 g = _____
c. 4.973 g ÷ 5 mL = _____
d. 12.9 m × 66.899 m = _____

Answers

a. 9.9 mL   b. 1800 g   c. 1 g/mL   d. 863 m²

1-5 Metric-Metric Conversions

In this section you will apply the factor-label method using significant figures.

Exercise 1-8

Exercise 1-8a

Convert 150 cm to m.

Solution

Since 1 m = 100 cm,

$$150 \, \cancel{cm} \times \frac{1 \, m}{100 \, \cancel{cm}} = 1.50 \, m$$

Note that the answer contains three significant figures, as does the original number. The definition 1 m = 100 cm is exact and has as many significant figures as needed.

Exercise 1-8b

Convert 3.76 L to mL.

Solution

$$3.76 \, \cancel{L} \times \frac{1000 \text{ mL}}{1 \, \cancel{L}} = 3760 \text{ mL}$$

$$= 3.76 \times 10^3 \text{ mL (to three significant figures)}$$

Exercise 1-8c

Convert 512 mg to kg.

Solution

Since 1 g = 1000 mg and 1 kg = 1000 g, then

$$512 \, \cancel{\text{mg}} \times \frac{1 \, \cancel{\text{g}}}{1000 \, \cancel{\text{mg}}} \times \frac{1 \text{ kg}}{1000 \, \cancel{\text{g}}} = 5.12 \times 10^{-4} \text{ kg}$$

Self-test

A hot-air balloon has a volume of 1.47×10^3 kiloliters. How many milliliters is this?

Answer

1.47×10^9 mL

1-6 English–SI Conversion Factors

Although English–English conversion units contain exact numbers (e.g., by definition, there are exactly 12 inches in 1 foot), the conversion factors from English units to SI units and vice versa depend on the number of significant figures used, as shown below.

English–SI conversion factors

1 in = 2.54 cm	(three significant figures)
1 lb = 454 g	(three significant figures)
2.2 lb = 1 kg	(two significant figures)
1.06 qt = 1 L	(three significant figures)

Exercise 1-9

a. 775 g = _____ lb   b. 2.50 mi = _____ km

Solution

a. $775 \, \cancel{\text{g}} \times \dfrac{1 \text{ lb}}{454 \, \cancel{\text{g}}} = 1.71$ lb (three significant figures)

b. In this example we can convert mi to yd, to ft, to in, to cm, to m, and finally to km. All of this may be done in one overall set of conversion factors, as shown below.

$$2.50 \text{ mi} \times \frac{1760 \text{ yd}}{1 \text{ mi}} \times \frac{3 \text{ ft}}{1 \text{ yd}} \times \frac{12 \text{ in}}{1 \text{ ft}} \times \frac{2.54 \text{ cm}}{1 \text{ in}} \times \frac{1 \text{ m}}{100 \text{ cm}} \times \frac{1 \text{ km}}{1000 \text{ m}} = 4.02 \text{ km}$$

Self-test

What is the volume in quarts of a 5.00×10^2 mL glucose solution?

Answer

0.530 qt or 5.30×10^{-1} qt

1-7 Density

The density of an object helps to characterize it physically and is determined by dividing its mass (or weight) by its volume. Density can be expressed by the formula

$$D = \frac{M}{V}$$

where D is the density, M is the mass, and V is the volume. Density must have units of mass per unit volume such as g/mL, g/cm³, lb/ft³, or mg/L.

Urine Density

The density of normal human urine ranges from 1.003 g/mL to 1.030 g/mL. Values outside this range are of diagnostic value. For example, the urine of a person suffering from diabetes mellitus contains large amounts of sugar and thus will have a density greater than normal. Conversely, a person suffering from diabetes insipidus will have a urine density close to 1.000 g/mL because of the large amount of water being excreted (see page 512).

Osteoporosis and Bone Density

Osteoporosis is a disease that decreases bone mass as a result of a deficiency of calcium compounds. The bones of an individual with osteoporosis are brittle and susceptible to fractures. Bone density changes are measured by using radiographic or

a 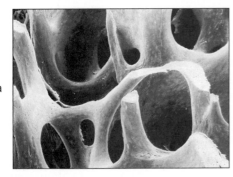 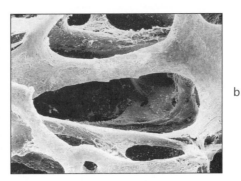 b

FIGURE 1-7
Photomicrographs of (a) normal bone tissue, and (b) osteoporitic bone tissue showing loss of bone mass.

ultrasound imaging techniques to determine the extent of calcium loss. Older individuals, particularly women, are more likely to develop osteoporosis, and adding a daily calcium supplement to the diet is highly recommended.

Exercise 1-10

Exercise 1-10a

What is the density of the mercury in a thermometer if 31.2 g of it occupies 2.29 mL?

$$D = \frac{M}{V}$$

$$= \frac{31.2 \text{ g}}{2.29 \text{ mL}} = 13.6 \frac{\text{g}}{\text{mL}} \text{ (three significant figures)}$$

Exercise 1-10b

Alcohol has a density of 0.80 g/mL. How much will 100 mL of it weigh?

$$D = \frac{M}{V}$$

$$M = D \times V$$

$$= 0.80 \frac{\text{g}}{\text{mL}} \times 100 \text{ mL} = 80 \text{ g (two significant figures)}$$

Self-test

A car radiator contains 8.0 L of antifreeze (ethylene glycol). If the density of ethylene glycol is 1.12 g/mL, how many grams of antifreeze are in the radiator?

Answer

9.0×10^3 g (two significant figures)

1-8 Temperature Scales

Old-fashioned thermometers rely on the expansion of mercury to indicate temperature. As the mercury is warmed, it expands and rises in the thermometer tube—a relatively slow process. Modern thermometers measure the infrared (heat) energy radiated by the tympanic membrane and surrounding tissue in the ear. Infrared thermometers accurately measure body temperature in a matter of seconds. (Photograph used with permission of Thermoscan, Inc., San Diego, Calif.)

Temperature is a measure of the availability of heat or cold or, in simpler terms, of how hot or cold a substance is. It can be measured by means of a thermometer. When we speak of body temperature as being 98.6 degrees or when we say that the outside temperature is 80 degrees or 40 degrees or even 2 degrees below zero, we are usually speaking in terms of degrees Fahrenheit (°F), even though we may not express the unit itself. In chemistry, the Celsius (C) and Kelvin (K) temperature scales are commonly used. Let us see first how the Celsius and Fahrenheit temperature scales compare with one another.

If a Fahrenheit and a Celsius thermometer are placed in a mixture of ice and water, the reading on the Fahrenheit scale will be 32° and the reading on the Celsius scale 0° (see Figure 1-8)—that is, 32° F corresponds to 0° C. These temperatures indicate the freezing point of water. They also indicate the melting point of ice.

Next, if the same two thermometers are placed in boiling water at 1 atmosphere pressure, the reading on the Fahrenheit scale will be 212° and the reading on the Celsius scale will be 100°. These temperatures indicate the boiling point of water.

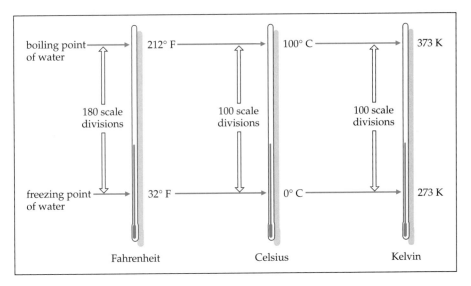

FIGURE 1-8

Comparison of Fahrenheit, Celsius, and Kelvin temperature scales.

Thus there is a 180-degree difference between the boiling and the freezing points of water on the Fahrenheit scale (212° to 32°) and a 100-degree difference between the boiling and freezing points of water on the Celsius scale (100° to 0°). The difference between the freezing points of water on the Fahrenheit and Celsius scales is 32° (32° − 0°).

This information can be combined into the following formulas:

$$°F = \frac{9}{5}°C + 32° \quad \text{and} \quad °C = \frac{5}{9}(°F - 32°)$$

The first formula is used to change Celsius temperatures to the corresponding Fahrenheit readings; the second formula is used to do the reverse, to change Fahrenheit readings to Celsius. Note that the first formula can be changed to the second by subtracting 32° from both sides and then dividing by $\frac{9}{5}$.

Exercise 1-11

Exercise 1-11a

Change 80.0° C to °F.

Solution

Using the formula

$$°F = \tfrac{9}{5}°C + 32°,$$
$$°F = \tfrac{9}{5}(80.0) + 32°$$
$$= 144° + 32°$$
$$= 176° F$$

Thus,

$$80.0° C = 176° F$$

Exercise 1-11b

Change 50° F to °C.

Solution

Using the formula

$$°C = \tfrac{5}{9}(°F - 32°),$$
$$°C = \tfrac{5}{9}(50° - 32°)$$
$$= \tfrac{5}{9}(18°)$$
$$= 10° C$$

Thus,

$$50° F = 10° C$$

A Kelvin thermometer will indicate the freezing point of water as 273 K and the boiling point of water as 373 K. Temperatures on the Kelvin scale are given as numbers only, without the degree sign. Why pick 273 K for the freezing point of water? Why not some even number? This temperature was selected so that zero on the Kelvin scale would represent the lowest temperature it is possible to reach. This temperature, 0 K, is also called absolute zero.

There are 100 units between the boiling and freezing points of water on the Kelvin scale (373 K − 273 K), just as there are on the Celsius scale. The difference between the freezing points of water on the Kelvin and Celsius scales is 273 (273 − 0). This information can be combined into the following formula.

$$K = °C + 273$$

Exercise 1-12

Change 37° C to K.

Solution

Using the formula

$$K = °C + 273,$$
$$K = 37 + 273$$
$$= 310$$

Thus,

$$37° C = 310 K$$

Self-test

Liquid oxygen boils at −183° C. What is oxygen's boiling point in degrees Fahrenheit? Kelvin?

Answer

−297° F, 90 K

Summary

The SI units of length and mass (weight) are the meter (m) and the kilogram (kg). Common units of length, weight, and volume are the meter (m), the gram (g), and the liter (L), respectively. Prefixes in common use are micro (1/1,000,000), milli (1/1000), centi (1/100), deci (1/10), kilo (1000), and mega (1,000,000).

Significant figures are those that are known to be reliable.

Density is defined as mass divided by volume, or $D = M/V$.

Although Fahrenheit and Celsius temperature scales are in common use, the Celsius and Kelvin temperature scales are the ones used almost exclusively in scientific measurements.

To convert Fahrenheit temperatures to Celsius, use the formula $°C = \frac{5}{9}(°F - 32°)$. To convert Celsius temperatures to Fahrenheit, use the formula $°F = \frac{9}{5}°C + 32°$. To convert Celsius temperature to Kelvin temperature, use the formula $K = °C + 273$.

Questions and Problems

1. What is the SI unit of mass? length? time? temperature?

2. What do the following abbreviations indicate: mg, dL, cm, kg, m, µL?

3. What do the following prefixes indicate: micro-, centi-, kilo-, mega-, milli-?

4. A patient has a body temperature of 104° F. What is this temperature in Celsius?

5. How many significant figures are present in the following numbers?

    a. 1.56   b. 0.0951

    c. 13.007   d. 3.29×10^{-2}

6. Which patient has a higher fever, one with a temperature of 102.2° F or one with a temperature of 39.0° C?

7. Which is longer, a yard or a meter?

8. Compare the following units: mL, cc, cm³.

9. If water in a pressure cooker boils at 110° C, what is the Fahrenheit reading? the Kelvin reading?

10. At the top of a certain mountain, water boils at 88° F. What is the Celsius boiling point? the Kelvin boiling point?

11. Convert the following:

    a. 1.58 L = _____ mL
    b. 122.6 mg = _____ g
    c. 0.058 cm = _____ m
    d. 1.25 mg = _____ µg
    e. 408 mL = _____ L
    f. 12.66 cm³ = _____ mL
    g. 2.04 kg = _____ g
    h. 2.00 kg = _____ mg
    i. 187 dg = _____ g
    j. 100 mm = _____ km

12. Convert the following (give answers to one decimal place):

    a. 125 cm = _____ in
    b. 110 lb = _____ kg
    c. 5 ft 8 in = _____ cm

d. 125 g = _____ lb
e. 40.5 in = _____ cm
f. 62.4 kg = _____ lb
g. 40 cm = _____ in
h. 60.6 kg = _____ lb
i. 240 g = _____ lb
j. 240 lb = _____ kg

13. The following information was recorded for a patient: height, 160 cm; weight, 70 kg; temperature, 38.6° C. What would the readings be in inches, pounds, and degrees Fahrenheit?

14. What is the density of a medication if the contents of a filled 2.00 mL syringe weigh 2.75 g?

15. A patient is 5 ft 2 in tall and weighs 120 lb. What is the patient's height in centimeters and weight in kilograms?

16. Normal urine has a density of 1.020 g/mL. What is the weight of a 100 mL sample of urine?

17. Three 10 mL samples of urine weigh 10.01 g, 10.20 g, and 10.36 g, respectively. Which sample represents normal urine in terms of density? See problem 16.

18. Mercury has a density of 13.6 g/mL. What is the weight of 100 mL of mercury?

19. A rectangular object is 20 cm tall, 18 cm wide, and 15 cm deep. What is its volume in cm³, mL, L, and kL?

20. Three patients have temperatures of 99° F, 39° C, and 313 K. Which patient has the highest fever?

21. An infant weights 3.0 kg at birth. What is the infant's birth weight in grams? pounds?

22. Compare mass and weight.

23. An order for a medication reads: Give 0.05 mL per kilogram of body weight. How much medication should be given to a patient weighing 160 lb?

24. An order for a medication reads: Give 1.0 mg per kilogram of body weight. How much medication should be given to a patient weighing 180 lb? Give your answer to the closest whole number of milligrams.

25. What do the following mean in terms of known units: microseconds, kilowatts, milliamperes, megacuries?

26. Consider the following food label:

Nutrition Facts
Serving Size 2 Tsp (5g)
Servings Per Container about 45

Amount Per Serving

Calories 20	Calories from Fat 15

	% Daily Value*
Total Fat 1.5g	**2%**
Saturated Fat 1g	**5%**
Cholesterol 5mg	**2%**
Sodium 85mg	**4%**
Total Carbohydrate 0g	**0%**
Dietary Fiber 0g	**0%**
Sugars 0g	
Protein 2g	

Vitamin A 0%	•	Vitamin C 0%
Calcium 6%	•	Iron 0%

* Percent Daily Values are based on a 2,000 calorie diet. Your daily values may be higher or lower depending on your calorie needs:

	Calories:	2,000	2,500
Total Fat	Less than	65g	80g
Sat Fat	Less than	20g	25g
Cholest	Less than	300mg	300mg
Sodium	Less than	2,400mg	2,400mg
Total Carb		300g	375g
Fiber		25g	30g

a. How many milligrams of total fat are present in each serving?
b. How many grams of cholesterol?
c. How many grams of sodium?

27. Consider the following label on a vial of medication:

Cefizox
Pediatric dosage
200 mg/kg/day

a. How much medication should be administered per day to an infant who weighs 13.2 lb?
b. If the medication is to be administered three times per day, how much medication should be administered per dose?

28. Consider the following label on a vial of medication:

Erythromycin ethylsuccinate
Pediatric dosage
30 mg/kg/day

a. How much medication per day should be administered to a child weighing 26.4 lb?
b. If the medication is to be administered four times per day, how much medication should be administered per dose?

29. The dosage for many antineoplastic medications is calculated based on the surface area of the patient's body. The following nomogram indicates the amount of surface area according to the patient's height (in centimeters) and weight (in kilograms).

To use the nomogram, place one end of a straight edge at the patient's height and the other end at the patient's weight. Where the edge crosses the center line, read the body surface area in square meters (m^2).

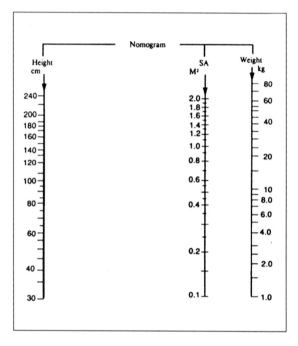

Haycock nomogram for calculating body surface area of adults. (From Haycock GB, Schwartz GJ, Wisofsky DH: Geometric method for measuring body surface area: A height-weight formula validated in adults.)

Consider the following medication order:

Name: Lebaron, Mary	**Room-Bed:** 216-B
Height: 5'4"	
Weight: 120 lb.	
Medication Order Dactinomycin IV qd for 5 days	
Dosage Instruction 400 mcg/m²	

a. What is the patient's height in centimeters?
b. What is the patient's weight in kilograms?
c. What is the patient's body surface area?
d. Calculate the amount of this medication to be given to the patient.

30. Consider the following medication order:

Name: McDonald, Arthur	Room-Bed: 100-A
Height: 5'8"	
Weight: 150 lb.	
Medication Order Doxorubicin HCl IV q21d	
Dosage Instruction 60 mg/m²	

  a. What is the patient's height in centimeters?
  b. What is the patient's weight in kilograms?
  c. What is the patient's body surface area?
  d. Calculate the amount of this medication to be given to the patient.

31. Calculate the volume of a 10.0-carat diamond (1 carat = 200.0 mg). The density of diamond is 3.51 g/mL.

32. A hyperbaric chamber with a volume of 8.0×10^3 L is filled with oxygen (density of oxygen is 1.43 g/L). Calculate the number of grams of oxygen in the chamber.

33. At what temperature is the numerical value of the Fahrenheit and Celsius scales the same?

34. Aspirin has a density of 1.40 g/mL. What is the volume of one aspirin tablet containing 250 mg of pure aspirin? What is the weight of exactly 1 L of aspirin?

Practice Test

1. 40° C = _____ K
  a. 104   b. 233   c. 313   d. 400

2. An object has a mass of 200 g and a volume of 125 mL. What is its density in g/mL?
  a. 0.620   b. 0.750   c. 1.60   d. 3.25

3. 125 cm = _____ m
  a. 0.0125   b. 0.125   c. 1.25   d. 12.5

4. 12.5 mg = _____ µg
  a. 0.0125   b. 125   c. 1250   d. 12,500

5. 45.8 cm³ = _____ mL
  a. 0.0458   b. 0.458   c. 4.58   d. 45.8

6. An object has a density of 1.25 g/mL and a volume of 180 mL. What is its mass?
  a. 14.4 g   b. 22.5 g   c. 144 g   d. 225 g

7. 2 ft 11 in = _____ cm
  a. 12   b. 62   c. 77   d. 89

8. 100 kg = _____ lb
  a. 45   b. 100   c. 220   d. 254

9. How many watts in 1.58 megawatts?
  a. 1580   b. 158,000
  c. 1,580,000   d. 1,580,000,000

10. 100.8° F = _____ °C
  a. 38.0   b. 38.2   c. 38.6   d. 39.0

2

Energy and Matter

Objectives

- To distinguish among various forms of energy
- To study the properties of matter
- To understand the states of matter
- To calculate the energy requirements for several processes
- To distinguish among and to compare elements, compounds, and mixtures

The manifold properties of matter give rise to the natural diversity of the world around us. In this view of Mono Lake, in California, solid, liquid, and gaseous phases of matter can be seen.

2-1 Energy

Energy is defined as the ability to do work. The muscles in our bodies get their energy from the chemical reactions that take place in the muscle cells. The heat energy necessary to keep our bodies at a temperature of 98.6° F or 37° C comes from the oxidation of the foods we eat. The electrical energy we use in our homes comes from burning a fuel or from atomic energy. Energy exists in several forms—heat, light, electric, mechanical, sound, chemical, and atomic. Energy can be classified into two categories: kinetic energy and potential energy.

Kinetic Energy

Kinetic energy is energy of motion; that is, energy that is doing something now, such as heat energy obtained from burning wood, light energy from an incandescent lightbulb, mechanical energy from a motor, and atomic energy from a nuclear reactor.

Potential Energy

Potential energy is stored energy, energy not associated with motion. Examples of potential energy are a dry cell (which can supply electrical energy when it is connected to something), food (which supplies energy to our bodies when it is metabolized), and water at the top of a waterfall (which becomes kinetic energy that can supply mechanical energy as it falls to the bottom).

Chemical energy is a form of potential energy. Most chemical reactions involve changes in heat energy. If heat is given off during a chemical reaction, that reaction is said to be exothermic. If heat is absorbed during a chemical reaction, that reaction is said to be endothermic.

Transformation of Energy

Energy can be transformed from one form into another. Thus, burning a piece of coal changes its potential energy into heat (kinetic) energy. The heat energy thus produced might be used to boil water, which produces large amounts of steam. The steam might be used to drive a generator to produce electrical energy. In turn, this electrical energy might be used to drive a motor (mechanical energy), produce light in a fluorescent lamp (light energy), operate a radio (sound energy), or operate a toaster (heat energy).

The sun produces energy by nuclear reactions and radiates this energy to the Earth. Plants on the Earth pick up the light energy from the sun during the process of photosynthesis and produce compounds that contain chemical energy. When humans eat these compounds in the plants, their bodies convert the chemical energy into heat energy and mechanical energy.

Conservation of Energy and Matter

The law of conservation of energy states that energy is neither created nor destroyed during a chemical reaction. Energy can be changed from one form to another, but the total amount of energy remains the same regardless of what form the energy is changed into.

The law of conservation of mass states that during a chemical reaction mass is neither created nor destroyed. This means that the total weights of substances before they react should be the same as the weights of the products after the reaction. For example, if a candle is placed in a sealed container, a certain weight will be obtained for both the candle and the container. If the candle is lit and allowed to burn (inside the sealed container) and if it is weighed as it continues to burn, the weight will be found to remain the same even though part of the candle is disappearing and several gaseous products are being produced. The same is true for a camera flashbulb before and after it is fired. Other experiments performed under very carefully controlled conditions have produced similar results—that is, the sum of the weights after a reaction is the same as the sum of the weights before the reaction.

In the early twentieth century Albert Einstein stated that mass and energy were interchangeable. That is, under certain conditions, mass could be changed into energy or energy into mass. These changes do not occur under the conditions of an ordinary chemical reaction, and so the laws of conservation of energy and conservation of mass are still used. However, these laws can be combined into one overall law, which states that mass and energy cannot be created or destroyed but they can be converted from one to the other.

Measurement of Energy

Heat is the most common form of energy; all other forms of energy can be converted into heat energy. The unit of heat energy is the calorie (cal), which is defined as the amount of heat required to raise the temperature of 1 g of water 1 K or 1° C. The calorie is abbreviated as cal. It is a rather small unit of heat.

A larger unit of heat, the kilocalorie, is equal to 1000 cal. The kilocalorie is abbreviated as kcal. The kilocalorie is used when measuring the heat energy of the body and for nutritional values of foods. The kilocalorie is also called a "large Calorie," abbreviated Cal. The caloric values of foods are listed in Cal.

Another unit of heat is the joule (abbreviated J); 1 cal equals 4.18 J. Calories are used primarily as a unit of measurement for medical work, whereas joules (and kilojoules, kJ) are used in chemical work.

Exercise 2-1

It takes 2.26 kJ to convert 1.00 g of liquid water into steam. Express this value in kcal.

Solution

- **Step 1** Convert kJ into J $\qquad$ $2.26 \text{ kJ} \times \dfrac{1000 \text{ J}}{1 \text{ kJ}} = 2260 \text{ J}$

- **Step 2** Convert J into calories $\qquad$ $2260 \text{ J} \times \dfrac{1 \text{ cal}}{4.18 \text{ J}} = 541 \text{ cal}$

- **Step 3** Convert cal into kcal $\qquad$ $541 \text{ cal} \times \dfrac{1 \text{ kcal}}{1000 \text{ cal}} = 0.541 \text{ kcal}$

Self-test

8.4×10^2 J are required to convert 1.00 g of liquid ethyl alcohol into vapor. Express this value in calories.

Answer

2.0×10^2 cal

Table 2-1 Specific Heat Values (in cal/g° C)					
Gases		**Liquids**		**Solids**	
Ammonia	0.502	Alcohol	0.587	Aluminum	0.215
Chlorine	0.114	Chloroform	0.231	Calcium	0.156
Oxygen	0.219	Ether	0.555	Copper	0.092
Nitrogen	0.249	Water	1.00	Iron	0.106

2-2 Specific Heat

When heat is added to a substance, its temperature rises; when heat is removed, its temperature falls. The amount of heat required to raise the temperature of 1 g of a substance 1 K or 1° C is called the specific heat capacity or the specific heat. From the definition of the calorie, the specific heat of water is 1 cal/g° C.

The specific heat values of various substances are listed in Table 2-1. Note that water has an abnormally high specific heat value when compared with other substances.

The relatively high specific heat of water indicates that it takes more heat to raise the temperature of water than it does for an equal amount of other substances. Conversely, a given amount of heat will increase the temperature of almost any substance more than it will raise the temperature of an equal mass of water.

The heat from sunshine will warm the air around an outdoor swimming pool more than it will warm the water, so the water feels cooler than the air. If the poolside chairs are made of metal, they will feel much hotter than those made of plastic, whose specific heat is greater than that of metals.

The amount of heat required to change the temperature of a substance is given by the following formula:

$$\text{heat gain (or loss)} = \text{mass} \times \text{specific heat} \times \text{temperature change}$$

or

$$\text{calories} = m \times sp\ ht \times \Delta t$$

where m is the mass in grams, $sp\ ht$ is the specific heat in cal/g° C, and Δt is the change in temperature in degrees Celsius.

Exercise 2-2

How many calories are required to change the temperature of 1.0×10^2 g water from 18.0° C to 19.5° C?

Solution

$$\text{Using calories} = m \times sp\ ht \times \Delta t$$

$$\text{calories} = 1.0 \times 10^2\ g \times 1.00\ cal/g° C \times 1.5° C$$

$$\text{calories} = 1.5 \times 10^2$$

Self-test

A pan containing 1.0×10^3 g of water is heated from $20.0°$ C to $100°$ C. Calculate the number of calories needed to heat the water.

Answer

8.0×10^4 cal

Exercise 2-3

500 calories are added to 100 g copper at $25.0°$ C.
a. What will the final temperature be?
b. If the same amount of heat was added to an equal amount of water, what would be the final temperature?

Solution

Using calories $=$ m $\times$ sp ht $\times$ Δt

a. $\Delta t = \dfrac{\text{calories}}{\text{m} \times \text{sp ht}} = \dfrac{500 \text{ cal}}{100 \text{ g} \times 0.092 \text{ cal/g}° \text{ C}} = 54.3°$ C

So, final temperature $= 25.0°$ C $+ 54.3°$ C $= 79.3°$ C.

b. $\Delta t = \dfrac{\text{calories}}{\text{m} \times \text{sp ht}} = \dfrac{500 \text{ cal}}{100 \text{ g} \times 1.00 \text{ cal/g}° \text{ C}} = 5.0°$ C

So, final temperature of water $= 25.0°$ C $+ 5.0°$ C $= 30.0°$ C.

Self-test

A 2.0 kg iron pan (temperature is $250°$ C) is taken off the stove and placed into a sink of water at $50°$ C. If the final temperature of the water is $70°$ C, how much water was in the sink?

Answer

1.9 L or 1.9×10^3 mL

2-3 Calorimetry

A calorimeter is a device used to measure the number of Cal produced by the oxidation (metabolism) of food. Essentially, a calorimeter consists of one container inside another, the two being separated by insulation. The inner container holds a given amount of water, a combustion device, a stirrer, and a thermometer.

The caloric values of the three main food types are as follows (see Table 2-2):

Carbohydrate	4 Cal/g
Fat	9 Cal/g
Protein	4 Cal/g

FIGURE 2-1

Calorimeter. The caloric value of foods may be obtained by using such a device.

Table 2-2 Caloric Values of Some Foods

Food	Portion	Kilocalories
Milk, whole	1 cup	160
Cheese, cheddar	1 ounce	115
Hamburger, broiled	3 ounces	245
Almonds, shelled, whole kernels	1 cup	850
Carrots, cooked	1 cup	91
Potato chips, 2 inch diameter	10 chips	115
Orange juice, fresh	1 cup	115
Bread, white, toasted	1 slice	70
Brownies, with nuts	1 brownie	95
Cola drink	12 ounces	145

Exercise 2-4

2.0 g of glucose, a carbohydrate, are burned in a calorimeter containing 500 g of water at 20.0° C. The final temperature is 36.0° C. How much heat in Cal is produced per gram of carbohydrate burned (oxidized)?

Solution

Using calories = m × sp ht × Δt

calories = 500 g × 1.00 cal/g° C × 16.0° C

calories = 8000

$$8000 \; \cancel{cal} \times \frac{1 \, Cal}{1000 \; \cancel{cal}} = 8.0 \, Cal$$

calories = 8.0 Cal for 2.0 g or 4.0 Cal/g

Therefore, each gram of carbohydrate yields 4.0 Cal.

Self-test

A patient was given 1000 mL of 30% protein hydrolysate intravenously. How many Calories did she receive?

Answer

1200 Cal

Table 2-3

Nutritional Information per Serving for a Loaf of Bread

Serving size	1 slice
Servings per package	20
Calories	89
Protein	3 g
Carbohydrate	17 g
Fat	1 g
Sodium	180 mg
Total dietary fiber	2 g

Table 2-4

Energy Used in Various Activities

Activity	Cal/Hr
Running at 8.5 mph	870
Swimming at 2 mph	540
Walking at 4 mph	230
Bicycling slowly	170
Gymnastics	168
Driving a car	61

Many people are becoming more calorie conscious and limiting their fat intake. Package labeling with a chart listing the carbohydrate, fat, and protein values and calorie content of the food product is an example of how the food industry has responded to this trend of consumer awareness.

In addition to the energy required for normal metabolic body processes, more energy is needed for various activities (Table 2-4).

Exercise 2-5

How long would it take to walk off the calories from two slices of bread?

Solution

From Table 2-3, note that one slice of bread provides 89 Cal, and from Table 2-4, note that walking uses 230 Cal per hour. Therefore,

$$89 \text{ Cal/slice} \times 2 \text{ slices} \times 1 \text{ hr}/230 \text{ Cal} = 0.77 \text{ hr or } 46 \text{ min}$$

Self-test

A chocolate candy bar provides 420 Cal. How long must you swim to use up this amount of Calories?

Answer

0.77 hr or 46 min

2-4 What is Matter?

Matter is anything that occupies space and has weight. Everything we see or feel is matter, as well as many things we cannot see or feel. Such things as trees, food, machinery, and soil are examples of matter we can see and feel. Air is an example of matter we cannot see, yet we know that it is all around us. Not all matter is of the same type. Matter can be classified as solid, liquid, or gas.

2-5 States of Matter

A piece of iron is an example of matter in the solid state. A bar of iron has a definite shape, a shape that cannot be easily changed. The volume of the piece of iron (the amount of space that it occupies) is also definite and cannot be easily changed. Also, the piece of iron does not flow.

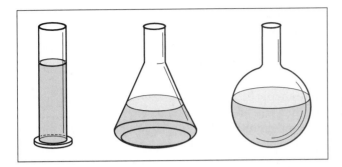

FIGURE 2-2

The same quantity (1 pint or approximately 500 mL) of water in variously shaped containers.

When a pint of water is poured from a container of one shape into a container of another shape, the water (a liquid) assumes the shape of the new container as far as it fills it. However, the volume of the water remains the same—1 pint—regardless of the shape of the container into which it is poured (see Figure 2-2).

When air is pumped into an empty bottle, the air occupies all of the space and also takes on the shape of the container. Forcing more air into the bottle will increase the pressure, but the air will still occupy all of the space and take the shape of the container. If the air is transferred to another bottle of different shape and size, again the air will occupy all of the space and take the shape of the new container.

In most solids, the particles are closely adhering and tightly packed in a highly ordered system. The motion of the particles is highly restricted (Figure 2-3a). These factors account for the fact that solids have a definite shape and resist changes in that shape. Tight packing also accounts for the incompressibility of solids. Heating allows the particles in a solid to move about slightly. Therefore, most solids expand when heated.

In liquids, the particles are moderately ordered and farther apart (but still in contact with each other) than the particles in solids (see Figure 2-3b). This loose structure accounts for the fact that liquids flow and have no definite shape. The particles in a liquid are close enough to resist compression. Liquids expand slightly when heated.

In gases, the motion of the particles is unrestricted so that the particles are independent of one another and are relatively very far apart (see Figure 2-3c). This independence allows gases to assume any shape or volume, to be compressed or expanded.

Some solids have a high density (gold, 19.3 g/mL), whereas others have a low density (cork, 0.2 g/mL). Many liquids have a relatively low density (water, 1 g/mL; gasoline, 0.66 g/mL), but some, such as mercury (13.6 g/mL), have a high density. All gases have a very low density, expressed in the unit of grams per liter (g/L) (e.g., air, 1.3 g/L; hydrogen, 0.09 g/L). In summary:

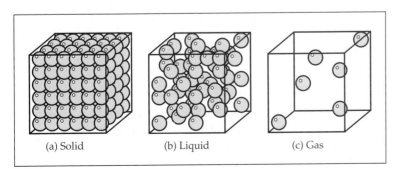

(a) Solid   (b) Liquid   (c) Gas

FIGURE 2-3

Rough representation of the particles of matter in (a) solid, (b) liquid, and (c) gas.

1. Solids have a definite shape and a definite volume, do not flow, and have particles that are closely adhering (interacting) and tightly packed in a highly ordered system. The motion of the particles in solids is highly restricted so that solids generally are incompressible. Most solids expand slightly when heated and may exhibit high or low density.

2. Liquids have no definite shape, do have a definite volume, flow, have particles that are relatively close to one another and are moderately ordered with some interaction, and may have high or low density. Liquids are incompressible and expand slightly when heated.

3. Gases have no definite shape and no definite volume, flow, and have particles whose motion is unrestricted, so they are independent and relatively far apart. Gases have a low density, are highly compressible, and expand greatly when heated.

Physical and Chemical Changes

Matter can usually be changed from one state to another merely by changing its temperature. Such changes are called physical changes. A physical change is one in which no new substance is produced, although there may be a change of state or density, or both.

However, not all substances can be changed from solid to liquid to gas merely by changing the temperature. For example, a piece of wood bursts into flame when heated sufficiently; it does not become liquid. Dry ice changes from solid to gas without becoming a liquid.

Dissolving sugar in water, breaking a sheet of glass, and freezing water also bring about physical changes. No new substance is produced, and each substance retains its own properties. Note that when water freezes there is a change in both state and density, since the density of water is $1.00\,\text{g/mL}$ and that of ice $0.917\,\text{g/mL}$.

The other type of change that matter can undergo is called a chemical change. A chemical change is one in which one or more substances disappear and a new substance or new substances are formed. The new substance or substances have entirely different properties from those of the original substance or substances.

Burning a piece of wood causes a chemical change. New substances—ash and smoke—are produced. These substances have properties different from those of the original piece of wood. Chemical changes are also involved in cooking food and in digestion.

Exercise 2-6

Classify each of the following observations as either a physical or a chemical change.

a. dissolving salt in water
b. boiling water
c. burning a candle
d. brewing coffee
e. rusting of iron

Solution

Examples **a, b,** and **d,** are all physical changes because no new substances were produced; examples **c** and **e** are chemical changes since new substances were produced.

2-6 Changes of State

Consider a piece of solid material. It consists of particles in a definite arrangement, as indicated in the preceding section. These particles are not motionless—they vibrate about fixed points. As heat is added to the solid, the temperature rises. Because of the increased temperature, the rate of vibration increases and the particles move slightly farther apart. Eventually, a point is reached when the vibrating particles can no longer retain their orderly arrangement. When this occurs, the solid begins to melt and becomes a liquid. The temperature at which this occurs is called the melting point of the solid. As more heat is added, more of the solid melts. All of the heat being added is used to change the state from solid to liquid. There is a constant temperature while the solid is changing to a liquid (melting).

If the temperature of a liquid is lowered, the same process takes place in reverse. The particles slow down and soon regain their original orderly arrangement. Such a process is called freezing, that is, freezing is the reverse of melting. The freezing point (FP) of a substance is the same as its melting point (MP). The freezing point of water is 0° C, and the melting point of ice is also 0° C. Every substance that undergoes melting and freezing has a characteristic MP and FP.

When heat is added to a solid (ice), it soon melts and becomes a liquid (water). Then when additional heat is added, the temperature of the liquid rises. The particles move faster and also move farther apart. Soon a point is reached at which the particles move far enough apart to become independent of one another. When this occurs, the liquid changes to a gas; it boils. The temperature at which this occurs is called the boiling point (BP) of the liquid. As more heat is added, more of the liquid boils, but all of the heat being added is used to change the state from liquid to gas. Thus the temperature remains constant during boiling.

However, water in a humidifier soon disappears into the air; it changes from a liquid to a gas even though no heat is applied. Such a process is called evaporation. What causes this process?

The water molecules are in random motion in all directions. Some of the faster-moving molecules at the surface may escape into the air. As this process continues, the slower moving molecules remain behind in the liquid. These slower-moving molecules have less energy and so the temperature of the water decreases. Therefore, evaporation is a cooling process.

The evaporation of perspiration from the skin cools the body by this same method. When a person has a high fever, a water sponge bath is used to lower the body's temperature by removing heat through the process of evaporation. An alcohol sponge bath does the same thing much more rapidly because alcohol evaporates more quickly than water (see Figure 2-4).

The reverse process, a change from the gaseous state back to the liquid state, is called condensation. As steam or water vapor molecules cool, they lose energy and change back into liquid water. Such a process can be observed by looking at a win-

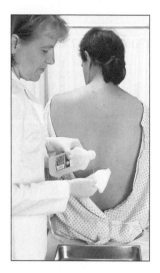

FIGURE 2-4
Alcohol sponge bath.

dowpane when the room temperature drops. The humid air (air that contains a large amount of water vapor) condenses on the cooler surface of the glass.

Some substances, when heated, change from the solid state directly into the gaseous state without any liquid being formed. Such a process is called **sublimation.** For instance, "dry" ice, or solid carbon dioxide, when allowed to stand in air, soon *sublimes* and becomes carbon dioxide gas; it does not melt. Ordinary ice, at a temperature below freezing, does not melt but can sublime. This process is used in the preparation of freeze-dried foods: the ice in frozen foods is allowed to sublime from the crystals.

Sublimation can also occur in reverse. The change from gas directly to the solid state is also called condensation. Iodine vapor, when cooled, condenses into iodine crystals.

2-7 Properties of Matter

One portion of matter can be distinguished from another by means of its properties. These distinguishing properties of matter can be classified into two main types: **physical properties** and **chemical properties.**

Physical Properties

Physical properties include color, odor, taste, solubility in water, density, hardness, melting point, and boiling point. These physical properties can serve to identify a substance, although not all of these properties may be necessary for the identification. For example, when we say that the color of a substance is white, we automatically eliminate all substances that are not white. Next, if we say that the white substance is odorless, we can eliminate all white objects that have an odor, leaving a smaller number of substances that are both white and odorless. If we continue to eliminate in this manner by using additional physical properties, such as density, hardness, melting point, and boiling point, eventually only one substance will fit all of these properties—the substance we are trying to identify.

Chemical Properties

Properties such as reacting (or not reacting) in air, reacting (or not reacting) with an acid, or burning (or not burning) in a flame are chemical properties. A substance can be identified by means of its chemical properties, but it is usually much simpler to do so by means of the physical properties.

FIGURE 2-5
Humidifier/nebulizer.

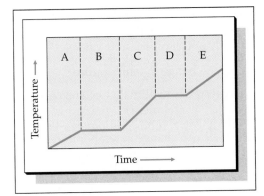

FIGURE 2-6

Temperature-time diagram. Section A indicates heating of the solid; section B, a change of state from solid to liquid; section C, heating of the liquid; section D, a change of state from liquid to gas; section E, heating of the gas.

Comparison of Physical and Chemical Properties

A physical property tells what a substance *is*—it is white or it is green; it is odorless or it has a sharp odor; it is hard or it is soft. A chemical property tells what a substance *does*—it burns or it does not burn; it reacts with an acid or it does not react with an acid; and so on.

2-8 Energy Involved in Changes of State

Figure 2-6 illustrates the various steps involved in changing a solid to a liquid and then to a gas (vapor). In each step there are certain energy requirements. The temperatures and specific heats necessary to change the state depend on the nature of the substance being heated. The following exercise describes the process for water. Other substances have similar heat requirements.

Exercise 2-7

How much heat is required to change 100 g ice at $-20°$ C to steam at $140°$ C?

Solution

- **Step 1** Heating a solid: the ice is heated from $-20°$ C to its melting point, $0°$ C. The formula is:

  calories = m $\times$ sp ht $\times$ change in temp

  where the specific heat of ice is 0.500 cal/g° C and the change in temperature is $20°$ C.

  calories = 100 g $\times$ 0.500 cal/g° C $\times$ 20° C = 1000 cal

- **Step 2** Changing state from solid to liquid: involves a change of state but no change in temperature. All of the heat applied is used to change the ice to water at the melting point.

  The amount of heat required to change 1 g of a solid to a liquid at its melting point is called the *heat of fusion*. The heat of fusion of ice is 80 cal/g. For the conversion of solid to liquid,

  calories = m $\times$ heat of fusion

  $\qquad$ = 100 g $\times$ 80 cal/g = 8000 cal

- **Step 3** Heating a liquid: the water is heated from its freezing point (0° C) to its boiling point (100° C), a change of 100° C. The specific heat of water is 1.00 cal/g° C. So,

$$\text{calories} = \text{mass} \times \text{sp ht} \times \text{change in temperature}$$

$$= 100 \text{ g} \times 1.00 \text{ cal/g° C} \times 100° \text{ C} = 10,000 \text{ cal}$$

- **Step 4** Change of state from liquid to gas: the water is boiled (changing in state from liquid to gas) with no change in temperature. All of the heat applied is used to change the state. The amount of heat required to change 1 g of a liquid to a gas at its boiling point is called the *heat of vaporization*. The heat of vaporization of water is 540 cal/g.

$$\text{amount of heat required} = \text{mass} \times \text{heat of vaporization}$$

$$\text{calories} = 100 \text{ g} \times 540 \text{ cal/g} = 54,000 \text{ cal}$$

- **Step 5** Heating a gas: the gas (steam) is heated from 100° C to 140° C, a change in temperature of 40° C. The specific heat of steam is 0.5 cal/g° C. So,

$$\text{calories} = \text{m} \times \text{sp ht} \times \text{change in temperature}$$

$$= 100 \text{ g} \times 0.5 \text{ cal/g° C} \times 40° \text{ C} = 2000 \text{ cal}$$

The total amount of heat required in all five steps is:

Step 1	1,000 cal
Step 2	8,000 cal
Step 3	10,000 cal
Step 4	54,000 cal
Step 5	2,000 cal
Total	74,000 cal = 74.0 Cal (three significant figures)

Note that when water at its boiling point is changed to steam at the same temperature, 540 cal are required for each gram changed. Conversely, for each gram of steam converted back to water at its boiling point, 540 cal are liberated. That is why a steam burn is so damaging to living tissue.

Self-test

Calculate the amount of heat removed when 5.0×10^2 mL of water at 18° C are converted into ice cubes at −15° C.

Answer

5.4×10^4 cal

2-9 Composition of Matter

All matter can be divided into three classes, depending on the properties of the material being considered. These three classes are **elements**, **compounds**, and **mixtures** (see Figure 2-7).

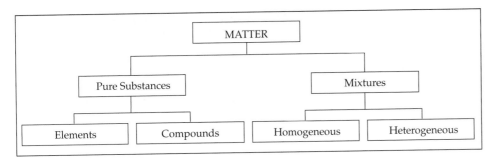

FIGURE 2-7
Matter flowchart.

Elements

Elements are the building blocks of all matter. An element can be defined as a substance that cannot be broken down into any simpler substance by ordinary chemical means; it contains only one type of substance. There are more than 100 known elements. The names of some of these elements, such as oxygen, hydrogen, iron, carbon, copper, mercury, and uranium, are probably familiar. The names of some of the rarer elements, such as cesium, einsteinium, molybdenum, and xenon, may not be familiar. Many elements are in common industrial use. Consider iron in steel, germanium in transistors, and tungsten in incandescent light bulbs.

Elements can be classified into two main types: metals and nonmetals. Each has its own specific properties.

Classification Metals conduct heat and electricity. They have a luster (a shiny surface) similar to that of silver and aluminum. Metals reflect light. Some metals are ductile (they can be drawn out into a thin wire). Some metals are malleable (they can be pounded out into thin sheets). Some metals have a high tensile strength. Metals such as iron, copper, and silver are solid at room temperature, but mercury is a liquid.

Nonmetals usually do not conduct heat and electricity very well. They have little luster and seldom reflect light. Nonmetals frequently are brittle. They cannot be pounded into thin sheets (they are not malleable) and cannot be drawn into thin wires (they are not ductile). Nonmetals may be solid at room temperature (carbon and sulfur), liquid at room temperature (bromine), or gaseous at room temperature (oxygen and nitrogen). There are many more metallic elements than nonmetallic; see the periodic table inside the front cover. Among the nonmetals are the noble gases, which are gases at room temperature. They are relatively unreactive. Among the noble gases are the elements helium, neon, and krypton.

Symbols for the Elements Each element can be identified by a symbol that represents that element. The symbol C stands for the element carbon, S for sulfur, and O for oxygen. In these instances, the symbol is the first letter of the name of the element. Because conflicts would arise if the first letter were to be used for every element, the first two letters are used for some elements, such as Ca for calcium and Al for aluminum. Or the symbol may use the first letter and one other letter to suggest a sound that is apparent in the name, such as Zn for zinc and Cl for chlorine. Note that when two letters are used to form a symbol, the first letter is capitalized and the second letter is not.

Table 2-5 Symbols for Some Elements

Symbol Using First Letter of Name	Symbol Using First Two Letters of Name	Symbol Using First Letter and One Other Letter in the Name	Symbol Based on Latin Name
C (carbon)	Al (aluminum)	As (arsenic)	Ag (silver—argentum)
H (hydrogen)	Br (bromine)	Cd (cadmium)	Au (gold—aurum)
I (iodine)	Ca (calcium)	Cl (chlorine)	Cu (copper—cuprum)
N (nitrogen)	Co (cobalt)	Cr (chromium)	Fe (iron—ferrum)
O (oxygen)	Ne (neon)	Mg (magnesium)	Na (sodium—natrium)
P (phosphorus)	Ni (nickel)	Mn (manganese)	Pb (lead—plumbum)
S (sulfur)	Si (silicon)	Zn (zinc)	Sn (tin—stannum)

The symbols for some elements are based on their Latin names. Ag, the symbol for silver, comes from the Latin word *argentum*. Fe, the symbol for iron, comes from the Latin word *ferrum*. Table 2-5 lists the symbols and names of various elements based on these categories.

Elements Present in the Human Body Table 2-6 lists the elements necessary for life, the symbols for these elements, and their functions in the body.

Compounds

When an electric current is passed through a container of water, the water is decomposed into two gases, each having its own set of properties. One gas is hydrogen and the other is oxygen (see Figure 2-8). If sugar is placed in a test tube and heated intensely, drops of water will collect at the top of the test tube while pieces of carbon will remain at the bottom. Both of these changes are chemical changes; that is, the original substance has disappeared and some new substances have been formed. Both sugar and water are classified as compounds, as are such other substances as salt, boric acid, carbon dioxide, and ether. Both water and sugar can be broken down into other substances by chemical means. Compounds, then, are substances that can be broken down into simpler substances by chemical means. Compare this definition with that of elements, which cannot be broken down into simpler substances by ordinary means.

Take a crystal of sugar from a sugar bowl. Examine it carefully. List its physical and chemical properties. Take another crystal of sugar from the sugar bowl and again list its physical and chemical properties. The properties of the second crystal are identical to those of the first, as are the pieces of sugar. That is, the sugar in a bowl of sugar is homogeneous; it is the same throughout.

An analysis of a sample of sugar shows that it contains carbon, hydrogen, and oxygen. The percentage of carbon in a portion of sugar can be calculated by weighing the piece of sugar as it is and again after heating it until nothing is left but carbon. Repeating this procedure with another portion of sugar will yield the same percentage of carbon. Likewise, analysis of several portions of sugar would yield identical results for the percentage of oxygen and for the percentage of hydrogen. The analysis of many samples of water gives a constant percentage of oxygen and hydrogen. The result of examining many different compounds for the percentages of their components can be stated as the law of definite proportions—

Table 2-6 Elements of Human Life

Element	Symbol	Function
Major components of molecules found in humans		
Oxygen	O	Required for water and organic compounds
Carbon	C	Required for organic compounds
Hydrogen	H	Required for water and organic compounds
Nitrogen	N	Required for many organic compounds and for all proteins
Sulfur	S	Required for some proteins and some organic compounds
Nutritionally important elements required in amounts greater than 100 mg/day		
Calcium	Ca	Required for bones and teeth; necessary for certain enzymes, nerve muscle function, hormonal action, cellular motility, and clotting of the blood
Phosphorus	P	Required for bones and teeth; necessary for high-energy compounds, nucleoproteins, nucleic acids, phospholipids, and some proteins
Magnesium	Mg	Required for many enzymes; necessary for energy reactions requiring adenosine triphosphate (ATP)
Sodium	Na	Principal positive extracellular ion
Potassium	K	Principal positive intracellular ion
Chlorine	Cl	Principal negative ion
Trace elements		
Iodine	I	Required for thyroid hormones
Fluorine	F	Required for bones and teeth; inhibitor of certain enzymes
Iron	Fe	Required for hemoglobin and many enzymes
Copper	Cu	Required for many oxidative enzymes, for the synthesis of hemoglobin, and for normal bone formation
Zinc	Zn	Required for many enzymes; related to action of insulin; essential for normal growth and reproduction and for nucleic acid metabolism
Manganese	Mn	Required for some enzymes acting in the mitochondria; essential for normal bone structure, reproduction, and for normal functioning of central nervous system
Cobalt	Co	Required for vitamin B_{12}
Molybdenum	Mo	Required for some enzymes; essential for purine metabolism
Chromium	Cr	Related to action of insulin
Selenium	Se	Essential for the action of vitamin E

compounds have a definite proportion or percentage by weight of the substances from which they were made.

It has already been mentioned that compounds have properties that are entirely different from those of the substances from which they were made. That is, water, a compound, has entirely different properties from the oxygen and hydrogen of which it is composed.

Compounds, then, have the following characteristics:

1. They can be separated into their component substances by chemical means.
2. They are homogeneous in composition.

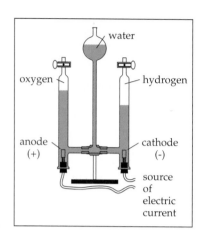

FIGURE 2-8
Electrolysis of water.

3. They have a definite proportion by weight of the substances from which they were made.
4. They have different properties from those of the substances from which they were made.

Mixtures

If a few crystals of salt are dissolved in a cup full of water, a mixture of salt and water—or salt water—is produced. If a few more crystals of salt are added to that same salt water, a little stronger salt solution is produced. If a teaspoonful of salt is added to the cup and stirred until it is dissolved, an even stronger salt solution will be produced. The salt and water form a mixture when they are placed together and stirred. The salt waters made by dissolving different amounts of salt in water will have different compositions, depending on how much salt was added to the water. Thus, one property of a mixture is that it can have a variable composition or variable proportions. More water, more salt, or both salt and water can be added to change the strength of the mixture, but in each case a salt water (saline) is produced.

The property of variable proportions is characteristic an any mixture. If sugar and sand are stirred together, a mixture is produced regardless of the amounts of each used. Likewise, sugar and iron filings can be mixed, and again, the proportion of iron filings to sugar does not alter the fact that it is a mixture.

The ingredients of any mixture can be separated from each other by such physical processes as evaporation and filtering. Salt can be separated from a salt–water mixture merely by evaporating the water. The sugar–sand mixture can be separated by placing it in water, stirring to dissolve the sugar, filtering the solution, and recovering the sand. The sugar can then be recovered from the filtered solution by evaporating the water. A sugar–iron filings mixture can be separated by passing a magnet over the mixture. The magnet will attract the iron filings, leaving the sugar behind. Mixtures can also be separated by a process known as chromatography.

Evaporation, chromatography, and separation by means of a magnet are examples of a physical change—one in which no new substances is produced. In each of the mixtures, the individual substances retain their own properties. There is no evidence of a chemical reaction because no new substance is produced. Thus, in the salt–water mixture, the salt retains its own properties, as does the water. In the sugar–sand mixture each retains its own properties. The sugar and the sand can be recognized separately under a microscope.

Exercise 2-8

Classify the following as an element, a compound, or a mixture.

a. hydrogen   b. gasoline
c. granite   d. gold
e. quartz   f. glucose

Solution

Hydrogen and gold are elements. Glucose and quartz are compounds. Gasoline and granite are mixtures.

Self-test

Classify the following as an element, a compound, or a mixture.

a. blood   b. saline solution
c. vitamin C (ascorbic acid)   d. zinc
e. sterling silver   f. diamond

Answers

Zinc and diamond are elements. Ascorbic acid is a compound. Blood, sterling silver, and saline solution are mixtures.

In summary, then, mixtures have the following characteristics.

1. They have no definite proportion or composition.
2. They can be separated into their component substances by physical means.
3. They retain the properties of the individual substances from which they were made.

Mixtures can be either homogeneous or heterogeneous in composition. **Homogeneous** means the same composition throughout. **Heterogeneous** means different composition throughout. Let us consider two of the mixtures already discussed, the salt–water and the sugar–sand mixtures.

Samples of a salt–water mixture are found to be identical from the top, from the middle, and from the bottom of a container of salt water. That is, the salt–water mixture is homogeneous. It is of the same composition throughout.

Now consider a mixture of sugar and sand. Samples from different parts of the mixture, when examined carefully under a microscope or with a magnifying glass, would not appear to have the same composition. That is, a mixture of sugar and sand is heterogeneous.

A mixture of two or more solids is heterogeneous. One solid can be distinguished from another, even if they are ground together in making the mixture. On the other hand, a solution is a homogeneous mixture. When a substance is dissolved in a liquid such as water, the mixture becomes the same throughout; it is homogeneous.

Although both the salt–water and sugar–sand mixtures contain two different substances, mixtures can be made from any number of substances. Note the word *substances*. It can mean either elements or compounds. That is, a mixture can be composed of two (or more) elements (powdered iron and powdered sulfur), two (or more) compounds (sugar and salt), or both elements and compounds (iodine and water).

Summary

Matter is anything that occupies space and has weight. Matter can exist in the solid, the liquid, or the gaseous state. These states of matter are interchangable and depend primarily on temperature. Physical properties of matter describe what a substance is; chemical properties describe what a substance does.

Energy is the ability to do work. The two types of energy are kinetic (the energy of motion) and potential or stored energy. The law of conservation of energy states that energy can be transformed from one type to another but is not created or destroyed. The law of conservation of mass states that mass is not created or destroyed. These two laws can be combined into one stating that mass and energy are not created or destroyed but can be converted from one to the other.

The calorie is the unit of heat energy and is the amount of heat required to raise the temperature of 1 g of water 1° C. One kilocalorie (kcal or Cal) equals 1000 cal.

Matter can be divided into three categories: elements, compounds, and mixtures.

Elements are substances that cannot be broken down into simpler substances by ordinary means. Elements are homogeneous in composition. Elements are either metals or nonmetals. Each element can be represented by a symbol.

Compounds can be separated into their component substances by chemical means, are homogeneous in composition, have a definite proportion by weight of their component substances, and have properties different from those of the substances from which they were made.

Mixtures have no definite proportions or composition, can be separated into their component substances by physical means, and retain the properties of the individual substances from which they were made. Mixtures can be made from two or more substances. Mixtures of solids are heterogeneous; mixtures made by dissolving a substance in a liquid are homogeneous in composition.

Questions and Problems

1. Which of the following is an element, which is a compound, and which is a mixture?
   a. iron
   b. air
   c. table salt
   d. zinc oxide
   e. water
   f. sugar
   g. boric acid
   h. calcium

2. Which of the following properties are physical and which are chemical?
   a. color
   b. odor
   c. density
   d. flammability
   e. boiling point
   f. taste
   g. reactivity
   h. solubility

3. Compare the properties of solids, liquids, and gases as to
   a. compressibility
   b. density
   c. definite volume
   d. definite shape
   e. motion of particles
   f. expansion on heating

4. Which of the following processes involve a physical change and which involve a chemical change?
   a. distilling water
   b. chopping wood
   c. digesting food
   d. souring milk
   e. breaking glass

5. Which *must* be homogeneous: elements, compounds, or mixtures? Which *may* be?

6. Which elements are represented by the following symbols?
   a. H   b. C   c. Fe   d. O   e. N   f. S
   g. Al   h. Na   i. Mg   j. Cr   k. Mn   l. Cl

7. A patient is given 1 L of glucose solution intravenously. If the solution contains 100 g glucose, a carbohydrate, how many kilocalories would the metabolism of that glucose produce?

8. Distinguish between potential energy and kinetic energy, and between endothermic and exothermic.

9. State the laws of conservation of energy, conservation of mass, and conservation of mass and energy.

10. What are the symbols for the following elements?
    a. sodium
    b. carbon
    c. silver
    d. bromine
    e. aluminum
    f. mercury
    g. potassium
    h. chlorine
    i. helium
    j. iron
    k. copper
    l. boron

11. Which element(s) is (are) required for the following?
    a. synthesis of hemoglobin
    b. formation of bones and teeth
    c. action of insulin
    d. synthesis of thyroid hormones

12. Can all substances be changed from solid to liquid? from liquid to gas? Explain.

13. Which particular elements(s) is (are) required for each of the following?
    a. blood clotting
    b. vitamin B_{12}
    c. formation of all protein

14. Why are gases compressible? Why are solids noncompressible?

15. Which expands more when heated: a solid, a liquid, or a gas? Why?

16. Give one example of an exothermic reaction and one of an endothermic reaction.

17. Compare the properties of metals and nonmetals.

18. The energy value of foods is measured in terms of what unit?

19. What happens to the particles when a solid melts? when a liquid boils?

20. Why does the temperature of a solid remain constant while it is melting?

21. Which is larger, a calorie or a joule?

22. Look at the label in Chapter 1, question 26. If a gram of both carbohydrate and protein produce 4 Cal and a gram of fat produces 9 Cal, is the total number of Calories produced correct?

23. A patient is given 1000 mL of 5% (5 g/100 mL) glucose (a carbohydrate) solution intravenously. How many Calories did the patient receive?

24. Define the following terms: calorie; joule; Calorie; kilocalorie; specific heat; heat of fusion; heat of vaporization; sublimation; evaporation; condensation.

25. Why does the temperature remain constant during a change of state?

26. How many calories are required to change:
    a. 50 g water from 15° C to 25° C?
    b. 120 g iron from 25° C to 140° C?
    c. 75 g water at 0° C to ice at 0° C?
    d. 200 g water at 100° C to steam at 100° C?
    e. 150 g ice at −30° C to steam at 200° C?

27. Why is evaporation a cooling process? Why is an alcohol sponge bath used instead of a water sponge bath?

28. Why is it more damaging to tissue to receive a burn from steam than from hot water?

29. How are freeze-dried foods prepared? What processes are involved?

30. When iron is heated, it changes from a solid to a liquid and eventually to a vapor. Would its change-of-state curve be similar to that of water? How would it differ?

31. What is the reverse of freezing called? The reverse of boiling? The reverse of evaporation?

32. The insensible evaporation of water results from the continual diffusion of water through the skin and respiratory tract regardless of body temperature. The amount of water loss per day is about 6.0×10^2 mL. If the water being lost has a heat of vaporization of 5.40×10^2 cal/g, calculate the number of Cal consumed by insensible evaporation per day.

33. A 21-year-old woman with anorexia nervosa for 3 years is admitted to a hospital. The patient is 5 ft 5 in tall and weighs 76 lb (about 60% of her ideal weight). She is given a nasogastric feeding of a nutrient solution containing the following: fat = 42 g/L; protein = 32 g/L; carbohydrates = 125 g/L. If she is fed 3.0 L of solution each day, how many Calories will she receive?

34. Calculate the final temperature of a mixture of 150 g of water at 52° C and 150 g of ice at −12° C.

35. Calculate the amount of time necessary to lose 1 lb of fat by performing each of the following activities: running; walking; swimming; and biking.

36. In a nuclear power plant, radioactive uranium breaks apart (fissions) and releases heat. The heat converts water into steam and the steam drives a turbine that generates electricity. Identify the four forms of energy involved and determine whether a chemical or physical change occurs in each process.

Practice Test

1. An example of a chemical property is
   a. color   b. odor   c. reactivity   d. taste

2. Gases have
   a. definite shape   b. definite volume
   c. compressibility   d. all of these

3. An example of a physical property is
   a. weight   b. reactivity
   c. taste   d. height

4. An example of an element is
   a. sulfur   b. water   c. zinc oxide   d. sugar

5. The metabolism of 50 g of protein will produce how many kilocalories?
   a. 50   b. 100   c. 200   d. 450

6. An example of a mixture is
   a. sugar   b. salt–water
   c. magnesium oxide   d. powdered zinc

7. The symbol for sodium is
   a. S   b. So   c. Na   d. No

8. What is the symbol for iron?
   a. I   b. Ir   c. Fe   d. K

9. Which must be homogeneous?
   a. elements   b. compounds
   c. mixtures   d. both a and b

10. The element necessary for the formation of vitamin B_{12} is
    a. Na   b. Co   c. Mo   d. Fe

3

Structure of Matter

Objectives

- To draw the structure of an atom and to know its fundamental particles
- To distinguish among isotopes and to diagram their structures
- To diagram the electronic structure of an atom
- To understand the importance and use of the periodic table

The properties of matter arise from atomic and molecular structure. This example shows the crystalline arrangement of carbon atoms in diamond.

3-1 The Atom

The symbol of an element not only represents that element, it also represents one atom of that element. But what is an atom? We have all heard of atoms in connection with atomic bombs, "splitting the atom," and atomic power.

Consider a bar of iron. Iron is an element. It has certain properties. Cutting the bar in half produces two pieces of iron. Both pieces have the same properties as the original bar. Continued cutting procedures smaller and smaller pieces, all with identical properties. In time, we would theoretically arrive at the smallest piece of iron attainable. This smallest piece of iron is an atom—an atom of iron. If this atom of iron were cut in two, particles with different properties would be produced. It would no longer be iron. Thus, an **atom** can be defined as the smallest portion of an element that retains all of the properties of the element.

A piece of iron is made up of many atoms of iron; a piece of copper, of many atoms of copper; and a piece of silver, of many atoms of silver. The atoms of one element differ from those of another and so give characteristic properties of each element. Atoms are called the building blocks of the universe. A chemist uses different kinds of atoms to build chemical compounds, just as we all use the different letters of the alphabet to form words. Since there are more than 100 elements, there are more than 100 different kinds of atoms.

3-2 Dalton's Atomic Theory

In 1808 John Dalton, an English school teacher, proposed his Atomic Theory to account for the then-known properties of matter. His theory stated:

1. All matter is composed of tiny, indivisible particles called atoms.
2. All atoms of an element are the same, but they differ from atoms of other elements.
3. Atoms of two or more elements combine to form compounds in ratios of simple whole numbers.
4. A chemical reaction involves a rearrangement of atoms.
5. Atoms cannot be created or destroyed.

Dalton's theory has been updated as more data were discovered, but his ideas in general are still valid. This theory can still be used to explain many of the properties of matter, with the following clarifications:

1. We now know that atoms can be "broken down" into simpler substances. They are not indivisible.
2. All atoms of the same element are not identical (see Section 3-8, Isotopes).
3. When atoms of one element react with atoms of another or other elements, they do so as individual units. That is, they react in the ratio of whole numbers, as stated in the law of definite proportions (see pages 36 and 37).
4 and 5. These sections explain the law of conservation of mass. For example, when hydrogen atoms combine with oxygen atoms to form water, the atoms are rearranged but there is no change in mass and matter is neither created nor destroyed.

We know that atoms are very small, but exactly how small are they and what do they weigh?

Size

Although an atom is extremely small, its size can be accurately measured. An atom has a diameter of approximately one hundred-millionth of a centimeter (1/100,000,000 cm). Because an atom is so small, one hundred trillion of them (100,000,000,000,000) could be placed on the head of a pin.

Mass

An atom, as we have seen, is extremely small; therefore, it is not surprising that a single atom weighs very little. In fact, it would take 18 million billion billion (18,000,000,000,000,000,000,000,000) hydrogen atoms to have a mass of 1 oz.

3-3 Inside the Atom

An atom is composed of a small, heavy nucleus with particles surrounding it at relatively great distances. Thus, an atom is composed mostly of empty space, that space being between the nucleus an the surrounding particles.

   If the nucleus of an atom could be expanded so that it was about 400 ft in diameter, the closest surrounding particle would be 4000 miles away. Between the nucleus and the closest particle would be empty space—4000 miles it. Actually, more than 99.9 percent of the volume of an atom is empty space.

3-4 Fundamental Particles

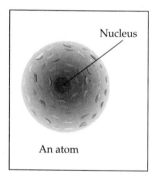

Nucleus

An atom

Atoms are made primarily of three fundamental particles: the proton, the electron, and the neutron.[†] The proton (p) has a charge of positive one (+1) and mass (weight) of approximately one atomic mass unit (amu).[‡] Protons are located inside the nucleus of the atom. The electron (e) has a negative charge (−1) and a mass (weight) of 1/1837 amu. Electrons are located outside the nucleus of the atom. The neutron (n) has no charge; it is neutral, as the name implies. It has a mass (weight) of approximately 1 amu. Neutrons are located inside the nucleus (Table 3-1). In addition to these three fundamental particles there are many, many more particles—among them the positron, the meson, and the neutrino—but a discussion of these is beyond the scope of this book.

3-5 Atomic Number

Each element has a given atomic number that represents that element and no other. The atomic number indicates the number of protons in the nucleus of an atom of that element. However, since all atoms are electrically neutral, there must be as many electrons (negative charges) as protons (positive charges). Therefore, the atomic number also tells the number of electrons in the atom, these electrons being located outside the nucleus.

[†]Theoretical physicists now believe that electrons are fundamental particles but that protons and neutrons are made of even smaller particles called quarks. However, in this discussion we shall assume that protons, neutrons, and electrons are all fundamental particles.

[‡]One atomic mass unit (amu) has a mass of 1.6605×10^{-24} g. A proton has a mass of 1.6726×10^{-24} g or 1.0073 amu. A neutron has a mass of 1.67495×10^{-24} g or 1.0087 amu. An electron has a mass of 9.1094×10^{-28} g or 5.5×10^{-4} amu. The actual masses of the fundamental particles are cumbersome to deal with, so we use the amu. For simplicity, as indicated in Table 3-1, we will assume that the proton and the neutron each have a mass of 1 amu and that the electron has a negligible mass.

Table 3-1	Fundamental Particles			
Particle	Symbol	Charge	Approximate Mass/ Weight in amu	Location in the Atom
Proton	p or p$^+$	+1	1	Inside nucleus
Electron	e or e$^-$	−1	1/1837	Outside nucleus
Neutron	n or n^0	0	1	Inside nucleus

The atomic number is written as a subscript to the left of the symbol. Thus, $_6$C indicates that the carbon atom has an atomic number of 6.

3-6 Atomic Mass

The mass of one atom cannot be detected even on the most sensitive device. However, the masses of individual atoms can be determined accurately by measuring large numbers of them. As would be expected, the mass of an individual atom is infinitesimal. But the chemist is not as interested in the exact masses of atoms as in their relative masses. The relative mass of an atom is called its atomic mass. The chemist uses atomic masses rather than exact masses. Atomic masses are easier to use than the exact masses, and they are just as accurate because they can be determined very precisely.

What does the term *relative mass* mean? The chemist has arbitrarily given the carbon-12 atom (see Section 3-9) a mass of 12.0000 amu. The masses of atoms of all other elements can then be compared with this mass. Thus, if an atom of an element has exactly twice the mass of the carbon-12 atom, that element is assigned a relative mass, or atomic mass, of 24.0000 amu. For our purposes, we will usually use the atomic masses as whole numbers, ignoring the decimal values. Thus, the atomic mass of carbon-12 is 12, that of oxygen is 16, and that of sodium is 23. For precise work, however, the exact atomic masses must be used.

The atomic masses of all the elements are listed in the periodic table inside the front cover and also in the chart inside the back cover.

3-7 Mass Number

The mass number of a nucleus is equal to the total number of protons and neutrons in that nucleus. The mass number is always a whole number. The mass number is written as a superscript to the left of the symbol. Thus, $_6^{12}$C indicates a carbon atom with an atomic number of 6 and a mass number of 12.

3-8 Number of Neutrons in the Nucleus

Because the mass of the electron is quite small (1/1837 that of the proton or the neutron) and because only the electrons are located outside the nucleus of the atom, practically all the mass of the atom is located in its nucleus. The atomic number of an element indicates the number of protons in its nucleus. The mass number indicates the total number of protons and neutrons in that nucleus. *The number of neutrons can be found by subtracting the atomic number of an element from its mass number.*

Exercise 3-1

Refer to the periodic table and obtain the following information for each element below: atomic number; atomic mass; number of protons, electrons, and neutrons.

a. helium   b. sodium

Solution

a. From the periodic table the atomic number of helium is 2 and the mass number is 4. The number of protons and electrons is equal to the atomic number, so protons $= 2$ and electrons $= 2$. Neutrons $=$ mass number $-$ atomic number, therefore, $4 - 2 = 2$.

b. From the periodic table the atomic number is 11 and the mass number is 23. Protons and electrons are equal to atomic number 11. Neutrons $=$ mass number $-$ atomic number; therefore, $23 - 11 = 12$.

Self-test

Obtain the atomic number, atomic mass, and number of protons, electrons, and neutrons for the following elements: iron, iodine, and potassium.

Answers

Element	Atomic number	Mass number	Protons	Electrons	Neutrons
Fe	26	56	26	26	30
I	53	127	53	53	74
K	19	39	19	19	20

3-9 Isotopes

The periodic chart at the front of the book indicates that the element chlorine has an atomic number of 17 and an atomic mass of 35.5 (to one decimal place). According to the discussion in the previous paragraphs, the chlorine atom should have 17 protons in its nucleus, 17 electrons outside its nucleus, and 18.5 neutrons ($35.5 - 17$) in its nucleus. However, a neutron is a fundamental particle so there can never be a fraction of a neutron. How can this problem be resolved?

The answer is that there are two types of chlorine atoms. One chlorine atom has a mass number of 35 and the other has a mass number of 37. (Chlorine with a mass number of 36 does not exist in nature.) Both of these types of atoms of chlorine have an atomic number of 17. These two varieties of chlorine are termed isotopes. Isotopes are defined as atoms of an element having the same atomic numbers but different mass numbers. The first isotope of chlorine—atomic number 17 and mass number 35—has 17 protons in its nucleus, 17 electrons outside its nucleus, and 18 neutrons ($35 - 17$) in its nucleus. The second isotope of chlorine— atomic number 17 and mass number 37—has 17 protons in its nucleus, 17 electrons outside its nucleus, and 20 neutrons ($37 - 17$) in its nucleus. The two isotopes of chlorine can be represented as follows:

$$\begin{array}{cccc} \begin{array}{c} 17\,\text{p} \\ 18\,\text{n} \end{array} & 17\,\text{e} & \begin{array}{c} 17\,\text{p} \\ 20\,\text{n} \end{array} & 17\,\text{e} \end{array}$$

$${}^{35}_{17}\text{Cl} \qquad\qquad {}^{37}_{17}\text{Cl}$$

The circle represents the nucleus of the atom, and the letter *p* the protons, *e* the electrons, and *n* the neutrons.

The atomic mass is the average mass of all the isotopes. It the two isotopes of chlorine, mass numbers 35 and 37, were present in equal amounts, the atomic (average) mass would be 36. However, since the atomic mass is listed as 35.5, the isotope of mass number 35 must be the predominant one because the atomic mass is closer to 35 than to 37.

The element carbon—atomic number 6—has three isotopes. Their mass numbers are 12, 13, and 14. They all have atomic number 6, which means that they all have six protons in their nucleus and six electrons outside their nucleus. The isotope of mass number 12 has six neutrons in its nucleus; the isotope of mass number 13 has seven neutrons in its nucleus; and the isotope of mass number 14 has eight neutrons in its nucleus. Carbon-12 is the most abundant, since the atomic mass of carbon is 12.011, including small amounts of the other isotopes.

$$\begin{array}{ccc} \overset{6\,p}{\underset{6\,n}{\bigcirc}}\ 6\,e & \overset{6\,p}{\underset{7\,n}{\bigcirc}}\ 6\,e & \overset{6\,p}{\underset{8\,n}{\bigcirc}}\ 6\,e \\ {}^{12}_{6}C & {}^{13}_{6}C & {}^{14}_{6}C \end{array}$$

three isotopes of carbon

Isotopes have been defined as atoms of an element having the same atomic number but different mass numbers; therefore isotopes of an element must have the same number of protons and electrons but different numbers of neutrons.

Most of the known elements have isotopes. Some have only two, whereas others have many more. In addition to the naturally occurring isotopes, there are many more artificially prepared isotopes. These isotopes will be discussed in the chapter on radioactivity (Chapter 13). In general, isotopes have identical chemical properties because they contain the same number of electrons as well as the same number of protons. However, isotopes have different physical properties.

The weighted average atomic mass of the naturally occurring isotopes of an element is called its atomic weight. The atomic weight of an element is usually a decimal number rather than a whole number, as indicated in the charts on the front and back covers of this book.

Exercise 3-2

Carbon occurs naturally as two isotopes. Calculate the atomic mass for carbon given the following information.

Isotope	Mass (amu)	Abundance (%)
C-12	12.0000	98.89
C-13	13.0034	1.11

Solution

The mass contribution of each isotope equals the atomic mass of an average carbon atom. To calculate the mass contribution, multiply the atomic mass by the decimal form of the percentage and add the two results.

C-12 $12.0000 \times 0.9889 = 11.87$ amu

C-13 $13.0034 \times 0.0111 = \underline{\ 0.144}$ amu

$\overline{12.01}$ amu

Self-test

Iron has four naturally occurring isotopes. Calculate the atomic mass for iron given the following data (give the answer to two decimal places).

Isotope	Mass (amu)	Abundance (%)
Fe-54	53.9396	5.92
Fe-56	55.9349	91.72
Fe-57	56.9354	2.11
Fe-58	57.9333	0.282

(Note that percentages have been rounded off; they do not add up to exactly 100%.)

Answer

55.85 amu

3-10 Structure of the Atom

What does the atom look like? The simplest one, the hydrogen atom, has the atomic number 1 and a mass number of 1. The atomic number indicates one proton inside the nucleus of this atom; it also indicates one electron outside that nucleus. There are no neutrons in the nucleus of a hydrogen atom. The hydrogen atom can be represented as follows:

1 p
0 n   1 e

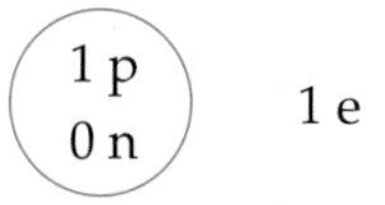

hydrogen, $_1^1H$

The element helium—$_2^4He$—has two protons in the nucleus, two electrons outside that nucleus, and two neutrons in the nucleus. The helium atom can thus be represented as follows.

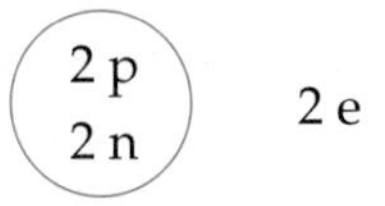

2 p
2 n   2 e

helium, $_2^4He$

The potassium atom—$_{19}^{39}K$—has 19 protons in its nucleus, 19 electrons outside its nucleus, and 20 neutrons in its nucleus.

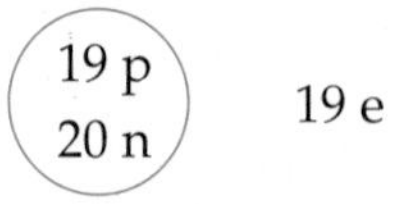

19 p
20 n   19 e

potassium, $_{19}^{39}K$

The gold atom—$_{79}^{197}Au$—has in its nucleus 79 protons and 118 neutrons. Outside the nucleus are 79 electrons.

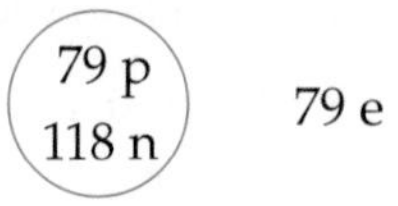

79 p
118 n   79 e

gold, $_{79}^{197}Au$

3-11 Arrangement of the Electrons in the Atom

There is a definite order to the arrangement of electrons in atoms. The electrons are located in energy levels. An energy level represents a volume occupied by an electron cloud (Figure 3-1). For our purposes, we shall use the terms electron cloud and energy level interchangeably.

The maximum number of electrons in each energy level can be calculated from the formula $2n^2$, where n is the number of the energy level counting out from the nucleus. Thus, the first energy level holds a maximum of two electrons (if $n = 1$, $2n^2 = 2$); the second energy level holds a maximum of eight electrons (if $n = 2$, $2n^2 = 8$), and the third energy level has a maximum of 18 electrons. However, as will be discussed later in this chapter, the maximum number of electrons in any outer energy level is always eight.

The first energy level must be completely filled before electrons can begin filling the second energy level; the second energy level must be completely filled before electrons can begin filling the third energy level.

The element hydrogen—$^1_1$H—has one proton and no neutrons in its nucleus and one electron outside of its nucleus. This one electron goes into the first energy level, so the hydrogen atom can be represented as follows. (The curved line indicates the first energy level.)

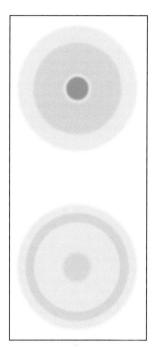

FIGURE 3-1
Electron clouds.

$$\begin{matrix} 1\,p \\ 0\,n \end{matrix} \quad 1\,e$$

$$^1_1\text{H}$$

The element helium—$^4_2$He—has two protons and two neutrons in its nucleus. The helium atom has two electrons outside its nucleus. These two electrons go into the first energy level, which can hold a maximum of two electrons, as shown.

$$\begin{matrix} 2\,p \\ 2\,n \end{matrix} \quad 2\,e$$

$$^4_2\text{He}$$

The lithium atom—$^7_3$Li—has three protons and four neutrons in its nucleus. Outside of its nucleus it has three electrons. Two of these electrons can go into the first energy level. The third electron must go into the second energy level because the first energy level can hold only two electrons.

$$\begin{matrix} 3\,p \\ 4\,n \end{matrix} \quad 2\,e \quad 1\,e$$

$$^7_3\text{Li}$$

The element sodium—$^{23}_{11}$Na—has 11 protons and 12 neutrons in its nucleus and 11 electrons outside its nucleus. The first energy level can hold two electrons and the second energy level eight electrons, so the one remaining electron must go into the third energy level. Therefore, the structure of the sodium atom is as follows.

$$\begin{matrix} 11\,p \\ 12\,n \end{matrix} \quad 2\,e \quad 8\,e \quad 1\,e$$

$$^{23}_{11}\text{Na}$$

			Electron Arrangement		
Element	Symbol	Atomic Number	First Energy Level	Second Energy Level	Third Energy Level
Hydrogen	H	1	1		
Helium	He	2	2		
Lithium	Li	3	2	1	
Beryllium	Be	4	2	2	
Boron	B	5	2	3	
Carbon	C	6	2	4	
Nitrogen	N	7	2	5	
Oxygen	O	8	2	6	
Fluorine	F	9	2	7	
Neon	Ne	10	2	8	
Sodium	Na	11	2	8	1
Magnesium	Mg	12	2	8	2
Aluminum	Al	13	2	8	3
Silicon	Si	14	2	8	4
Phosphorus	P	15	2	8	5
Sulfur	S	16	2	8	6
Chlorine	Cl	17	2	8	7
Argon	Ar	18	2	8	8

Table 3-2 Electron Arrangements for the First 18 Elements

Table 3-2 lists the progression of atomic numbers and the electron arrangement in the first 18 elements.

3-12 Energy Sublevels

An energy level is composed of sublevels, differing from one another in their spatial arrangement. The sublevels are composed of atomic orbitals, also called orbitals. The rules relating to sublevels are

1. Each energy level has one *s* sublevel containing one *s* orbital.
2. Beginning with the second energy level, each energy level contains a *p* sublevel, which consists of three *p* orbitals.
3. Beginning with the third energy level, each energy level contains a *d* sublevel, which consists of five *d* orbitals.
4. Beginning with the fourth energy level, each energy level contains an *f* sublevel, which consists of seven *f* orbitals.
5. Each orbital can contain no more than two electrons.
6. Each group of orbitals must be completely filled before electrons can begin to fill the next one.

This can be represented as follows:

Sublevels	s	p	d	f
Number of orbitals	1	3	5	7
Total number of electrons possible	2	6	10	14

If only the s sublevel is filled, there is a total of 2 electrons.
If the s and p sublevels are filled, there is a total of 8 electrons.
If the s, p, and d sublevels are filled, there is a total of 18 electrons.
If the s, p, d, and f sublevels are filled, there is a total of 32 electrons.

Recall that the maximum number of electrons in the first, second, third, and fourth energy levels is 2, 8, 18, and 32, respectively.

The orbitals are filled in a very definite sequence, as indicated in the following chart.

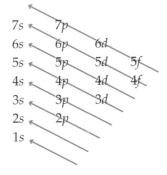

The order of filling the orbitals can be found by reading diagonally upward in the direction indicated by the arrows. Thus, the orbitals are filled in the order $1s$, $2s$, $2p$, $3s$, $3p$, $4s$, $3d$, $4p$, and so on.

Hydrogen, $_1$H, has only one electron. This electron must fill the lowest energy sublevel, the $1s$ orbital, so the hydrogen atom is indicated as $1s^1$, which means one electron in the s orbital of the first energy level.

Helium, $_2$He, has two electrons. Both of these electrons can go into the $1s$ orbital since this orbital can hold a total of two electrons. Thus, helium may be indicated as $1s^2$, meaning two electrons in the s orbital of the first energy level.

Lithium, $_3$Li, has a total of three electrons. The $1s$ orbital can hold only two electrons, or $1s^2$. The remaining electron must go into the next available orbital, the $2s$ orbital, so the electron arrangement for lithium is $1s^2 2s^1$, where the $2s^1$ indicates one electron in the s orbital of the second energy level.

For beryllium, $_4$Be, there is a total of four electrons. The first orbital, the $1s$, can hold two electrons, or $1s^2$. The next orbital, the $2s$, can also hold two electrons, or $2s^2$, so the electron arrangement for beryllium is $1s^2 2s^2$.

For boron, $_5$B, there is a total of five electrons. The first orbital, the $1s$, can hold two electrons, or $1s^2$. The second orbital, the $2s$, can also hold two electrons, or $2s^2$. The fifth electron must go to the next available orbital, the $2p$, or $2p^1$, so the electron arrangement for boron is $1s^2 2s^2 2p^1$.

Exercise 3-3

Describe the electron arrangement for the following elements.
a. Fluorine ($_9$F)   b. Potassium ($_{19}$K)   c. Cobalt ($_{27}$Co)

Solution

a. Fluorine ($_9$F) has a total of nine electrons. The first orbital can hold two electrons, or $1s^2$. The second orbital can also hold two electrons, or $2s^2$. The remaining five electrons go into the $2p$ orbitals since there are three of these $2p$ orbitals and each orbital can hold two electrons. Thus, the electron arrangement for fluorine is $1s^2 2s^2 2p^5$.

b. Potassium ($_{19}$K) has a total of 19 electrons. The $1s$ orbital can hold two electrons and the $2s$ orbital can also hold two electrons. The $2p$ orbitals can hold six electrons, making a total of 10 electrons so far. The next orbital to be filled is the $3s$, which can hold two electrons. Then comes the $3p$ orbitals, which can hold a total of six electrons, making a grand total of 18 electrons so far. The one remaining electron must go into the next orbital in order, the $4s$, so the electron arrangement for potassium is $1s^2 2s^2 2p^6 3s^2 3p^6 4s^1$.

c. Cobalt, ($_{27}$Co) has a total of 27 electrons. The $1s$ orbital can hold two electrons, $1s^2$. The $2s$ orbital can also hold two electrons, or $2s^2$, and $2p$ orbitals can hold six electrons, or $2p^6$. The $3s$ orbital can hold two electrons, $3s^2$, and the $3p$ orbitals can hold six electrons, or $3p^6$. The total number of electrons placed in orbitals so far is 18, with nine more to be placed. The next orbital in order is the $4s$, which can hold two electrons, or $4s^2$, followed by the $3d$ orbitals, which can hold up to 10 electrons. Therefore, the remaining seven electrons can go into the $3d$ orbitals, or $3d^7$, so the electron arrangement for cobalt is $1s^2 2s^2 2p^6 3s^2 3p^6 4s^2 3d^7$.

Self-test

Write electron configurations for the following elements.
a. iron   b. oxygen   c. iodine

Answers

a. iron: $1s^2 2s^2 2p^6 3s^2 3p^6 4s^2 3d^6$
b. oxygen: $1s^2 2s^2 2p^4$
c. iodine: $1s^2 2s^2 2p^6 3s^2 3p^6 4s^2 3d^{10} 4p^6 5s^2 4d^{10} 5p^5$

3-13 The Periodic Table

Toward the end of the nineteenth century, chemists noted many similarities among elements and tried to group them into certain families with similar properties and reactions. This was the beginning of the periodic table.

The modern periodic table places the elements according to the number and arrangement of the electrons in the atom. Look at the abbreviated periodic table shown in Figure 3-2 and the complete periodic table inside the front cover of this book. Note that the table is divided into horizontal rows and vertical columns. Each box in the table represents one element. There are over 100 known elements; thus, there are over 100 different kinds of atoms, each having its own place on the periodic table. The vertical columns are called (main) **groups** and are labeled with Roman numerals. The horizontal rows are called **periods** and are indicated by Arabic numerals.

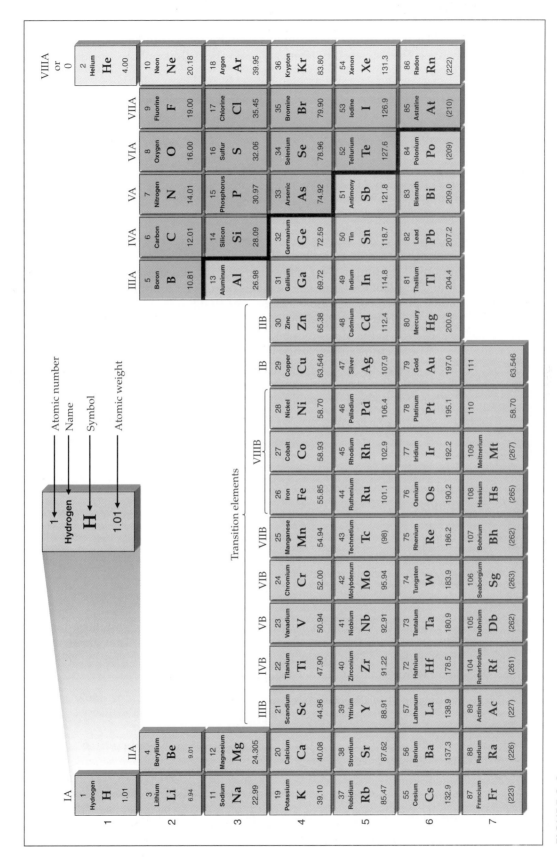

FIGURE 3-2
Abbreviated periodic table.

Each box on the chart contains a symbol with a number above and below that symbol. The name of the element is given above the symbol. Consider the box with the symbol O in it,

8
Oxygen
O
16.00

The symbol O represents the element oxygen. The atomic number of the element (8) is given above the name. Below the symbol is the atomic weight, 16.00.

Groups

Group IA elements are called **active metals.** This group consists of the elements lithium, (Li), sodium (Na), potassium (K), rubidium (Rb), cesium (Cs), and francium (Fr). All of the Group IA elements (except for hydrogen) are metals and have similar properties: they are soft, shiny metals; they are good conductors of heat and electricity; they react vigorously with water. All of the elements in group IA have one electron in the highest (outermost) energy level.

The electrons in the highest energy level are called **valence electrons.** The elements in a group have similar chemical properties because they have the same number of valence electrons. We will see that elements in the same group form similar compounds and frequently can substitute for each other.

Arrangement of Electrons in Group IA Elements

	Energy Level						
Element	*1*	*2*	*3*	*4*	*5*	*6*	*7*
Hydrogen	1						
Lithium	2	1					
Sodium	2	8	1				
Potassium	2	8	8	1			
Rubidium	2	8	18	8	1		
Cesium	2	8	18	18	8	1	
Francium	2	8	18	32	18	8	1

Group IIA elements, beryllium (Be), calcium (Ca), strontium (Sr), barium (Ba), and radium (Ra), are called **alkaline earth elements.** They are all metals similar to those in group IA but are less active. All of the elements in group IIA have two electrons in the outermost energy level.

Arrangement of Electrons in Group IIA Elements

	Energy Level						
Element	*1*	*2*	*3*	*4*	*5*	*6*	*7*
Beryllium	2	2					
Magnesium	2	8	2				
Calcium	2	8	8	2			
Strontium	2	8	18	8	2		
Barium	2	8	18	18	8	2	
Radium	2	8	18	32	18	8	2

The elements in group IIIA, boron (B), aluminum (Al), gallium (Ga), indium (In), and thallium (Tl), all have three electrons in their outermost energy level.

Group VIIA elements, fluorine (F), chlorine (Cl), bromine (Br), iodine (I), and astatine (At), are called **halogens.** These elements are nonmetals, and all have seven electrons in their outermost energy level.

Arrangements of Electrons in Group VIIA Elements

Energy Level

Element	1	2	3	4	5	6	7
Fluorine	2	7					
Chlorine	2	8	7				
Bromine	2	8	18	7			
Iodine	2	8	18	18	7		
Astatine	2	8	18	32	18	7	

Group VIIIA elements, also known as group 0 elements, are called the **noble gases.** All (except for helium) have eight electrons in their outermost energy level. These elements are also called *inert gases* because they are usually unreactive. Eight electrons in the outer energy level corresponds to filled s and p orbitals, which in turn leads to great stability. Table 3-3 compares the electron arrangements of the noble gases.

Thus, in general, the A group number indicates the number of electrons in the outermost energy level. The B group elements are called the **transition elements.** They are all metals and usually have two electrons in the outer energy level. They are located at the center and at the bottom of the periodic chart.

Exercise 3-4

Find an element with chemical properties similar to:
a. Fe   b. Mg   c. O

Solution

Any element in the same group will have similar properties.
a. Ru, Os   b. Be, Ca, Sr, Ba   c. S, Se, Te, Po

Self-test

Find an element with chemical properties similar to:
a. Cu   b. Si   c. As

Answers

a. Ag, Au   b. C, Ge, Sn Pb   c. N, P, Sb, Bi

Periods

Reading horizontally (see page 53), in period 1 there are only two elements, hydrogen and helium. Both of these elements have an electron or electrons in the first energy level only. In period 2 and period 3 there are eight elements, all of which have one or more electrons in their highest energy level, the second and third energy levels, respectively. In period 4 there are 18 elements, all having electrons in their fourth energy level.

Table 3-3 The Noble Gases		Electron Arrangement by Energy Level					
Symbol	Atomic Number	1	2	3	4	5	6
He	2	2					
Ne	10	2	8				
Ar	18	2	8	8			
Kr	36	2	8	18	8		
Xe	54	2	8	18	18	8	
Rn	86	2	8	18	32	18	8

Another generalization, then, is that the period corresponds to the number of energy levels in the atom. Thus, the element oxygen—group VIA and period 2—has six electrons in its highest energy level (from group VIA) and two energy levels (from period 2). The element vandium (symbol V, group VB) is in period 4. Vanadium has two electrons in its highest energy level (transition elements usually have two electrons in their highest energy level) and four energy levels (from period 4).

Melting and Boiling Points of Elements

Look at the melting and boiling points of some metals in group IA.

Elements in Group IA

	Melting Point (°C)	Boiling Point (°C)
Lithium	181	1342
Sodium	98	883
Potassium	63	760
Rubidium	39	686
Cesium	28	669

Note that the melting and boiling points decrease down the group. This is true in general for the metals in other groups as well.

Now consider the elements in group VIIA (the nonmetals).

Elements in Group VIIA

	Melting Point (°C)	Boiling Point (°C)
Fluorine	−220	−188
Chlorine	−101	−35
Bromine	−7	59
Iodine	114	184

Note that for nonmetals, the melting and boiling points increase down the group.

Table 3-4 provides the melting and boiling points of the elements in two periods of the periodic chart.

Note that in general, across a period the melting and boiling points of those elements that are solids at room temperature increase and that the reverse is true for those elements that are gases.

Note also that the metals have higher melting and boiling points than nonmetals do.

Table 3-4 Melting and Boiling Points of Some Elements (°C)							
Period 2	**Li**	**Be**	**B**	**C**	**N**	**O**	**F**	**Ne**
Melting point	181	1278	2300	Sublimes 3652	−210	−218	−220	−249
Boiling point	1342	2970	2550	—	−196	−183	−188	−246
Period 3	**Na**	**Mg**	**Al**	**Si**	**P**	**S**	**Cl**	**Ar**
Melting point	98	649	660	1410	44	119	−101	−189
Boiling point	883	1107	2467	2355	280	445	−35	−186

3-14 Metals, Nonmetals, Metalloids

Note the zigzag line on the periodic chart on page 53 and also inside the front cover. Elements to the left of that line and those at the bottom of the chart are metals. Those are the upper right are the nonmetals. Recall that all transition elements are metals; that is, most of the elements are metals and just a few are nonmetals.

Usually, metals are shiny, are good conductors of heat and electricity, have a relatively high density and high melting and boiling points, are ductile (can be formed into thin wires), malleable (can be pounded into thin sheets), and are solids at room temperature (except for mercury). The names of many metals end in the letters "ium," except for such metals as silver, gold, and iron.

Usually, nonmetals are not shiny and are poor conductors of heat and electricity. They have relatively low densities and low melting and boiling points and may be solids (sulfur), liquids (bromine), or gases (oxygen) at room temperature.

Along the dividing line between metals and nonmetals are some elements that exhibit characteristics of both metals and nonmetals. Such elements are called metalloids. Selenium (Se) and germanium (Ge) are examples of metalloids. Metalloids have properties between those of metals and nonmetals. They are semiconductors of electricity, which makes them useful in the manufacture of electronic equipment such as computer chips and solar batteries.

Pentium chip.

Exercise 3-5

Classify the following as a metal, metalloid, or nonmetal.
a. Sn   b. B   c. Ca   d. Br

Solution

a. Sn is a metal because it is below the dividing line for metals/nonmetals.
b. B is a metalloid because it is on the dividing line.
c. Ca is a metal because it is one the far left of the periodic table.
d. Br is a nonmetal because it is on the far right of the periodic table and it is above the dividing line.

Self-test

Classify the following as metal, metalloid, or nonmetal.
a. Cr   b. As   c. Ne   d. Cs

Answers

Cr and Cs are metals; As is a metalloid; Ne is a nonmetal

FIGURE 3-3

A scanning electron microscope in use. By its use of a highly focused electron beam in place of light, the electron microscope allows researchers to view objects at magnifications that far exceed those of light microscopes. In this photograph an immunologist photographs white blood cells caught in the process of attacking pathogens.

3-15 Electron Microscopy

The electron microscope uses the wavelike properties of electrons to see extremely small images. Because electrons are charged particles, they can be easily focused by an electrical field. Recent advances in the electron microscope have allowed scientist to see atoms for the first time. The electron microscope has become one of the most powerful tools in chemical, biologic, and material science research (see Figure 3-3).

Summary

Atoms have an extremely small size and mass. The masses of all atoms are compared with that of the carbon-12 atom, which has been assigned a mass of 12.0000 amu. These relative masses are called atomic masses.

The atom is composed of nucleus containing protons and neutrons and of electrons surrounding that nucleus. Both protons and neutrons each have a mass of 1 amu; electrons have practically no mass. Almost the entire mass of the atom is therefore in its nucleus.

The atomic number of an element indicates the number of protons inside the nucleus of an atom of that element and also the number of electrons outside that nucleus.

The number of neutrons can be found by subtracting the atomic number of an element from its mass number.

Isotopes are atoms having the same atomic number but different mass numbers. Most elements have isotopes; some have two, many have several.

The electrons are located in electron clouds or energy levels. The first energy level holds a maximum of 2 electrons, the second a maximum of 8 electrons, and the third energy level a maximum of 18 electrons.

The first electron energy level must be filled before electrons can begin filling the second. The second energy level must be filled before electrons can begin filling the third energy level.

Electron energy levels are composed of sublevels, which in turn consist of orbitals. Each orbital can hold a maximum of two electrons. The order of filling the orbitals is $1s$, $2s$, $2p$, $3s$, $3p$, $4s$, $3d$, and so on.

The noble gases each have eight electrons in their highest energy level with the exception of helium, which has only two.

The periodic table lists all of the elements in the order of their atomic numbers. The vertical columns are called groups and the horizontal rows are called periods. The A group number indicates the number of electrons in the highest energy level. The transition element are all metals and usually have two electrons in their highest energy level. The period indicates the number of energy levels.

Questions and Problems

1. Diagram the following atoms showing protons, electrons, and neutrons in each energy level and sublevel:

Symbol	Atomic Number	Mass Number
H	1	1
C	6	12
Ne	10	20
S	16	32
Li	3	7
Al	13	27

2. Diagram the structures of the following isotopes showing electrons in each energy level:

Symbol	Atomic Number	Mass Number
H	1	1,2,3
Na	11	22,23,24
N	7	13,14,15
C	6	12,13,14
Mg	12	24,25,26
O	8	16,18

3. Using the periodic table inside the front cover of this book, predict the number of electrons in the highest energy level and the number of energy levels in the following elements:
   a. barium (Ba)
   b. selenium (Se)
   c. chlorine (Cl)
   d. scandium (Sc)
   e. silicon (Si)
   f. boron (B)
   g. radium (Ra)
   h. zinc (Zn)
   i. lanthanum (La)
   j. manganese (Mn)
   k. arsenic (As)
   l. krypton (Kr)

4. What is an atom?

5. What are the vertical columns in the periodic table called? the horizontal rows?

6. In general, the elements in the periodic chart are listed in order of increasing atomic masses, but there are exceptions. Give two examples.

7. Where is the mass of the atom concentrated?

8. Name the three fundamental particles. Indicate their masses, charges, and location in the atom.

9. How do isotopes differ from one another with respect to chemical properties? physical properties?

10. Why do elements in the same group have similar chemical properties?

11. Are all transition elements metals? Explain.

12. What are noble gases? What are their properties?

13. In which part of the periodic table are metals located? nonmetals? noble gases?

14. Magnesium has two isotopes of mass numbers 24 and 25. If the isotopes were present in equal amounts, what would the atomic mass of magnesium be? If 67 percent

of the lighter isotope and 33 percent of the heavier isotope were present, what would the atomic mass be?

15. As new elements are prepared synthetically, will they fit into the existing periodic table? Explain.

16. If element number 109 is produced, will it be a transition element? Explain.

17. An atom of an element contains 14 protons and 15 neutrons.
   a. How many electrons does it have?
   b. What is its mass number?
   c. What is its atomic number?
   d. In which group is it?
   e. In which period is it?
   f. Is it a metal or nonmetal?
   g. What is its symbol?

18. An element is located in group VA, period 3. What can you tell about that element?

19. Why are atomic weights not whole numbers?

20. Are isotopes of an element identical in all respects? Explain.

21. What can the atomic mass indicate about the abundance of various isotopes in nature?

22. What is meant by the term *electron cloud*?

23. Suggest a reason why ingested radioactive radium would accumulate in the bones.

24. Compare the terms *atomic number* and *mass number*.

25. How does the group number aid in determining the electron structure of the elements?

26. Are most of the elements in nature metals or nonmetals?

27. Given the group number and period, predict the electronic structure of the following elements: IIA, 3; VA, 4; IIIA, 7.

28. How does the boiling point and melting point of a metal vary in a group? a nonmetal?

29. How does the BP and MP of an element vary in a period?

30. What are metalloids? How are they used? Where are they located on the periodic chart?

31. What are halogens, alkali metals, alkaline earth metals, transition elements?

32. What are the special properties of group VIIIA elements?

33. The next member of the oxygen family should be a metalloid. What will the atomic number of this element be?

34. Write electron configurations for As, Ba, Br, Xe.

Practice Test

1. Electrons are found:
   a. inside the nucleus
   b. outside the nucleus
   c. either inside or outside the nucleus

2. Isotopes differ in the number of _____.
   a. electrons   b. neutrons
   c. protons   d. charged particles

3. Most of the mass of the atom is located in the _____.
   a. electrons   b. protons
   c. nucleus   d. neutrons

4. The fundamental particle with a negative charge is the _____.
   a. electron   b. proton
   c. neutron   d. none of these

5. All noble gases except helium have _____ electrons in their outermost energy level.
   a. 2   b. 8   c. 18   d. 32

6. The $^{14}_{6}C$ atom contains _____ protons.
   a. 6   b. 8   c. 14   d. 20

7. Transition elements usually have _____ electrons in their highest energy level.
   a. 0   b. 2   c. 8   d. 18

8. The maximum number of electrons in the fourth energy level is _____.
   a. 8   b. 18   c. 32   d. 64

9. The atomic number of an element indicates the number of _____ in the atom.
   a. electrons and neutrons
   b. neutrons only
   c. protons and electrons
   d. protons and neutrons

10. The relative mass of an atom is called its _____.
   a. isotopic mass   b. nuclear mass
   c. atomic mass   d. all of these

Chemical Bonding

Objectives

- To become acquainted with symbols and formulas
- To write electron-dot structures
- To understand the formation of ions and of ionic bonds
- To name ionic and covalent binary compounds

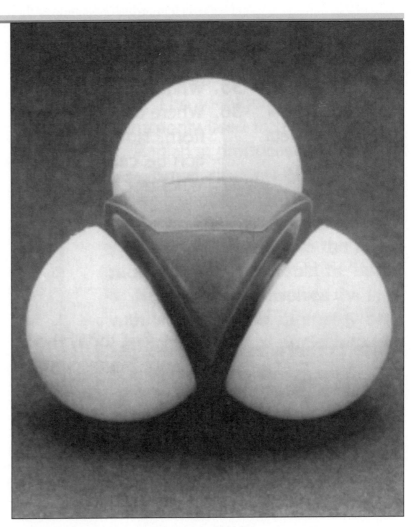

Molecules are held together tightly by chemical bonding between their constituent atoms, as exemplified by this space-filling model of an ammonia molecule. Ammonia molecules consist of three hydrogen atoms (white) bound to a central nitrogen atom.

4-1   Molecules

A molecule is a combination of two or more atoms. These atoms may be of the same elements, as in the oxygen molecule (O_2), or of different elements, as in the hydrogen chloride molecule (HCl). A more complicated molecule is that of glucose, $C_6H_{12}O_6$. What holds the atoms together in a molecule? Atoms are held together by bonds that can be classified into two main types: ionic and covalent.

4-2   Stability of the Atom

Most atoms are considered stable (nonreactive) when their highest (outer) energy level has eight electrons in it and is therefore filled. The noble gases neon, argon, krypton, xenon, and radon all have eight electrons in their highest energy level. They are stable. One exception to this rule of eight (the octet rule) is the lightest noble gas, helium, which is stable even though it has only two electrons in its highest energy level because that energy level is the first and can hold only two electrons.

Atoms that do not have eight electrons in their highest energy level may lose, gain, or share their valence electrons with other atoms in order to reach a more stable structure with lower chemical potential energy. This process of rearrangement of the valence electrons is responsible for chemical reactions between atoms.

4-3   Symbols and Formulas

A symbol not only identifies an element but also represents one atom of that element. Thus, the symbol Cu designates the element copper and also indicates one atom of copper (the number 1 being understood and not written). Two atoms of copper are designated as 2 Cu.

A formula consists of a group of symbols that represent the elements present in a substance. It also indicates one molecule of that substance. Thus the formula NaCl indicates that the compound (sodium chloride) consists of one atom of sodium (Na) and one atom of chlorine (Cl).

If there is more than one atom of an element present in a compound, numerical subscripts are used to indicate how many atoms of each element are present. In the compound HNO_3 (nitric acid) there are one atom of hydrogen (H), one atom of nitrogen (N), and three atoms of oxygen (O), all of which make up one molecule of HNO_3. In the compound $K_2Cr_2O_7$ (potassium dichromate) there are two atoms of potassium (K), two atoms of chromium (Cr), and seven atoms of oxygen making up one molecule.

To designate more than one molecule of that substance, a number (a coefficient) is placed in front of the formula for that substance. For example, 2 HNO_3 indicates two molecules of HNO_3; 6 $K_2Cr_2O_7$ indicates six molecules of $K_2Cr_2O_7$.

The formula O_2 indicates one molecule of oxygen, with the 1 being understood. This molecule consists of two atoms of oxygen. The formula H_2 indicates one molecule of hydrogen, which consists of two atoms of hydrogen. Both O_2 and H_2 are called diatomic molecules because they are each made up of two atoms. Other examples of diatomic molecules are N_2, F_2, Cl_2, Br_2, and I_2. Water (H_2O) is a triatomic molecule—it contains two atoms of hydrogen and one atom of oxygen.

Elements may be monatomic; that is, they consist of only one atom. Examples of monatomic elements are neon (Ne) and argon (Ar). Molecules of other elements such as sulfur, S_8, are polyatomic; they contain several atoms in their molecules.

Be very careful in distinguishing between 2 O and O_2. The 2 O represents two atoms of oxygen that are not combined; they are separate, independent atoms; O_2

represents one molecule of oxygen, which consists of two atoms of oxygen that are chemically combined with a covalent bond between them. (Two molecules of oxygen would shown as 2 O_2.) This note of caution applies to other diatomic molecules as well.

Exercise 4-1

Give the composition in atoms of the following compounds:
a. vitamin C, $C_6H_8O_6$ b. $KMnO_4$

Solution

Vitamin C contains six atoms of carbon, eight atoms of hydrogen, and six atoms of oxygen. $KMnO_4$ has one atom of potassium, one atom of manganese, and four atoms of oxygen.

Self-test

Give the composition in atoms for the following compounds:
a. aspirin, $C_9H_8O_4$ b. chalk, $CaCO_3$

Answers

a. Nine atoms of carbon, eight atoms of hydrogen, and four atoms of oxygen
b. One atom of calcium, one atom of carbon, and three atoms of oxygen

---

4-4 Electron-Dot Structures

The electron-dot structure of an atom (also called a Lewis structure) is an abbreviated representation for the structure of that atom. In this system, the nucleus and all of the energy levels except the highest one are represented by the symbol for that element. Each valence electron is indicated by a dot. For example, the element sodium (symbol Na, atomic number 11) has its nucleus surrounded by 11 electrons—2 in the first energy level, 8 in the second energy level, and 1 in the third (highest) energy level. The electron-dot structure for the sodium atom is Na · , with the dot representing the one valence electron and the symbol Na representing the remainder of the atom. Carbon, atomic number 6, has the electron configuration 2e) 4e). That electron-dot representation for carbon is

$$\cdot \overset{\textstyle \cdot}{\underset{\textstyle \cdot}{C}} \cdot$$

Argon, atomic number 18, has the electron configuration of 2e) 8e) 8e). The electron-dot structure for argon is

$$: \overset{\textstyle ..}{\underset{\textstyle ..}{Ar}} :$$

4-5 Formation of Ions

Figure 4-1 shows electron-dot structures of the elements in the first three periods of the periodic chart. Consider the sodium atom with the electron structure 2e) 8e) 1e). If the sodium atom loses its one outer electron, it will reach a noble gas structure of eight electrons in its outer energy level. A noble gas structure has great stability.

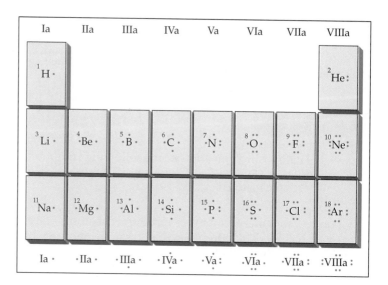

FIGURE 4-1

Electron-dot structures of the first 18 elements.

When a sodium atom loses an electron, it becomes a positively charges particle called a sodium ion. This reaction may be written as

$$Na\cdot - e^- \longrightarrow Na^+$$

or, more appropriately,

$$Na \longrightarrow Na^+ + e^-$$

where the positive sign indicates a charge of +1 on the sodium ion. (Note that the number 1 is understood and not written.) The charge on the sodium ion is positive because other sodium ion still has eleven protons in its nucleus but now has only ten electrons outside that nucleus.

Likewise, the aluminum atom, which has the electron structure 2e) 8e) 3e), loses all three outer electrons when it forms an aluminum ion with a charge of +3 (written to the upper right of the symbol as 3+).

$$\cdot Al\cdot \longrightarrow Al^{3+} + 3\,e^- \quad \text{or simply} \quad Al \longrightarrow Al^{3+} + 3\,e^-$$

A metal that has one valence electron forms an ion with a 1+ charge; a metal with two valence electrons forms an ion with 2+ charge, and so on. The positive charge on a metallic ion is equal to the number of electrons lost by the metal. Also, the positive charge on a metallic ion is equal to its A group number.

Elements that have six or seven electrons in their highest energy level tend to gain electrons to reach a stable configuration of eight. Such elements are called nonmetals. (Most elements having four or five outer electrons are also nonmetals. These will be discussed separately under covalent bonds.)

Consider the element chlorine, 2e) 8e) 7e). The chlorine atom will be tend to gain one electron to bring its highest energy level to eight, reaching a stable (noble gas) structure. Chlorine will thus form an ion with a charge of 1−, or, omitting the dots,

$$Cl + e^- \longrightarrow Cl^-$$

Since the ion has one more electron than the atom, it will have a charge of 1−, again with the 1 being understood and not written.

Likewise, the sulfur atom, 2e) 6e), can gain two electrons to form an ion with a charge of 2−.

$$S + 2e^- \longrightarrow S^{2-}$$

The S^{2-} ion has eight electrons in the highest energy level, a noble gas structure.
The negative charge on a nonmetallic ion is equal to the A group number minus 8.

Exercise 4-2

a. What is the charge on a radium ion. Radium is in group IIA.
b. What is the charge on an iodine ion? Iodine is in group VIIA.

Solution

a. For metal ions, the charge of an ion equal to its A group number, so the charge on radium ion is +2.
b. For nonmetals, the charge of an ion is equal to its A group number − 8, so the charge on an iodine ion is 7 − 8 or −1.

Self test

What are the normal charges for ions of the following elements?
a. K   b. Ca   c. S

Answers

a. K = 1   b. Ca = 2   c. S = −2

An atom that has either lost or gained electrons in its highest energy level is called an **ion.** Ions formed from a metal will have a positive charge equal to the number of electrons lost. Ions formed from nonmetals will have a negative charge equal to the number of electrons gained.

Positive ions are attracted toward a negatively charged electrode called a cathode. Such ions are called *cations.[†] Likewise, negative ions are attracted towards a positively charge electrode, an anode. These ions are called *anions. Common cations in body fluids are the sodium ion, Na^+, the potassium ion, K^+, and the calcium ion, Ca^{2+}. The chloride ion, Cl^-, is the most common anion in body fluids.

4-6 Size of Ions

When a metal loses an electron (or electrons), the positive charge on the nucleus is greater than the negative charge in the electron energy levels, so the nucleus pulls in the electrons and thus decreases the size of the ion. That is, for metals, the ionic radius is less than the atomic radius.

When a nonmetal gains an electron (or electrons), the positive charge on the nucleus is less that the negative charge in the electron energy levels; thus the nucleus cannot hold the electrons as tightly as before. Therefore, for nonmetals, the ionic radius is greater than the atomic radius.

Figure 4-2 Illustrates the relative sizes of several atoms and their corresponding ions.

---

†See the glossary for starred terms.

	IA	IIA	IIIA	IVA	VA	VIA	VIIA	0
1	H ○							He
2	Li / Li⁺	Be / Be²⁺	B	C	N	O / O²⁻	F / F⁻	Ne
3	Na / Na⁺	Mg / Mg²⁺	Al / Al³⁺	Si	P	S / S²⁻	Cl / Cl⁻	Ar

FIGURE 4-2

Atomic and ionic sizes of some elements.

4-7 Ionic Bonds

When a sodium atom (Na) combines with a chlorine atom (Cl) to form a sodium chloride molecule (NaCl), the sodium atom loses one electron to form a positively charged sodium ion (Na^+). At the same time the chlorine atom gains that one electron to form a negatively charged chloride ion (Cl^-).

The reaction is

$$Na \cdot + \cdot \ddot{\underset{..}{Cl}} : \longrightarrow Na^+ + :\ddot{\underset{..}{Cl}}:^- \qquad or \qquad Na + Cl \longrightarrow Na^+ + Cl^-$$

The positively charged sodium ion and the negatively charged chloride ion will be attracted to each other and will be held together by the electrostatic attraction of their charges (opposite charges attract each other). This type of bonding is called an **ionic bond.** An ionic bond results from the transfer of an electron or electrons from one atom to another with the formation of ions that attract one another.

Another example of a transfer of electrons from a metal to a nonmetal is in the reaction between magnesium (Mg) and two chlorine atoms,

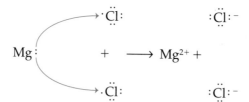

or

$$Mg + 2\,Cl \longrightarrow Mg^{2+} + 2\,Cl^-$$

in which the positively charged magnesium ion and the negatively charged chloride ions are held together by ionic bonds. (Again, each ion has a completed highest energy level of eight.)

4-8 Polyatomic Ions

A group of atoms that stay together and act as a unit in a chemical reaction is called a **polyatomic ion.** A polyatomic ion acts as if it were a simple ion. Table 4-1 indicates the name, formula, and charge of several common polyatomic ions.

4-9 Naming Ionic Compounds

Compounds that contain ions are called ionic compounds or electrolytes. As will be discussed later, ionic compounds fall into three categories: acids, bases, and salts.

Ionic compounds that contain only two types of elements are called **binary** compounds. To name binary compounds, the following system is used.

name of positive ion followed by **stem of negative ion** + *ide*

Note that the names of all binary compounds end in the letters *-ide*. The stem is the first part of the name of the element forming the negative ion. Stems for some common elements are listed in Table 4-2.

To name ionic compounds containing polyatomic ions, the following system is used.

name of positive ion followed by **name of polyatomic ion**

In NH_4Cl, the NH_4 polyatomic ion is treated as if it were a simple positive ion. Thus, the compound NH_4Cl is treated as if it were a binary compound, and it is called ammonium chloride.

Table 4-1	
Common Polyatomic Ions	
Name	**Formula and Charge**
Sulfate	SO_4^{2-}
Nitrate	NO_3^-
Phosphate	PO_4^{3-}
Carbonate	CO_3^{2-}
Hydroxide	OH^-
Bicarbonate	HCO_3^-
Ammonium	NH_4^+

Table 4-2	
Stems for Some Common Elements	
Element	**Stem**
Oxygen	ox
Chlorine	chlor
Bromine	brom
Iodine	iod
Nitrogen	nitr
Sulfur	sulf
Carbon	carb
Phosphorus	phosph

Exercise 4-3

a. What if the name of the compound $CaBr_2$?
b. What if the name of the compound KCl?
c. Name the compound $MgSO_4$.

Solution

a. The name of the positive ion, which is always written first in an ionic compound, is calcium. The negative ion comes from Br, the element bromine. The stem for bromine is *brom*; the ending is *ide*. Thus the name of $CaBr_2$ is calcium bromide.
b. The name of the positive ion is potassium. The stem of the negative ion, from the element chloride, Cl, is *chlor*; the ending is *ide*. The name of KCl is potassium chloride.
c. The positive ion is magnesium. The name of the SO_4 polyatomic ion, from Table 4-1, is *sulfate*. Thus, $MgSO_4$ is called magnesium sulfate.

Self-test

Name the following compounds:
a. $Al(OH)_3$ b. Na_2CO_3 c. NH_4NO_3

Answers

a. aluminum hydroxide   b. sodium carbonate   c. ammonium nitrate

4-10 Covalent Bonds

Ionic bonding results from the loss or gain electrons. However, there is another method by which atoms can be bonded together: by the sharing of electrons (covalent bonding).

In the chloride molecule, Cl_2, each of the two chlorine atoms has seven outer electrons. In this case both atoms will share electrons so that each will have a completed outer energy level of eight electrons. Diagram (4-1) shows two chlorine atoms with their electrons so situated that each has eight electrons around it.

$$:\ddot{Cl}\cdot\ +\ \cdot\ddot{Cl}:\ \longrightarrow\ \left(:\ddot{Cl}\ddot{:}\ddot{Cl}:\right) \tag{4-1}$$

Note that each atom has a noble gas structure, with eight outer electrons. Each of the chlorine atoms shares one electron with the other. The bond that holds these two atoms together is called a covalent bond. Note that in the chlorine molecule, Cl_2, there has been no electron loss or gain and so there are no chloride ions present. This is one of the primary differences between ionic and covalent bonds. In compounds containing ionic bonds, ions are present, whereas in compounds containing covalent bonds no ions are present.

The covalent bond between the two chlorine atoms can be indicated by a short line joining the atoms, Cl—Cl, with the electrons being understood and not written.

Covalent bonds can also be formed between atoms of different elements. In compounds containing covalent bonds, each atom usually has eight electrons around it, since eight electrons in the outer energy level represent a stable structure. An exception is hydrogen, which in compounds has only two electrons around it. (Recall that the first energy level can hold only two electrons.)

To satisfy the requirement of eight electrons around the nitrogen and only two electrons around each hydrogen the structure of NH_3 can be represented as

$$
\begin{array}{ccc}
& & H \\
\begin{array}{c} H \\ H:N:H \end{array} & \text{or} & \begin{array}{c} | \\ H\!-\!N\!-\!H \end{array}
\end{array}
\tag{4-2}
$$

In ammonia there are three covalent bonds. In carbon tetrachloride these are four covalent bonds, one between each of the chlorines and the carbon.

Since metals tend to lose electrons, they usually do not form covalent compounds. Thus, we can say that most covalent compounds are formed between nonmetals.

The compound carbon dioxide, CO_2, can be represented as

$$(O::C::O)\quad \text{or}\quad O\!=\!C\!=\!O \tag{4-3}$$

There are two double covalent bonds present in carbon dioxide. In the previous examples, single covalent bonds were present, representing one shared pair of electrons. Single covalent bonds are also called single bonds. A double covalent bond (also called a double bond) represents two shared pairs of electrons. A triple covalent bond (triple bond) represents three shared pairs of electrons, as in the compound nitrogen, N_2.

$$(:N:::N:)\quad \text{or}\quad N\!\equiv\!N \tag{4-4}$$

Note again that in CO_2 and in N_2 there are eight electrons around each atom.

Exercise 4-4

Exercise 4-4a

Draw the electron-dot structure for carbon tetrachloride, CCl_4, where carbon is the central atom.

Solution

Carbon, atomic number 6, has four electrons in its valence shell and may be represented as

$$\cdot \overset{\displaystyle \cdot}{\underset{\displaystyle \cdot}{C}} \cdot$$

Chlorine, atomic number 17, has seven electrons in its valence shell and may be represented as

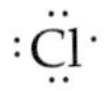

$$:\overset{\displaystyle \cdot\cdot}{Cl}\cdot$$

To satisfy the requirements of eight electrons around each atom, the following structure may be drawn:

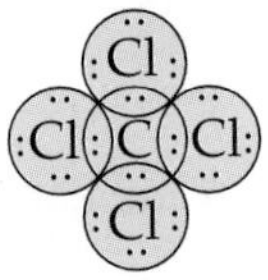

This structure may also be drawn as

$$\begin{array}{c} Cl \\ | \\ Cl-C-Cl \\ | \\ Cl \end{array}$$

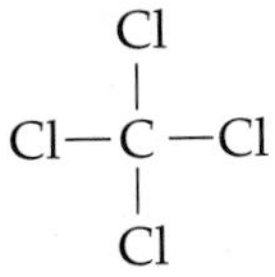

where a dash indicates a pair of shared electrons.

Exercise 4-4b

Draw the electron-dot structure for ammonia, NH_3, where the nitrogen is the central atom.

Solution

Nitrogen, atomic number 7, has five electrons in its valence shell and may be represented as

$$\cdot \overset{\displaystyle \cdot}{\underset{\displaystyle \cdot\cdot}{N}} \cdot$$

To satisfy the requirement of eight electrons around the nitrogen atom and two electrons around each hydrogen atom, the following structure may be drawn:

This structure may also be drawn as

$$\begin{array}{c} H \\ | \\ H-N-H \end{array}$$

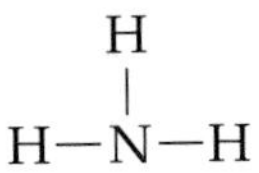

Exercise 4-4c

Draw the electron-dot structure for carbon dioxide, CO_2.

Solution

Carbon has four valence electrons and can be represented as

$$\cdot \ddot{C} \cdot$$

Oxygen has six valence electrons and can be represented as

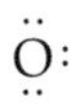

$$\ddot{\underset{\cdot\cdot}{O}} :$$

Oxygen needs two more electrons and carbon needs four more electrons. Therefore, carbon shares two electrons with oxygen and each oxygen shares four electrons with carbon.

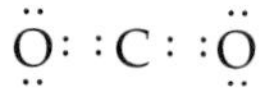

$$\ddot{\underset{\cdot\cdot}{O}} : \, : C : \, : \ddot{\underset{\cdot\cdot}{O}}$$

Exercise 4-4d

Draw the electron-dot structure for molecular nitrogen, N_2.

Solution

Nitrogen has five valence electrons and can be represented as

$$\cdot \, \ddot{N} \, \cdot$$

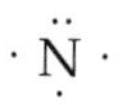

Each nitrogen needs three more electrons; therefore each nitrogen shares three electrons with the other nitrogen.

$$: N \vdots\vdots N :$$

Exercise 4-4e

Draw the electron-dot structure for the sulfate ion (SO_4^{2-}).

Solution

Structures for ions are similar to compounds except we must add or subtract electrons to account for charge. Sulfur has six valence electrons and can be written as

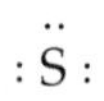

$$: \ddot{S} :$$

Oxygen has six valence electrons and can be written as

$$: \ddot{O} :$$

The ionic charge of -2 means that there are two extra electrons that can be placed around the sulfur.

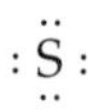

$$: \ddot{S} :$$

Sulfur now has eight electrons and needs no more electrons. Each oxygen needs two more electrons and sulfur provides them as shown.

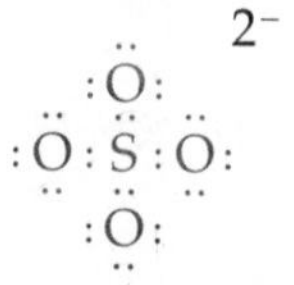

$$\begin{array}{c} 2^- \\ : \ddot{O} : \\ : \ddot{O} : \ddot{S} : \ddot{O} : \\ : \ddot{O} : \end{array}$$

Self-test

Draw the electron-dot structure for each of the following:

a. H_2S b. CH_2O c. PCl_3

d. NH_4^+ e. PO_4^{3-}

Answers

a.
$$H:\overset{..}{\underset{..}{S}}:H$$

b.
$$H:C::\overset{..}{\underset{..}{O}}:$$
$$H$$

c.
$$:\overset{}{\underset{}{Cl}}:\overset{..}{P}:\overset{}{\underset{}{Cl}}:$$
$$:\overset{}{\underset{..}{Cl}}:$$

d.
$$\overset{+}{}$$
$$H$$
$$H:\overset{..}{N}:H$$
$$H$$

e.
$$\left[\ :\overset{..}{\underset{}{O}}: \atop :\overset{..}{\underset{..}{O}}:\overset{}{P}:\overset{..}{\underset{..}{O}}: \atop :\overset{..}{\underset{..}{O}}:\ \right]^{3-}$$

4-11 Nonpolar and Polar Covalent Bonds

Consider the compound Cl_2 or Cl—Cl. There is a single bond representing a pair of shared electrons between the two chlorine atoms. The two chlorines are identical, and so the electrons should be shared equally between them. Such a type of bond is called a **nonpolar covalent bond**.

In the compound HCl, or H—Cl, the single bond again represents a shared pair of electrons. However, this pair of electrons is not shared equally. Let us see why.

Electronegativity is the attraction of an atom for electrons. The greater the electronegativity, the greater the attraction for electrons; the lower the electronegativity the less the attraction for electrons. Table 4-3 indicates the electronegativities of some elements.

Again consider the HCl molecule. Chlorine is more electronegative than hydrogen and so attracts the shared electrons more strongly. Thus, the shared pair of electrons will be closer to the chlorine than to the hydrogen, or

$$H:\overset{..}{\underset{..}{Cl}}: \qquad \text{which can also be represented as} \qquad \overset{\delta+}{H}\ \overset{\delta-}{Cl}$$

where the δ sign (Greek delta) indicates a partial charge. So, one end of the HCl molecule has a partial positive charge and the other end has a partial negative charge. Note, however, that there are no ions formed, only partial charges. Such a type of bond is called a **polar covalent bond**. Thus, with equally shared electrons, as between like atoms or between atoms of equal electronegativity, nonpolar covalent bonds are formed. When atoms of different electronegativity form a covalent bond, the bond is always polar.

Another example of a compound containing a polar covalent bond is water, H_2O. Its structure is

$$H:\overset{..}{\underset{..}{O}}: \qquad \text{or} \qquad \overset{\delta+}{H}-\overset{\delta-}{O}$$
$$H \qquad\qquad\qquad\qquad\qquad H^{\delta+}$$

Because oxygen is more electronegative that hydrogen, it becomes the negative side of the molecule. The hydrogens are at the positive side. The polar nature of the water molecule will be discussed in more detail in Chapter 9.

Table 4-3	
Electronegativities of Various Elements	
F	4.0
O	3.5
N, Cl	3.1
Br	2.8
C, S	2.5
I	2.4
H	2.1
Ca	1.0
Na	0.9

However, a nonpolar molecule may contain polar bonds. Consider the compound CCl_4 whose partial charges are indicated in structure (4-5).

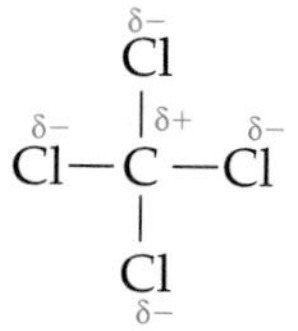

$$
\overset{\delta-}{\underset{\underset{\delta-}{Cl}}{\overset{Cl}{\underset{|}{\overset{\delta-}{Cl}-\overset{\delta+}{C}-\overset{\delta-}{Cl}}}}}
$$

(4-5)

The bonds between the carbon and each chlorine are polar. But there is no negative end to the molecule and no positive end. Rather, the outer part is negative and the inner part positive (partially), so the molecule is nonpolar. In general, symmetrical molecules are nonpolar even though they may contain polar bonds.

4-12 Resonance

Sometimes more than one electron-dot structure can be drawn for the same group of atoms. Consider the compound SO_3. Three different electron-dot structures can be drawn, all of which satisfy the requirements of eight electrons around each atom.

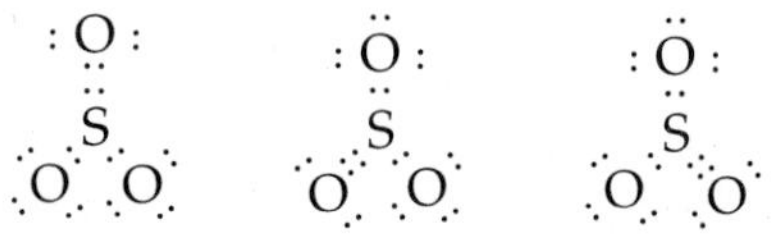

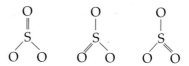

or

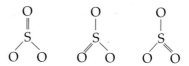

According to these structures there are two different types of bonds between the sulfur and the oxygen atoms—single covalent bonds and a double covalent bond. However, experimental evidence shows that all the bonds in the SO_3 molecule are the same.

The above three electron-dot structures are said to be in **resonance** and are so indicated by a double-headed arrows, as follows:

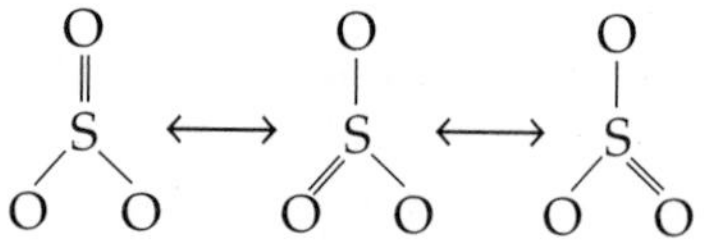

Resonance occurs when more than one electron-dot structure can be drawn for a given molecule or ion. The resulting structures are called **resonance structures**. Resonance occurs because electrons are not fixed objects near a given atom. Instead they move around the entire molecule. Therefore the correct structure lies somewhere between the various resonance structures. For most purposes in inorganic chemistry, the electron-dot structures are satisfactory, but for organic compounds based on benzene (see Chapter 16), resonance structures are important and must always be considered.

4-13 Naming Covalent Compounds

Covalent binary compounds have names ending in -*ide*, as do ionic binary compounds. To name covalent binary compounds, the following system is used.

prefix + name of first element followed by
prefix + stem of second element + *ide*

Note that this system is similar to that for ionic compounds except that prefixes are used for covalent compounds, whereas no prefixes are used in naming ionic compounds. The prefixes in common use are

mono-, 1   di-, 2   tri-, 3   tetra-, 4   penta-, 5

However, the prefix mono- is usually understood and not written.

Exercise 4-5

Exercise 4-5a

Name the covalent compound CCl_4.

Solution

The name of the first element is carbon. There is only one carbon atom indicated in the formula, so the prefix is mono- (understood and not written). The second element is chlorine, stem *chlor* (see Table 4-2). Since there are four chlorines indicated in the formula, the prefix is *tetra-*. The ending as with all binary compounds is -*ide*; so the name is carbon tetrachloride.

Exercise 4-5b

Name the covalent compound P_2O_3.

Solution

By following the previous example and using the prefixes *di-* and *tri-*, respectively, the name is found to be *diphosphorus trioxide*.

Self-test

Name the following compounds
a. SF_6 b. N_2O_4 c. ClO_2

Answers

a. sulfur hexafluoride   b. dinitrogen tetroxide   c. chlorine dioxide

4-14 Strengths of Bonds

Consider Table 4-4 of the melting points of several ionic and covalent compounds. In general, compounds containing ionic bonds have higher melting points than compounds containing covalent bonds. That is, it takes more energy (heat) to separate the particles in ionic compounds than it takes to separate those in covalent compounds. Although the bonds holding ionic compounds together are generally weaker than those holding covalent compounds together, ionic compounds contain many more bonds than covalent compounds. However, covalent substances,

Table 4-4

Melting Points of
Various Substances

	Melting Point (°C)
Ionic compound	
Sodium chloride, NaCl	800
Calcium chloride, $CaCl_2$	782
Zinc oxide, ZnO	1975
Covalent compound	
Glucose, $C_6H_{12}O_6$	146
Carbon tetrachloride, CCl_4	−23
Urea, NH_2CONH_2	133

such as diamond, that contain a network of covalent bonds also have extremely high melting points.

Water solutions of ionic compounds conduct electricity (they are electrolytes) because they contain ions. Water solutions of covalent compounds do not contain ions and do not conduct electricity.

Most ionic compounds are soluble in polar solvents such as water. Most covalent compounds are insoluble in polar solvents.

Most ionic compounds are insoluble in nonpolar solvents such as benzene. Many covalent compounds are soluble in nonpolar solvents.

Molten ionic compounds conduct electricity because they contain ions. Molten covalent compounds contain no ions and do not conduct electricity.

4-15 Shapes of Molecules: The VSEPR Theory

Molecules have three-dimensional shapes even though they are commonly represented on paper as two-dimensional.

The valence shell electron-pair repulsion (VSEPR) theory is based on the idea that electron pairs in the valence shell of an atom repel one another. A molecule (or ion) will have the lowest potential energy (the greatest stability) when the electron pairs are in a geometric position that minimizes their repulsions. That is, the electrons try to get as far from each other as possible. The relative locations of electrons and of atoms accounts for the molecular geometry of the molecules (or ions). The VSEPR theory is useful in predicting the approximate shape of molecules (or ions) formed from nonmetals.

When using the VSEPR theory, the following rules apply:

1. Draw the electron-dot structure of the compound.
2. Count the number of atoms bonded to the central atom.
3. Count the number of nonbonded electron pairs on the central atom.
4. Add the number of atoms bonded to the central atom to the number of nonbonded electron pairs. This total will indicate the shape, as shown in Table 4-5.

Table 4-5 Bonding and Shape of Molecules and Ions

Number of Atoms Bonded to Central Atom	Number of Pairs of Nonbonded Electrons on Central Atom	Total	Shape of Molecule or Ion
2	0	2	Linear 
2	2	4	Bent 
3	0	3	Trigonal 
3	1	4	Pyramidal 
4	0	4	Tetrahedral 
4	2	6	Square planar
5	0	5	Trigonal bipyramidal 
6	0	6	Octahedral 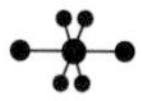

Exercise 4-6

Exercise 4-6a

Determine the shape of CH_4.

Solution

- **Step 1** The electron-dot structure is

$$
\begin{array}{c}
H \\
\ddot{} \\
H : C : H \\
\ddot{} \\
H
\end{array}
$$

- **Step 2** The number of atoms bonded to the central atom is 4.
- **Step 3** The number of nonbonded electron pairs on the central atom is 0.
- **Step 4** The total of Steps 2 and 3 is 4.

Referring to Table 4-5, the predicted shape should be tetrahedral.

Exercise 4-6b

Determine the structure of NH_3.

Solution

- **Step 1** The electron-dot structure is

$$
\begin{array}{c}
H \\
\ddot{} \\
H : N : H \\
\ddot{}
\end{array}
$$

- **Step 2** The number of atoms bonded to the central atom is 3.
- **Step 3** The number of nonbonded electron pairs on the central atom is 1.
- **Step 4** The total of Steps 2 and 3 is 4 (3 + 1), which corresponds to a predicted pyramidal shape, or

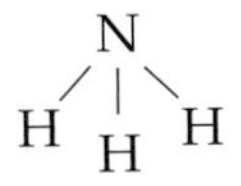

$$
\begin{array}{c}
N \\
\diagup \mid \diagdown \\
H \quad H \quad H
\end{array}
$$

Exercise 4-6c

Determine the shape of H_2O

Solution

- **Step 1** The electron-dot structure is

$$
H : \ddot{O} : H
$$

- **Step 2** The number of atoms bonded to the central atom is 2.
- **Step 3** The number of nonbonded electron pairs on the central atom is 2.
- **Step 4** The total of Steps 2 and 3 is 4 (2 + 2), which corresponds to a predicted shape of bent, or

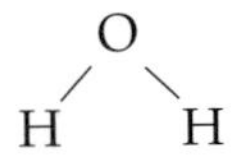

$$
\begin{array}{c}
O \\
\diagup \quad \diagdown \\
H \qquad H
\end{array}
$$

Exercise 4-6d

Determine the shape of CO_2.

Solution

■ **Step 1** The electron-dot structure is

$$O::C::O$$

■ **Step 2** The number of atoms bonded to the central atom is 2.

■ **Step 3** The number of nonbonded electron pairs on the central atom is 0.

■ **Step 4** The total of step 2 and 3 is 2, which predicts a linear shape, or

$$O=C=O$$

Self-test

Predict the shape of the following:

a. H_2S b. HCN   c. SO_3
d. $SiCl_4$ e. SF_6 f. NO_3^-

Answers

a. bent   b. linear   c. trigonal
d. tetrahedral   e. octahedral   f. trigonal

Summary

Molecules are combinations of two or more atoms. Atoms are held together in molecules by ionic or covalent bonds.

A symbol for an element not only identifies that element but also represents one atom of that element. A formula consists of a group of symbols that represent the elements present in a substance.

The number of outer electrons determines the chemical properties of the atom. Atoms are most stable when they have eight electrons in their highest (outer) energy level.

The electron-dot structure of an element uses the symbol of that element to represent the nucleus and all of the electrons except those in the highest energy level. Each electron in that highest energy level is represented by a dot placed near the symbol.

Metals have one, two, or three electrons in their highest energy level and tend to lose all those electrons to form positively charged ions. Nonmetals with six or seven electrons in their highest energy level tend to gain electrons to bring that energy level to eight, thereby forming ions with a negative charge. Positively charged ions are called cations; negatively charged ions are called anions.

The ionic radius of a metal is less than that of the corresponding atom; the ionic radius of a nonmetal is greater than that of the corresponding atom.

When a metal loses an electron to form a positively charged ion and a nonmetal gains that electron to form a negatively charged ion, these ions are held together by the attraction of their charges. This type of bonding is called ionic bonding.

A polyatomic ion is a group of atoms that acts as a unit in a chemical reaction.

Ionic binary compounds are named by writing the name of the positive ion and then the stem of the negative ion with the ending *-ide*. Ionic compounds containing polyatomic ions are named by writing the name of the positive ion and then the name of the polyatomic ion.

Nonmetals may also share electrons to complete their highest energy level with eight. Such a bond is called a covalent bond. In a covalent bond no ions are formed. When a covalent bond is formed, each element has eight electrons around it, except for hydrogen, which has only two.

Covalent compounds in which a pair of electrons is shared either between two identical atoms or between two atoms of equal electronegativity contain nonpolar covalent bonds. Covalent compounds containing bonds between atoms of different electronegativity contain polar covalent bonds.

Covalent compounds are named by writing a prefix and the name of the first element and then writing a prefix, the stem of the second element, and the ending *-ide*.

The VSEPR (valence shell electron pair repulsion) theory aids in the prediction of the shape of a molecule or ion, depending on the number of atoms bonded to a central atom and the number of pairs of nonbonded electrons on the central atom.

Questions and Problems

1. What is an atom? a molecule?

2. Give an example of a monatomic molecule, a diatomic molecule, and a polyatomic molecule.

3. What are valence electrons? What effect do they have on the properties of an atom?

4. Give the electron-dot structures for the following elements (use the periodic table).
   - a. sulfur
   - b. nitrogen
   - c. carbon
   - d. sodium
   - e. helium
   - f. oxygen

5. Compare the size of a metal atom and its ion. Do the same for a nonmetal. Explain the difference.

6. What electron configuration does an element usually need to reach maximum stability? What are the exceptions to this rule?

7. What is an anion? a cation? Give two examples of each.

8. What is an ion? What type of elements form positively charged ions? negatively charged ions?

9. What determines the charge of an ion?

10. Compare ionic and covalent bonding.

11. Name to following ionic compounds:
    - a. NaBr
    - b. $MgSO_4$
    - c. CaS
    - d. KNO_3
    - e. K_2S
    - f. AlI_3

12. Draw the electron-dot structures for the following covalent compounds:
    - a. H_2S
    - b. H_2
    - c. NH_3
    - d. CH_3Cl
    - e. O_2
    - f. HBr

13. What is a polyatomic ion? Give two examples.

14. Give an example of a single covalent bond, a double covalent bond, and a triple covalent bond.

15. Name the following covalent compounds:
    - a. N_2O_5
    - b. IBr
    - c. CS_2
    - d. PBr_3
    - e. SO_3
    - f. As_2O_3

16. What is a polar covalent bond? a nonpolar covalent bond? Give an example of each.

17. What do the following symbols or formulas indicate? O; O_2; O_3; CO; CO_2; S_8

18. Do water solutions of ionic compounds conduct electricity? of covalent compounds? Explain the difference.

19. What is the most electronegative element?

20. Can a nonpolar molecule contain polar bonds? Explain.

21. Why is the charge of magnesium ion +2 rather than some other number?

22. Is the first element in a binary compound always a metal? What effect will this have on the naming of the binary compound?

23. Is a covalent bond always polar? nonpolar? How can you tell?

24. Where can resonance occur?

25. Predict the shape of the following compounds or ions.
    - a. BCl_3
    - b. PBr_5
    - c. CO_3^{2-}
    - d. SCl_2
    - e. $BeCl_2$
    - f. $AsBr_5$
    - g. SO_3^{2-}
    - h. PO_3^{3-}
    - i. CI_4
    - j. AsF_6^-

26. Name the following. Draw their electron-dot structure and predict their shape.
    - a. BF_3
    - b. CH_4
    - c. PF_3
    - d. OF_2
    - e. XeO_4

Practice Test

1. The electron-dot structure for magnesium is _____ .
   a. Mg   b. Mg :   c. Mg^{2+} d. Mg ·

2. The formula for calcium iodide is _____ .
   a. CaI   b. CaI_2 c. CaIO   d. Ca_2I

3. In a double covalent bond, how many electrons are shared?
   a. 2   b. 4   c. 6   d. 8

4. The name of NCl_3 is _____ .
   a. nitrogen chloride   b. nitrogen trichloride
   c. trinitrogen chloride   d. nitrogen trichlorine

5. Nonmetals sometimes gain electrons to form _____ .
   a. positively charged ions
   b. negatively charged ions
   c. ions that are smaller than the original atom
   d. ions that are polyatomic

6. In general, atoms reach greatest stability when they have _____ electrons in their highest energy level.
   a. 6   b. 8   c. 2   d. 4

7. The symbol ⟷ indicates
   a. resonance   b. equilibrium
   c. a reversible reaction   d. none of these

8. An example of a polar compound is _____ .
   a. H_2 b. CO_2 c. CCl_4 d. H_2O

9. The VSEPR theory predicts the shape of a molecule or ion, depending on
   a. the number of atoms bonded to a central atom
   b. the number of nonbonded electron pairs on the central atom
   c. the sum of a and b
   d. the difference betwee a and b

10. Which of the following is an anion?
    a. Cl^- b. NH_4^+ c. Mg^{2+} d. O_2

Chemical Formulas and Reactions

Objectives

- To balance chemical equations
- To interpret chemical reactions
- To understand equilibrium reactions and the factors that can affect them

Chemical reactions—the recombining of molecules into new compounds—account for many of the physical changes occurring in nature. These processes can be isolated and studied in the laboratory.

5-1 Molecular Mass

The molecular mass or formula mass of any compound is the sum of the atomic masses[†] of all the atoms present in one molecule of that compound. The molecular mass of sodium bromide, NaBr, is 103, which represents the sum of the atomic mass of sodium (23) plus that of bromine (80). (See Table of Atomic Masses, inside back cover.)

Calculating the Molecular Mass (Formula Mass) of a Compound

To find the molecular mass of a compound, add the atomic masses of all of the atoms that are present in that compound. In the compound H_2O, the molecular mass can be calculated by adding the mass of two atoms of hydrogen and one atom of oxygen.

$$
\begin{array}{r}
\text{2 hydrogen atoms (atomic mass 1) } 2 \times 1 = 2 \\
\text{1 oxygen atom (atomic mass 16) } 1 \times 16 = \underline{16} \\
\text{molecular mass} = 18
\end{array}
$$

The molecular mass of glucose, $C_6H_{12}O_6$, can be calculated as follows:

$$
\begin{array}{r}
\text{6 carbon atoms (atomic mass 12) } 6 \times 12 = 72 \\
\text{12 hydrogen atoms (atomic mass 1) } 12 \times 1 = 12 \\
\text{6 oxygen atoms (atomic mass 16) } 6 \times 16 = \underline{96} \\
\text{molecular mass} = 180
\end{array}
$$

The molecular mass of calcium phosphate, $Ca_3(PO_4)_2$, can be calculated as follows:

$$
\begin{array}{r}
\text{3 calcium atoms (atomic mass 40) } 3 \times 40 = 120 \\
\text{2 phosphorus atoms (atomic mass 31) } 2 \times 31 = 62 \\
\text{8 oxygen atoms (atomic mass 16) } 8 \times 16 = \underline{128} \\
\text{molecular mass} = 310
\end{array}
$$

Exercise 5-1

Calculate the molecular mass of ascorbic acid, $C_6H_8O_6$.

Solution

The molecular mass is calculated as follows

$$
\begin{array}{r}
\text{6 carbon atoms (atomic mass 12) } 6 \times 12 = 72 \\
\text{8 hydrogen atoms (atomic mass 1) } 8 \times 1 = 8 \\
\text{6 oxygen atoms (atomic mass 16) } 6 \times 16 = \underline{96} \\
\text{molecular mass} = 176
\end{array}
$$

Self-test

Calculate the molecular mass of sodium carbonate, Na_2CO_3.

Answer

106

[†]Atomic masses are usually rounded off to the closest whole number.

Percentage Composition

The percentage composition of a compound can be calculated from the relative atomic masses of the elements present in that compound. Consider the compound $Ca_3(PO_4)_2$, calcium phosphate, whose molecular mass was found to be 310. Of this mass, 120 is calcium, 62 phosphorus, and 128 oxygen. Then

$$\% \text{ Ca} = \frac{\text{mass of calcium in compound}}{\text{mass of compound}} \times 100 = \frac{120}{310} \times 100 = 38.7\%$$

$$\% \text{ P} = \frac{\text{mass of phosphorus in compound}}{\text{mass of compound}} \times 100 = \frac{62}{310} \times 100 = 20.0\%$$

$$\% \text{ O} = \frac{\text{mass of oxygen in compound}}{\text{mass of compound}} \times 100 = \frac{128}{310} \times 100 = 41.3\%$$

Exercise 5-2

Calculate the percentage composition of Na_2CO_3.

Solution

$$\% \text{ Na} = \frac{\text{mass of sodium in } Na_2CO_3}{\text{mass of } Na_2CO_3} \times 100 = \frac{46}{106} = 43\%$$

$$\% \text{ C} = \frac{\text{mass of carbon in } Na_2CO_3}{\text{mass of } Na_2CO_3} \times 100 = \frac{12}{106} = 11\%$$

$$\% \text{O} = \frac{\text{mass of oxygen in } Na_2CO_3}{\text{mass of } Na_2CO_3} \times 100 = \frac{48}{106} = 45\%$$

The percentages do not total 100% because of rounding off.

Self-test

Calculate the percentage composition of ascorbic acid, $C_6H_8O_6$.

Answer

41 percent C; 4.5 percent H; 55 percent O

The Mole

We measure distances on Earth in terms of miles. However, this unit is far too small for distances to the stars. We need a much larger unit, the light-year. So, too, chemists need a larger unit to weigh molecules, since individual molecules are too small to measure even with the most sensitive equipment. Such a unit, the **mole** (abbreviated mol), is defined as the number of atoms in 12.000 g of ^{12}C. One mole[†] of anything—molecules, atoms, ions, electrons—always contains the same number of particles, 6.02×10^{23} (602,000,000,000,000,000,000,000). This large number is called **Avogadro's number.**

---

†Note that mole is a "counting word" like dozen.

One mole of any substance has a mass, in grams, equal to its molecular (or atomic) mass.

1 mol of water, H_2O,   contains 6.02×10^{23} molecules and has a mass of 18 g
1 mol of carbon, C,   contains 6.02×10^{23} atoms   and has a mass of 12 g
1 mol of glucose, $C_6H_{12}O_6$, contains 6.02×10^{23} molecules and has a mass of 180 g

Note that H_2O and $C_6H_{12}O_6$ are molecules, so we deal with Avogadro's number of molecules. Carbon, C, represents an atom, so we deal with Avogadro's number of atoms.

Exercise 5-3

Exercise 5-3a

What is the mass of 1 mol of NH_3, ammonia? How many molecules are present in 1 mol of ammonia?

Solution

The molecular mass of ammonia is 17 (N = 14 and each H = 1). So 1 mol of NH_3 has a mass of 17 g. By definition, 1 mol of any substance contains Avogadro's number of particles, so 1 mol of NH_3 contains 6.02×10^{23} molecules.

Exercise 5-3b

What is the mass of 2 mol of NH_3?

Solution

If 1 mol of NH_3 has a mass of 17 g, 2 mol will have a mass of 34 g. We can write this mathematically as follows, using the conversion factor 1 mol NH_3 = 17 g:

$$2 \text{ mol NH}_3 \times \frac{17 \text{ g NH}_3}{1 \text{ mol NH}_3} = 34 \text{ g NH}_3$$

Exercise 5-3c

A container holds 45 g of sugar, $C_6H_{12}O_6$ (molecular mass 180). How many moles of sugar are present?

Solution

Since the molecular mass of sugar is 180, 1 mol of sugar has a mass of 180 g. Using this latter figure as a conversion factor,

$$45 \text{ g sugar} \times \frac{1 \text{ mol sugar}}{180 \text{ g sugar}} = 0.25 \text{ mol sugar}$$

Exercise 5-3d

How many molecules are present in 27.0 g of water, H_2O (molecular mass 18)?

Solution

We know that 1 mol of any substance contains Avogadro's number of particles, 6.02×10^{23}. So first we have to convert grams of water to moles of water (1 mol of water has a mass of 18.0 g) and then moles of water to molecules of water (1 mol of water contains 6.02×10^{23} molecules).

$$27.0 \, \text{g H}_2\text{O} \times \frac{1 \, \text{mol H}_2\text{O}}{18.0 \, \text{g H}_2\text{O}} = 1.50 \, \text{mol H}_2\text{O}$$

$$1.50 \, \text{mol H}_2\text{O} \times \frac{6.02 \times 10^{23} \, \text{molecules H}_2\text{O}}{1 \, \text{mol H}_2\text{O}} = 9.03 \times 10^{23} \, \text{molecules H}_2\text{O}$$

Self-test

Calculate the number of moles and molecules in
a. 10.0 g of water   b. 10.0 g of carbon dioxide

Answers

a. 0.556 mol, 3.34×10^{23} molecules   b. 0.227 mol, 1.37×10^{23} molecules

5-2 Empirical and Molecular Formulas

An **empirical (simplest) formula** represents the relative number of each type of atom present in each molecule of a given compound.

A **molecular formula** represents the actual number of atoms present in each molecule of a given compound.

The empirical formula for both acetylene and benzene is CH, indicating one atom of carbon for each atom of hydrogen in both compounds. The molecular formula for acetylene is C_2H_2, and that of benzene is C_6H_6. Note that in both of these compounds the ratio of carbons to hydrogens is 1:1.

The empirical formula does not always represent the actual number of atoms present and so cannot represent the molecular mass. The molecular formula is always a simple integral multiple (1, 2, 3, etc.) of the empirical formula.

Exercise 5-4

Exercise 5-4a

A compound contains 11.2 percent hydrogen and 88.8 percent oxygen. What is its empirical formula? If its molecular mass is 18, what is its molecular formula?

Solution

- **Step 1** *Assume that 100 g of the compound is present.* If there is 100 g of compound present, there will be 11.2 g of hydrogen (11.2 percent of 100 g) and also 88.8 g of oxygen (88.8 percent of 100 g).
- **Step 2** *Convert grams to moles.*

$$\text{For hydrogen: } 11.2 \, \text{g} \times \frac{1 \, \text{mol}}{1.0 \, \text{g}} = 11.2 \, \text{mol}$$

$$\text{For oxygen: } \quad 88 \, \text{g} \times \frac{1 \, \text{mol}}{16 \, \text{g}} = 5.6 \, \text{mol}$$

■ **Step 3** *Find the empirical formula by dividing by the smaller number of moles.*

$$\text{For hydrogen: } \frac{11.2 \text{ mol}}{5.6 \text{ mol}} = 2$$

$$\text{For oxygen: } \frac{5.6 \text{ mol}}{5.6 \text{ mol}} = 1$$

The ratio of 2 mol of hydrogen to 1 mol of oxygen gives the empirical formula of H_2O.

■ **Step 4** *Find the molecular formula by dividing the mass of the empirical formula into the molecular mass and multiplying that number by the empirical formula.*

The mass of $H_2O = 18$ ($H = 1$, $O = 16$). The molecular mass is 18, so the ratio of $18/18 = 1$, and thus in this case the empirical formula is also the molecular formula.

Exercise 5-4b

A compound of molar mass 56 contains 85.6 percent carbon and 14.4 percent hydrogen. What is its empirical formula? its molecular formula?

Solution

■ **Step 1**   Carbon: $85.6 \text{ g} \times \dfrac{1 \text{ mol}}{12 \text{ g}} = 7.1 \text{ mol}$

Hydrogen: $14.4 \text{ g} \times \dfrac{1 \text{ mol}}{1 \text{ g}} = 14.4 \text{ mol}$

■ **Step 2**   For carbon: $\dfrac{7.1 \text{ mol}}{7.1 \text{ mol}} = 1$

For hydrogen: $\dfrac{14.4 \text{ mol}}{7.1 \text{ mol}} = 2$

Note that the number of moles is rounded off to whole numbers because in step 1 the exact atomic masses are not used.

■ **Step 3** The empirical formula is CH_2.
■ **Step 4** The mass of CH_2 is 14. The molecular mass was given as 56. Since $56/14 = 4$, the molecular formula is $(CH_2)_4$ or C_4H_8.

Exercise 5-4c

A compound of molar mass 270 contains 17.0 percent sodium, 47.4 percent sulfur, and 35.6 percent oxygen. What is its molecular formula?

■ **Step 1**   Sodium: $17.0 \text{ g} \times \dfrac{1 \text{ mol}}{23 \text{ g}} = 0.74 \text{ mol}$

Sulfur: $47.4 \text{ g} \times \dfrac{1 \text{ mol}}{32 \text{ g}} = 1.5 \text{ mol}$

Oxygen: $35.6 \text{ g} \times \dfrac{1 \text{ mol}}{16 \text{ g}} = 2.2 \text{ mol}$

■ **Step 2**   For sodium: $\dfrac{0.74 \text{ mol}}{0.74 \text{ mol}} = 1$

For sulfur: $\dfrac{1.5 \text{ mol}}{0.74 \text{ mol}} = 2$

For oxygen: $\dfrac{2.2 \text{ mol}}{0.74 \text{ mol}} = 3$

■ **Step 3** The empirical formula is NaS_2O_3.

■ **Step 4** The mass of the empirical formula is 135 ($23 + 2 \times 32 + 3 \times 16$). The given molar mass is 270. $270/135 = 2$, so the molecular formula is $Na_2S_4O_6$.

Self-test

An analysis of nicotine gave the following percent composition: 74.0 percent C, 8.65 percent H, and 17.3 percent N. What is the empirical formula for nicotine? If the molar mass of nicotine is 81, what is the molecular formula for nicotine?

Answers

C_5H_7N; the same

5-3 Chemical Equations

When an electrical current (energy) is passed through water (a process known as electrolysis), hydrogen gas and oxygen gas are produced. The chemist uses symbols and formulas in a chemical equation to describe this chemical reaction.

$$H_2O_{(l)} + \text{energy} \longrightarrow H_{2(g)} + O_{2(g)}$$

The arrow is used instead of an equal sign and is read as "yields" or "produces." The plus sign on the right-hand side of the equation is read as "and." A plus sign on the left-hand side of the equation is read as "reacts with." The materials that react are called the reactants. The reactants are written on the left-hand side of the equation. The substances that are produced are called the products. They are written on the right-hand side of the equation. The (l) indicates a liquid, (g) a gas. A solid, or precipitate (an insoluble solid), is indicated by (s). The energy involved in the reaction can be written in words on the left (or right) side of the equation; it may be indicated above the arrow; or it may be indicated by the symbol Δ (Greek delta). The use of a catalyst can be shown above or below the arrow.

A chemical equation must be balanced before any specific interpretation can be made about that reaction. For example, if we wish to know how much oxygen will be required to metabolize a given amount of a particular carbohydrate, we must first set up a balanced equation for the reaction.

Exercise 5-5

Write a chemical equation for the following reactions.
a. Calcium carbonate is heated to produce calcium oxide and carbon dioxide.
b. Hydrochloric acid reacts with sodium hydroxide to produce sodium chloride and water.

Solution

To write a chemical equation we must provide a chemical formula for each reactant and product.
a. $CaCO_3 + \Delta \text{ (heat)} \longrightarrow CaO + CO_2$
b. $HCl + NaOH \longrightarrow NaCl + H_2O$

Self-test

Write a chemical equation for the following reactions.
a. Sulfur burns in air (oxygen) to produce sulfur dioxide.
b. Copper oxide reacts with hydrogen to produce copper and water.

Answers

a. $S + O_2 \longrightarrow SO_2$
b. $CuO + H_2 \longrightarrow Cu + H_2O$

Balancing Chemical Equations

Note that the chemical equation shown below does not contain the same numbers of hydrogen and oxygen atoms on both sides of the arrow. To be balanced, a chemical equation must contain the same number of atoms of each element on both sides. Thus, equation (5-1) is not balanced.

$$H_2O_{(l)} + \text{energy} \longrightarrow H_{2(g)} + O_{2(g)} \quad \text{(unbalanced)} \quad (5\text{-}1)$$

In balancing a chemical equation, you must not change the subscripts (the small numbers to the right of the symbols) because doing so would change either the reactants or the products, thus changing the meaning of the reaction. Instead, place coefficients in front of the symbols and formulas to indicate how many atoms or molecules of each are needed.

In the equation $H_2O \longrightarrow H_2 + O_2$, there are two hydrogen atoms on each side of the equation. There are one oxygen atom on the left side and two oxygen atoms on the right side of the equation. To get two atoms of oxygen on the left side of the equation (to balance the two on the right side), place a 2 in front of the H_2O. The 2 cannot be placed as a subscript after the O in H_2O because then another substance would be represented, not water. The 2 cannot be placed between the H and the O because this also would change the meaning of the formula. Therefore, place the 2 in front of the H_2O.

$$2\,H_2O_{(l)} + \text{energy} \longrightarrow H_{2(g)} + O_{2(g)} \quad \text{(unbalanced)}$$

However, now there are four hydrogen atoms (two H_2's) on the left side of the equation. To get four hydrogen atoms on the right side of the equation, place a 2 in front of the H_2. There are already two oxygen atoms on each side of the equation. Thus, the balanced equation is as follows:

$$2\,H_2O_{(l)} + \text{energy} \longrightarrow 2\,H_{2(g)} + O_{2(g)} \quad \text{(balanced)}$$

This balanced equation now shows that two molecules of water, on electrolysis, yield two molecules of hydrogen gas and one molecule of oxygen gas.

When aluminum metal reacts with sulfuric acid, the products are hydrogen gas, H_2, and aluminum sulfate, $Al_2(SO_4)_3$. The unbalanced equation for this reaction is

$$Al + H_2SO_4 \longrightarrow Al_2(SO_4)_3 + H_{2(g)} \quad \text{(unbalanced)} \qquad (5\text{-}2)$$

Equation (5-2) is not balanced because there are more aluminum atoms on the right side of the equation than on the left side. The same is true for the sulfur and oxygen atoms. To balance an equation of this type, pick out the most complicated-looking formula and assume that one molecule of it is present. The most complicated-looking formula in equation (5-2) is $Al_2(SO_4)_3$. Assuming that one molecule of it is produced, then there are two atoms of aluminum on the right side of the equation. To balance this, place a 2 in front of the Al on the left side of the equation. (For simplicity, we will omit the Δ sign from all the equations that follow.)

$$2\,Al + H_2SO_4 \longrightarrow Al_2(SO_4)_3 + H_{2(g)} \quad \text{(unbalanced)}$$

Next note that there are three SO_4 groups in the molecule of $Al_2(SO_4)_3$. There must then be three SO_4 groups on the left side of the equation. To get these three groups place a 3 in front of the H_2SO_4.

$$2\,Al + 3\,H_2SO_4 \longrightarrow Al_2(SO_4)_3 + H_{2(g)} \quad \text{(unbalanced)}$$

To complete the equation, note that there are now six hydrogen atoms on the left side of the equation (in the three H_2's). Therefore, there must be six hydrogen atoms on the right side, so place a 3 in front of the H_2.

$$2\,Al + 3\,H_2SO_4 \longrightarrow Al_2(SO_4)_3 + 3\,H_{2(g)} \quad \text{(balanced)}$$

Now the equation is balanced. There are 2 aluminums, 6 hydrogens, 3 sulfurs, and 12 oxygens (or 3 SO_4's) on each side of the equation.

When sulfur is burned in excess oxygen, sulfur trioxide is produced. This reaction, written in equation form, is shown in equation (5-3).

$$S_{(s)} + O_{2(g)} \longrightarrow SO_{3(g)} \quad \text{(unbalanced)} \qquad (5\text{-}3)$$

Following the balancing procedure, pick out the most complicated compound and take one molecule of it. Thus, in equation (5-3), take one molecule of SO_3. This molecule contains one atom of sulfur. There is already one atom of sulfur on the left side of the equation. There are three atoms of oxygen on the right side of the equation and only two on the left side. However, there is no *whole* number that can be placed in front of the O_2 to make three oxygen atoms on that side of the equation. If a 2 is placed there, there will be four atoms of oxygen. In this case, then, instead of selecting one molecule of the most complicated compound, select two molecules of it.

$$S_{(s)} + O_{2(g)} \longrightarrow 2\,SO_{3(g)} \quad \text{(unbalanced)}$$

Then, to balance two sulfur atoms on the right side of the equation, start with two sulfur atoms on the left side.

$$2\,S_{(s)} + O_{2(g)} \longrightarrow 2\,SO_{3(g)} \quad \text{(unbalanced)}$$

Next, the right side of the equation contains six oxygen atoms and so must the left side. Place a 3 in front of the O_2 to have six oxygen atoms on that side of the equation. The equation then is balanced.

$$2\,S_{(s)} + 3\,O_{2(g)} \longrightarrow 2\,SO_{3(g)} \quad \text{(balanced)}$$

Exercise 5-6

Balance the following reaction.

$$N_2 + H_2 \longrightarrow NH_3$$

Solution

The most complicated compound is NH_3. Assuming one molecule of NH_3, we have one atom of nitrogen on the right side but two atoms of nitrogen on the left side. To balance the nitrogen, we place a 2 in front of the NH_3.

$$N_2 + H_2 \longrightarrow 2\,NH_3$$

Now we have six hydrogen atoms on the right; we need six hydrogen atoms on the left. We place a 3 in front of H_2 to get six hydrogens on the left.

$$N_2 + 3\,H_2 \longrightarrow 2\,NH_3$$

The equation is now balanced. There are two nitrogen and six hydrogen atoms on both sides of the equation.

Self-test

Balance this equation.

$$C_6H_{12}O_6 + O_2 \longrightarrow CO_2 + H_2O$$

Answer

$$C_6H_{12}O_6 + 6\,O_2 \longrightarrow 6\,CO_2 + 6\,H_2O$$

5-4 Reaction Rates

Some chemical reactions proceed at a slow rate. Iron, for example, rusts very slowly. Wood takes years to decay. On the other hand, some chemical reactions proceed more rapidly. Coal burns steadily and quickly. Concrete begins to set within a few hours. Some chemical reactions not only occur rapidly, they take place almost instantaneously. Consider the violent explosion of dynamite. Within a fraction of a second, the complete reaction has taken place.

For a chemical reaction to occur between two substances, their molecules must collide (interact) with sufficient energy to overcome any force of repulsion between the electron clouds.

The minimum amount of energy required for a collision to produce a successful reaction is called the *activation energy (Figure 5-1).

What determines the speed of a chemical reaction? The speed of a chemical reaction depends on several factors: (1) the nature of the reacting substances, (2) the temperature, (3) the concentration of the reacting substances, (4) the presence of a catalyst, and (5) the surface area and the intimacy of contact of the reacting substances.

Nature of Reacting Substances

When a solution of sodium sulfate (Na_2SO_4) is mixed with a solution of barium chloride ($BaCl_2$), a white precipitate of barium sulfate ($BaSO_4$) is formed immediately.

$$Na_2SO_4 + BaCl_2 \longrightarrow 2\,NaCl + BaSO_{4(s)} \tag{5-4}$$

Equation (5-4) can be rewritten to show the ions of which these salts consist.

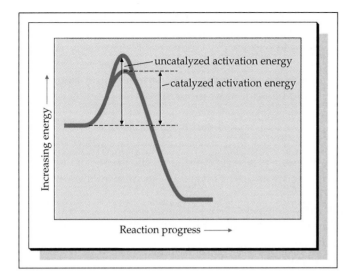

FIGURE 5-1

A catalyst increases the rate of a reaction by lowering the activation energy so more reactant molecules collide with enough energy to react.

$$2\,Na^+ + SO_4^{2-} + Ba^{2+} + 2\,Cl^- \longrightarrow 2\,Na^+ + 2\,Cl^- + BaSO_{4(s)}$$

Next, as in any algebraic equation, cancel the sodium ions and the chloride ions from both sides of equation (5-4), leaving the net equation

$$Ba^{2+} + SO_4^{2-} \longrightarrow BaSO_{4(s)}$$

This is an example of an ionic reaction—the reaction between ions. Many of the reactions taking place in the body are of this type.

Consider, however, the reaction between hydrogen (H_2) and oxygen (O_2) to form water (H_2O). This reaction proceeds very slowly, even at a temperature of 200° C, unless a spark is introduced into the mixture.

$$2\,H_2 + O_2 \xrightarrow{spark} 2\,H_2O$$

In this reaction, it is necessary for the bonds between the hydrogen atoms in the hydrogen molecules to be broken. Also, the bonds between the oxygen atoms must be broken before the reaction can occur. This is an example of a reaction in which covalent bonds must be broken and new ones formed. In general, such reactions proceed much more slowly than ionic reactions.

Temperature

As the temperature rises, the speed of a chemical reaction increases because at higher temperatures the molecules move faster and therefore collide more often. Thus, even a slight change in temperature can affect the speed of a reaction with noticeable results. Every 10° C rise in temperature doubles the rate, or speed, of the reaction. In other words, the reaction would occur in half the time. For a 20° C rise, the reaction would occur four times faster, or in one-fourth the time.

A patient who has a fever of only a few degrees has an increased pulse rate and also an increased respiratory rate. Reactions taking place throughout the body proceed at an accelerated rate.

When the temperature of the human body drops, the various metabolic processes slow down considerably. This fact is of great importance, for example, during open-heart surgery when the temperature of the body is lowered considerably.

A drop in body temperature of 2 to 3° F causes uncontrolled shivering as the body tries to generate heat by muscle activity. Death from exposure occurs when the body temperature drops below 78° F. Victims who fall overboard into water whose temperature is near freezing may survive only a short time unless they are rescued or are wearing the proper type of insulated coverings. A person who has been "chilled" in such an accident should be given dry clothing and warmed as soon as possible. Warm liquids are helpful in increasing body temperature, as are "sugars." Contrary to popular belief, alcohol is not helpful in such a situation because it dilates blood vessels, allowing cold blood to flow more rapidly to the vital organs.

Concentration

The concentration of a reactant is the amount present in a given unit of volume. The more of a given material present in a certain volume, the greater its concentration. Greater concentration produces faster reactions because there are more molecules that can react.

A patient with a respiratory disease can breathe more easily when using a nasal catheter with oxygen because the concentration of the oxygen in the lungs is increased. This increased concentration increases the speed of oxygen uptake, making breathing easier for the patient.

Catalyst

When a protein substance is placed in water and heated, a hydrolysis reaction (see page 205) proceeds at an extremely slow speed. If a strong acid is added to the mixture, the reaction proceeds at a much faster rate. The acid is not used up (that is, it is not changed chemically). Its presence merely increases the speed of the reaction. Any substance that increases the speed of a reaction without itself being changed chemically is called a catalyst.

Catalysts function to increase the rate of a reaction because they provide a pathway with a lower activation energy (see Figure 5-1). Many of the chemical reactions used in industry would not be practical without a catalyst. They would take too long to be of commercial use.

The body uses catalysts to enable its chemical reactions to proceed at a rapid pace. Those catalysts present in the body are called enzymes. During digestion, for example, the food undergoes many chemical changes, each under the influence of a specific enzyme. There are also inhibitors that slow down rather than speed up chemical reactions.

Surface Area

The speed of a chemical reaction also depends on the amount of surface area present in the reacting substances. Although a pile of flour is quite harmless, the same flour in the form of dust can cause a dangerous explosion. This effect results from the tremendous amount of surface area of the dust. This large surface area can react rapidly with the oxygen in the air to cause an explosion.

Many medications are given in the form of finely divided suspended solids. In this manner, more surface area means more rapid absorption in the body.

Increasing the concentration of the reacting substances can also be considered as increasing the amount of surface area.

5-5 Equilibrium Reactions

Often, when two or more reactants unite to form a certain number of products, these products themselves unite to re-form the original reactants. Reactions of this type are called reversible reactions. They are indicated by double arrows $\rightleftharpoons$ showing that the reaction may proceed in either direction, depending on the conditions that exist.

If we start with a mixture of N_2 and H_2, at a given temperature and pressure (with a catalyst), we will soon have some NH_3 formed. As more NH_3 is formed, it will begin to decompose into N_2 and H_2, or

$$N_2 + 3H_2 \rightleftharpoons 2NH_3$$

When the rates of formation and decomposition become equal, a chemical equilibrium exists. This does not mean that all reaction has stopped; it merely means that the rate of decomposition is the same as the rate of formation so the composition remains constant. An equilibrium can be defined as a dynamic state in which the rate of the forward reaction is equal to the rate of the reverse reaction.

Two examples of equilibrium reactions in the body are

$$HCO_3^- + H^+ \rightleftharpoons H_2CO_3 \rightleftharpoons CO_2 + H_2O$$

and

$$\text{hemoglobin} + \text{oxygen} \rightleftharpoons \text{oxyhemoglobin}$$

Equilibrium Constant

Consider the general equilibrium reaction:

$$A + B \rightleftharpoons C + D$$

The law of mass action states that the rate of a chemical reaction is proportional to the concentration of the reacting substances. So, for the above reaction, rate forward $= k_1[A][B]$, where k_1 is a proportionality constant and the brackets [ ] indicate concentrations in moles per liter. Likewise, the rate of the reverse reaction equals $k_2[C][D]$, where k_2 is another proportionality constant.

At equilibrium, the rate of the forward reaction is equal to the rate of the reverse reaction, so

$$k_1 \times [A] \times [B] = k_2 = [C] \times [D]$$

from which we have

$$\frac{[C] \times [D]}{[A] \times [B]} = \frac{k_1}{k_2} = K_{eq}$$

where K_{eq} is the equilibrium constant (since the ratio of two constants k_1/k_2 is itself another constant).

In general, the equilibrium constant, K_{eq}, equals the product of the concentrations of the products divided by the product of the concentrations of the reactants, each concentration raised to the power indicated by its coefficient in the equation. So, for the reaction $4A + 3C \rightleftharpoons 2D + F$, we have

$$K_{eq} = \frac{[D]^2[F]}{[A]^4[C]^3}$$

Exercise 5-7

In the conversion of glucose to vitamin C, the following equilibrium reaction takes place:

$$C_5H_{11}O_5COOH \rightleftharpoons H^+ + C_5H_{11}O_5COO^-$$

$$\quad\text{gluconic acid} \qquad\qquad \text{hydrogen ion} \qquad \text{gluconate ion}$$

If the equilibrium concentrations in moles per liter are $C_5H_{11}O_5COOH$, 0.10; H^+, 3.7×10^{-3}, $C_5H_{11}O_5COO^-$, 3.7×10^{-3}, calculate the value of K_{eq}.
Using the formula for K_{eq}, we have

$$K_{eq} = \frac{[H^+][C_5H_{11}O_5COO^-]}{[C_5H_{11}O_5COOH]}$$

$$= \frac{(3.7 \times 10^{-3})(3.7 \times 10^{-3})}{0.10}$$

$$= 1.4 \times 10^{-4}$$

Exercise 5-8

In the manufacture of wood alcohol (page 284), the equilibrium reaction $CO + 2\,H_2 \rightleftharpoons CH_3OH$ occurs. At equilibrium, the concentrations in moles per liter are CO, 0.025; H_2, 0.050; CH_3OH, 0.12. Calculate the value of K_{eq}.

$$K_{eq} = \frac{[CH_3OH]}{[CO][H_2]^2} = \frac{0.12}{(0.025)(0.050)^2} = 1.9 \times 10^3$$

In general, a large value of K_{eq} indicates an equilibrium that has been shifted far to the right, whereas a small value indicates one shifted to the left.

Le Châtelier's Principle

Le Châtelier's principle states that **if a stress is applied to a reaction at equilibrium, the equilibrium will be displaced in such a direction as to relieve that stress.** Thus, if we apply a stress such as a change in concentration or temperature, we should be able to predict the results of such a stress upon the given equilibrium.

Effect of Concentration

Let us see what the effect on the equilibrium and on the equilibrium constant will be if we add more A to a mixture of A, B, C, and D. The reaction is

$$A + B \rightleftharpoons C + D$$

Increasing the concentration of a reactant increases the number of collisions between reactant molecules so that the rate of the forward reaction is greater than the rate of the reverse reaction. The system is no longer in equilibrium. However, as the reactant molecules are used up, the rate of the forward reaction will decrease and the rate of the reverse reaction will increase until a new equilibrium is established. That is, the addition of more A will cause a stress that the reaction will tend to oppose by using up more A, thus shifting the equilibrium to the right. Since the equilibrium constant depends on temperature only, the addition of more A will have no effect upon it. Thus, the addition of more A increases the amount of products; likewise for the addition of more B to the equilibrium mixture. In accordance with Le Châtelier's principle, removel of a product (or products) should also shift the equilibrium to the right.

Consider the enzyme-catalyzed reaction

$$CO_2 + H_2O \rightleftharpoons H_2CO_3$$

Carbon dioxide is a waste product produced in the cells. As the CO_2 flows into the blood, its concentration increases and so, according to Le Châtelier's principle, the equilibrium shifts toward the right, producing more H_2CO_3. In the lungs, as carbon dioxide is exhaled, the equilibrium is shifted to the left, again according to Le Châtelier's principle.

Effect of a Catalyst

The addition of a catalyst will speed up the forward reaction but will speed up the reverse reaction equally. Thus, all that is accomplished is that the system reaches equilibrium much sooner. The equilibrium reached is the same as would have been reached if no catalyst had been used. Again we note that temperature is the only thing that affects the equilibrium constant, so naturally a catalyst will have no effect on it.

Effect of Temperature and Pressure

In the reaction

$$4\,HCl + O_2 \rightleftharpoons 2\,H_2O + 2\,Cl_2 + heat$$

heat is liberated when the reaction proceeds to the right and is absorbed when it proceeds to the left.

If we raise the temperature of this reaction mixture at equilibrium, the reaction will tend to go in a direction to relieve this stress. That is, as the temperature is raised, the equilibrium will shift to the left, favoring the reaction that tends to absorb this heat. Conversely, as the temperature is lowered, the reaction speeds more to the right for the production of more heat to relieve the new stress.

Increasing the pressure increases the concentration of all the gases involved in an equilibrium reaction; decreasing the pressure has the opposite effect. The side with the most moles of gases is the side most sensitive to pressure changes. If we assume that all reactants and products in the example just given are gases, then the left side has 5 mol of gases (4 mol HCl and 1 mol O_2) and the right side has 4 mol gas (2 mol H_2O and 2 mol Cl_2). The left side is more pressure sensitive. An increase in pressure would drive the reaction to the right to relieve the stress, and an decrease in pressure would drive the reaction to the left.

Exercise 5-9

Predict the direction of equilibrium shift for this reaction.

$$H_2O_{(l)} + CO_{2(g)} \rightleftharpoons H_2CO_{3(l)} + heat$$

a. Increase in the CO_2 level
b. Increase in temperature
c. Adding a catalyst
d. Increasing the pressure

Solution

a. Increasing the level of CO_2 will force the reaction to the side away from CO_2. The reaction will shift to the right.

b. Increasing the temperature will drive the reaction away from the side with the heat. The reaction will shift to the left.
c. A catalyst will have no effect on an equilibrium.
d. There is only one gas in this reaction, CO_2. Increasing the pressure will force the reaction away from the CO_2. Reaction will shift to the right.

Self-test

Predict the direction of equilibrium shift for the following reaction. All compounds are gases

$$N_2 + 3H_2 \rightleftharpoons 2NH_3 + heat$$

a. Decrease in temperature
b. Increase in pressure
c. Addition of nitrogen
d. Removal of ammonia

Answers

All four will shift the equilibrium to the right.

5-6 Interpreting Chemical Equations

A balanced chemical equation specifies a great deal of quantitative information. Consider the balanced equation:

$$2H_2 + O_2 \longrightarrow 2H_2O \tag{5-5}$$

Equation (5-5) indicates that two molecules of H_2 react with one molecule of O_2 to produce two molecules of H_2O. Note that the numbers are the same as the coefficients (numbers in front of the substances) in the balanced equation. That is why the equation must be balanced before any type of calculation can be carried out.

Equation (5-5) can also be interpreted in terms of moles: 2 mol of H_2 react with 1 mol of O_2 to produce 2 mol of H_2O.

We can substitute molecular masses in grams for moles in equation (5-5). Thus, the equation can be represented as

$$2H_2 + O_2 \longrightarrow 2H_2O$$

2 molecules + 1 molecule $\longrightarrow$ 2 molecules

2 mol + 1 mol $\longrightarrow$ 2 mol

2 (2 g) + 32 g $\longrightarrow$ 2 (18 g)

Note that the law of conservation of mass must be satisfied.

Exercise 5-10

Exercise 5-10a

Given the balanced equation

$$C_6H_{12}O_6 + 6O_2 \longrightarrow 6CO_2 + 6H_2O + 686 \text{ kcal}$$

how many moles of CO_2 will be produced from 3.5 mol of $C_6H_{12}O_6$?

Solution

We note from the balanced equation that 1 mol of $C_6H_{12}O_6$ yields 6 mol of CO_2. Using this as a conversion factor,

$$3.5 \text{ mol } C_6H_{12}O_6 \times \frac{6 \text{ mol } CO_2}{1 \text{ mol } C_6H_{12}O_6} = 21 \text{ mol } CO_2$$

Exercise 5-10b

Using the balanced equation in Example 5-10a, how many grams of water will be produced from the reaction of 0.50 mol of $C_6H_{12}O_6$?

Solution

The balanced equation indicates that 1 mol of $C_6H_{12}O_6$ yields 6 mol of H_2O, so

$$0.50 \text{ mol } C_6H_{12}O_6 \times \frac{6 \text{ mol } H_2O}{1 \text{ mol } C_6H_{12}O_6} = 3 \text{ mol } H_2O$$

Then, recalling that 1 mol of H_2O has a mass of 18 g, we obtain

$$3 \text{ mol } H_2O \times \frac{18 \text{ g } H_2O}{1 \text{ mol } H_2O} = 54 \text{ g } H_2O$$

Example 5-10c

Using the balanced equation in Example 5-10a, how much energy will be produced from the complete combustion of 0.5 mol of $C_6H_{12}O_6$?

Solution

The balanced equation indicates that complete oxidation of 1 mol of $C_6H_{12}O_6$ yields 686 kcal, so

$$0.50 \text{ mol } C_6H_{12}O_6 \times \frac{686 \text{ kcal}}{1 \text{ mol } C_6H_{12}O_6} = 343 \text{ kcal}$$

Self-test

One of the active ingredients in antacids is calcium carbonate. Calcium carbonate reacts with stomach acid according to the following equation:

$$2 \text{ HCl} + CaCO_3 \longrightarrow CaCl_2 + H_2O + CO_2$$

Assuming 10 g of $CaCO_3$, how stomach acid can be neutralized and how much CO_2 is produced?

Answer

7.3 g of HCl and 4.4 g of CO_2

Summary

The molecular mass of a compound is equal to the sum of the atomic masses of the atoms present in that compound. One mole of any substance contains Avogadro's number (6.02×10^{23}) of particles.

An empirical (simplest) formula represents the relative number of each type of element present in a given compound.

A molecular formula represents the actual number of atoms present in each molecule of a given compound.

A chemical equation uses symbols and formulas to represent a chemical reaction. The substances on the left side of the equation are called reactants and those on the right side products. A (g) indicates a gas; an (s) indicates a precipitate or solid and (l) a liquid.

To balance a chemical equation, pick out the most complicated-looking compound and assume that one molecule of it is present. Then proceed back and forth adding coefficients in front of the reactants and products until the number of atoms of each type is the same on both sides of the equation.

The speed of a chemical reaction depends upon the nature of the reacting substances, the temperature, the concentration of the reacting substances, the presence of a catalyst, and the surface area of the reacting substances.

An equilibrium reaction is a dynamic state in which the rate of the forward reaction is equal to the rate of the reverse reaction.

Le Châtelier's principle states that if a stress is applied to a reaction at equilibrium, the equilibrium will be displaced in such a direction as to relieve that stress.

Questions and Problems

1. Balance the following equations:
   a. $S + O_2 \longrightarrow SO_3$
   b. $Mg + HCl \longrightarrow MgCl_2 + H_2$
   c. $C + O_2 \longrightarrow CO_2$
   d. $Al + O_2 \longrightarrow Al_2O_3$
   e. $Zn(OH)_2 + H_2SO_4 \longrightarrow ZnSO_4 + H_2O$
   f. $Ca + AgNO_3 \longrightarrow Ca(NO_3)_2 + Ag$
   g. $N_2 + O_2 \longrightarrow N_2O_5$
   h. $K_2SO_4 + BaCl_2 \longrightarrow BaSO_4 + KCl$

2. Interpret question 1 in terms of moles, molecules, and grams.

3. Calculate the percentage composition for each compound in question 1(a), (c), and (d).

4. What is a mole? What is the mass of 1 mol of $C_6H_{10}O_5$?

5. How many moles are in 200 g of NaOH? how many molecules?

6. For the balanced equation $Zn + 2HCl \longrightarrow ZnCl_2 + H_2$, how many moles of H_2 can be produced from 3.75 mol of Zn?

7. How many grams of $ZnCl_2$ will be produced according to the equation in question 6?

8. Calculate the value for K_{eq} for the reaction $A + 3B \rightleftharpoons 2C$ if the molar concentrations are A, 1.0; B, 2.0; and C, 1.5.

9. What factors determine the rate of a chemical reaction? Indicate at least one practical application for each factor.

10. What does Le Châtelier's principle state? Do all reactions exist at equilibrium?

11. What will be the mass of 6.02×10^{23} atoms of S?

12. How many moles of $AlCl_3$ are in 5.00×10^{-2} g?

13. How many molecules are in 2 mol of CO_2? how many atoms?

14. Must an equation be balanced before mass calculations are made? Explain.

15. How many grams of N_2 will react with oxygen to produce 2.5 mol of N_2O_3?

$$2N_2 + 3O_2 \longrightarrow 2N_2O_3$$

16. Calculate K_{eq} for the reaction $N_2 + 3H_2 \rightleftharpoons 2NH_3$ + heat if the molar concentrations are N_2, 4.0: H_2, 3.0; and NH_3, 60.0.

17. How may the reaction in question 16 be shifted to the left? to the right?

18. What effect does a catalyst have on an equilibrium? Why?

19. Is a reaction at equilibrium static or dynamic? Explain.

20. What effects does a fever have on the rate of body reactions?

21. Calculate the molecular mass of each of the following compounds (use whole numbers for atomic masses).
    a. $KClO_3$ b. $C_6H_{10}O_5$
    c. $FeSO_4$ d. P_2O_5
    e. $K_2Cr_2O_7$ f. $Al_2(SO_4)_3$

22. Which of the following is an empirical formula? a molecular formula? How can you tell?
    CH_4; CH; C_2H_2; C_6H_6; $C_6H_{12}O_6$; CO_2

23. Can an empirical formula and a molecular formula be the same? Explain.

24. Calculate the empirical formula for a compound containing 63.2 percent manganese and 36.8 percent oxygen.

25. Cisplatin, an anticancer agent used in the treatment of solid tumors, is prepared by the following reaction:

$$K_2PtCl_4 + 2NH_3 \longrightarrow \underset{\text{cisplatin}}{Pt(NH_3)_2Cl_2} + 2KCl$$

How many grams of ammonia and K_2PtCl_4 are needed to make 100.0 g of cisplatin?

26. Lithium hydroxide is used on the space shuttle to remove exhaled carbon dioxide.

$$LiOH + CO_2 \longrightarrow LiHCO_3$$

If the average astronaut produces 930 g of CO_2 per day, how much LiOH is needed for a 10-day mission of five astronauts?

27. Which of the following fertilizers contains the highest percentage of nitrogen?
a. CH_4N_2O (urea), b. NH_3, c. NH_4NO_3, d. $NaNO_3$

28. Hemoglobin, the oxygen-carrying protein in red blood cells, has four iron atoms per molecule. If hemoglobin is 0.34 percent iron, calculate the molecular mass of hemoglobin.

29. Verapamil is used in the treatment of coronary heart disease. It acts by blocking the transport of calcium, thus preventing an accumulation of excess calcium in the muscle tissue of the heart. The formula for verapamil is $C_{27}H_{38}O_4N_2$. Determine the following:
a. The molecular mass for verapamil
b. The number of moles in one 120 mg tablet
c. The number of molecules in one 120 mg tablet

30. A packet of white powder, thought to be cocaine, is seized by the police. The formula for cocaine is $C_{17}H_{21}NO_4$. A preliminary analysis indicates that the powder is 3.82 percent N. Is the powder cocaine?

31. The U.S. Recommended Daily Allowances for several vitamins are as follows: 60.0 mg vitamin C, $C_6H_8O_6$; 400 µg of folic acid, $C_{19}H_{19}N_7O_6$; 1.5 mg of vitamin A, $C_{20}H_{30}O$. Calculate the number of moles and molecules in each allowance.

32. The fermentation of sugar to produce ethyl alcohol is
$$\underset{\text{sugar}}{C_6H_{12}O_6} + yeast \longrightarrow \underset{\text{ethyl alcohol}}{2\,C_2H_6O} + 2\,CO_2$$
How many grams of ethyl alcohol can be produced from 1.00 kg of sugar?

33. Brain cells "burn" glucose ($C_6H_{12}O_6$) for energy.
$$C_6H_{12}O_6 + 6\,O_2 \longrightarrow 6\,CO_2 + 6\,H_2O + 11{,}800 \text{ kcal}$$
If 1.0 g of glucose is "burned," calculate the following:
a. Amount of heat produced
b. Amount of CO_2 produced
c. Amount of O_2 consumed

34. For spaceflight, the energy content per gram of fuel needs to be as large as possible. Which of the following has the highest energy content per gram?

Compound	Heat produced (kcal/mol)
Octane C_8H_{18}	22,900
Methyl alcohol CH_4O	3030
dimethylhydrazine $C_2H_8N_2$	7080

Practice Test

Use the following equation for questions 1 and 2.
$$Al + H_2SO_4 \longrightarrow Al_2(SO_4)_3 + H_{2(g)}$$

1. The (g) indicates _____.
a. a solid   b. a liquid   c. a gas   d. a gel

2. When the equation is balanced, the number in front of H_2SO_4 is _____.
a. 1   b. 2   c. 3   d. 4

Use the following balanced equation for questions 3-6.
$$2\,N_2 + 5\,O_2 \rightleftharpoons 2\,N_2O_5 + heat \quad \text{(all gases)}$$

3. An increase in N_2O_5 will shift the equilibrium _____.
a. to the right   b. to the left   c. not at all

4. An increase in pressure will shift the equilibrium _____.
a. to the right   b. to the left   c. not at all

5. A decrease in temperature will _____.
a. shift the equilibrium to the right
b. shift the equilibrium to the left
c. have no effect on the equilibrium

6. The addition of catalyst will speed up _____.
a. the forward reaction
b. the reverse reaction
c. both reactions

Use the following balanced equation for questions 7-9.
$$2\,H_2 + O_2 \longrightarrow 2\,H_2O$$

7. How many moles of water will be produced from 4 mol of H_2?
a. 2   b. 4   c. 6   d. 8

8. How many grams of water will be produced from 3 mol of H_2?
a. 18   b. 36   c. 54   d. 72

9. How many moles of H_2 are required to react completely with 1.5 mol of O_2?
a. 1.5   b. 2   c. 2.5   d. 3.5

10. An equilibrium reaction is an example of a(n) _____ reaction.
a. exothermic   b. reversible
c. endothermic   d. oxidative

Oxidation-Reduction Reactions

Objectives

- To understand the significance of oxidation and reduction
- To balance oxidation-reduction reactions
- To become familiar with some breath tests

Much of the chemical change occurring around us is caused by oxidation-reduction reactions. An example is the corrosion of metals. The iron gears pictured here have been oxidized by atmospheric oxygen to produce rust.

6-1 Oxidation Numbers

For ionic compounds, the oxidation number, sometimes called the *charge* of an element, is equal to the number of electrons lost or gained and therefore is the same as the charge on the ion. That is, in sodium chloride, Na^+Cl^-, the oxidation number of sodium is +1 and that of chlorine is −1. In the compound $MgBr_2$, where the magnesium ion has a charge of 2+ and each bromide ion a charge of 1−, the oxidation number of magnesium is +2 and that of each bromine −1.

For covalent compounds, where electrons are shared and not transferred, oxidation numbers are assigned to elements using the following rules.

1. All elements in their free state have an oxidation number of zero.
2. The oxidation number of oxygen is −2 (except in peroxides, where it is −1).
3. The oxidation number of hydrogen is +1 (except in metal hybrides, where it is −1).
4. The sum of oxidation numbers in all compounds must equal zero. (That is, all compounds are electrically neutral.)
5. All elements in group IA have an oxidation number of +1.
6. All elements in group IIA have an oxidation number of +2.

Table 6-1 lists the oxidation numbers of various elements.

Table 6-1 Oxidation Numbers of Some Elements and Ammonium Ion

Name and Symbol		Oxidation Number[†]	Name and Symbol		Oxidation Number
Positive oxidation numbers			**Negative oxidation numbers**		
Hydrogen	H^+	+1	Chloride	Cl^-	−1
Sodium	Na^+	+1	Bromide	Br^-	−1
Potassium	K^+	+1	Iodide	I^-	−1
Silver	Ag^+	+1	Sulfide	S^{2-}	−2
Ammonium	NH_4^+	+1	Oxide	O^{2-}	−2
Calcium	Ca^{2+}	+2			
Magnesium	Mg^{2+}	+2			
Aluminum	Al^{3+}	+3			
Iron	Fe^{2+} and Fe^{3+}	+2 and +3			
Copper	Cu^+ and Cu^{2+}	+1 and +2			
Tin	Sn^{2+} and Sn^{4+}	+2 and +4			

†Note that some elements, such as copper, tin, and iron, have more than one oxidation number.

Calculating Oxidation Numbers From Formulas

As has been previously mentioned, the sum of the oxidation numbers in any compound must equal zero. Let us find the oxidation number of zinc in zinc oxide, ZnO. Note that the oxidation number of oxygen, as listed in Table 6-1, is -2. Writing the formula of the compound with the known oxidation number above,

$$\underset{\text{Zn}}{?} \quad + \underset{\text{O}}{-2} = 0$$

we see that the oxidation number of the Zn must be $+2$ for the sum of the oxidation numbers to be zero.

Next let us find the oxidation number of Mn in potassium permanganate, $KMnO_4$. From the table we see that the oxidation number of K is $+1$ and that of O is -2. Therefore, four oxygens will have a total oxidation number of $4(-2)$, or

$$\underset{\text{K}}{+1 +} \quad \underset{\text{Mn}}{(?)} \quad \underset{\text{O}_4}{+4(-2) = 0}$$

For the sum of the oxidation numbers to be zero, the oxidation number of Mn must be $+7$.

Now consider the compound diarsenic trisulfide, As_2S_3. The oxidation number of S from the table is -2. Therefore, three S's will have a total oxidation number of $3(-2)$.

$$\underset{\text{As}_2}{?} \quad + \underset{\text{S}_3}{3(-2) = 0}$$

For the sum of the oxidation numbers to be zero, the total oxidation number of the As atoms must be $+6$. However, this value of $+6$ applies to two As atoms, so the oxidation number of each As is $+3$.

Exercise 6-1

Calculate the oxidation number for carbon in each of these compounds.
a. glucose, $C_6H_{12}O_6$ b. CO
c. CO_2 d. $NaHCO_3$

Solution

a. For $C_6H_{12}O_6$, hydrogen is $+1$, 12 hydrogens $= +12$; O is -2, $6 \times -2 = -12$. $12 - 12 = 0$; therefore, carbon is 0.
b. For CO, oxygen is -2; therefore, carbon must be $+2$.
c. For CO_2, oxygen is -2, $2 \times 2- = -4$; therefore, carbon must be $+4$.
d. For $NaHCO_3$, sodium is $+1$, hydrogen is $+1$, oxygen is -2; $3 \times -2 = -6$, $1 + 1 - 6 = -4$; therefore, carbon must be $+4$.

Self-test

Calculate the oxidation numbers for chlorine in each of these compounds.
a. Cl_2 b. KCl
c. Cl_2O d. $KClO_3$
e. HOCl

Answers

a. 0   b. -1 c. $+1$ d. $+5$ e. $+1$

Writing Formulas From Oxidation Numbers

To write the formula of a compound formed between calcium and chlorine, look up the oxidation numbers of these elements in Table 6-1 and write them above the symbols

$$\overset{+2-1}{CaCl}$$

Note that the sum of the oxidation numbers is not zero, so this is not the correct formula for the compound. An easy method for obtaining the correct formula is to use a system of crisscrossing the oxidation numbers:

$$\overset{+2}{Ca}\quad\overset{-1}{Cl}$$

Thus, the formula of the compound between calcium and chlorine is $CaCl_2$. (Note that the subscript 1 is always understood and never written.) In this compound the sum of the oxidation numbers (+2 for the calcium and −2 for the two chlorines) does equal zero.

To write the formula for the compound formed between magnesium and the phosphate ion, we first write the oxidation numbers above the symbols and then crisscross them

$$\overset{+2}{Mg}\quad\overset{-3}{PO_4}$$

so that the formula is $Mg_3(PO_4)_2$, where the sum of the oxidation numbers now equals zero (+6 from three Mg's and −6 from two PO_4's). The parentheses around the PO_4^{3-} ion indicate that the ion occurs more than once in the formula. If the ion occurs only once in a compound, parentheses are not necessary. Thus, in the compound between the sodium ion (oxidation number +1) and the nitrate ion (oxidation number −1), the formula is simply written as $NaNO_3$.

When the positive and negative oxidation numbers are equal, the formula is correctly written without subscripts. The compound formed between calcium (oxidation number +2) and the sulfate ion (oxidation number −2) is $CaSO_4$ since the sum of the oxidation numbers is already zero.

Occasionally, when both positive and negative oxidation numbers are even numbers, the formula of the compound can be simplified by dividing by 2. Thus, the compound formed between tin (oxidation number +4) and the sulfate ion (oxidation number −2) can be written as

$$\overset{+4}{Sn}\quad\overset{-2}{SO_4}$$

or $Sn_2(SO_4)_4$, which should be simplified to $Sn(SO_4)_2$.

6-2 Oxidation

Oxidation can be defined as "a loss of electrons." Consider the electron-dot structures in reaction (6-1). The sodium atom has one outer electron. When the sodium atom loses this one electron, it forms a sodium ion with a +1 charge. This loss of an electron is defined as oxidation. Therefore, the sodium atom was oxidized.

$$Na\cdot + \cdot\ddot{\underset{..}{Cl}}: \longrightarrow Na^+ + :\ddot{\underset{..}{Cl}}:^- \qquad (6\text{-}1)$$

A second definition of oxidation states that it is an increase in oxidation number. Consider reaction (6-2).

$$2\,Na + Cl_2 \longrightarrow 2\,NaCl \qquad (6\text{-}2)$$

An uncombined element has an oxidation number of zero. (See Section 5-15 for a discussion of oxidation numbers.) The oxidation number of sodium in sodium chloride (NaCl) is +1, and that of chlorine is −1. Therefore, the reaction (6-2) can be written as follows:

$$\overset{0}{2\,Na} + \overset{0}{Cl_2} \longrightarrow \overset{+1}{2\,Na} + \overset{-1}{2\,Cl}$$

where the upper numbers indicate the respective oxidation numbers of the substances. The sodium has changed in oxidation number from zero to +1, a gain. This is oxidation. The sodium atom was oxidized.

The cells in the body "burn" glucose, producing carbon dioxide, water, and energy.

$$\overset{0}{C_6H_{12}O_6} + \overset{0}{6\,O_2} \longrightarrow \overset{+4-2}{6\,CO_2} + \overset{-2}{6\,H_2O} + energy$$
glucose

The oxidation number of each carbon atom in glucose is zero. The oxidation number of the carbon atom in carbon dioxide (CO_2) is +4.[†] Therefore, the carbon atom increased in oxidation number. A gain in oxidation number is oxidation; therefore, the carbon atom in glucose was oxidized, or it can be said that the glucose, which contains the carbon atom, was oxidized.

A third definition of oxidation is "addition of oxygen." Consider reaction (6-3), which involves the oxidation of formaldehyde to formic acid (this type of reaction will be discussed in Chapter 18).

$$2\,HCHO + O_2 \longrightarrow 2\,HCOOH \qquad (6\text{-}3)$$
formaldehyde   formic acid

Note that this reaction involves the addition of oxygen.

A fourth definition of oxidation involves the "removal of hydrogen." In reaction (6-4),

$$2\,CH_3CH_2OH + O_2 \longrightarrow 2\,CH_3CHO + 2\,H_2O \qquad (6\text{-}4)$$
ethanol   acetaldehyde

ethanol is changed to acetaldehyde. This process is called oxidation because it involves the loss of hydrogen.

The following oxidation reactions take place in the body. They will be discussed in detail in the appropriate chapters on carbohydrates, fats, and proteins.

$$carbohydrate + O_2 \longrightarrow CO_2 + H_2O + energy$$
$$fat + O_2 \longrightarrow CO_2 + H_2O + energy$$
$$protein + O_2 \longrightarrow CO_2 + H_2O + urea + energy$$

[†]Refer to Section 6-1.

$$\underset{C_6}{6(0)} + \underset{H_{12}}{12(+1)} + \underset{O_6}{6(-2)} = 0 \quad \text{and} \quad \underset{C}{+4} + \underset{O_2}{2(-2)} = 0$$

Thus, oxidation can be defined as

1. An increase in oxidation number
2. A loss of electrons
3. A gain of oxygen
4. A loss of hydrogen

6-3 Reduction

Oxidation is defined as a loss of electrons and also as an increase in oxidation number. Reduction is the opposite of oxidation—a gain of electrons and, therefore, a decrease in oxidation number. **Oxidation can never take place without reduction** because something must be able to pick up the electrons lost by the oxidized atom, ion, or compound. Free electrons cannot exist by themselves for very long.

In the reactions of sodium with chlorine,

$$\overset{0}{2\,Na} + \overset{0}{Cl_2} \longrightarrow \overset{+1}{2\,Na^+} + \overset{-1}{2\,Cl^-}$$

the sodium increased in oxidation number from 0 to $+1$. It was oxidized. At the same time, the chlorine decreased in oxidation number from 0 to -1. Therefore, the chlorine was reduced. Consider the structures in reaction (6-1). The chlorine atom has seven outer electrons. It gains one electron to form the chloride ion with a charge of -1. This gain of an electron is called reduction. Thus, by either definition, the chlorine was reduced.

In the reaction of glucose with oxygen,

$$\overset{0}{C_6H_{12}O_6} + \overset{0}{6\,O_2} \longrightarrow \overset{+4\,-2}{6\,CO_2} + \overset{-2}{6\,H_2O} + energy$$

glucose is oxidized because the carbon atoms change in oxidation number from 0 to $+4$. However, at the same time, the oxygen changes from 0 to -2, a decrease in oxidation number. Therefore, the oxygen is reduced.

Since reduction is the opposite of oxidation, reduction can also be defined as a loss of oxygen or as a gain of hydrogen. Equations (6-5) and (6-6) illustrate reduction reactions that will be discussed in Chapter 18.

$$CH_3COOH + H_2 \longrightarrow CH_3CHO + H_2O \qquad (6\text{-}5)$$
$$\text{acetic acid} \qquad\qquad\quad \text{ethanol}$$

$$CH_3COCH_3 + H_2 \longrightarrow CH_3CH(OH)CH_3 \qquad (6\text{-}6)$$
$$\text{acetone} \qquad\qquad\quad \text{isopropyl alcohol}$$

Reaction (6-5) is reduction because it involves a loss of oxygen; reaction (6-6) is reduction because it involves a gain of hydrogen.

Thus, reduction is

1. A decrease in oxidation number
2. A gain of electrons
3. A loss of oxygen
4. A gain of hydrogen

6-4 Oxidizing Agents and Reducing Agents

In reaction (6-7),

$$H_2 + PbO \longrightarrow Pb + H_2O \qquad\qquad (6-7)$$

$$\overset{0}{H_2} + \overset{+2\,-2}{PbO} \longrightarrow \overset{0}{Pb} + \overset{2(+1)\,-2}{H_2O}$$

the lead decreases in oxidation number from +2 to 0. Therefore, the lead was reduced. What reduced it? What supplied the electrons that it had to gain to decrease in oxidation number? The answer is that the hydrogen reduced it. The hydrogen supplied the electrons so that the lead could be reduced; therefore, hydrogen is called the **reducing agent.** The substance that causes the reduction of an element or compound is known as a reducing agent.

At the same time, the hydrogen gained in oxidation number from 0 to +1. Therefore, the hydrogen was oxidized. What oxidized it? What picked up the electrons that the hydrogen must have lost in being oxidized? The answer is that the PbO picked up these electrons; thus, the PbO is called the **oxidizing agent.** An oxidizing agent is defined as a substance that causes the oxidation of some element or compound.

Another way of stating this is that **whatever is oxidized is the reducing agent** and **whatever is reduced is the oxidizing agent.**

Rewriting equation (6-7) and indicating what was oxidized, what was reduced, what the oxidizing agent was, and what the reducing agent was, we have the following:

$$\underset{\substack{\text{oxidized}\\\text{(reducing}\\\text{agent)}}}{H_2} + \underset{\substack{\text{reduced}\\\text{(oxidizing}\\\text{agent)}}}{PbO} \longrightarrow Pb + H_2O$$

Exercise 6-2

For the following reaction find:
a. the substance oxidized
b. the substance reduced
c. the oxidizing agent
d. the reducing agent

$$2\,C_3H_6 + 9\,O_2 \longrightarrow 6\,H_2O + 6\,CO_2$$

Solution

■ **Step 1** Assign oxidation numbers for each element.

For C_3H_6: $C = -2;\quad H = +1$

For O_2: $O = O$

For H_2O: $H = +1;\quad O = -2$

For CO_2: $C = +4;\quad O = -2$

■ **Step 2** Look for changes in the oxidation numbers.
H remained at +1.

C changed from −2 to +4. Since the oxidation number increased, the C in C_3H_6 was oxidized and acts as the reducing agent.

O changed from 0 to -2. Since the oxidation number decreased, O_2 was reduced and acts as the oxidizing agent.

Self-test

For the following reaction determine:
a. the substance oxidized;   b.   the substance reduced
c. the oxidizing agent;   d.   the reducing agent

$$Zn + CuSO_4 \longrightarrow ZnSO_4 + Cu$$

Answers

a. and d. Zn   b. and c. Cu in $CuSO_4$

Balancing Oxidation-Reduction Reactions

To balance a redox equation use the following rules:

1. Write the complete equation listing the oxidation numbers of all the elements (see Table 6-1), and locate those elements that are changing their oxidation number.
2. Calculate the change in oxidation number, and indicate this change below the equation.
3. Multiply the change in oxidation number for each atom by the total number of those atoms that are changing in oxidation number.
4. Find a least common multiple (LCM) for the total increase and decrease in oxidation number.
5. Equate the increase and decrease in oxidation numbers, since they must always be equal.
6. Complete the remainder of the equation by adding the proper coefficients to make sure that there are the same number of each type of atom on both sides of the equation.
7. Note that any element in the free state, that is, uncombined, has an oxidation number of zero.

Exercise 6-3

Exercise 6-3a

Balance $H_2 + O_2 \longrightarrow H_2O$.

Solution

- **Step 1** *Write in all the oxidation numbers.*

$$\overset{0}{H_2} + \overset{0}{O_2} \longrightarrow \overset{2(+1)-2}{H_2O}$$

(Since the molecules of hydrogen and oxygen are uncombined, their atoms have an oxidation number of zero; from Table 5-5 we know that the oxidation number of oxygen in water is -2 and that of each hydrogen is $+1$.)

■ **Step 2** *Show the increase and decrease in oxidation numbers.*

Since the hydrogen is changing from an oxidation number of 0 to +1, there is an increase of 1 in oxidation number. The oxygen, in changing from an oxidation number of 0 to -2, decreases by 2 in oxidation number. These changes can be indicated in the equation as

$$
\overset{0}{H_2} + \overset{0}{O_2} \longrightarrow \overset{2(+1)-2}{H_2O}
$$

$$\uparrow \text{1 for each atom} \qquad \downarrow \text{2 for each atom}$$

where the upward-pointing arrow indicates an increase in oxidation number and the downward-pointing arrow indicates a decrease in oxidation number.

■ **Step 3** *Find the total change in oxidation number for each of the elements that is changing.*

The total increase in oxidation number for the hydrogen is 2 (1 for each of the hydrogens). The total decrease in oxidation number for the oxygen is 4 (2 for each oxygen). These total changes in oxidation number can be indicated as

Total changes:
$$
\overset{0}{H_2} + \overset{0}{O_2} \longrightarrow \overset{2(+1)-2}{H_2O}
$$

$$\uparrow 2 \qquad \downarrow 4$$

■ **Step 4** *The LCM for an increase of 2 and a decrease of 4 is 4.*

■ **Step 5** *Equate the increase in oxidation number with the decrease.*

The increase in oxidation number for the hydrogen (2) goes into the LCM twice, so we place a 2 in front of the H_2. The decrease in oxidation number for the O_2 (4) goes into the LCM once, so we place a 1 in front of the O_2 (no number in front of a formula designates 1). Thus, we have

$$2\,H_2 + O_2 \longrightarrow H_2O$$

where the total increase in oxidation number is now equal to the total decrease.

■ **Step 6** *Complete the equation.*

Since we now have a total of four H's on the left side of the equation, we must have four H's on the right side, so we place a 2 in front of the H_2O and then we have four H's on each side of the equation, or

$$2\,H_2 + O_2 \longrightarrow 2\,H_2O$$

We note that there are also two O's on each side of the equation, and it is therefore completely balanced.

Exercise 6-3b

Balance $KClO_3 \longrightarrow KCl + O_2$.
Listing the oxidation numbers we have

$$
\overset{\displaystyle 3\times}{\underset{KClO_3}{\overset{+1+5-2}{}}} \longrightarrow \overset{+1-1}{KCl} + \overset{0}{O_2}
$$

Note that the Cl and the O are changing in oxidation number. The Cl, in changing from $+5$ to -1, exhibits a decrease of 6 in oxidation number. The O, in changing from -2 to 0, exhibits an increase of 2 in oxidation number of each of the O atoms present. Since there are three O atoms present, the total increase in oxidation number is 6. These changes may be indicated as

$$
Total\ changes: \quad \overset{\displaystyle 3\times}{\underset{KClO_3}{\overset{+5-2}{}}} \longrightarrow \overset{-1}{KCl} + \overset{0}{O_2}
$$

$$
\downarrow 6 \quad \uparrow 6
$$

Because the increase in oxidation number here does equal the decrease, we need no LCM and so leave the left side of the equation as is.

Now to complete the equation. Since we have one $KClO_3$ on the left side of the equation we can have only one KCl on the right because the number of K's on each side of the equation must be the same and so must the number of Cl's. Since we have three O's on the left side of the equation we must have three O's on the right side; we get these by placing a $1\frac{1}{2}$ in front of the O_2 ($1\frac{1}{2} \times O_2 = 3\,O$). Thus, we have

$$
KClO_3 \longrightarrow KCl + 1\tfrac{1}{2}\,O_2
$$

However, we can never have fractions of atoms or molecules in a chemical equation, so we multiply the whole equation by 2 to get

$$
2\,KClO_3 \longrightarrow 2\,KCl + 3\,O_2
$$

which is balanced because it contains the same number of each type of atom on each side of the equation.

What was reduced? An error made by many students is the statement that the chlorine was reduced because it decreased in oxidation number. It is true that a decrease in oxidation number is reduction, but the chlorine did not decrease in oxidation number, the Cl^{+5} decreased. This makes quite a difference as we shall see in the following paragraph.

What was oxidized? The O^{2-} was oxidized because it increased in oxidation number. If we had said that the oxygen was oxidized, that would have been incorrect because oxygen is a product of the reaction and not one of the original reactants.

Also the Cl^{+5}, since it was reduced, is the oxidizing agent, and the O^{2-}, which was oxidized, is the reducing agent.

Exercise 6-3c

Balance

$$
FeSO_4 + K_2Cr_2O_7 + H_2SO_4 \longrightarrow Fe_2(SO_4)_3 + Cr_2(SO_4)_3 + K_2SO_4 + H_2O
$$

Since the equation is a little more complex than those we have studied so far, we will solve it carefully and slowly to see why we take each step. First, we will list the oxidation numbers, giving the whole (SO_4) polyatomic ion a charge of -2 instead of giving the separate oxidation numbers of the S and the O because we see that SO_4 occurs unchanged on both sides of the equation. So we have

$$\underset{+2-2}{FeSO_4} + \underset{\substack{2\times2\times7\times \\ +1+6-2}}{K_2Cr_2O_7} + \underset{\substack{2\times \\ +1-2}}{H_2SO_4} \longrightarrow \underset{\substack{2\times\ 3\times \\ +3\ \ -2}}{Fe_2(SO_4)_3} + \underset{\substack{2\times\ 3\times \\ +3\ \ -2}}{Cr_2(SO_4)_3} + \underset{\substack{2\times \\ +1\ -2}}{K_2SO_4} + \underset{\substack{2\times \\ +1-2}}{H_2O}$$

Note that the Fe is changing from $+2$ to $+3$, which is an increase of 1, and the Cr is changing from $+6$ to $+3$, which is a decrease of 3 in oxidation number for each Cr or a total decrease of 6 in oxidation number for the two Cr's, or

$$\underset{+2}{FeSO_4} + \underset{2(+6)}{K_2Cr_2O_7} + H_2SO_4 \longrightarrow \underset{2(+3)}{Fe_2(SO_4)_3} + \underset{2(+3)}{Cr_2(SO_4)_3} + K_2SO_4 + H_2O$$

$\uparrow 1 \qquad\qquad \downarrow 6$

The LCM for an increase of 1 and a decrease of 6 in oxidation number is 6. Therefore, an increase of 1 (by the Fe) goes into the LCM six times, so we place a 6 in front of the $FeSO_4$; a decrease of 6 in oxidation number (by the Cr's in $K_2Cr_2O_7$) goes into the LCM once, so we place a 1 (understood) in front of the $K_2Cr_2O_7$, or

$$6\ FeSO_4 + K_2Cr_2O_7 + (\ )\ H_2SO_4 \longrightarrow$$
$$(\ )\ Fe_2(SO_4)_3 + (\ )\ Cr_2(SO_4)_3 + (\ )\ K_2SO_4 + (\ )\ H_2O$$

Now to complete the rest of the equation. Since there are six Fe's on the left side, there must be six on the right; we get these by placing a 3 in front of the $Fe_2(SO_4)_3$ (three Fe_2 = six Fe). Since there are two K's on the left side of the equation, there must be two on the right, and we get these by taking one K_2SO_4. Since there are two Cr's on the left side, there must be two on the right, and so we place a 1 in front of the $Cr_2(SO_4)_3$, or

$$6\ FeSO_4 + K_2Cr_2O_7 + (\ )\ H_2SO_4 \longrightarrow$$
$$3\ Fe_2(SO_4)_3 + Cr_2(SO_4)_3 + K_2SO_4 + (\ )\ H_2O$$

Now let us count the SO_4 groups on the right side of the equation. There are 9 in the 3 $Fe_2(SO_4)_3$ plus 3 in the $Cr_2(SO_4)_3$ plus 1 in the K_2SO_4, or a total of 13 SO_4's. We started with six SO_4's in the 6 $FeSO_4$, so we need seven more; we get these by taking seven H_2SO_4. These seven H_2's on the left side must give seven H_2's on the right, so we have seven H_2O's, or

$$6\ FeSO_4 + K_2Cr_2O_7 + 7\ H_2SO_4 \longrightarrow$$
$$3\ Fe_2(SO_4)_3 + Cr_2(SO_4)_3 + K_2SO_4 + 7\ H_2O$$

To check whether the equation is correctly balanced, let us add up the O's on each side and see if the totals are the same.

Left Side		*Right Side*	
6 $FeSO_4$	24 O's	3$Fe_2(SO_4)_3$	36 O's
1 $K_2Cr_2O_7$	7 O's	$Cr_2(SO_4)_3$	12 O's
7 H_2SO_4	28 O's	K_2SO_4	4 O's
		7 H_2O	7 O's
	59 O's		59 O's

And so we see that the equation is correctly balanced.

What is the oxidizing agent? We know that the oxidizing agent is reduced, which means that it decreases in oxidation number. The Cr^{6+}, in going from $+6$ to $+3$, decreases in oxidation number, therefore, it is the oxidizing agent; or we might say that the $K_2Cr_2O_7$ is the oxidizing agent, since the Cr^{6+} is a part of this compound and the rest of it does not change in oxidation number. Likewise, the reducing agent is the Fe^{2+} or the $FeSO_4$ (which is oxidized).

Oxidation and reduction reactions produce the energy the body needs to carry out its normal functions. Oxidation-reduction in the body involves either oxygen or hydrogen, or both.

Enzymes involved in oxidation-reduction reactions in the body are called oxidoreductases (see page 428). Many of these enzymes are present in the mitochondria (page 429).

Self-test

Balance

$$Fe_2O_3 + CO \longrightarrow Fe + CO_2$$

Answer

$$Fe_2O_3 + 3\,CO \longrightarrow 2\,Fe + 3\,CO_2$$

6-5 Importance of Oxidation-Reduction

Antiseptic Effects

Because they are oxidizing agents, many *antiseptics have the property of killing bacteria. Among these is chlorine, which oxidizes organic matter and bacteria and so is used in the treatment of water to make it potable (see page 156). Calcium hypochlorite, $Ca(OCl)_2$, another commonly used oxidizing agent and bleaching powder, is used as a disinfectant for clothes and hospital beds. Table 6-2 lists some of the common antiseptics that are oxidizing agents.

Formaldehyde and sulfur dioxide are two reducing agents used in disinfecting rooms formerly occupied by patients with contagious diseases.

Effects on Hair Protein

Oxidizing and reducing agents denature protein by affecting the disulfide bonds (see page 289) of the amino acid cysteine. Use is made of this effect in "home permanents." Hair protein is primarily keratin, and keratin contains a large amount of cysteine. During the treatment, a reducing agent is used first. This substance breaks the disulfide bonds in the hair protein. The hair is then shaped with rollers. The new shape is "set" by using an oxidizing agent, which forms new disulfide bonds in the desired places. The hair will retain its new shape only until new hair grows out. Then the entire process has to be repeated.

Black and White Photography

In black and white photography (as in the making of x-ray films), oxidation-reduction reactions occur.

Photographic film has an emulsion containing silver bromide (AgBr), which is highly sensitive to light. Exposure to light (or radiation) activates some of the silver ions in the emulsion.

Table 6-2 Antiseptic Agents		
Formula	Name	Use
3% H_2O_2	Hydrogen peroxide	Minor cuts and scratches
$KMnO_4$	Potassium permanganate	Treatment of infection in urethra and bladder
$KClO_3$	Potassium chlorate	Treatment of sore throat
I_2 in H_2O	Lugol's solution	Treatment of minor cuts
NaOCl	Sodium hypochlorite (Dakin's solution)	Treatment of wounds
$(C_6H_5CO)_2$	Benzoyl peroxide	Treatment of acne
Cl_2	Chlorine	Treatment of water
O_3	Ozone	Treatment of water
$Ca(OCl)_2$	Calcium hypochlorite	Disinfection of hospital beds
$C_6H_5 - C_6H_5OH$	O-phenylphenol (Lysol)	Household disinfectant
C_6H_5OH	Phenol and its derivatives	Mouthwashes

When the film is developed, the activated silver ions react with the developer, hydroquinone, which is a reducing agent. The activated silver ions are reduced to black metallic silver.

$$Ag^+ \longrightarrow Ag \downarrow$$

The next step in the process, called fixing, uses sodium thiosulfate ($Na_2S_2O_3$) to dissolve the unactivated silver bromide. If the remaining AgBr were not removed, it would gradually darken when later exposed to light.

After washing and drying, the negative is ready for use. Areas of the film that were exposed to the greatest amount of light (or radiation) are darkest, whereas those exposed to the least amount of light are the lightest.

That is, in a negative, the dark and light areas of the object are reversed.

A print, called a positive, is made by shining light through the negative onto a piece of photographic paper containing a silver bromide emulsion. The same processes as in developing the negative are used. But this time, the dark areas of the negative (which come from the light areas of the object) appear light and the light areas on the negative appear dark (see below).

Photographic positive

Photographic negative

Breath-Alcohol Analyzer

Reactions involving oxidation-reduction are used to measure the amount of alcohol in a driver's breath. A sample of the driver's breath is blown through an orange-colored solution of acidified potassium dichromate. If alcohol, which is a reducing agent, is present, it causes the following reaction to take place.

$$3\ C_2H_5OH\ +\ 2\ K_2Cr_2O_7\ +\ 8\ H_2SO_4\ \longrightarrow$$

alcohol   potassium   sulfuric
             dichromate   acid

$$3\ CH_3COOH\ +\ 2Cr_2(SO_4)_3\ +\ 2\ K_2SO_4\ +\ 11\ H_2O$$

      acetic   chromic   potassiuim   water
       acid   sulfate   sulfate

The chromic sulfate thus produced is green. The greater the amount of alcohol in a driver's breath, the greater the change from orange to green. The actual alcohol content can be determined by comparing the color produced with that of a standardized chart.

6-6 Other Breath Tests

Analysis of human breath has confirmed the presence of nearly 400 gaseous compounds. One of these tests, for alcohol, as indicated in the previous paragraph, has been widely used. However, the detection of other gases in the breath can be of diagnostic value.

The detection of acetone in the breath (see page 483) is an indication of uncontrolled diabetes mellitus.

Some breath analysis tests require an individual to consume large quantities of a specific precursor to a volatile compound. The disease may reveal itself by the presence of certain breakdown products that appear in the breath.

In the test for malabsorption syndrome, a patient is given an oral dose of xylose, a carbohydrate. The appearance of large quantities of hydrogen in the breath in the succeeding few hours is a confirmation of this syndrome.

Damage to the pancreas can be detected by the administration of rice starch orally, and then testing the breath for hydrogen.

Pancreatic disease can also be detected by using radioactive carbon, ^{14}C. The amount of radioactive carbon dioxide can be an indication of pancreatic malfunction.

The use of galactose containing ^{14}C can be used to test for liver damage by testing for the amount of radioactive carbon dioxide.

Another test for liver disease involves the presence of dimethyl sulfide in the breath.

Summary

Oxidation is defined as a loss of electrons. Oxidation is also an increase in oxidation number, a combination with oxygen, or a loss of hydrogen.

Reduction is a gain of electrons. Reduction is also a decrease in oxidation number, a gain of hydrogen, or a loss of oxygen. Oxidation can never take place without reduction, and vice versa.

Whatever is oxidized is called a reducing agent.

Whatever is reduced is called an oxidizing agent.

Oxidizing and reducing agents are useful as antiseptics and also as stain removers.

Questions and Problems

1. In the following equations, indicate the oxidizing agent and the reducing agent.
   a. $S + O_2 \longrightarrow SO_2$
   b. $Mg + 2 HCl \longrightarrow MgCl_2 + H_2$
   c. $Fe + CuSO_4 \longrightarrow FeSO_4 + Cu$
   d. $Cl_2 + 2 NaBr \longrightarrow 2 NaCl + Br_2$

2. In the following equations, what was oxidized and what was reduced?
   a. $2 Al + 3 O_2 \longrightarrow Al_2O_3$
   b. $Zn + H_2SO_4 \longrightarrow ZnSO_4 + H_2$
   c. $2 N_2 + 3 O_2 \longrightarrow 2 N_2O_3$
   d. $C_6H_{12}O_6 + 6 O_2 \longrightarrow 6 CO_2 + 6 H_2O$
   e. $3 Fe + 2 O_2 \longrightarrow Fe_3O_4$

3. How does a breath analyzer work?

4. Why can oxidation never take place without reduction?

5. Balance the following oxidation-reduction reactions:
   a. $FeCl_3 + SnCl_2 \longrightarrow FeCl_2 + SnCl_4$
   b. $P + HNO_3 + H_2O \longrightarrow NO + H_3PO_4$
   c. $C_{12}H_{22}O_{11} + O_2 \longrightarrow CO_2 + H_2O$
   d. $Al + H_2SO_4 \longrightarrow Al_2(SO_4)_3 + H_2$

6. In each equation in question 5, indicate what was oxidized, what was reduced, the oxidizing agent, and the reducing agent.

7. Give two examples of antiseptic agents and indicate their use.

8. Give an example of reduction involving (a) gain of electrons, (b) loss of oxygen, and (c) gain of hydrogen.

9. Consider the reaction $2 N_2 + 5 O_2 \rightleftharpoons 2 N_2O_5$. Does it involve oxidation-reduction in both directions? Are all oxidation-reduction reactions reversible?

10. Discuss oxidation-reduction in terms of disulfide bonds.

11. Determine the substance oxidized, the substance reduced, the oxidizing agent, and the reducing agent in the breath-alcohol test.

12. The following reaction takes place when light strikes a photographic film.
$$2 AgBr + light \longrightarrow 2 Ag + Br_2$$
What is oxidized and what is reduced in this reaction?

13. The overall reaction in a car battery is
$$Pb + PbO_2 + H_2SO_4 \longrightarrow PbSO_4 + H_2O$$
Balance this equation. What is oxidized? reduced?

14. Chlorine is manufactured by the following process:
$$NaCl + H_2O \xrightarrow[\text{current}]{\text{electrical}} Cl_2 + NaOH + H_2$$

Balance this equation. What is oxidized? reduced?

Practice Test

1. Oxidation _____.
   a. always involves hydrogen
   b. always involves oxygen
   c. occurs in nuclear reactions
   d. can never take place without reduction

2. Oxidation may involve _____.
   a. a gain of oxygen
   b. a gain of electrons
   c. a gain of hydrogen
   d. a decrease in oxidation number

Use the following balanced oxidation-reduction equation for questions 3 and 4.
$$K_2Cr_2O_7 + 3 H_2S + 4 H_2SO_4 \longrightarrow$$
$$K_2SO_4 + Cr_2(SO_4)_3 + 3 S + 7 H_2O$$

3. The oxidizing agent is _____.
   a. $K_2Cr_2O_7$ b. H_2S c. H_2SO_4 d. S

4. The substance oxidized is _____.
   a. $K_2Cr_2O_7$ b. H_2S c. H_2SO_4 d. S

5. Oxidizing and reducing agents denature protein by affecting what type of bond?
   a. triphosphate   b. oxygen
   c. disulfide   d. hydrogen

6. In the reaction $C_6H_{10}O_5 + 6 O_2 \longrightarrow$ $6 CO_2 + 5 H_2O$ the oxidation number of carbon _____.
   a. increases by 2   b. increases by 4
   c. decreases by 2   d. decreases by 4

7. The use of a breath-alcohol analyzer is based on the fact that alcohol acts as a(n) _____.
   a. analgesic   b. antiseptic
   c. reducing agent   d. oxidizing agent

8. The reaction $2 S + 3 O_2 \longrightarrow 2 SO_3$ is an example of _____.
   a. oxidation
   b. reduction
   c. oxidation and reduction
   d. neither oxidation nor reduction

9. Oxidizing agents can act as _____.
   a. antipyretics   b. analgesics
   c. antiseptics   d. all of these

10. An oxidizing agent _____.
    a. is oxidized   b. is reduced
    c. loses electrons   d. always gains hydrogen

7

The Gaseous State

Objectives

- To understand the kinetic molecular theory
- To use the various gas laws
- To use molar gas volumes in calculations

The presence of boron in the incandescent sample at center is revealed by a green-colored flame. A flame is a gaseous substance whose atoms, excited by the intense heat, emit light of a color characteristic of the chemical composition of the gas.

7-1 General Properties

In Chapter 2 we saw that gases have no definite shape, no definite volume, and a low density. Gases have a much greater volume than an equal mass of solid or liquid. For example, 1 g of liquid water occupies 1 mL and 1 g of solid water (ice) occupies nearly 1 mL, but 1 g of water vapor (at 0° C and 1 atm pressure) occupies nearly 1250 mL. These and other properties of gases can be explained in terms of the kinetic molecular theory.

7-2 The Kinetic Molecular Theory

The principal assumptions of the kinetic molecular theory are as follows.

1. Gases consist of tiny particles called molecules.
2. The distance between the molecules of gas is very great compared to the size of the molecules themselves (that is, the volume occupied by a gas is mostly empty space).
3. Gas molecules are in rapid motion and move in straight lines, frequently colliding with each other and with the walls of the container.
4. Gas molecules do not attract each other.
5. When molecules of a gas collide with each other or with the walls of the container, they bounce back with no loss of energy. Such collisions are said to be perfectly elastic.
6. The average kinetic energy of the molecules is the same for all gases at the same temperature. The average kinetic energy increases as the temperature increases and decreases as the temperature decreases.

Let us see how the various properties of gases can be explained in terms of this theory.

1. If the distance between molecules is very great compared to the size of the molecules themselves, the molecules will occupy only a small fraction of that volume so the density will be very low.
2. If the molecules are moving rapidly in all directions, they can fill any size container. They can keep on moving until they hit a wall of the container or until they hit each other and bounce back; that is, the gas will have no definite volume and no definite shape.
3. If the molecules of a gas strike the walls of a container, they should exert a pressure on each wall. And, since the molecules are moving in all directions, they should exert a pressure equally in all directions. Gases do just this.
4. If a bottle of ether is opened in a room, the odor is soon apparent in all parts of that room; that is, the molecules of ether gas diffuse into the air (a mixture of gases). According to the kinetic molecular theory, there is a great deal of empty space between the molecules of a gas, so the ether molecules (or any other gas) can diffuse into the spaces between the air molecules.
5. We can show that the collisions must be perfectly elastic by means of a reverse type of reasoning. Suppose that the collisions between molecules were not perfectly elastic; that is, suppose some energy was lost upon collision with other molecules and with the walls of the container. Eventually, the gas molecules would have so little energy left that they would settle to the bottom of the container. However, gases never settle. Therefore, the collisions between the molecules themselves and with the walls of the container must be perfectly elastic.

6. Since gas molecules are so far apart, it should be possible to force them closer together by increasing the pressure; that is, gases should be compressible, as indeed they are.

7-3 Measurement of Pressure

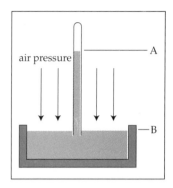

FIGURE 7-1
Pressure is measured by a barometer.

Air is all around us and consists of many molecules moving rapidly in all directions. We do not feel the pressure exerted by the air molecules on our body because we experience the same pressure from the air molecules within our body. That is, the pressures are equal.

At high altitudes, the atmospheric pressure is lower because there are fewer air molecules there. Conversely, the atmospheric pressure increases greatly with increasing depth in the ocean.

If we change altitude rapidly, we can feel the sudden change in pressure. When riding in a rapidly ascending elevator, our ears may "pop." Similarly, deep-sea divers must breathe air under high pressure to compensate for the increased pressure exerted by the water. The farther down they go, the greater the pressure.

Pressure is measured by a barometer, one form of which is shown in Figure 7-1. A glass tube about 1 yard long is filled with mercury and placed open end down in a dish of mercury. The mercury falls in the tube, leaving a vacuum above it, until the pressure exerted by the air just balances the mercury column in the tube. Thus, the atmospheric pressure is expressed as being equal to so many millimeters of mercury (mm Hg) or the height of the column from A to B.

One unit of pressure is the torr, named after the Italian scientist Torricelli, who invented the barometer. One torr is the pressure exerted by 1 mm of mercury at sea level. The SI unit of pressure is the pascal (Pa).

Standard Temperature and Pressure

Standard pressure is 1.00 atm and will support a column of mercury 760 mm tall. Standard pressure can be expressed in many different units. Among these are the following:

760 mm Hg	29.92 in/Hg	1.013×10^5 pascals (Pa)
76.0 cm Hg	14.7 lb/in^2	101.3 kilopascals (kPa)
760 torr		

Standard temperature is 0° C or 273 K. Standard temperature and pressure is abbreviated STP.

7-4 The Gas Laws

Boyle's Law

If the volume of a gas is reduced, the molecules will have less space in which to move. Therefore, they will strike the walls of the container more often and cause a greater pressure.

The relationship between the volume of a given quantity of a gas and its pressure is expressed by Boyle's law, which states that the volume occupied by a gas is inversely proportional to the pressure if the temperature remains constant.

Figure 7-2 shows a graph of pressure versus volume. Note that as the pressure increases the volume of the gas decreases, and as the pressure decreases the volume of the gas increases.

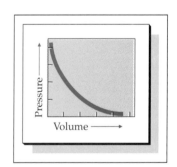

FIGURE 7-2
Graph illustrating Boyle's law; pressure and volume are related inversely.

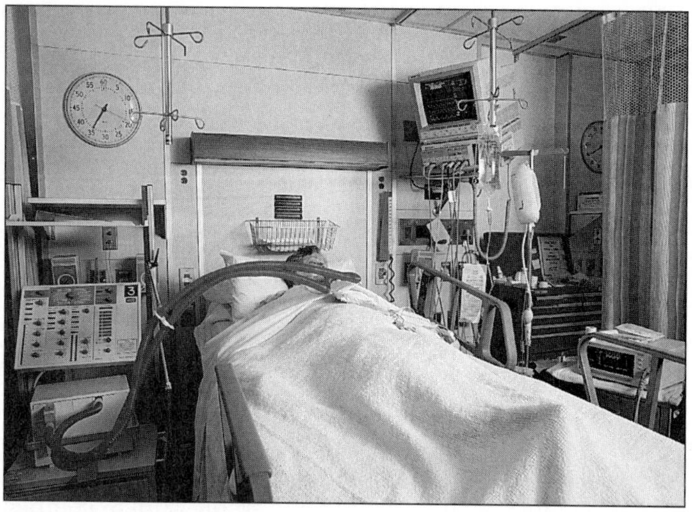

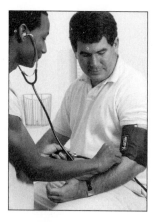

FIGURE 7-3

A chest respirator is a mechanical application of Boyle's law that is used to aid patients with respiratory difficulties.

FIGURE 7-4

A sphygmomanometer being used to monitor blood pressure.

In normal breathing, the diaphragm moves downward, allowing the lungs to expand. The increased volume in the lungs causes the pressure in the lungs to drop slightly. Since air always flows from an area of higher pressure to one of lower pressure, air flows into the lungs. When the diaphragm moves upward, the volume of the lungs decreases and the pressure of the air inside the lungs increases. Now the air flows out of the lungs (again from an area of high pressure to one of lower pressure).

A direct application of Boyle's law is seen in a chest respirator, a machine used in the treatment of patients with respiratory difficulties. When the pressure inside the respirator is decreased, the air in the lungs expands, forcing the diaphragm down. When the pressure in the respirator is increased, the volume of air in the lungs is decreased, allowing the diaphragm to move upward again. This alternate increase and decrease in pressure enables the patient to breathe even though he or she cannot control the movement of the diaphragm muscles.

Another example of Boyle's law is in the use of a sphygmomanometer, a device used to measure blood pressure. When the rubber bulb is squeezed, the volume of air in that bulb is decreased and its pressure is increased. This increased pressure is transmitted to the cuff (see Figure 7-4).

Boyle's law can be stated mathematically as

$$P_1V_1 = P_2V_2 \qquad \text{(at constant temperature)}$$

where P_1 and P_2 are the initial and final pressures, and V_1 and V_2 are the initial and final volumes, respectively. Pressures are usually expressed in millimeters of mercury (mm Hg). Recall that 760 mm Hg equals 1 atm pressure.

Exercise 7-1

What volume will 500 mL of gas initially at 25° C and 750 mm Hg occupy when conditions change to 25° C and 650 mm Hg?

Solution

Note first that the temperature is constant, so we can use Boyle's law. Note also that P_1 is 750 mm Hg, V_1 is 500 mL, and P_2 is 650 mm Hg. Then, using $P_1V_1 = P_2V_2$, we have

$$750 \text{ mm Hg} \times 500 \text{ mL} = 650 \text{ mm Hg} \times V_2$$

Dividing by 650 mm Hg, we have

$$\frac{750 \text{ mm Hg} \times 500 \text{ mL}}{650 \text{ mm Hg}} = V_2 \quad \text{and} \quad V_2 = 577 \text{ mL}$$

Self-test

A gas exerts a pressure of 858 torr when confined in a 5.00 L container. What will be the pressure if the gas is confined in a 10.0 L container at constant temperature?

Answer

$P_2 = 429$ torr

Charles's Law

When gases are heated, they expand; when gases are cooled, they contract. The relationship between volume and temperature is expressed by Charles's law, which states that the volume of a fixed quantity of gas is directly proportional to its Kelvin temperature if the pressure remains constant. Recall that Kelvin (absolute) temperature is Celsius temperature plus 273 (see page 18).

Charles's law can be explained in terms of the kinetic molecular theory. As the temperature of a gas increases, the molecules move faster and strike the walls of the container more often. However, if the pressure is kept constant, then the volume must increase; that is, at constant pressure, the higher the temperature, the greater the volume of a gas, and vice versa.

A direct application of Charles's law can be seen in such equipment as an incubator. When air comes into contact with the heating element, it expands and becomes lighter. This lighter air rises, causing a circulation of warm air throughout the incubator (see Plate 1).

Charles's law can be expressed mathematically as

$$\frac{V_1}{T_1} = \frac{V_2}{T_2} \qquad \text{(at constant pressure)}$$

where V_1 and V_2 are the initial and final volumes and T_1 and T_2 the initial and final Kelvin temperatures, respectively.

Exercise 7-2

A sample of gas occupies 368 mL at 27° C and 600 mm Hg. What will be the volume of that gas at 127° C and 600 mm Hg?

Solution

We note that pressure is constant, so we can use Charles's law. Also note that V_1 is 368 mL, T_1 is 27 + 273, or 300 K, and T_2 is 127 + 273, or 400 K. Then, substituting into:

$$\frac{V_1}{T_1} = \frac{V_2}{T_2}$$

we have

$$\frac{368 \text{ mL}}{300 \text{ K}} = \frac{V_2}{400 \text{ K}}$$

$$V_2 = 491 \text{ mL}$$

Self-test

At constant pressure 200 mL of gas at 35° C is cooled to −20° C. What will be its new volume?

Answer

$V_2 = 164 \text{ mL}$

Gay-Lussac's Law

Gay-Lussac's law expresses the relationship between the pressure exerted by a gas and its temperature. These two factors are directly proportional. That is, as the temperature of a gas increases, the pressure increases, and vice versa, if the volume remains constant.

Gay-Lussac's law may be explained in terms of the *kinetic molecular theory*. In a fixed volume of gas (such as in a rigid container), the molecules move at random. As the temperature increases, the molecules move faster and have greater kinetic energy. When these faster moving molecules strike the walls of the container, they exert a greater pressure. Conversely, as the temperature decreases, the molecules move more slowly and so the pressure drops.

Gay-Lussac's law may be expressed mathematically as

$$\frac{P_1}{T_1} = \frac{P_2}{T_2} \qquad \text{(at constant pressure)}$$

where P_1 and P_2 are the initial and final pressures, and T_1 and T_2 are the initial and final Kelvin temperatures, respectively.

Exercise 7-3

Exercise 7-3a

What will be the final pressure of a given volume of a gas at 140° C and 740 mm Hg pressure if the temperature is increased to 175° C?

Solution

Using

$$\frac{P_1}{T_1} = \frac{P_2}{T_2}$$

where P_1 is 740 mm Hg, T_1 is 140° C or 140 + 273 = 413 K, and T_2 is 175° C or 175 + 273 = 448 K.

$$\frac{740 \text{ mm}}{413 \text{ K}} = \frac{P_2}{448 \text{ K}}$$

$$P_2 = 803 \text{ mm Hg}$$

Exercise 7-3b

500 mL steam at 100° C and 1.00 atm pressure is heated until the pressure becomes 1.05 atm (volume remaining constant). What will the new temperature be in degrees Celsius?

Solution

Using

$$\frac{P_1}{T_1} = \frac{P_2}{T_2}$$

where P_1 is 1.00 atm, P_2 is 1.05 atm, T_1 is 100° C or 100 + 273 = 373 K,

$$\frac{1.00 \text{ atm}}{373 \text{ K}} = \frac{1.05 \text{ atm}}{T_2}$$

$$T = 392 \text{ K}$$

$$T_2 = 392 - 273 = 119° \text{ C}$$

Self-test

An automobile tire has a pressure of 24.2 lb/in² at 20.4° C.

After driving for an hour the tire pressure has increased to 28.3 lb/in².

Assuming that the tire's volume did not change, what is the tire's temperature now? Note that tire pressures are always measured in excess of atmospheric pressure (14.7 lb/in²).

Answer

51.3° C

A common application of Gay-Lussac's law is the autoclave, a device used in hospitals for sterilization (see Figure 7-5). The normal temperature of steam is 100° C, but in an autoclave it can rise as high as 120° C because of the increased pressure. This higher temperature is sufficient to destroy any microorganisms that may exist in the material being autoclaved.

FIGURE 7-5

An autoclave uses pressurized steam to sterilize medical instruments. The increased pressure raises the temperature of the steam to a level that is sufficiently high to kill most microorganisms.

Combined Gas Laws

Boyle's law refers to the volume of a fixed quantity of a gas at constant temperature; Charles's law refers to the volume of such a gas at constant pressure. However, frequently neither of these factors is constant. In such a case we use the combined gas laws, which can be stated mathematically as

$$\frac{P_1 V_1}{T_1} = \frac{P_2 V_2}{T_2}$$

Exercise 7-4

What volume will 250 mL of gas at 27° C and 800 mm Hg occupy at STP?

Solution

Using

$$\frac{P_1 V_1}{T_1} = \frac{P_2 V_2}{T_2}$$

we have

$$\frac{800 \text{ mm Hg} \times 250 \text{ mL}}{300 \text{ K}} = \frac{760 \text{ mm Hg } V_2}{273 \text{ K}}$$

$$V_2 = 239 \text{ mL}$$

Self-test

A high-altitude balloon has a volume of 2.43 L at 755 torr and 20.1° C. What is the volume at 1.00 torr and −10.1° C?

Answer

1.64×10^3 L

Dalton's Law

Dalton's law refers to a mixture of gases rather than to a pure gas. Dalton's law states that in a mixture of gases, each gas exerts a partial pressure proportional to its concentration. For example, if air contains 21 percent oxygen, then 21 percent of the total air pressure is exerted by the oxygen. Normal air pressure of 1 atm will support a mercury column 760 mm high. The partial pressure of the oxygen in the air would be 21 percent of 760 mm Hg, or $0.21 \times 760 = 160$ mm Hg.

Dalton's law can be explained in terms of the kinetic molecular theory. Since there is no attraction between gas molecules, each kind of molecule strikes the walls of the container the same number of times per second as if it were the only kind of molecule present. That is, the pressure exerted by each gas (its partial pressure) is not affected by the presence of other gases. Each gas exerts a partial pressure proportional to the number of molecules of that gas (proportional to its concentration).

Gases always diffuse from an area of higher partial pressure to one of lower partial pressure. An example of the diffusion of gases caused by a difference in partial pressures is found in our own bodies (see Figure 7-6). The partial pressure of oxygen in the inspired air is 158 mm Hg. The partial pressure of oxygen in the *alveoli is 104 mm Hg. Therefore, oxygen passes from the lungs into the alveoli

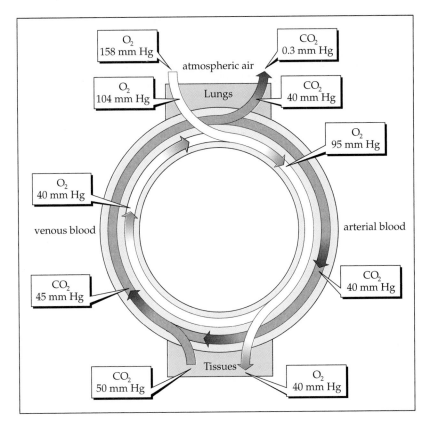

FIGURE 7-6
Diagram showing oxygen and carbon dioxide (CO_2) flow and partial pressures in body.

(from a higher partial pressure to a lower one). From the alveoli the oxygen diffuses into the venous blood (from a partial pressure of 104 mm Hg to one of 40 mm Hg). This diffusion of oxygen into the venous blood in the lungs changes the venous blood into arterial blood, in which the partial pressure of oxygen is 95 mm Hg. When the arterial blood reaches the tissues, where the partial pressure of oxygen is 40 mm Hg, oxygen diffuses to those tissues (again from a higher partial pressure to a lower one). When the arterial blood loses oxygen to the tissues, its oxygen partial pressure drops to 40 mm Hg, and it becomes venous blood, which returns to the lungs to begin the cycle anew.

Conversely, in the tissues the partial pressure of the carbon dioxide is 50 mm Hg, and in the arterial blood it is 40 mm Hg, so that carbon dioxide diffuses out of the tissues into the blood. When the arterial blood picks up carbon dioxide (and at the same time loses oxygen) it becomes venous blood with a carbon dioxide partial pressure of 45 mm Hg. This venous blood, in turn, loses carbon dioxide to the alveoli, where the carbon dioxide partial pressure is 40 mm Hg. From the alveoli the carbon dioxide passes into the lungs (from a partial pressure of 40 mm Hg to one of 0.3 mm Hg) and then is exhaled.

Graham's Law

If a container of ammonia and a container of ether are opened simultaneously, a person standing at the far end of the room will notice the odor of ammonia before that of the ether. Both gases will diffuse through the air, but ammonia has a molecular mass of 17 and ether a molecular mass of 74. Lighter molecules diffuse faster than heavier ones. This relationship was stated by Graham as follows: "The rates of diffusion of gases are inversely proportional to the square roots of the molecular masses" (or inversely proportional to the square roots of the densities).

$$\frac{\text{rate of diffusion gas}_1}{\text{rate of diffusion gas}_2} = \sqrt{\frac{\text{molecular mass gas}_2}{\text{molecular mass gas}_1}}$$

Exercise 7-5

Compare the relative rates of diffusion of CH_4 and O_2.

Solution

$$\frac{\text{rate of diffusion } CH_4}{\text{rate of diffusion } O_2} = \frac{\sqrt{\text{molecular mass of } O_2}}{\sqrt{\text{molecular mass of } CH_4}}$$

$$\frac{\text{rate of diffusion } CH_4}{\text{rate of diffusion } O_2} = \frac{\sqrt{32}}{\sqrt{16}} = \frac{5.6}{4.0} = \frac{1.4}{1}$$

so that Ch_4 will diffuse 1.4 times as rapidly as O_2.

Self-test

An unknown gas will diffuse 2.92 times more slowly than H_2; what is the molar mass of this gas?

Answer

17 g/mol

Henry's Law

Carbonated drinks are made by dissolving carbon dioxide, under pressure, in a given liquid. When the container is opened, the pressure is released and bubbles of carbon dioxide escape.

This relationship is expressed in terms of Henry's law, which states, "the solubility of a gas in a liquid at a given temperature is proportional to the pressure of that gas." That is, more gas dissolves at higher pressures and less at lower pressures.

Oxygen should be more soluble in blood at pressures above normal. Hyperbaric chambers, in which air pressures are two to three times normal, are used in the treatment of cancer (see Section 13-13) and for victims of scuba diving accidents (see Figure 7-7). Decompression sickness can occur in any individual who has been breathing compressed gas at depths greater than 30 ft. Flying within 24 hr of diving can precipitate illness after an otherwise safe decompression.

Deep-sea divers are subject to pressures far above normal and so can have greater amounts of dissolved gases in their bloodstream. This condition might lead to the "bends" (see page 136).

Molar Gas Volumes

The volume occupied by 1 mol of any gas at STP (standard *t*emperature and *pres*sure) is 22.4 L. However, this volume may be used only at STP. If a gas is at some other temperature or pressure, its volume must first be converted to STP before the molar gas volume can be used.

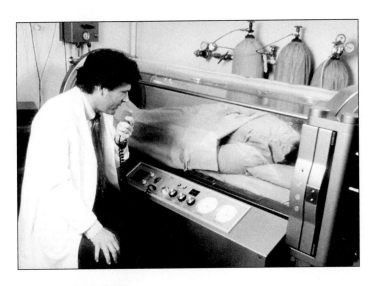

FIGURE 7-7
A hyperbaric chamber in use.

Exercise 7-6

Exercise 7-6a

How many moles of O_2 gas are present in 5.6 L at STP?
Since 1 mol of a gas at STP occupies 22.4 L, then

$$5.6 \, \cancel{L} \times \frac{1 \, \text{mol}}{22.4 \, \cancel{L}} = 0.25 \, \text{mol} \, O_2$$

Exercise 7-6b

What volume will 1.50 mol of CH_4 gas occupy at STP?

Solution

$$1.50 \, \cancel{\text{mol}} \times \frac{22.4 \, \text{L}}{1 \, \cancel{\text{mol}}} = 33.6 \, \text{L at STP}$$

Exercise 7-6c

At STP, 2.50 g of a gas occupies 500 mL. What is the molecular mass of that gas?

Solution

First we must convert the volume to liters.

$$\frac{2.50 \, \text{g}}{500 \, \cancel{\text{mL}}} \times \frac{1000 \, \cancel{\text{mL}}}{1 \, \text{L}} = \frac{5.00 \, \text{g}}{1 \, \text{L}}$$

Then using the conversion factor that 1 mol occupies 22.4 L,

$$\frac{5.00 \, \text{g}}{1 \, \cancel{\text{L}}} \times \frac{22.4 \, \cancel{\text{L}}}{1 \, \text{mol}} = \frac{112 \, \text{g}}{1 \, \text{mol}}$$

That is, 1 mol of the gas has a mass of 112 g, so the molecular mass is 112.

Self-test

Dry ice sublimes to gaseous CO_2. What volume will 1.00 g of dry ice occupy at STP?

Answer

0.509 L

Exercise 7-7

Calculate the density of O_2 gas at STP.

Solution

Since 1 mol of any gas at STP occupies 22.4 L and since the molecular mass of O_2 is 32.0,

$$D = \frac{M}{V} = \frac{32.0 \text{ g}}{22.4 \text{ L}} = 1.43 \text{ g/L}$$

Molar gas volumes may also be used in calculations involving chemical reactions.

Exercise 7-8

Given the balanced reaction

$$MnO_2 + 4 HCl \qquad MnCl_2 + Cl_2 + 2 H_2O$$

Calculate
a. the number of moles of Cl_2 produced from 2.00 moles of MnO_2
b. the number of liters of Cl_2 at STP produced from 2.00 moles of MnO_2

Solution

a. number of moles of Cl_2 = 2.00 $\cancel{\text{mol } MnO_2} \times \dfrac{1 \text{ mol } Cl_2}{1 \cancel{\text{mol } MnO_2}}$ = 2.00 mol Cl_2

b. number of liters of Cl_2 =

$$2.00 \cancel{\text{mol } MnO_2} \times \frac{1 \cancel{\text{mol } Cl_2}}{1 \cancel{\text{mol } MnO_2}} \times \frac{22.4 \text{ L } Cl_2}{1 \cancel{\text{mol } Cl_2}} = 44.8 \text{ L } Cl_2$$

Self-test

a. Calculate the molar mass of a gas that has a density of 1.342 g/L at STP.
b. Pure oxygen was first prepared by heating mercuric oxide (HgO). What volume of O_2 can be prepared at STP by heating 20.0 g of HgO?

$$2 HgO \qquad 2 Hg + O_2$$

Answers

a. 30.1 g/mol   b. 1.03 L

The Ideal Gas Law

The ideal gas law indicates the relationship between the pressure, temperature, volume, and the number of moles of a gas. It may be expressed as

$$PV = nRT$$

where

> P is the pressure in atmospheres
>
> V is the volume in liters
>
> n is the number of moles
>
> T is the temperature measured in Kelvins
>
>   and R is a constant 0.0281 L $\cdot$ atm/mol $\cdot$ K

Exercise 7-9

Exercise 7-9a

Calculate the volume occupied by 2.00 moles of CO_2 gas at 35.0° C and 1.20 atm pressure.

Solution

$$PV = nRT$$

$$V = \frac{nRT}{P} \quad (\text{where } T = 35.0 + 273 = 308 \text{ K})$$

$$V = \frac{2.00 \text{ mol} \times \dfrac{0.0281 \text{ L} \cdot \text{atm}}{\text{mol K}} \times 308 \text{ K}}{1.20 \text{ atm}}$$

$$V = 42.1 \text{ atm}$$

Exercise 7-9b

Using the ideal gas law, calculate the volume occupied by 0.500 mol helium gas in a 5.00 L container at 100° C.

$$PV = nRT$$

$$P = \frac{nRT}{V} \quad (\text{where } T = 100 + 273 = 373 \text{ K})$$

$$P = \frac{0.500 \text{ mol} \times \dfrac{0.0821 \text{ L} \cdot \text{atm}}{\text{mol K}} \times 373 \text{ K}}{5.00 \text{ L}}$$

$$P = 3.06 \text{ atm}$$

Self-test

A compressed air tank carried by scuba divers has a volume of 8.0 L and a pressure of 140 atm at 20.1° C. How many moles of gas are in the tank?

Answer

46.5 mol

Summary

The properties of gases can be explained in terms of the kinetic molecular theory, which states that (1) gases are composed of tiny particles called molecules; (2) the distances between molecules of a gas are very great compared to the size of the molecules themselves; (3) gas molecules move rapidly in straight lines; (4) gas molecules do not attract each other; (5) collisions between molecules and between the molecules and the walls of the container are perfectly elastic; and (6) the average kinetic energy of the molecules is the same for all gases at the same temperature.

Pressure is measured by means of a barometer. Standard pressure is 1 atm and will support a column of mercury 760 mm tall.

Boyle's law states that the volume of a fixed quantity of a gas is inversely proportional to the pressure if the temperature remains constant. Boyle's law can be expressed mathematically as $P_1V_1 = P_2V_2$.

Charles's law states that the volume of a fixed quantity of a gas is directly proportional to its Kelvin temperature if the pressure remains constant. Charles's law can be stated mathematically as $V_1/T_1 = V_2/T_2$.

Gay-Lussac's law states that pressure exerted by a gas is directly proportional to its Kelvin temperature if the volume remains constant.

The combined gas laws can be stated mathematically as $P_1V_1/T_1 = P_2V_2/T_2$.

Dalton's law states that in a mixture of gases each gas exerts a partial pressure proportional to its concentration.

Gases diffuse from an area of higher partial pressure to one of lower partial pressure.

Graham's law states that the rates of diffusion of gases are inversely proportional to the square roots of their molecular masses.

The molar gas volume is 22.4 L at STP. It is the volume occupied by 1 mol of any gas under those conditions.

Questions and Problems

1. Explain in terms of the kinetic molecular theory (a) why gases are compressible, (b) why gases have no definite shape, (c) why gases diffuse into each other, (d) why gases have a low density, (e) why gases do not settle, (f) Charles's law, and (g) Boyle's law.

2. What volume will 400 mL of argon gas initially at 30° C and 725 mm Hg occupy at 30° C and 650 mm Hg?

3. Calculate the pressure at which 2.50 L of hydrogen at 25° C and 625 mm Hg will occupy 1.40 L at 25° C.

4. What will be the volume of 100 mL of neon at 127° C and 450 mm Hg when the temperature rises to 227° C at 900 mm Hg?

5. When 4.00 ft³ of carbon dioxide gas at 30° C and 1.25 atm pressure is cooled to −10° C and 1.25 atm pressure, what is its volume?

6. What volume will 120 mL of oxygen gas at 25° C and 800 mm Hg occupy at STP (0° C and 760 mm Hg)?

7. Why does carbon dioxide gas pass from the cells to the bloodstream instead of vice versa?

8. Explain the flow of oxygen from the air into the cells in terms of partial pressures.

9. Explain how a chest respirator works.

10. Explain how an autoclave works in terms of the appropriate gas law.

11. Which gas will diffuse faster? carbon dixoide (CO_2) or nitrogen dioxide (NO_2)?

12. Arrange the following gases in order of decreasing rates of diffusion: Cl_2, H_2, CO_2, CH_4, CO, SO_2, Ar, F_2.

13. Explain in terms of the kinetic molecular theory why gases exert a pressure.

14. What is meant by the term *elastic collisions* in terms of gases? What consequence does this have on the property of a gas?

15. Explain how a barometer works.

16. What does a falling barometer indicate? Why?

17. How does a sphygmomanometer work? What is it used for?

18. Explain, in terms of partial pressure, the use of oxygen for patients with emphysema.

19. Define Henry's law. What use is made of this law in the treatment of cancer?

20. Explain why gases exert a pressure equally in all directions.

21. How many moles of CO are present in 4.5 L at STP?

22. What volume will 1.50 mol of C_2H_4 gas occupy at STP?

23. At STP, 0.50 g of a gas occupies 448 mL. What is the molecular mass of the gas?

24. Which represents the highest pressure: 800 torr, 15 lb/in^2, 30.00 in Hg, 81.0 cm Hg, or 120 kPa?

25. At what temperature would 1 mol of an ideal gas occupy 1 L at 1 atm pressure?

26. Gas pressure in interstellar space is approximately 10^{-17} torr at a temperature of 100 K. How many molecules are there in 1 L of interstellar space? If we assume that all the molecules are hydrogen, (H_2), what is the density of interstellar space?

27. A small cylinder of helium used for filling balloons has a volume of 2.30 L at 22.7° C. The helium pressure inside the cylinder is 80.4 atm. How many balloons with a volume of 2.0 L at a pressure of 2.0 atm and a temperature of 22.7° C can be filled by the cylinder?

28. Ammonium nitrate is a fertilizer and can also be used as an explosive. When heated, it explodes by this reaction:

$$2\ NH_4NO_3 \longrightarrow 2\ N_2 + 4\ H_2O + O_2$$

If all the products are gases, how many liters of gaseous products at 770° C and 1.00 atm of pressure are produced by the explosion of 100.0 kg of ammonium nitrate?

29. The surface temperature of Venus is about 1050 K and the air pressure is about 75 atm. Assuming that this represents STP on Venus, what is the volume of 1 mol of O_2 on the surface of Venus? density?

30. The main engines on the space shuttle rocket use the following reaction:

$$2\ H_2 + O_2 \longrightarrow 2\ H_2O$$

Assuming that H_2 and O_2 diffuse into the reaction chamber through identical holes, what must the ratio of hydrogen holes to oxygen holes be to maintain a 2:1 hydrogen-to-oxygen ratio?

Practice Test

1. Gases are compressible because the molecules
   a. have elastic collisions
   b. do not attract one another
   c. are in rapid motion
   d. are far apart

2. The molar gas volume at STP, in liters, is _____.
   a. 11.2   b. 22.4   c. 44.8   d. 67.2

3. Blood pressure can be measured with a _____.
   a. hygrometer   b. sphygmomanometer
   c. thermometer   d. barometer

4. As the pressure of a gas decreases, its solubility in water _____.
   a. increases   b. decreases   c. is unaffected

5. Charles's law applies when _____ remain(s) constant.
   a. pressure   b. temperature
   c. volume   d. a, b, and c

6. Boyle's law applies when _____ remains constant.
   a. pressure   b. temperature
   c. volume   d. all of these

7. The partial pressure of oxygen is higher _____.
   a. in the tissues than in the alveoli
   b. in the air than in the lungs
   c. in the tissues than in the bloodstream
   d. in venous blood than in arterial blood

8. 100 mL of a gas at 0° C and 600 mm Hg occupy what volume at 0° C and 1200 mm Hg?
   a. 50 mL   b. 100 mL
   c. 150 mL   d. 200 mL

9. 1000 mL of O_2 at 127° C and 800 mm Hg occupy what volume at −73° C and 1200 mm Hg?
   a. 333 mL   b. 750 mL
   c. 1333 mL   d. 3000 mL

10. How many moles of NO_2 gas are in 5.6 L at STP?
    a. 0.25   b. 0.5   c. 2   d. 4

8

Oxygen and Other Gases

Objectives

- To become acquainted with the uses and properties of oxygen

- To become familiar with problems associated with various noxious gases

The gaseous phase is unique among the three physical states in that a small change in ambient temperature or pressure strongly affects its density. In this photo, a balloonist fills his craft with hot gases from its burner prior to ascent. This heated exhaust is lower in density than the surrounding air. This phenomenon will cause the balloon to rise once it has been filled.

8-1 Oxygen

Occurrence

Oxygen is the most abundant element on the Earth's surface. Air is about 21 percent free oxygen. The oceans and lakes on the Earth's surface consist of about 80 percent oxygen in the combined state. The Earth's crust consists of about 50 percent oxygen, combined mostly with silicon. Oxygen compounds comprise most of the weight of plants and animals.

Properties

Physical Properties   At room temperature oxygen is a colorless, odorless, tasteless gas. It is slightly heavier than air and is slightly soluble in water.

The method of preparation illustrated in Figure 8-1 shows that oxygen does not dissolve appreciably in water. If it were soluble in water, it could not be collected by this method. However, a small amount of oxygen does dissolve in water. This amount, though small, is of very definite importance to marine life. It is this small amount of dissolved oxygen that enables fish and other aquatic animals to "breathe."

The density of oxygen is 1.43 g/L. The density of air is 1.29 g/L. Therefore, oxygen is slightly heavier than air. When oxygen is collected in the laboratory, it is kept in covered bottles with the mouth upward.

When cooled sufficiently, oxygen forms a pale blue liquid that boils at $-182.5°$ C. Further cooling produces a pale blue solid when the oxygen freezes at $-218.4°$ C.

Chemical Properties   Oxygen is a moderately active element at room temperature but is extremely active at higher temperatures. It combines with almost all elements to produce a class of compounds called oxides.

$$2\,Mg_{(s)} + O_{2(g)} \longrightarrow 2\,MgO_{(s)}$$
$$\text{magnesium} \quad \text{oxygen} \qquad \text{magnesium oxide}$$

$$C_{(s)} + O_{2(g)} \longrightarrow CO_{2(g)}$$
$$\text{carbon} \quad \text{oxygen} \qquad \text{carbon dioxide}$$

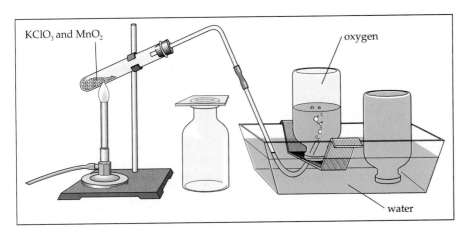

FIGURE 8-1

Laboratory preparation of oxygen. The reaction is

$$2\,KClO_{3(s)} + \frac{MnO_2}{\text{heat}} \longrightarrow 2\,KCl_{(s)} + 3\,O_{2(g)}$$

The reaction between oxygen and some other substance is an example of oxidation. Common examples of oxidation are the rusting of iron, the burning of a candle, and the decay of wood. Oxidation also occurs in living plant and animal tissues. These oxidation reactions are able to occur rapidly at relatively low temperatures because of the presence of specific catalysts called enzymes. An example of such a reaction occurring in the human body is the oxidation of glucose, a simple sugar, to carbon dioxide and water.

$$C_6H_{12}O_{6(aq)} + 6\,O_{2(g)} \xrightarrow{\text{enzymes}} 6\,CO_{2(g)} + 6\,H_2O_{(1)} + \text{energy}$$

$$\underset{\text{glucose}}{} \quad \underset{\text{oxygen}}{} \qquad \underset{\substack{\text{carbon} \\ \text{dioxide}}}{} \quad \underset{\text{water}}{}$$

(This reaction is greatly oversimplified. As will be discussed later, when glucose is oxidized to carbon dioxide and water in the body, there are many intermediate steps involved, each with its own particular enzyme. The subscript (aq) indicates an aqueous solution.)

Combustion

Wood, coal, and gas burn, that is, undergo combustion, in the presence of oxygen. Combustion can be defined as a rapid oxidation in which heat and light are produced, usually accompanied by a flame. Oxygen supports combustion (that is, substances burn in oxygen), but oxygen itself does not burn.

Combustion is usually thought of in terms of a reaction with oxygen. However, oxygen is not absolutely necessary for a combustion reaction. When powdered iron and sulfur are heated together in a test tube, a rapid reaction occurs in which heat and light are given off. This is also a combustion reaction even though no oxygen is involved.

Spontaneous Combustion   When iron combines with the oxygen of the air, it rusts, continuously liberating a small amount of heat as the reaction continues. The total amount of liberated heat, however, will be the same as that which would have been liberated had the iron been burned in oxygen. That is, the total amount of heat produced is the same regardless of whether the reaction proceeds rapidly or slowly.

This is the principle underlying spontaneous combustion, which can be defined as a slow oxidation that develops by itself into combustion. How is such a process possible? If rags containing a "drying oil" such as boiled linseed oil are placed in an open dry container without adequate ventilation, the oil will slowly combine with the oxygen in the air, liberating a small amount of heat during the process. Without ventilation the heat will not be dissipated, especially since rags are such poor conductors of heat. As the oxidation proceeds, more and more heat will be liberated until a sufficient amount accumulates to start the rags burning. This is spontaneous combustion.

If such rags are placed in a closed container, the oxidation will not continue after the oxygen supply is used up. Likewise, if the rags are hung in a place where freely circulating air can carry away the heat, no spontaneous combustion can occur.

Fire Prevention and Control   Care should be taken in handling and storing flammable liquids such as ether and alcohol. Where smoking is allowed, non-flammable receptacles should be provided for the butts. One frequent cause of fires is lighted cigarettes thrown into a wastebasket. Oily rags and mops should be stored in a well-ventilated fireproof locker to avoid the danger of spontaneous combustion.

When a fire does occur, how can it be extinguished? There are two methods of putting out a fire: removing the oxygen from the burning material or lowering the temperature of the burning substance below its kindling point.

One type of fire extinguisher contains carbon dioxide gas under pressure (Figure 8-2). This type of extinguisher has several advantages. The carbon dioxide is extremely cold and lowers the temperature of the burning substance. A large amount of carbon dioxide, which is heavier than oxygen, surrounds the burning area, keeping out the oxygen. For these reasons the carbon dioxide should be directed at the base of the flame. This type of extinguisher is recommended for electrical fires because carbon dioxide does not conduct electricity. It is also used for oil and gasoline fires because it shuts off the oxygen supply and at the same time lowers the temperature. When a fire is extinguished, the carbon dioxide disappears into the air, so it does not have to be cleaned up along with the material that was burning. A cylinder of carbon dioxide holds a large volume of the gas under pressure and thus has a great capability in fire fighting.

Carbon tetrachloride (CCl_4) was formerly used in fire extinguishers, but its use for this purpose has been outlawed because (1) its vapors are toxic and (2) a hot fire converts it to phosgene ($COCl_2$), a poisonous gas.

FIGURE 8-2

Fire extinghisher containing pressurized carbon dioxide gas. When released from an extinguisher over a fire, carbon dioxide gas displaces the oxygen needed for combustion, effectively smothering the fire.

Exercise 8-1

Complete and balance the following reaction of oxygen.
a. Fe + O_2 $\longrightarrow$
b. Ca + O_2 $\longrightarrow$

Solution

a. Fe reacts with O_2 to produce iron oxide, Fe_2O_3.

$$Fe + O_2 \longrightarrow Fe_2O_3$$

To balance we need 4 Fe; 3 O_2; and 2 Fe O:

$$4\,Fe + 3\,O_2 \longrightarrow 2\,Fe_2O_3$$

b. The product is calcium oxide, CaO:

$$Ca + O_2 \longrightarrow CaO$$

To balance we need 2 Ca and 2 CaO:

$$2\,Ca + O_2 \longrightarrow 2\,CaO$$

Self-test

Complete and balance these reactions.
a. Si + O_2 $\longrightarrow$
b. Al + O_2 $\longrightarrow$

Answers

a. Si + O_2 $\longrightarrow$ SiO_2 b. 4 Al + 3 O_2 $\longrightarrow$ 2 Al_2O_3

Preparation

Laboratory Methods   One very common method for preparing oxygen in the laboratory is by heating potassium chlorate, $KClO_3$ (see Figure 8-1). The reaction is

$$2\ KClO_{3(s)} \xrightarrow{\text{heat}} 2\ KCl_{(g)} + 3\ O_{2(g)}$$

It takes a considerable amount of heat to produce oxygen by this method because the potassium chlorate must be heated to its melting point (370° C) before it gives off oxygen.

However, when manganese dioxide (MnO_2) is added to the potassium chlorate, oxygen is evolved from the heated mixture at a much lower temperature and at a more rapid rate. The manganese dioxide acts as a catalyst in this reaction; a catalyst increases the speed of the reaction but does not take part in it. The presence of a catalyst in a reaction is indicated over the arrow in the equation for that reaction.

$$2\ KClO_{3(s)} \xrightarrow[\text{heat}]{MnO_2} 2\ KCl_{(g)} + 3\ O_{2(g)}$$

Oxygen can also be produced in the laboratory by the electrolysis of water.

$$2\ H_2O_{(1)} \xrightarrow{\text{electricity}} 2\ H_{2(g)} + O_{2(g)}$$

Commercial Method   The commercial source of oxygen is the air—an inexhaustible supply as long as we have trees and other green plants, both on land and in the water. When air is cooled to a low enough temperature under compression, it becomes a liquid. Liquid air, like ordinary air, consists mostly of nitrogen and oxygen. Liquid nitrogen boils at $-196°$ C. Liquid oxygen boils at $-182.5°$ C. When liquid air is allowed to stand, the nitrogen boils off first (because of its lower boiling point), leaving almost pure oxygen behind. Liquid oxygen is stored in steel cylinders under high pressure.

Uses

Medical Uses   Oxygen is necessary to life. When oxygen is taken into the lungs, it combines with the hemoglobin of the blood to form a compound called oxyhemoglobin (see Section 30-12). The blood carries the oxyhemoglobin to the tissues where oxygen is released. This oxygen then reacts with the food products in the cells, producing energy. At the same time, carbon dioxide is produced and is carried back to the lungs, where it is exhaled. Blood going to the tissues (arterial blood) contains oxyhemoglobin and has a characteristic bright red color. Blood coming from the tissues (venous blood) does not contain oxyhemoglobin and has a characteristic reddish purple color.

Patients with lung diseases such as pneumonia frequently do not have enough functioning lung tissue to pick up sufficient oxygen from the air. These patients are given a mixture of oxygen and air to breathe instead of air alone. Then, because of the higher partial pressure of oxygen, a small functioning area of the lung can pick up more oxygen than it could if air alone were breathed in. This enables the patient to breathe and live until the diseased area is cured and returns to normal. The oxygen can be administered by nasal cannula or mask (Figures 8-3

FIGURE 8-3
Nasal cannula used by a
woman with a respiratory
disease.

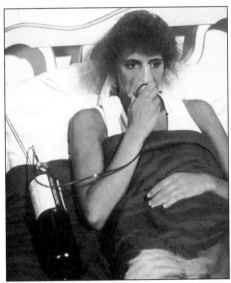

FIGURE 8-4
Oxygen mask in use.

and 8-4). Oxygen is given to patients with lung cancer to help them adjust to a decreased lung area.

Living tissues require oxygen. Without it they soon die. However, if the temperature is lowered sufficiently, tissues can survive with very little oxygen. At normal human body temperature, 98.6° F (37° C), the brain is extremely sensitive to a lack of oxygen. However, it too can live for a longer period of time without oxygen if its temperature is lowered. But how low is low when we are talking about body temperature?

In one form of surgery used to repair ruptured blood vessels in the brain, the body temperature is dropped from normal to 86° F (30° C) by means of an ice bath.

At this temperature the heart can be stopped for about 15 minutes without damage to the body. However, brain surgery requires more time than the 15 minutes allowed. After the body temperature is lowered to 86° F and after the heart is stopped, a saltwater solution at 32° F (0° C) is pumped directly into the main artery that feeds the brain. This cold solution lowers the brain's temperature to approximately 60° F; at this temperature brain surgery can be completed without damage to either the brain or the body. After surgery, the heart is restarted and blood (carrying oxygen) again flows to the brain.

Hypothermia is also used for open-heart surgical procedures. The patient is placed on a plastic pad through which an iced solution is run to lower body temperature to 68° or 77° F (20° or 25° C). However, the main cooling is done through the bypass machine during the surgery. The heart itself is cooled by perfusion of coronary arteries with an iced solution or packed in an ice solution. The temperature of the heart can be lowered to 41° F (5° C). Hypothermia of the organs is also necessary during transplant surgery.

Hyperthermia is a technique of using microwave radiation to raise the temperature of target cancer cells without harming nearby normal tissue.

Since oxygen supports combustion, an object such as a candle, which burns slowly in air, will burn very vigorously in oxygen. Therefore, patients and others must not smoke or use matches in a room in which oxygen is in use. The following precautions should also be taken.

1. Electrical devices such as radios, televisions, electric razors, and so on are likewise banned because of the danger that a spark from the equipment could cause a fire.
2. Electric signal cords should be replaced by a hand bell because of the danger of a spark.
3. Patients should not be given backrubs with alcohol or oil because of the danger of a fire. Instead, lotion or powder should be used.
4. Oil or grease should never be applied to any part of the oxygen equipment. Nurses should take care not to have oil on their hands when manipulating the regulator of the oxygen tank.

*Hypoxia is a condition in which the body does not receive enough oxygen. In cases of hypoxia, oxygen must be administered to permit the body to function normally. Newborn babies who have difficulty breathing are given oxygen containing a small amount of carbon dioxide. The carbon dioxide stimulates the respiratory center of the brain so that the rate of breathing increases and the oxygen is picked up more rapidly and carried to the tissues.

During dental surgery, when nitrous oxide is used as the anesthetic, oxygen must be administered along with the nitrous oxide to prevent asphyxiation (lack of oxygen). Firefighters who breathe large quantities of smoke may suffer from asphyxiation. They are treated by breathing from an oxygen mask. Death by smoke inhalation during a fire is often the result of breathing in a noxious material such as hydrogen cyanide, HCN, or phosgene produced by the burning of plastics.

When a person has been under water for several minutes and becomes unconscious because of lack of oxygen, oxygen is administered in an effort to build up the oxygen concentration in the blood rapidly. Pilots and astronauts breathe oxygen through a face mask because of the decrease or lack of oxygen in the atmosphere around them, as do miners, deep-sea divers, and firefighters.

In the hospital, oxygen under pressure (hyperbaric treatment) is administered in the treatment of cancerous tissues (see Section 13-13), gangrene, and carbon monoxide poisoning.

Another use of oxygen in the hospital is in determining an individual's basal metabolic rate (BMR). Basal metabolism tests measure the energy production of the body by measuring the amount of oxygen a patient breathes during a specified period of time (usually 6 minutes). The BMR has been used as an indicator of thyroid function, but its use for this purpose has been largely supplanted by tests using [131]I (radioactive iodine).

One measure of the body's use of oxygen (pulmonary function) is the MET, derived from the word metabolism. One MET corresponds to the consumption of 1 mL of O_2 per minute for each kilogram of body weight (Figure 8-5). Table 8-1 compares MET values for various activities.

Formerly, all premature infants were routinely given oxygen until respiratory sufficiency had been established (Figure 8-6). Now it is known that when the concentration of oxygen rises above 40 percent in the inspired air, premature infants develop retrolental fibroplasia, a disease that affects the eyes. This disease produces complete or nearly complete blindness as a result of separation and fibrosis of the retina.

Table 8-1	
MET Values	
	$\dfrac{mL}{min.kg}$
At rest	1
Walking 2 mph	3
Running 7 mph	12
Walking upstairs	15
Shoveling snow	7

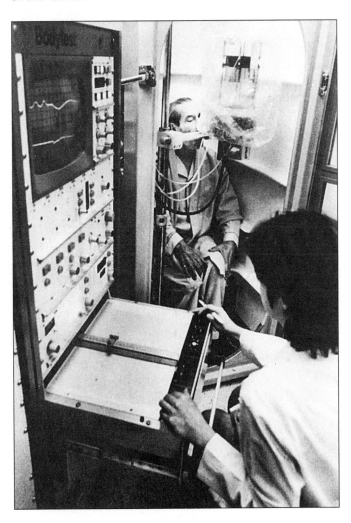

FIGURE 8-5

MET test being administered to a patient in a pulmonary function laboratory.

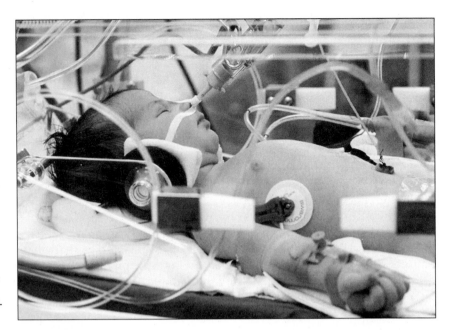

FIGURE 8-6
Infant in Isolette at a pediatric intensive care facility. Formerly, all premature infants were given oxygen until they reached respiratory competence. This practice has been abandoned since the discovery of associated deleterious effects.

Exercise 8-2

How many milliliters of oxygen are consumed by a 60 kg woman running for 30 minutes at 7 mph?

Solution

Running has a MET value of $\dfrac{12 \text{ mL}}{\text{min.kg}}$

$$\frac{12 \text{ mL}}{\text{min.kg}} \times 60 \text{ kg} \times 30 \text{ min} = 2.2 \times 10^4 \text{ mL}$$

Self-test

Calculate the milliliters of O_2 consumed by an 80 kg (176 lb) person shoveling snow for 20 minutes.

Answer

1.1×10^4 mL of O_2

Commercial Uses Deep-sea divers who work at great depths formerly were supplied air under pressure to enable them to breath. However, this process introduced a difficulty in that the nitrogen in the air became much more soluble in the blood under the higher pressure. When the diver was taken out of the water too rapidly, this dissolved nitrogen became less soluble because of the decrease in pressure. When this happened, bubbles of nitrogen formed in the blood, causing a condition called "the bends." Unless the diver could be placed immediately in a compression chamber under high pressure, which then was gradually reduced to normal pressure, the results were generally fatal. Recall the similar effect of tiny bubbles being formed in a liquid when a bottle of warm carbonated beverage is opened.

A method enabling the diver to breathe a mixture of oxygen and helium is now used. Helium is less soluble in the blood than nitrogen and so decreases the chances of the bends occurring.

A mixture of helium and oxygen can flow more easily through a partially obstructed air passageway than can a mixture of nitrogen and oxygen. In small children, the airways are extremely tiny, so even a partial obstruction can cause major problems. Administration of a He-O_2 mixture to an infant suffering from an infection that results in the swelling of the air passageways allows the passage of oxygen to the body until the antibiotics can take effect and cure the infection.

The He-O_2 mixture is also useful in buying time for adult patients who have a tumor partially blocking the air passageways and so cannot get enough oxygen into their bodies. Normally, such patients cannot survive long enough for radiation to shrink the tumor. With the He-O_2 mixture, they can. They are then gradually weaned off the He-O_2 mixture. Such a treatment is not a cure but merely a method of buying time until the breathing difficulty can be corrected.

Industrially, oxygen is used in the operation of furnaces for production of iron and steel. It is also used in welding torches. Liquid oxygen (LOX) is used in rockets for launching satellites.

8-2 Hydrogen Peroxide

Another product besides water (H_2O) is possible between the elements hydrogen and oxygen. This compound is hydrogen peroxide, H_2O_2, produced by means of the following reactions:

$$Ba_{(s)} + O_{2(g)} \longrightarrow BaO_{2(s)}$$

$$BaO_{2(s)} + H_2SO_{4(aq)} \longrightarrow BaSO_{4(s)} + H_2O_{2(aq)}$$

Hydrogen peroxide is a pale blue oily liquid that boils at 150° C and freezes at −0.41° C. It is very unstable and decomposes according to the following reaction:

$$2\,H_2O_{2(aq)} \longrightarrow 2\,H_2O_{(1)} + O_{2(g)}$$

This reaction is induced by light, and therefore hydrogen peroxide is usually stored in dark-colored bottles. A 3 percent solution of H_2O_2 is used as an antiseptic; a 30 percent solution is used as a bleach and as an oxidizing agent in chemistry laboratories.

The oxidation state of oxygen in water and in oxides is minus −2. The oxidation state of oxygen in the free state is 0. In peroxides, the oxidation state of the oxygen (the oxidation number) is −1.

Hydrogen peroxide is an intermediate product in the reduction of O_2 to H_2O in many cellular reduction reactions. Since hydrogen peroxide is toxic to living cells, it must be removed quickly before it can cause harm. The enzyme that catalyzes such a reaction is catalase (see page 428).

8-3 Superoxides

When a metal reacts with oxygen, an oxide is formed.

$$2\,Al + 3\,O_2 \longrightarrow 2\,Al_2O_3 \quad \text{(aluminum oxide)}$$

When an active metal reacts with oxygen, a peroxide is formed.

$$2\,Na + O_2 \longrightarrow Na_2O_2 \quad \text{(sodium peroxide)}$$

When a very active metal reacts with oxygen, a superoxide is formed.

$$K + O_2 \longrightarrow KO_2 \quad \text{(potassium superoxide)}$$

In superoxides, the oxidation number of oxygen is $-\frac{1}{2}$. Superoxide ions exist in the human body and in some microorganisms. In the production of uric acid (page 513), superoxide ions are produced. Superoxide ions are formed during the oxidation of Fe^{2+} to Fe^{3+} in the hemoglobin molecule. Superoxide ions are also involved in the immune response and possibly in the aging process.

Both the superoxide ion (O_2^-) and hydrogen peroxide (H_2O_2) are very reactive substances and extremely dangerous to living cells; therefore, their removal is of great importance to the well-being of the organism. The superoxide ion is rapidly converted to O_2 and H_2O_2 by the enzyme superoxide dismutase. The H_2O_2 thus produced is decomposed into H_2O and O_2 by the enzyme catalase (page 428).

$$2\,H_2O_2 \xrightarrow{\text{catalase}} 2\,H_2O + O_2$$

The reaction of both peroxides and superoxides with carbon dioxide is of use in designing a breathing apparatus for astronauts, submariners, and firefighters. The purpose of such an apparatus is to remove exhaled CO_2 and replace it with (hopefully) an equal volume of O_2. The reactions are

$$2\,Na_2O_{2(s)} + 2\,CO_{2(g)} \longrightarrow 2\,Na_2CO_{3(s)} + O_{2(g)}$$

$$4\,KO_{2(s)} + 2\,CO_{2(g)} \longrightarrow 2\,K_2CO_{3(s)} + 3\,O_{2(g)}$$

According to the first reaction, 2 volumes of CO_2 are converted into 1 volume of O_2. But in the second reaction, 2 volumes of CO_2 are converted into 3 volumes of O_2. While the superoxides regenerate more O_2 than the peroxides do, they are less stable and much more difficult to handle.

8-4 Ozone

Oxygen occurs most frequently as the diatomic molecule O_2. However, it also exists as a triatomic molecule, O_3, called ozone (Figure 8-7). These two different forms of the element oxygen are called allotropic forms. Allotropic forms of an element have different physical and chemical properties. Other elements that eixst in allotropic forms are carbon, sulfur, phosphorus, tin, and lead.

Preparation

When air or oxygen is passed between two electrically charged plates, the volume decreases and a pale blue gas with a strong odor is formed. This gas is ozone. An equation for the formation of ozone is

$$3\,O_2 \rightleftharpoons 2\,O_3$$

The double arrow indicates an equilibrium reaction, one that proceeds in both directions.

Ozone is formed by the action of ultraviolet light from the sun on the oxygen in the air. It is also formed in the air by electrical discharges (lightning). Ozone can also be found around high-voltage machinery, where sparks convert some of the oxygen of the air into ozone.

The ozone layer in the upper atmosphere helps protect the Earth from the harmful effect of the sun's ultraviolet radiation (see Section 8-9).

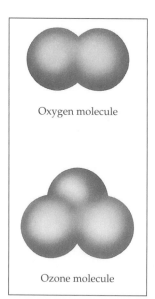

Oxygen molecule

Ozone molecule

FIGURE 8-7

Models of oxygen and ozone molecules.

Properties and Uses

Ozone is a colorless (or sometimes slightly blue) gas at room temperature. It has a very pungent odor similar to that of garlic. It is heavier than oxygen and more soluble in water. Liquid ozone is blue. Ozone is a powerful oxidizing agent. The use of ozone is rather limited because it is poisonous and unstable. Ozone is irritating to the mucous membranes, and it is quite toxic to the body except in extremely small amounts.

Ozone is not tolerated in industrial establishments in concentrations of more than 1 part per million (ppm), whereas carbon monoxide (CO) can be tolerated in strengths as high as 100 ppm. In other words, ozone is 100 times as poisonous as carbon monoxide. It is also 100 times as poisonous as hydrogen sulfide (H_2S) and approximately 10 times as poisonous as hydrogen cyanide (HCN).

Ozone is not used medically for any purpose because it reaches a fatal concentration long before it can act in any useful way. However, the industrial use of ozone is increasing rapidly. It is a more powerful oxidizing agent than oxygen itself. Its usefulness as a *bactericide, a decolorizer, and a deodorizer is great when it is employed carefully and in sufficient strengths. Ozone is used in the rapid aging of wood, for rapid drying of varnishes and inks, in the treatment of water, and in the disinfection of swimming-pool water, but it should only be handled with great care by trained personnel.

8-5 Nitrous Oxide

Nitrous oxide (N_2O) is frequently used as a general anesthetic. It is a colorless gas with almost no odor or taste. It is heavier than air and has a relatively low solubility in blood. N_2O does not combine with hemoglobin and is carried dissolved in the bloodstream. It is excreted, unchanged, primarily through the lungs, but a small fraction may escape through the skin. It has been found that nitrous oxide can be useful in relieving the pain in heart attack victims by reducing the strain on the heart muscles.

Although N_2O is not flammable, it does support combustion, as does oxygen, so care must be taken with its use.

The highest concentration of N_2O that can be safely given for the maintenance of anesthesia is 70 percent. Above this concentration hypoxia develops. However, at a concentration of 70 percent or less, N_2O is not potent enough for most patients, so additional drugs such as halothane or methoxyflurane are used to complete the anesthesia.

Nitrous oxide can react with the benzpyrene (page 279) in the air to produce *mutagenic substances. Large doses of vitamin E seem to protect mice and rats from these substances; however, humans do not benefit in the same way.

8-6 Nitric Oxide

Nitric oxide (NO) is one of the products of the combustion of petroleum in gasoline engines. It is rapidly converted into nitrogen dioxide, NO_2, by oxygen in the air. Both NO and NO_2 are constituents of smog and acid rain (see Section 8-8).

Recent evidence has shown that NO is also produced in the body and is found in cells in the brain, the pancreas, the liver, the gastrointestinal tract, and the blood vessels. It is produced from the amino acid arginine (see page 420) and is involved in such functions as blood pressure, penile erection, and inhibition of blood clotting. It also acts as a neurotransmitter (see page 595).

Nitric oxide has also been implicated in such disorders and diseases as septic shock, diabetes mellitus, Alzheimer's disease, and Huntington's chorea.

NO is also found in the sinuses, the air-filled cavities in the skull surrounding the nose. There the NO, which is lethal to bacteria and viruses even in small doses, keeps the sinuses sterile even though they are a bacteria-friendly environment.

The NO is produced in the epithelial cells of the membranes of the sinuses. NO acts as a bacteriocide and a *virucide by binding to enzymes that these organisms need to grow and reproduce.

NO has another role in the human body; it is a powerful vasodilator. It has been used to open up the blood vessels of patients receiving ventilatory therapy.

8-7 Noxious Gases

Carbon monoxide (CO) is a colorless, odorless, tasteless gas. It is nonirritating to the body but has a great affinity for hemoglobin. CO is poisonous because the hemoglobin forms such a strong bond with the gas that the blood is unable to carry sufficient oxygen to the tissues.

Chlorine is a greenish yellow gas with a pungent odor. In neutral or acid solutions chlorine acts not only as a bactericide but also as a virucide and *amoebicide. It is less effective as a germicide in basic solution. Chlorine kills microorganisms in water and so is used for water purification. In large amounts, chlorine vapors are toxic because they react with and destroy lung tissues.

Sulfur dioxide is released into the air during the combustion of fossil fuels. Catalytic converters in automobiles change SO_2 to SO_3, which can then form sulfuric acid and other sulfates that are extremely harmful to the health of humans and aquatic creatures. Sulfur dioxide has been used as a preservative.

Hydrogen sulfide (H_2S) is characterized by its distinctive odor of rotten eggs. H_2S is extremely poisonous, even more so than hydrogen cyanide (HCN), but its odor warns a person of its presence long before its concentration reaches lethal proportions. This is not true for HCN.

8-8 Air Pollution

Air pollutants can be gases such as sulfur dioxide, nitrogen oxides, ozone, hydrocarbons, and carbon monoxide, or they can be particulate matter such as smoke particles, asbestos, and lead *aerosols.

Concentrations of pollutants in the air are often expressed as parts per million (ppm), where 1 ppm corresponds to 1 part pollutant to 1 million parts air.

Concentrations of pollutants at levels far below 1 ppm can have an adverse effect upon human life. For example, 0.2 ppm sulfur dioxide in the atmosphere leads to an increased death rate, and 0.02 ppm peroxybenzoyl nitrate (a constituent of smog) causes severe eye irritation.

The word *smog* is derived from the words *smoke* and *fog*. Smog is characterized by air that contains lung and eye irritants, along with reduced visibility. Smog is formed frequently during a thermal inversion. Normally, warm air near the ground surface rises and carries away pollutants. However, during a thermal inversion, the air near the surface is cooler than the air above it and so remains at the surface, keeping the air pollutants down at that level.

Gases

The principal source of three gaseous pollutants—carbon monoxide, hydrocarbons, and nitrogen oxides—is the automobile. Tobacco smoke is a source of carbon monoxide as well as particulate matter (ash); fossil-fuel-powered electrical generating stations are also a major source of air pollutants such as sulfur dioxide.

Each pollutant poses a different threat to human life.

1. **Carbon monoxide** is a deadly poison that interferes with the transportation of oxygen by the blood by competing with oxygen for the iron-binding sites in hemoglobin. Low concentrations of this gas, such as are found in automobiles, garages, downtown streets during rush hours, and space-heated rooms, cause impairments of judgement and vision. Evidence has shown that intermittent exposure to carbon monoxide at low levels of concentration can cause strokes and hypertension in susceptible individuals. High concentrations of carbon monoxide cause headache, drowsiness, coma, respiratory failure, and death.

2. **Nitrogen oxides** are just as dangerous a pollutant as carbon monoxide, even though environmental groups do not stress them equally. The first effects of nitrogen oxides upon humans is an irritation of the eyes and respiratory passages. Concentrations of 1.6 to 5 ppm of nitrogen dioxide for a 1-hour exposure cause increased airway resistance and diminish diffusing capacity of the lungs. Concentrations of 25 to 100 ppm cause acute but reversible bronchitis and *pneumonitis. Concentrations above 100 ppm are usually fatal, with death resulting from pulmonary *edema.

   Nitrogen dioxide, NO_2, can react with water to form nitric acid, a constituent of "acid rain."

3. **Oxides of sulfur** cause acute airway spasm and poor airway clearance in all people and exert a deadly effect on patients already disabled by lung disease. Concentrations of 8 to 10 ppm cause immediate throat irritation, and concentrations of 20 ppm cause immediate coughing.

   Particulate matter absorbs sulfur dioxide, and the resulting tiny particles can enter the small air passages in the lungs, causing spasms and destruction of cells. Particles of the smallest size penetrate deepest into the lungs and remain there the longest.

   Sulfur dioxide, SO_2, is oxidized in the air to sulfur trioxide, SO_3. Sulfur trioxide can react with the moisture in the air to form sulfuric acid, another constituent of acid rain.

   Why is acid rain so detrimental? It affects trees and fruits and vegetables with considerable commercial consequences. Also it can react with calcium carbonate, which is found in marble and in mortar. Many famous marble sculptures around the world are rapidly deteriorating because of the acid in the air. The mortar between the bricks is also affected by acid rain and is slowly being dissolved, leaving the bricks with very little to hold them together.

4. **Tobacco smoke** is one of the most dangerous types of air pollution. Smokers inhale large amounts of carbon monoxide, tars, and particulate matter. People around smokers are also exposed to these same pollutants. Evidence has shown that expectant mothers who smoke have smaller babies and babies with a higher infant mortality rate than those of nonsmokers. Smokers also have a greater chance of developing lung cancer, *emphysema, and cardiovascular disease (see Plate 25).

5. **Ozone** is formed by the action of sunlight on oxygen and is normally present in the atmosphere in extremely small amounts. Ozone reacts with hydrocarbon emissions and with oxides of nitrogen to form peroxyacyl nitrates, which are the eye irritants of smog. Since ozone is produced by the action of sunlight, its levels in the air are usually lower at night than during the day. Concentrations as low as 0.15 ppm can, within 1 hour, adversely affect such plants as tomatoes and corn. While low concentrations of ozone cause eye irritation, higher concentrations can cause pulmonary edema and hemorrhaging.

Los Angeles–type smog is primarily a photochemical phenomenon because many of the changes that take place in the air pollutants are due to the action of sunlight upon the substances mentioned previously.

8-9 The Ozone Layer

Ozone is found in a layer 20 to 50 km (12 to 30 miles) above the Earth's surface. Ozone is formed by the reaction of the sun's high-energy ultraviolet radiation on oxygen molecules in the upper air.

$$3\,O_2 \underset{\text{ultraviolet radiation}}{\rightleftharpoons} 2\,O_3$$

This layer, called the *ozone layer*, filters out most of the sun's high-energy ultraviolet radiation. This type of radiation is detrimental to life, and therefore the ozone layer serves a vital function. If the ozone layer were reduced or eliminated, the effect on life would be catastrophic. The two principal threats to the ozone layer are

1. Chlorofluorocarbons, used in aerosol spray cans, in refrigeration, and in air conditioners
2. Oxides of nitrogen

Chlorofluorocarbons were widely used in air-conditioning and refrigeration units and in aerosol spray cans because they are *volatile, are chemically unreactive, and liquefy under pressure. In the atmosphere, they gradually diffuse upward into the stratosphere where they can react with the ozone layer and reduce the amount of ozone present. Even if the ozone layer were depleted by only a small amount, the result would be a marked increase in skin cancer, caused by the dangerous ultraviolet radiation from the sun reaching the Earth's surface. For these reasons, the use of chlorofluorocarbons has been banned or strictly limited.

Oxides of nitrogen such as NO and NO_2 are formed in the exhaust of automobiles and planes. In the lower levels of the atmosphere, although these compounds are detrimental to human health, they pose no threat to the ozone layer because they do not remain in the atmosphere long enough to rise into the stratosphere. However, supersonic and military planes do inject large amounts of oxides of nitrogen into the ozone layer because they fly at such great altitudes. This is one reason the use of supersonic planes has not been greatly expanded.

The Greenhouse Effect

The infrared energy coming from the sun heats the Earth and in turn the Earth reradiates some of that energy back into space. Nature balances heat gain and heat loss to maintain a liveable temperature on the Earth's surface. If the amount of infrared

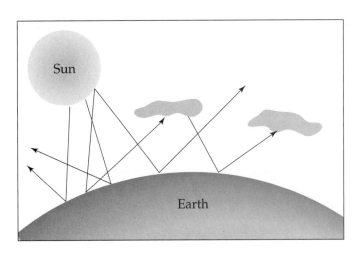

FIGURE 8-8

Carbon dioxide in the air also absorbs infrared radiation and prevents it from flowing back into space.

energy reaching the Earth is less than the amount being reradiated outward, an ice age may occur. If, however, the amount of incoming energy is greater than that being lost, global warming might happen.

Such a global warming, called the "greenhouse effect," results primarily from the occurrence of certain gases in the earth's atmosphere. One of these gases is water vapor, both in the air and in the clouds around the Earth. Water vapor absorbs infrared radiation from the Earth's surface. An example of such an absorption can be noted in the fact that on cloudy nights the earth loses less heat by radiation than it does on cloudless nights (Figure 8-8).

Carbon dioxide in the air also absorbs infrared radiation and prevents it from flowing back into space.

Other gases such as methane and chlorofluorocarbons act similarly. Even though these latter gases are present in much smaller quantities than carbon dioxide, they exert a relatively large effect because they are more efficient in trapping infrared energy.

The amount of carbon dioxide in the atmosphere has been increasing rapidly since the beginning of the Industrial Revolution as more and more fossil fuels are burned to provide energy.

A global warming trend of even a few degrees could melt the polar ice cap and cause worldwide flooding along coastal areas.

Even though mankind may attempt to regulate the amount of carbon dioxide being released into the air, we cannot control the release of carbon dioxide (and volcanic ash) being produced by volcanic activity. Volcanic ash also hinders the reradiation of infrared energy.

Summary

Oxygen is the most abundant element on the Earth's surface. It is colorless, odorless, tasteless, slightly heavier than air, and very slightly soluble in water.

Oxygen combines with most elements to form a class of compounds called oxides. The reaction between oxygen and some other substances is an example of oxidation. Oxidation also occurs in plant and animal tissues in the presence of specific catalysts called enzymes.

Combustion is a rapid reaction with oxygen, one in which heat and light are produced, usually accompanied by a flame. Spontaneous combustion is a slow oxidation that develops by itself into combustion.

Fire extinguishers accomplish their work either by removing the oxygen from the burning substance or by lowering its temperature below its kindling point, or both.

Oxygen can be prepared in the laboratory by heating such compounds as potassium chlorate ($KClO_3$) or by the electrolysis of water.

Oxygen is prepared commercially from liquid air.

Oxygen is used medically in the treatment of various respiratory diseases. Oxygen is inhaled and carried to the tissues by the hemoglobin in the blood. Care must be taken to prevent fires in the room of a patient receiving oxygen therapy. This means no smoking or use of electric devices, no alcohol or oil backrubs.

Asphyxiation (hypoxia) can be overcome by the administration of oxygen. Oxygen is used in the treatment of cancer for some patients and in the determination of basal metabolic rate.

Industrially, oxygen is used in welding, in the production of steel, and in rockets.

When oxygen reacts with an active metal, a peroxide is formed; when oxygen reacts with a very active metal, a superoxide is formed.

Ozone (O_3) is an allotropic form of oxygen (O_2). Ozone is colorless gas with a pungent odor. It is very toxic to the body. Ozone is not used medically but has many industrial uses; it must be employed under carefully controlled conditions by trained personnel.

Nitrous oxide is used as a general anesthetic, usually in conjunction with another drug such as halothane or methoxyflurane.

Among the noxious gases are carbon monoxide, the vapor of carbon tetrachloride, chlorine, sulfur dioxide, and hydrogen sulfide.

Air pollutants may be gases such as carbon monoxide, ozone, nitrogen oxides, and sulfur oxides, and they also may be particulate matter such as smoke particles, asbestos, and lead aerosols. Each pollutant poses a different threat to human life.

Questions and Problems

1. List several physical properties of oxygen.
2. Compare combustion with spontaneous combustion.
3. How can oxygen be prepared in a laboratory? commercially?
4. How is oxygen carried to the tissues?
5. How can ozone be prepared? Does it have any medical uses? Explain.
6. What is hypoxia? How may it be treated?
7. How can carbon dioxide be removed from the air and replaced with oxygen (a) during photosynthesis and (b) in a submarine?
8. What causes retrolental fibroplasia? How can it be prevented?
9. Discuss medical uses of oxygen and also the precautions to be taken during its use.
10. What is the oxidation state of oxygen in each of the following compounds?

    Na_2O; KO_2; Li_2O; H_2O_2; O_2
11. Why is hydrogen peroxide stored in dark bottles?
12. Where are superoxides formed in the body? How are they formed?
13. Why is carbon monoxide dangerous? carbon tetrachloride?

14. Why must nitrous oxide be used at concentrations less than 70 percent? Why is it used in conjunction with other drugs?
15. List the general properties of nitrous oxide. What precautions should be taken with its use?
16. What is an oxide?
17. Why is carbon dioxide sometimes given along with oxygen to newborn babies?
18. How is hydrogen peroxide removed from living cells?
19. Compare the reactions of metals, active metals, and very active metals with oxygen.
20. What is meant by the term MET? How does it vary during various activities?
21. Discuss the medical uses of hypothermia.
22. Do all oxidation reactions involve oxygen? Explain.
23. Compare the reaction involving the oxidation of glucose with photosynthesis.
24. Why are fire extinguishers containing CCl_4 no longer in use?
25. Compare the use of a nasal catheter with the use of an oxygen tent.
26. What causes the red color of blood?
27. Compare the poisonous nature of hydrogen sulfide with that of hydrogen cyanide.

28. Explain the use of an He-O$_2$ mixture in the treatment of obstructions of the air passageways.
29. What is hyperbaric treatment? Where is it used?
30. Name three greenhouse gases.
31. Explain how the ozone layer protects us.
32. What threats to life are caused by (a) ozone, (b) particulate matter in the air, (c) oxides of sulfur?

33. What are the hazards of cigarette smoking?
34. Why is the ozone layer important? What might affect it? What might be done to protect it?
35. What is acid rain and why is it harmful?
36. What is smog? What are its principal constituents?
37. What gases are used in the breathing apparatus of deep-sea divers? Why?

Practice Test

1. Commercially, oxygen is produced from _____.
   a. KClO$_3$
   b. water
   c. air
   d. metal peroxides

2. Oxygen can be mixed with _____ gas for use in diving equipment.
   a. N$_2$ b. He   c. H$_2$ d. CO$_2$

3. A gas frequently used for general anesthesia is _____.
   a. carbon tetrachloride   b. carbon dioxide
   c. nitrous oxide   d. methane

4. The molecular formula for ozone is _____.
   a. O   b. O$_2$ c. O$_3$ d. O$_4$

5. If a patient is using oxygen, which of the following precautions should be taken?
   a. ban all electrical devices
   b. avoid smoking
   c. remove electrical signal cord
   d. all of these

6. An example of a noxious gas is _____.
   a. N$_2$ b. Cl$_2$ c. H$_2$ d. CO$_2$

7. Spontaneous combustion can be prevented by _____.
   a. the addition of oxygen
   b. poor ventilation
   c. use of a closed container
   d. none of these

8. Oxides are compounds formed by the reaction of oxygen with _____.
   a. metals only
   b. nonmetals only
   c. either metals or nonmetals

9. Oxygen is _____.
   a. heavier than air and insoluble in water
   b. heavier than air and slightly soluble in water
   c. lighter than air and slightly soluble in water
   d. lighter than air and insoluble in water

10. Oxygen can be produced in the laboratory by _____.
    a. electrolysis of water
    b. heating of any metal oxide
    c. heating any nonmetal oxide
    d. all of these

Water

Objectives

- To understand the properties and reactions of water
- To draw the structure of the water molecule
- To understand the nature of hydrogen bonding
- To become familiar with methods of water purification
- To become familiar with processes of water softening
- To understand the causes of water pollution

Water, a compound present in great abundance on Earth, is essential for living processes.

9-1 The Importance of Water

Water is one of the most important chemicals known. Without it neither animal nor plant life would exist. Humans can live for a few weeks without food but for only a few days without water. Almost three-fourths of the Earth's surface is covered by water. Rain, snow, sleet, hail, fog, and dew are manifestations of the water vapor present in the air.

Water is present on the solid part of the Earth's surface as lakes, streams, waterfalls, and glaciers. The human body is approximately 50 percent water. Water is essential in the processes of digestion, circulation, elimination, and the regulation of body temperature. Indeed, every activity of every cell in the body takes place in a water environment. Normally, in the body, water intake equals water output. If water intake is greater than water output, a condition known as *edema results (see Figure 9-1). If water intake is less than output, dehydration occurs.

Water is important as a solvent. Many substances dissolve in water; sugar, salt, and alcohol are examples.

9-2 Physical Properties of Water

Pure water is colorless, odorless, and tasteless. Tap water owes its taste to dissolved gases and minerals. Large bodies of water such as lakes and oceans appear blue owing to the presence of finely divided solid material and also to the reflection of the sky (see Table 9-3).

When water at room temperature is cooled, its volume contracts. However, water is an unusual liquid in that, after it is cooled to 4° C, further cooling causes expansion in volume. When water freezes at 0° C, it expands even more, increasing in volume by almost 10 percent and decreasing in density. These changes on freezing explain why ice floats and why pipes may burst when the water in them freezes. Expansion of water during frostbite explains the damage to the cells.

The density of water changes with the temperature. At 4° C water has its smallest volume and its maximum density, 1.00 g/mL. However, for all practical purposes, the density of water is given as 1 g/mL regardless of temperature, since the variation in density between 0° and 100° C is quite small.

Pure water boils at 100° C at 1 atm pressure. At lower pressures it boils at a lower temperature. In certain mountainous localities, water boils at 80° C because of the lower pressure. If more heat is applied to the water, it merely boils faster. The boiling point does not increase. When water boils at 80° C, there may not be sufficient heat to cook food. In this case, the food must be heated in a pressure cooker. This device increases pressure and so increases the boiling point of the water. The same principle holds in the autoclave, where the increased pressure raises the boiling point.

For water to evaporate, a certain amount of heat is necessary. This amount of heat, called the heat of vaporization, is 540 cal/g. When water is placed on the skin, the heat it needs to make it evaporate comes from that skin. Therefore, the skin loses heat and so is cooled (see Section 9-6). The evaporation of perspiration also cools the skin.

For ice to melt, the amount of heat required is 80 cal/g. When an ice pack is placed on the skin, the heat needed to melt the ice comes from the body, thus lowering the temperature of the body.

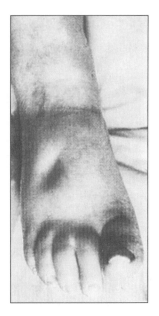

FIGURE 9-1

Patient's leg showing pitting edema.

9-3   Physical Constants Based on Water

The freezing point of water (0° C or 32° F) and the boiling point of water at 1 atm pressure (100° C or 212° F) are the standard reference points for the measurement of temperature.

The mass of 1 mL of water at 4° C (its maximum density) is 1 g.

The calorie† is defined as the amount of heat required to change the temperature of 1 g of water by 1° C.

9-4   Structure of the Water Molecule

Pure water does not conduct electricity. This indicates that water is a covalent compound. In covalent compounds atoms share electrons. Thus, each hydrogen atom of the water molecules shares its one electron with the oxygen atom.

We might expect the oxygen and the hydrogen atoms in the water molecule to be arranged in a straight line such as HOH. However, laboratory evidence indicates that atoms in the water molecule are arranged in a nonlinear manner with the angle between the hydrogen atoms being approximately 105 degrees (see page 71).

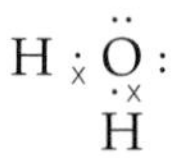

$$H \overset{..}{\underset{\cdot \times}{\times} O} :$$
$$H$$

The x's represent the electrons of the hydrogen atom, and the dots the electrons from the oxygen atom. However, the oxygen atom has a greater attraction for electrons than the hydrogen atom. Therefore, the electrons will spend more of their time closer to the oxygen atom than to the hydrogen atom. This shifting of the electrons toward the oxygen atom will tend to give the oxygen atom a partial negative charge, whereas the hydrogen atoms have a partial positive charge. The structure and distribution of relative charges in the water molecule can be represented

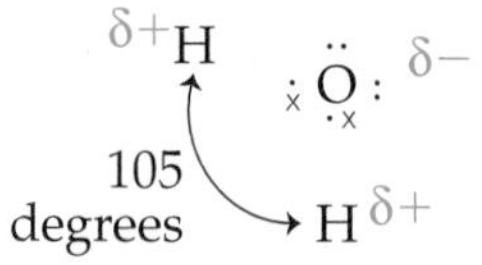

Molecules in which there is an unequal or uneven *distribution* of charges are called polar molecules. Water is a polar molecule. The polar nature of the water molecule is responsible for its property of dissolving many materials. When the ionic compound sodium chloride (NaCl) is placed in water, it dissolves partly because of the water molecule attracting the ions and pulling them apart (see Figure 9-2). Sodium chloride is a polar compound because of the uneven distribution of charges, Na^+Cl^-, within the molecule. A general rule is that polar compounds dissolve in polar liquids, and nonpolar compounds dissolve in nonpolar liquids. Nonpolar compounds generally do not dissolve in polar liquids. Carbon tetrachloride (CCl_4) and benzene (C_6H_6) are nonpolar liquids. Water is a polar liquid. As should be expected, carbon tetrachloride dissolves in benzene but not in water. Likewise, benzene does not dissolve in water (see page 164).

---

†The SI unit of heat is the joule, but we will use the more familiar unit of the calorie.

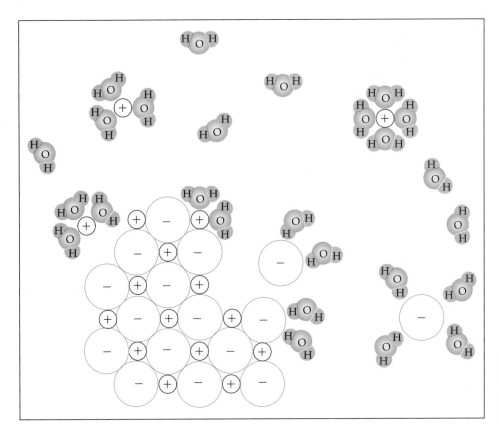

Model of the dissolution of an ionic solid such as sodium chloride (Na^+Cl^-) in water.

Exercise 9-1

Would you expect HCl or oil to be more water soluble?

Solution

HCl is a polar molecule, whereas oil is not; therefore, HCl will be more water soluble.

Self-test

Would O_2 be more soluble in water or gasoline (nonpolar)?

Answer

Gasoline

9-5 Hydrogen Bonding

Since water molecules are polar, we might expect the positive end of one dipole to attract the negative end of another dipole. This is exactly what is observed. This type of attraction is called hydrogen bonding. Hydrogen bonding is a weak inter-molecular attraction between an electropositive hydrogen atom and a nonbonded electron pair of an electronegative atom of another water molecule or another

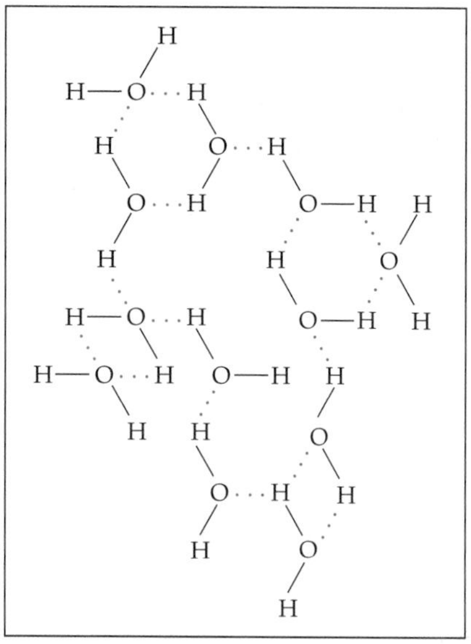

FIGURE 9-3
Hydrogen bonding in water.

polar molecule. Hydrogen bonding plays an important role in determining the globular shape of proteins and in stabilizing the DNA double helix.

Hydrogen bonds also account for many of the unusual properties of water. In the following section we will see how this type of bonding can account for the abnormally low vapor pressure of water.

9-6 Evaporation

The particles in a liquid, such as water, are not held as tightly as those in a solid (see Figure 2-3), so they have some freedom of motion. When some of the surface molecules escape completely from the liquid, the process is called evaporation. As the temperature of the liquid rises, the molecules move faster, so more of them can escape from the surface. Thus, the rate of evaporation increases as the temperature increases.

Evaporation requires energy. When water evaporates, it takes energy (540 cal/g) either from its surroundings or from the remaining water. When water is placed on the skin, it evaporates by taking heat from the body. That is, evaporation is a cooling process. Alcohol evaporates faster than water, so it has a greater cooling effect when placed on the skin. Thus, alcohol sponge baths lower body temperature faster than water sponge baths.

Consider two containers of water, one open to the air and the other sealed (Figure 9-4). The water in the open container will continue to evaporate until none is left. However, the situation is different in the sealed container. As more and more of the surface molecules escape into the air above the liquid, some of those molecules will return to the liquid. Soon the rate of evaporation will equal the rate at which the gaseous molecules return to the liquid. This condition is known as an equilibrium. The pressure exerted by the gaseous form of the liquid under equilibrium conditions is known as the **vapor pressure** of the liquid.

Since the rate of evaporation increases as the temperature increases, the vapor pressure also increases as the temperature increases. Table 9-1 indicates the vapor pressure of water at various temperatures. Table 9-2 indicates the vapor pressures of various liquids at room temperature (20° C).

Because alcohol has a higher vapor pressure than water at the same temperature, alcohol will evaporate faster than water. Therefore, medications containing

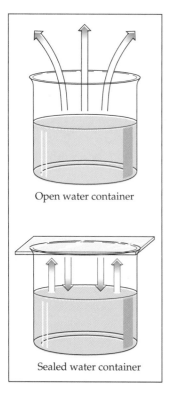

Open water container

Sealed water container

FIGURE 9-4
Evaporation takes place from the open container; equilibrium is established in the closed container, after which the amounts of evaporation and condensation become equal.

Table 9-1	Vapor Pressure of Water		
Temperature (°C)	Pressure (mm Hg)	Temperature (°C)	Pressure (mm Hg)
20	18	70	234
30	32	80	355
40	55	90	526
50	93	100	760
60	149		

Table 9-2	
Vapor Pressures of Various Liquids at Room Temperature (20° C)	
Liquid	Pressure (mm Hg)
Water	18
Alcohol (ethyl)	44
Benzene	75
Acetone	177
Ether (ethyl)	442

alcohol should be tightly closed when stored. If not, some of the alcohol will evaporate, thus changing the strength of that medication.

Water has a relatively low vapor pressure. Why? Recall that water molecules are held together by hydrogen bonds. For water to evaporate, not only must the surface molecules have enough energy to escape but they must also have enough energy to break the hydrogen bonds holding them together. Therefore, the tendency to evaporate is low, and hence the vapor pressure is low.

9-7 Boiling Point

Both boiling and evaporation involve a change from the liquid to the gaseous state. In boiling, however, heat is applied directly, whereas in evaporation heat is taken from the surroundings.

The boiling point of a liquid is defined as the temperature at which the vapor pressure of the liquid is equal to atmospheric pressure. For water, the temperature at which the vapor pressure equals 1 atm (760 mm Hg) is 100° C (see Table 9-1). Thus, at 1 atm pressure water boils at 100° C. If the atmospheric pressure is lower than 1 atm, water will boil at a correspondingly lower temperature. Conversely, at pressures above 1 atm, water boils at a temperature higher than 100° C. An autoclave (see page 119) uses this principle.

9-8 Chemical Properties

Electrolysis

When water undergoes electrolysis—that is, when an electric current is passed through it—hydrogen gas (H_2) and oxygen gas (O_2) are formed. The volume of hydrogen produced is twice that of oxygen.

$$2\,H_2O \xrightarrow{\text{electric current}} 2\,H_{2(g)} + O_{2(g)}$$

Stability

When water is heated to 100° C at 1 atm pressure, it boils and turns into a gas, steam. Even when the steam is heated to a high temperature, it does not decompose. The water molecule is extremely stable. There is very little decomposition of the water molecule even at a temperature of 1600° C.

Reaction with Metal Oxides

Water reacts with soluble metal oxides to form a class of compounds called **bases** (see Chapter 12). For example,

$$CaO + H_2O \longrightarrow Ca(OH)_2$$

calcium   calcium hydroxide,
oxide   a base

$$Na_2O + H_2O \longrightarrow 2\,NaOH$$

sodium   sodium hydroxide,
oxide   a base

Reaction with Nonmetal Oxides

Water reacts with soluble nonmetal oxides to form a class of compounds called acids (see Chapter 12). For example,

$$CO_2 + H_2O \longrightarrow H_2CO_3$$

carbon   carbonic
dioxide   acid

$$SO_3 + H_2O \longrightarrow H_2SO_4$$

sulfur   sulfuric
trioxide   acid

Reaction with Active Metals

When an active metal such as sodium or potassium is placed in water, a vigorous reaction takes place, with the rapid evolution of hydrogen gas. At the same time, a base is formed.

$$2\,Na + 2\,H_2O \longrightarrow 2\,NaOH + H_{2(g)}$$

sodium   sodium hydroxide

Exercise 9-2

Complete and balance the following reactions:
a. $K_2O + H_2O \longrightarrow$
b. $N_2O_5 + H_2O \longrightarrow$
c. $Ca + H_2O \longrightarrow$

Solution

a. The product is KOH:
   The balanced equation is $K_2O + H_2O \longrightarrow 2\,KOH$
b. The product is HNO_3:
   The balanced equation is $N_2O_5 + H_2O \longrightarrow 2\,HNO_3$
c. The products are $Ca(OH)_2$ and H_2
   The balanced equation is $Ca + 2\,H_2O \longrightarrow Ca(OH)_2 + H_2$

Self-test

Complete and balance the following reactions:
a. $Li + H_2O \longrightarrow$
b. $P_2O_5 + H_2O \longrightarrow$
c. $MgO + H_2O \longrightarrow$

Answers

a. $2\,Li + H_2O \longrightarrow 2\,LiOH$
b. $P_2O_5 + 3\,H_2O \longrightarrow 2\,H_3PO_4$
c. $MgO + H_2O \longrightarrow Mg(OH)_2$

Formation of Hydrates

When water solutions of some soluble compounds are evaporated, the substances separate as crystals that contain the given compound combined with water in a definite proportion by weight. Crystals that contain a definite proportion of water as part of their crystalline structure are called hydrates. The water contained in a hydrate is called water of hydration or water of crystallization.

Barium chloride crystallizes from solution as a hydrate containing two molecules of water. The formula for this hydrate is $BaCl_2 \cdot 2\,H_2O$. It is called barium chloride dihydrate. The dot in the middle of the compound indicates a loose association between two molecules of water and the molecule of barium chloride. When the water of hydration is removed from a hydrate, the resulting compound is said to be anhydrous (without water). When barium chloride dihydrate is heated, the water of hydration is driven off, leaving anhydrous barium chloride.

$$BaCl_2 \cdot 2\,H_2O \xrightarrow{\text{heat}} BaCl_2 + 2\,H_2O_{(g)}$$

barium chloride dihydrate   anhydrous barium chloride

Substances that lose their water of hydration on exposure to air are said to be efflorescent. An example of an efflorescent hydrate is washing soda ($Na_2CO_3 \cdot 10\,H_2O$), sodium carbonate decahydrate. Substances that pick up moisture from the air are said to be hygroscopic. An example of a hygroscopic compound is calcium chloride ($CaCl_2$). A hygroscopic compound can be used as a drying agent because it will pick up and remove moisture from the air.

Table 9-3 Properties of Water	
Physical	**Chemical**
Colorless, except in deep layers, when it is greenish blue	Stable compound, not easily decomposed; at 1600° C about 0.3% dissociates
Odorless	
Tasteless	Reacts violently with sodium and potassium
At 1 atm, water freezes and ice melts at 0° C or 32° F	Reacts with metal oxides (e.g., CaO) to form bases
At 1 atm, water boils and steam condenses at 100° C or 212° F	Reacts with nonmetal oxides (e.g., SO_2) to form acids
Water expands when it freezes to ice	When crystals are formed from aqueous solutions of certain substances, water and the solute combine to build crystals, which are called hydrates
Density at 4° C, 1.0 g/mL	
Heat of vaporization, 540 cal/g	
Heat of fusion, 80 cal/g	

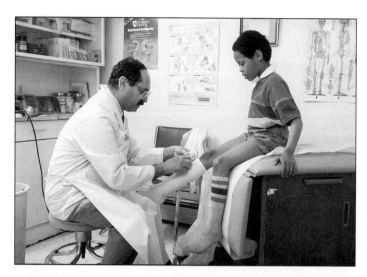

FIGURE 9-5
Cast being applied.

A hydrate of particular importance in the medical field is plaster of paris. When plaster of paris is mixed with water, it forms a hard crystalline compound called gypsum.

$$(CaSO_4)_2 \cdot H_2O + 3 H_2O \longrightarrow 2(CaSO_4 \cdot 2 H_2O)$$

plaster of paris (soft)   gypsum (hard)

Plaster of paris was formerly used extensively in preparing surgical casts (see Figure 9-5). It was spread on crinoline to form a bandage. When the bandage was to be used, it was placed in water, wrung out, and quickly applied. Plaster of paris expands on setting, so the cast may be comfortable when applied but become too tight as it sets. Therefore, it is very important for the nurse to check the circulation to the body part to which a plaster cast has been applied.

The current method of preparing surgical casts involves the use of knitted fiberglass fabric impregnated with a water-activated urethane resin. Exposure to moisture or water initiates a chemical reaction that results in a rigid tape within a few minutes. These casts have the advantages that they are lighter than comparable plaster casts, they can be immersed in water, they are porous and allow free air circulation, and they are resistant to breakage.

Other hydrates are $MgSO_4 \cdot 7 H_2O$, magnesium sulfate heptahydrate (commonly known as Epsom salts), and $Na_2SO_4 \cdot 10 H_2O$, sodium sulfate decahydrate (commonly known as Glauber's salts). Both of these hydrates are used as *cathartics.

9-9 Hydrolysis

When some salts are placed in water, hydrolysis occurs. Hydrolysis is the reaction of a compound with water. When ammonium chloride, a salt, is placed in water, some hydrolysis occurs according to the following equilibrium reaction.

$$NH_4Cl + H_2O \rightleftharpoons NH_4OH + HCl$$

ammonium   water   ammonium   hydrochloric
chloride   hydroxide   acid

Note that hydrolysis in this example is the reverse of neutralization.

Hydrolysis occurs during the process of digestion of foods. For example, sucrose is hydrolyzed into glucose and fructose through the action of an enzyme.

$$C_{12}H_{22}O_{11} + H_2O \xrightarrow{\text{enzyme}} C_6H_{12}O_6 + C_6H_{12}O_6$$

sucrose   glucose   fructose

Note that both glucose and fructose have the same formula, $C_6H_{12}O_6$. How two different compounds can have the same formula will be discussed in Chapter 20.

The hydrolysis of fats in the body yields fatty acids and glycerol. The hydrolysis of proteins yields amino acids. Specific enzymes are necessary for these hydrolytic reactions. They will be discussed in the appropriate chapters on fats and protein.

9-10 Purification of Water

Impurities Present in Water

Natural water contains many dissolved and suspended materials. Rainwater contains dissolved gases—oxygen, nitrogen, and carbon dioxide—plus air pollutants (see Section 8-8), suspended dust particles, and other particulate matter. Groundwater contains minerals dissolved from the soil through which the water has passed. It also contains some suspended materials. Seawater contains over 3.5 percent dissolved matter, the principal compound being sodium chloride. Both seawater and groundwater also contain dissolved and undissolved pollutants.

Lake water or river water may appear clear when a glass full of it is held up to the light, or it may at first contain suspended clay or mud, which tends to settle slowly, leaving what appears to be pure water. However, either of these "clear" waters may contain bacteria and other microorganisms that can be quite harmful to the body. Their destruction or removal is necessary for the proper purification of water. Water can be purified by several processes. The most common are distillation, boiling, filtration, and aeration.

Distillation

Distillation is a process of converting water to steam and then changing the steam back to water again. In the laboratory, a "still" is used to prepare distilled water. In the still shown in Figure 9-6, impure water is placed in the flask at the left and then heated to boiling. As the water boils and changes into steam, it passes into a condenser, which consists of a long glass tube surrounded by another glass tube through which cold water runs. The cold water, by absorbing heat, causes the steam to condense back into liquid water, which then runs out of the end of the tube into the receiving vessel at the right. The suspended and dissolved solids (including the bacteria) that were present in the impure water remain behind in the flask. They do not pass over into the condenser with the steam. The dissolved gases originally present in the impure water, however, do pass over with the steam. The usual practice is to discard the first few milliliters of distilled water coming from the condenser, since they will contain most of the dissolved gases.

Although distilled water is pure water, it is too expensive and the process too slow for large-scale use. The principle use of distilled water in the hospital is in the preparation of sterile solutions.

Boiling

Groundwater or contaminated water can usually be made safe for drinking by boiling it for at least 15 minutes. The boiling does not remove the dissolved impurities, but it does kill any bacteria that might be present. Freshly boiled water has a flat taste because of the loss of dissolved gases. The taste may be brought back to

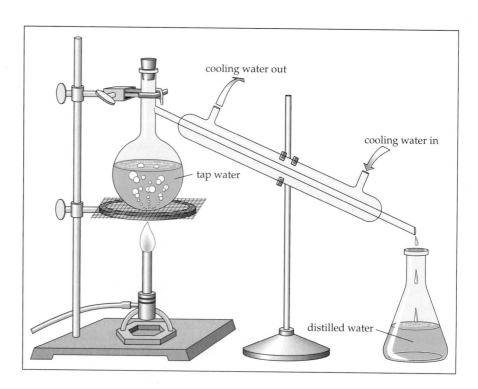

FIGURE 9-6
A laboratory distillation apparatus.

normal by pouring the water back and forth from one clean vessel to another. This process, called aeration, allows air to dissolve in the water again.

Sedimentation and Filtration

For large-scale use, water is first allowed to stand in large reservoirs where most of the suspended dirt, clay, and mud settle out. This process is called sedimentation. However, sedimentation is a very slow process. Put some finely divided clay in a graduated cylinder full of water, shake it, and see how slowly the clay settles out. Commercially, a mixture of aluminum sulfate and lime is added to the water. These two chemicals combine to form aluminum hydroxide which precipitates as a gelatinous (sticky) substance. As the sticky aluminum hydroxide settles out, it carries down with it most of the suspended material. The main advantage of this material is that it settles much more rapidly than does the suspended material by itself and so increases the rate of sedimentation.

After sedimentation, the water is filtered through several beds of sand and gravel to remove the rest of the suspended material. The water then is essentially free of suspended material. However, this process does not remove the dissolved material or much of the bacteria originally present. The water must then be treated with chlorine to kill the bacteria before it is fit to drink.

Aeration

Water can be purified by exposing it to air for a considerable period of time. The oxygen in the air dissolves in water and destroys the bacteria by the process of oxidation. The oxygen also oxidizes the dissolved organic material in the water so that the bacteria have no source of food. However, this process is slow and expensive because of the length of time involved in exposing water to the air. Commercially, aeration is accomplished by spraying filtered chlorinated water into the air. This additional process also removes objectionable odors from the water.

9-11 Hard and Soft Water

When a small amount of soap is added to soft water, it forms copious suds. When a small amount of soap is added to hard water, it forms a precipitate or scum and no lather. What is the difference between soft water and hard water? Hard water contains dissolved compounds of calcium and magnesium. Soft water may contain other dissolved compounds, but these compounds do not cause hardness. The calcium and magnesium compounds (which cause the hardness) react with soap to form a precipitate, thus removing the soap from the water. More and more soap must be added until all the hardness-causing compounds are removed. Only then will the soap cause a lather. The reactions involved are as follows:

$$Ca^{2+} + Na(soap) \longrightarrow Ca(soap)_{(s)} + 2\,Na^+$$

$$\underset{\text{hardness}}{Mg^{2+}} + Na(soap) \longrightarrow \underset{\substack{\text{hardness} \\ \text{(a precipitate)}}}{Mg(soap)_{(s)}} + \underset{\substack{\text{no} \\ \text{hardness}}}{2\,Na^+}$$

The precipitated soap adheres to washed materials, making them rough and irritating to tender skin, or to washed hair, making it sticky and gummy (see Figure 9-7). Food cooked in hard water is likely to be tougher than that cooked in soft water because of the presence of additional minerals. When hard water is boiled, some of the salts form a deposit on the inside of the container in which it is heated. Look inside an old teakettle at home and see the "boiler scale." Hard water must never be used to sterilize surgical instruments because the precipitated salts will dull the cutting edges. (If iron compounds are present in water, they also will cause hardness.)

Detergents have replaced soaps for washing clothes because they do not precipitate in hard water.

Home Methods for Water Softening

Temporary hardness is hardness that can be removed by boiling. It is caused by the bicarbonates of calcium and magnesium. Boiling converts these to insoluble carbonates and carbon dioxide gas, thus removing part of the hardness.

$$Ca(HCO_3)_{2(aq)} \xrightarrow{\text{heat}} CaCO_{3(s)} + CO_{2(g)} + H_2O$$

Other soluble compounds of calcium and magnesium cause permanent hardness. Permanent hardness is not affected by boiling.

Ammonia (ammonium hydroxide) is frequently used to soften water used in washing clothes and windows because it precipitates all the ions that cause temporary hardness and some of those that cause permanent hardness. Borax, sodium tetraborate ($Na_2B_4O_7$), has been used in the home as a laundry water softener. Its effect is similar to that of ammonia. Washing soda (Na_2CO_3), frequently used as a home-laundry water softener, removes both temporary and permanent hardness from the water. Trisodium phosphate or TSP (Na_3PO_4), which is another home-laundry water softener, has an action similar to that of washing soda. However, TSP is no longer recommended for laundry use because of the effects of phosphates on our lakes and streams.

9-12 Fluoridation of Water

Approximately 1 ppm of fluoride ion added to drinking water in the form of NaF reduces the incidence of dental caries. One of the major constituents of bones and teeth is hydroxyapatite ($Ca_5[PO_4]_3OH$). The presence of the hydroxide ion (OH^-)

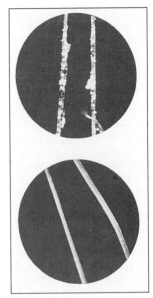

FIGURE 9-7

When hair is shampooed in hard water, curd clings to the hair strands, dulling their natural luster and interfering with their ability to reflect light. The hair strands at top are stringy and not clean because of the clinging hard water curd. Those at bottom, washed in soft water, are radiant and clean.

makes this substance susceptible to attack by acidic substances such as soft drinks with a low pH (see Section 12-5). Replacement of the hydroxide ion with a fluoride makes teeth more resistant to acid attack and hence more resistant to decay.

Fluorides are also used in toothpastes and gels in the form of stannous fluoride (SnF_2).

9-13 Water as a Moderator

Many nuclear reactors use water as a moderator, a substance that slows down neutrons. Why slow down neutrons? Because slower-moving neutrons can react with ^{235}U more effectively than faster-moving neutrons (see page 233).

9-14 Water Pollution

What is polluted water? Strictly speaking, it is any water that is not pure. However, tap water contains many dissolved and suspended substances. It is not pure, yet it is not called polluted water. Any substance that prevents or prohibits the normal use of water is termed a pollutant. The signs of polluted water are usually quite obvious— oil and dead fish floating on the surface of a body of water or deposited along the shores, a bad taste to drinking water, a foul odor along a waterfront, unchecked growth of aquatic weeds along the shore, or tainted fish that cannot be eaten.

Water pollutants can be classified into several categories.

1. Oxygen-demanding wastes. Dissolved oxygen is required for both plant and animal life in a body of water. Anything that tends to decrease the supply of this vital element endangers the survival of the life forms. Oxygen-demanding wastes are acted upon by bacteria in the presence of oxygen, thus leading to a depletion of the dissolved oxygen. Oxygen-demanding pollutants include sewage and wastes from papermills, food-processing plants, and other industrial processes that discharge organic materials into the water.

2. Disease-causing agents. Among the diseases that can be caused by pathogenic microorganisms present in polluted water are typhoid fever, cholera, infectious hepatitis, and poliomyelitis.

3. Radioactive material. Low-level radioactive wastes from nuclear power plants are sealed in concrete and buried underground. High-level radioactive wastes are initially stored as liquids in large underground tanks and later converted into solid form for burial in concrete. In either case, leakage can lead to pollution of nearby water supplies.

4. Heat. Although heat is not normally considered a pollutant of waterways, it does have a detrimental effect on the amount of dissolved oxygen. Thermal pollution results when water is used as a coolant for industrial plants and nuclear reactors and then returned to its source. In addition to decreasing the amount of dissolved oxygen, thermal pollution also causes an increase in the rate of chemical reactions. The metabolic processes of fish and microorganisms are speeded up, increasing their need for oxygen,

at a time when the supply of oxygen is diminishing. Higher water temperatures can also be fatal to certain forms of marine life.

5. Plant nutrients. Nutrients stimulate the growth of aquatic plants. This may lead to lower levels of dissolved oxygen. It may also lead to disagreeable odors when the large amount of plant material decays. Excessive plant growth is often unsightly and interferes with recreational use of water. Excess phosphorus in sewage comes from phosphate detergents and is one cause of this type of pollution.

6. Synthetic organic chemicals. In this category of pollutants are such substances as surfactants in detergents, pesticides, plastics, and food additives.

7. Inorganic chemicals and minerals. These pollutants come from industrial wastes as well as from runoff water from urban areas. One example of such a pollutant is mercury. It was once believed that metallic mercury was inert and settled to the bottom of a lake. It is now known that anaerobic bacteria in bottom muds are capable of converting this mercury into compounds that are poisonous. Another pollutant in this category is sulfuric acid, which is formed by the reaction of sulfur-containing ores with water and oxygen in the air. Salt is also a pollutant, which can occur when brine from oil wells is released into freshwater.

Summary

Water is one of the most important chemicals known. Pure water is a colorless, odorless, flat-tasting liquid that freezes at 0° C and boils at 100° C at 1 atm pressure. Water has its maximum density at 4° C. When water freezes, its volume increases by almost 10 percent.

Water is used as a solvent for many substances. Many chemical reactions take place only when the reactants are dissolved in water. The evaporation of water or perspiration from the skin is a cooling process because the skin provides (loses) the heat required to change the liquid to the vapor state. Water is the standard of reference for such physical constants as the temperature scale and the calorie.

The water molecule is a covalent one with a hydrogen-oxygen-hydrogen angle of about 105 degrees. Water is a polar liquid. Polar liquids usually dissolve polar compounds, and nonpolar liquids dissolve nonpolar compounds.

The electrolysis of water yields hydrogen and oxygen gases. Water does not otherwise appreciably decompose into these gases, even when heated to 1600° C.

Water reacts with certain metal oxides to form bases; water reacts with certain nonmetal oxides to form acids. Water reacts with sodium and potassium to form hydrogen gas and a base. Water reacts with certain salts to form hydrates. Hydrates that lose their water of hydration on standing are said to be efflorescent. Substances that pick up moisture from the air are said to be hygroscopic.

Hydrolysis of a salt is the reaction of that salt with water whereby the water molecule is split.

Water can be purified by distillation, by boiling, by sedimentation and filtration, and by aeration.

Hardness in water is caused by ions of calcium and magnesium (and iron). Temporary hardness is caused by bicarbonates of calcium and magnesium. Materials causing permanent hardness cannot be removed by boiling. Hard water can be softened by using ammonia, washing soda, trisodium phosphate, or lime-soda.

Water pollution can be caused by oxygen-demanding wastes, disease-causing agents, radioactive materials, heat, plant nutrients, synthetic organic chemicals, and inorganic chemicals and minerals.

Questions and Problems

1. Define the term *calorie*. What is the SI unit of heat?
2. How much heat is required to evaporate 30 g of water from the skin? Where does this heat come from?
3. Why does ice float?
4. Water is used as a standard of reference for which physical constants?
5. Why is the water molecule polar? Explain by means of its structure.
6. Why use a pressure cooker at high altitudes? What effect does pressure have on the boiling point of a liquid?
7. What type of compound is produced when water reacts with a metal oxide? with a non-metal oxide?
8. Why purify water?
9. What causes temporary hardness in water? permanent hardness? How can they be removed?
10. What diseases can be caused by polluted water?
11. What is thermal pollution? What are its effects?
12. Name three synthetic organic pollutants and indicate where they could come from.
13. What is a hydrate? What happens when a hydrate is heated?
14. What is an efflorescent compound? a hygroscopic compound?
15. Describe the preparation of distilled water. What types of impurities are removed during the distillation process?
16. Will benzene, a nonpolar compound, dissolve in water? Explain.
17. Why does salt dissolve in water?
18. What effect does boiling have on the purification of water?
19. What effect does hydrogen bonding have on the boiling point of water? Explain.
20. What causes the vapor pressure of a liquid?
21. Of what importance is hydrolysis in digestive processes?
22. Is hard water safe to drink? Explain.
23. Explain the relationship of vapor pressure and boiling point of a liquid.
24. Explain the use of water as a moderator in nuclear reactors.
25. How are the following chemicals used in water purification? Cl_2, CaO (lime), and $Al_2(SO_4)_3$ (alum)
26. Explain why alcohols like ethyl alcohol (C_2H_5OH) are water soluble.

Practice Test

1. Hardness in water is caused by the presence of _____.
   a. K^+
   b. Cu^+
   c. Ca^{2+}
   d. all of these
2. Water reacts with nonmetal oxides to form _____.
   a. acids
   b. bases
   c. salts
   d. hydrates
3. Water used as a moderator
   a. activates neutrons
   b. slows down neutrons
   c. absorbs neutrons
   d. has no effect on neutrons
4. The reaction of water with a salt is called _____.
   a. efflorescence
   c. neutralization
   b. acidification
   d. hydrolysis
5. An example of a cooling process is _____.
   a. distillation
   b. condensation
   c. evaporation
   d. hydration
6. Water
   a. is a polar compound
   b. has a maximum density at 4° C
   c. boils at 100° C at 1 atm pressure
   d. all of these
7. A process for purification of water is _____.
   a. solvation
   b. evaporation
   c. condensation
   d. distillation
8. As the atmospheric pressure increases, the boiling point of a liquid _____.
   a. increases
   b. decreases
   c. is unaffected
9. Water has a relatively low vapor pressure because of _____.
   a. dissolved gases
   b. ionic bonds
   c. hydrogen bonds
   d. dissolved minerals
10. What types of compounds would be expected to dissolve in water?
    a. covalent
    b. nonpolar
    c. polar
    d. metallic

10

Liquid Mixtures

Objectives

- To understand the importance and properties of solutions

- To express strength of solutions in different ways

- To understand the uses and properties of suspensions and emulsions

- To describe the properties of colloids and to compare these properties with those of solutions, suspensions, and emulsions

- To describe osmosis and dialysis

Most liquids are not pure; they often contain dissolved solutes or particulate matter. In this photo, a cube of sugar dissolves in water to produce a solution of sucrose.

10-1 Solutions

Liquid mixtures can be divided into four types: solutions, suspensions, colloids, and emulsions. Each type has its own specific properties and uses.

General Properties

A solution is a homogeneous mixture of two or more substances evenly distributed in each other. A liquid solution consists of two parts: the solid, liquid, or gaseous material that has dissolved—the solute—and a liquid material in which it has dissolved—the solvent.

When a crystal of salt is placed in water, which is then stirred, the crystal dissolves and a clear solution is formed. When more salt is added to this salt-water solution, it too dissolves, making the solution more concentrated than the previous one. Even more salt can be dissolved in the water to make it a much more concentrated salt solution. Thus, one of the properties of solutions is that they have a variable composition. Varying amounts of salt and water can be mixed to form various concentrations of saltwater.

When salt is dissolved in water to make salt water, the solution formed is clear and colorless. When sugar is dissolved in water, again a clear, colorless solution is formed. When copper sulfate is dissolved in water, it also forms a clear solution. However, the solutions formed with salt and sugar are colorless, whereas that formed with copper sulfate is blue. Solutions are always clear. They may or may not have a color. Clear merely means that the solution is transparent to light.

When a salt solution is examined under a high-power microscope, it appears to be homogeneous. The same is true for sugar solution. In general, all solutions are homogeneous. The solute cannot be distinguished from the solvent in a solution.

When a solution is allowed to stand undisturbed for a long period of time, no crystals of solute settle out, provided the solvent is not allowed to evaporate. This is another property of solutions—the solute does not settle out.

The salt in a salt-water solution can be recovered by allowing the water to evaporate; the same is true of the sugar in a sugar solution. In general, solutions can be separated by physical means.

If a solution (such as saltwater) is poured into a funnel containing a piece of filter paper, the solution will pass through unchanged. That is, the particles in solution must be smaller than the openings in the filter paper.

The properties of solutions are summarized as follows. Solutions:

1. Consist of a soluble material or materials (the solute) dissolved in a liquid (the solvent)
2. Have a variable composition
3. Are clear
4. Are homogeneous
5. Do not settle
6. Can be separated by physical means
7. Pass through filter paper

Solvents Other Than Water

Solvents other than water are also used. One common solvent used in hospitals is alcohol. An alcohol solution used medicinally is called a *tincture. Tincture of iodine contains iodine dissolved in alcohol. Tincture of green soap contains potassium soap dissolved in alcohol. Ether is an excellent solvent for fats and oils.

Factors Affecting Solubility of a Solute

Temperature Most solid solutes are more soluble in hot water than in cold water. Figure 10-1 shows that KNO_3 becomes much more soluble as the temperature increases; however, $Ce_2(SO_4)_3$ becomes less soluble with an increase in temperature, and NaCl shows little change in solubility. Gases such as HCl and SO_2 become less soluble with increasing temperature. The solubility of Br_2, a liquid, is practically unaffected by temperature.

Pressure A change in pressure has no noticeable effect on the solubility of a solid or liquid solute in a given solvent but will affect the solubility of a gaseous solute. The greater the pressure, the greater the solubility of a gas in a liquid.

Surface Area Although surface area does not affect the amount of solute that will dissolve, it does affect the rate of dissolution. The greater the amount of surface area, the quicker a solute will dissolve in a solvent. Thus, to make a solid solute dissolve faster we frequently powder it, thereby increasing the surface area.

Stirring The rate at which a solute dissolves can also be increased by stirring the mixture. The process of stirring brings fresh solvent into contact with the solute and so permits more rapid solution.

Therefore, to dissolve most solid solutes rapidly, the solute should be powdered and the mixture should be heated while it is stirred.

Nature of Solvent In general, polar liquids dissolve polar compounds and nonpolar liquids dissolve nonpolar compounds. Water is a polar liquid (see Section 9-2) and dissolves polar compounds such as sodium chloride (NaCl).

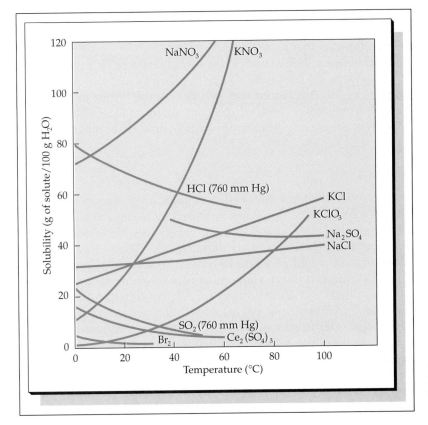

FIGURE 10-1

Variation in the solubilities of several substances in water as a function of temperature.

Table 10-1	
Some Polar and Nonpolar Liquids	
Name	Formula
Polar liquids	
Water	H_2O
Methyl alcohol	CH_3OH
Ethyl alcohol	C_2H_5OH
Nonpolar liquids	
Benzene	C_6H_6
Carbon tetrachloride	CCl_4
Ether	$C_2H_5OC_2H_5$

Sometimes, in spite of all attempts, a substance does not appreciably dissolve in water. If the substance is nonpolar, then it should not be expected to be soluble in water. But it should be soluble in a nonpolar liquid. Sometimes polar materials do not dissolve in water. These are called insoluble salts (see Section 12-6).

Table 10-1 lists several polar and nonpolar liquids. Using this table we could predict that potassium iodide (KI), a polar compound, should be soluble in water and that most likely it would not be soluble in a nonpolar liquid such as benzene. Likewise we could predict that oils and waxes, which are nonpolar compounds, should be insoluble in water but soluble in ether (or other nonpolar liquids).

In addition to liquid solutions, there are also solid solutions. They usually consist of one metal dissolved in another while both are in the molten (liquid) state. Such a solid solution is called an *alloy. Some alloys are used in preparing replacements for such body parts as hip bones and knee joints.

A special type of alloy consists of a metal dissolved in mercury, a liquid metal. An alloy containing mercury is called an amalgam. If silver is dissolved in mercury, a silver amalgam is produced. This substance was used in dental work to fill a cavity in a tooth.

Importance of Solutions

During digestion, foods are changed to soluble substances so that they can pass into the bloodstream and be carried to all parts of the body. At the same time the waste products of the body are dissolved in the blood and carried to other parts of the body where they can be eliminated. Plants obtain minerals from the groundwater in which those minerals have dissolved.

Many chemical reactions take place in solution. When solid silver nitrate is mixed with solid sodium chloride, no reaction takes place because the movement of the ions in the solid state is highly restricted. However, when a solution of silver nitrate is mixed with a solution of sodium chloride, a precipitate of silver chloride is formed instantaneously. This reaction occurs because the ions in the solution are free to move and react with other ions.

Many medications are administered orally, subcutaneously, or intravenously as solutions.

Drugs must be in solution before they can be absorbed from the gastrointestinal (GI) tract. As you might expect, when drugs are taken in solution, such as syrups and elixirs, they are absorbed more rapidly than drugs in a solid form, such as tablets and capsules.

Strength of Solutions

Dilute and Concentrated
When a few crystals of sugar are placed in a beaker of water, a *dilute* sugar solution is produced. As more and more sugar is added to the water, the solution becomes more *concentrated*. But when does the solution change from dilute to concentrated? There is no sharp dividing line. "Dilute" means that the solution contains a small amount of solute in relation to solvent. "Concentrated" merely means that the solution contains a large amount of solute in relation to solvent.

However, both dilute and concentrated are relative terms. A dilute sugar solution may contain 5 g of sugar per 100 mL of solution, whereas 5 g (the same amount) of boric acid per 100 mL of solution will produce a concentrated boric acid solution. That is, the terms dilute and concentrated usually have no specific quantitative meaning and so are are not generally used for medical applications.

Saturated
Suppose a small amount of salt is placed in a beaker of water. When the mixture is stirred, all the salt will dissolve. If more and more salt is added with

stirring, a point will soon be reached where some of the salt settles to the bottom of the beaker. This excess salt does not dissolve even upon more rapid agitation. This type of solution is called a *saturated solution.* Some of the crystals are continuously dissolving and going into the solution, but at the same time, the same amount of solute crystallizes out of the solution. This type of interchange, as you may remember, is called equilibrium.

Thus, a saturated solution can be defined as one in which there is an equilibrium between the solute and the solution. A saturated solution can also be defined as one that contains all the solute that it can hold under the given conditions.

In the bloodstream, if the concentration of uric acid exceeds its solubility, the excess undissolved uric acid crystallizes. When these crystals accumulate in the great toe or in the earlobes, the condition is called gout (Figure 10-2). If the uric acid crystallizes in the gallbladder, gallstones can be formed. Kidney stones may also be caused by crystallized uric acid.

Unsaturated   An unsaturated solution contains less of a solute than it could hold under normal conditions. Suppose that a saturated solution of glucose in a certain amount of water contains 25 g of glucose. An unsaturated glucose solution would be one that contained less than 25 g of glucose in the same amount of water. In an unsaturated solution, no equilibrium exists because there is no undissolved solute.

Like dilute and concentrated, the terms saturated and unsaturated are relative terms. The same amount of two different solutes may produce entirely different types of solutions. For glucose, 5 g in 100 mL of water produces an unsaturated solution, whereas 5 g of boric acid in 100 mL of water produces a saturated solution. Therefore, the terms *saturated solution* and *unsaturated solution* are not used for medical applications.

Supersaturated   Under certain conditions a solvent can be made to dissolve more solute than its saturated solution can hold under the same conditions. A supersaturated solution can be prepared by adding excess solute to a saturated solution, heating that mixture, filtering off the excess solute, and then allowing the liquid to cool slowly. If this is done carefully, some excess solute will remain dissolved, thus forming a supersaturated solution. However, such a solution is very unstable. If one crystal of solute is added, or if the liquid is shaken, the excess solute will crystallize immediately and a saturated solution will remain.

Percentage Solutions   The **weight-volume method** expresses the weight of solute in a given volume of solvent, usually water. A 10 percent glucose solution will contain 10 g of glucose per 100 mL of solution. A 0.9 percent saline solution will contain 0.9 g of sodium chloride per 100 mL of solution. The percentage indicates the number of grams of solute per 100 mL of solution.

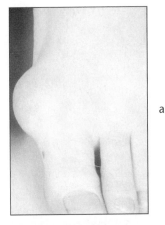

a

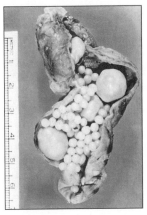

b

FIGURE 10-2
Gout in large toe (a); gallstones (b).

Exercise 10-1

Exercise 10-1a

Prepare 500 mL of 2 percent citric acid solution.

Solution

First let us calculate how much citric acid will be required. A 2 percent citric acid solution contains 2 g citric acid per 100 mL of solution. Therefore, in 500 mL of solution there should be

$$500 \text{ mL} \times \frac{2 \text{ g citric acid}}{100 \text{ mL}} = 10 \text{ g citric acid}$$

To prepare the solution, proceed as follows:

1. Weigh out exactly 10 g of citric acid.
2. Dissolve the 10 g of citric acid in a small amount of water contained in a 500 mL graduated cylinder.
3. Add water to the 500 mL mark and stir.

Note that the 10 g of citric acid was dissolved in water and then diluted to the required volume—500 mL. It was *not* dissolved directly in 500 mL of water. (If it had been, the final volume would have been more than 500 mL.)

Exercise 10-1b

A patient is given 1000 mL of 0.9 percent NaCl intravenously. How many grams of NaCl did the patient receive?

Solution

$$0.9\% \text{ means } \frac{0.9 \text{ g NaCl}}{100 \text{ mL solution}}$$

$$1000 \text{ mL} \times \frac{0.9 \text{ g NaCl}}{100 \text{ mL}} = 9 \text{ g NaCl}$$

Self-test

How many milliliters of a 2.00% KCl solution are needed to provide a patient with 5.00 g of KCl?

Answer

250 mL

In clinical work involving dilute solutions, concentrations are sometimes expressed in terms of milligram percent (mg %), which indicates the number of milligrams of solute per 100 mL of solution. Milligram percent is also referred to as milligrams per deciliter (mg/dL).

Parts per Million   Low concentrations may be expressed in the units *milligrams per liter* (mg/L). Another unit for expressing low concentrations is parts per million (ppm). One part per million is equivalent to 1 mg/L. That is, if a solution has a concentration of 40 mg/L, its concentration can also be expressed as 40 ppm.

Parts per million are used to indicate the hardness of water and also to show the concentration of both common substances and pollutants in water and in air.

Extremely low concentrations of pollutants are expressed in the units parts per billion, which is equivalent to milligrams per 1000 L (mg/1000 L).

Ratio Solutions   Another method of expressing concentration is a ratio solution. A 1:1000 merthiolate solution contains 1 g of merthiolate in 1000 mL of solution. A 1:10,000 KMnO$_4$ solution contains 1 g of KMnO$_4$ in 10,000 mL of solution. The first number in the ratio indicates the number of grams of solute, and the second number gives the number of milliliters of solution. As with percentage solutions, the solute is dissolved in a small amount of solvent (water) and then diluted to the desired volume.

Percentage and ratio solutions are frequently used by doctors, nurses, and pharmacists.

Molar Solutions   Molar solutions are used most frequently by chemists. A molar solution is defined as one that contains 1 mol (see Section 5-1) of solute per liter of solution. A 1 molar (1 M) solution of glucose ($C_6H_{12}O_6$) will contain 1 mol of glucose (180 g) in 1 L of solution. As before, the solute is dissolved in a small amount of water and then diluted to the desired volume.

Exercise 10-2

Exercise 10-2a

Prepare 3 L of 2 M (2 molar) KCl (molecular mass 74.5).

Solution

The problem calls for the preparation of 3 L of 2 M KCl or 3 L × 2 M KCl. Recall that molarity means moles per liter.

$$\text{Moles of substance} = \text{volume} \times \text{concentration}$$

$$\text{Moles of KCl needed} = 3 \, \cancel{L} \times \frac{2 \, \text{mol KCl}}{1 \, \cancel{L}} = 6 \, \text{mol KCl}$$

Then, since 1 mol KCl has a mass of 74.5 g,

$$\text{Moles KCl required} = 6 \, \cancel{\text{mol KCl}} \times \frac{74.5 \, \text{g KCl}}{1 \, \cancel{\text{mol KCl}}} = 447 \, \text{g KCl}$$

So, we take 447 g KCl, dissolve it in water, and dilute to a total of 3 L.

Exercise 10-2b

Prepare 500 mL of 0.1 M NaOH (molecular mass 40).

Solution

The problem calls for 500 mL of 0.1 M NaOH, or 500 mL × 0.1 M NaOH. Changing molarity to moles per liter and also changing milliliters to liters, we have

$$500 \, \cancel{\text{mL}} \times \frac{1 \, \cancel{L}}{1000 \, \cancel{\text{mL}}} \times \frac{0.1 \, \text{mol NaOH}}{1 \, \cancel{L}} = 0.05 \, \text{mol NaOH}$$

Then, changing moles of NaOH to grams of NaOH,

$$0.05 \, \cancel{\text{mol NaOH}} \times \frac{40 \, \text{g NaOH}}{1 \, \cancel{\text{mol NaOH}}} = 2 \, \text{g NaOH}$$

Thus, we should dissolve 2 g of NaOH in water and dilute to 500 mL.

Exercise 10-2c

How many grams of glucose are present in 0.5 L of 2.0 M glucose solution? The molecular mass of glucose is 180.

Solution

Changing molarity to moles per liter, we have

$$0.5 \, \cancel{L} \times \frac{2.0 \text{ mol glucose}}{1 \, \cancel{L}} = 1 \text{ mol glucose}$$

Then,

$$1 \, \cancel{\text{mol glucose}} \times \frac{180 \text{ g glucose}}{1 \, \cancel{\text{mol glucose}}} = 180 \text{ g glucose}$$

Self-test

The average concentration of sodium ions in the blood serum is 0.34 percent. What is the molarity of Na in blood serum?

Answer

0.15 M

Osmolarity   The osmotic pressure of a solution is expressed in terms of the osmolarity of a solution. *Osmolarity (osmol) is a measure of the number of particles in solution and will be discussed later in this section under Osmotic Pressure.

Dilution   Solutions are diluted by adding water to a more concentrated solution. Because the amount of solute does not change, a simple equation allows us to calculate the new concentration:

$$C_1 \times V_1 = C_2 \times V_2$$

where C_1 and C_2 indicate the initial and final concentrations, respectively, and V_1 and V_2 are the initial and final volumes, respectively.

Exercise 10-3

Calculate the final concentration of 100 mL of 5.0 percent glucose diluted to 600 mL.

Solution

Using the equation

$$C_1 \times V_1 = C_2 \times V_2$$

where

$$C_1 = 5.0\%, V_1 = 100 \text{ mL, and } V_2 = 600 \text{ mL}$$
$$5.0\% \times 100 \text{ mL} = C_2 \times 600 \text{ mL}$$
$$C_2 = 0.83\%$$

Self-test

How many milliliters of a 10.0 percent saline solution are needed to make 500 mL of 0.90 percent saline.

Answer

45 mL

Special Properties of Solutions

Effect of Solute on Boiling Point and Freezing Point Whenever a nonvolatile solute is dissolved in a solvent, the boiling point of the solution thus prepared is always greater than that of the pure solvent. Pure water boils at 100° C at 1 atm pressure. A solution of salt in water or a solution of sugar in water will boil at a temperature above 100° C.

Likewise, when a nonvolatile solute is dissolved in a solvent, the freezing point of the solution is always less than the freezing point of the pure solvent. The freezing point of water is 0° C. The freezing point of salt solution or sugar solution is always less than 0° C.

Use is made of these facts in the cooling system of automobiles. Antifreeze is added to the water in the car radiator to lower the freezing point so that the liquid will not freeze when the temperature drops below 32° F. The same material is used in some areas to raise the boiling point of the liquid in the automobile radiator—to prevent it from boiling over when the temperature rises.

Surface Tension Consider a water molecule in the center of a beaker of water (see Figure 10-3). This water molecule will be attracted in all directions by the water molecules around it. Next, consider a molecule at the surface of the water. This molecule is attracted sideways and downward, but it is not attracted very much by the air molecules above it. Therefore, there is a net downward attraction on the surface water molecules. This downward pull on the surface molecules causes them to form a surface film. Surface tension can be defined as the force that causes the surface of a liquid to contract. Surface tension also is the force necessary to break this surface film. All liquids exhibit surface tension; the surface tension of water is higher than that of most liquids.

Surface tension is responsible for the formation of drops of water on a greasy surface. The surface film holds the drop in a spherical shape rather than letting it spread over the surface as a sheet of water.

Some medications designed for use on the tissues in the throat contain a very special surface-active agent—one that will reduce the surface tension of the water. This surface-active agent, called a surfactant, lowers the surface tension of the liquid so that it spreads rapidly over the tissues rather than collecting in the form of droplets with less "active" surface area. Soaps and detergents are *surfactants. Bile, which is secreted by the liver, contains bile salts, which act as surfactants. These surfactants help in the digestion of fats (page 443).

A lack of a surfactant in the lungs of premature infants causes respiratory distress syndrome (RDS), formerly called hyaline membrane disease. The surfactant is necessary to form a coating on the inner lining of the small air sacs (alveoli) in the lungs. If the surfactant is present in low amounts or is not present at all, the surface tension in the alveoli rises, causing portions of the lung to collapse and producing respiratory distress.

The attraction of like molecules for each other is known as cohesion; the attraction of unlike molecules is known as adhesion. Water rises in a capillary tube because of surface tension and also because of adhesion of the water molecules to the glass. This effect, known as capillary action, is used when drawing blood samples. Capillary action also accounts for the absorption of water by absorbent cotton and by paper towels.

Viscosity Some liquids flow readily (water) whereas others (molasses) do not. A measure of the resistance to flow is called *viscosity (see page 441). Table 10-2

FIGURE 10-3
Surface tension of a liquid.

Table 10-2	
Relative Viscosities of Various Liquids at 20° C	
Alcohol (ethyl)	1.20
Blood, male	4.71
Blood, female	4.46
Ether (ethyl)	0.23
Glycerol	1490
Water	1.00

indicates the viscosities of several liquids compared to water, which is assigned a value of 1.00. Note the viscosity of blood compared with that of water.

As the temperature increases, liquids tend to flow more readily, and so the viscosity decreases.

Diffusion   Diffusion, also known as passive transport, is the process whereby a substance moves from an area of its higher concentration to a region where it is less concentrated. That is, the molecules of the substance move from an area where they are crowded together and where molecular collisions are frequent to a region where they are less crowded and collisions will occur less often. The greater the difference in concentration between the two areas (the greater the concentration gradient), the faster will be the rate of diffusion. During diffusion, no external source of energy is required.

For example, when a crystal of copper sulfate pentahydrate (a blue crystalline substance) is dropped into a cylinder of water, the blue color is soon observed in the water surrounding the crystal. After a while, the blue color can be seen extending upward from the crystal. The liquid at the bottom of the cylinder will be darker blue, and the liquid above it will be lighter blue. After several hours the entire contents of the cylinder will be uniformly blue. That is, the solute particles from the crystal are uniformly distributed into all parts of the solution.

Gases will diffuse into one another. When a bottle of ether is opened the odor can soon be detected at a distance. Diffusion into a gas takes place more rapidly than diffusion into a liquid.

Another example of diffusion is the loss of perspiration from the body. The moisture flows from an area of high concentration (the skin) to one of lower concentration (the air). The higher the moisture content of the air (the lower the concentration gradient), the slower the rate of evaporation of moisture from the skin.

Passive transport (diffusion) occurs along a concentration gradient from an area of higher concentration to one of lower concentration with no energy other than kinetic molecular energy being required.

Active transport occurs when a substance is moved against the concentration gradient, that is, from an area of low concentration to one of higher concentration. In this process energy is required. Active transport is responsible for the high concentration of potassium ions inside the cells compared to a much lower potassium ion concentration outside the cells. Active transport is also responsible for the low concentration of sodium ions inside the cells compared with a much higher sodium ion concentration outside the cells. The amount of energy required for active transport could be as much as 35 percent of the energy output of a resting cell.

Osmosis   When the diffusing substance is water and when the diffusion takes place through a *semipermeable (properly, selectively permeable) membrane, the process is called osmosis.

Osmosis can be defined as the diffusion of water (solvent) through a semipermeable membrane from a weaker solution (one containing less dissolved solute) to a stronger solution (one containing more dissolved solute). Osmosis can also be defined as the diffusion of water (solvent) down a concentration gradient from an area of high solvent concentration (a weak solution) to a region of low solvent concentration (a stronger solution).

Consider two salt solutions (one dilute and the other concentrated) separated by a semipermeable membrane. The two solutions will tend to equalize in concentration. That is, the dilute one will tend to become more concentrated, whereas the concentrated one will tend to become more dilute. How can they do this? There are two possibilities. First, osmosis can take place. That is, the solvent, water, can diffuse through the membrane from the weaker to the stronger solution (or from an area of high solvent concentration to a region of lower solvent concentration). This process will continue until the two solutions have the same concentration.

The second possibility is that of diffusion. The solute will diffuse through the membrane from the stronger to the weaker solution (down the concentration gradient) until the two solutions have equal concentrations. Both osmosis and diffusion can occur at the same time, but not at the same rate. In general, osmosis occurs more rapidly than diffusion.

An example of osmosis that is quite common in the home can be observed by placing a dried prune in water (Figure 10-4). The skin of the prune acts as a semipermeable membrane. Inside the prune are rather concentrated juices. The water surrounding the prune is certainly dilute in comparison to the juices inside. Thus there are two different concentrations of a solution separated by a semipermeable membrane, and osmosis can take place. In which direction? In osmosis the diffusion of solvent is from the weaker to the stronger solution. Therefore, the water will diffuse into the prune, causing it to swell.

Another example of osmosis can be seen when a cucumber is placed in a strong salt solution (Figure 10-5). The skin of the cucumber acts as a semipermeable membrane. The liquid inside the cucumber is quite dilute in comparison to that of the salt solution. Therefore, osmosis takes place between solutions of two different concentrations. The flow of water, again from dilute to concentrated, is from the cucumber into the solution. Thus the cucumber shrinks and becomes a pickle, again by the process of osmosis.

Sailors lost at sea die of dehydration if they drink salt water. In this case, the seawater has a higher concentration of salts than the body fluids, so water diffuses out of the tissues and dehydration results.

Osmotic Pressure Consider a beaker of water into which is placed a thistle tube with a semipermeable membrane over its end as in Figure 10-6. Assume that the thistle tube contains sugar solution. In which direction will osmosis take place? Osmosis will take place with the water diffusing into the thistle tube (from weaker to stronger solution). As the water diffuses into the thistle tube, the water level in the tube will rise. The rising water in the thistle tube will exert a certain amount of pressure, as does any column of liquid. The pressure exerted during osmotic flow is called osmotic pressure. This concept is of great importance in the regulation of fluid and electrolyte balance in the body (see Section 30-14).

The osmotic pressure of a solution can be expressed in terms of *osmolarity and depends upon the number of particles in solution. The unit of osmolarity is the osmol. For dilute solutions, osmolarity is expressed in milliosmoles (mOsmol). Body fluids have an osmolarity of about 300 mOsmol.

Osmolarity can be calculated from molarity as follows:

$$\text{osmolarity} = \text{molarity} \times \frac{\text{number of particles}}{\text{molecule of solute}}$$

Exercise 10-4

Exercise 10-4a

What is the osmolarity of a 1 M glucose solution?

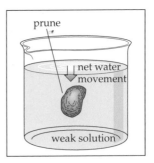

FIGURE 10-4
Osmosis: prune in water.

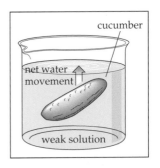

FIGURE 10-5
Osmosis: cucumber in salt water.

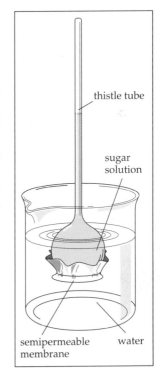

FIGURE 10-6
Demonstration of osmotic pressure.

Solution

Glucose is a nonelectrolyte (see Section 11-1); it is undissociated. That is, it produces no ions in solution. Therefore, one molecule of glucose yields one particle in solution. Thus,

$$\text{osmolarity} = \text{molarity} \times \frac{\text{number of particles}}{\text{molecule of solute}}$$

$$= 1\,\text{M} \times 1 = 1\,\text{osmol}$$

Exercise 10-4b

What is the osmolarity of a 1 M NaCl solution?

Solution

Each NaCl yields two ions, Na^+ and Cl^-, in solution (see Section 11-3). Thus,

$$\text{osmolarity} = 1\,\text{M} \times 2 = 2\,\text{osmol}$$

Therefore, we should expect a 1 M NaCl solution to have twice the osmotic pressure of a 1 M glucose solution, as it does.

Exercise 10-4c

What is the osmolarity of a 0.1 M Na_2SO_4 solution?

$$Na_2SO_4 \longrightarrow 2\,Na^+ + SO_4^{2-}$$

Solution

Since each molecule of Na_2SO_4 yields three ions in solution,

$$\text{osmolarity} = 0.1\,\text{M} \times 3 = 0.3\,\text{osmol}$$

Self-test

a. What is the osmolarity of physiologic saline solution with a concentration of 0.9 percent NaCl? (0.9 percent NaCl is equivalent to 0.15 M NaCl.)
b. What is the osmolarity of a 5.5 percent glucose solution? (A 5.5 percent glucose solution is 0.305 M.)

Answers

a. 300 mOsmol
b. 305 mOsmol

A cardioplegic solution is used for cardiac instillation during open-heart surgery. This buffered solution contains the following electrolytes.

	mEq/L[†]		*mEq/L*
Ca^{2+}	2.4	Na^+	120
Mg^{2+}	32	Cl^-	160
K^+	16	HCO_3^-	10

The osmolar concentration of this solution is 280 mOsmol/L, and the pH[‡] is approximately 7.8.

[†]mEq/L will be discussed in Chapter 30.
[‡]pH will be discussed in Chapter 13.

The osmotic pressure, π, of a solution is related to the molar concentration of solute M by the equation:

$$\pi = MRT$$

where R is the gas constant and T is the Kelvin temperature. There is a similarity between this equation for the osmotic pressure and the equation for an ideal gas. Ionic compounds produce greater effects as a result of ionization. For ionic substances the molarities of all the ions must be added together to get a total molarity. Thus, a 2 molar NaCl solution will have 2 M Na^+ and 2 M Cl^- for a total of 4 M ions.

Exercise 10-5

The osmolar concentration of normal plasma is 0.290 molar. What is the osmotic pressure of normal blood plasma?

Solution

$= MRT$ $\quad$ M $=$ 0.290 mol/L
$\quad$ T $=$ normal body temperature in K $=$ 310 K
$\quad$ R $=$ 0.082 $\dfrac{\text{L atm}}{\text{mol K}}$

$= 0.290 \, \text{moles}/\text{L} \times 310 \, \text{K} \times \dfrac{0.082 \, \text{L atm}}{\text{mol K}} = 7.4 \, \text{atm}$

Self-test

Calculate the molarities of glucose solution and salt solution that will produce an osmotic pressure equal to that of normal blood plasma.

Answer

0.290 M glucose and 0.145 M NaCl

Reverse Osmosis As was indicated in the preceding paragraphs, when water and a salt solution are separated by a semipermeable membrane, osmosis takes place. There is net water diffusion from the weaker to the stronger solution with the result that osmotic pressure is produced. However, if a pressure greater than the osmotic pressure is applied to the salt solution side of the membrane (Figure 10-7), the entire process will be reversed. That is, water will diffuse from the salt solution side to the water side. This process is called reverse osmosis and is used in the desalination of seawater, which is not drinkable because of its high salt content.

Isotonic Solutions

Two solutions that have the same solute concentration are said to be isotonic. The normal salt concentration of the blood is approximately equal to that of a 0.9 percent sodium chloride solution. The common name for 0.9 percent sodium chloride solution is physiologic saline solution. The blood and physiologic saline solution are isotonic—they have the same salt concentration. A 5.5 percent glucose solution is also approximately isotonic with body fluids.

Suppose that a red blood cell were placed in a small amount of physiologic saline solution. What would happen? The red blood cell is surrounded by its cell

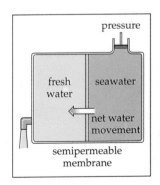

FIGURE 10-7
Reverse osmosis.

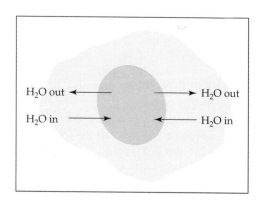

FIGURE 10-8

Red blood cell in isotonic solution. There is no change in the shape of a cell since water enters and leaves at the same rate.

wall, which acts as a semipermeable membrane. Will osmosis take place? The answer is no, because there is no difference in concentration on either side of the semipermeable membrane. Actually, the water molecules move in both directions equally with no net change (see Figure 10-8). Thus, a physiologic saline solution can be given intravenously to a patient without any effect on the red blood cells (see Figure 10-9). A physiologic saline solution is administered under the following conditions:

1. When the patient has become dehydrated
2. When the patient has lost considerable fluid, as in the case of hemorrhage
3. To prevent postoperative shock

Hypotonic Solutions

A hypotonic solution is one that contains a lower solute concentration than that of another solution. Distilled water and tap water are hypotonic compared with blood.

Suppose that a red blood cell is placed in water (a hypotonic solution). What will happen? The salt concentration in the red blood cell is higher than that of the

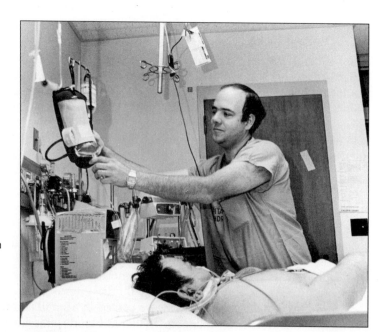

FIGURE 10-9

A nurse readies saline solution for intravenous infusion into a patient's circulatory system. This solution is similar to blood plasma with respect to the types and concentrations of solutes it contains.

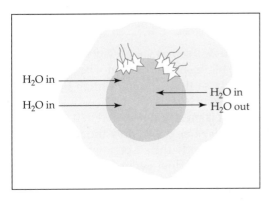

FIGURE 10-10

Red blood cell in hypotonic solution. The rate of water passing into the cell exceeds the rate of loss of water. As a result the cell distends and may burst.

water. Therefore, osmosis will take place, with the water diffusing into the red blood cell (from dilute to concentrated solution) (see Figure 10-10). The red blood cell thus enlarges until it bursts. This bursting of a red blood cell because of a hypotonic solution is called *hemolysis. During hemolysis the blood is said to be *laked. Thus, a hypotonic solution is *not* usually used for transfusions.

Hypertonic Solutions

A *hypertonic solution is one that contains a higher solute concentration than that of another solution. A 5 percent sodium chloride solution or a 10 percent glucose solution is an example of a hypertonic solution when compared with blood.

Suppose a red blood cell is placed in a hypertonic solution. What will happen? The salt concentration in the red blood cell is less than in the hypertonic solution. Therefore, osmosis will take place with the water diffusing out of the red blood cell (from dilute to concentrated) (see Figure 10-11). The red blood cell thus shrinks. This shrinking of the red blood cell in a hypertonic solution is called *plasmolysis.

Usually only isotonic solutions can be safely introduced into the bloodstream. Hypotonic solutions can cause hemolysis, and hypertonic solutions can cause plasmolysis.

Saline *cathartics such as magnesium sulfate, milk of magnesia, and magnesium citrate are absorbed from the large intestine slowly and incompletely. When these substances are ingested, a hypertonic solution is produced in the large intestine and water will diffuse from the tissue spaces into the intestinal tract until the solution is again isotonic with the body fluids. This additional water in the large intestine produces a watery stool that is easily evacuated.

Therefore, the continual use of cathartics may cause a patient to become dehydrated. One the same basis, cathartics have been used to rid the body of excess fluid, although diuretics are now preferred for this purpose.

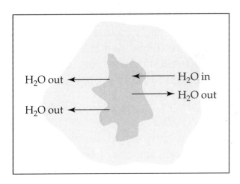

FIGURE 10-11

Red blood cell in hypertonic solution. The rate at which water leaves the cell exceeds the rate of water entering the cell. This causes the cell to shrink.

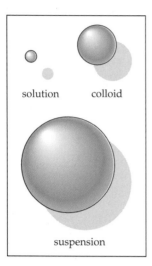

FIGURE 10-12

Relative sizes of solution, colloid, and suspension particles. Particles in solution are ions or molecules; colloidal particles consist of small clumps of molecules; suspension particles consist of large clumps of molecules.

10-2 Suspensions

Suppose that some powdered clay is placed in water and vigorously shaken. A suspension of clay in water will be produced. This suspension will not be clear; it will be opaque. On standing, the clay will slowly settle. The composition of the suspension is actually changing as the clay settles out, so it is a heterogeneous mixture.

The clay is not dissolved in the water; it is merely suspended in it. When the suspension is poured into a funnel lined with filter paper, only the water passes through the filter paper; the clay does not. Evidently the suspended clay particles are too large to pass through the holes in a piece of filter paper. Undoubtedly they will also be too large to pass through a membrane that has even finer openings (Figure 10-12).

Properties of suspensions are summarized as follows. Suspensions

1. Consist of an insoluble substance dispersed in a liquid
2. Are heterogeneous
3. Are not clear
4. Settle
5. Do not pass through filter paper
6. Do not pass through membranes

Some medications, such as milk of magnesia, are administered as a suspension. Many bottles of medication state on the label "shake before using." Most suspensions use water as the suspending medium, but procaine penicillin G, for example, is usually administered as an oil suspension.

A mist is a suspension of a liquid in a gas. Water droplets suspended in air are one example of a mist.

Patients with a decreased water content in their lungs usually have thickened bronchial secretions. To increase the water content of the lungs requires a higher-than-normal water content in the inspired air. A nebulizer is a device that generates an aerosol mist consisting of large water particles that can penetrate into the trachea and large bronchi. An ultrasonic nebulizer produces a supersaturated mist by means of ultrasonic sound waves. Such a mist is very effective in inhalation therapy because the particle size is small enough to reach the smallest *bronchioles. Also, the concentration of the mist is greater than that produced by other means.

Care must be taken in the use of a nebulizer to avoid (1) bacterial contamination in the water reservoir of the equipment (such bacteria will go directly into the bronchi) and (2) excess water being added to the patient's lungs.

10-3 Colloids

The third class of liquid mixtures, called colloids, consists of tiny particles suspended in a liquid. We might ask, if these particles are suspended in a liquid, why aren't they suspensions? The answer is that these colloids behave quite differently from ordinary suspensions and have an entirely different set of properties.

Size

When a colloid (or a colloidal dispersion as it is frequently called) is poured into a funnel lined with filter paper, the colloid passes through the filter paper. This indicates that the colloidal particles are smaller than the openings in a filter paper.

When a colloid is placed in a membrane, the colloidal particles do not pass through. Thus, colloidal particles are larger than the openings in the membrane.

We know that solutions pass through filter paper. They can also pass through certain types of membranes, whereas suspensions pass through neither. Therefore, colloidal particles must be intermediate in size between solution particles and suspension particles. Colloidal particle sizes are measured in nanometers (nm) (1 nanometer is 0.000000001 m or one-millionth of a millimeter). Colloidal particles range in size from 1 to 100 nm. Solution particles are smaller than 1 nm, whereas suspension particles are larger than 100 nm (see Figure 10-12).

Colloids have a tremendous amount of surface area because they consist of so many tiny particles. Consider a cube 1 cm on an edge. The volume of such a cube is 1 cubic centimeter (cm^3) and the surface area is 6 square centimeters (cm^2). (There are six faces to a cube, and each face has an area of 1 cm^2.)

If the cube is cut in half, the total volume of both pieces is still 1 cm^3. However, the total surface area has now increased to 8 cm^2 (cutting introduces two new faces each having an area of 1 cm^2). Cutting in half again still retains the volume of 1 cm^3. The surface area now is 10 cm^2. Continued cutting further increases the surface area.

When the size of the individual particle reaches colloidal size—approximately 10 nm—the total amount of surface area is about 60,000,000 cm^2. This tremendous surface area gives colloids one of their most important properties—adsorption.

Adsorption

Adsorption is defined as the property of holding substances to a surface. Colloidal charcoal will adsorb tremendous amounts of gas. It has a selective adsorption, as do most colloids. Coconut charcoal is used in gas masks because it selectively adsorbs poisonous gases from the air. It does not adsorb ordinary gases from the air.

Several commonly used medications owe their use to their adsorbent properties. Charcoal tablets are administered to patients to aid digestion. Kaolin, a finely divided aluminum silicate, is administered for the relief of diarrhea. Colloidal silver adsorbed on protein (Argyrol) has been used as a germicide.

Electrical Charge

Almost all colloidal particles have an electrical charge—either positive or negative. Why? Colloids selectively adsorb ions on their surface. If a colloid selectively adsorbs negative ions, it becomes a negatively charged colloid; it adsorbs positive ions, it becomes a positively charged colloid.

Consider two negatively charged colloidal particles suspended in water. What would happen if these particles were to come close together? They would repel each other because of the repulsion of their like charges. Therefore, these colloidal particles will have little tendency to form large particles that would then settle out.

How can colloids be made to settle? Colloidal particles have an electrical charge that will repel all other similarly charged colloidal particles. However, these charged particles *will* attract particles of opposite charge. Therefore, a negative colloid can be made to coagulate (begin to settle out) by adding to it positively charged particles.

Bichloride of mercury ($HgCl_2$) is a poisonous substance. When swallowed, it forms a positive colloid in the stomach. The antidote for this type of poisoning is egg white, which is a negative colloid. These two oppositely charged colloids neutralize each other and coagulate in the stomach. The stomach must then be pumped out to remove the coagulated material. If this is not done, the stomach will digest the egg white, exposing the body once again to the poisonous substance.

Protein can also be coagulated by heat. Egg white—a colloidal substance—is quickly coagulated when heated.

The charge of a colloid can be determined by placing it in a U tube containing two electrodes. When a current is passed through the U tube, each electrode will attract particles of opposite charge. A negative colloid will begin to accumulate around the positive electrode, and a positive colloid around the negative electrode. The movement of electrically charged suspended particles toward an oppositely charged electrode is called electrophoresis. Electrophoresis is a slow process because the charged particles are not soluble and are much larger than the particles in solution (see page 531).

Tyndall Effect

When a strong beam of light is passed through a colloidal dispersion, the beam becomes visible because the colloidal particles reflect and scatter the light. This phenomenon is called the Tyndall effect (see Figure 10-13). The Tyndall effect can also be observed when a beam of sunlight passes through a darkened room. The dust particles in the air scatter the light so that the sun's rays become visible. The blue color of the sky is the result of scattered light, as is the blue color of the ocean. However, we actually do not see the colloidal particles; we merely see the light scattered by them.

When a strong beam of light is passed through a solution, no Tyndall effect is observed because the solution particles are too small to scatter the light. Thus, the Tyndall effect is a way of distinguishing between solutions and colloids.

Brownian Movement

When colloidal particles are observed with a transmission electron microscope, the particles are seen to move in a haphazard irregular motion called Brownian movement (see Figure 10-14).

What causes this irregular motion? It cannot be caused by the vibration of the slide on the microscope stand because the same motion can be observed when the microscope is mounted on a concrete pillar sunk deep into the earth. Brownian movement can be observed during the day or night, in the city and the country, in warm weather and cold weather, at high altitudes and low altitudes. The

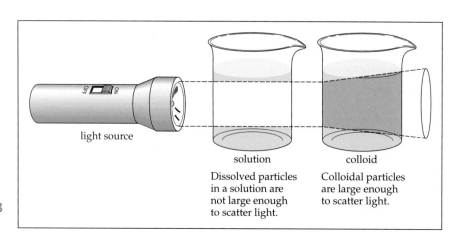

light source

solution
Dissolved particles in a solution are not large enough to scatter light.

colloid
Colloidal particles are large enough to scatter light.

FIGURE 10-13
Tyndall effect.

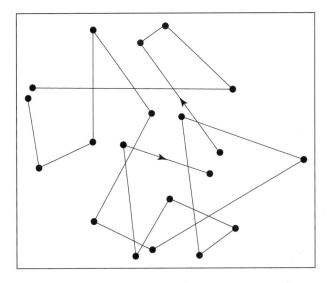

FIGURE 10-14
Brownian movement.

strangest characteristic of all is that this particlar motion *never* ceases. It can be observed in containers that have been sealed for years. It has also been observed in liquid *occlusions in quartz samples that have been undisturbed for perhaps millions of years.

Careful investigation by scientists established the fact that Brownian movement is not caused by the colloidal particles themselves but rather to their bombardment by the molecules of the suspending medium. The molecules of the suspending medium are in continuous random motion. When these rapidly moving molecules of the medium strike the colloidal particles, they cause these particles to have the random motion characteristic of Brownian movement.

Gels and Sols

Colloidal dispersions can be subdivided into two classes. Where there is a strong attraction between the colloidal particle and the suspending liquid (water), the system is said to be *hydrophilic (water loving). Systems of this type are called gels. Gelatin in water is an example of such a system. Gels are semisolid and semi-rigid; they do not flow easily.

A colloidal system where there is little attraction between the suspended particles and the suspending water is *hydrophobic (water hating). Systems of this type are called sols. They pour easily. A small amount of starch in water forms a sol. A hydrosol is a colloidal dispersion in water, while an aerosol is a colloidal dispersion in air.

If a gel, such as gelatin, is heated, it turns into a sol but returns to its original gel state on cooling. Protoplasm has the ability to change gel (in membranes) into sol, and vice versa.

Dialysis

Dialysis is the separation of solute particles from colloidal particles by means of a semipermeable membrane. Recall that solute particles can pass through semipermeable membranes but colloidal ones cannot. Suppose that a colloidal starch suspension and sodium chloride solution are placed inside that type of membrane, which in turn is placed in a beaker of water. The starch is a colloid and cannot pass through the membrane. This is an example of diffusion. The salt will continue to pass through the membrane until the salt concentration inside the membrane is the same as that in the water surrounding the membrane.

When this happens, no more salt will be removed from the mixture inside the membrane bag. However, if the bag is suspended in running water, soon all the salt will be removed from the inside, leaving behind only the starch—the colloid. This is dialysis—the separation of a solute from a colloid by means of a membrane (Figure 10-15). *Antitoxins are prepared by this method. The impure material is placed inside a container made of membrane suspended in running water. The soluble impurities diffuse out through the membrane, leaving the pure antitoxin behind. The same process is used to prepare low-sodium milk. Milk enclosed in a membrane is suspended in running water. The soluble salt (sodium chloride) leaves the milk, so the remaining liquid is practically free of sodium compounds.

In the body the membranes in the kidneys allow the soluble waste material to pass through. The same membranes do not allow protein to pass through, since proteins are colloids. In an average 150-pound adult, approximately 180 L of blood is purified daily.

Hemodialysis

Hemodialysis refers to the removal of soluble waste products from the bloodstream by means of a membrane. Purification of the blood can be accomplished in this way because soluble particles can diffuse through dialyzing membranes, whereas blood cells and plasma proteins cannot. When a patient has problems related to renal excretion, an artificial kidney machine can be used (Figure 10-16). This machine applies the principles of hemodialysis.

The artifical kidney machine consists of a long cellophane or plastic tube wrapped around itself to form a coil and immersed in a temperature-controlled solution whose chemical composition is carefully regulated according to the needs of the patient. The patient's blood is pumped through the coil, and the soluble end products of protein catabolism, water, and *exogenous poisons are removed from the blood. At the same time the blood cells and plasma proteins remain in the

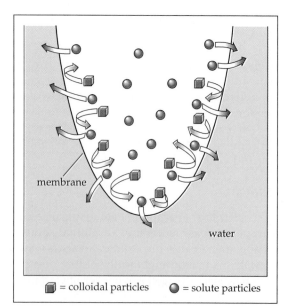

FIGURE 10-15
Dialysis.

◨ = colloidal particles   ● = solute particles

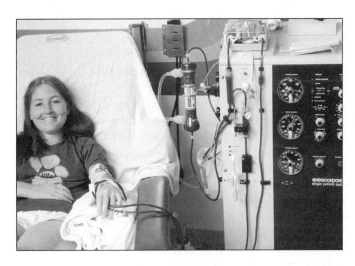

FIGURE 10-16

A woman undergoes renal dialysis on an artifical kidney machine.

blood (recall that proteins, which are colloids, cannot pass through a membrane, and here the cellophane acts as the membrane). If the solution in which the coil is immersed has the same concentration of sodium ions as the blood, no net diffusion of sodium ions will take place. The same applies to other soluble substances in the bloodstream. That is, by regulating the composition of the solution, the waste products and unwanted material can be removed from the bloodstream.

A patient may remain on the artificial kidney machine 4 to 7 hours or even more. During this time the solution must be changed at intervals to avoid accumulation of waste products. Otherwise, the waste products would no longer continue to diffuse out of the blood. After passing through the coil, the blood is returned to the patients's veins.

As we might expect, the dialysate is mostly water. The kind of water used in its preparation is of utmost importance to the patient. If the water has been softened by the zeolite process, it may contain too high a concentration of sodium ions (and so add sodium to the blood; see Section 30-15). Improper levels of calcium and magnesium ions in the water can lead to metabolic bone diseases. Traces of copper in the water (from copper tubing in the equipment, copper pipes in the water supply, or copper sulfate used to kill algae in the water treatment plant) have caused fatalities in dialysis patients.

10-4 Emulsions

When two liquids are mixed, either they dissolve in each other or they do not. Two liquids that are soluble in each other are **miscible.** Two liquids that do not dissolve in each other are **immiscible.**

Suppose two immiscible liquids such as oil and water are poured together and then shaken vigorously. The oil forms tiny drops, which are suspended in the water. After a while the tiny drops come together to form larger drops; these soon rise to the top and separate from the water. This type of liquid mixture is called a **temporary emulsion**—emulsion because it consists of a liquid colloidally suspended in a liquid, and temporary because it separates.

What are the properties of a temporary emulsion? First, as was just noted, it separates. Since it separates, it must be heterogeneous; its composition continues to change. Temporary emulsions, such as oil in water, are not clear. If placed in a funnel containing a piece of filter paper, they do not pass through the filter paper. Neither do they pass through any membrane.

A temporary emulsion such as oil and water separates because the oil drops attract one another to form larger drops, which soon rise to the top. However, suppose that each drop of oil were given a negative charge. Then the drops would not attract one another. They would not settle out; they would form a permanent emulsion. To give the oil drops an electric charge we need an emulsifying agent. An emulsifying agent is a protective colloid that coats the suspended oil drops and prevents them from coming together. Thus the surface area of oil is greatly increased by the addition of an emulsifying agent. Permanent emulsions are not clear; they are homogeneous. They do not settle; they do not pass through either filter paper or membranes.

An example of a permanent emulsion is mayonnaise, which is an emulsion of oil in vinegar with egg yolk as an emulsifying agent. Milk is an emulsion of butter fat in water, with casein acting as the emulsifying agent. Soap acts as an emulsifying agent on grease and oils in water. Many medications are given in the form of emulsions, gum acacia being the most common emulsifying agent for the dispersion of fats.

Summary

Solutions consist of a solute and a solvent. Solutions are clear and homogeneous, have a variable composition, do not settle, may be separated by physical means, and pass through filter papers.

Factors affecting the solubility of a solute are temperature, pressure, surface area, agitation, and nature of the solvent.

Many chemical reactions take place in solution. Many medications are administered as solutions.

Solutions may be labeled as dilute or concentrated, as saturated or unsaturated, all of which are relative terms and do not indicate any definite amount of solute and solvent.

A percentage solution indicates the number of grams of solute per 100 mL of solution. A ratio solution indicates the number of grams of solute and the number of milliliters of solution. A molar solution indicates the number of moles of solute per liter of solution.

When a nonvolatile solute is dissolved in a solvent, the boiling point is elevated and the freezing point is depressed.

All liquids exhibit surface tension, which causes the surface molecules to form a surface film.

The movement of solute into a solvent or through a solution is called diffusion. The flow of solvent through a semipermeable membrane is called osmosis. The pressure exerted during osmosis is called osmotic pressure. Osmotic pressure is expressed in terms of the osmolarity of the solution.

An isotonic solution has the same salt concentration as blood and is used for transfusions.

A hypotonic solution has a solute concentration less than that of blood. If injected into the bloodstream, a hypotonic solution can cause hemolysis—the bursting of the red blood cells.

A hypertonic solution has a solute concentration greater than that of blood. If injected into the bloodstream, a hypertonic solution may cause plasmolysis—the shrinking of the red blood cells.

Suspensions consist of a nonsoluble solid suspended in a liquid medium. Suspensions are not clear; they settle out; they are heterogeneous; they do not pass through a filter paper; and they do not pass through a membrane.

Colloids consist of tiny particles suspended in a liquid. Colloids do not settle; they pass through filter paper but not through membranes; they adsorb (hold) particles on their surface; they have electric charges, owing to the adsorption of charged particles (ions); they exhibit the Tyndall effect and Brownian movement.

Colloidal dispersions can be subdivided into two classes—sols and gels.

Dialysis is the separation of a solute from a colloid by means of a semipermeable membrane.

Hemodialysis refers to the removal of soluble waste products from the bloodstream by means of a membrane. When a patient has problems related to renal excretion, an artificial kidney machine may be used.

An emulsion consists of a liquid suspended in a liquid. An emulsion that settles is called a temporary emulsion. When an emulsifying agent is added to a temporary emulsion, it becomes a permanent emulsion.

Questions and Problems

1. Give specific directions for the preparation of
   a. 250 mL of 4 percent boric acid solution
   b. a 1:1,000 $KMnO_4$ solution
   c. 1.5 L of 2 M KCl
   d. 100 mL of 0.9 percent NaCl solution

2. List the general properties of solutions, suspensions, colloids, and emulsions.

3. Define solvent, dialysis, diffusion, osmosis, and isotonic solution.

4. Why must an isotonic solution be used during a blood transfusion?

5. What is the Tyndall effect? Brownian movement?

6. What is surface tension?

7. What factors affect the rate of solution of a solid solute?

8. Compare a temporary emulsion with a permanent emulsion. How can the former be changed to the latter?

9. How does an artifical kidney machine function? Why must the solution in such a machine be changed at regular intervals?

10. How can colloids be made to settle? What use is made of this process?

11. Why do colloids pass through filter paper but not through membranes?

12. What is a supersaturated solution? How can it be prepared? Is it stable?

13. Compare a sol with a gel. Give an example of each.

14. How can you tell whether a solution is saturated, unsaturated, or supersaturated?

15. What is a nebulizer? For what purposes is it used in a hospital? What precautions should be taken with its use?

16. What is a tincture? Give two examples.

17. A patient is given 1000 mL of 5 percent glucose solution intravenously in a 12-hour period. How many grams of glucose did the patient receive per hour?

18. How many grams of $C_{12}H_{22}O_{11}$ (molar mass 342) are present in 100 mL of 0.75 M solution?

19. Does pressure always affect the amount of a solute that will dissolve in a given solvent? Explain.

20. Why is the control of water important in hemodialysis? Would you expect the control of temperature also to be important? Explain.

21. From Figure 10-1 calculate the number of grams of $NaNO_3$ that will dissolve in 1000 mL of water at 20° C.

22. What is a surfactant? Explain why surfactants are important in the body.

23. What is viscosity, adhesion, cohesion, capillary action?

24. For what purpose(s) is a blood extender used?

25. How would you prepare 200 mL of 0.5 M $C_6H_{12}O_6$?

26. Calculate the osmolarity of the following solutions:
    a. 1.5 M sucrose solution
    b. 0.5 M H_2SO_4 solution
    $$(H_2SO_4 \longrightarrow 2\,H^+ + SO_4^{2-})$$
    c. 5 percent glucose solution (0.27 M)

27. Explain how drinking salt water can cause dehydration.

28. How does reverse osmosis work? What use is made of this process?

29. Compare the terms *percent solution* and *milligram percent*.

30. On what basis can the solubility of a solute in water be predicted? Explain.

31. Using Figure 10-1, answer the following:
    a. Which compound has the greatest solubility at 0° C? at 40° C?
    b. Which compound shows the least change in solubility over the temperature range? the greatest change over the temperature range?
    c. What can be stated about gas solubilities that cannot be said about solid solubilities?
    d. Calculate the grams of Br_2 that can dissolve in 500 mL of water at 20° C.
    e. Calculate the grams of $NaNO_3$ that can dissolve in 600 mL of water at 20° C.

32. Most states consider a blood-alcohol level of 0.10 percent to be the legal threshold of intoxication. What volume of beer (4.8 percent alcohol) is necessary to reach this limit for a person with a blood volume of 7.0 L? Assume all the alcohol goes directly into the blood. A fatal blood-alcohol level is about 0.70 percent. What volume of beer is necessary to reach this level? What volumes of vodka (32 percent alcohol) are necessary to reach the intoxicated and lethal levels?

33. Ringer's solution, used in the treatment of burns and wounds, is prepared by dissolving 4.30 g of NaCl, 0.150 g of KCl, and 0.165 g of $CaCl_2$ in water and diluting to 500 mL at 37° C. What is the percent composition for each component? What is the molarity of Na^+, K^+, Ca^{2+} and Cl^-? Calculate the osmotic pressure of this solution.

34. The concentration of glucose (molar mass = 180 g/mol) in normal blood is 90 mg/100 mL at body temperature. What is the molarity of glucose at this level? What osmotic pressure does this concentration exert? Diabetics often have blood glucose levels of 240 mg/mL or higher. Calculate the osmotic pressure of the elevated blood glucose level.

35. What volume of 5.0 M NaCl is needed to provide 10.0 grams of NaCl?

36. Gold sodium thiomalate (Myochrysine) is a gold-containing drug used in the treatment of rheumatoid arthritis. One milliliter of a 10 percent solution of Myochrysine is injected into a patient with a blood volume of 7.0 L. What is the final blood concentration of Myochrysine? Assume all of the drug remains in the bloodstream.

Practice Test

1. A solution that has a higher salt concentration than the blood is said to be _____.
    a. isotonic   b. hypotonic
    c. hypertonic   d. normal

2. The separation of a solution from a colloid by means of a semipermeable membrane is called _____.
    a. electrolysis   b. dialysis
    c. osmosis   d. neutralization

3. How many grams of $C_6H_{12}O_6$ (molecular mass 180) are needed to prepare 2 L of a 0.1 M solution?
    a. 18   b. 36   c. 180   d. 360

4. When a nonvolatile solute is dissolved in water, the boiling point _____.
    a. increases
    b. decreases
    c. does not change

5. Colloids _____.
    a. settle
    b. pass through filter paper
    c. pass through membranes
    d. all of these

6. A 5 percent solution (molecular mass 180) contains how many grams of glucose per 100 mL?
    a. 0.5   b. 2.5   c. 5   d. 9

7. Colloids exhibit _____.
    a. Brownian movement
    b. the Tyndall effect
    c. both of these effects
    d. neither of these effects

8. A solution that can cause plasmolysis is _____.
    a. isotonic   b. hypotonic
    c. hypertonic   d. normal

9. An alcohol solution used medicinally is called a(n) _____.
    a. activator   b. tincture
    c. antiseptic   d. analgesic

10. When glucose is dissolved in water
    a. glucose is the solute and water is the solvent
    b. glucose is the solvent and water is the solute
    c. glucose is the solute and water is the solution
    d. glucose is the solvent and water is the solution

11

Ionization

Objectives

- To distinguish between electrolytes and nonelectrolytes
- To become familiar with the theory of ionization
- To distinguish between strong and weak electrolytes
- To understand the importance of ions in body chemistry

Salts ionize as they dissolve in water, resulting in an aqueous solution with different physical properties than pure water. Because a salt solution has a lower melting point than water, salt will melt ice on an icy sidewalk.

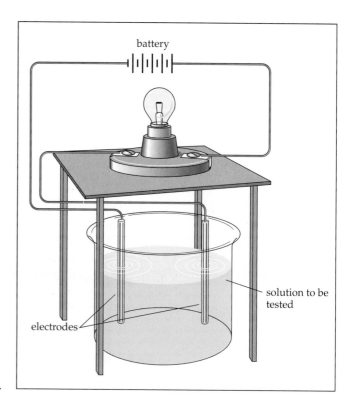

FIGURE 11-1
Conductivity apparatus.

11-1 Conductivity of Solutions

Figure 11-1 illustrates an apparatus designed to show whether a liquid will conduct electricity or not. When two metal plates, called electrodes, are immersed in a liquid and the current is turned on, the bulb will light if the liquid conducts a current. If a liquid does not conduct a current, the bulb will not glow at all.

It can be shown experimentally that the only water solutions that conduct electricity are those of acids, bases, and salts. The substances whose water solutions conduct electricity are called **electrolytes.** Substances whose water solutions do not conduct electricity are called nonelectrolytes. Pure water is a nonelectrolyte, as is alcohol. Sugar solutions are also nonelectrolytes.

11-2 Effect of Electrolytes on Boiling Points and Freezing Points

Electrolytes have an unusual effect on the boiling point and the freezing point of a solution compared with the boiling point and the freezing point of the solvent itself. When a solid compound is dissolved in water, the resulting solution has a boiling point above 100° C: that is, solid solutes raise the boiling point of water. However, the results are considerably exaggerated if the solute is an electrolyte rather than a nonelectrolyte.

Consider a 1 M solution of an electrolyte such as sodium chloride and a 1 M solution of a nonelectrolyte such as sugar. The increase in the boiling point of the water containing the electrolyte sodium chloride will be approximately twice that for the water containing the nonelectrolyte sugar.

Likewise, the freezing point of the sodium chloride solution will be lowered approximately twice as much as that of the sugar solution.

Why do electrolytes conduct electricity? Why do they have an unusual effect on the boiling point and freezing point of the solution?

11-3 Theory of Ionization

In 1887 a Swedish chemist, Svante Arrhenius, proposed a theory to explain the behavior of electrolytes in solution. The main points of his theory are as follows.

1. When electrolytes are placed in water, the molecules of the electrolyte break up into particles called ions. This process he called ionization.
2. Some of the ions have a positive charge; others have a negative charge.
3. The sum of the positive charges is equal to the sum of the negative charges; that is, the original molecules were neutral, so the sum of the charges making up this molecule must also be neutral.
4. The conductance of electricity by solutions of electrolytes is the result of the presence of ions.
5. Nonelectrolytes do not conduct electricity because of the absence of ions.
6. The effect of electrolytes on the boiling point and on the freezing point of a solution is the result of the increased number of particles (ions) present in the solution.

Arrhenius proposed that an equilibrium exists between the ions and the nonionized molecules. This equilibrium can be represented by the equation

$$NaCl \rightleftharpoons Na^+ + Cl^-$$

| sodium chloride molecule | sodium ion | chloride ion |

Arrhenius also proposed that the ionization should increase as the solution becomes more dilute. That is, in extremely dilute solutions, the electrolyte molecule would be almost completely dissociated (ionized).

Arrhenius's theory had to be modified as more information became available. Yet the modern theory of ionization retains most of his principles.

Arrhenius believed that the molecules of an electrolyte such as sodium chloride broke up into ions when placed in water. Later evidence proved conclusively that sodium chloride exists as ions, even in the solid state. When sodium chloride is formed from its elements the sodium atom loses one electron to form a positively charged sodium ion. At the same time the chlorine atom gains one electron to form a negatively charged chloride ion. Thus, sodium chloride exists in an ionic state even in the solid state. When sodium chloride is dissolved in water, the ions are free to move around.

Solid sodium chloride consists of sodium ions and chloride ions bonded together by ionic bonds in a definite crystalline structure. These ions are held tightly in place and cannot move about to any great extent. Thus, they are not free to conduct an electric current, and solid sodium chloride does not conduct electricity. If sodium chloride is heated until it melts (about 800° C), the resulting liquid will conduct electricity because the ions have some freedom to move around.

11-4 Conductivity of Solutions of Electrolytes

When sodium chloride crystals are placed in water, the ions present in the crystal are free to move about in the solution. These positively and negatively charged ions will be attracted to oppositely charged electrodes. Thus, if two electrodes are connected to a battery and are placed in a sodium chloride solution, the solution will conduct electricity.

The positively charged sodium ions will be attracted to the negative electrode, the cathode. Ions attracted to a cathode are called cations. At the same time, the negatively charged chloride ions will be attracted to the positive electrode, the anode. Ions attracted toward an anode are called anions. This movement of ions through the solution consists of a flow of current. If a nonelectrolyte is placed in water, no ions are formed and no conductance takes place.

11-5 Strong and Weak Electrolytes

Acids are electrolytes; however, not all acids behave the same when placed in water. A dilute solution of hydrochloric acid (HCl) is a strong electrolyte. When a beaker containing it is placed under a conductivity apparatus, the bulb glows brightly. However, when a beaker containing a dilute solution of acetic acid ($HC_2H_3O_2$) is placed under the conductivity apparatus, the bulb glows dimly. This indicates that the acetic acid is a weak electrolyte. Likewise, sodium hydroxide (NaOH) solution is a strong electrolyte, whereas ammonium hydroxide (NH_4OH) solution is a weak electrolyte.

How can one acid or base be a strong electrolyte and another one be weak? What accounts for the difference between the types of electrolytes?

When hydrochloric acid or sodium hydroxide is placed in water, it dissociates (breaks up) almost completely into ions, as indicated by the arrows pointing in one direction only.

$$HCl \longrightarrow H^+ + Cl^-$$
$$NaOH \longrightarrow Na^+ + OH^-$$

Because these substances are just about completely ionized, they are called strong electrolytes.

However, some acids and bases ionize only to a limited extent when they are placed in water. They remain primarily as nonionized molecules. These substances maintain an equilibrium between the nonionized molecule and the ions, with the equilibrium being far to the left, as indicated by the arrows.

$$HC_2H_3O_2 \rightleftharpoons H^+ + C_2H_3O_2^-$$
$$NH_4OH \rightleftharpoons NH_4^+ + OH^-$$

These substances are called weak electrolytes.

Thus, the original Arrhenius theory is true for weak electrolytes but not for strong ones.

11-6 Other Evidence of Ionization

It has already been mentioned that the conductivity of solutions of electrolytes indicates the presence of ions. Likewise, the effect of electrolytes on the boiling point and the freezing point can be explained by the presence of ions. The change in the boiling point or freezing point depends on the number of particles present in solution. Sugar ($C_6H_{12}O_6$) contributes only one particle per molecule because it is not ionized. Sodium chloride (NaCl) contributes two particles—the two ions— and so should have twice the effect on the boiling point.

Another factor indicating the presence of ions is the instantaneous reaction of solutions of electrolytes. When a solution of sodium chloride (NaCl) is mixed with a solution of silver nitrate ($AgNO_3$), a white precipitate of silver chloride (AgCl) is formed instantaneously.

$$Na^+ + Cl^- + Ag^+ + NO_3^- \longrightarrow AgCl_{(s)} + Na^+ + NO_3^-$$

The silver chloride is written as a molecule rather than as ions. This is because that substance precipitates, thus removing the ions from the solution.

When an acid reacts with a base, a salt and water are formed. If potassium hydroxide, KOH, is reacted with nitric acid, HNO_3, potassium nitrate and water are formed. The ionic reaction is

$$K^+ + OH^- + H^+ + NO_3^- \longrightarrow K^+ + NO_3^- + H_2O$$

Potassium nitrate is a soluble salt, an electrolyte, which is ionized. Water is a non-electrolyte; it is not ionized and is therefore written as a molecule.

Hydrolysis of salts (see Section 12-7) can also be explained on the basis of ionization. When sodium acetate, $NaC_2H_3O_2$, is placed in water, the acetate ion reacts with the water according to the equation

$$\underset{\text{acetate ion}}{Na^+ + C_2H_3O_2^-} \rightleftharpoons \underset{\text{acetic acid}}{HC_2H_3O_2} + Na^+ + OH^- \qquad (11\text{-}1)$$

The acetic acid, being a weak acid, does not ionize appreciably to furnish any hydrogen ions and so is written as a molecule in equation (11-1). If the ions common to both sides of equation (11-1) (the sodium ions) are eliminated, the net reaction becomes

$$C_2H_3O_2^- + H_2O \rightleftharpoons HC_2H_3O_2 + OH^- \qquad (11\text{-}2)$$

That is, the acetate ion (which came from a weak acid) hydrolyzes (reacts with water) to form noniodized acetic acid, a weak electrolyte, leaving hydroxide ions in solution. Therefore, when sodium acetate is placed in water, hydroxide ions are formed, causing the solution to be basic. Note that ions from a strong acid or a strong base (see Section 14-7) do not hydrolyze. Thus, the sodium ion in reaction (11-1) does not react with water and so can be eliminated from both sides of the equation, giving equation (11-2).

11-7 Importance of Ions in Body Chemistry

Ions play the chief role in the various processes that take place in the body. Many of the body's vital processes take place in the ionic state within the cell. Table 11-1 lists some of the more important ions that are found in the body.

In addition, ions are necessary as part of the blood buffer system. They cause osmotic pressure in the cells and are necessary to control the contraction and relaxation of muscles. Ions are necessary to carry nerve impulses and help regulate the digestive processes (see Sections 30-13 and 30-15).

Table 11-1	Ions Found in the Body	
Calcium ion	Ca^{2+}	Necessary for clotting of the blood; for formation of milk curd during digestion in the stomach; for formation of bones and teeth; for action of muscle, including heart
Iron ion	Fe^{2+}	Necessary for formation of hemoglobin and cytochromes
Sodium ion	Na^+	Principal extracellular positive ion
Potassium ion	K^+	Principal intracellular positive ion
Chloride ion	Cl^-	Intracellular and extracellular negative ion
Bicarbonate ion	HCO_3^-	Extracellular negative ion and blood buffer
Iodide ion	I^-	Present in thyroid hormones
Ammonium ion	NH_4^+	Plays a role in maintaining body's acid-base balance
Phosphate ion	PO_4^{3-}	Plays an important role, along with calcium ions, in the formation of bones and teeth
Magnesium ion	Mg^{2+}	An important activator for many enzyme systems

11-8 Milliequivalents

Milliequivalents per Liter

Concentrations of ions in body fluids are frequently expressed in the units milliequivalents per liter (mEq/L) (see Table 30-1). The number of equivalents of an ion is determined by multiplying the number of moles of ions by the value of the charge the ion carries. That is, 1 mol of sodium ions (Na^+) contains 1 Eq of sodium ions. One mole of calcium ions Ca^{2+}, contains 2 Eq of calcium ions, and 1 mol of CO_3^{2-} ions contains 2 Eq of carbonate ions.

One milliequivalent (mEq) is 1/1000 of an equivalent.

For example, 1 mol KCl contains 1 mol K^+ and 1 mol CL^- and so has 1 Eq K^+ and 1 Eq Cl^-. Likewise, 1 mol $MgSO_4$ contains 2 Eq Mg^{2+} and 2 Eq SO_4^{2-}.

To change grams to milliequivalents the following conversion factor is used.

$$\# \, mEq \;=\; \# \, g \text{ substance} \times \frac{1 \text{ mol}}{\text{mol mass}} \times \frac{\# \, Eq}{1 \text{ mol}} \times \frac{1000 \text{ mEq}}{1 \text{ Eq}}$$

Exercise 11-1

Exercise 11-1a

A patient is given 3 g KCl (molecular mass 75). How many mEq K^+ will she receive?

Solution

$$\# \, mEq \;=\; 3 \text{ g} \times \frac{1 \text{ mol}}{75 \text{ g}} \times \frac{1 \text{ Eq } K^+}{1 \text{ mol}} \times \frac{1000 \text{ mEq}}{1 \text{ Eq}}$$

$$\# \, mEq \;=\; 40$$

Exercise 11-1b

How many mEq Ca^{2+} are present in each 0.6 g calcium lactate tablet if the molecular mass is 388?

Solution

$$\# \text{mEq} = 0.6 \text{ g} \times \frac{1 \text{ mol}}{388 \text{ g}} \times \frac{2 \text{ Eq Ca}^{2+}}{1 \text{ mol}} \times \frac{1000 \text{ mEq}}{1 \text{ Eq}}$$

$$\# \text{mEq Ca}^{2+} = 3.9$$

Self-test

Calculate the number of milliequivalents of each ion in a solution containing 5.0 g of NaCl and 5.0 g of KCl.

Answer

86 mEq of Na^+; 67 mEq of K^+; and 153 mEq of Cl^-

Summary

Substances whose water solutions conduct electricity are called electrolytes. Soluble acids, bases, and salts are electrolytes. Solutions that do not conduct electricity are nonelectrolytes.

Electrolytes have an effect on the boiling and freezing points of a solution. This effect is caused by the presence of ions.

Ionization (the formation of ions) accounts for the electrical conductivity of a solution. When a substance ionizes, the sum of the positive charges equals the sum of the negative charges. The positive ions, cations, are attracted toward the cathode (negative electrode), whereas the negative ions, the anions, are attracted toward the anode (positive electrode).

The modern theory of ionization is that ions are already present in a crystalline salt and that these ions are released to move about when that substance is placed in solution.

Most salts are strong electrolytes because they are completely (strongly) ionized. Acids and bases that are strong electrolytes are highly ionized. Acids and bases that are poor electrolytes are weakly ionized.

The presence of ions is of great importance in maintaining the electrolyte balance of body fluids.

Questions and Problems

1. What is a strong electrolyte? Give two examples.
2. What is a weak electrolyte? Give two examples.
3. What effect does an electrolyte have on the freezing point of a solution? Why?
4. What effect does an electrolyte have on the boiling point of a solution? Why?
5. What effect does a nonelectrolyte have on the freezing point of a solution? on the boiling point?
6. How can you tell whether a solution contains an electrolyte or a nonelectrolyte?
7. What is an anion? a cation? Give two examples of each.
8. List several ions necessary for the proper functioning of the body and indicate the purpose each serves.
9. State the main ideas of Arrhenius's theory of ionization.
10. Compare the osmolarity of equal molar concentrations of sucrose and sodium chloride. What effect will these solutions have on the freezing point?
11. Explain why solid NaCl does not conduct electricity but molten NaCl does.
12. If a salt is placed in water, will the resulting mixture conduct electricity? Explain.

Practice Test

1. An ion necessary for the formation of bones and teeth is _____.
   a. HCO_3^- b. CO_3^{2-} c. Cl^- d. PO_4^{3-}

2. An ion necessary for the thyroid hormone is _____.
   a. Cl^- b. I^- c. K^+ d. Mg^{2+}

3. A solution conducts electricity because of the presence of _____.
   a. hydrates
   b. covalent bonds
   c. hydrogen bonds
   d. ions

4. A solution that conducts electricity is called a(n) _____.
   a. molar solution   b. isotonic solution
   c. electrolyte   d. neutral solution

5. An ion necessary for the formation of bones and teeth is _____.
   a. Fe^{2+} b. Ca^{2+} c. Mg^{2+} d. Al^{3+}

6. Ions attracted toward a negatively charged electrode are called _____.
   a. anions   b. cations
   c. electrolytes   d. positrons

7. A soluble nonelectrolyte has what effect on the freezing point of a solvent.
   a. increases it   b. decreases it
   c. has no effect

8. An ion necessary for the formation of hemoglobin is _____.
   a. K^+ b. Ca^{2+} c. Mg^{2+} d. Fe^{2+}

9. Ions attracted toward a positively charged electrode are called _____.
   a. electrons   b. anions
   c. cations   d. positrons

10. The principal extracellular positive ion is _____.
    a. Na^+ b. K^+ c. Ca^{2+} d. Mg^{2+}

Acids, Bases, and Salts

Objectives

- To become familiar with the properties and uses of acids
- To become familiar with the properties and uses of bases
- To understand the ionization of water
- To become familiar with acid-base titration
- To understand the use of pH in expressing strengths of acids and bases

Oranges are fruits replete with citric acid, a weak organic acid that gives orange juice its tart taste.

12-1 Acids

Acids can be defined as compounds that yield or donate hydrogen ions (H^+) in a water solution. Brønsted, a Danish chemist, defined an acid as a substance that donates protons. Note that a hydrogen ion results when a hydrogen atom loses its only electron, leaving a single proton, so both definitions say essentially the same thing; that is, a hydrogen ion is a proton.

It is the hydrogen ions that are responsible for the particular properties of acids. Most people think of acids as liquids. However, there are many solid acids; boric acid and citric acid are two common examples. The names and formulas of the most common acids are given in Table 12-1.

Table 12-1	
Common Acids	
HCl	Hydrochloric acid
H_2SO_4	Sulfuric acid
HNO_3	Nitric acid
H_2CO_3	Carbonic acid
H_3PO_4	Phosphoric acid
$HC_2H_3O_2$	Acetic acid

Properties of Acids

Acids yield hydrogen ions when placed in water solution.

$$HCl \longrightarrow H^+ + Cl^-$$
hydrocholric acid → hydrogen ion + chloride ion

$$HNO_3 \longrightarrow H^+ + NO_3^-$$
nitric acid → hydrogen ion + nitrate ion

$$HC_2H_3O_2 \rightleftharpoons H^+ + C_2H_3O_2^-$$
acetic acid ⇌ hydrogen ion + acetate ion

Recall that strong acids are almost completely ionized in solution, whereas weak acids are only partially ionized.

Hydrochloric, nitric, and acetic acids are called *monoprotic acids* because each molecule yields one hydrogen ion in solution.

Sulfuric acid (H_2SO_4) is an example of a *diprotic acid*; it yields two hydrogen ions per molecule.

$$H_2SO_4 \longrightarrow H^+ + HSO_4^-$$
sulfuric acid → hydrogen ion + bisulfate ion

$$HSO_4^- \rightleftharpoons H^+ + SO_4^{2-}$$
bisulfate ion ⇌ hydrogen ion + sulfate ion

Sulfuric acid is a strong electrolyte, as indicated by the single arrow, which shows complete ionization. HSO_4^- is a weak acid, as indicated by its equilibrium reaction.

Phosphoric acid (H_3PO_4) is a *triprotic acid*, as the following equations indicate.

$$H_3PO_4 \rightleftharpoons H^+ + H_2PO_4^-$$
phosphoric acid ⇌ hydrogen ion + dihydrogen phosphate ion

$$H_2PO_4^- \rightleftharpoons H^+ + HPO_4^{2-}$$
dihydrogen phosphate ion ⇌ hydrogen ion + monohydrogen phosphate ion

$$HPO_4^{2-} \rightleftharpoons H^+ + PO_4^{3-}$$
monohydrogen phosphate ion ⇌ hydrogen ion + phosphate ion

Hydrogen ions are too reactive to exist in solution by themselves. They react with water to form *hydronium ions* (H_3O^+).

$$H^+ + H_2O \rightleftharpoons H_3O^+$$

While this is the more correct representation of the presence of hydrogen ions in solution, for simplicity we will use the term H^+ in this book.

Solutions of acids have a sour taste. Lemon juice and grapefruit juice owe their sour taste to citric acid. Vinegar owes its sour taste to acetic acid. Sour milk owes its taste partly to lactic acid.

When acids react with certain compounds, these compounds change in color. Substances that change in color in the presence of acids are called indicators. One of the most common indicators for acids is litmus. Blue litmus turns red in the presence of an acid (in the presence of hydrogen ions). Another common indicator—phenolphthalein—turns from red to colorless in the presence of an acid.

Acids react with metal oxides and hydroxides to form water and a salt. For example,

$$\underset{\text{acid}}{2\,HCl} + \underset{\text{metal oxide}}{MgO} \longrightarrow \underset{\text{water}}{H_2O} + \underset{\text{salt}}{MgCl_2}$$

$$\underset{\text{acid}}{H_2SO_4} + \underset{\substack{\text{metal}\\\text{hydroxide}}}{2\,NaOH} \longrightarrow \underset{\text{water}}{2\,H_2O} + \underset{\text{salt}}{Na_2SO_4}$$

The reaction of acids with certain hydroxides (called bases) is called neutralization. That is, acids neutralize bases to form water and a salt.

If we rewrite the above neutralization reaction in ionic form, we have

$$2\,H^+ + SO_4^{2-} + 2\,Na^+ + 2\,OH^- \longrightarrow 2\,H_2O + 2\,Na^+ + SO_4^{2-}$$

Canceling ions that appear in the same quantities on both sides of the equation:

$$2\,H^+ + \cancel{SO_4^{2-}} + \cancel{2\,Na^+} + 2\,OH^- \longrightarrow 2\,H_2O + \cancel{2\,Na^+} + \cancel{SO_4^{2-}}$$

we have the equation

$$2\,H^+ + 2\,OH^- \longrightarrow 2\,H_2O$$

which simplifies to

$$H^+ + OH^- \longrightarrow H_2O$$

This final reaction is called a *net equation*. It indicates that a hydrogen ion from any acid reacts with a hydroxide ion (OH^-) from any base to form water.

The activity series of metals (Table 12-2) lists metals in order of decreasing activity. Note that hydrogen is classified with the metals. Any metal above hydrogen in this series will displace the hydrogen from an acid. The farther above hydrogen a metal is in the activity series, the greater will be its tendency to displace hydrogen from an acid.

Acids react with any metal above hydrogen in the activity series to produce hydrogen gas and a salt. For example,

$$Zn + H_2SO_4 \longrightarrow \underset{\text{salt}}{ZnSO_4} + H_{2(g)}$$

$$Mg + 2\,HCl \longrightarrow \underset{\text{salt}}{MgCl_2} + H_{2(g)}$$

Table 12-2	
The Activity Series of the Metals	
K	Potassium
Ca	Calcium
Na	Sodium
Mg	Magnesium
Al	Aluminum
Zn	Zinc
Fe	Iron
Sn	Tin
Pb	Lead
H	Hydrogen
Cu	Copper
Hg	Mercury
Ag	Silver
Au	Gold

These are examples of single-replacement reactions—the metal replaces the hydrogen in the acid. Thus, acids cannot be stored in containers made of these active metals. Iron is above hydrogen in the activity series and should replace hydrogen in an acid. Therefore, acids should not be allowed to come into contact with surgical or dental instruments, which are frequently made of stainless steel. Acids are usually stored in glass or plastic containers. Reactions involving acids are usually carried out in glass or plastic vessels.

Since any metal above hydrogen in the activity series will replace the hydrogen from an acid, any metal below hydrogen in the series should not be able to replace a hydrogen from an acid. That is, we should expect the mixing of copper with hydrochloric acid to produce no hydrogen, and it does not.

$$Cu + HCl \longrightarrow \text{no reaction}$$

Acids react with carbonates and bicarbonates to form carbon dioxide, water, and salts. For example,

$$\underset{\text{acid}}{2\,HCl} + \underset{\text{carbonate}}{CaCO_3} \longrightarrow \underset{\text{salt}}{CaCL_2} + \underset{\text{carbonic acid}}{H_2CO_3}$$

and

$$H_2CO_3 \longrightarrow CO_{2(g)} + H_2O$$

$$\underset{\text{acid}}{HNO_3} + \underset{\text{bicarbonate}}{NaHCO_3} \longrightarrow \underset{\text{salt}}{NaNO_3} + \underset{\text{carbonic acid}}{H_2CO_3}$$

and

$$H_2CO_3 \longrightarrow H_2O + CO_{2(g)}$$

The stomach normally secretes hydrochloric acid (HCl), which is required for the digestion of protein (Section 25-6). Emotional stress can lead to hyperacidity, too great an acid concentration in the stomach. Commerical **antacids** are available to react with the excess stomach acid. Among these are

Tums, which contains calcium carbonate

Maalox, which contains magnesium hydroxide (Mg[OH]$_2$) and aluminum hydroxide (Al[OH]$_3$) in suspension

Rolaids, which contains calcium carbonate (CaCO$_3$) and magnesium hydroxide (Mg [OH]$_2$)

All of these antacid substances can react with the hydrochloric acid present in the stomach.

New studies indicate that ingested aluminum compounds may have a detrimental effect on brain function. Soluble aluminum compounds found in fruit juices and wine coolers seems to have a more rapid effect on brain function than the insoluble aluminum compounds found in some antacids.

In Alzheimer's disease, unusually large amounts of aluminum ions are found in abnormal brain cells. It is not clear whether these ions cause the disease or are a result of it.

Alka-Seltzer, another commercial antacid, contains sodium bicarbonate (NaHCO$_3$), citric acid, and aspirin. When placed in water, the bicarbonate ions react with the acidic components to produce carbon dioxide gas (CO_2), which provides the "fizz," or effervescence. The net ionic reaction is

$$H^+ + HCO_3^- \longrightarrow CO_{2(g)} + H_2O$$

(The prefix *bi-* is sometimes used to indicate hydrogen in a compound. Sodium bicarbonate is also known as sodium hydrogen carbonate.) Sodium bicarbonate ($NaHCO_3$) can also be used to remove excess stomach acidity. However, its continued use may interfere with the normal digestive processes in the stomach. In addition, the continued use of $NaHCO_3$ adds a considerable amount of sodium ions to the body.

Strong acids will attack clothing. Vegetable fibers such as cotton and linen, animal fibers such as wool and silk, and synthetic fibers are rapidly destroyed by strong acids. All these effects are actually the result of the hydrogen ions present in the acids.

Strong acids (acids with a high hydrogen ion concentration) also have an effect on tissues. Concentrated nitric acid (HNO_3) and concentrated sulfuric acid (H_2SO_4) are extremely corrosive to the skin; therefore, great care must be exercised in handling them. The yellowing of the skin by nitric acid is a test that is specific for protein. If a strong acid is spilled on the skin, a serious burn may result. The area should be washed copiously with water. Then it should be treated with sodium bicarbonate to neutralize any remaining acid. Dilute acids are not as corrosive to the tissues. A few may even be used internally in the body.

Drugs are absorbed from the gastrointestinal tract more rapidly when they are nonionized. If a weakly acidic drug is swallowed, it will remain nonionized the stomach because of the low pH (high acid content; see Section 12-5) of the stomach. Since the drug is nonionized, it will be absorbed rapidly. Conversely, a weakly basic drug will be highly ionized in the stomach and thus will be poorly absorbed there.

Aspirin, or acetylsalicylic acid (see page 310), is more than 90 percent nonionized in a strongly acidic solution, whereas it is about 1 percent nonionized in a neutral solution. Thus, we should expect aspirin to be absorbed rapidly from the stomach, where the pH is near 2, and more slowly from the small intestines, where the pH is above 7.

Uses of Acids

Acids such as hydrochloric acid (HCl, commercially known as muriatic acid) are used in industry and in laboratory work in large amounts. Several acids are also used for medical purposes; among these are hydrochloric, nitric, hypochlorous, boric, acetylsalicylic, and ascorbic acids.

Hydrochloric acid, normally found in the gastric juices, is necessary for the proper digestion of proteins in the stomach. Patients who have a lower than normal amount of hydrochloric acid in the stomach, a condition called hypoacidity, are given dilute hydrochloric acid orally before meals to overcome this deficiency.

Nitric acid (HNO_3) is used to test for the presence of albumin in urine because it will coagulate protein. Nitric acid has been used to remove warts, but dichloroacetic acid (bichloracetic acid) and trichloroacetic acid are now commonly used for this purpose.

Hypochlorous acid ($HClO$) is used as a disinfectant for floors and walls in the hospital.

Boric acid (H_3BO_3) has had extensive use as a germicide. Although boric acid has been used in eyewashes, its use in solutions or as a powder on extensive inflamed surfaces or in body cavities is now practically obsolete. Containers of boric acid should have a label reading "poison."

Acetylsalicylic acid (aspirin) is widely ised as an *analgesic and as an *antipyretic. Aspirin is frequently taken by people with a cold to relieve headache, muscle pain, and fever. However, the aspirin does not remove the source of infection or effect a cure. Aspirin also may interfere with the normal clotting of the blood and may cause bleeding in the stomach of some individuals (see page 310).

Ascorbic acid (vitamin C) is normally found in citrus fruits and is used in the prevention and treatment of scurvy.

12-2 Bases

Bases can be defined as substances that yield hydroxide (OH^-) ions in a water solution. That is, bases increase the hydroxide ion concentration in water. If we write the ionization reactions (see Chapter 11) of the bases sodium hydroxide and potassium hydroxide, we have

$$NaOH \longrightarrow Na^+ + OH^-$$

$$KOH \longrightarrow K^+ + OH^-$$

A more general definition of a base is that of Brønsted: **a base is a substance that accepts protons.** According to this definition, substances that yield hydroxide ions in solution are bases, as are substances that yield bicarbonate ions, because these ions are then free to combine with or accept protons (hydrogen ions) as indicated in the following reactions:

$$OH^- + H^+ \longrightarrow H_2O$$

$$\underset{\text{base}}{HCO_3^-} + \underset{\text{proton}}{H^+} \longrightarrow H_2CO_3$$

Strong bases are highly ionized and have a great attraction for protons. Weak bases are slightly ionized and have a weak attraction for protons. Ammonia (NH_3) is a base according to the Brønsted definition because it accepts a proton from water or acids.

$$NH_3 + H_2O \rightleftharpoons NH_4^+ + OH^-$$

C_2H_5OH (ethyl alcohol) is not a base because it does not accept a proton; it does not ionize in water.

Table 12-3 indicates several commonly used bases. Note that they consist of a metal ion ionically bonded to an OH^- ion. The only exception to this rule is the base ammonium hydroxide (NH_4OH), in which the ammonium ion (NH_4^+) is considered to act as a metal ion. Bases are produced when metallic oxides are dissolved water. For example,

$$\underset{\substack{\text{metal} \\ \text{oxide}}}{CaO} + \underset{\text{water}}{H_2O} \longrightarrow \underset{\substack{\text{calcium hydroxide,} \\ \text{a base}}}{Ca(OH)_2}$$

Table 12-3

Commonly Used Bases

Name	Formula
Sodium hydroxide	NaOH
Ammonium hydroxide	NH_4OH
Potassium hydroxide	KOH
Calcium hydroxide	$Ca(OH)_2$
Magnesium hydroxide	$Mg(OH)_2$

Properties of Bases

Solutions of bases have a slippery, soapy feeling and a biting, bitter taste. Like acids, bases also react with indicators. Bases turn red litmus blue, turn methyl orange from red to yellow, and turn phenolphthalein from colorless to red.

Bases neutralize acids to form water and a salt. For example,

$$Ca(OH)_2 + H_2SO_4 \longrightarrow 2 H_2O + CaSO_4$$

     base   acid   water   salt

Strong bases react with certain metals to produce hydrogen gas. For example,

$$2 Al + 6 NaOH + 6 H_2O \longrightarrow 3 H_{2(g)} + 2 Na_3Al(OH)_6$$

aluminum   sodium hydroxide,   water   hydrogen   sodium aluminate,
             a strong base   a soluble compound

Thus, a strong base such as lye (NaOH) should never be used or stored in an aluminum container because it will rapidly react with and dissolve the container.

Strong bases have a high hydroxide ion concentration. They have a corrosive effect on tissues because of their ability to react with proteins and fats. If a strong base is spilled on the skin, a serious burn may result. The procedure in this case is to apply copious amounts of water.

Strong laundry soaps are quite basic and should not be used for washing woolen clothing because the hydroxide ion will attack the fibers and cause them to shrink. Particular care must be taken not to use strong soap on diapers because, if it is not thoroughly removed, the basic soap can cause severe sores on the tender skin of a baby.

Uses of Bases

Sodium hydroxide (NaOH), commonly known as lye, is used to remove fats and grease from clogged drains. It is quite caustic, and care must be exercised handling this substance. Sodium hydroxide is also used in the conversion of fat to soap (see page 365).

Calcium hydroxide solution ($Ca[OH]_2$), commonly known as lime water, is used to overcome excess acidity in the stomach. It is also used for medical purposes as an antidote for oxalic acid poisoning because it reacts with the oxalic acid to form an insoluble compound, calcium oxalate.

Magnesium hydroxide ($Mg[OH]_2$) is commonly known as milk of magnesia. In dilute solutions it is used as an antacid for the stomach. In the form of a suspension of magnesium hydroxide in water, it is used as a laxative.

Spirits of ammonia, which contains ammonium hydroxide (NH_4OH) and ammonium carbonate ($[NH_4]_2CO_3$), is used as a heart and respiratory stimulant. Ammonium hydroxide, also known as household ammonia, is used as a water softener for washing clothes.

12-3 Acid-Base Titration

When an acid reacts with a base (neutralization), a salt and water are produced. This reaction can be carried out in the laboratory by a process called titration (Figure 12-1). In this process, a buret is filled with an acid of known concentration. A buret is a cylindrical glass tube with graduated markings, so that the exact volume of liquid withdrawn can be easily determined. A flask placed below the buret contains a measured volume of a base of unknown concentration. An indicator is added to the flask. The indicator is one that will change in color when the acid has reacted with all of the base present. This is called the *endpoint* or equivalence point of the titration.

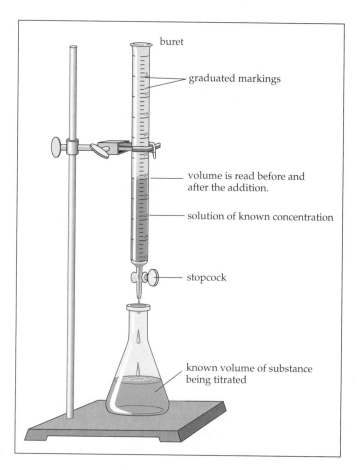

FIGURE 12-1
Titration of acid with base.

12-4 Ionization of Water

Even though pure water does not conduct electricity, very exact measurements show that a very slight amount of ionization does take place. The reaction is

$$H_2O \rightleftharpoons H^+ + OH^-$$

At 25° C, only one water molecule out of 10 million ionizes (dissociates). The concentration of hydrogen ions in pure water is 10^{-7} M, as is the hydroxide ion concentration. Since the hydrogen and hydroxide ion concentrations are equal, pure water is neutral.

The equilibrium constant for water (Section 5-5) can be written as

$$K_{eq} = \frac{[H^+][OH^-]}{[H_2O]}$$

where the brackets indicate concentration in moles per liter. However, the concentration of water remains constant, so the equilibrium constant can be rewritten as

$$K_{eq} \times [H_2O] = [H^+][OH^-] = K_w$$

where K_w, the ion product of water, is itself the product of two constants, K_{eq} and $[H_2O]$.

At equilibrium at 25° C, pure water contains 10^{-7} M H^+ and 10^{-7} M OH^-, so

$$K_w = [H^+][OH^-] = 10^{-7} \times 10^{-7} = 10^{-14}$$

Since $[H^+][OH^-]$ always equals 10^{-14}, if $[H^+]$ increases, the $[OH^-]$ must decrease, and vice versa. A solution in which the $[H^+]$ is greater than the $[OH^-]$ is acidic, and a solution in which the $[H^+]$ is less than the $[OH^-]$ is basic. For example, in 0.01 M HCl, $[H^+]$ is 10^{-2}. Therefore, since $[H^+][OH^-] = 10^{-14}$, $(10^{-2})[OH^-] = 10^{-14}$ and $[OH^-] = 10^{-12}$. The solution is acidic because the $[H^+]$ is greater than $[OH^-]$; likewise in a 0.1 M NaOH solution, $[OH^-] = 10^{-1}$, so $[H^+] = 10^{-13}$. The solution is basic since $[H^+]$ is less than $[OH^-]$.

Exercise 12-1

A solution has an $[H^+]$ of 3.75×10^{-4} M. What is its OH^- concentration?

Solution

Since $[H^+][OH^-] = 10^{-14}$

$$[OH^-] = \frac{10^{-14}}{[H^+]} = \frac{10^{-14}}{3.75 \times 10^{-4}} = 2.67 \times 10^{-11} \text{ M}$$

Self-test

What is the $[H^+]$ and $[OH^-]$ of the following solutions
a. 6 M HCl   b. 2 M NaOH

Answers

a. $[H^+] = 6$ M and $[OH] = 1.7 \times 10^{-15}$ M
b. $[H^+] = 5 \times 10^{-15}$ M and $[OH] = 2$ M

12-5 pH

Placing a few drops of concentrated hydrochloric acid in water produces a dilute acid solution. A few more drops produce another solution, still dilute but a little stronger than the previous one. If a piece of blue litmus paper is placed in either of these two solutions it will turn red, indicating that the solution is acidic. However, it will not tell which one is more strongly acidic. Likewise, if a piece of red litmus paper turns blue when placed in a solution, this merely indicates that the solution is basic or alkaline, it does not indicate how strongly basic the solution is. The term pH is used to indicate the exact strength of an acid or a base. The pH indicates the hydrogen ion concentration in a solution.

Mathematically, pH is defined as the negative logarithm (log) of the hydrogen ion concentration, or

$$pH = -\log [H^+]$$

A logarithm is an exponent. Therefore, the logarithm of 10^{-2} is -2 and log 10^{-12} is -12. Thus, a solution that has $[H^+] = 10^{-4}$ has a pH of 4; that is,

$$pH = -\log [H^+] = -\log 10^{-4} = -(-4) = 4$$

A pH of 7 indicates a neutral solution because $[H^+] = [OH^-]$. pH values below 7 indicate an acidic solution; pH values between 5 and 7 indicate a weakly acidic solution; values between 2 and 5 a moderately acidic solution; and pH values between 0 and 2 a strongly acidic solution.

Table 12-4 pH Values			
pH	Strength of Acid ($[H^+]$ in mol/L)	Strength of Base ($[OH^-]$ in mol/L)	
0	10^0	10^{-14}	Strong acid
1	10^{-1}	10^{-13}	
2	10^{-2}	10^{-12}	Moderate acid
3	10^{-3}	10^{-11}	
4	10^{-4}	10^{-10}	
5	10^{-5}	10^{-9}	Weak acid
6	10^{-6}	10^{-8}	
7	10^{-7}	10^{-7}	Neutral
8	10^{-8}	10^{-6}	Weak base
9	10^{-9}	10^{-5}	
10	10^{-10}	10^{-4}	Moderate base
11	10^{-11}	10^{-3}	
12	10^{-12}	10^{-2}	
13	10^{-13}	10^{-1}	Strong base
14	10^{-14}	10^0	

Likewise, pH values above 7 indicate a basic solution: pH values between 7 and 9 indicate a weakly basic solution; those between 9 and 12 a moderately basic solution; and pH values between 12 and 14 a strongly basic solution. This is summarized in Table 12-4.

Exercise 12-2

A solution has $[H^+]$ of 2.43×10^{-5} M. What is its pH? Its pOH?

$$pH = -\log [H^+] = -\log 2.43 \times 10^{-5}$$

Solution

To find a log, a calculator can be used to determine the following:

Step	Press	Display	
1. Enter the coefficient	2.43	2.43	
2. Access the exponent mode	EE or EXP	2.43	00
3. Enter the exponent	5	2.43	05
4. To get negative exponent	$+/-$	-2.43	05
5. Find the log	log	-4.6144	
6. To obtain negative log	$+/-$	4.6144	

So $-\log 2.45 \times 10^{-5}$ is 4.6144 or 4.61 (to two decimal places).

Since $pH + pOH = 14$, $pOH = 14 - pH$

$$pOH = 14 - 4.61 = 9.39$$

Self-test

Calculate the pH of 0.02 M hydrochloric acid (HCl) solution and the pH of a 0.015 sodium hydroxide (NaOH) solution.

Answers

The HCl solution has a pH of 1.7; the NaOH solution has a pH of 12.2.

Table 12-5	
pH Values of Body Fluids and Common Household Substances	
	pH Range
Fluid	
Blood	7.35-7.45
Gastric juices	1.6-1.8
Bile	7.8-8.6
Urine	5.5-7.5
Saliva	6.2-7.4
Substance	
Black coffee	4.8-5.2
Eggs	7.6-8.0
Lemon juice	2.8-3.4
Milk	6.3-6.6
Tap water	6.5-8.0

The pH values of some common body fluids and some household substances are listed in Table 12-5. From these values it can be seen that blood is a slightly basic liquid, the gastric juices are strongly acidic, bile is weakly basic, and urine and saliva are both weakly acidic or weakly basic. The pH of pure water is 7.0.

A difference of 1 in pH value represents a tenfold difference in strength. That is, an acid of pH 4.5 is 10 times as strong as one of pH 5.5. Likewise, a base of pH 10.7 is 10 times as strong as one of pH 9.7, and 100 times as strong (10 × 10) as one of pH 8.7. Therefore, a small change in pH indicates a definite change in acid or base strength.

To measure pH in a laboratory, a pH meter (Figure 12-2) can be used. This instrument is standardized by placing the electrodes into a solution of known pH to see that it is functioning and recording properly. Then the electrodes are placed in a solution of unknown pH and the pH is determined by reading the value on the pH meter. A quicker but less accurate method is to touch a drop of the liquid to a specially prepared piece of indicator paper and then determine the pH by comparison with pH color scale.

FIGURE 12-2
Digital pH meter.

Exercise 12-3

Two solutions have pHs of 10.84 and 12.84, respectively.
a. Are these solutions acidic or basic?
b. Which is stronger?
c. How much stronger?

Solution

a. Since the pH values are above 7, the solutions are basic.
b. The higher the pH of a basic solution, the stronger the solution, so the solution with a pH of 12.84 is the stronger solution.
c. A difference of 1 in pH corresponds to a tenfold difference in strength. So a solution with a pH of 12.84 is 10 times as strong as a solution with a pH of 11.84; a solution with a pH of 11.84 is 10 times as strong as a solution with a pH of 10.84. Therefore, the solution with a pH of 12.84 is 100 (10×10) times as strong as the solution with a pH of 10.84.

Self-test

Compare the acidity of a solution with a pH of 4.0 with that of a solution with a pH of 8.0.

Answer

The solution with a pH of 4.0 is 10^4 times more acidic than the solution with a the pH of 8.0.

Salts

Acids have one ion in common, the hydrogen ion (H^+). Consider the dissociation of the following salts:

$$NaCl \longrightarrow Na^+ + Cl^-$$
$$K_2SO_4 \longrightarrow 2\,K^+ + SO_4^{2-}$$
$$Mg(NO_3)_2 \longrightarrow Mg^{2+} + 2\,NO_3^-$$

Salts have no common ion. Salts in solution yield a positive ion and a negative ion. Salts are formed by the reaction of an acid and a base. For example,

Acid		*Base*		*Salt*		*Water*
HCl	+	KOH	$\longrightarrow$	KCl	+	H_2O
H_2SO_4	+	$Mg(OH)_2$	$\longrightarrow$	$MgSO_4$	+	$2\,H_2O$
$2\,HNO_3$	+	$Zn(OH)_2$	$\longrightarrow$	$Zn(NO_3)_2$	+	$2\,H_2O$

Recall that the reaction of an acid with a base is called neutralization.

12-6 Solubility of Salts

Some salts are quite soluble in water. Others are classified as slightly soluble or insoluble. Table 12-6 indicates the solubility of most common salts, and Table 12-7 indicates the solubilities in water of selected common salts as predicted by using Table 12-6.

Table 12-6 Solubility of Common Salts and Bases

Soluble	Insoluble
Sodium salts	Carbonates (except sodium, potassium, ammonium
Potassium salts	
Ammonium salts	Phosphates (except sodium, potassium, ammonium)
Acetates	
Nitrates	Sulfides (except sodium, potassium, ammonium)
Chlorides (except silver, lead, and mercury +1)	
	Hydroxides (except sodium, potassium, ammonium)
Sulfates (except calcium, barium, and lead)	

Table 12-7 Water Solubility of Some Salts

Name of Salt	Formula	Solubility
Sodium chloride	$NaCl$	Soluble
Silver chloride	$AgCl$	Insoluble
Sodium suflate	Na_2SO_4	Soluble
Zinc nitrate	$Zn(NO_3)_2$	Soluble
Barium sulfate	$BaSO_4$	Insoluble
Calcium phosphate	$Ca_3(PO_4)_2$	Insoluble
Magnesium carbonate	$MgCO_3$	Insoluble

12-7 Reactions of Salts

Hydrolysis

Hydrolysis is the reaction of a compound with the hydrogen ion or the hydroxide ion derived from water. Ions derived from a weak acid or base hydrolyze (react with water) to form the corresponding acid or base. Ions derived from a strong acid or base do not hydrolyze. For simplicity we will assume that the acids and bases listed in Table 12-8 are strong and that all other commonly used acids and bases are weak (For a discussion of strong and weak acids and bases see Section 12-1.)

Since salts are produced by the reaction of an acid with a base, salts must contain parts of each. The positive ion of a salt is derived from a base, and its negative ion is derived from an acid.

Let us consider the hydrolysis of the salt sodium cyanide (NaCN). We first write the formula of the salt in ionic form, sodium ion (Na^+) and cyanide ion (CN^-). Directly below these ions we write the ionized formula for water, with the

Table 12-8
Strong Acids and Bases in Common Use

Strong acids
HCl	Hydrochloric acid
HNO_3	Nitric acid
H_2SO_4	Sulfuric acid

Strong bases
NaOH	Sodium hydroxide
KOH	Potassium hydroxide
$Ca(OH)_2$	Calcium hydroxide

OH$^-$ ion below the positive ion of the salt and the H$^+$ ion below the negative ion of the salt, or

$$\begin{array}{c|c} Na^+ & CN^- \\ OH^- & H^+ \end{array} \qquad (12\text{-}1)$$

The sodium ion is derived from one of the strong bases (NaOH) and does not hydrolyze. The CN$^-$ ion is not derived from one of the strong acids and does hydrolyze. Reading upward on the right side of the line in setup (12-1), we have the essentially nonionized compound HCN. Thus, the products of hydrolysis of NaCN are the two ions on the left side of the line—Na$^+$ and OH$^-$—and the non-ionized compound on the right side of the line—HCN. The hydrolysis reaction for NaCN can be written as

$$Na^+ + CN^- + H_2O \rightleftharpoons Na^+ + OH^- + HCN$$

Since the hydrolysis reaction produces OH$^-$ ions, the solution will be basic. This is what we should expect from the hydrolysis of NaCN, which is derived from a strong base and a weak acid.

Now consider the hydrolysis of NH$_4$Cl. If we write this compound in ionic form, with water in its ionized form below it, and place the OH$^-$ ion below the positive ion of the salt and the H$^+$ ion of the water below the negative ion of the salt, we have setup (12-2).

$$\begin{array}{c|c} NH_4^+ & Cl^- \\ OH^- & H^+ \end{array} \qquad (12\text{-}2)$$

The NH$_4^+$ ion is not derived from one of the strong bases and does hydrolyze. The nonionized compound formed, as indicated by reading down on the left side of the line, is NH$_4$OH. The Cl$^-$ ion is derived from one of the strong acids and does not hydrolyze. The right side of the line in (12-2) indicates the ions remaining in solution, Cl$^-$ and H$^+$. Thus, the hydrolysis reaction is

$$NH_4^+ + Cl^- + H_2O \rightleftharpoons NH_4OH + H^+ + Cl^-$$

The resulting solution will be acidic because of the presence of hydrogen ions. This is what should be expected, since the compound NH$_4$Cl is derived from a weak base and a strong acid.

A prediction of the pH of the solution produced by the hydrolysis of a salt is shown in Table 12-9.

Table 12-9　pH of Solution Produced During Hydrolysis of a Salt		
Substances from which Salt was Derived	**Predicted pH**	**Example**
Strong acid, weak base	Below 7	$Al_2(SO_4)_3$
Weak acid, strong base	Above 7	KCN
Weak acid, weak base	Not easily predicted	$NH_4C_2H_3O_2$

Reaction with Metals

Some metals react with salt solutions to form another salt and a different metal. Refer to the activity series of metals (Table 12-2) and recall that any metal can replace a metal ion below it in the activity series. Thus zinc can replace the copper from the copper sulfate solution because the zinc is higher in the activity series than copper. This is an example of a single replacement reaction.

$$Zn + CuSO_4 \longrightarrow ZnSO_4 + Cu$$

Reaction with Other Salts

Two different salts in solution can react by double displacement. Consider reactions (12-3) and (12-4):

$$K_2SO_4 + Ba(NO_3)_2 \longrightarrow 2\,KNO_3 + BaSO_{4(s)} \qquad (12\text{-}3)$$

$$K_2SO_4 + Zn(NO_3)_2 \longrightarrow \text{no reaction} \qquad (12\text{-}4)$$

Why does reaction (12-3) proceed, whereas (12-4) does not? Review the solubility rules (Section 12-6). Note that one of the products formed in reaction (12-3), $BaSO_4$, is insoluble in water. Therefore, reaction (12-3) will proceed. In reaction (12-4), the products, if formed, would be $ZnSO_4$ and KNO_3. Both of these salts are soluble and ionized; therefore, no reaction will take place. For a reaction to take place between solutions of two salts, at least one of the products must be insoluble in water, a weak electrolyte, or a nonionized compound.

Reaction with Acids and Bases

Salts react with acids or bases to form other salts and other acids and bases. These reactions will proceed if one of the products is insoluble in water or is an insoluble gas.

$$CaCO_3 + 2\,HCl \longrightarrow CaCl_2 + H_2O + CO_{2(g)}$$

$$AlCl_3 + 3\,NaOH \longrightarrow 3\,NaCl + Al(OH)_{3(s)}$$

12-8 Uses of Salts

Salts are necessary for the proper growth and metabolism of the body. Iron salts are necessary for the formation of hemoglobin; iodine salts for the proper functioning of the thyroid gland; calcium and phosporus salts for the formation of bones and teeth; sodium and potassium salts help regulate the acid-base balance of the body. Salts regulate the irritability of nerve and muscle cells and the beating of the heart. Salts help maintain the proper osmotic pressure of the cells.

Many salts have specific uses. Barium sulfate ($BaSO_4$) is used for x-ray work. Although barium compounds are poisonous, barium sulfate is insoluble in body fluids and so has no harmful effect on the body. Barium sulfate is opaque to x-rays and, when swallowed, it can be used to outline the gastrointestinal (GI) system for x-ray photographs (Figure 12-3).

Table 12-10 lists some common salts and their specific uses in medicine.

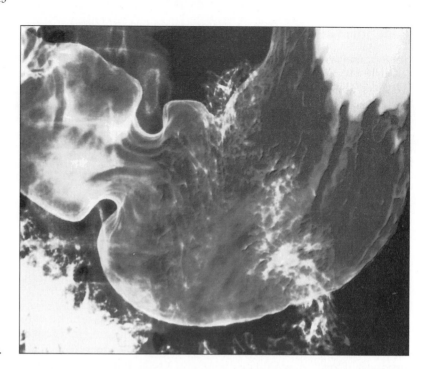

FIGURE 12-3
X ray of the stomach.

Table 12-10   Common Salts and Their Uses

Classification	Formula	Chemical Name	Common Name
Antacids	$CaCO_3$	Calcium carbonate	Precipitated chalk
	$NaHCO_3$	Sodium bicarbonate	Baking soda
Cathartics	Na_2SO_4	Sodium sulfate	Glauber's salt
	$MgSO_4 \cdot 7\,H_2O$	Magnesium sulfate	Epsom salts
	$MgCO_3$	Magnesium carbonate	
	$MgHC_6H_5O_7 \cdot 5\,H_2O$	Magnesium citrate	Citrate of magnesia
	$KNaC_4H_4O_6 \cdot 4\,H_2O$	Potassium sodium tartrate	Rochelle salt
Diuretic	NH_4Cl	Ammonium chloride	Sal ammoniac
Expectorants	NH_4Cl	Ammonium chloride	Sal ammoniac
	KI	Potassium iodide	
Germicide	$AgNO_3$	Silver nitrate	Lunar caustic
Miscellaneous uses			
X-ray work	$BaSO_4$	Barium sulfate	Barium
Caries reduction	NaF	Sodium fluoride	
	SnF_2	Stannous fluoride	
For casts	$(CaSO_4)_2 \cdot H_2O$	Calcium sulfate hydrate	Plaster of paris
Treatment of anemia	$FeSO_4$	Ferrous sulfate	
Decrease of blood clotting time	$CaCl_2$	Calcium chloride	
Physiologic saline solution used as an irrigant and as IV replacement fluid	$NaCl$	Sodium chloride	Table salt
Thyroid treatment	KI	Potassium iodide	
	NaI	Sodium iodide	
Prevention of clotting of stored blood	$Na_3C_6H_5O_7$	Sodium citrate	

12-9 Buffer Solutions

The pH of pure water (a neutral solution) is 7.0. If an acid is added to water, the pH goes down. How far below 7.0 it goes depends on how much acid and how strong an acid is added. When a base is added to pure water, the pH rises above 7.0.

However, when small amounts of acid or base are added to a buffer solution, the pH does not change appreciably. A buffer solution is defined as a solution that will resist changes in pH on the addition of small amounts of either acid or base.

Buffer solutions, or buffers, are found in all body fluids and are responsible for helping maintain the proper pH of those fluids. The normal pH range of the blood is 7.35 to 7.45. Even a slight change in pH can cause a very definite pathologic condition. When the pH falls below 7.35, the condition is known as *acidosis. *Alkalosis is the condition under which the pH of the blood rises above 7.45 (see Section 30-13).

What does a buffer solution consist of, and how does it work? A buffer system usually consists of a weak acid and a salt of a weak acid. There are several buffer systems in the blood. One of these consists of carbonic acid (H_2CO_3), a weak acid, and sodium bicarbonate ($NaHCO_3$), the salt of a weak acid.

Suppose that an acid such as hydrochloric acid (HCl) enters the bloodstream. The HCl reacts with the $NaHCO_3$ part of the buffer according to the reaction

$$HCl + NaHCO_3 \longrightarrow NaCl + H_2CO_3$$

The NaCl produced is neutral; it does not hydrolyze (see Section 12-7). The H_2CO_3 produced is part of the original buffer system and is only slightly ionized. In the body, acids (hydrogen ions) are produced by various metabolic processes. When these acids enter the bloodstream, they are removed by this reaction or a similar reaction with other buffers.

A base such as sodium hydroxide (NaOH) would react with the carbonic acid part of the buffer,

$$NaOH + H_2CO_3 \longrightarrow H_2O + NaHCO_3$$

forming water, a harmless neutral normal metabolite, and sodium bicarbonate, part of the original buffer. Thus, when either acid or base is added to the buffer, something neutral (NaCl or water) is formed plus more of the buffer system (H_2CO_3 or $NaHCO_3$). Therefore, the pH of the blood should not change because the buffer system has reduced the number of free hydrogen or hydroxide ions.

The principal intracellular buffer is the phosphate buffer system, which consists of a mixture of monohydrogen phosphate ions (HPO_4^{2-}) and dihydrogen phosphate ions ($H_2PO_4^-$). Excess acid (H^+) reacts with the monohydrogen phosphate ions,

$$H^+ + HPO_4^{2-} \longrightarrow H_2PO_4^-$$

Excess base (OH^-) reacts with the dihydrogen phosphate ions,

$$OH^- + H_2PO_4^- \longrightarrow HPO_4^{-2} + H_2O$$

In each case, more buffer is produced and the pH remains constant.

In addition to the carbonate and phosphate buffers there are also several organic buffer systems. These will be discussed in Chapter 30.

If there is an overproduction of acid in the tissues and if these acids cannot be excreted rapidly enough, a condition known as acidosis results during which the buffers are unable to handle the excess acid. Acidosis can occur in certain diseases, such as diabetes mellitus, and during starvation (see Section 30-13).

Prolonged vomiting can result in alkalosis because of the continued loss of the acid contents of the stomach.

Summary

Acids are compounds that yield hydrogen ions or protons in solution. The general properties of acids are sour taste, turn blue litmus red, reaction with metal oxides and hydroxides (neutralization), reaction with metals to yield hydrogen gas, reaction with carbonates and bicarbonates to produce carbon dioxide gas, effect on clothing, and effect on tissues.

Antacids are used to neutralize excess acidity in the stomach.

Acids and bases are used industrially, pharmaceutically, and in the hospital and laboratory.

Bases are compounds that yield hydroxide (OH^-) ions in solution. Bases are also compounds that accept protons. Bases have a slippery, soapy feeling; they turn red litmus blue, neutralize acids, react with certain metals to produce hydrogen, and affect tissue and clothing.

Titration is a method for determining the strength of an unknown acid (or base) by using a buret containing a base (or acid) of known concentration and an indicator.

The ion product of water is $K_w = [H^+] \times [OH^-] = 10^{-14}$.

The pH of a solution indicates numerically the acid (or base) strength of the solution in terms of its hydrogen ion concentration. A pH of 7 indicates a neutral solution; a pH below 7, an acid solution; a pH above 7, a basic solution.

Salts are formed by the reaction of an acid with a base. Salts yield ions other than hydrogen or hydroxide.

Some salts are soluble and others are insoluble in water. The solubility of most common salts can be determined from the solubility rules.

Hydrolysis is a double displacement reaction in which water is a reactant.

Salts react with some metals to yield other salts and other metals. Salts can react with other salts by a double displacement reaction. Salts can react with acids or bases to form other salts and other acids or bases.

Salts serve a definite purpose in the various metabolic processes of the body.

Buffer solutions do not change in pH when small amounts of acid or base are added. Buffer solutions help maintain the proper pH of the body fluids.

Questions and Problems

1. List the general properties of acids; of bases.
2. What effect does an acid have on litmus paper? What effect does a base have?
3. Define neutralization, pH, acid, base, and titration.
4. How can pH be measured? What is the pH of blood; saliva; urine? Are these liquids acidic, basic, or netural?
5. Name three bases used in hospitals and give one medical use for each.
6. Name three acids used in hospitals and give one medical use for each.
7. What substances can be used to neutralize stomach acid?
8. Why are acids usually stored in glass containers rather than metal ones?
9. What treatment should be given if a strong acid is spilled on the skin? a strong base?
10. Name five soluble salts; five insoluble salts.
11. Define hydrolysis; buffer solution.
12. Indicate whether the hydrolysis of the following salts will produce an acid, basic, or neutral solution:
    a. KCl
    b. NH_4NO_3
    c. $ZnSO_4$
    d. MgI_2
    e. Na_2CO_3
13. Give a reaction of a salt with an acid; with a base; with another salt; with a metal.
14. What is acidosis; alkalosis?
15. Name five salts in common use in a hospital and indicate what each is used for.
16. Why is the pH of drinking water usually not 7.0?

17. Why is $Ca(OH)_2$ a base and $C_2H_4(OH)_2$ not a base?
18. How do antacids work?
19. Why is boric acid not in medical use today?
20. What is acetylsalicylic acid? Where and for what purposes is it used?
21. Why is lime water used as an antidote for oxalic acid poisoning?
22. What is meant by the term *endpoint of a titration*?
23. Compare the ionization of strong and weak acids; strong and weak bases.
24. Explain how a buffer solution works.
25. When the following compounds are placed in water, will the resulting solutions have a pH of 7? above 7? below 7?
    a. $CaCl_2$
    b. NH_4NO_3
    c. $NaCN$
    d. $MgSO_4$
    e. $KClO$
26. Why do ions from weak acids or bases hydrolyze whereas those from strong acids or bases do not?

27. Why are buffer solutions important in the body?
28. Urine normally has a pH of about 6. If a person excretes 1700 mL of urine in 1 day, how many grams of hydrogen are lost?
29. The ionization constant of water, K_i is 2.42×10^{-14} at 37° C (normal body temperature). What is the pH of neutral water at this temperature?
30. Calculate the pH of the following solutions: 60 g of NaOH in 500 mL of solution; 100 g HCl in 2.0 L of solution.
31. Normal venous blood has a pH of about 7.4. Blood pH is controlled by the buffer system of H_2CO_3 and HCO_3^-. Which would you expect to have a higher concentration? Why?
32. Compare the Brønsted definitions of acids and bases with the general definitions.
33. If a solution has a hydrogen ion concentration of 10^{-5} M, what is its pH? Is the solution acid or basic?
34. Compare the strengths of bases of pH 8.47 and 10.47.

Practice Test

1. The principal intracellular buffer is a _____.
   a. chloride
   b. carbonate
   c. phosphate
   d. sulfate
2. A salt used in caries reduction is _____.
   a. NaF   b. $AgNO_3$ c. $MgSO_4$ d. $CaCl_2$
3. An example of an insoluble salt is _____.
   a. $CaCl_2$ b. $CaCO_3$ c. NaCN   d. $Al_2(SO_4)_3$
4. An example of a soluble salt is _____.
   a. KCl   b. AgCl   c. $MgCO_3$ d. $FeCO_3$
5. Prolonged vomiting can lead to _____.
   a. acidosis
   b. alkalosis
   c. hydrolysis
   d. hemolysis
6. A salt used in x-ray work is _____.
   a. $MgSO_4$ b. $AgNO_3$ c. KI   d. $BaSO_4$
7. Diabetes mellitus can cause a condition known as _____.
   a. alkalosis
   b. hypoxia
   c. acidosis
   d. hypervitaminosis
8. Which of the following pH values could indicate alkalosis?
   a. 7.55   b. 7.35   c. 7.15   d. 7.0
9. Potassium iodide can be used as a(n) _____.
   a. antacid
   b. cathartic
   c. thyroid treatment
   d. aid to prevent blood clotting
10. A solution used to maintain a constant pH is called a(n) _____.
    a. buffer
    b. substrate
    c. hydrolysate
    d. acid-base solution

11. Which of the following is not a strong acid?
    a. nitric
    b. sulfuric
    c. acetic
    d. hydrochloric
12. An example of a diprotic acid is _____.
    a. H_2O
    b. H_2SO_4
    c. C_2H_2
    d. $Ca(OH)_2$
13. Which pH indicates a strong base?
    a. 1.2   b. 5.4   c. 8.2   d. 11.4
14. An example of an antacid is _____.
    a. $Mg(OH)_2$
    b. $CaCO_3$
    c. $AlNa(OH)_2CO_3$
    d. all of these
15. An acid is a substance that _____.
    a. donates electrons
    b. donates protons
    c. donates neutrons
    d. none of these
16. Acids react with _____.
    a. all bases
    b. all salts
    c. all metals
    d. all of these
17. The reaction of an acid and a base is called _____.
    a. oxidation
    b. hydrolysis
    c. equilibrium
    d. neutralization
18. An example of a strong base is _____.
    a. NH_4OH
    b. C_2H_5OH
    c. KOH
    d. $Fe(OH)_3$
19. An acid found in the stomach is _____.
    a. HCl
    b. HNO_3
    c. $HC_2H_3O_2$
    d. H_2SO_4
20. The pH of blood is _____.
    a. 6.55-7.05
    b. 7.35-7.45
    c. 8.20-9.20
    d. 6.5-8.0

Radioactivity

Objectives

- To describe the types of radiation produced during radioactive decay
- To balance nuclear reactions
- To be aware of the units of radiation and of the methods used to detect radiation
- To understand the importance of half-life
- To become acquainted with various radioisotopes used in medicine
- To become aware of the biologic effects of radiation
- To understand nuclear fission and fusion

A physician studies a patient's x-ray. X rays, in addition to other types of radiation, are used to probe the internal structure of the human body.

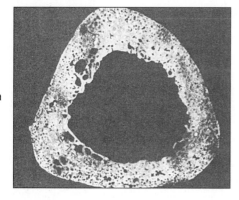 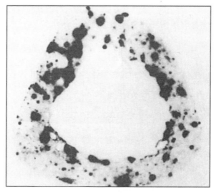

a

b

FIGURE 13-1

A section of bone from the body of a former radium watch dial painter who, to maintain a fine tip of his paintbrush, habitually touched it to his tongue. (a) Photograph of the bone section in normal light, showing dark regions damaged by radiation from the ingested radium. (b) An autoradiograph (a photo in which radioactivity emitted from the bone section exposed a piece of photographic film placed in contact with it) of the same bone section. Note that areas of high radioactivity generally correspond to the areas of greatest damage in (a).

13-1 Discovery of Radioactivity

In 1896 a French physicist, Henri Becquerel (1852-1908), found that uranium crystals had the property of "fogging" a photographic plate that had been placed near those crystals. This fogging took place even though the photographic plate was wrapped in black paper. By placing crystals of uranium on a photographic plate covered with black paper and then developing the plate, he obtained a self-photograph of the crystals. Becquerel concluded that the uranium gave off some kind of radiation or rays that affected the photographic plate. Figure 13-1 illustrates an *autoradiograph of a radioactive bone section.

Substances like uranium that spontaneously give off radiation are said to be radioactive. Radioactivity is the property that causes an element to emit radiation. This radiation comes from the nucleus of the atom.

13-2 Types of Radiation Produced by a Radioactive Substance

The following experiment was performed to study the radiation produced by a radioactive element (Figure 13-2). A piece of radium was placed at the bottom of a thick lead well. The purpose of the lead was to absorb all the radiation except that going directly upward. The escaping radiation was allowed to fall on a photographic plate. When the radiation was passed through a strong electrostatic field, three different areas showed up on the photographic plate. This indicated that there were actually three different kinds of radiation. These were called alpha, beta, and gamma.

Alpha Particles

Alpha particles (α particles) are attracted toward the negative electrostatic field, which indicates that they are positively charged. Alpha particles consist of positively charged helium nuclei; that is, they consist of the nuclei of helium atoms

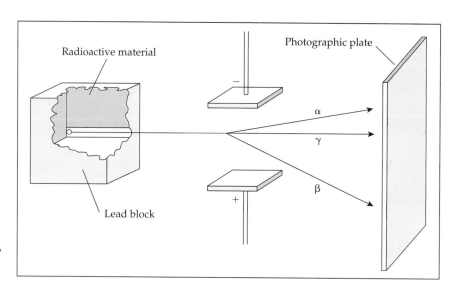

FIGURE 13-2

Radium emits radiation that an electrostatic field separates into alpha (α) particles, beta (β) particles, and gamma (γ) rays.

(each of which contains two protons and two neutrons) and so they have a charge of +2. Alpha particles have a very low penetrating power. They can be stopped by a piece of paper or by a thin sheet of aluminum foil. Alpha particles are relatively harmless when they strike the body because they do not penetrate the outer layer of the skin. However, if a source of alpha particles is inhaled or ingested or gets into the body through an open wound, then those particles can cause damage to the cells and to the internal organs.

Alpha particles result from the radioactive decay of heavy elements such as uranium and radium.

Beta Particles

Beta particles (β particles) are attracted toward the positive electrostatic field, which indicates that they consist of negatively charged particles. Beta particles consist of high-speed electrons that travel in excess of 100,000 miles per second. Note that beta particles (electrons) are produced in the nucleus by the transformation of a neutron into a proton and an electron. The electron is emitted as a beta particle and the proton remains in the nucleus. Beta particles (electrons) have a charge of −1.

Beta particles have a slight penetrating power. They pass through a sheet of paper but can be stopped by heavy clothing. When beta particles strike the body, they penetrate only a few millimeters and do not reach any vital organs. If a source of beta particles should be inhaled or ingested, those particles could also cause internal damage to the body cells and organs.

Note in Figure 13-2 that beta particles are deflected by the electrostatic field to a much greater extent than are the alpha particles. This indicates that beta particles have a much smaller mass than alpha particles.

Gamma Rays

Gamma rays (γ rays) are not affected by an electrostatic field because they have no charge. They are not particles at all; they have no mass. Gamma rays are a form of electromagnetic radiation similar to x rays. Gamma rays are very penetrating; they will pass through the body, causing cellular damage as they travel through (see Section 13-15). Gamma rays are often emitted along with alpha or beta particles. Gamma rays originate from unstable atoms releasing energy to gain stability.

The "gamma knife" focuses a dose of gamma radiation to a precise target point in the brain. It is used to treat deep-seated brain tumors that were previously considered inoperable. There is no incision in the scalp and no need for general anesthesia.

Other Types of Radiation

X rays are a form of electromagnetic radiation usually produced by machines, whereas gamma rays are emitted by radioactive substances.

Neutrons are released from elements that undergo spontaneous fission. Their relatively large mass gives them great energy, and because they have no charge they readily penetrate the body. Neutrons are used in the treatment of cancer.

13-3 Nuclear Reactions

When the nucleus of an atom emits a ray, its atomic number and mass number may change. Such changes in the nucleus are called nuclear reactions.

In writing equations for nuclear reactions the following symbols are used.

1. The atomic number and the mass number are written to the left of the symbol of the element.

$$\text{mass number} \longrightarrow {}^{238}_{92}\text{U}$$
$$\text{atomic number}^{\dagger} \nearrow$$

2. The alpha particle, which consists of a helium nucleus, is indicated as ${}^{4}_{2}\text{He}$.
3. The beta particle, which is an electron, is indicated as ${}^{0}_{-1}\text{e}$. (The beta particle has arbitrarily been assigned an atomic number of -1.)
4. The gamma ray, which is a radiation only, is indicated as γ.
5. The neutron is indicated as ${}^{1}_{1}\text{n}$.
6. The proton is indicated by ${}^{1}_{1}\text{H}$.
7. The deuteron, which is the nucleus of a heavy hydrogen atom, is indicated by ${}^{2}_{1}\text{H}$.
8. The positron, which is a positively charged electron, is indicated by ${}^{0}_{1}\text{e}$.

The following rules must be observed when writing nuclear reactions.

1. The sum of the atomic numbers on both sides of the equation must be the same.
2. The sum of the mass numbers on both sides of the equation must be the same.

When a uranium atom—${}^{238}_{92}\text{U}$—loses an alpha particle, the nuclear reaction can be written as follows:

$$\quad {}^{238}_{92}\text{U} \longrightarrow \alpha + ? \quad \text{or} \quad {}^{238}_{92}\text{U} \longrightarrow {}^{4}_{2}\text{He} + X$$

where X is the product other than an alpha particle produced during the decomposition. The atomic number on the left side of the equation is 92 and on the right side it is 2, so the atomic number of X must be 90. Likewise, the mass number on

---

†Note that in the periodic table (inside front cover) the atomic number is above the symbol.

the left side of the equation is 238 and on the right side it is 4, so the mass number of X must be 238 − 4 or 234. Therefore, the reaction is written as

$$^{238}_{92}\text{U} \longrightarrow \ ^{4}_{2}\text{He} + \ ^{234}_{90}\text{X}$$

Element 90 (from the periodic table in the front of the book) is thorium, symbol Th. The complete reaction therefore becomes

$$^{238}_{92}\text{U} \longrightarrow \ ^{4}_{2}\text{He} + \ ^{234}_{90}\text{Th}$$

indicating that when uranium gives off an alpha particle, it changes into a new element, thorium. If a thorium atom should emit an alpha particle, the reaction could be written as

$$^{234}_{90}\text{Th} \longrightarrow \ ^{4}_{2}\text{He} + \text{Z}$$

The atomic number of element Z must be 88 for the sum of the atomic numbers to be the same on both sides of the equation. Likewise, the mass number of element Z must be 230, so the equation would be

$$^{234}_{90}\text{Th} \longrightarrow \ ^{4}_{2}\text{He} + \ ^{230}_{88}\text{Z}$$

The element of atomic number 88 is radium, symbol Ra, so the actual reaction is

$$^{234}_{90}\text{Th} \longrightarrow \ ^{4}_{2}\text{He} + \ ^{230}_{88}\text{Ra}$$

Thus, when an atom emits an alpha particle, its mass number decreases by 4 and its atomic number decreases by 2.

If a thorium atom, $^{234}_{90}\text{Th}$, emits a beta particle, $^{0}_{-1}\text{e}$, the following reaction takes place:

$$^{234}_{90}\text{Th} \longrightarrow \ ^{0}_{-1}\text{e} + \ ^{234}_{91}\text{Pa}$$

Therefore, when a thorium atom emits a beta particle, an isotope of protactinium (Pa) is produced. Note that the sum of the atomic numbers is the same on both sides of the equation, as is the sum of the mass numbers.

When an atom emits a beta particle, its mass number remains the same, but its atomic number increases by 1.

When an atom emits a gamma ray, there is no change in the atomic number or mass number because the gamma ray is not a particle. It has no mass and no charge. However, even though the same element is produced in the reaction, it has a slightly lower energy because of the energy carried away by the gamma ray,

$$^{238}_{92}\text{U} \longrightarrow \gamma + \ ^{238}_{92}\text{U}^*$$

where the * indicates a slightly lower energy.

Another way of indicating a loss of a gamma ray and the resulting change in energy is

$$^{99\text{m}}_{43}\text{Tc} \longrightarrow \ ^{99}_{43}\text{Tc} + \gamma$$

where the "m" indicates metastable (unstable) form of technetium (Tc). Thus, the unstable form of technetium emits a gamma ray and becomes a more stable (less energetic) isotope of technetium. $^{99\text{m}}\text{Tc}$ is used in various types of scans (see page 224).

Exercise 13-1

Write a balanced nuclear equation for the following radioactive nuclear decay reactions:

a. Radon-222 (^{222}Rn) decays by alpha emission.

b. Technetium-99 m (^{99m}Tc) decays by gamma and beta emission.

Solution

To write a balanced nuclear equation we must (1) have an equal sum of the atomic numbers and (2) have an equal sum for the atomic masses.

a. Write the equation for the decay of radon-222 as follows:

$$^{222}_{86}\text{Rn} \longrightarrow {}^{4}_{2}\text{He} + ?$$

Since the atomic numbers must add up, the missing atomic number must be $86 - 2 = 84$. The missing atomic mass must be $222 - 4 = 218$.

The element with atomic number 84 is polonium (Po).

$$^{222}_{86}\text{Rn} \longrightarrow {}^{4}_{2}\text{He} + {}^{218}_{84}\text{Po}$$

b. The equation is $\quad {}^{99m}_{43}\text{Tc} \longrightarrow \gamma + {}^{0}_{-1}\text{e} + ?$

The missing atomic number is $43 - (-1) = 44$. The missing atomic mass is $99 - 0 = 99$. The element with atomic number 44 is ruthenium (Ru).

$$^{99m}_{43}\text{Tc} \longrightarrow \gamma + {}^{0}_{-1}\text{e} + {}^{99}_{44}\text{Ru}$$

Self-test

Find the identity of X in the following nuclear reactions:

a. ${}^{22}_{11}\text{Na} \longrightarrow {}^{0}_{1}\text{e} + \text{X}$

b. ${}^{14}_{6}\text{C} \longrightarrow {}^{14}_{7}\text{N} + \text{X}$

c. ${}^{131}_{53}\text{I} \longrightarrow {}^{0}_{-1}\text{e} + \text{X}$

Answers

a. ${}^{22}_{10}\text{Ne}$ b. ${}^{0}_{-1}\text{e}$ c. ${}^{131}_{54}\text{Xe}$

13-4 Natural and Artificial Radioactivity

Some elements, such as uranium and radium, are naturally radioactive. Natural radioactivity can be defined as the spontaneous change of one element into another. Many of the heavier elements are naturally radioactive, whereas most of the lighter elements are naturally nonradioactive.

However, normally nonradioactive lighter elements can be changed into radioactive ones by bombardment with such particles as protons (${}^{1}_{1}\text{H}$), neutrons (${}^{1}_{0}\text{n}$), electrons (${}^{0}_{-1}\text{e}$), or alpha particles (${}^{4}_{2}\text{He}$). The radioactive substances so produced are said to be artificially radioactive. They may be produced in a nuclear reactor. If nitrogen (${}^{14}_{7}\text{N}$) is bombarded with neutrons (${}^{1}_{0}\text{n}$), the following nuclear reaction takes place.

$$^{14}_{7}\text{N} + {}^{1}_{0}\text{n} \longrightarrow {}^{14}_{6}\text{C} + {}^{1}_{1}\text{H}$$

The products are a radioactive isotope of carbon, ^{14}C, and a proton. Note that again the sum of the atomic numbers on both sides of the equation is the same; likewise the sum of the mass numbers.

13-5　Units of Radiation

Radiation is measured in terms of several different units, depending on whether the measurement relates to a physical or a biologic effect.

The physical unit of radiation is a measure of the number of nuclear disintegrations occurring per second in a radioactive source. The standard unit is the *curie, which is defined as the number of nuclear disintegrations occurring per second in 1 g of radium; 1 curie (1 Ci) equals 37 billion disintegrations per second. Smaller units are the millicurie (1 mCi = 37 million disintegrations per second) and the microcurie (1 µCi = 37,000 disintegrations per second). These smaller units are frequently used in describing an amount of radioactive fallout. The curie is not useful in biologic work because it simply indicates the number of disintegrations per second regardless of the type of radiation and regardless of the effect of that radiation upon tissue.

The *roentgen (abbreviated R) is a unit of radiation generally applied to x rays and gamma rays only. X rays and gamma rays produce ionization in air and also in tissue. The roentgen is defined as the intensity of x rays or gamma rays that produces 2 billion ion pairs (see page 231) in 1 mL of air. This is not the same for tissue as it is for air, so that the roentgen does not accurately indicate the amount of radiation on tissue.

The *rad (radiation absorbed dose) refers to the amount of radiation energy absorbed by tissue that has been radiated. One rad corresponds to the absorption of 100 ergs of energy per gram of tissue. An erg is a very small unit of energy. More than 40 million ergs are required to equal 1 cal. However, even though the erg is an extremely small unit of energy, the effect of 1 rad (100 ergs per gram) is important because of the ionization that the radiation produces in the cells. The SI unit for absorbed dosage of radiation is the *gray* (Gy). 1 Gy = 100 rad.

Different types of radiation cause different biologic damage to cells. This difference in biologic effectiveness may be expressed in terms of the *relative biologic effectiveness* (RBE) of radiation.

The standard for RBEs is the gamma radiation from ^{60}Co. The RBE of a radiation is defined as

$$RBE = \frac{\text{dose of gamma radiation from cobalt–60}}{\text{dose of radiation required to produce the same biologic effect}}$$

For alpha particles RBE = 10; for beta particles RBE = 1.

The *rem (radiation equivalent, man) represents the amount of radiation absorbed by a human being. This unit of measure takes into consideration the difference in energy for various radioactive sources. A rem is the amount of ionizing radiation that, when absorbed by a human, has an effect equal to the absorption of 1 R. A smaller unit is the millirem, mrem.

Rem = rad × RBE, with the RBE being larger for types of radiation that produce greater numbers of ions along the path of the radiation. The production of a greater number of ions implies more damage to the individual cell.

13-6　Detection and Measurement of Radiation

The problem of detecting and measuring radiation is very important in medical work, particularly in the protection of personnel. One device used to detect radiation is the Geiger counter. This device consists of a glass tube containing a gas at low pressure through which runs a wire connected to a high-voltage power supply.

When the device is brought close to a radioactive substance, the radiation causes a momentary current to flow through the tube. A speaker is usually placed in the circuit to produce a click, indicating a momentary flow of current. Sometimes a counting device is connected to the tube to indicate the amount of radiation.

Scanners (see page 221 and Figure 13-3) use another type of device, called a scintillation counter, to detect and measure radiation. This detector consists of a crystal of sodium iodide containing a small amount of thallium iodide. When the crystal is hit by radiation, it gives off a flash of light, a scintillation; hence the name.

A counting device records these scintillations, and the result is produced as a "scan."

X-ray technicians and others who work around radiation usually are required to wear film badges (Figure 13-4). These badges indicate the accumulated amount of radiation to which they have been exposed. They contain a piece of photographic

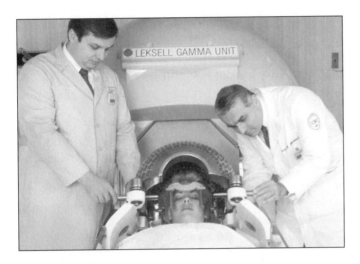

FIGURE 13-3

Use of a "gamma knife" for the treatment of a deep-seated brain tumor.

FIGURE 13-4

Radiologists and other personnel who work in areas where radiation exposure is a potential danger are required to wear film badges. These badges, which contain a strip of photographic film, are checked regularly for evidence of exposure to radiation.

film whose darkening is directly proportional to the amount of radiation received. This film must be checked frequently to see how much radiation has been absorbed.

13-7   Radioisotopes

The isotopes produced artificially by bombardment with one of the various particles are called radioactive isotopes, or radioisotopes. Radioisotopes have the same chemical properties as nonradioactive isotopes of the same element because chemical properties are based upon electrons only, and isotopes of an element have identical electron structures. Thus, an organism cannot distinguish between normal carbon (^{12}C) and radioactive carbon (^{14}C).

If a plant is exposed to carbon dioxide containing ^{14}C, the resulting carbohydrates will contain radioactive carbon. The movement of the radioactive atoms as they proceed through the plant can be followed by means of a Geiger counter. Radioisotopes introduced into living organisms are called "tagged" atoms because their path can be followed readily as they move through the organism.

A radioisotope is usually indicated by its symbol and mass number only. The atomic number is not necessary because the symbol itself serves to identify the element. Some radioisotopes commonly used in medicine and in biochemistry are ^{131}I, ^{60}Co, ^{99m}Tc, ^{14}C, and ^{59}Fe (see Sections 13-11 and 13-12).

Radioisotopes are used medicinally in the diagnosis and treatment of various disorders of the human body. A medical tool called a scanner helps to locate malignancies (see Figure 13-3). A patient is given a selected radioisotope that will accumulate in the body area being studied. The scanner moves back and forth across that site and detects the difference in uptake between healthy and abnormal tissue. This information is fed to a receiver and recorder, which then produces a picture called a scan (Figure 13-5). Proper interpretation of a scan can tell not only if there is a malignancy but also where it is located and its exact dimensions.

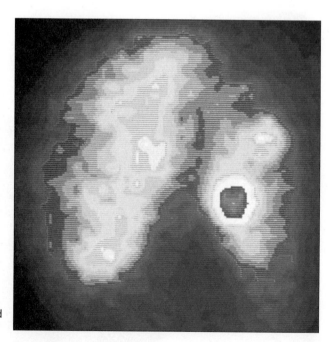

FIGURE 13-5

Scintillation scan of multi-nodular goiter of the thyroid gland (front view).

13-8 Positron Emission Tomography (PET)

Some synthetic radioactive substances give off positrons from their nuclei. Positrons are positively charged electrons and are produced as indicated in the following reaction.

$$\underset{\text{protein}}{{}^{1}_{1}p} \longrightarrow \underset{\text{neutron}}{{}^{1}_{0}n} + \underset{\text{positron}}{{}^{0}_{1}e}$$

The positron thus emitted can exist for only a very short time before colliding with an electron. When this occurs, the electron (negatively charged) and the positron (positively charged) annihilate each other and in so doing convert their masses into two bursts of gamma rays.

$$\underset{\text{electron}}{{}^{0}_{-1}e} + \underset{\text{positron}}{{}^{0}_{1}e} \longrightarrow \underset{\text{gamma rays}}{2\gamma}$$

Positron-emitting elements such as ${}^{11}C$, ${}^{15}O$, ${}^{13}N$, and ${}^{18}F$ are made part of a molecule that, when administered to a patient, will travel to the section of the body being studied. There it will produce gamma rays inside the tissue (as opposed to radiation treatment from outside the body). Positron-emitting elements have a short half-life (see following section), ranging from 2 minutes to 2 hours. Because of this short half-life, large amounts of these materials can be given with relatively small exposure to the patient. In addition, because of the short half-life, repeated measurements can be made effectively.

The PET method allows glucose, made partially from positron-emitting ${}^{11}C$ instead of the normal ${}^{12}C$, to be traced through the sensory sections of the brain. In one test, when volunteers were stimulated visually in only one eye, the region of the brain on the opposite side showed increased glucose usage. Similar tests are being performed for other senses, such as sound and smell. Not only can the change in glucose usage be determined but also the location of that usage can be pinpointed. PET is also being used in detecting such disorders as epilepsy, heart disease, stroke, Parkinson's disease, and mental illnesses.

13-9 Half-life

When a radioactive element gives off a particle, it changes or decays into another element. The rate of decay of all radioactive elements is not the same. Some elements decay rapidly, whereas others decay at an extremely slow rate. The half-life of a radioactive element is defined as the amount of time required for half of the atoms in a given sample to decay. Some radioactive elements have a half-life measured in terms of billions of years, whereas others are measured in fractions of a second.

The radioisotope ${}^{131}I$ has a half-life of approximately 8 days. Consider a 60 mg sample of ${}^{131}I$. After 8 days (one half-life period of time) only half as much, or 30 mg, would be left. After another half-life period (a total of 16 days), only half of that amount, or 15 mg, of ${}^{131}I$ would be left. During every half-life period, half of the remaining amount decays.

This can be shown diagrammatically in a decay curve in which the amount of radioactive material is plotted against elapsed time (Figure 13-6).

For medical work, a radioisotope must have a half-life such that it will remain in the body long enough to supply the radiation needed and yet not expose the

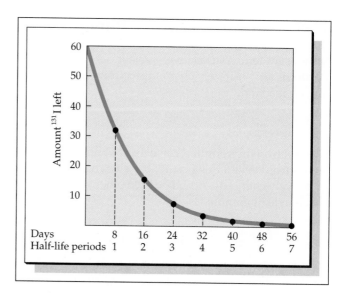

FIGURE 13-6

Decay curve for a radioactive substance, showing amount of ^{131}I versus time and half-life periods.

body to excess radiation. This requires either a relatively short half-life or rapid elimination from the body.

The half-life periods of several commonly used radioisotopes are listed in Table 13-1.

Radioisotopes of long half-life are very dangerous to the body. For example, radium has a half-life of 1590 years; therefore, if it is taken into the body, it continues to give off its radiation during the lifetime of that person. Radium and strontium-90 are members of group IIa in the periodic chart. This indicates that they act like calcium. They are easily taken up in the bones and are not readily excreted.

Biologic half-life is the time taken for one-half of an administered radioactive substance to be lost through biologic processes.

Exercise 13-2

The half-life of $^{99m}_{43}$Tc is 6.0 hours. How much remains of a 100 g sample after 1 day?

Solution

A half-life is the time that it takes for $\frac{1}{2}$ of a sample to decay. One day is 24 hours, which is 4 half-lifes.

$$100 \text{ g} \times \tfrac{1}{2} \times \tfrac{1}{2} \times \tfrac{1}{2} \times \tfrac{1}{2} = 6.25 \text{ g}$$

Self-test

A vial containing 400 mg of $^{131}_{53}$I is misplaced on a shelf. After 80 days, the vial is found. How much of the original sample of $^{131}_{53}$I remains? (See Table 13-1 for the half-life of $^{131}_{53}$I.)

Answer

0.39 mg

Table 13-1	Half-Life Periods of Several Common Radioisotopes	
Radioisotope	Half-life	Radiation
^{99m}Tc	6.0 hr	γ
^{59}Fe	45 days	$\beta + \gamma$
^{198}Au	2.7 days	$\beta + \gamma$
^{131}I	8.0 days	$\beta + \gamma$
^{123}I	13 hr	$\beta + \gamma$
^{32}P	14.3 days	β
^{60}Co	5.3 yr	$\beta + \gamma$
^{14}C	5760 yr	β
^{3}H	12 yr	β
^{67}Ga	78 hr	γ
^{51}Cr	27.8 days	γ
^{24}Na	15.0 hr	$\beta + \gamma$
^{111}In	2.8 days	γ

13-10 Boron Neutron Capture Therapy

A promising new procedure is boron neutron capture therapy (BNCT). In this procedure, a stable isotope of boron, ^{10}B, is irradiated with a beam of low-energy neutrons. These low-energy neutrons have little effect on normal cells and tissues, but cause the boron to emit alpha particles. These alpha particles destroy cancerous cells but do not reach nearby healthy cells or tissues. For this method to work, the boron is carried into tumor cells by drugs, monoclonal antibodies, or derivatives of naturally occurring compounds. For some unknown reason, ^{10}B concentrates in tumor cells; hence the use of this method.

13-11 Radioisotopes in Medicine

When a radioisotope is to be used for diagnostic purposes, it must meet several criteria. Among them are the following: The radioactive element must be contained in a compound that will tend to concentrate in the area under study or in certain abnormal tissues. Since the presence of radiation is usually determined by an external counter or by a scan, a radioisotope emitting alpha particles is not generally used because such particles have too low a penetrating power to be detected outside the body. Beta-emitting radioisotopes must be located very close to the skin to be detected. Radioisotopes with gamma radiation are preferred.

The radioisotope selected should have a short half-life and should be in the form of a compound that will be eliminated from the body shortly after its diagnostic use is completed. Thus, the body will receive a minimum amount of radiation after the test is completed. In addition, the amount of radioisotope used should be as small as practicable.

When a radioisotope is to be used for therapy, external measurement is not so necessary, so alpha as well as beta and gamma emitters can be used. In radiation therapy, selected cells or tissues are to be destroyed without damage to nearby healthy tissues. Thus, the given radioisotope should have the property of concentrating in the desired area and, preferably, should emit alpha or beta particles because these have limited penetrating power and will not damage adjacent tissues.

A radioimmunoassay technique measures the blood levels of an enzyme called CK BB. The blood levels of this enzyme increase after damage to the nervous system. This method allows rapid assessment of brain damage that can occur during a stroke.

Some of the many radioisotopes in common medical use are discussed in the sections that follow.

Iodine-131, Iodine-123

Iodine-131 and iodine-123 are used in the diagnosis and treatment of thyroid conditions. The thyroid gland requires iodine to function normally. A patient suspected of having a thyroid disorder is given a drink of water containing a small amount of ^{131}I (or ^{123}I) in the form of sodium iodide. If the thyroid is functioning normally, it should take up about 12 percent of the radioactive iodine within a few hours. This iodine uptake can be measured with scanning equipment. If less than the normal amount of ^{131}I is taken up, the patient may have a hypothroid condition. If the amount of ^{131}I taken up is greater than normal, the patient may have a condition known as hyperthyroidism.

For a typical thyroid scan, a dose of 50 μCi is administered orally. This amount of radiation gives 50 rad to the thyroid gland and 0.02 rad to the total body. Another method for thyroid scanning involves the use of metastable technetium-99 (^{99m}Tc). This radioisotope has the following advantages: (1) 5 mCi of ^{99m}Tc administered intravenously gives a dose of only 1 rad to the thyroid gland and 0.02 rad to the whole body, and (2) the scan can be performed 20 minutes after injection rather than after a 24-hour wait, as when ^{131}I is used.

In addition, ^{131}I and ^{123}I are used for scans of the adrenal glands and to assess renal function. ^{131}I is used to screen for impaired fat digestion and decreased intestinal absorption of fats. A normal result implies good pancreatic function. ^{131}I is employed *therapeutically in the treatment of hyperthyroidism and in cancer of the thyroid gland.

Iodine-125 (^{125}I) injected intravenously in the form of radioisotope-labeled fibrinogen is being used to detect deep *thromboses.

Technetium-99m

Metastable technetium-99 (^{99m}Tc) is one of the most widely used radioisotopes for various types of scans. The term "metastable technetium-99m" indicates that the element is highly energetic and loses its extra energy as gamma radiation to yield a more stable radioisotope, $^{99m}Tc \longrightarrow {}^{99}Tc + \gamma$. ^{99m}Tc as sodium pertechnate is used for brain scans and thyroid scans (see Figure 13-5); technetium sulfur colloid, for liver scans and bone marrow scans; technetium macroaggregates of albumin, for lung-perfusion scans; technetium pyrophosphate, for myocardial infarction scans; technetium diethylenetriaminepentaacetic acid (DTPA), for renal scans; technetium-labeled red blood cells, for pericardial studies, heart function studies, and bleeding studies; technetium phosphate compounds, for bone scans; and technetium-IDA (iminodiacetic acid) analogues, for hepatobiliary imaging. Technetium albuminate is used to scan the placenta for bleeding.

^{99m}Tc is also used to measure blood volume. A sample of this radioisotope is injected into the bloodstream. A short while later a sample of blood is withdrawn and its radioactivity is measured. By comparing the radioactivity of the ^{99m}Tc and the radioactivity present in the sample withdrawn, an accurate determination of the total blood volume can be obtained.

The use of scanners (CT, PET, and MRI) for the detection of tumors and cancerous tissue has greatly reduced the use of ^{99m}Tc for these purposes.

Cobalt-60, Cobalt-57

Cobalt-60 (^{60}Co) gives off powerful gamma rays as well as beta particles. For treatment purposes, the beta particles are shielded out and only the gamma rays are used. ^{60}Co has a half-life of 5.3 years and is used as a substitute for radium because it is much cheaper and easier to handle. This radioisotope is employed in the treatment of many different types of cancer. Hospitals use large machines containing a cobalt "bomb" to supply gamma radiation directly to the cancer site (Figure 13-7).

Cobalt-57 (^{57}Co) is used in the diagnosis of vitamin B_{12} malabsorption, a condition that might occur in pernicious anemia.

Other Radioisotopes

Among the many other radioisotopes in medical use are krypton-79 (^{79}Kr) for evaluation of the cardiovascular system, selenium-75 (^{75}Se) for the determination and size of the pancreas, gadolinium-153 (^{153}Gd) for the determination of bone density, and mercury-197 (^{197}Hg) for the evaluation of spleen function and for brain scans. Copper-64 (^{64}Cu) is used in the diagnosis of Wilson's disease (see page 412), and gold-198 (^{198}Au) is used for the assessment of kidney functions. ^{3}H is used to determine the total amount of water present in the body; ^{14}C has been employed to study the path of carbohydrates, fats, and proteins in the body and their conversion into other substances; and gold-198 (^{198}Au) has been employed for the treatment of pleural and peritoneal *metastases.

Unlike ^{99m}Tc, which concentrates in damaged heart cells, thallium-201 (^{201}Tl) concentrates in normal heart muscle but not in abnormal tissue. Thus, damaged heart muscle areas show up on a scan as "cold spots." ^{201}Tl is also used to detect hyperparathyroidism.

Iron-59 (^{59}Fe) is used to measure the rate of disappearance of iron from the plasma, the plasma iron turnover rate, bone marrow function, and the utilization of iron in red blood cell production and to determine anemias caused by iron deficiency or chronic infection. The iron is usually injected as ferrous citrate, with blood samples being withdrawn at various intervals depending on the diagnostic test.

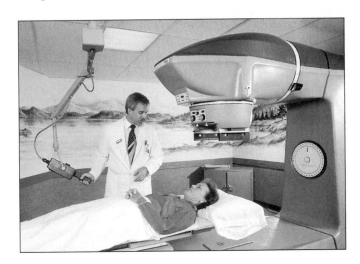

FIGURE 13-7

Radiotherapy: a linear accelerator focuses a beam of subatomic particles on a tumor in order to destroy it.

Chromium-51 (^{51}Cr), in the form of sodium chromate that has been tagged to albumin, is used to determine protein loss through the gastrointestinal tract. It is also used to determine the size, shape, and location of the spleen and can be used in blood studies including plasma volume, red blood cell mass, and red blood cell survival time.

Phosphorus-32 (^{32}P) provides a useful means of determining eye tumors. It is not possible to perform a biopsy of the eye without loss of vision and destruction of the eye. Since ^{32}P uptake is greater in ocular tumors than in other areas of the eye, the location of a tumor can be pinpointed. ^{32}P is also used in the treatment of leukemia, and in the detection of skin cancer or cancer in tissue exposed by surgery.

Indium-111 (^{111}In) is used for the detection of infection and neurologic disorders such as normal-pressure hydrocephaly. ^{111}In, when labeled to monoclonal antibodies (MoAb), is used for cancer detection.

13-12 Radioimmunoassay

Radioimmunoassay (RIA) is a method of measuring the concentration of substances that are present in very small amounts in the blood. The procedure is based upon the body's ability to provide immunity against disease.

When a foreign substance enters the bloodstream, the body produces antibodies to react with and neutralize the invading material. The foreign body, which causes the production of antibodies, is called an *antigen. The reaction is

$$\text{antigen + antibody} \longrightarrow \text{antigen-antibody complex}$$

For each antigen there is a specific antibody. An antibody that protects against one disease will, in general, have no effect on other diseases.

Suppose an RIA is ordered for renin, an enzyme of the kidney. The laboratory will produce an artificially radioactive form of renin and also the specific antibody for that substance. As indicated in the previous paragraph, the antigen reacts with the antibody to form an antigen-antibody complex. However, in this case there are two antigens, the one present in the blood and the artificially radioactive one. Both can react with the antibody. The two antigen-antibody complexes are separated from the rest of the material. When the ratio of the radioactive complex to the nonradioactive complex is measured, the concentration of the antigen being determined can be calculated.

This method is capable of determining protein concentration as low as 10^{-9} g/mL. Concentrations of lower molecular mass compounds can be determined to within 10^{-12} g/mL.

Dying heart muscle releases MB isoenzyme, which is different from that produced by other types of cells. Radioimmunoassay for this isoenzyme can help diagnose heart attack in persons with chest pains.

13-13 X-ray Therapy

X rays are a penetrating type of radiation, similar to gamma rays but of a lower energy. The amount of radiation and how deep it penetrates the tissue are adjustable in an x-ray machine, whereas these factors are generally fixed in a radioactive source such as Ra or ^{60}Co. X rays can be used for treatment of superficial skin conditions by adjusting the voltage of the machine so that it produces a "soft" or nonpenetrating radiation. Although more penetrating x rays have been

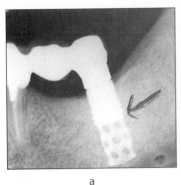
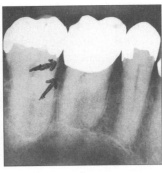
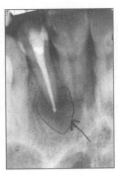
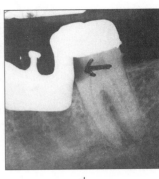

a   b   c   d

FIGURE 13-8

Dental x rays reveal pathologies not visible on the surface, as well as evidence of past restorative work: (a) root canal; (b) bone destruction with calculus formation (irregular mineral deposition) on root surfaces; (c) periapical abscess; and (d) carious lesion ("cavity").

used for the treatment of deep-seated malignancies, the cobalt machine (Figure 13-7) is now widely used for the treatment of different types of cancer.

Another form of x-ray therapy employs oxygen at pressures higher than atmospheric. This treatment is based on the fact that cancer cells are almost three times as sensitive to destruction by x rays when the tissues are under 3 atmospheres (atm) pressure as they are at normal atmospheric pressure. This is not true for normal cells. The patient is placed in a chamber containing air at 3 atm pressure and then given x-ray treatments. This type of treatment is called hyperbaric cancer radiation.

X rays will not pass through bone and teeth as easily as through tissue, so dental x rays show the presence of cavities, advanced bone destruction, and abscesses as well as the positions of normal and impacted teeth (Figure 13-8).

Radiopaque substances are compounds that absorb x rays, thereby allowing body parts to become visible on film. One such radiopaque substance, barium sulfate ($BaSO_4$), is used to detect abnormalities in the stomach and esophagus. Barium sulfate, which is highly insoluble in water and is also nontoxic, is administered orally in the form of a suspension.

13-14 Diagnostic Instrumentation

X-ray Scanners

In addition to x-ray machines, another type of instrument used is the CT (computed tomography) scanner. This type of scanner rotates in a circle around the body and makes sharp, detailed records of narrow strips of a cross section of the body, each record yielding thousands of bits of information. All of the information is fed into a computer, which then produces a picture of the body section scanned. The results are available in minutes on television screens in black and white or in color (see Plate 29). Even though the scanner makes many passes, the patient is exposed to about the same amount of radiation as with traditional x-ray equipment.

The use of the CT scanner makes possible a detailed examination of any part of a patient's body for the detection of tumors of the brain, breast, kidney, lung, or pancreas. It can also detect abnormal cavities in the spinal cord and enlargements of such organs as the liver, spleen, and heart. In addition, this equipment allows examinations to be performed on an outpatient basis, since no preparation is necessary. The examination can even be performed with the patient wearing clothing.

CT scanners are also used to distinguish between benign and malignant lung tumors. A lung nodule that contains calcium absorbs more x radiation than the surrounding tissues and so appears darker on the "scan." It has been found that calcium-containing lung nodules are almost always benign.

Xeroradiography

In xeroradiography, conventional x-ray equipment is used to make the exposure; however, the image is produced by an entirely different process. In x-ray work, the image is produced on a piece of film by a photochemical process. In xeroradiography the image is produced on opaque paper by a photoelectric process. In contrast to x-ray methods, the xeroradiographic process is dry and requires no darkroom.

Xeroradiography is applied principally to *mammography for early detection of breast cancer. It produces better resolution and makes the interpretation of soft tissue studies easier and more accurate (see Plate 24). Xeroradiography is also useful in the search for foreign bodies, such as plastic, wood, and glass, that may not show up on a regular x ray.

MRI

MRI (magnetic resonance imaging) is a noninvasive (nonsurgical) method of following biochemical reactions in both cells and entire organs under normal physiological conditions. The principle of MRI, in general, is that a sample is placed in a strong magnetic field and subjected to radio waves of a frequency appropriate to the nucleus of the element being studied. For biochemical work, ^{31}P and ^{13}C are frequently used. In chemical work, this method is called NMR, nuclear magnetic resonance.

With MRI it is possible to follow changes in phosphate compounds during skeletal muscle contraction and even in heart muscle itself. Changes in heart muscle caused by oxygen deprivation can now be studied. For example, CO competes with O_2 for binding sites on hemoglobin, causing CO poisoning as a result of oxygen deprivation.

MRI is used to scan for brain damage in newborns that have suffered asphyxia or similar problems. MRI can detect brain damage due to strokes and heart attacks within minutes after an attack rather than the hours or days needed in a CT scan. However, MRI poses a potential hazard to patients with pacemakers and artificial metal joints.

Magnetic resonance can also be used to examine blood flow in the brain without the use of dyes injected into the bloodstream. Such a procedure is called MR angiography (Figure 13-9).

In addition to MRI and PET scans (Section 13-8), there are also super-conducting quantum interference devices (SQUID) and single-photon emission computed tomography (SPECT). The former picks up magnetic fields in the brain and indicates brain activity (see Plate 22), the latter tracks blood flow in the brain.

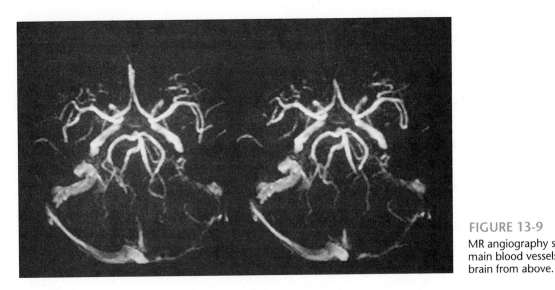

FIGURE 13-9

MR angiography showing main blood vessels in the brain from above.

13-15 Biologic Effects of Radiation

Externally, alpha and beta particles are relatively harmless to humans, for they have slight penetrating power (Figure 13-10). Gamma rays, with their great penetrating power, have a very definite effect upon the body. If a radioactive substance is taken inside the body, it is the alpha particles that are most harmful.

Protection

Shielding, distance, and limiting exposure are the only effective preventive methods against radiation exposure.

Exposure to external radiation can be controlled by increasing the distance between the body and the source of the radiation. The amount of radiation received varies inversely as the square of the distance; therefore, doubling the distance from a radioactive source permits the body to receive only one fourth as much radiation. Shielding material, such as lead, when placed between the body and the radioactive source will also protect the body against radiation.

FIGURE 13-10

Radiation burns suffered from the Chernobyl accident.

This is the reason that dentists and dental assistants stand behind a lead shield when performing x-ray evaluations. They also should shield patients by using lead aprons across their bodies. Clothing or plastic will offer protection against beta emitters.

Exercise 13-3

A nurse receives an exposure of 20 mrem when standing 3 feet from a radioactive source. What will be the exposure at a distance of (a) 6 feet? (b) 10 feet?

Solution

Since exposure to radiation varies inversely with the square of the distance, we can use the relationship

$$\frac{\text{exposure at distance 1}}{\text{exposure at distance 2}} = \frac{(\text{distance 2})^2}{(\text{distance 1})^2}$$

Exposure at distance 1 is 20 mrem; distance 1 is 3 ft and distance 2 is 6 ft. Exposure at distance 2 is unknown, or x.

$$\frac{20 \text{ mrem}}{x} = \frac{(6 \text{ ft})^2}{(3 \text{ ft})^2}$$

$$\frac{20 \text{ mrem}}{x} = \frac{36 \text{ ft}^2}{9 \text{ ft}^2}$$

Canceling ft² and cross-multiplying, we have

$$36x = 180 \text{ mrem}$$

$$x = 5 \text{ mrem}$$

Distance 2 is now 10 ft, so

$$\frac{20 \text{ mrem}}{x} = \frac{(10 \text{ ft})^2}{(3 \text{ ft})^2}$$

$$\frac{20 \text{ mrem}}{x} = \frac{100 \text{ ft}^2}{9 \text{ ft}^2}$$

$$100x = 180 \text{ mrem}$$

$$x = 1.8 \text{ mrem}$$

Self-test

An x-ray source has a radiation level of 500 mrem at a distance of 1 foot. How far away do you need to be to reduce this exposure to 5 mrem?

Answer

10 feet

Radiologists and others working with radioactive material must wear protective gloves and handle the material with long tongs to avoid the material's directly contacting the body.

Another factor to consider is the length of time of the exposure. The shorter the time that any individual is exposed to radiation, the smaller the dose that will be absorbed.

Radioactive wastes must be disposed of in specially marked containers and should be handled with care.

Persons in regular contact with radioactive material or radiation must wear film badges or other types of detectors (see Section 13-6), and these must be checked periodically for exposure to radiation.

Radiation can strike the molecules within a cell. Since water is the most abundant cellular molecule, we will consider the effects of radiation on a water molecule. Radiation may knock an electron from a water molecule, or it may remove a hydrogen ion, as indicated in the following equations:

$$H_2O \xrightarrow{\text{radiation}} e^- + H_2O^+$$

$$H_2O \xrightarrow{\text{radiation}} H^+ + OH^-$$

In each case, a pair of ions, an ion pair, is produced (see page 218). That is, radiation causes ionization within the cells. Ionization within the cell can disrupt the chemical processes going on inside that cell. Ionization can alter DNA. As a result, some cells may die or fail to multiply, but other cells will continue to live and reproduce. Because of the alteration in DNA, there may be genetic changes (mutations) that can show up in future generations. Large amounts of radiation can also produce cataracts, sterility, and leukemia. Exposure to radiation lessens life expectancy. Rapidly dividing tissue is highly susceptible to radiation. For this reason, unnecessary irradiation of growing children and pregnant women should be avoided. However, since cancerous cells are also rapidly dividing, they too should be highly susceptible to radiation. Hence irradiation is the form of treatment for some cancers.

Exposure to large amounts of radiation can cause "radiation sickness." The symptoms are gastrointestinal disturbances (nausea, vomiting, diarrhea, general body weakness), a drop in red and white blood cell counts, loss of hair, extensive skin damage, and ulcerative sores that are difficult to heal. Extremely large doses of radiation (more than 5000 rad) are fatal.

Table 13-2 indicates the biologic effects of increasing amounts of radiation on rats, and Table 13-3 indicates the symptoms in humans.

Table 13-2	Effects of Radiation on Rats	
Number of Roentgens	Time Period	Results
10		Genetic changes
50		Shortened life span
100	30 days	Cataracts; leukemia
500	10-14 days	Destruction of spleen, sternum, bone marrow; disappearance of leukocytes and platelets; decrease in gamma globulin
1000	4-6 days	Failure of intestinal system; loss of intestinal mucosa; bacterial invasion; no antibodies present; no immunity
10,000	1-2 days	Severe body burns
100,000	1-2 hr	Destruction of central nervous system

Table 13-3	Symptoms in Humans Resulting From Exposure to Acute Whole-Body Radiation
Dose (rad)	**Symptoms (for 50% of population)**
50-100	Decrease in circulating lymphocytes
120	Anorexia
170	Nausea
210	Vomiting
242	Diarrhea

The National Council on Radiation Protection and Measurement and the International Commission on Radiological Protection have set the following radiation standards:

1. A dose not exceeding 0.5 rem (500 mrem) per year of whole-body exposure for individual members of the general population
2. An average dose to the general population not exceeding 0.17 rem (170 mrem) per year of whole-body radiation
3. A dose not exceeding 5 rem per year, whole-body exposure for radiation workers

It is also recommended that exposures to radiation be kept as low as possible since the results are cumulative.

13-16 Sources of Radiation

The body receives radiation externally from three principal sources: natural background radiation, medical radiation, and radioactive wastes. Background radiation comes from space and from radioactive material present in the soil, in the air, in water, and in building materials. The average natural background radiation exposure received by each individual in the United States is 120 to 150 mrem per year.

The amount of diagnostic medical radiation varies with the type and frequency of medical treatment. The average in the United States is 70 mrem per year. However, when x-ray films are made of specific parts of the body that body part may receive a very high dose of radiation. Table 13-4 indicates the average exposure of radiation received during selected types of medical treatment.

Table 13-4	Amount of Radiation Received During Various Types of Medical Treatment
Type of Treatment	**Average Exposure (in mrem)**
Chest x-ray	27
Abdominal x-ray	620
Diagnostic upper GI series	1970
Dental x-ray	910
^{131}I thyroid scan	10,000-20,000 (to thyroid)
^{99m}Tc thyroid scan	1500 (to thyroid)
Radium implant treatment	3000-8000

13-17 Irradiation of Food

Irradiation consists of exposing food to some form of ionizing radiation, such as gamma rays or x rays, to kill insects and microorganisms and also to halt the ripening of fruit. ^{60}Co is most commonly used for this purpose. Irradiation lengthens the shelf life of the food and reduces the need for preservatives, some of which have toxic effects. Table 13-5 gives FDA-allowed limits for irradiation of foods.

13-18 Nuclear Energy

Some people believe that nuclear power is the answer to the world's energy problems; others say that it is too dangerous to use and should be outlawed. What is nuclear power? How can the nucleus of an atom produce such tremendous amounts of energy? We shall discuss these topics under two general headings: nuclear fission and nuclear fusion.

Nuclear Fission

When bombarded with neutrons, the nuclei of several heavy elements split into smaller pieces. This process is call *fission. ^{235}U can be split by a neutron into smaller pieces, such as strontium and xenon, accompanied by the release of more neutrons and a tremendous amount of energy. When this reaction takes place, the sum of the masses of the products is less than the masses of the reactants. That is, nuclear reactions do not obey the law of conservation of mass. They do, however obey the combined law of conservation of mass and energy. The amount of mass that disappears is converted into an equivalent amount of energy. This amount of energy can be calculated by using Einstein's equation

$$E = mc^2$$

where E is the amount of energy, m is the loss in mass, and c is a constant, the speed of light. When m is in kilograms and c is in meters per second, E is in joules.

Consider the fission of ^{235}U, written in equation form, with masses of reactants and products also being indicated.

$$^{1}_{0}n \; + \; ^{235}_{92}U \longrightarrow \; ^{94}_{38}Sr \; + \; ^{139}_{54}Xe \; + \; 3\,^{1}_{0}n \; + \; \text{energy}$$
$$\text{(kg) } 1.0087 \quad 234.9934 \qquad\qquad 93.9154 \qquad 138.9179 \qquad 3(1.0087)$$

Total mass of reactants	236.0021 Kg
Total mass of products	235.8594 Kg
Loss in mass	0.1427 Kg

Using Einstein's equation, $E = mc^2$, we find

$$E = 0.1427 \text{ kg} \times (3.00 \times 10^8 \text{ m/s})^2$$

$$= 1.28 \times 10^{16} \text{ joules}$$

This is equivalent to 3 million million Cal, a tremendous amount of energy.

Chain Reactions Note in the above reaction that when a neutron strikes a uranium nucleus, three more neutrons are produced. If each of these three neutrons strikes another uranium nucleus, then nine more neutrons will be produced. This process keeps on building up as more and more uranium nuclei undergo fission. Such a reaction, called a *chain reaction, can sustain itself, producing more and

Table 13-5
Radiation Limits for Foods Set by FDA

Food	Dose Limit kGy)*
Fruits and vegetables	1
Dehydrated herbs, seeds	30
Pork	0.3-1
White potatoes	1
Wheat, wheat flour	1

*1 kilogray = 100,000 rad.

more energy (Figure 13-11). However, a minimum amount of uranium nuclei (called the critical mass) must be present for a chain reaction to occur.

In an atomic bomb, the chain reaction is uncontrolled, producing a tremendous explosion. If the chain reaction is controlled, large amounts of useful energy are available. Devices that control nuclear chain reactions are called reactors.

Nuclear Reactors A nuclear reactor uses a fissionable isotope such as ^{235}U or ^{239}Pu as its fuel. Control rods made of cadmium or boron steel—both of which absorb neutrons—are used to control the reaction. When inserted into the reactor, these rods absorb neutrons and slow down the rate of the reaction. When pulled partway out, they allow the rate of the reaction to increase by absorbing fewer neutrons. Surrounding the fissionable material is a moderator such as water or graphite, which slows down the neutrons produced during fission.

The energy produced during fission is primarily in the form of heat. A heat transfer system removes heat from the core of the reactor and transfers this heat to a steam-generating unit.

Naturally, because of the intense radiation in the reactor, there must be sufficient shielding to protect personnel from radiation. Also, safety equipment is designed to shut down the reactor in case of a malfunction or radiation leakage.

Nuclear Fusion

The sun and other stars produce their energy by a process called *fusion. In this process, small nuclei are combined to form larger ones. In the sun, the overall reaction is the combination of hydrogen to form helium, with the subsequent release of incredible amounts of energy. Note that fusion is the opposite of fission.

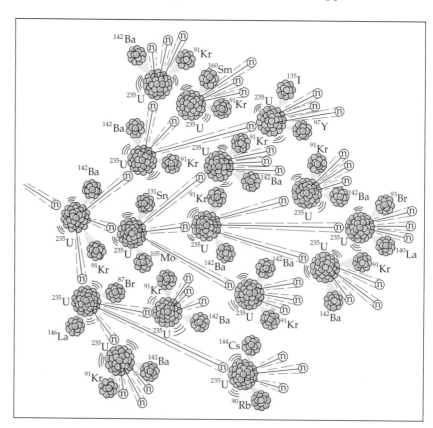

FIGURE 13-11
Chain reaction.

Fusion is the combination of small nuclei to make larger ones; fission is the splitting of a larger nucleus into smaller ones. It has been said that fusion is the energy source of the future. Fusion produces more energy than a comparable fission reaction, and with a much smaller amount of radioactive by-products to be disposed of. The fuel for a fusion reaction is deuterium, an isotope of hydrogen, which comes from seawater and is relatively plentiful. What has prevented the production of fusion reactors?

The answer lies in the fact that in order for a fusion reaction to work, temperatures of hundreds of millions of degrees Celsius must be produced. Also, suitable containers for the fusion reaction have not been designed in a workable form.

When a fusion reaction is produced in the laboratory, it will probably be the following:

$$\underset{\text{isotopes of}\atop\text{hydrogen}}{{}^2_1\text{H} + {}^3_1\text{H}} \longrightarrow {}^4_2\text{He} + {}^1_0\text{n} + \text{energy}$$

The explosion of a hydrogen bomb is an example of an uncontrolled fusion reaction.

13-19 Disposal of Radioactive Waste

Waste disposal in the United States has become a problem because of the diminishing amount of landfill available. A larger problem is the disposal of radioactive materials, not only from nuclear power plants and nuclear weapons production but also from medical applications.

From the beginning of the nuclear age until 1982, there was no permanent method of disposal of radioactive waste. It was all placed in temporary storage. Much of the waste will remain very radioactive for 20,000 years or longer. The potential problems of genetic mutations and cancer resulting from radioactive material make the safe permanent disposal of such substances a matter of immediate concern.

The Nuclear Waste Policy Act of 1982 mandated the choosing and preparing of deep underground disposal sites for nuclear waste materials. Of great importance in the site selection are the stability of the rock formation in which the waste is to be placed; the absence of volcanic or earthquake activity; and the presence of rocks impermeable to water. With proper disposal, the amount of radioactivity will gradually diminish with little or no effect on the environment.

Summary

Radioactivity is the property of emitting radiation from the nucleus of an atom. The three types of radiation are alpha, beta, and gamma. Alpha particles are positively charged helium nuclei. Beta particles are high-speed electrons and are negatively charged. Gamma rays are a high-energy form of electromagnetic radiation and have no charge or mass.

In nuclear reactions both the sum of the atomic numbers and the sum of the mass numbers are the same on both sides of an equation. In addition to naturally occurring radioactive substances, artificially radioactive substances can be prepared by bombardment with particles such as protons, neutrons, and alpha particles.

A scanner helps to locate malignancies by moving back and forth across the site being studied and detecting the radiation in each area over which it travels. The radiation comes from radioisotopes administered to the patient, which are selected to accumulate at the desired body part.

The half-life of a radioisotope is the amount of time required for half of its atoms to decay. For medical work, a radioisotope must have a half-life long enough to give the body part the radiation it needs and short enough that it will not give the patient too much radiation during the period it remains in the body.

Radioisotopes are used in the diagnosis and treatment of various disorders: ^{131}I for thyroid conditions, ^{32}P for eye tumors, ^{99m}Tc for scans, and ^{60}Co for radiation therapy.

X rays are a type of radiation similar to gamma rays. The penetrating powers of x rays can be controlled, whereas those of gamma rays cannot. x-ray treatment combined with high-pressure oxygen has had some success in the treatment of cancer.

The units of radiation are the curie, the gray, the roentgen, the rad, and the rem. The roentgen applies primarily to x rays. The rem is the unit most commonly used in relation to the body.

Radiation produces ionization within the cells, causing some type of damage. Small amounts of radiation produce genetic changes; larger amounts cause a shortened life span; very large amounts of radiation can cause death within a short period of time. X rays are harmful to the body because of the effects produced by the radiation, so care should be taken to avoid unnecessary exposure.

Nuclear fission involves the splitting of a large nucleus into smaller ones with large amounts of energy being produced. Nuclear fusion involves the combining of small nuclei into larger ones, with the release of tremendous amounts of energy.

The disposal of radioactive waste is an ongoing problem, with the effect upon the environment being an item of top priority.

Questions and Problems

1. Are alpha particles harmful if they strike the body from external sources? from internal sources? Explain.
2. Are beta particles harmful if they strike the body from external sources? from internal sources? Explain.
3. Are gamma rays harmful if they strike the body from external sources? from internal sources? Explain.
4. What are alpha particles? beta particles? gamma rays?
5. Balance the following equations:
   a. $^{218}_{84}Po \longrightarrow \alpha + ?$
   b. $^{7}_{3}Li + ^{2}_{1}H \longrightarrow ^{8}_{4}Be + ?$
   c. $^{14}_{7}N + \alpha \longrightarrow ^{1}_{1}H + ?$
   d. $^{10}_{5}B + n \longrightarrow \alpha + ?$
6. Compare natural and artifical radioactivity.
7. Define the term *half-life*.
8. What are radioisotopes? Give one use for each of the following:
   a. ^{131}I b. ^{14}C c. ^{75}Se d. ^{197}Hg
9. For medical usage, should radioisotopes have a long or short half-life? Explain.
10. What is a roentgen? a rad? a rem?
11. What is meant by the term *RBE*? How is it defined?
12. If 1.00 mg of ^{51}Cr is present today, how much will be present approximately 1 month from now? 6 months from now?
13. Describe the use of a scanner. Name two radioisotopes used in scans and indicate where they can be used.
14. What is PET and where is it used?
15. Explain how ^{10}B can be used in the treatment of cancer.

16. How can radiation be detected?
17. What are the units of radiation?
18. What are x rays used for?
19. What is xeroradiography? What is it used for?
20. From what source does the body receive external radiation?
21. How can the body receive internal radiation?
22. List some of the physiologic effects of radiation.
23. What can be done to minimize the effects of radiation on the body?
24. Compare the criteria for selection of radioisotopes for diagnostic and therapeutic use.
25. What changes take place in the atomic number when an atom emits an alpha particle? a beta particle? a gamma ray?
26. What changes take place in the mass number when an atom emits an alpha particle? a beta particle? a gamma ray?
27. What is an MRI? Explain its use.
28. What is SQUID? SPECT?
29. What effect does irradiation have on food?
30. What is a chain reaction?
31. Compare nuclear fission with nuclear fusion.
32. If a nurse standing 4 feet from a patient with a radium implant receives an exposure of 40 mrem, what would be the exposure level 12 feet away?
33. How does radiation produce ion pairs? What effects do ion pairs have on cells?

34. What is radioimmunoassay? What is it used for?

35. What is a scintillation counter?

36. Where does the energy of nuclear fission come from? How can the energy of this type of reaction be calculated?

37. What does the "m" in ^{99m}Tc stand for?

38. Sodium-24 in the form of NaCl is given by injection to measure sodium balance in a patient. If 80 mg of sodium-24 is given, how much will remain after 2.5 days (60 hours)?

39. The carbon-14 radioactivity from wood found in an ancient fireplace is 12.5 percent of a current reference sample. What is the age of the sample? The half-life of carbon-14 is 5700 years.

Practice Test

1. The charge on a beta particle is _____.
   a. +1   b. −1   c. +2   d. −2

2. The type of radiation carrying a positive charge is the _____.
   a. gamma ray   b. beta particle
   c. alpha particle   d. none of these

3. The most penetrating type of external radiation is _____.
   a. alpha   b. beta   c. gamma   d. delta

4. In the reaction
$$^{234}_{92}U \longrightarrow ^{0}_{-1}e + X$$
   the atomic number of element X is _____.
   a. 89   b. 91   c. 92   d. 93

5. The mass number of element X in the preceding question is _____.
   a. 230   b. 232   c. 233   d. 234

6. An element has a half-life of 2 weeks. If 100 mg is present today, how many milligrams will be present 6 weeks from now?
   a. 50   b. 33   c. 25   d. 12.5

7. The unit of radiation relating directly to human tissue is the _____.
   a. rem   b. rad   c. roentgen   d. curie

8. Radiation
   a. may cause nausea
   b. affects rapidly dividing cells
   c. may decrease life expectancy
   d. all of these

9. If a person receives a certain amount of radiation from a source at a distance of 8 feet, what fraction of that radiation will that person receive at a distance of 16 feet?
   a. 3/4   b. 1/2   c. 1/4   d. 0

10. A device used to detect radiation is the _____.
    a. film badge   b. Geiger counter
    c. scintillation counter   d. all of these

14

Introduction to Organic Chemistry

Objectives

- To recognize the importance of organic compounds
- To compare properties of organic and inorganic compounds
- To become familiar with the method of indicating covalent bonding in organic compounds
- To understand and be able to draw structural formulas
- To draw structures of isomers
- To visualize the three-dimensional arrangement of bonds around the carbon atom

The molecules of life encompass a wide range of organic molecules. Indeed, life is based on them. A short segment of DNA, the genetic component of living cells, is illustrated in this figure. DNA, itself an organic molecule, is made up of a repeating series of smaller organic molecule arranged in a double-helical pattern.

In the eighteenth century it was believed that a "vital force" was needed to make the compounds produced by living cells, which were classified as organic compounds. However, this belief was overthrown by a German chemist, Friedrich Wöhler, in 1828. He prepared urea, a compound normally found in the blood and urine, by heating a solution of ammonium cyanate, an inorganic compound.

$$NH_4CNO \xrightarrow{\text{heat}} NH_2 - \overset{\displaystyle O}{\overset{\displaystyle \|}{C}} - NH_2$$

ammonium cyanate   urea

After Wöhler's work many other organic compounds were produced in the laboratory. This led to the subdivision of chemistry into two parts—inorganic and organic. Organic chemistry was defined as the chemistry of carbon compounds. Why have one category for the element carbon and place all other elements in the other category? Carbon is unique in that it forms covalent bonds to other carbon atoms as well as to other elements. Although there are tens of thousands of inorganic compounds known today, millions of organic compounds are known.

14-1 Importance of Organic Chemistry

Organic chemistry touches our lives in countless ways. It is the chemistry associated with all living matter in both plants and animals. Carbohydrates, fats, proteins, vitamins, hormones, enzymes, and many drugs are organic compounds. Wool, silk, cotton, linen, and such synthetic fibers as nylon, rayon, and Dacron contain organic compounds. So do perfumes, dyes, flavors, soaps, detergents, plastics, gasolines, and oils.

Comparison of Organic and Inorganic Compounds

Most organic compounds differ from inorganic compounds in many ways, as shown in Table 14-1.

Table 14-1 Comparison of Properties of Most Organic and Inorganic Compounds		
Property	Organic	Inorganic
Flammable	Yes	No
Melting point	Low	High
Boiling point	Low	High
Solubility in water	No (for most)	Yes
Solubility in nonpolar liquids	Yes	No
Types of bonding	Covalent	Ionic
Reactions occur between	Molecules	Ions
Atoms per molecule	Many	Few
Structure	Complex	Simpler
Electrolyte	No	Yes

14-2 Bonding

Organic compounds—compounds of carbon—are held together by covalent bonds. Recall that covalent bonds are formed by sharing electrons. In organic chemistry the term bond is used to designate a shared pair of electrons. Carbon has four electrons, $\cdot\overset{\cdot}{\underset{\cdot}{C}}\cdot$; this means that carbon can form a maximum of four covalent bonds. Bonds are usually represented by a short, straight line connecting the atoms, with each bond representing a shared pair of electrons.

Each carbon atom in the following compounds forms four bonds.

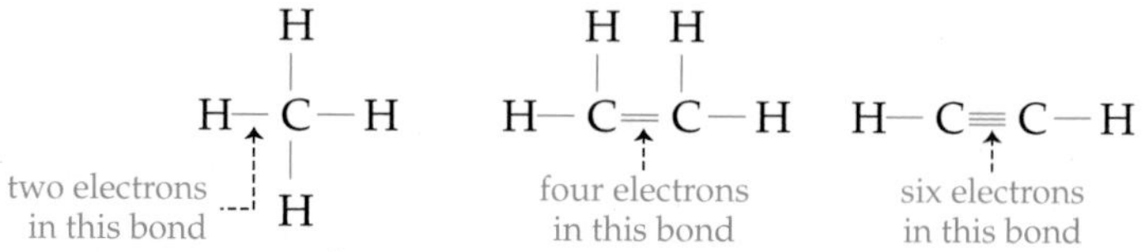

two electrons in this bond

four single bonds to carbon

four electrons in this bond

double bond between carbons

six electrons in this bond

triple bond between carbons

Likewise, since the oxygen atom has an oxidation number of −2, it forms two covalent bonds, as shown in the following compounds.

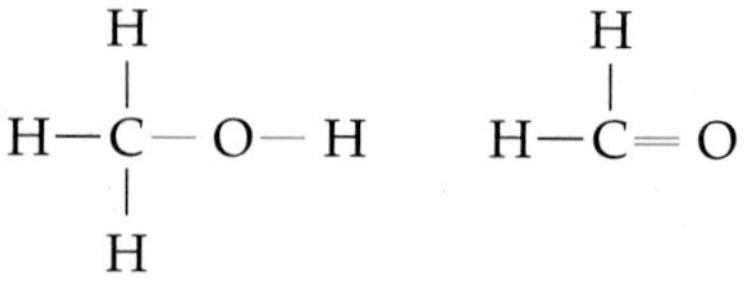

two single bonds to oxygen

double bond to oxygen

Nitrogen, with an oxidation number of −3, forms three covalent bonds.

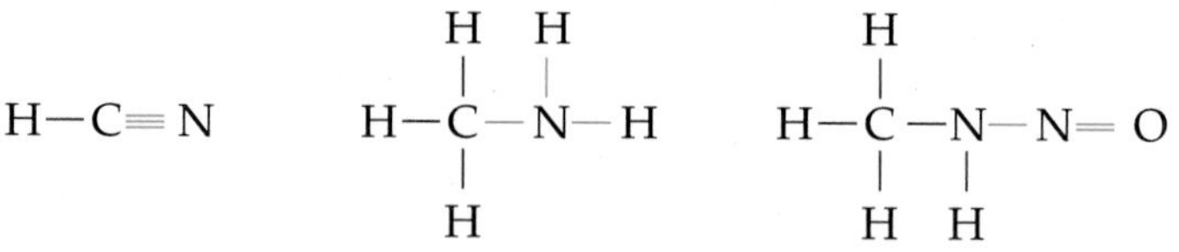

triple bond to nitrogen

three single bonds to nitrogen

one single and one double bond to nitrogen

Note that in all the preceding examples, hydrogen, with an oxidation number of +1, forms only one covalent bond. The halogens (fluorine, chlorine, bromine, and iodine), all with an oxidation number of −1, also form only one bond. Recall that metals lose electrons to form positive ions and nonmetals gain electrons to form negative ions.

14-3 Structural Formulas

As will be seen later, organic compounds are often written using a structural rather than a molecular formula. What is the difference between a structural formula and a molecular formula? And why use the former and not the latter?

A structural formula shows the exact way in which the atoms are connected to each other, but a molecular formula does not. You cannot tell the structure of a molecule from its molecular formula.

If the carbons, hydrogens, and oxygens in C_2H_6O are arranged in such a manner that each carbon atom has four bonds attached to it, each hydrogen atom has one bond, and the oxygen has two bonds, there are two possible structures, both having the formula C_2H_6O:

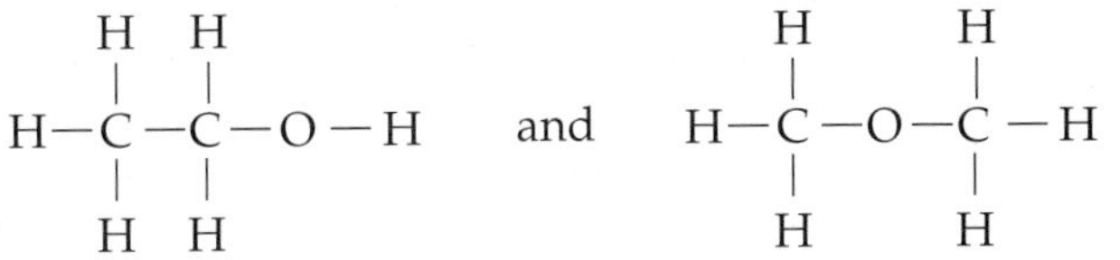

$$
\begin{matrix}
& H & H & & & & H & & H \\
& | & | & & & & | & & | \\
H- & C- & C- & O-H & \text{and} & H- & C- & O- & C-H \\
& | & | & & & & | & & | \\
& H & H & & & & H & & H
\end{matrix}
\tag{14-1}
$$

The two compounds shown above have different structures and different properties and are actually different compounds. They are called isomers. They have the same molecular formulas but different connectivity of the atoms or groups. The first compound is called ethyl alcohol, the second dimethyl ether. This difference in the structure of compounds having the same molecular formula illustrates the importance of using structural rather than molecular formulas for organic compounds.

In diagram (14-1) the lines represent bonds or shared electrons. Using electron dots, these structures are

$$
\begin{matrix}
& H & H & & & & H & & H \\
& \cdot\cdot & \cdot\cdot & & & & \cdot\cdot & & \cdot\cdot \\
H: & C: & C: & O:H & \text{and} & H: & C: & O: & C:H \\
& \cdot\cdot & \cdot\cdot & & & & \cdot\cdot & & \cdot\cdot \\
& H & H & & & & H & & H
\end{matrix}
$$

It is simpler to write a structural formula using the bond notation.

Exercise 14-1

Draw the structural formulas for C_5H_{12}.

Solution

There are three possibilities for the arrangement of five carbon atoms.

1. Five carbons in a row:

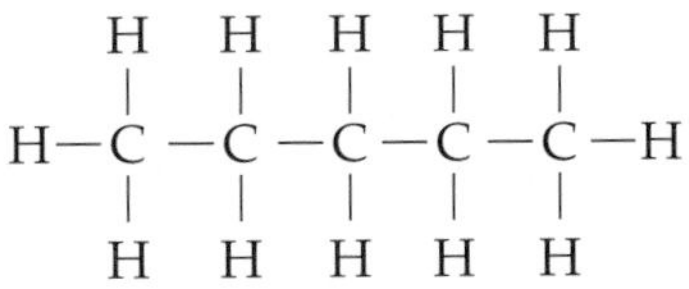

$$
\begin{matrix}
H & H & H & H & H \\
| & | & | & | & | \\
H-C- & C- & C- & C- & C-H \\
| & | & | & | & | \\
H & H & H & H & H
\end{matrix}
$$

2. Four carbons in a row, with the fifth carbon attached to one of the middle carbons:

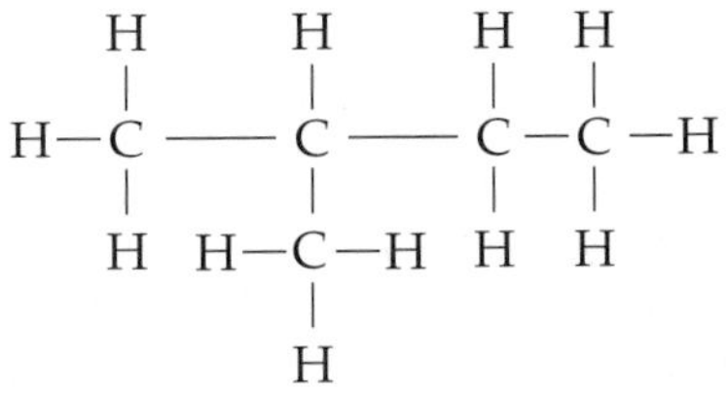

$$
\begin{matrix}
H & & H & & H & H \\
| & & | & & | & | \\
H-C & — & C & — & C-C-H \\
| & & | & & | & | \\
H & & H-C-H & & H & H \\
& & | & & & \\
& & H & & &
\end{matrix}
$$

3. Three carbons in a row, with the other two carbons attached to the middle carbon:

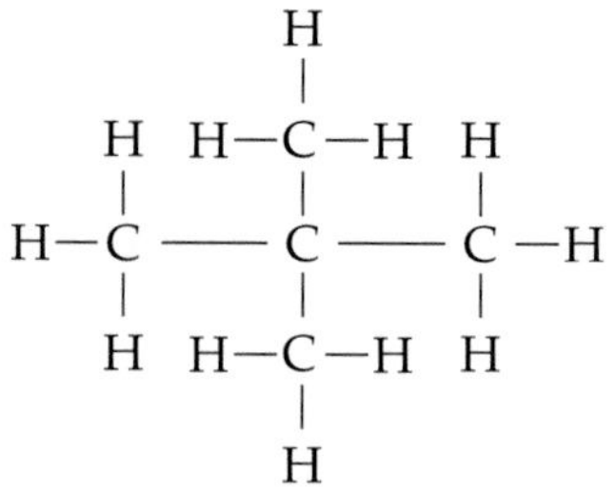

$$
\begin{array}{ccccc}
 & & H & & \\
 & & | & & \\
 & H & H-C-H & H & \\
 & | & | & | & \\
H-C & - & C & - & C-H \\
 & | & | & | & \\
 & H & H-C-H & H & \\
 & & | & & \\
 & & H & &
\end{array}
$$

Self-test

Draw the structural formulas for C_3H_8O.

Answer

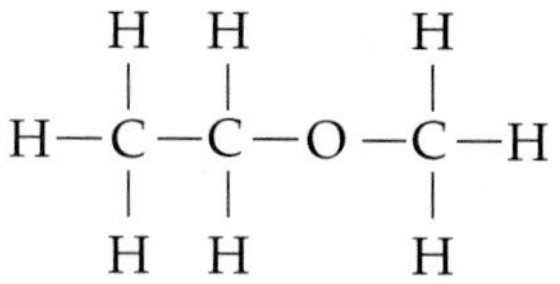

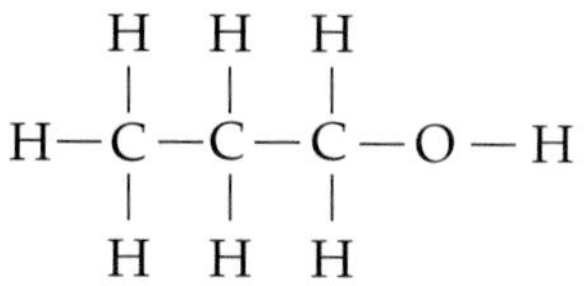

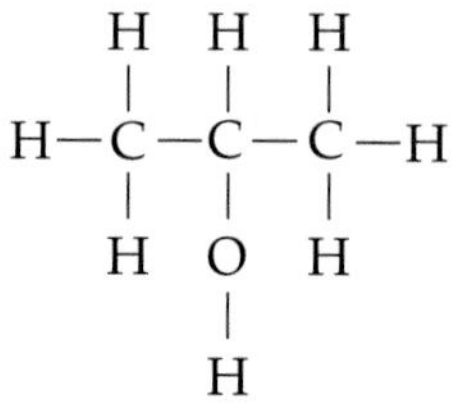

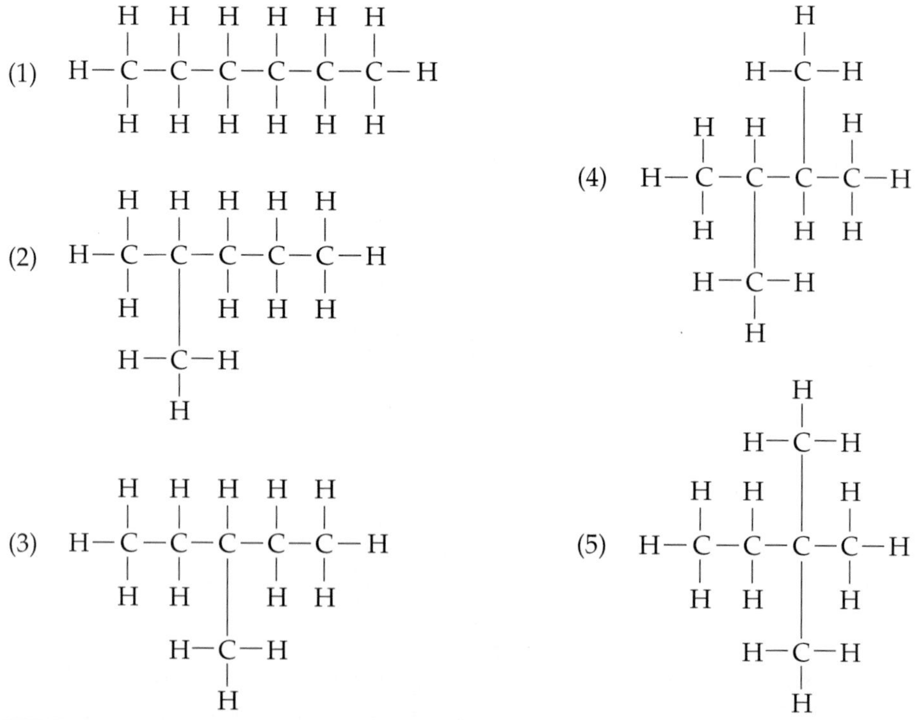

$$
\begin{array}{ccccccc}
H & H & & H & \\
| & | & & | & \\
H-C-C-O & -C-H & \\
| & | & & | & \\
H & H & & H &
\end{array}
\qquad
\begin{array}{cccc}
H & H & H & \\
| & | & | & \\
H-C-C-C-O-H \\
| & | & | & \\
H & H & H &
\end{array}
$$

$$
\begin{array}{ccc}
H & H & H \\
| & | & | \\
H-C-C-C-H \\
| & | & | \\
H & O & H \\
 & | & \\
 & H &
\end{array}
$$

---

(1)
$$
\begin{array}{cccccc}
H & H & H & H & H & H \\
| & | & | & | & | & | \\
H-C-C-C-C-C-C-H \\
| & | & | & | & | & | \\
H & H & H & H & H & H
\end{array}
$$

(2)
$$
\begin{array}{ccccc}
H & H & H & H & H \\
| & | & | & | & | \\
H-C-C-C-C-C-H \\
| & | & | & | & | \\
H & | & H & H & H \\
 & H-C-H & & & \\
 & | & & & \\
 & H & & &
\end{array}
$$

(3)
$$
\begin{array}{ccccc}
H & H & H & H & H \\
| & | & | & | & | \\
H-C-C-C-C-C-H \\
| & | & | & | & | \\
H & H & | & H & H \\
 & & H-C-H & & \\
 & & | & & \\
 & & H & &
\end{array}
$$

(4)
$$
\begin{array}{cccc}
 & & H & \\
 & & | & \\
 & & H-C-H & \\
 & H & H & | & H \\
 & | & | & | & | \\
H-C-C-C-C-H \\
 & | & | & | & | \\
 & H & | & H & H \\
 & & H-C-H & \\
 & & | & \\
 & & H &
\end{array}
$$

(5)
$$
\begin{array}{cccc}
 & & H & \\
 & & | & \\
 & & H-C-H & \\
 & H & H & | & H \\
 & | & | & | & | \\
H-C-C-C-C-H \\
 & | & | & | & | \\
 & H & H & | & H \\
 & & & H-C-H & \\
 & & & | & \\
 & & & H &
\end{array}
$$

FIGURE 14-1
Isomers of C_6H_{14}.

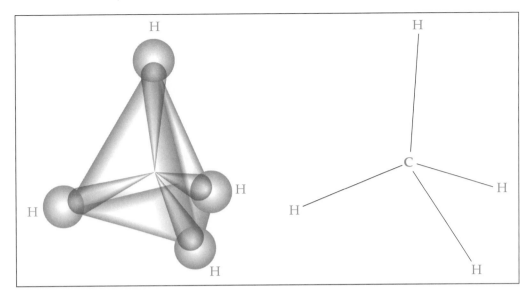

FIGURE 14-2
Tetrahedral structure methane, CH_4.

14-4 Isomers

Isomers are defined as compounds having the same molecular formula but different arrangement of atoms. Thus, the compound C_2H_6O has two isomers. The compound C_6H_{14} has five isomers, as illustrated in Figure 14-1. $C_6H_{12}O_6$ is usually called glucose, but this molecular formula actually represents 16 different compounds, or isomers, each of which is a different sugar (see page 337).

14-5 Three-Dimensional Arrangement of the Bonds in the Carbon Atom

It should be realized that the compounds represented by structural formulas are three-dimensional and not planar. For example, each carbon atom has four bonds attached to it. If these bonds are symmetrically arranged, the planar representation of the structure is

$$-\overset{\displaystyle |}{\underset{\displaystyle |}{C}}-$$

What does it actually look like in three dimensions? The four bonds of the carbon atom are arranged in a *tetrahedral shape. The carbon atom lies at the center of the tetrahedron; the angle between the bonds is 109.5 degrees. Therefore the compound CH_4, methane, can be represented as shown in Figure 14-2.

14-6 Bonding Ability of Carbon

What is unique about the element carbon that it forms so many compounds? The other elements form relatively few compounds. The answer is that the carbon atom has the ability to bond other carbon atoms to itself to form very large and complex molecules. Carbon atoms can join together to form continuous or branched chains of carbon atoms. Compounds of this type are called aliphatic compounds (Figure 14-3, a). Carbon compounds can also bond together in the shape of rings to form cyclic (also called aromatic) compounds (Figure 14-3, b). A third type of organic

FIGURE 14-3
Types of carbon compounds.

compounds are the heterocyclic compounds, which also have a ring structure. However, this ring structure contains some element other than carbon in the ring (Figure 14-3, c). In this particular case the other element is nitrogen.

14-7 Carbon Compounds with Other Elements

Carbon also forms compounds with other elements besides hydrogen and oxygen. If a chlorine atom is bonded to a carbon atom, the compound chloromethane or methyl chloride is formed.

$$H-\overset{\displaystyle H}{\underset{\displaystyle H}{\overset{|}{\underset{|}{C}}}}-Cl$$

Examples of carbon bonded to oxygen and nitrogen appear earlier in this chapter.

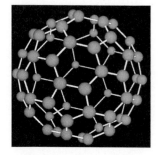

Fullerene.

14-8 Allotropic Forms of Carbon

Carbon has long been known in two allotropic forms. Graphite is a soft, dark black solid with good electrical conduction. Diamond is formed over long geologic time periods, when graphite is subjected to extreme underground pressures. Diamond is clear and is the hardest natural substance known. Recently a third allotrope of carbon has been discovered. Fullerenes, the most common being C_{60}, have shapes similar to soccer balls. Scientists are just beginning to investigate the many potential uses for this substance. The Nobel Prize in chemistry for 1996 was awarded for the discovery of fullerene.

Soccer ball.

Summary

Organic chemistry is defined as the chemistry of carbon compounds. Most organic compounds differ from inorganic compounds as follows: they are combustible; they have lower melting points; they are insoluble in water; reaction takes place between molecules rather than between ions; the molecules contain many atoms; the molecules have a complex structure.

In organic chemistry the term bond is used rather than oxidation number. A bond is indicated by a short line and represents a pair of shared electrons. The carbon atom always has four bonds associated with it; the oxygen atom, two; the hydrogen atom, one.

Structural formulas are used for organic compounds rather than molecular formulas because the same molecular formula can often represent more than one structural formula.

Isomers are compounds having the same molecular formula but different arrangement of atoms.

The four single bonds of the carbon atom are arranged in three dimensions in a tetrahedral structure with the carbon atom at the center and each bond pointing to a corner.

Organic compounds can be divided into three categories: aliphatic compounds, aromatic compounds, and heterocyclic compounds. Heterocyclic compounds contain elements other than carbon in the ring.

Carbon also forms compounds with other elements besides hydrogen and oxygen.

Questions and Problems

1. What are the three types of organic compounds?
2. Compare the terms *bond* and *covalent bond*.
3. Compare the properties of organic and inorganic compounds in terms of solubility in water, types of bonding, types of structure, flammability, boiling and melting points, and numbers of atoms in the molecules.
4. Draw the structures of the three isomers of C_5H_{12}.
5. Compare structural and molecular formulas. Why are the former used in organic chemistry?

6. Do all organic compounds have isomers? Explain.
7. How are the bonds arranged around the central carbon atom in methane (CH_4)?
8. How many bonds should be attached to hydrogen; sulfur; nitrogen; oxygen; carbon?
9. Is it possible for a heterocyclic compound to contain hydrogen in the ring? Explain.
10. Phosphorus is in group Va of the periodic chart. How many bonds should be attached to the phosphorus atom?

11. Decide whether the following compounds are aliphatic, cyclic, or heterocyclic.

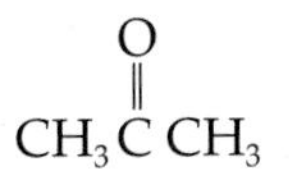

a. acetone

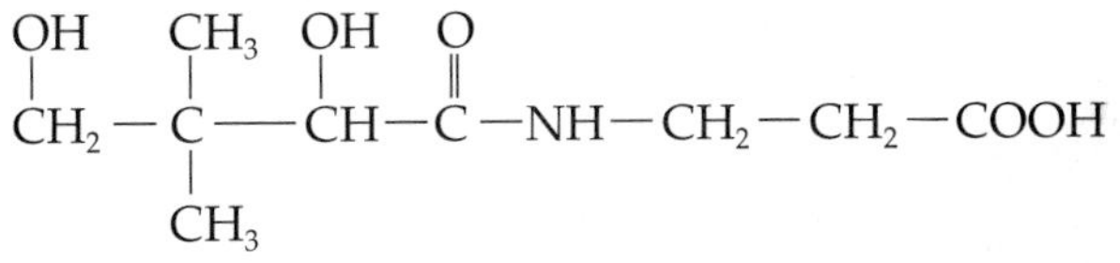

b. pantothenic acid

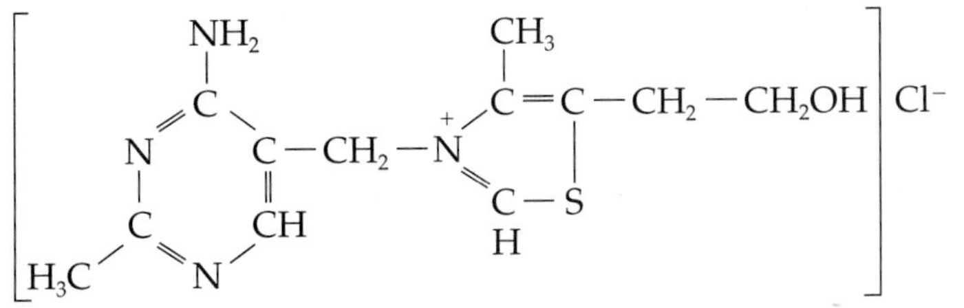

c. thiamine hydrochloride

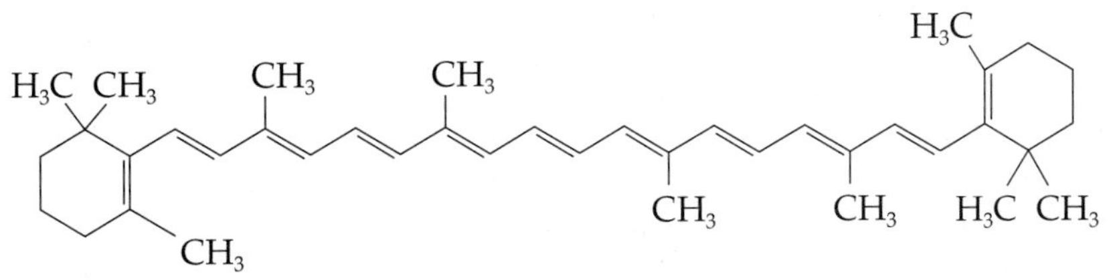

d. β-carotene

11. Decide whether the following compounds are aliphatic, cyclic, or heterocyclic—cont'd.

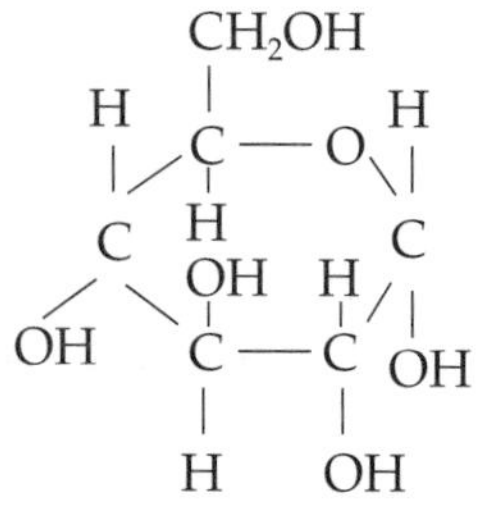

e. glucose

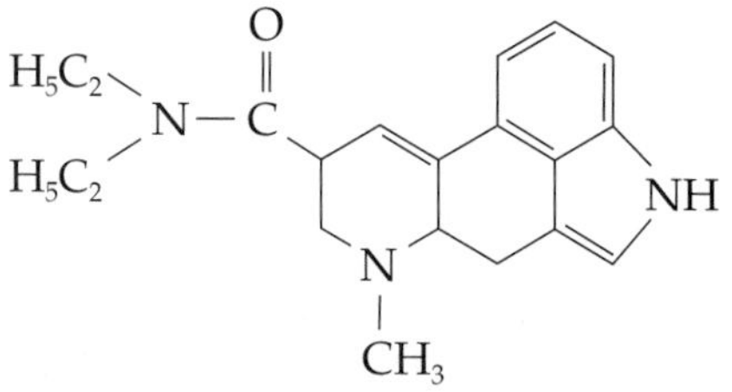

f. lysergic acid diethylamide (LSD)

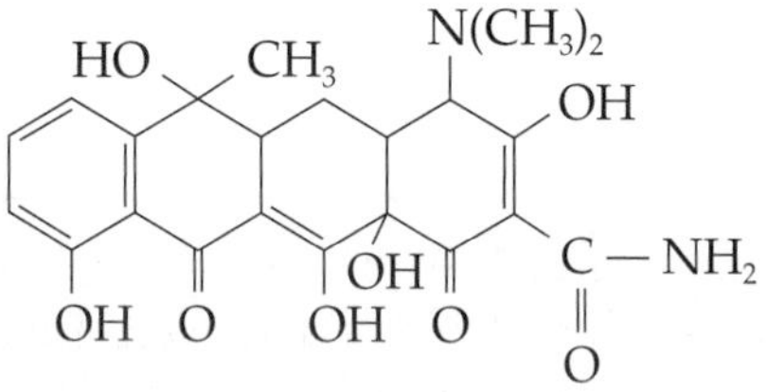

g. tetracycline

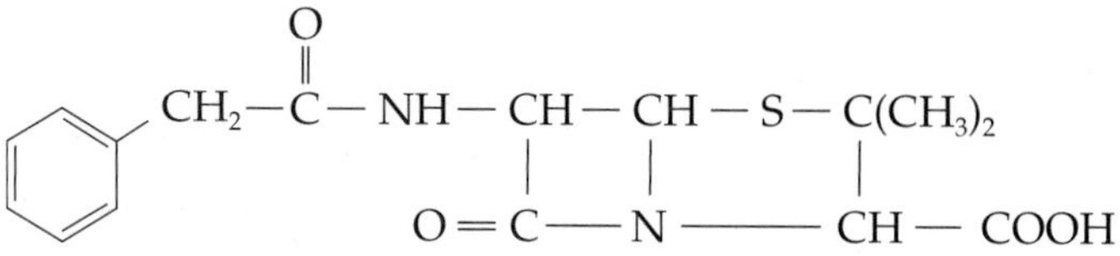

h. penicillin G

12. Vitamin E (tocopherol) contains all three types of organic compounds: aliphatic, aromatic, and heterocyclic. Label the regions in vitamin E that correspond to each.

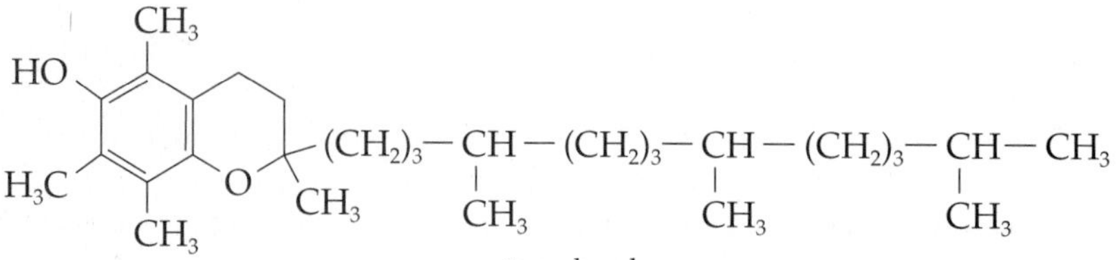

α-tocopherol

Practice Test

1. In what shape are the bonds around the central carbon atom in methane, CH_4, arranged?
   a. planar
   b. tetrahedral
   c. octahedral
   d. trigonal

2. Organic compounds consisting of straight chains of saturated carbon atoms are called _____.
   a. aliphatic
   b. aromatic
   c. isomers
   d. heterocyclic

3. Carbon can bond to _____.
   a. C
   b. O
   c. N
   d. All of these

4. Carbon atoms form how many bonds?
   a. 4   b. 3   c. 2   d. 1

5. Oxygen can form how many bonds?
   a. 4   b. 3   c. 2   d. 1

6. An organic bond represents _____.
   a. a transfer of electrons
   b. a pair of shared electrons
   c. a pair of shared protons
   d. a transfer of protons

7. Organic compounds having the same molecular formula but different structural formulas are called _____.
   a. polymers
   b. isomers
   c. heterocyclics
   d. carbohydrates

8. Organic compounds, when compared with inorganic compounds, have _____.
   a. lower melting points
   b. many more atoms
   c. a more complex structure
   d. all of these

9. Most organic compounds are _____.
   a. soluble in water
   b. nonflammable
   c. held together by covalent bonds
   d. all of these

10. An aromatic compound consists of a _____.
    a. ring of carbon atoms
    b. a chain of carbon atoms
    c. a compound containing chlorine
    d. none of these

Saturated Hydrocarbons: The Alkanes and Their Halogen Derivatives

Objectives

- To draw structures for and name some alkanes and cycloalkanes
- To understand the properties and reaction of alkanes
- To understand the sources, properties, and uses of hydrocarbons and their halogen derivatives

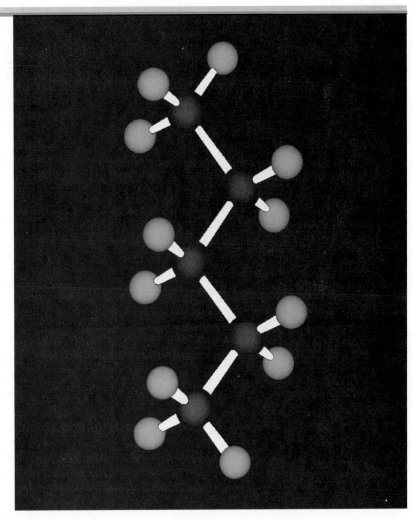

Hydrocarbons are found in a variety of places, from the fatty acids chains of lipids to synthetic polymers. This computer model illustrates the structure of a pentane hydrocarbon molecule.

15-1 Alkanes

As the name implies, hydrocarbons are compounds that contain carbon and hydrogen only.

Consider the hydrocarbon with only one carbon atom. Since a carbon atom must have four bonds, four hydrogen atoms can be attached to that carbon atom. The hydrocarbon thus formed is called methane and has the structural formula shown in Figure 15-1.

If two carbon atoms are bonded together, six hydrogen atoms can be joined to them. This hydrocarbon is called ethane. The molecular formula is C_2H_6, and the structure is shown in Figure 15-2.

The hydrocarbon of three carbon atoms needs eight hydrogen atoms to satisfy all the bonds. This compound, C_3H_8, is called propane.

These compounds are called alkanes, and they are said to be saturated; that is, they have single covalent bonds between carbon atoms.

Table 15-1 lists several hydrocarbons. Note that the names of all alkanes end in *-ane*. The names of the first four compounds are not systematic; they were named before a system of nomenclature was devised. Beginning with the hydrocarbon containing five carbon atoms, however, the names follow a definite pattern. The name *pentane* was derived from the prefix *penta-*, meaning five; pentane contains five carbon atoms. The name *hexane* is derived from the prefix *hexa-*, which means six; and so on through the series.

Structural formulas can be condensed as indicated in Table 15-1 and discussed in the next sections.

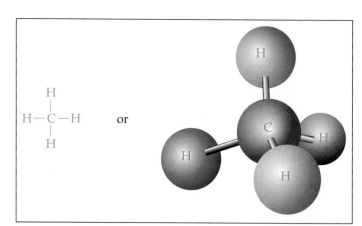

FIGURE 15-1
Methane.

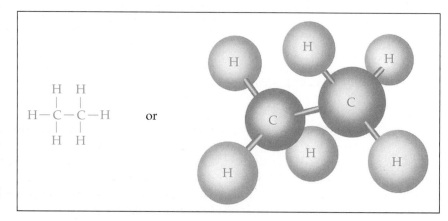

FIGURE 15-2
Ethane.

Table 15-1 Alkanes

Number of Carbon Atoms	Name	Molecular Formula	Structural Formula	Condensed Structural Formula
1	Methane	CH_4	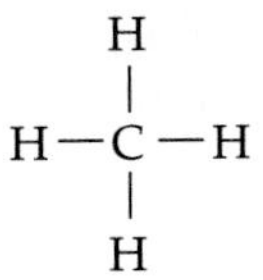 $$H-\overset{\displaystyle H}{\underset{\displaystyle H}{C}}-H$$	CH_4
2	Ethane	C_2H_6	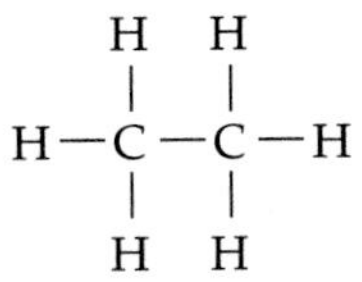 $$H-\overset{\displaystyle H}{\underset{\displaystyle H}{C}}-\overset{\displaystyle H}{\underset{\displaystyle H}{C}}-H$$	CH_3-CH_3 or CH_3CH_3
3	Propane	C_3H_8	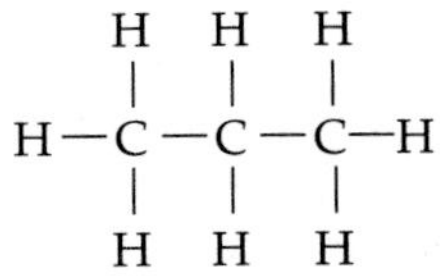 $$H-\overset{\displaystyle H}{\underset{\displaystyle H}{C}}-\overset{\displaystyle H}{\underset{\displaystyle H}{C}}-\overset{\displaystyle H}{\underset{\displaystyle H}{C}}-H$$	$CH_3-CH_2-CH_3$ or $CH_3CH_2CH_3$
4	Butane	C_4H_{10}	$$H-\overset{\displaystyle H}{\underset{\displaystyle H}{C}}-\overset{\displaystyle H}{\underset{\displaystyle H}{C}}-\overset{\displaystyle H}{\underset{\displaystyle H}{C}}-\overset{\displaystyle H}{\underset{\displaystyle H}{C}}-H$$ and $$H-\overset{\displaystyle H}{\underset{\displaystyle H}{C}}-\overset{\displaystyle H}{\underset{\displaystyle \,}{C}}-\overset{\displaystyle H}{\underset{\displaystyle H}{C}}-H$$ $$\underset{\displaystyle H}{|}$$ $$H-\overset{}{C}-H$$ $$\underset{\displaystyle H}{|}$$	$CH_3-CH_2-CH_2-CH_3$ or $CH_3(CH_2)_2CH_3$ and $CH_3-\underset{\displaystyle CH_3}{CH}-CH_3$ or $CH_3CH(CH_3)CH_3$
5	Pentane	C_5H_{12}	3 isomers[†]	
6	Hexane	C_6H_{14}	5 isomers	
7	Heptane	C_7H_{16}	9 isomers	
8	Octane	C_8H_{18}	18 isomers	
9	Nonane	C_9H_{20}	35 isomers	
10	Decane	$C_{10}H_{22}$	75 isomers	

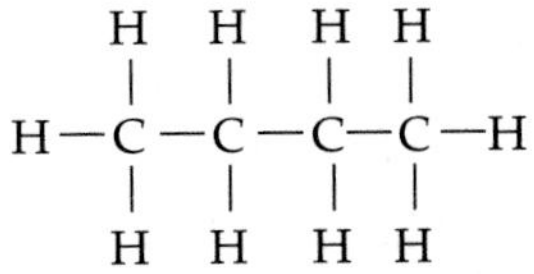

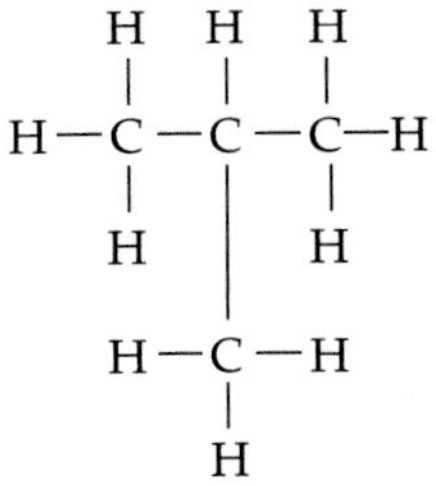

[†]*The number of isomers increases rapidly as the number of carbons in the compounds increases. $C_{40}H_{82}$ has 62,491,178,805,831 possible isomers.*

General Formula

The general formula for alkanes is

$$C_nH_{2n+2}$$

where n represents the number of carbon atoms. That is, if a compound contains n carbon atoms, the number of hydrogen atoms is twice n plus two more.

Butane contains four carbon atoms. According to the formula the number of hydrogen atoms should be $(2 \times 4) + 2$ or 10, which agrees with the molecular formula for butane listed in Table 15-1. Likewise, octane has eight carbon atoms and so should have $(2 \times 8) + 2$, or 18, hydrogen atoms, which it does.

Continuing with a larger number of carbon atoms, the alkane containing 16 carbon atoms would have $(2 \times 16) + 2$, or 34, hydrogen atoms, giving it the formula $C_{16}H_{34}$.

Carbon atoms connected by single bonds exhibit *free rotation* about those bonds. That is, the carbon atoms can rotate with respect to each other. Thus the molecular formula C_3H_8, which represents a continuous chain of three carbon atoms, can be diagrammed in several ways, such as

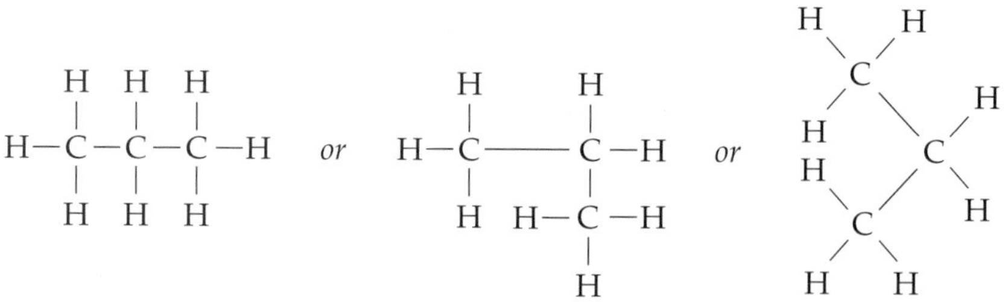

However, they all represent the same compound, which consists of three carbon atoms connected in one continuous chain.

In larger molecules, bonding between carbon atoms usually produces twisted shapes because of molecular interactions. However, for simplicity we will consider such compounds as having straight chains.

In the case of butane, C_4H_{10}, we can draw several models for four carbon atoms connected continuously:

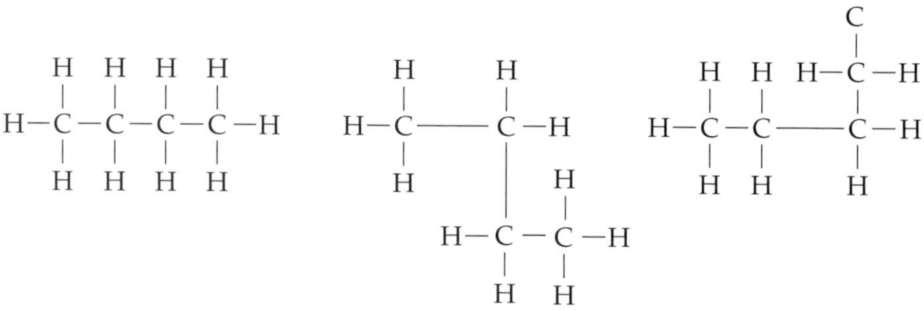

However, because of free rotation of the carbon-to-carbon bonds, they all represent the same compound, consisting of four continuously connected carbon atoms.

But the structure

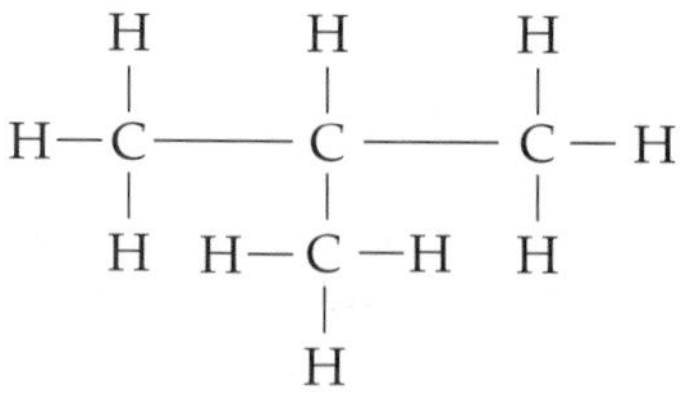

which also has the molecular formula C_4H_{10}, cannot rotate its bonds to form a continuous chain of four carbon atoms. Therefore, it is an isomer—it has the same molecular formulas as the preceding compound but a different structural formula.

Condensed Structural Formulas

It is usually too cumbersome to write a structural formula showing all the bonds and all the individual hydrogen atoms. So we frequently condense the formula, writing the hydrogens attached to a carbon atom immediately after that carbon atom. There are several different types of condensed structural formulas that can be used, depending on preference. For example,

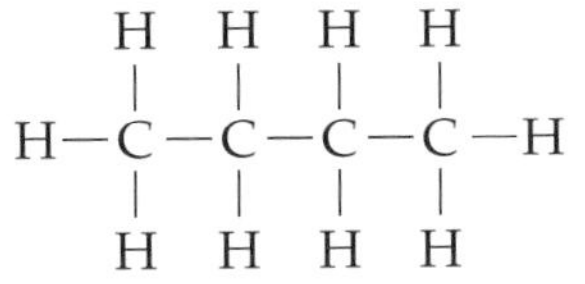

$$H-\underset{\underset{H}{|}}{\overset{\overset{H}{|}}{C}}-\underset{\underset{H}{|}}{\overset{\overset{H}{|}}{C}}-\underset{\underset{H}{|}}{\overset{\overset{H}{|}}{C}}-\underset{\underset{H}{|}}{\overset{\overset{H}{|}}{C}}-H$$

can be semi-condensed to

$$CH_3-CH_2-CH_2-CH_3$$

It can be condensed even further by omitting the bonds and assuming that they are present, or

$$CH_3CH_2CH_2CH_3$$

By grouping identical arrangements of atoms and indicating the number of these arrangements with a subscript, another condensed formula for the same compound can be written as

$$CH_3(CH_2)_2CH_3$$

For a branched compound, any group attached to the chain is indicated by parentheses placed around that group following the carbon to which it is attached. For example, the formula of the compound

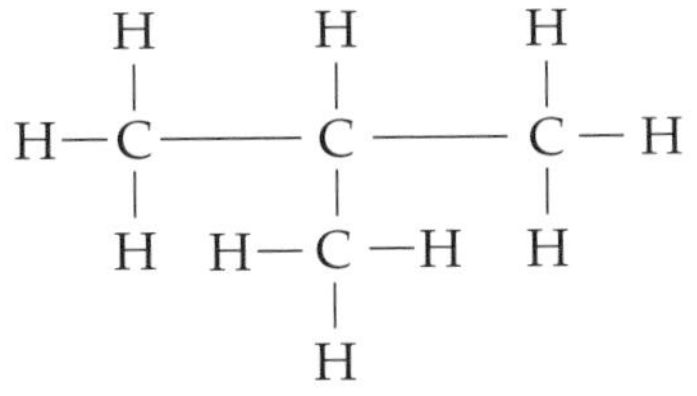

$$H-\underset{\underset{H}{|}}{\overset{\overset{H}{|}}{C}}\text{------}\underset{\underset{H-\underset{\underset{H}{|}}{\overset{\overset{|}{}}{C}}-H}{|}}{\overset{\overset{H}{|}}{C}}\text{------}\underset{\underset{H}{|}}{\overset{\overset{H}{|}}{C}}-H$$

can be condensed to

$$CH_3-CH(CH_3)-CH_3 \quad \text{or} \quad CH_3CH(CH_3)CH_3$$

Sometimes carbon atoms are grouped to form a condensed structural formula. Thus

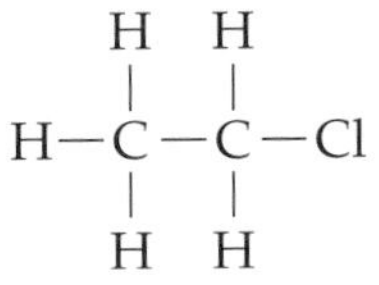

$$H-\underset{\underset{H}{|}}{\overset{\overset{H}{|}}{C}}-\underset{\underset{H}{|}}{\overset{\overset{H}{|}}{C}}-Cl$$

can be condensed to

$$CH_3-CH_2-Cl \quad \text{or} \quad CH_3CH_2Cl \quad \text{or} \quad C_2H_5Cl$$

Exercise 15-1

Write condensed structural formulas for the isomers of pentane, C_5H_{12}.

Solution

a. Five carbons in a row:

$$CH_3CH_2CH_2CH_2CH_3$$

b. Four carbons in a row, with one carbon attached in the middle:

$$CH_3CH(CH_3)CH_2CH_3$$

c. Three carbons in a row, with two carbons attached to the middle carbon:

$$CH_3C(CH_3)_2CH_3$$

Self-test

Write condensed structural formulas for the four isomers of $C_3H_6Cl_2$.

Answer

$CH_3CH_2CHCl_2$
$CH_3CCl_2CH_3$
$CH_3CHClCH_2Cl$
$CH_2ClCH_2CH_2Cl$

Alkyl Groups

Whey a hydrogen atom is removed from an alkane, an alkyl group is formed. The names of the alkyl groups are obtained by changing the ending of the name from *-ane* to *-yl*. The alkyl group of one carbon atom formed from the alkane methane is called the methyl group. The alkyl group of two carbon atoms formed from ethane is called the ethyl group. The alkyl group of three carbon atoms formed from propane is called propyl (see Table 15-2).
The compound CH_3Cl, or

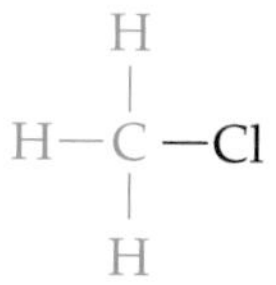

$$\begin{array}{c} H \\ | \\ H-C-Cl \\ | \\ H \end{array}$$

is made up of a methyl group (CH_3—) attached to a chlorine atom. It is called methyl chloride. The compound C_2H_5I consists of an ethyl group (C_2H_5—) bonded to an iodine. It is called ethyl iodide.

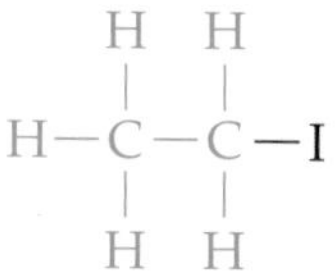

$$\begin{array}{cc} H & H \\ | & | \\ H-C-C-I \\ | & | \\ H & H \end{array}$$

Naming Alkanes

An international system of nomenclature for organic compounds has been devised and is recognized and used by chemists all over the world. This system was devised and approved by the International Union of Pure and Applied Chemistry and is frequently designated by the initials *IUPAC. The rules of the IUPAC system are

1. Pick out the longest continuous chain of carbon atoms.
2. Identify that chain as an alkane.
3. Pick out the alkyl groups or substituents attached to but not part of that chain.
4. Number the carbons in the chain, starting at whichever end of the chain will give the smallest numbers to the carbons to which the alkyl groups are attached. Continue the numbering of this carbon chain in the same direction from one end to the other.
5. List the numbers and the names of the alkyl groups or substituents in alphabetical order.
6. Use commas between numbers and a hyphen between a number and a letter.

Table 15-2 Simple Alkyl Groups	
Name of Alkyl Group	Condensed Structural Formula
Methyl	CH_3-
Ethyl	CH_3-CH_2- or C_2H_5-
Propyl	$CH_3-CH_2-CH_2-$ or C_3H_7-
Butyl	$CH_3-CH_2-CH_2-CH_2$ or C_4H_9-
Isopropyl	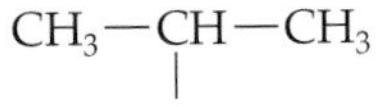 $CH_3-CH-CH_3$ $\quad\quad\quad\mid$ $\quad\quad\quad CH_3$
Tertiary butyl	$CH_3-\underset{\displaystyle \mid}{\overset{\displaystyle CH_3}{\underset{\displaystyle CH_3}{C}}}-CH_3$

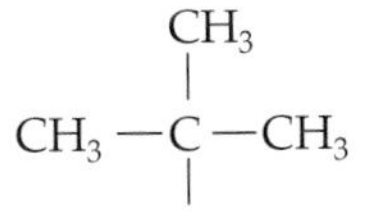

Exercise 15-2

Name the following compound.

$$CH_3-\underset{\displaystyle \underset{\displaystyle CH_3}{\mid}}{CH}-CH_2-CH_2-CH_2-CH_3$$

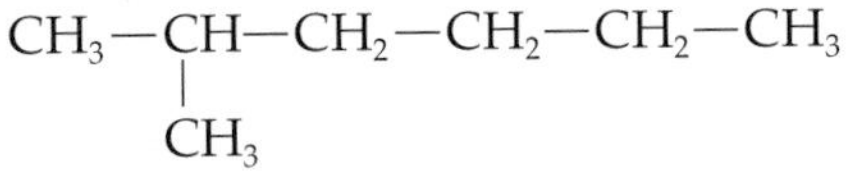

The longest chain contains six carbon atoms; therefore, this compound is a hexane. To identify the alkyl group attached to the chain and also to identify the chain itself, it is sometimes easier to highlight the chain. Then whatever is attached to the chain, the alkyl groups, will be outside the shading and can be easily noticed.

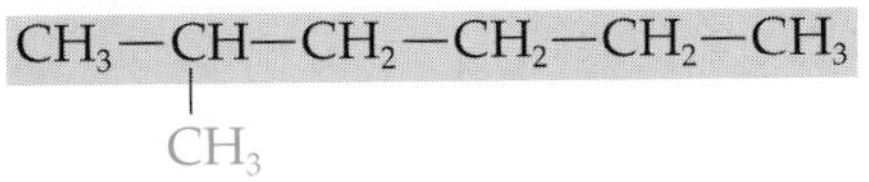

$$CH_3-\underset{\displaystyle \underset{\displaystyle CH_3}{\mid}}{CH}-CH_2-CH_2-CH_2-CH_3$$

Attached to the chain (sticking out) is a substitutent. In this case the substituent is an alkyl group of one carbon atom, the CH_3- or methyl group. Thus, this compound is a methylhexane.

The next step calls for numbering the carbons in the chain. They can be numbered in either direction, from left to right or from right to left. It should be observed that the methyl group is on the second carbon atom from the left end or on the fifth carbon atom from the right end. The rule states that the numbering should be such that the carbon to which the alkyl group is attached has the smallest number. Therefore, the correct name of this compound is *2-methylhexane,* which indicates that there is a methyl group on the second carbon from the end in a chain consisting of six carbon atoms. If the methyl group had been pointing upward instead of down, the compound would still be the same and so would the name.

Exercise 15-3

Name the following compound.

$$CH_3-CH_2-\underset{\displaystyle \underset{\displaystyle CH_3}{\mid}}{CH}-\overset{\displaystyle \overset{\displaystyle CH_3}{\mid}}{CH}-CH_3$$

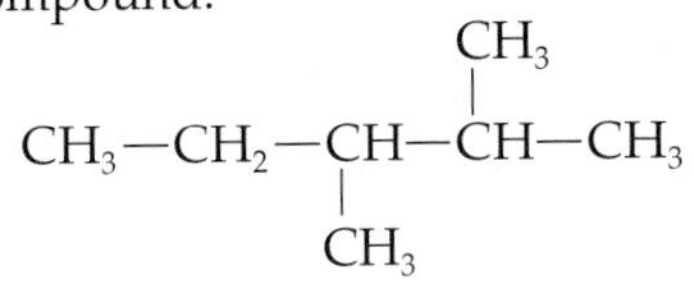

Solution

The longest chain contains five carbon atoms. This compound is a pentane. There are two methyl groups attached to the chain, one on the second carbon and one on the third carbon. This time the numbering is from right to left in order to obtain the lowest numbers for the alkyl groups.

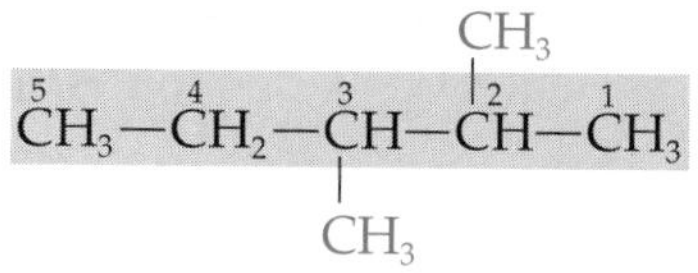

$$
\begin{array}{ccccc}
& & & \overset{\displaystyle CH_3}{\underset{\displaystyle |}{}} & \\
\overset{5}{CH_3} - \overset{4}{CH_2} - \overset{3}{CH} - \overset{2}{CH} - \overset{1}{CH_3} \\
& & \underset{\displaystyle CH_3}{\underset{\displaystyle |}{}} & &
\end{array}
$$

This compound is called *2,3-dimethylpentane*, where the prefix *di-* indicates that there are two identical groups. Dimethyl means two methyl groups, and the numbers tell us on which carbon atoms they are located.

Whenever an alkyl group appears more than once in a compound, a prefix is used to designate how many of these alkyl groups are present in that compound. Following are the most commonly used prefixes and their meanings:

di-	two
tri-	three
tetra-	four
penta-	five

Exercise 15-4

Name the following compound.

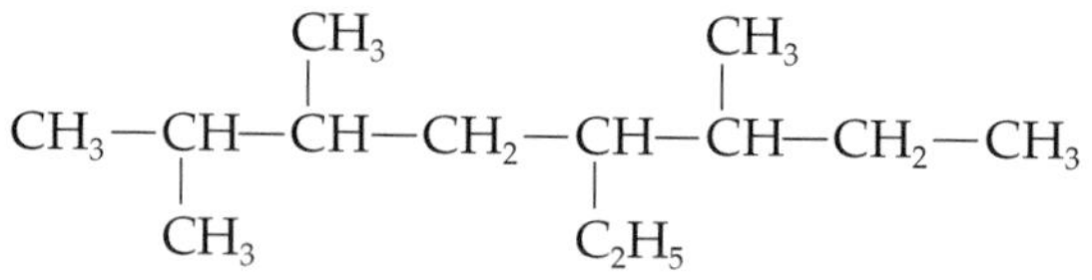

$$
\begin{array}{ccccccc}
& & CH_3 & & & CH_3 & \\
& & | & & & | & \\
CH_3 - CH - CH - CH_2 - CH - CH - CH_2 - CH_3 \\
& | & & & | & & \\
& CH_3 & & & C_2H_5 & &
\end{array}
$$

Solution

We first highlight the longest chain. This chain contains eight carbon atoms, so the compound is an octane.

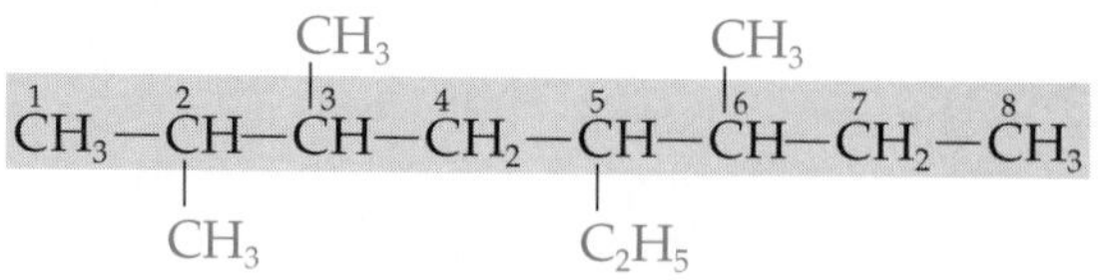

$$
\begin{array}{cccccccc}
& & CH_3 & & & CH_3 & & \\
& & | & & & | & & \\
\overset{1}{CH_3} - \overset{2}{CH} - \overset{3}{CH} - \overset{4}{CH_2} - \overset{5}{CH} - \overset{6}{CH} - \overset{7}{CH_2} - \overset{8}{CH_3} \\
& | & & & | & & & \\
& CH_3 & & & C_2H_5 & & &
\end{array}
$$

Attached to the chain are three methyl groups and one ethyl group. The chain should be numbered from left to right to obtain alkyl groups of the lowest numbers. There are methyl groups on carbons numbered 2, 3, and 6 and an ethyl group on carbon number 5 of the eight-carbon chain.

The correct name of this compound is *5-ethyl-2,3,6-trimethyloctane*. Note that the alkyl groups are named in alphabetical order (ethyl before methyl). Also note that the prefix *tri-* is used to indicate three alklyl groups of the same type.

Exercise 15-5

Name the following compound.

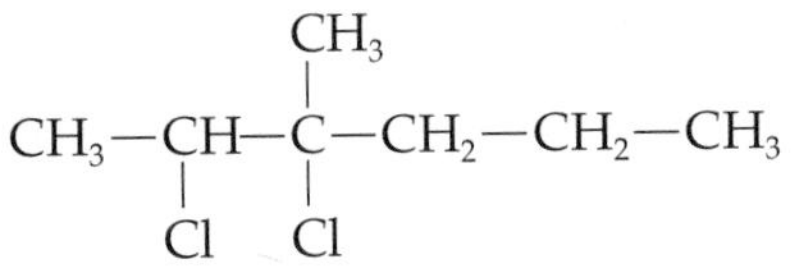

$$CH_3-\underset{\underset{Cl}{|}}{CH}-\underset{\underset{Cl}{|}}{\overset{\overset{CH_3}{|}}{C}}-CH_2-CH_2-CH_3$$

Solution

The longest chain contains six carbon atoms; it is a hexane. The numbering is from left to right to give the lowest numbers to the added groups. There are two chlorine atoms attached to the chain, one on carbon number 2 and one on carbon number 3. Also, there is a methyl group on carbon number 3.

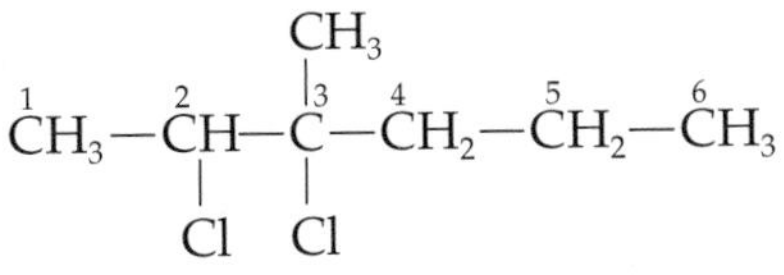

$$\overset{1}{CH_3}-\underset{\underset{Cl}{|}}{\overset{2}{CH}}-\underset{\underset{Cl}{|}}{\overset{\overset{CH_3}{|}}{\overset{3}{C}}}-\overset{4}{CH_2}-\overset{5}{CH_2}-\overset{6}{CH_3}$$

The compound is called *2,3-dichloro-3-methylhexane.*

Drawing Structural Formulas

If the IUPAC name of a compound is known, its structure can be drawn. Consider the following name: *3,4-diethyl-5-methyloctane.*

>   *octane* indicates a chain of eight carbon atoms.
>
>   *ethyl* indicates a group of two carbon atoms.
>
>   *diethyl* indicates two groups of two carbon atoms.
>
>   *3,4-diethyl* indicates that the two ethyl groups are on carbons 3 and 4 of the chain.
>
>   *methyl* indicates a group of one carbon atom.
>
>   *5-methyl* indicates that the methyl group is on carbon number 5.

The chain can be numbered from either end, but it is usually easier to number from left to right:

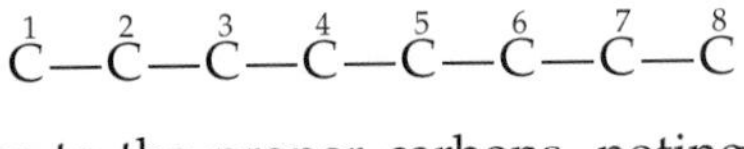

$$\overset{1}{C}-\overset{2}{C}-\overset{3}{C}-\overset{4}{C}-\overset{5}{C}-\overset{6}{C}-\overset{7}{C}-\overset{8}{C}$$

After we add the groups to the proper carbons, noting that it does not matter whether the groups are attached above or below the structure, and add hydrogen to complete the bonding we have:

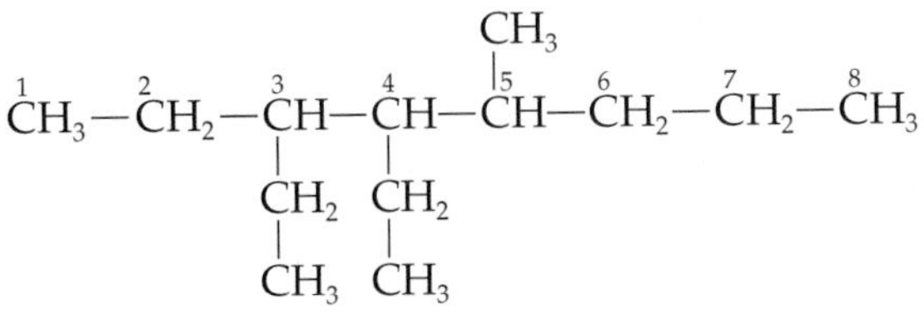

$$\overset{1}{CH_3}-\overset{2}{CH_2}-\underset{\underset{\underset{CH_3}{|}}{CH_2}}{\overset{3}{CH}}-\underset{\underset{\underset{CH_3}{|}}{CH_2}}{\overset{4}{CH}}-\underset{\underset{CH_3}{|}}{\overset{\overset{\overset{CH_3}{|}}{5}}{CH}}-\overset{6}{CH_2}-\overset{7}{CH_2}-\overset{8}{CH_3}$$

Exercise 15-6

Draw the structure for 2,5-dichloro-3-ethylnonane.

Solution

Nonane contains nine carbon atoms, numbered as follows:

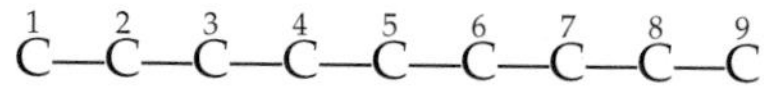

$$\overset{1}{C}-\overset{2}{C}-\overset{3}{C}-\overset{4}{C}-\overset{5}{C}-\overset{6}{C}-\overset{7}{C}-\overset{8}{C}-\overset{9}{C}$$

2,5-*dichloro* indicates a chlorine atom on carbons 2 and 5; *3-ethyl* indicates an ethyl group attached to carbon number 3. So, the structure is

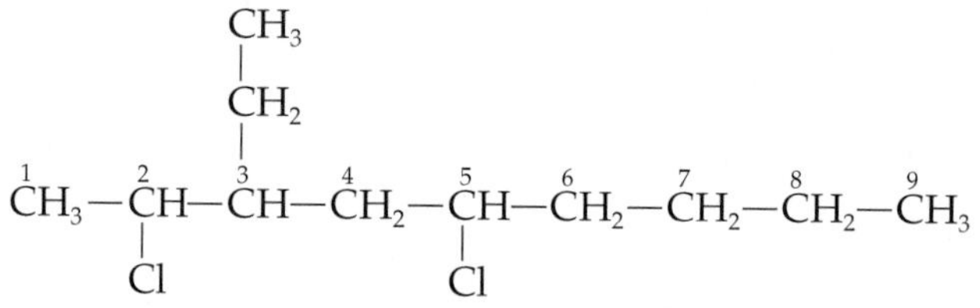

$$CH_3$$
$$|$$
$$CH_2$$
$$|$$
$$\overset{1}{CH_3}-\overset{2}{CH}-\overset{3}{CH}-\overset{4}{CH_2}-\overset{5}{CH}-\overset{6}{CH_2}-\overset{7}{CH_2}-\overset{8}{CH_2}-\overset{9}{CH_3}$$
$$|\qquad\qquad\quad|$$
$$Cl\qquad\qquad\;\;Cl$$

2,5-dichloro-3-ethylnonane

Self-test

Draw structures for the following compounds.
a. 2,6-dimethyloctane   b. 3-ethylpentane
c. 2,2,3,3-tetramethylhexane

Answers

a. 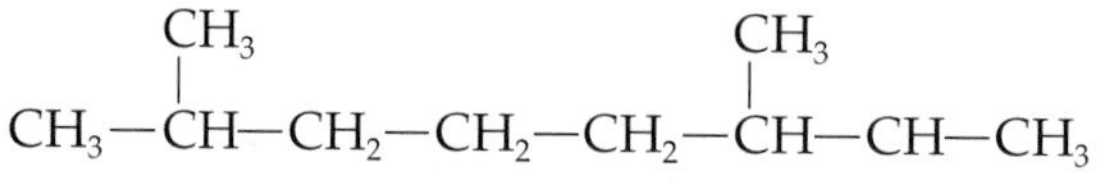
$$CH_3 \qquad\qquad\qquad CH_3$$
$$|\qquad\qquad\qquad\qquad\;\;|$$
$$CH_3-CH-CH_2-CH_2-CH_2-CH-CH-CH_3$$

b. 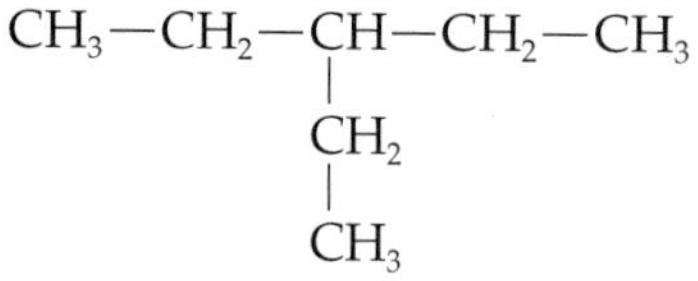
$$CH_3-CH_2-CH-CH_2-CH_3$$
$$|$$
$$CH_2$$
$$|$$
$$CH_3$$

c. 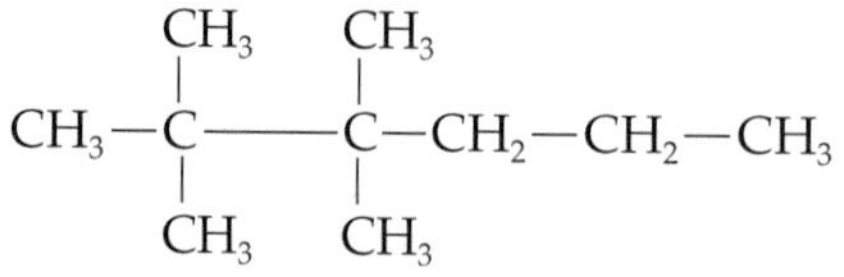
$$CH_3 \quad\; CH_3$$
$$|\qquad\; |$$
$$CH_3-C-----C-CH_2-CH_2-CH_3$$
$$|\qquad\; |$$
$$CH_3 \quad\; CH_3$$

Table 15-3   Names of Simple Hydrocarbons

Structure	IUPAC Name	Common Name
$CH_3-CH_2-CH_2-CH_2-CH_3$	*n*-Pentane	Pentane
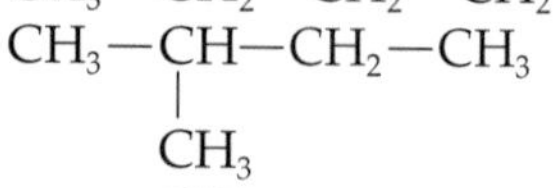 $CH_3-CH-CH_2-CH_3$ $\quad\;\;\mid$ $\quad\;\;CH_3$	2-Methylbutane	Isopentane
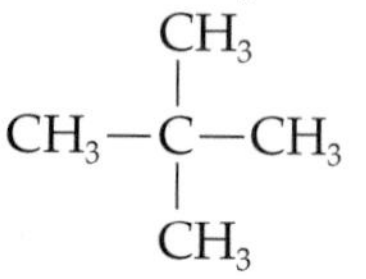 $\quad\quad CH_3$ $\quad\quad\mid$ CH_3-C-CH_3 $\quad\quad\mid$ $\quad\quad CH_3$	2,2-Dimethylpropane	Neopentane

Many simple hydrocarbons and hydrocarbon derivatives have common names that were in use before the IUPAC system was adopted. Many of these names are still in use, although IUPAC naming is preferred. Table 15-3 indicates the IUPAC name and the common name for the three isomers of pentane, C_5H_{12}. Note that the prefix *n-* (for normal) indicates a straight-chain compound.

15-2 Cycloalkanes

Alkanes also exist in the shape of a ring. Such structures are called cycloalkanes. They are named by placing the prefix *cyclo-* before the name of the corresponding continuous-chain alkane. Thus, the cyclic alkane of three carbons is called *cyclopropane*. Its structure is

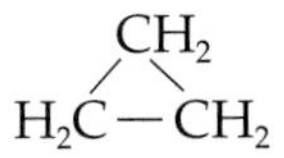

$$\begin{array}{c} CH_2 \\ \diagup \diagdown \\ H_2C - CH_2 \end{array}$$

Cyclopropane is used as a general anesthetic. Both induction and recovery time are short with its use. Muscular relaxation is greater than with nitrous oxide but less than ether. As with ether, the danger of explosion is great, so care must be taken in its use.

The structures of cycloalkanes can also be abbreviated by using a geometric shape in which each line represents a bond between carbon atoms and each corner a carbon atom and its hydrogens. The following structures represent cyclopropane and cyclobutane, respectively.

cyclopropane   cyclobutane

When alkyl groups are attached to a cycloalkane, the compound is numbered in such a manner as to give the groups the lowest numbers. When only one group is attached to a cycloalkane, the number 1 is understood and not written.

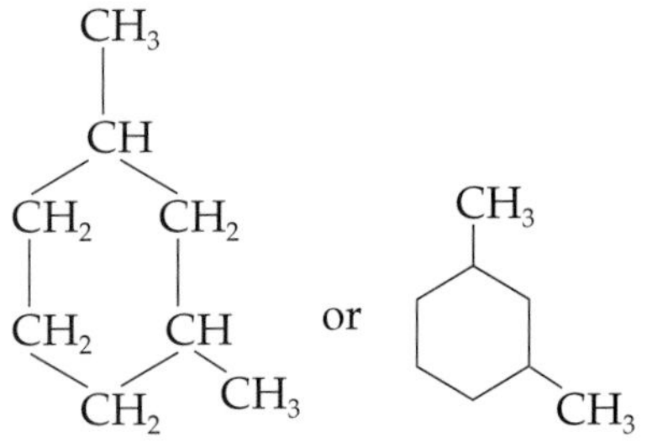

1,3-dimethylcyclohexane 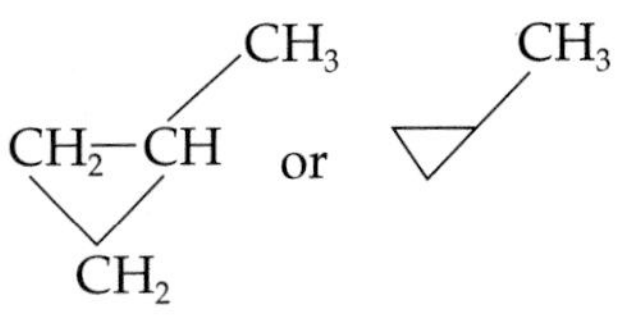   methylcyclopropane

Exercise 15-7

Name the following cyclocompounds:

a.

b.

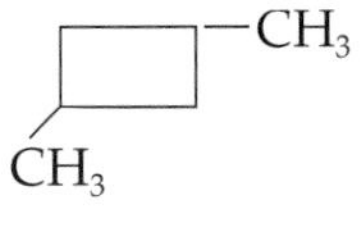

Solution

a. The ring has five corners, which indicates five carbons, so it is a *cyclopentane*. Therefore, the name of this compound is chlorocyclopentane. Note that no number is needed since all carbons in this ring are identical.
b. There are two methyl groups attached to the opposite sides of a four-carbon ring, which is a cyclobutane. Therefore, the name of the compound is 1,3-dimethylcyclobutane.

Self-test

Show the structural difference between 1,3-dichloropentane and 1,3-dichlorocyclopentane.

Answer

1,3-dichloropentane is

$$CH_2Cl—CH_2—CHCl—CH_2—CH_3$$

1,3-dichlorocyclopentane is

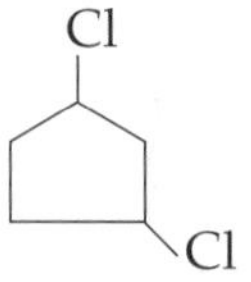

15-3 Sources of Hydrocarbons

The chief sources of hydrocarbons are petroleum and natural gas. Petroleum is a very complex mixture of solid, liquid, and gaseous hydrocarbons plus a few compounds of other elements. Natural gas is primarily a mixture of alkanes of one to four carbon atoms. In general, alkanes are gaseous if the compounds contain between 1 and 4 carbon atoms, liquid if they contain between 5 and 16 carbon atoms, and solid if they contain more than 16 carbon atoms. The various hydrocarbons present in petroleum are isolated by a process known as fractional distillation.

One of the chief products of the fractional distillation of petroleum is gasoline. One measure of the quality of a gasoline is octane number. The octane number of a gasoline refers to how smoothly it burns. A high-octane gasoline delivers power smoothly to the pistons. A low-octane fuel burns too rapidly and causes the engine to "knock." Isooctane, 2,2,4-trimethylpentane, which has excellent combustion properties, was arbitrarily assigned an octane rating of 100. Normal heptane, *n*-heptane (C_7H_{16}), which causes considerable engine knocking, was assigned an octane rating of 0. Mixtures of these two compounds were burned in test engines to establish an "octane scale." That is, if a sample of gasoline has the same amount of knock as an 87 percent isooctane—13 percent heptane mixture, then that gasoline has an octane rating of 87.

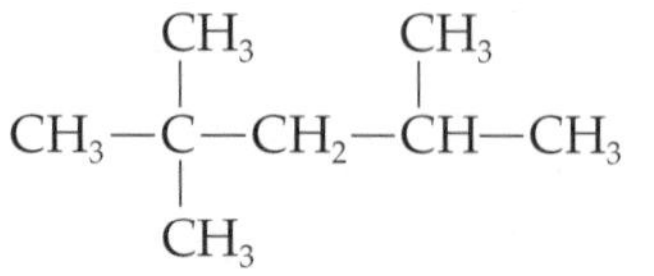

$$\begin{matrix} & CH_3 & & CH_3 \\ & | & & | \\ CH_3—C—CH_2—CH—CH_3 \\ & | \\ & CH_3 \end{matrix} \qquad CH_3—CH_2—CH_2—CH_2—CH_2—CH_2—CH_3$$

isooctane   *n*-heptane

Table 15-4	Comparison of Octane Ratings and Chain Branching	
Formula	Name	Octane Number
C_6H_{14}	Hexane	25
C_6H_{14}	2-Methylpentane	73
C_7H_{16}	Heptane	0
C_7H_{16}	2-Methylhexane	42
C_7H_{16}	2,2-Dimethylpentane	92

FIGURE 15-3
Oil refinery.

In general, increased branching of a chain increases the octane rating of a hydrocarbon, as indicated in Table 15-4.

Until the mid-1960s automotive gasolines commonly contained the additive tetraethyllead ($Pb[C_2H_5]_4$). This substance reduced knocking and increased the octane ratings of gasolines. However, it caused large amounts of lead compounds to be discharged into the air, with resulting hazards to human health. For that reason unleaded fuels must now be used in all modern cars (see Figure 15-3).

15-4 Properties of Hydrocarbons

In general, as the number of alkane carbon atoms increases, the boiling point and density increase. Table 15-5 indicates the changes in these properties with increasing numbers of carbon atoms. These changes are due to the large surface areas and increased intermolecular forces. Branching lowers the boiling point since the molecules become more compact and the intermolecular forces are lessened.

Table 15-5 Physical Properties of Some Alkanes

Condensed Structural Formula	Name	Boiling Point (°C)	Density (g/mL)
CH_4	Methane	-162	Gas
CH_3CH_3	Ethane	-89	Gas
$CH_3CH_2CH_3$	Propane	-42	Gas
$CH_3(CH_2)_2CH_3$	Butane	-0.5	Gas
$CH_3(CH_2)_3CH_3$	Pentane	36	0.626
$CH_3(CH_2)_4CH_3$	Hexane	69	0.659
$CH_3(CH_2)_5CH_3$	Heptane	98	0.684
$CH_3(CH_2)_6CH_3$	Octane	126	0.703
$CH_3(CH_2)_7CH_3$	Nonane	151	0.718
$CH_3(CH_2)_8CH_3$	Decane	174	0.730

Alkanes are nonpolar compounds and generally are soluble in nonpolar solvents such as carbon tetrachloride (CCl_4) and insoluble in polar compounds such as water.

Reactions of the Alkanes

Combustion Alkanes react with air or oxygen to produce carbon dioxide, water, and energy.

$$CH_4 + 2\,O_2 \longrightarrow CO_2 + 2\,H_2O + energy$$

$$C_3H_8 + 5\,O_2 \longrightarrow 3\,CO_2 + 4\,H_2O + energy$$

Incomplete combustion of alkanes can produce carbon monoxide (CO). This process can occur in an automobile engine when insufficient oxygen is present. Running a car engine in a closed garage can lead to CO poisoning and even death because of the accumulation of this deadly gas.

$$\underset{\text{isooctane}}{2\,C_8H_{18}} + 17\,O_2 \longrightarrow \underset{\text{carbon monoxide}}{16\,CO} + 18\,H_2O$$

Halogenation Alkanes react with halogens (F_2, Cl_2, Br_2, or I_2) by a process called substitution in which a halogen atom substitutes for a hydrogen atom. This type of reaction is called halogenation, and the products are haloalkanes (alkyl halides).

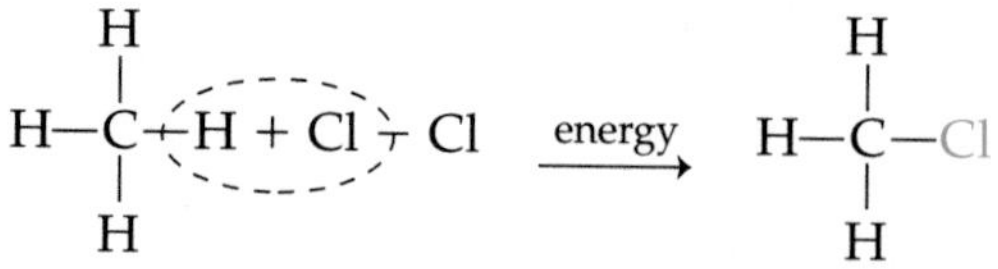

chloromethane
(methyl chloride)

The IUPAC name is given first, followed by the common name in parentheses. Further substitution of chlorines for hydrogen leads to

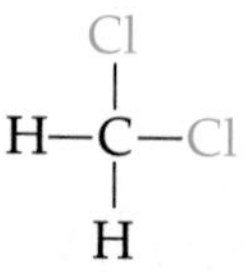

dichloromethane
(methylene chloride)

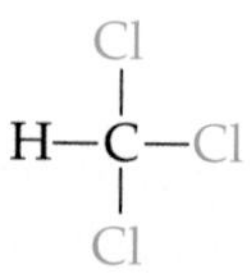

trichloromethane
(chloroform)

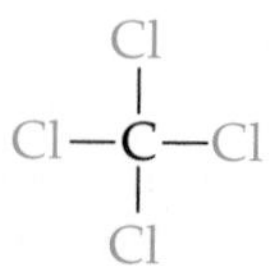

tetrachloromethane
(carbon tetrachloride)

Similar compounds can be formed with bromine and iodine. The common name for $CHCl_3$ is chloroform. Likewise, $CHBr_3$ is bromoform, and CHI_3 is iodoform.

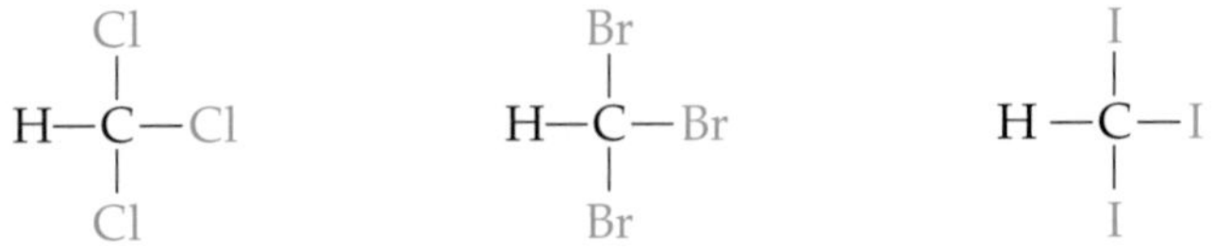

$$
\begin{array}{ccc}
\overset{\displaystyle Cl}{\underset{\displaystyle Cl}{H-C-Cl}} &
\overset{\displaystyle Br}{\underset{\displaystyle Br}{H-C-Br}} &
\overset{\displaystyle I}{\underset{\displaystyle I}{H-C-I}}
\end{array}
$$

trichloromethane (chloroform)	tribromomethane (bromoform)	triiodomethane (iodoform)

Since there is only one carbon atom, no numbers are necessary.

A hydrocarbon derivative used as an anesthetic is halothane, 2-bromo-2-chloro-1,1,1-trifluoroethane. (Note that the groups attached to the carbons are named in alphabetical order and lowest numbers are used.)

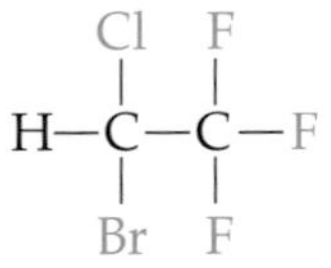

$$
\overset{\displaystyle Cl\ \ \ F}{\underset{\displaystyle Br\ \ \ F}{H-C-C-F}}
$$

Halothane's main advantage over other anesthetics formerly used is that it is non-flammable and is not irritating to the respiratory passages. Halothane is usually used in conjunction with nitrous oxide (see Section 8-5) and muscle relaxants to provide general anesthesia for surgery of all types, but it can also be administered alone.

A similar compound, chloral hydrate, induces sleep and prevents convulsions. A combination of ethyl alcohol and chloral hydrate is known as a "Mickey Finn" or "knockout drops."

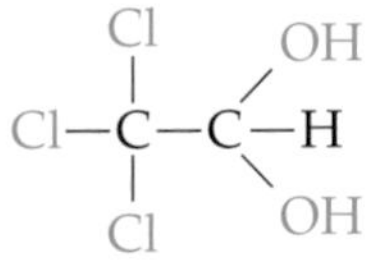

$$
\overset{\displaystyle Cl\ \ \ \ \ \ OH}{\underset{\displaystyle Cl\ \ \ \ \ \ OH}{Cl-C-C-H}}
$$

chloral hydrate

In veterinary medicine, chloral hydrate is used as a narcotic and anesthetic for cattle, horses, and poultry.

Other widely used haloalkanes include ethyl chloride, a fast-acting topically applied local anesthetic; Freon-12, formerly used in aerosol containers and in refrigeration and air conditioning units (see Section 8-9); Teflon and polyvinyl chloride (PVC), which are polymers. Teflon is used as a coating material for cooking utensils, and PVC is used in making plastic containers and pipes for plumbing. Their structures are as follows:

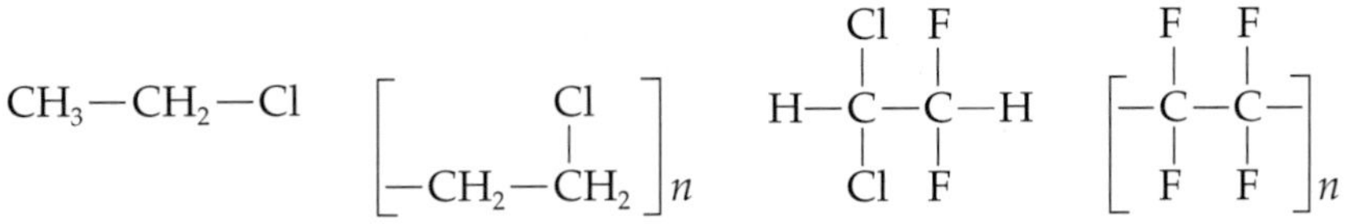

$$
CH_3-CH_2-Cl \qquad \left[\underset{\displaystyle -CH_2-CH_2-}{Cl} \right]_n \qquad \overset{\displaystyle Cl\ \ \ F}{\underset{\displaystyle Cl\ \ \ F}{H-C-C-H}} \qquad \left[\overset{\displaystyle F\ \ \ F}{\underset{\displaystyle F\ \ \ F}{-C-C-}} \right]_n
$$

ethyl chloride	polyvinyl chloride	Freon-12	Teflon

Exercise 15-5

Complete and balance the following reactions.

a. $CH_3-CH_3 + Cl_2 \xrightarrow{\text{energy}}$ b. $C_6H_{14} + O_2 \xrightarrow{\text{energy}}$

Solution

a. $CH_3-CH_3 + Cl_2 \longrightarrow CH_3-CH_2Cl + HCl$

This is a halogenation substitution reaction.

b. $2\,C_6H_{14} + 19\,O_2 \longrightarrow 12\,CO_2 + 14\,H_2O$

This is a combustion reaction.

Summary

Hydrocarbons are organic compounds that contain the elements carbon and hydrogen only. The simplest hydrocarbon, the hydrocarbon containing only one carbon atom, is methane (CH_4). The hydrocarbon of two carbon atoms is called ethane; that of three carbon atoms, propane. These compounds are called alkanes. The general formula for alkanes is C_nH_{2n+2}, where n is the number of carbon atoms. The names of all alkanes end in *-ane*. Beginning with pentane, the prefixes indicate the number of carbon atoms present.

When an alkane loses a hydrogen atom, it forms an alkyl group whose name ends in *-yl*. the alkyl group of one carbon atom, derived from methane, is called the methyl group (CH_3-). The alkyl group of two carbon atoms, derived from ethane, is called the ethyl group (C_2H_5-).

To identify and name a hydrocarbon compound, pick out the longest continuous chain of carbon atoms and name that chain. Then pick out the alkyl groups attached to that chain. Number the carbons, starting at whichever end of the chain will give alkyl groups with the smallest numbers. List the numbers and names of the alkyl groups, using prefixes to designate alkyl groups occurring more than once.

Alkanes that occur in the shape of a ring are called cycloalkanes.

Haloalkanes have one or more hydrogen atoms of an alkane replaced by a halogen.

Questions and Problems

1. Name an alkane of four carbon atoms; of six carbon atoms.
2. What is an alkyl group? Name and draw the structure of an alkyl group derived from ethane; propane.
3. Name the following compounds:

   a.
   $$CH_3-\underset{\underset{CH_3}{|}}{CH}-CH_2-CH_3$$

   b.
   $$CH_3-\underset{\underset{CH_3}{|}}{CH}-\overset{\overset{CH_3}{|}}{CH}-\underset{\underset{CH_3}{|}}{CH}-CH_2-CH_3$$

   c.
   $$CH_3-\overset{\overset{CH_3}{|}}{\underset{\underset{CH_3}{|}}{C}}-CH_3$$

   d.
   $$CH_3-CH_2-\underset{\underset{\underset{\underset{CH_3}{|}}{CH_2}}{|}}{CH}-CH_2-CH_2-CH_2-CH_3$$

4. List several uses for hydrocarbons.
5. What are the chief sources of hydrocarbons?
6. What is a cycloalkane? Give an example.
7. Draw the structure for:
   a. methylcyclopentane
   b. 1,3-dibromocyclobutane
8. Draw the structure for:
   a. 3-bromooctane
   b. 2-chloro-2-methylhexane
   c. 4,4-diethyldecane
   d. dibromomethane
9. What is the IUPAC name for isooctane? Write the equation for its complete oxidation.
10. Name the following compounds:

    a.
    $$CH_3-\overset{\overset{Cl}{|}}{\underset{\underset{Cl}{|}}{C}}-CH_2-CH_2-CH_3$$

    b.
    (cyclopentane ring with —Br and —Br substituents)

11. What problems are associated with the use of leaded gasoline?

12. Compare the complete and incomplete oxidation of alkanes.

13. Complete and balance the following reactions.

a. $C_8H_{18} + O_2 \xrightarrow[\text{combustion}]{\text{complete}}$

b. $CH_4 + Br_2 \xrightarrow{\text{energy}}$

14. Write condensed structural formulas for the compounds in question 3.

15. Name and draw all of the monochloroderivatives of the following:
    a. 2-methylbutane   b. cyclobutane

16. Write a balanced equation for the combustion of decane.

17. Draw structures for:
    a. cyclopropane   b. chloral hydrate
    c. carbon tetrachloride d. chloroform
    e. Freon-12   f. dichoromethane
    g. isooctane

18. Draw a representative structure for the following polymers. State what elements would be found in each.
    a. Teflon   b. polypropylene
    c. polyvinyl chloride

Practice Test

1. The name of the following compound is _____.

$$CH_3-CH-CH_2-CH-CH_2-CH_3$$
$$\quad\quad | \quad\quad\quad\quad |$$
$$\quad\quad CH_3 \quad\quad\quad CH_3$$

   a. dimethylhexane   b. 2,3-dimethylhexane
   c. 2,4-dimethylhexane d. 3,5-dimethylhexane

2. Alkanes react primarily by _____.
   a. addition   b. substitution
   c. oxidation   d. reduction

3. Alkanes of six or more carbon atoms are:
   a. soluble in water and insoluble in organic solvents
   b. insoluble in water and soluble in organic solvents
   c. soluble in water and soluble in organic solvents
   d. insoluble in water and insoluble in organic solvents

4. The structure △ represents _____.
   a. cyclomethane   b. cyclopropane
   c. heterocyclopropane   d. heterocyclobutane

5. The alkane of eight carbons is
   a. C_8H_{12} b. C_8H_{14} c. C_8H_{16} d. C_8H_{18}

6. Increased branching of a hydrocarbon has what effect on its octane rating?
   a. increases it   b. decreases it
   c. has no effect

Unsaturated Hydrocarbons and Their Halogen Derivatives: Alkenes, Alkynes, and Aromatic Compounds

Objectives

- To name and draw structures of alkanes and alkenes
- To distinguish between saturated and unsaturated hydrocarbons
- To understand reactions of alkenes and alkynes
- To recognize the structure of benzene
- To name various substitution products of benzene
- To recognize the structures of various aromatic compounds

Aromatic compounds are well known for their fragrance and are used in a variety of household cleansers and disinfectants.

Saturated hydrocarbons have single bonds between carbon atoms. Unsaturated hydrocarbons have multiple bonds between carbon atoms. Polyunsaturated hydrocarbons have many multiple bonds in their molecules.

Unsaturated hydrocarbons may be classified as follows:

- alkenes, which contain one or more double bonds between carbon atoms (compounds containing two double bonds are called *dienes* and those containing three double bonds are called *trienes*)
- alkynes, which contain one or more triple bonds between carbon atoms
- aromatics, which contain benzene rings that have multiple double bonds

16-1 Alkenes

Alkanes have a single bond between the carbon atoms. Alkenes have a double bond (two bonds) between two of the carbon atoms.

$$\underset{/}{\overset{\backslash}{C}}=\underset{\backslash}{\overset{/}{C}}$$

Consider two carbon atoms connected by a double bond. Since this double bond uses four electrons from both carbons, a total of only four hydrogen atoms will satisfy all of the remaining bonds. Recall that a single bond represents a pair of shared electrons; a double bond represents two pairs of shared electrons. This compound thus becomes

$$
\begin{array}{cc}
\text{H} & \text{H} \\
| & | \\
\text{H}-\text{C}=\text{C}-\text{H}
\end{array}
\qquad
\left(
\begin{array}{cc}
\text{H} & \text{H} \\
\text{H}:\ddot{\text{C}}:\ddot{\text{C}}:\text{H}
\end{array}
\right)
$$

and has the molecular formula C_2H_4. It is called *ethene*.

When three carbon atoms are arranged in a chain with a double bond between two of the carbon atoms, C=C—C, how many hydrogen atoms must be connected to these carbon atoms in order to satisfy all of the bond requirements. The answer is six, and the structure becomes

$$
\begin{array}{ccc}
\text{H} & \text{H} & \text{H} \\
| & | & | \\
\text{H}-\text{C}=\text{C}-\text{C}-\text{H} \\
& & | \\
& & \text{H}
\end{array}
$$

The molecular formula is C_3H_6. The name of this compound is *propene*.

Note that the names of these compounds end in *-ene*. This is true of all alkenes. Note that the names of these compounds are similar to those of the alkanes except for the ending, which is *-ene* instead of *-ane*.

Compare the structures of ethane (C_2H_6) and ethene (C_2H_4).

$$
\begin{array}{cc}
\text{H} & \text{H} \\
| & | \\
\text{H}-\text{C}-\text{C}-\text{H} \\
| & | \\
\text{H} & \text{H}
\end{array}
\qquad\qquad
\begin{array}{cc}
\text{H} & \text{H} \\
| & | \\
\text{H}-\text{C}=\text{C}-\text{H}
\end{array}
$$

$$\text{ethane} \qquad\qquad\qquad \text{ethene}$$

The three-carbon alkane is propane, and the three-carbon alkene is propene. Likewise, the four-carbon alkane is called butane, while the corresponding alkene is called butene. The names and formulas of some of the alkenes are listed in Table 16-1.

The general formula for alkenes is

$$C_nH_{2n}$$

There are twice as many hydrogen atoms as carbon atoms in every alkene. Thus, octene has eight carbon atoms and 16 hydrogen atoms, and the formula of an alklene of 15 carbons atoms would be $C_{15}H_{30}$.

The vinyl group has the structure

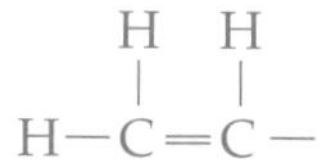

$$\begin{array}{cc} H & H \\ | & | \\ H-C & =C- \end{array}$$

Vinyl chloride is used in the manufacture of such products as floor tile, raincoats, fabrics, and furniture coverings. However, evidence has shown that several workers exposed to vinyl chloride during their work have died from a very rare form of liver cancer. In addition, exposure to vinyl chloride is suspected to be responsible for certain types of birth defects.

The structure of vinyl chloride is

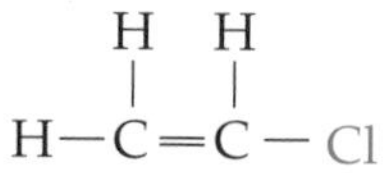

$$\begin{array}{cc} H & H \\ | & | \\ H-C & =C-Cl \end{array}$$

Vinyl chloride is used in the manufacture of polyvinyl chloride (PVC), a plastic.

Table 16-1   Some Simple Alkanes

Number of Carbon Atoms	Name	Molecular Formula	Structural Formula	Condensed Structural Formula
2	Ethene	C_2H_4	$\begin{array}{cc} H & H \\ | & | \\ H-C & =C-H \end{array}$	$CH_2=CH_2$
3	Propene	C_3H_6	$\begin{array}{ccc} & H & H & H \\ & | & | & | \\ H-C & -C & =C-H \\ & | \\ & H \end{array}$	$CH_3CH=CH_2$
4	Butene	C_4H_8	$\begin{array}{cccc} H & H & H & H \\ | & | & | & | \\ H-C & =C & -C & -C-H \\ & & | & | \\ & & H & H \end{array}$	$CH_2=CHCH_2CH_3$
			or	or
			$\begin{array}{cccc} H & H & H & H \\ | & | & | & | \\ H-C & -C & =C & -C-H \\ | & & & | \\ H & & & H \end{array}$	$CH_3CH=CHCH_3$

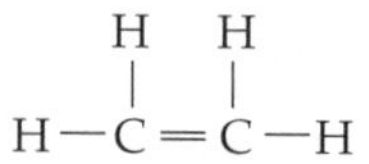

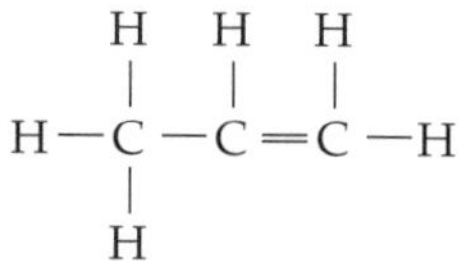

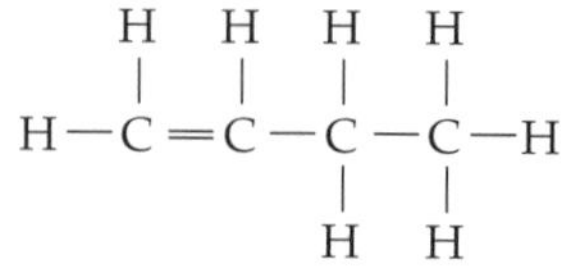

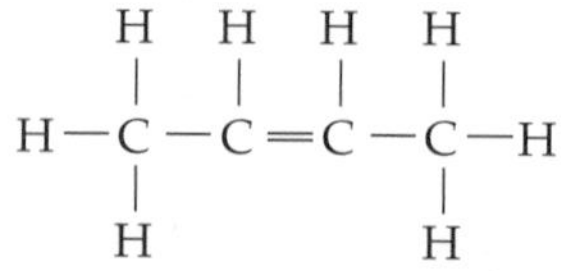

Naming Alkenes

The rules for naming alkenes according to the ILUPAC system are as follows:

1. Identify the longest chain containing the double bond.
2. Number the chain beginning at the end closest to the double bond so that the double bond will have the lowest number.
3. Locate and number the groups attached to the chain.
4. When naming, cite the lower number of the double-bonded carbon atoms. Note that all alkenes have names that end with *-ene*.

Exercise 16-1

Name the following compound.

$$CH_2{=}CH{-}CH{-}CH_3$$
$$|$$
$$CH_3$$

- **Step 1** The longest chain containing the double bond has four carbon atoms, so the compound is a *butene*.
- **Step 2** The chain is numbered as follows so that the *double bond has the lowest number*:

$$\overset{1}{C}H_2{=}\overset{2}{C}H{-}\overset{3}{C}H{-}\overset{4}{C}H_3$$
$$|$$
$$CH_3$$

- **Step 3** Attached to the chain is a methyl group at carbon number 3, so the name of the compound is

$$\overset{1}{C}H_2{=}\overset{2}{C}H{-}\overset{3}{C}H{-}\overset{4}{C}H_3$$
$$|$$
$$CH_3$$

3-methyl-1-butene

Note that 1-butene indicates a double bond starting at carbon number 1.
Similarly,

$$\overset{4}{C}H_3{-}\overset{3}{C}H{=}\overset{2}{C}{-}\overset{1}{C}H_3$$
$$|$$
$$CH_3$$

is called 2-methyl-2-butene.

If an alkene contains two double bonds, it is called a *diene*; a *triene* contains three double bonds.

1-3-butadiene is used in the manufacture of automobile tires, and leukotriene is involved in the body's allergic responses (see page 374).

Cycloalkenes

Cycloalkenes are named in a manner similar to that for alkenes.

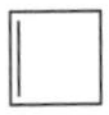

cyclobutene

1 is understood, so it is assumed that the double bond is between carbon atoms 1 and 2.

The following cycloalkene is numbered to show the double bond having the number 1 (understood) and the methyl group having the lowest possible number. If the numbering had been in the other direction, then the methyl group would have a larger number.

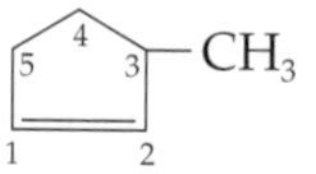

3-methylcyclopentene

Exercise 16-2

Name the following compounds.

a. $CH_2{=}CH{-}\overset{\overset{\displaystyle CH_3}{|}}{\underset{\underset{\displaystyle CH_3}{|}}{C}}{-}CH_2{-}CH_3$ b.

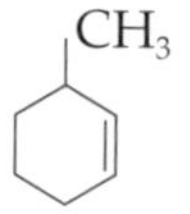

Solution

a. The longest chain containing a double bond has five carbon atoms, so the parent compound is a *pentene*. The methyl groups are both at carbon number 3, so the name of this compound is 3,3-dimethyl-1-pentene or 3,3-dimethylpentene, with the 1 being understood.
b. The ring contains six carbon atoms and a double bond, so it is a *hexene*. The methyl group is at the third carbon from the beginning of the double bond, so the name is 3-methylcyclohexene.

Drawing Structures of Alkenes

Exercise 16-3

Draw the structure of 3,4-diethyl-2-hexene.

Solution

Hexene indicates a six-carbon structure with one double bond. 2-hexene indicates a six-carbon chain with a double bond starting at carbon number 2:

$$\overset{1}{C}{-}\overset{2}{C}{=}\overset{3}{C}{-}\overset{4}{C}{-}\overset{5}{C}{-}\overset{6}{C}$$

3,4-diethyl indicates two alkyl groups of two carbons each, one on carbon number 3 and one on carbon number 4, or after adding hydrogen to complete the number of bonds,

$$CH_3{-}CH{=}\underset{\underset{\underset{\displaystyle CH_3}{|}}{\underset{\displaystyle CH_2}{|}}}{C}{-}\underset{\underset{\underset{\displaystyle CH_3}{|}}{\underset{\displaystyle CH_2}{|}}}{CH}{-}CH_2{-}CH_3$$

Exercise 16-4

Draw the structure of 4-methylcyclopentene. Cyclopentene indicates a five-carbon ring with one double bond:

Since no number is given for the double bond, it must be at carbon number 1.

4-methyl indicates a methyl group at carbon number 4, or

$$CH_3 - \text{(ring)}$$

Reactions of Alkenes

Ethene reacts with hydrogen (H_2) by a process known as *addition*; that is, the hydrogen atoms add to the double bond, making a single bond out of it. This reaction is called *hydrogenation*.

$$\underset{\text{ethene}}{H-\overset{\displaystyle H}{\underset{\displaystyle\:}{C}}=\overset{\displaystyle H}{\underset{\displaystyle\:}{C}}-H} \;+\; \underset{\text{hydrogen}}{H_2} \;\xrightarrow[\text{catalyst}]{\text{metal}}\; \underset{\text{ethane}}{H-\overset{\displaystyle H}{\underset{\displaystyle H}{C}}-\overset{\displaystyle H}{\underset{\displaystyle H}{C}}-H}$$

Cycloalkenes react similarly:

$$\underset{\text{cyclohexene}}{\bigcirc} \;+\; H_2 \;\xrightarrow{\text{catalyst}}\; \underset{\text{cyclohexane}}{\bigcirc}$$

Halogens also react with alkenes by addition. This reaction is called *halogenation*.

$$\underset{\text{propene}}{H-\overset{\displaystyle H}{\underset{\displaystyle H}{C}}-\overset{\displaystyle H}{\underset{\displaystyle\:}{C}}=\overset{\displaystyle H}{\underset{\displaystyle\:}{C}}-H} \;+\; \underset{\text{chlorine}}{Cl-Cl} \;\longrightarrow\; \underset{\text{1,2-dichloropropane}}{H-\overset{\displaystyle H}{\underset{\displaystyle H}{C}}-\overset{\displaystyle H}{\underset{\displaystyle Cl}{C}}-\overset{\displaystyle H}{\underset{\displaystyle Cl}{C}}-H}$$

$$\underset{\text{cyclopentene}}{\bigcirc} \;+\; Br_2 \;\xrightarrow{\text{catalyst}}\; \underset{\text{1,2-dibromocyclopentane}}{\bigcirc_{Br \quad Br}}$$

In general, the addition reaction takes place more rapidly and under milder conditions than the substitution reaction. Note that in these examples an unsaturated reactant has been changed into a saturated product. A practical application of this will be discussed in the hydrogenation of fats and oils (see page 364).

Polymers

When many small organic molecules that are alike or similar are joined together to form a much larger molecule, the process is polymerization. The product is called a polymer. Naturally occurring polymers include starch, cellulose, rubber, and protein. Synthetic polymers include plastics and such fibers as nylon, rayon, and Dacron (see page 286). Medical uses of polymers include synthetic heart valves and blood vessels, surgical mesh, disposable syringes, and drug containers (see Figure 16-1).

The equation that follows indicates the formation of polyethylene from ethene (ethylene). Note that an initiator is required. This substance starts the reaction, which continues until a long chain is obtained. The number of units of ethylene joined together varies between 100 and 1000.

$$CH_2{=}CH_2 \xrightarrow[\text{initiator}]{\substack{\text{heat,}\\ \text{pressure}}} -(CH_2-CH_2)-(CH_2-CH_2)- \text{ etc.} \quad \text{or} \quad (-CH_2-CH_2-)_n$$

ethene
(ethylene)   polythylene

If the starting material is propene (propylene), then the polymer produced is called polypropylene. If styrene is the starting material, polystyrene is produced. Likewise polyvinyl chloride (PVC) is produced from vinyl chloride.

Cis-trans Isomers

Cis-trans, or geometric isomers, are compounds with the same molecular formula that have different structural formulas because either a double bond or a ring system prevents the rotation necessary to change one into the other. That is, cis-trans isomers have the same sequence of atoms but different spatial arrangements. Cis-trans isomers have different physical properties, such as different boiling and freezing points.

If two carbon atoms are connected by a double bond, those two carbon atoms and the four groups attached to them all lie in a single plane. There are two possible isomeric structures, one having similar groups on the same "side" of the

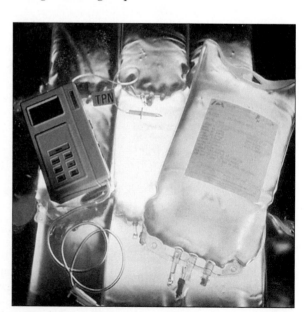

FIGURE 16-1
Plastic containers of intravenous solution.

double bond and one having similar groups on opposite "sides" of the double bond. Examples of cis-trans isomerism are

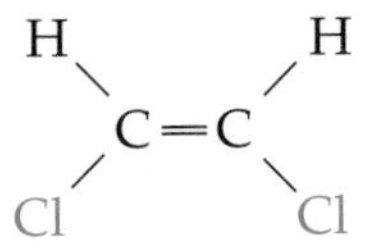

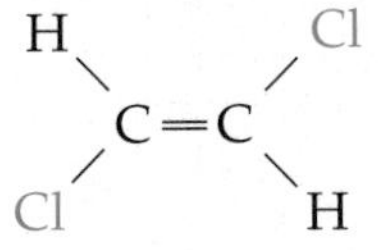

cis-1,2-dichloroethene   and   *trans*-1,2-dichloroethene

The prefix **cis-** means on the same side, and **trans-** means across or on opposite sides.

If the two groups or atoms attached to one of the doubly bonded carbon atoms are identical, only one structure is possible. The two groups or atoms attached to each carbon must be different for cis-trans isomerism to occur.

Cis-trans isomerism is also possible for ring structures. In cyclopropane, the three carbon atoms all lie in one plane. Substituents on adjacent carbons can be on the same side of the plane (cis) or on opposite sides of the plane (trans), such as

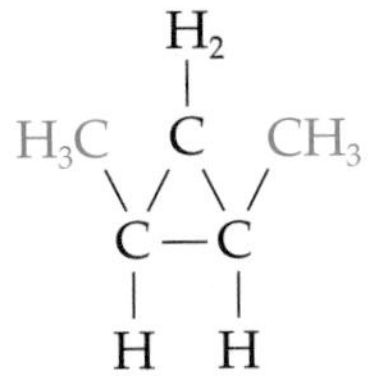

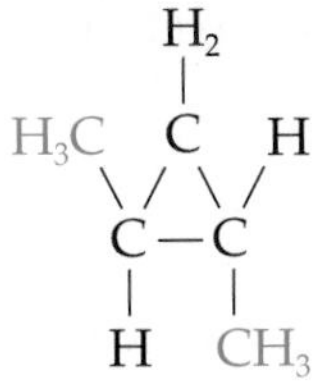

cis-1,2-dimethylcyclopropane   and   *trans*-1,2-dimethylcyclopropane

Cis-trans isomerism occurs in fatty acids (see Chapter 21) and is very important in the rhodopsin–vitamin A cycle (see page 555).

An example of a biologically active cis-trans isomer is cisplatin, a chemotherapeutic agent used in the treatment of cancer.

Exercise 16-5

Name the following compounds.

a.

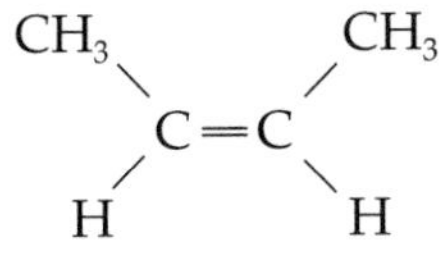

b.

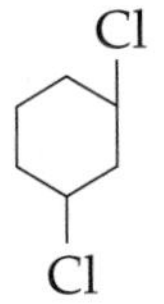

Solution

a. The compound contains a double bond between carbons 2 and 3 of a four-carbon compound, so it is a 2-butene. The two end carbons are on the same side of the double bond, so it is called *cis*-2-butene.
b. The compound contains a 6-membered ring, so it is a cyclohexane. The two chloro groups are on carbons 1 and 3, so it is called 1,3-dichlorocyclohexane. But the chlorines are on opposite sides of the ring, so the name is *trans*-1,3-dimethylcyclohexane.

Self-test

Draw the structures for the following.
a. *trans*-2-pentene   b. *cis*-1,2-dichlorocyclobutane.

Answers

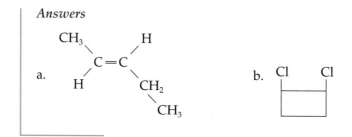

Examples of Biologically Important Alkenes

1. Ethene, a plant hormone responsible for ripening of fruits

$$CH_2=CH_2$$

2. Linolenic acid

$$CH_3CH_2-CH=CH-CH_2-CH=CH-CH_2-CH=CH-(CH_2)_7COOH$$

3. Arachidonic acid

$$CH_3(CH_2)_4-CH=CH-CH_2-CH=CH-CH_2-CH=CH-CH_2-CH=CH-(CH_2)_3COOH$$

4. β-carotene has eleven double bonds

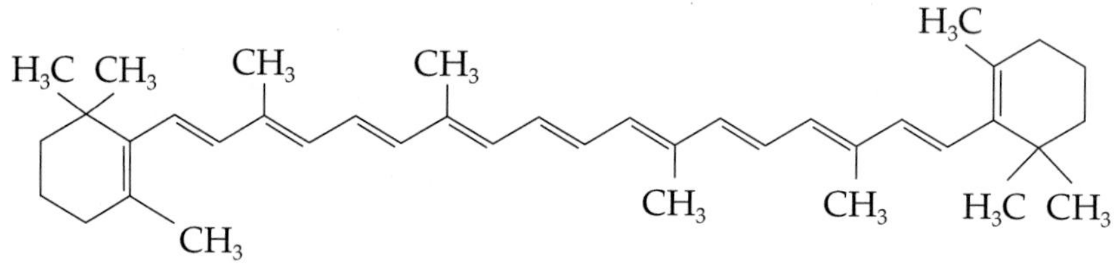

5. Vitamin A

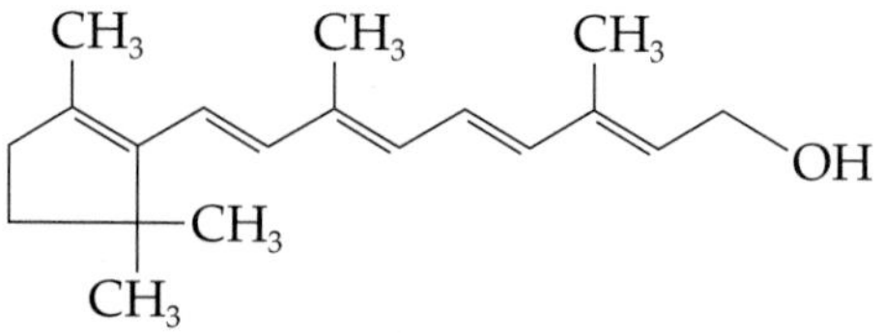

6. Isoprene, present in natural rubber

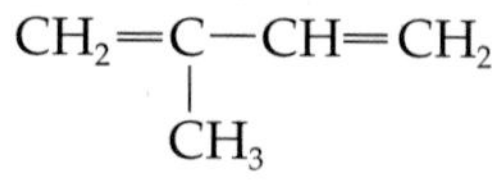

$$CH_2=C-CH=CH_2$$
$$\quad\;\; |$$
$$\quad\; CH_3$$

16-2 Alkynes

Consider two carbon atoms connected by a triple bond.

$$-C\equiv C-$$

How many hydrogen atoms must be connected to these two carbon atoms to satisfy all the bond requirements? The answer is two, so the molecular formula of this compound is C_2H_2. This compound is called ethyne. Its structure is

$$H-C\equiv C-H \qquad (H:C::C:H)$$

Ethyne is commonly called acetylene. However, this name is not preferred because the ending -*ene* denotes a double bond, whereas this compound actually has a triple bond between the carbon atoms.

If three carbon atoms are placed in a chain with a triple bond between two of them,

$$C\equiv C-C$$

only four hydrogen atoms can be placed around these carbons to satisfy all the bonds. The compound then becomes

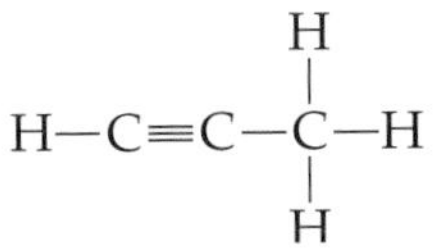

$$\begin{array}{c} H \\ | \\ H-C\equiv C-C-H \\ | \\ H \end{array}$$

with the molecular formula C_3H_4. This compound is called *propyne*.

These two compounds, ethyne and propyne, are **alkynes.** They have a triple bond between two of the carbon atoms. All their names end in -*yne*. The general formula for alkynes is

$$C_nH_{2n-2}$$

Thus hexyne, which has six carbon atoms, has the formula $C_6H_{(2\times6)-2}$, or C_6H_{10}. Likewise, octyne has the molecular formula C_8H_{14}.

Alkynes are relatively rare compounds and do not normally occur in the human body.

As with alkenes, alkynes are named with the triple bond having the smallest number.

$$HC\equiv C-CH_2-CH_2-CH_3 \qquad CH_3-C\equiv C-CH_2-CH_3$$

1-pentyne   2-pentyne

$$\begin{array}{c} CH_3-C\equiv C-CH-CH_2-CH_3 \\ | \\ CH_3 \end{array}$$

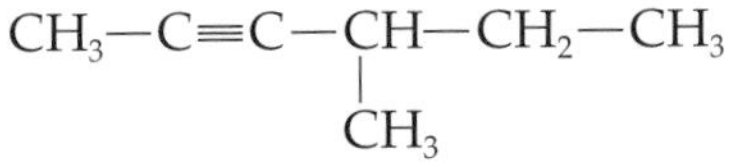

4-methyl-2-hexyne

There can be no hydrocarbon with four bonds between the carbon atoms because then there would be no bonds available for any hydrogen atoms. (Recall that hydrocarbons must contain both carbon and hydrogen.) Like alkenes, alkynes undergo addition reactions. Ethyne (acetylene) reacts with chlorine by a process of addition to form a single bond between the carbon atoms.

$$\begin{array}{c} \qquad\qquad\qquad\qquad\qquad\quad Cl \quad Cl \\ \qquad\qquad\qquad\qquad\qquad\quad | \quad\; | \\ H-C\equiv C-H + 2\,Cl-Cl \longrightarrow \quad H-C-C-H \\ \qquad\qquad\qquad\qquad\qquad\quad | \quad\; | \\ \qquad\qquad\qquad\qquad\qquad\quad Cl \quad Cl \end{array}$$

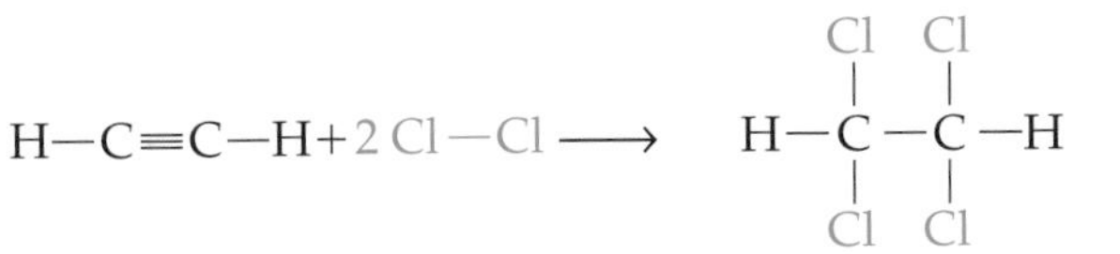

ethyne   chlorine   1,1,2,2-tetrachloroethane

16-3 Aromatic Compounds: Benzene

The term *aromatic* originally referred to certain compounds that had a pleasant odor and similar chemical and physcial properties. Further studies of these compounds showed that they all had a ring-shaped structure. However, many compounds have a pleasant odor and do not have this ring-shaped structure. The term aromatic is now usually used to designate compounds whose bonding has features in common with that of benzene.

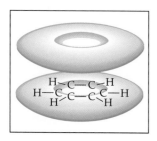

FIGURE 16-2

Electron-cloud picture of benzene.

Structure

Benzene has the formula C_6H_6. Since there is one hydrogen atom for every carbon atom, we might expect benzene to have several double bonds to fulfill the bonding requirements for carbon, thus making it unsaturated and predictably very reactive. On the contrary, however, benzene is quite stable. The structure for the benzene molecule was first deduced by Kekulé in 1865. He stated that the six carbon atoms were arranged in a ring with alternate single and double bonds, each carbon atom having one hydrogen atom attached to it. Kekulé also suggested that the position of the double and single bonds could change, producing two structures that represent benzene.

resonance structures of benzene

The actual structure of benzene is intermediate between these two Kekulé structures. This is indicated by the double-headed arrow. The two Kekulé structures are called *resonance structures. The electron-cloud picture (Figure 16-2) of the benzene molecule shows two continuous doughnut-shaped electron clouds, one above and the other below the plane of atoms.

The benzene structure is the basis of many thousands of organic compounds. Chemists have devised an abbreviated version of the benzene formula:

delocalized   localized
form   form

Abbreviated resonance structures for benzene are

In all these representations it is assumed that a hydrogen atom is present on each carbon or at each corner unless otherwise indicated.

If one of the hydrogen atoms is replaced by a methyl group (CH_3—), it is indicated as follows:

toluene

Properties

Benzene is a colorless liquid with a distinct gasolinelike odor. It is insoluble in water but soluble in alcohol and ether. Benzene is toxic when taken internally. Contact with the skin is harmful, and continued inhalation of benzene vapors decreases red and white blood cell counts. Benzene is now considered to be mildly carcinogenic, and care must be taken with its use. For this reason, the use of benzene in laboratory experiments by students has been, or should be, discontinued. Also, benzene, commonly known as naphtha, has been banned as an ingredient in most consumer products. In most experimental work, toluene can be used instead of benzene.

Halogen Derivatives

Although benzene has three double bonds, it does not easily undergo addition reactions as do alkenes. Instead, like alkanes, benzene usually undergoes substitution reactions. This difference results from resonance in the benzene structure. When benzene is treated with chlorine, chlorobenzene is produced.

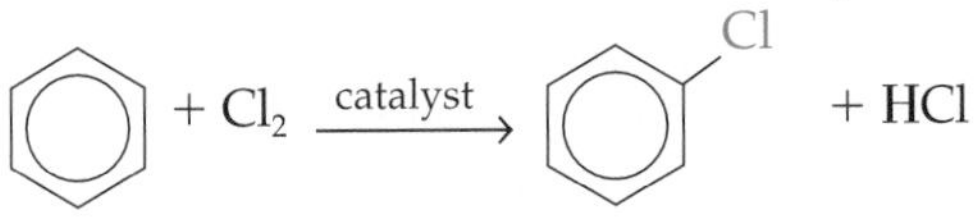

The structure of chlorobenzene indicates that one of the hydrogen atoms in the benzene ring has been replaced by a chlorine atom. Since all six carbon atoms and all six hydrogen atoms are equivalent in the benzene ring, there is only one possible monosubstitution product. That is, all of the structures shown in diagram (16-1) are identical—they represent the same compound, chlorobenzene.

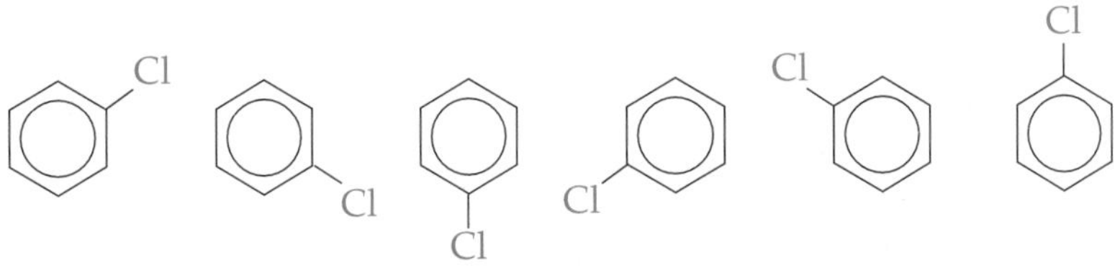

(16-1)

However, when two of the hydrogen atoms in the benzene ring are replaced, more than one possible *di*substitution product is possible. One of the disubstitution products obtained upon the reaction of chlorine with chlorobenzene is indicated as follows:

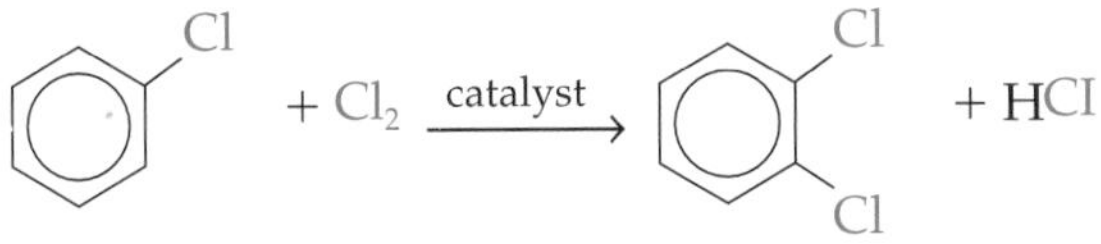

For the IUPAC name, the benzene ring is numbered from 1 to 6. The numbers must be such that the substituents attached to the ring have the lowest possible numbers, as shown in structure (16-2). The name of this compound is 1,2-dichlorobenzene, which indicates two chlorines on a benzene ring in positions 1 and 2.

(16-2)

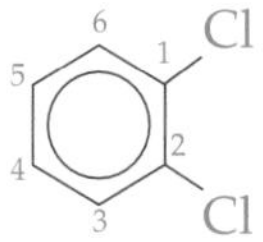

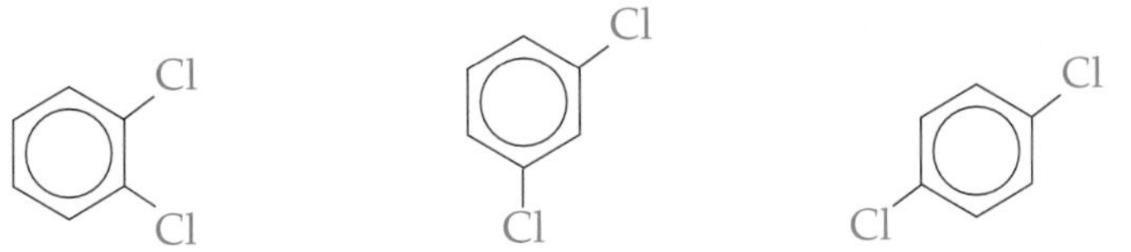

(16-3)

1,2-dichlorobenzene   1,3-dichlorobenzene   1,4-dichlorobenzene

Actually, when chlorine reacts with chlorobenzene, three different disubstitution products (isomers) are obtained. Their structures and names are indicated in (16-3).

If the disubstitution products on the benzene ring are different, the same system of naming may be used. Consider compound (16-4). It may be numbered with either the bromine atom at position 1 and the chlorine atom at position 2 or

(16-4)

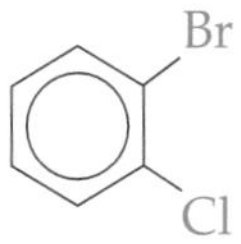

vice versa. Thus the compound may be named either 1-bromo-2-chlorobenzene or 1-chloro-2-bromobenzene. The preferred name lists the substituents in alphabetical order, but pharmaceutical companies frequently use several different systems for naming their drugs.

The common system for naming disubstituted benzene compounds is based upon the use of prefixes rather than numbers to designate positions in the benzene ring. The prefix *ortho-* indicates substances on the benzene ring in positions next to each other. The compound shown in (16-5a) is called *ortho*-dichlorobenzene, or simply *o*-dichlorobenzene.

a. 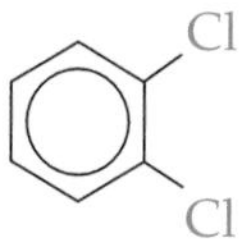   b.    (16-5)

The compound shown in (16-5b) may be called either *o*-chlorobromobenzene or *o*-bromochlorobenzene; the alphabetic sequence is preferred.

When substituents on the benzene ring are separated by one carbon atom (in positions 1 and 3), the prefix used is *meta-*. The compound shown in (16-6a) is called *meta*-dichlorobenzene or *m*-dichlorobenzene.

a. 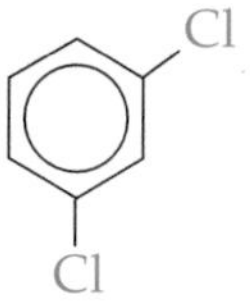   b. 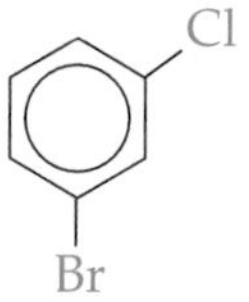   (16-6)

Structure (16-6b) is called *m*-bromochlorobenzene (note that the substituents are named in alphabetical order).

When two substituents on the benzene ring are separated by two carbon atoms (in positions 1 and 4), the prefix used is *para-*. The compound indicated in (16-7a) is called *para*-dichlorobenzene or *p*-dichlorobenzene. It is a modern version of mothballs.

a. 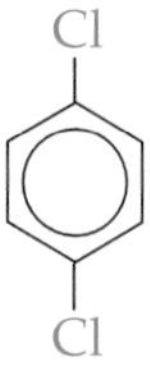   b. 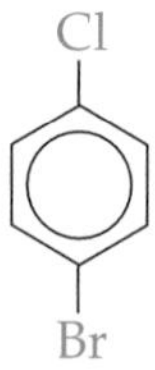   (16-7)

The structure of *p*-bromochlorobenzene is shown in (16-7b).

Other Derivatives

Toluene The methyl derivative of benzene is commonly called toluene. Its IUPAC name is methylbenzene. Its structure is

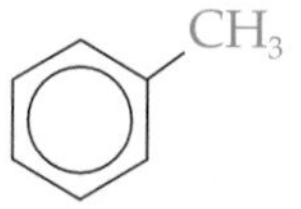

Toluene is a colorless liquid with a benzenelike odor. It is insoluble in water and soluble in alcohol and ether. Toluene is used as a preservative for urine specimens and in the preparation of dyes and explosives. Toluene is one ingredient in airplane glue; when "sniffed," it can produce blurred vision and a lack of coordination and may even be fatal. Toluene, however, is much less toxic than benzene and now is frequently being substituted for benzene. A derivative of toluene, trinitrotoluene (TNT), is a powerful explosive.

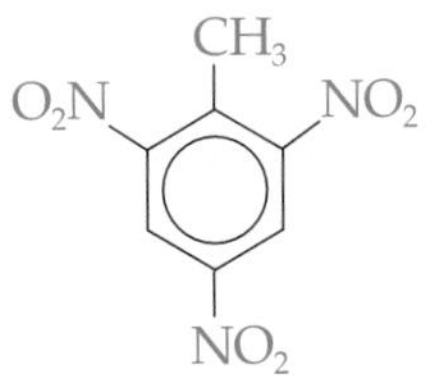

TNT

Exercise 16-6

Draw the structures for the following compounds.

a. 3-chlorotoluene
b. para-difluorobenzene

Solution

a. The chlorine is attached to carbon number 3 in the toluene molecule where the methyl group is assumed to be at carbon number 1.

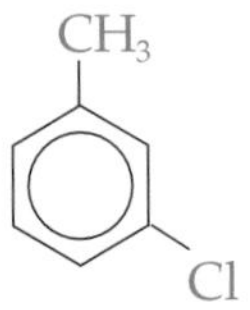

b. Para- indicates that the two fluorines are opposite one another on the benzene ring (at positions 1 and 4), or

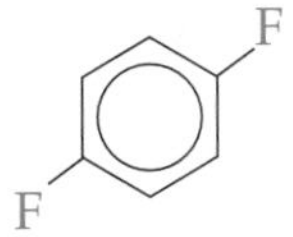

Self-test

Draw structures for the following compounds.

a. orthobromotoluene
b. 1,4-dibromobenzene

Answers

a.

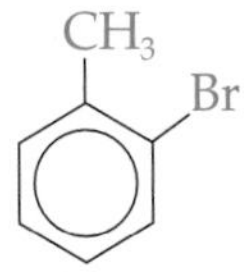

b.

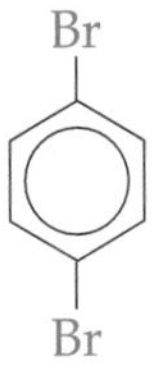

Xylene   Xylene is a dimethylbenzene. Again, with two substituents on the benzene ring, there are three possible structures of xylene.

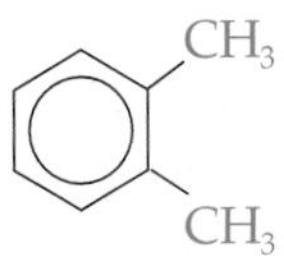

o-xylene 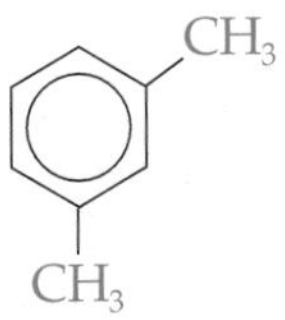   *m*-xylene 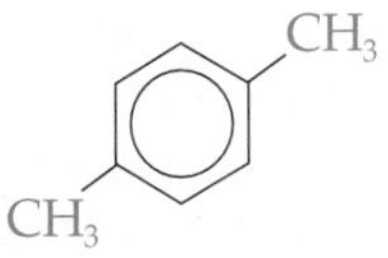   *p*-xylene

Xylenes are good solvents for oils and are used in cleaning lenses in microscopes.

Naphthalene   Naphthalene, $C_{10}H_8$, is an aromatic compound containing two benzene rings. These two rings are attached to each other in such a manner that they share two carbon atoms.

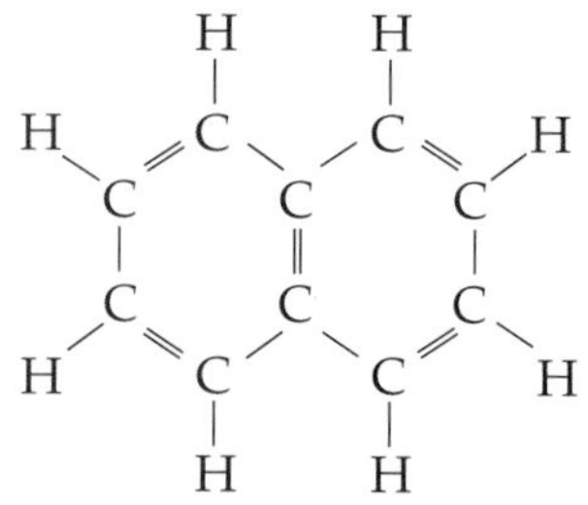

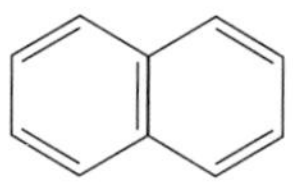

Naphthalene is a white crystalline solid obtained from coal tar. Naphthalene crystals were frequently used in the home under the name mothballs. Functional groups may be attached to naphthalene in either of two positions, called alpha (α) and beta (β). For example, the structures of α- and β-naphthols are

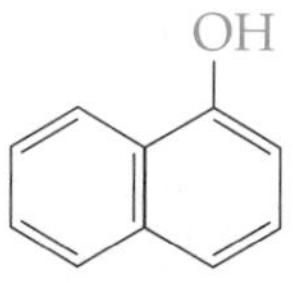

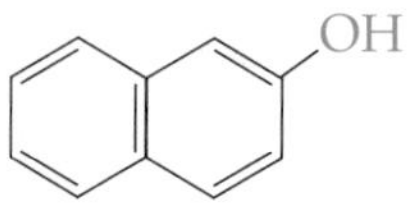

α-naphthol   β-naphthol

Note that naphthalene has four α positions and also four β positions.

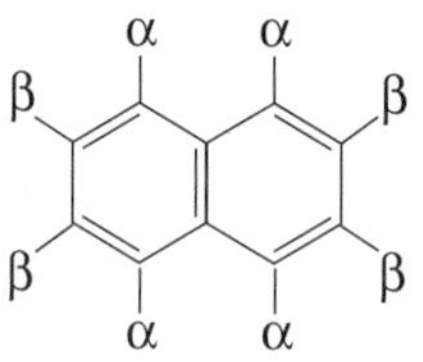

Anthracene and Phenanthrene

Anthracene and phenanthrene are aromatic compounds containing three benzene rings joined together. Their structures are

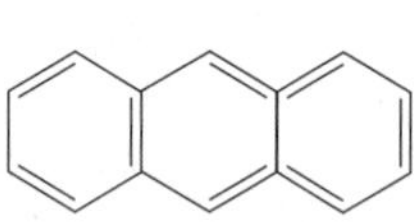

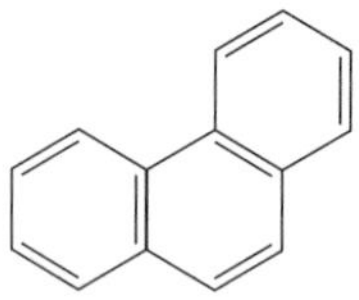

anthracene   phenanthrene

Anthracene is used commercially in the manufacture of dyes. Phenanthrene is an isomer of anthracene. It also contains three benzene rings but in a different structural arrangement. Phenanthrene has the basic structure of many biologically and medically important compounds. Among these are the male and female sex hormones, vitamin D, cholesterol, bile acids, and some alkaloids.

Other Polycyclic Aromatic Compounds

When coal or wood is burned, compounds containing several benzene rings are obtained. Such compounds are carcinogenic (cancer producing). One of the most active of this group of carcinogens is benzpyrene. In the seventeenth century, chimney sweeps in England were the first people known to develop cancer from such compounds, which are also found in the "tar" of cigarettes. In industrialized nations, large amounts of benzpyrene are emitted into the air. This dangerous substance is also found in well-done charbroiled meats and smoked fish.

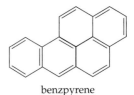

benzpyrene

Summary

Alkenes have a double bond between two of the carbon atoms. The names of all alkenes end in -*ene*. The general formula for alkenes is C_nH_{2n}.

Alkynes have a triple bond between two of the carbon atoms and have the general formula C_nH_{2n-2}. The names of all alkynes end in -*yne*.

A saturated hydrocarbon has only single bonds between the carbon atoms. An unsaturated hydrocarbon has double or triple bonds between its carbon atoms. Saturated hydrocarbons react by the process known as substitution. Unsaturated hydrocarbons react by the process known as addition. The addition reaction is usually much more rapid than the substitution reaction.

The chief sources of hydrocarbons are petroleum and natural gas.

Polymers consist of many small molecules joined together. Examples are polyethylene and polypropylene.

Since the carbon atoms in a double bond cannot rotate, the groups attached to that double bond are fixed on one side or the other. Such compounds are called cis-trans isomers. Cis-trans isomers can also exist in cyclic compounds.

Benzene (C_6H_6) is a symmetrical six-sided ring compound with a hydrogen at each corner. One or more of the hydrogen atoms of benzene may be replaced by a halogen yielding halogen derivatives.

There is only one monosubstitution product of benzene for any substituent because all six positions on the ring are equivalent.

There are three possible disubstitution products of benzene for any pair of substituents. These compounds can be named numerically. If both the substituents are chlorine, the disubstitution products are 1,2-dichlorobenzene, 1,3-dichlorobenzene, and 1,4-dichlorobenzene.

Prefixes are also used to designate positions on the benzene ring. The prefix *ortho*- corresponds to positions 1,2; *meta*- to 1,3; and *para*- to 1,4.

Naphthalene contains two benzene rings joined together. Anthracene and phenanthrene contain three benzene rings joined together.

Questions and Problems

1. Classify the following as alkanes, alkenes, or alkynes.
   a. C_5H_{10} b. C_9H_{16} c. C_7H_{14} d. C_3H_8
2. Name the following.
   a. An alkene of four carbon atons; of eight carbon atoms.
   b. An alkyne of three carbon atoms; of six carbon atoms.
3. What does the term *saturated* refer to in hydrocarbons? *unsaturated*?
4. How do unsaturated hydrocarbons react? Give an example.

5. Name the following compounds.

   a.
   $$CH_3-CH=CH-\overset{\overset{\displaystyle Cl}{|}}{CH_2}$$

   b.
   $$CH_3-\overset{\overset{\displaystyle }{|}}{\underset{\underset{\displaystyle CH_3}{|}}{CH}}-C\equiv C-CH_3$$

   c.
   $$\overset{\overset{\displaystyle Cl}{|}}{CH}=\overset{\overset{\displaystyle Cl}{|}}{C}-CH_3$$

6. Draw structures for the following compounds.
   a. 3-octene
   b. 3-methyl-2-heptene
   c. cyclohexene
   d. 2-pentyne
   e. 3,3-dimethylhexyne
7. Compare the complete and incomplete oxidations of alkenes.
8. How are polymers formed? Give examples of two.
9. What do the terms *cis* and *trans* refer to?
10. Why are cis-trans isomers important biologically?
11. Complete the following reactions and name the products.

   a. $CH_2{=}CH{-}CH_3 + Cl_2 \longrightarrow$

   b. 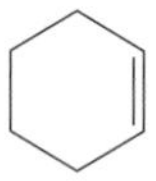 $+ \; HCl \longrightarrow$

   c. 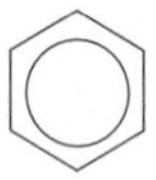 $+ \; Cl_2 \xrightarrow{\text{catalyst}}$

   d. $CH_3{-}C{\equiv}C{-}CH_3 + Br_2 \longrightarrow$

12. Draw the resonance structures for benzene; the abbreviated structure.
13. Draw the structure for chlorobenzene; 1,2-dibromobenzene; *m*-dichlorobenzene.
14. Why has the use of benzene in laboratories been discontinued?
15. Why is there only one monosubstitution product with any given substituent for benzene?

16. Why are there three and only three disubstitution products for benzene?
17. Give the IUPAC name for the following compounds.

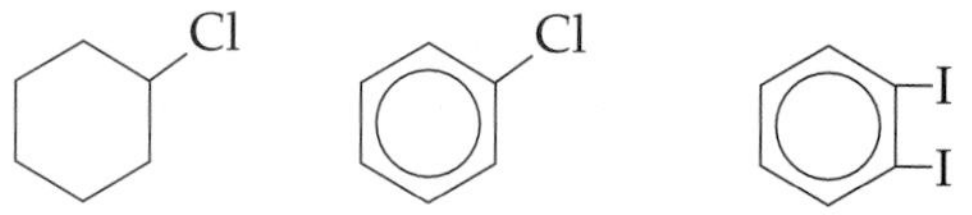

18. Draw the structures of naphthalene; anthracene; phenanthrene.
19. Why does naphthalene have only two monosubstitution products?
20. Why is the phenanthrene structure important biologically?
21. Benzpyrene contains how many rings? Where is it found? Why is it a dangerous compound?
22. Compare the structures of cyclohexane and cyclohexene; cyclohexene and benzene.
23. Name and draw all of the monosubstitution derivatives of the following with chlorine.
    a. benzene
    b. toluene
24. Write balanced equations for the combustion of benzene.
25. Draw structures for the following.
    a. 3-bromotoluene
    b. para-dichlorobenzene
    c. vinyl chloride
    d. acetylene
26. Write an equation for each of the following reactions and name the organic product(s).
    a. propene + chlorine
    b. propyne + chlorine
    c. propene + HCl
    d. propyne + HCl

Practice Test

1. Alkynes react primarily by:
   a. addition
   b. substitution
   c. polymerization
   d. hydrogenation
2. The structure $\triangle$ represents:
   a. cyclopropane
   b. cyclopropene
   c. propane
   d. propene
3. Which of the following is an alkene?
   a. C_2H_2
   b. C_6H_{12}
   c. $C_{12}H_{22}$
   d. none of these
4. Which of the following is saturated?
   a. an alkane
   b. an alkene
   c. an alkyne
   d. a cyclic compound
5. $HC{\equiv}CH$ is called _____.
   a. ethane
   b. ethene
   c. ethyne
   d. ethylene
6. Vitamin D and the sex hormones have the same basic structure as _____.
   a. anthracene
   b. naphthalene
   c. benzene
   d. phenanthrene

7. 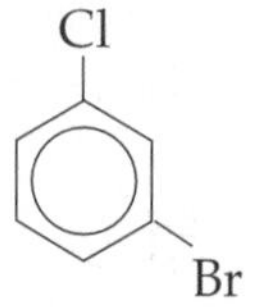 is called _____.
   a. chlorobromobenzene
   b. *o*-bromochlorobenzene
   c. *m*-bromochlorobenzene
   d. *p*-chlorobromobenzene
8. 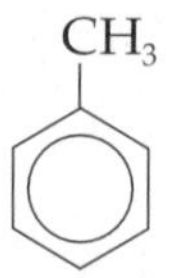 is called _____.
   a. benzocaine
   b. toluene
   c. anthracene
   d. methene
9. How may disubstitution products are possible for benzene?
   a. 1
   b. 2
   c. 3
   d. 4

Alcohols, Thiols, Phenols, and Ethers

Objectives

- To become familiar with the functional group present in alcohols

- To name various alcohols

- To become familiar with the uses of some alcohols

- To distinguish between primary, secondary, and tertiary alcohols

- To distinguish between alcohols and thiols

- To recognize the structure of and the importance of phenols and phenol derivatives

- To recognize the structure and the importance of alcohol and phenol derivatives

- To draw the structure of and name various ethers

- To write the reaction for the formation of an ether

- To become familiar with the uses of ethers

Ethylene glycol, an alcohol, is commonly used in antifreeze because it extends the temperature range of the liquid phase of water.

17-1 Alcohols

A functional group is a particular arrangement of a few atoms that imparts certain characteristic properties to an organic molecule. For example, alcohols are derivatives of hydrocarbons in which one or more of the hydrogen atoms has been replaced by a hydroxyl (—OH) functional group. Other functional groups will be discussed in Chapter 18 (also see the inside back cover).

In the IUPAC system, alcohols are named for the longest continuous chain containing the —OH group, with that functional group having the lowest numbers. The ending -*e* is changed to -*ol* to indicate the alcohol functional group. Thus, the alcohol derived from methane is methanol, and the alcohol derived from ethane is ethanol.

In the common system, alcohols are named by taking the name of the alkyl group and adding the word *alcohol*.

Examples of the names of simple alcohols in these two systems are

$$CH_3-OH \qquad CH_3-CH_2-OH \qquad CH_3-\overset{\displaystyle |}{\underset{\displaystyle OH}{CH}}-CH_3$$

methanol   ethanol   2-propanol

(methyl alcohol)   (ethyl alcohol)   (isopropyl alcohol)

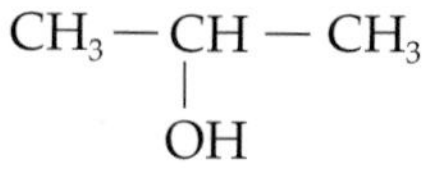

More complex alcohols are named according to the IUPAC systems.

Exercise 17-1

Name the following compound.

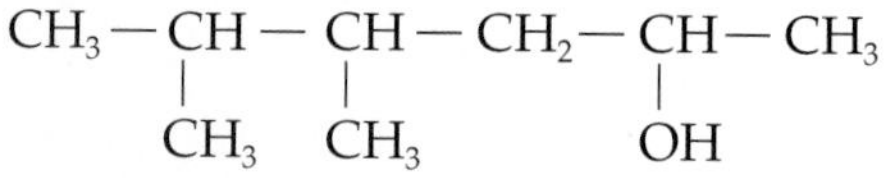

$$CH_3-\overset{\displaystyle |}{\underset{\displaystyle CH_3}{CH}}-\overset{\displaystyle |}{\underset{\displaystyle CH_3}{CH}}-CH_2-\overset{\displaystyle |}{\underset{\displaystyle OH}{CH}}-CH_3$$

Solution

The longest chain containing an —OH group is six carbon atoms, so this is a hexanol. The —OH group is on carbon number 2, numbering from right to left, so that the —OH group will have the lowest number.

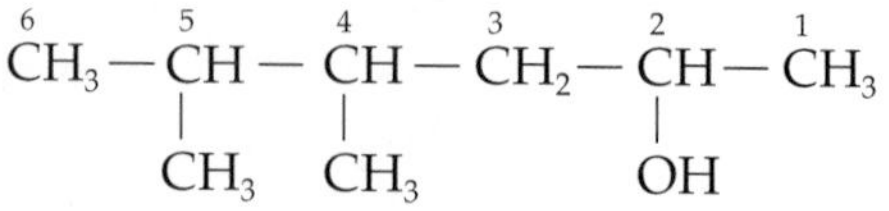

$$\overset{6}{C}H_3-\overset{5}{\overset{\displaystyle |}{\underset{\displaystyle CH_3}{CH}}}-\overset{4}{\overset{\displaystyle |}{\underset{\displaystyle CH_3}{CH}}}-\overset{3}{C}H_2-\overset{2}{\overset{\displaystyle |}{\underset{\displaystyle OH}{CH}}}-\overset{1}{C}H_3$$

Two methyl groups are attached to the chain, one at carbon number 4 and one at carbon number 5, so the name of this compound is 4,5-dimethyl-2-hexanol.

Exercise 17-2

Name

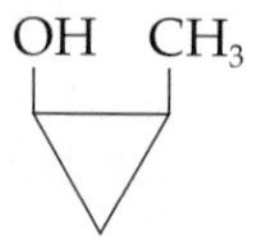

Solution

This is a cyclic alcohol of three carbon atoms, so it is a cyclopropanol. Because the —OH functional group is given the lowest number, it is assumed to be at carbon number 1 (understood). Therefore, the methyl group must be at carbon number 2. Thus, the name of the compound is 2-methylcyclopropanol.

Exercise 17-3

Name

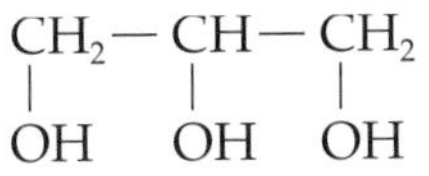

$$CH_2-CH-CH_2$$
$$|\quad\ |\quad\ |$$
$$OH\quad OH\quad OH$$

Solution

This alcohol contains three —OH groups and three carbon atoms, so it is a propanetriol. The —OH groups are at carbon numbers 1, 2, and 3. The name is 1,2,3-propanetriol. The common names for the compound are glycerol or glycerin.

Writing Structures of Alcohols

Exercise 17-4

Draw the structure of 3,5-dichloro-4-methyl-2-octanol.

Solution

2-octanol indicates an eight-carbon alcohol with the —OH group on carbon number 2:

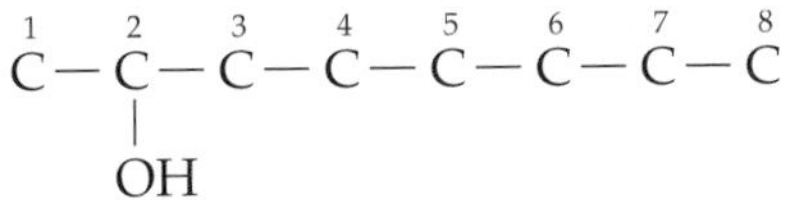

$$\overset{1}{C}-\overset{2}{C}-\overset{3}{C}-\overset{4}{C}-\overset{5}{C}-\overset{6}{C}-\overset{7}{C}-\overset{8}{C}$$
$$|$$
$$OH$$

3,5-dichloro indicates two chlorines, one at carbon number 3 and one at carbon number 5.

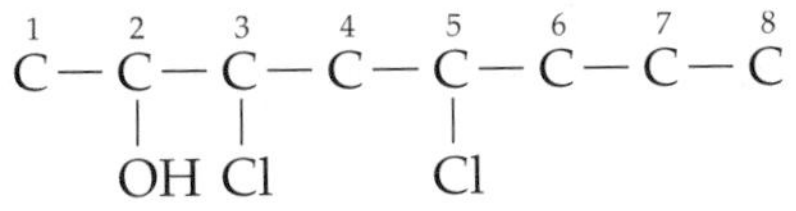

$$\overset{1}{C}-\overset{2}{C}-\overset{3}{C}-\overset{4}{C}-\overset{5}{C}-\overset{6}{C}-\overset{7}{C}-\overset{8}{C}$$
$$|\quad\ |\qquad\ |$$
$$OH\ Cl\qquad Cl$$

4-methyl indicates a methyl group at carbon number 4. Adding hydrogens to complete all the carbon bonds, we have

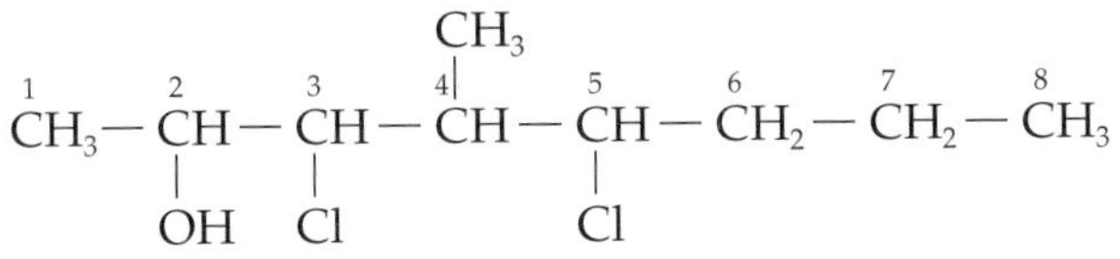

$$\qquad\qquad\qquad\overset{4}{\underset{|}{C}H_3}$$
$$\overset{1}{C}H_3-\overset{2}{C}H-\overset{3}{C}H-\overset{4}{C}H-\overset{5}{C}H-\overset{6}{C}H_2-\overset{7}{C}H_2-\overset{8}{C}H_3$$
$$\qquad\ |\qquad\ |\qquad\qquad |$$
$$\qquad OH\quad Cl\qquad\quad Cl$$

Self-test

Draw structures for these compounds.
a. 3-methyl-2-pentanol b. *cis*-1,3-cyclohexanediol

Answer

a. 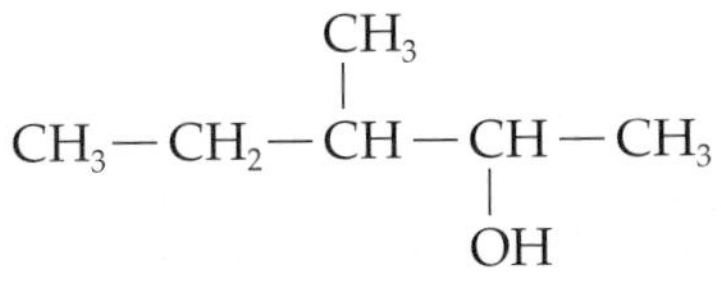
$$\qquad\qquad CH_3$$
$$\qquad\qquad\ |$$
$$CH_3-CH_2-CH-CH-CH_3$$
$$\qquad\qquad\qquad |$$
$$\qquad\qquad\qquad OH$$

b.

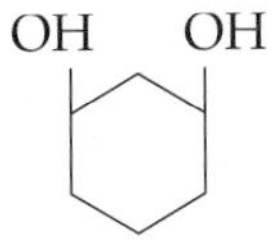

Although IUPAC nomenclature is preferred by chemists, medical personnel usually refer to alcohols by their common names. We will do so in the following sections.

The general formula for an alcohol is ROH, where the R signifies an alkyl group attached to an —OH functional group.

Because alcohols do not ionize, their reactions are much slower than those of inorganic bases, which contain a hydroxide ion (OH^-). Solutions of alcohols are nonelectrolytes; they are not bases. However, alcohols do react with acids to form compounds called esters, which will be discussed in Chapter 18.

Since oxygen is more electronegative than either carbon or hydrogen, alcohols are polar compounds. This polarity gives rise to hydrogen bonding between alcohol molecules and accounts for their relatively high boiling points.

The presence of an —OH group increases the ability of a substance to dissolve in water. Thus methyl alcohol and ethyl alcohol are quite soluble in water. However, as the length of the carbon chain increases, the molecules become more and more like alkanes and less and less like alcohols, so the solubility in water decreases.

Uses

Methyl Alcohol Methyl alcohol (methanol), CH_3OH, is commonly known as wood alcohol. It is used as a solvent in many industrial reactions. Methyl alcohol should never be applied directly to the body, neither should the vapors be inhaled because this substance can be absorbed both through the skin and through the respiratory tract. Ingestion of as little as 15 mL of methyl alcohol can cause blindness, and 30 mL can cause death.

Methy alcohol can be prepared from the distillation of wood. It is prepared commercially from carbon monoxide.

$$CO + 2\,H_2 \xrightarrow[\text{heat}]{\text{Pt catalyst}} CH_3OH$$

Ethyl Alcohol Ethyl alcohol (ethanol), CH_3CH_2OH, is known commonly as grain alcohol. In the hospital the word *alcohol* means ethyl alcohol.

One important property of ethyl alcohol is its ability to denature protein. Because of this property, ethyl alcohol is widely used as an antiseptic.

As an antiseptic, 70 percent alcohol is preferred to a stronger solution. It would seem that if 70 percent alcohol is a good antiseptic, then 100 percent alcohol would be even better; however, the reverse is true. The 70 percent alcohol is actually a better antiseptic than the 100 percent alcohol.

Pure alcohol coagulates protein on contact. Suppose that pure alcohol is poured over a single-celled organism. The alcohol will penetrate the cell wall of that organism in all directions, coagulating the protein just inside the cell wall, as shown in Figure 17-1. This ring of coagulated protein would then prevent the alcohol from penetrating farther into the cell, and no more coagulation would take place. At this time the cell would become dormant, but not dead. Under the proper conditions the organism could again begin to function. If 70 percent alcohol is poured over a single-celled organism, the diluted alcohol also coagulates the protein, but at a slower rate, so that it penetrates all the way through the cell before coagulation can block it. Then all the cell proteins are coagulated, and the organism dies (Figure 17-2).

Alcohol (ethyl) can also be used for sponge baths to reduce the fever of a patient. When alcohol is placed on the skin it evaporates rapidly. In order to evaporate, alcohol requires heat. This heat comes from the patient's skin. Thus an

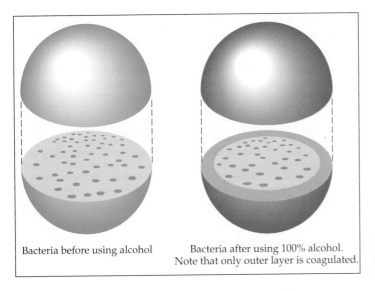

Bacteria before using alcohol

Bacteria after using 100% alcohol. Note that only outer layer is coagulated.

FIGURE 17-1
Effect of 100 percent alcohol on bacteria.

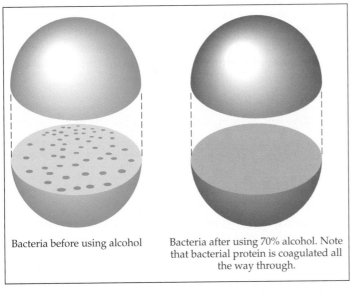

Bacteria before using alcohol

Bacteria after using 70% alcohol. Note that bacterial protein is coagulated all the way through.

FIGURE 17-2
Effect of 70 percent alcohol on bacteria.

alcohol sponge bath will remove heat from the patient's skin and so lower the body temperature. A water sponge bath will do the same thing, but water evaporates more slowly than alcohol so the heat is removed more slowly. However, water sponge baths are in common use in many hospitals because they are cheaper. For children it is kinder to allow them to play in a tepid bath. The alcohol sponge bath cools them too rapidly and is uncomfortable. Since alcohol is flammable it cannot be used in a room where oxygen is in use.

Alcohol is used as a solvent for many substances. Alcohol solutions are called tinctures. Tincture of iodine consists of iodine dissolved in alcohol.

Ethyl alcohol is also used as a beverage. The concentration of alcohol in alcoholic beverages is expressed as "proof." The proof is twice the percentage of alcohol in the solution. Thus a beverage marked "100 proof" contains 50 percent alcohol. Alcohol slows reaction time, so driving under the influence of alcohol can be very dangerous. Alcohol is not a stimulant; it actually depresses the nervous system and can remove an individual's normal inhibitions. Excessive use of alcohol may cause the destruction of the liver, a condition known as cirrhosis.

Alcohol (ethyl) can be prepared from the fermentation of blackstrap molasses, the residue that results from the purification of cane sugar. The principal constituent of molasses is sucrose ($C_{12}H_{22}O_{11}$). The fermentation reaction is brought about by the enzymes present in yeast.

$$C_{12}H_{22}O_{11} + H_2O \xrightarrow{\text{enzymes}} 4\,C_2H_5OH + 4\,CO_2$$
$$\text{sucrose} \qquad\qquad\qquad \text{ethyl alcohol}$$

The starches present in grains can be converted into sugar by malt. The sugar thus produced can be fermented under the influence of the enzymes in yeast to yield ethyl alcohol. Hence ethyl alcohol is also known as grain alcohol.

Ethyl alcohol can also be prepared synthetically from ethene.

$$CH_2{=}CH_2 + H_2O \xrightarrow{\text{catalyst}} C_2H_5OH$$
$$\text{ethene} \qquad\qquad\qquad \text{ethyl alcohol}$$

Isopropyl Alcohol   Isopropyl alcohol has the structural formula shown below. The IUPAC name for this compound is 2-propanol, indicating that the —OH functional group is on the second carbon of a three-carbon chain. Isopropyl alcohol is toxic and should not be taken internally. Since it is not absorbed through the skin, it is commonly used as rubbing alcohol and as an *astringent.

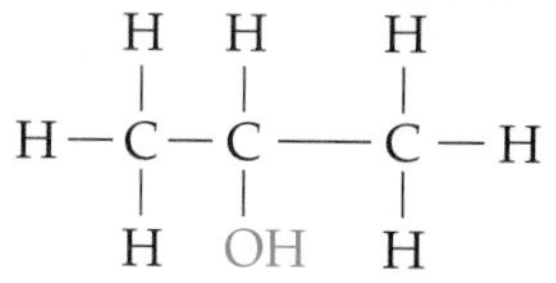

isopropyl alcohol
2-propanol

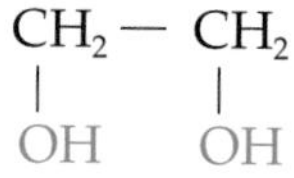

ethylene glycol
1,2-ethanediol

Ethylene Glycol   All of the previously mentioned alcohols have one —OH function group. They are called *monohydric alcohols*. Glycols are compounds with two —OH function groups. They are examples of *dihydric alcohols*.

Ethylene glycol has a high boiling point due to extensive hydrogen bonding. It is also completely miscible with water.

If taken internally, ethylene glycol (glycol) is extremely toxic because it is oxidized in the liver to oxalic acid. Oxalic acid, in turn, crystallizes as a calcium salt, calcium oxalate, in the kidneys, causing renal damage, which can lead to kidney failure and death.

The antidote for ethylene glycol poisoning (and also for methyl alcohol poisoning) is administration of high levels of ethyl alcohol. This causes the liver enzymes to oxidize the ethyl alcohol rather than the poisonous alcohol, which is present in smaller amounts.

Ethylene glycol is used in preparations to moisten the skin. It is also used as a permanent antifreeze in car radiators and as a raw material in the manufacture of the polymer Dacron. The IUPAC name for ethylene glycol is 1,2-ethanediol.

Glycerol   Gylcerol (sometimes called glycerin) is a trihydric or trihydroxy alcohol; it contains three —OH groups. The IUPAC name for glycerol is 1,2,3-propanetriol, since there is an — OH functional group on each carbon atom of the three-carbon chain.

$$H-\overset{\overset{\displaystyle H}{|}}{\underset{\underset{\displaystyle OH}{|}}{C}}-\overset{\overset{\displaystyle H}{|}}{\underset{\underset{\displaystyle OH}{|}}{C}}-\overset{\overset{\displaystyle H}{|}}{\underset{\underset{\displaystyle OH}{|}}{C}}-H$$

1,2,3-propanetriol/glycerol

Glycerol is an important alcohol in terms of body chemistry, especially as a constituent of fats (See Chapter 21). It is a by-product of the manufacture of soap (see page 365) and is used in the preparation of cosmetics and hand lotions and also in suppositories. Glycerol is used in the laboratory as a lubricant for rubber tubing and stoppers. When treated with nitric acid, glycerin forms nitroglycerin, an explosive. Medicinally, nitroglycerin is used to treat angina or heart pain. It causes a dilation of the coronary arteries, thus increasing the supply of blood to the heart muscles.

How does nitroglycerin work? It is only recently that the mechanism has been worked out. The process is as follows: nitroglycerin liberates nitric oxide from the endothelial cells of the blood vessels. The NO binds to the heme component of the enzyme guanylate cyclase, which in turn catalyzes the synthesis of cyclic guanosine monophosphate (c-GMP). This c-GMP acts as a signal that induces muscular relaxation in the blood vessels.

Other Alcohols   Menthol is an example of a cyclic alcohol. It has a cooling, refreshing feeling when rubbed on the skin and so is a frequently used ingredient in cosmetics and shaving lotions. Menthol is used in cough drops and nasal sprays.

menthol

Other alcohols of biologic importance are cholesterol, retinol (vitamin A), and tocopherol (vitamin E). These will be discussed later in appropriate chapters.

Types

Primary Alcohols   A primary ($1°$) alcohol is one that contains an —OH functional group attached to a carbon that has one or no carbon atoms attached to it. All the structures in diagram (17-1) are primary alcohols.

$$H-\overset{\overset{\displaystyle H}{|}}{\underset{\underset{\displaystyle H}{|}}{C}}-OH \qquad H-\overset{\overset{\displaystyle H}{|}}{\underset{\underset{\displaystyle H}{|}}{C}}-\overset{\overset{\displaystyle H}{|}}{\underset{\underset{\displaystyle H}{|}}{C}}-OH \qquad H-\overset{\overset{\displaystyle H}{|}}{\underset{\underset{\displaystyle H}{|}}{C}}-\overset{\overset{\displaystyle H}{|}}{\underset{\underset{\displaystyle H}{|}}{C}}-\overset{\overset{\displaystyle H}{|}}{\underset{\underset{\displaystyle H}{|}}{C}}-OH \qquad (17\text{-}1)$$

methyl alcohol   ethyl alcohol   propyl alcohol

Methyl alcohol (methanol) is an example of a primary alcohol in which the —OH functional group is attached to a carbon atom with no other carbons attached to it. Ethyl alcohol (ethanol) and propyl alcohol (propanol) are examples of primary alcohols in which the —OH function group is attached to a carbon atom with one carbon atom attached to it. Note that in a primary alcohol the functional group (—OH) is at the end of the chain.

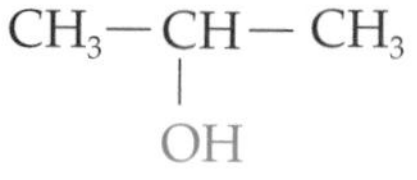

$$CH_3 - \underset{\underset{OH}{|}}{CH} - CH_3$$

isopropyl alcohol
(2-propanol)

Secondary Alcohols A secondary (2°) alcohol is one in which the —OH is attached to a carbon atom having two other carbon atoms attached to it. Isopropyl alcohol is an example of a secondary alcohol. Note that in a secondary alcohol the functional group is not at the end of the chain.

Tertiary Alcohols A tertiary (3°) alcohol is one in which the —OH is attached to a carbon atom that has three carbon atoms attached to it.

$$CH_3 - \underset{\underset{OH}{|}}{\overset{\overset{CH_3}{|}}{C}} - CH_3$$

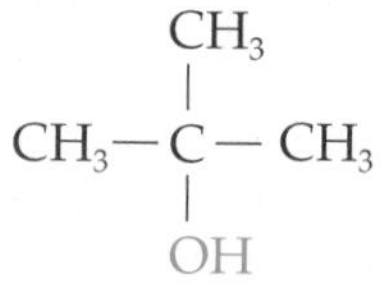

2-methyl-2-propanol, commonly known
as tertiary butyl alcohol

Reactions

Dehydration In the presence of a dehydrating agent, such as H_2SO_4, alcohols can be dehydrated to form alkenes.

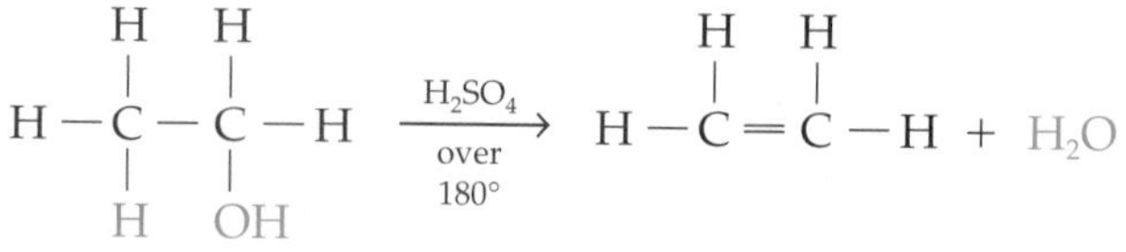

ethanol   ethene

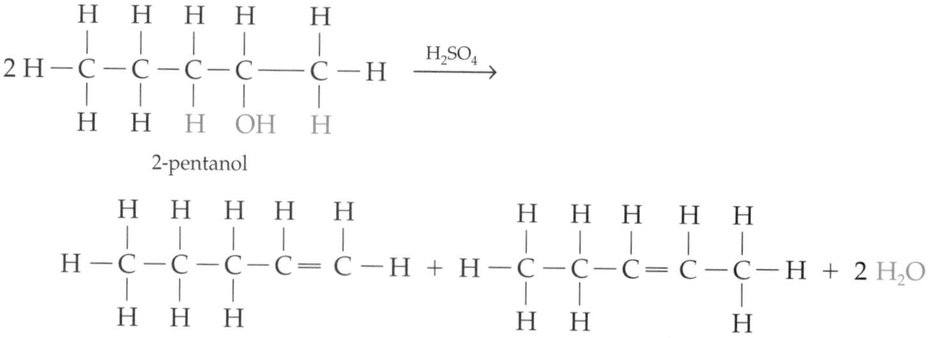

2-pentanol

1-pentene   2-pentene

Formation of Ethers As will be shown in Section 17-4, alcohols can also react in the presence of H_2SO_4 to form ethers.

Oxidation Primary alcohols can be oxidized to form aldehydes and then further oxidized to form acids and eventually CO_2 and H_2O. These reactions will be discussed in the next chapter. Some of these oxidation products are responsible for the toxicity of various alcohols.

Secondary alcohols can be oxidized to yield ketones (see Chapter 18). Tertiary alcohols cannot be oxidized under ordinary conditions. Alcohols also react with organic acids to yield compounds called esters (see Section 19-1).

17-2 Thiols

Thiols are sulfur analogs of alcohols and contain an —SH functional group in place of an —OH group. The IUPAC names of thiols are formed by adding the ending -*thiol* to the name of the parent hydrocarbon. Note that the -*e* ending of the parent compound is not deleted. The common names of thiols were formed by first naming the alkyl group and then adding the word mercaptan.

$$CH_3 - CH - CH_3$$
$$\hspace{2.7cm}|$$
$$CH_3 - SH \hspace{2.5cm} SH$$

methanethiol
(methyl mercaptan)

2-propanethiol
(isopropyl mercaptan)

Many thiols are found in nature, and they all have a disagreeable odor. When an onion is cut, 1-propanethiol is released; garlic owes its odor to the presence of thiols; and thiols are responsible for the odor given off by skunks. Since natural gas used for heating and cooking is odorless, thiols are added so that a gas leak can be easily detected.

Unlike alcohols, thiols do not exhibit hydrogen bonding. Therefore, they have lower boiling points than the corresponding alcohols. Also, since they do not exhibit hydrogen bonding, thiols are less soluble in water than alcohols with the same number of carbon atoms.

Thiols can be prepared by heating alkyl halides with sodium hydrogen sulfide, (NaHS).

$$CH_3 - CH_2 - CH_2 - Cl + NaHS \longrightarrow CH_3 - CH_2 - CH_2 - SH + NaCl$$

Thiols are easily oxidized to disulfides.

$$R - CH_2 - SH \xrightarrow{O_2} R - CH_2 - S - S - CH_2 - R$$

a thiol   a disulfide

An example of such an oxidation is the conversion of cysteine to cystine, a reaction that takes place when hair is given a "permanent" (see Section 6-5).

$$2\ H - \overset{\displaystyle COOH}{\underset{\displaystyle NH_2}{\overset{\displaystyle |}{\underset{\displaystyle |}{C}}}} - CH_2 - SH \underset{\text{reduction}}{\overset{\text{oxidation}}{\rightleftharpoons}} H - \overset{\displaystyle COOH}{\underset{\displaystyle NH_2}{\overset{\displaystyle |}{\underset{\displaystyle |}{C}}}} - CH_2 - S - S - CH_2 - \overset{\displaystyle COOH}{\underset{\displaystyle NH_2}{\overset{\displaystyle |}{\underset{\displaystyle |}{C}}}} - H$$

cysteine   cystine

Disulfide bonds are also involved in the formation of the tertiary structure of some proteins (see page 390).

Thiols (also called sulfhydryls) are also important in enzymes involved in carbohydrate metabolism.

17-3 Phenols

When an —OH group is attached to a benzene ring, a class of compounds known as *phenols* is formed. Generally, phenols are like alcohols but have been placed in a class by themselves because phenols are weak acids and alcohols are not.

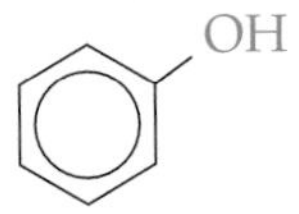

phenol

Phenol reacts with aqueous sodium hydroxide solution to form a water-soluble salt, sodium phenoxide. Alcohols such as ethanol and cyclohexanol do not react with basic solutions.

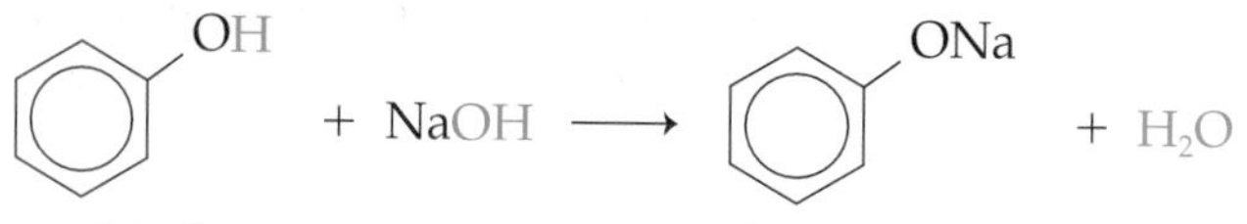

Pure phenol is a white crystalline solid with a low melting point, 41° C. However, on exposure to light and air phenol turns reddish. Phenol is poisonous if taken internally, and externally it causes deep burns and blisters on the skin. If phenol should accidentally be spilled on the skin, it should be removed as quickly as possible with 50 percent alcohol, glycerin, sodium bicarbonate solution, or water.

Phenol was the original Lister *antiseptic and is still used as a disinfectant for surgical instruments and utensils, clothing and bed linens, floors, toilets, and sinks. Phenol is used commercially in the manufacture of dyes and plastics.

Phenol is the standard of reference for germicidal activity of disinfectants; that is, their activity is compared with that of phenol. If a 1 percent solution of a germicide kills organisms in the same time that a 5 percent solution of phenol does, then that germicide is said to have a phenol coefficient of 5.

Exercise 17-5

Draw the structure for 2-chlorophenol.

Solution

Phenol is 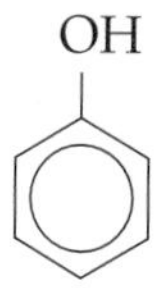 with the OH at position number 1 (understood),

so 2-chlorophenol is

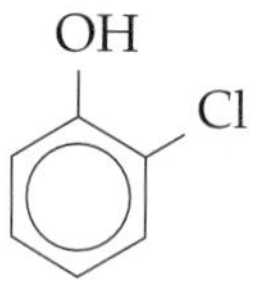

Self-test

Draw the structure for metabromophenol.

Answer

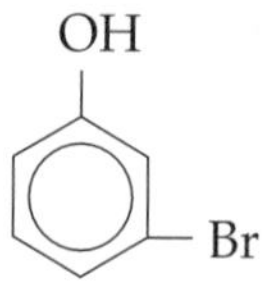

Phenol Derivatives

The methyl derivatives of phenol are called cresols. There are three different cresols—*ortho-*, *meta-*, and *para-*cresols.

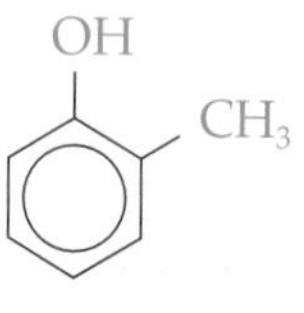

o-cresol 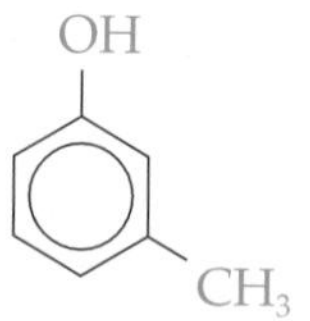   m-cresol 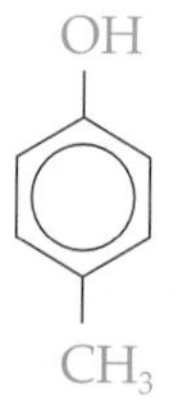   p-cresol

Usually, cresol is a mixture of all three of these isomers. Cresol is a better antiseptic than phenol and is also less toxic. Even though cresols are less toxic than phenol, they are still poisonous and should be used for external purposes only.

Resorcinol is *m*-dihydroxybenzene. It is also an antiseptic but is not as good as phenol.

A resorcinal derivative, hexylresorcinol, is a much better antiseptic and germicide than resorcinol. It is commonly used in mouthwashes. Its structure is

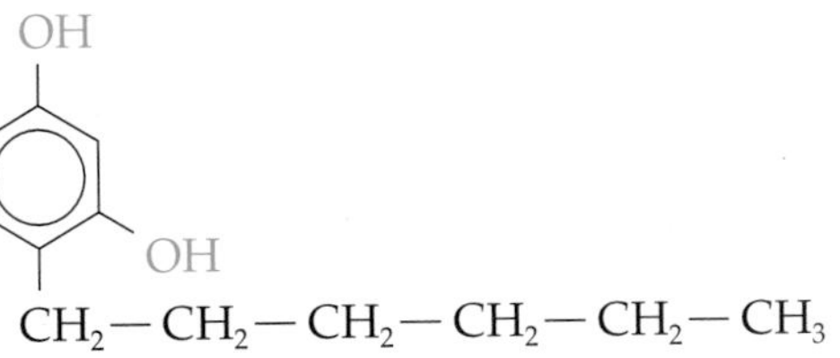

OH

OH

$$CH_2-CH_2-CH_2-CH_2-CH_2-CH_3$$
hexylresorcinol

Two compounds commonly found in foods (listed as food preservatives) are BHA (butylated hydroxyanisole) and BHT (butylated hydroxytoluene).

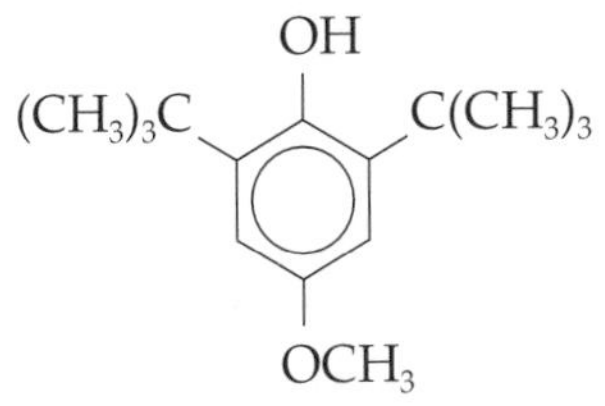

Hydroquinone is 1,4-dihydroxybenzene. Hydroquinone is easily oxidized to quinone. Quinone, in turn, is easily reduced to hydroquinone.

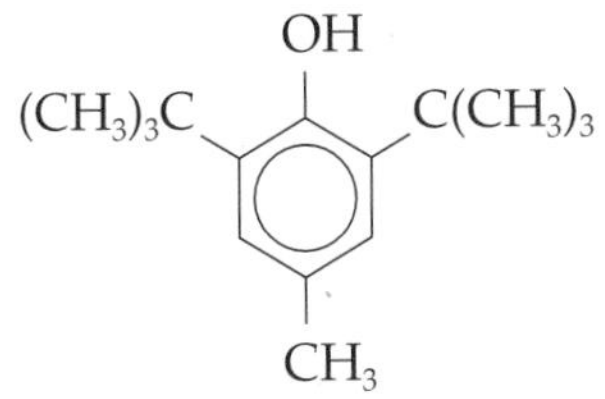

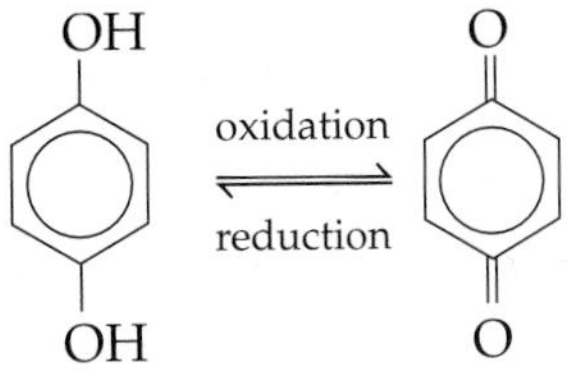

OH

$(CH_3)_3C$ $C(CH_3)_3$

OCH_3

BHA

OH

$(CH_3)_3C$ $C(CH_3)_3$

CH_3

BHT

OH → O
$\overset{oxidation}{\underset{reduction}{\rightleftharpoons}}$

hydroquinone   quinone

Hydroquinone and quinones are important in the respiratory system. One such compound, ubiquinone, is also known as coenzyme Q (see page 426).

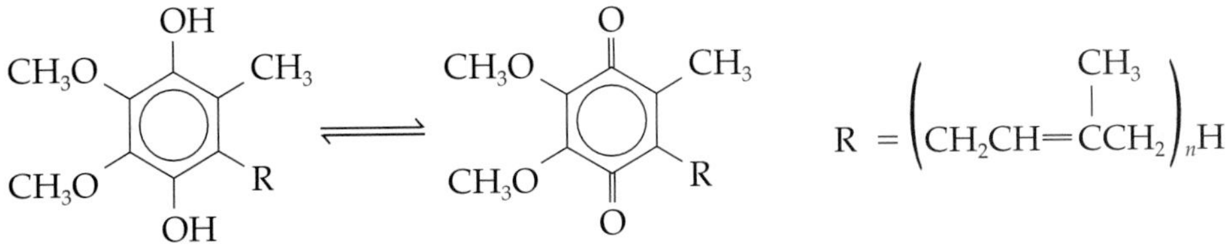

OH
CH_3O CH_3
CH_3O R
OH
$\rightleftharpoons$
O
CH_3O CH_3
CH_3O R
O

$$R = \left(CH_2CH=\overset{CH_3}{\overset{|}{C}}CH_2\right)_n H$$

ubiquinone (coenzyme Q)

reduced form   oxidized form

17-4 Ethers

An ether is formed during the dehydration of an alcohol. In this reaction the sulfuric acid can be considered a dehydrating agent that removes water from two molecules of alcohol. Consider the following reaction, where R indicates an alkyl group and hence ROH indicates an alcohol.

$$R-OH + HO-R \xrightarrow{H_2SO_4} R-O-R + H_2O$$
alcohol   alcohol   under 140°   ether

This equation indicates that two molecules of alcohol react in the presence of sulfuric acid to form water and an ether. When methyl alcohol is reacted with sulfuric acid, methyl ether (also called dimethyl ether) is formed.

$$CH_3\!-\!OH + HO\!-\!CH_3 \xrightarrow{H_2SO_4} CH_3\!-\!O\!-\!CH_3 + H_2O$$

methyl alcohol   methyl alcohol   methyl ether
(dimethyl ether)

When ethyl alcohol is treated with sulfuric acid, ethyl ether (also called diethyl ether) is formed.

$$CH_3\!-\!CH_2\!-\!OH + HO\!-\!CH_2\!-\!CH_3 \xrightarrow{H_2SO_4} CH_3\!-\!CH_2\!-\!O\!-\!CH_2\!-\!CH_3 + H_2O$$

ethyl alcohol   ethyl alcohol   ethyl ether
(diethyl ether)

$$CH_3\!-\!O\!-\!CH_2\!-\!CH_3$$
methoxyethane
(methyl ethyl ether)

$$CH_3\!-\!O\!-\!\underset{\underset{CH_3}{|}}{CH}\!-\!CH_3$$
2-methoxypropane
(methyl isopropyl ether)

The ether is named for the alcohol from which it is made; ethyl ether from ethyl alcohol and methyl ether from methyl alcohol. The general formula for an ether is ROR.

Under the IUPAC system, the $-OCH_3$ group is called methoxy and the $-OCH_2CH_3$ group is called ethoxy. Thus, methyl ether ($CH_3\!-\!O\!-\!CH_3$) is called methoxymethane and ethyl ether ($CH_3\!-\!CH_2\!-\!O\!-\!CH_2\!-\!CH_3$) is called ethoxyethane.

If two different alcohols are reacted, a mixed ether is formed. Mixed ethers are named as alkoxy derivatives of hydrocarbons, with the shorter chain being named as the alkoxy group and the longer chain as the alkane.

Under the common system, mixed ethers are named by listing the alkyl groups, followed by the word ether.

Exercise 17-6

Name the following ether under the IUPAC system and with the common name
$$CH_3\!-\!CH_2\!-\!O\!-\!CH_2\!-\!CH_2\!-\!CH_3$$

Solution

The shorter chain is named as the alkoxy group, so $CH_3\!-\!CH_2\!-\!O$ is called ethoxy. The longer chain has three carbon atoms, so it is a propane. Therefore, the IUPAC name is ethoxypropane. The common name is derived as follows: The alkyl group on the left has two carbon atoms—it is ethyl; the alkyl group on the right has three carbons—it is propyl. Therefore, the common name is ethylpropyl ether.

Self-test

Draw structures for (a) ethyl isopropyl ether and (b) methoxypentane.

Answers

a. $$CH_3\!-\!CH_2\!-\!O\!-\!\underset{\underset{CH_3}{|}}{\overset{\overset{CH_3}{|}}{CH}}$$

b. $$CH_3\!-\!O\!-\!CH_2\!-\!CH_2\!-\!CH_2\!-\!CH_2\!-\!CH_3$$

Recall from the previous section that alcohols react with H_2SO_4 to form alkenes. Here it is indicated that alcohols also react with H_2SO_4 to form ethers. What causes two different reactions to take place? When alcohols react with H_2SO_4 at a temperature of 140° C or lower, ethers are formed. At temperatures above 150° C, alkenes are formed.

Ethers have a low boiling point because the molecules do not form hydrogen bonds (see Section 9-5). Ethers are good solvents because they are inert—they do not react with the solute. Ethyl ether is frequently used to extract organic material from naturally occurring substances. The low boiling point of the ether allows it to be easily removed and recovered. However, ethyl ether is very flammable and must be used with care. Ethers that remain in a laboratory for a long period of time may contain organic peroxides that are extremely explosive. Care must be taken that ethers are not allowed to stand exposed to air and undisturbed for a long period of time.

Ether as an Anesthetic

Ethyl ether, commonly known as ether, has been used quite extensively as a general anesthetic. It is very easy to administer, is an excellent muscular relaxant, and has very little effect on the rate of respiration, blood pressure, or pulse rate. However, the disadvantages of ether outweigh its advantages. It is very flammable, it is irritating to the membranes of the respiratory tract, and it has an aftereffect of nausea. Today ether is infrequently employed as an inhalation anesthetic except in laboratory work. It has been replaced by such nonflammable anesthetics as nitrous oxide (see Section 8-5), halothane (see page 261), and the newer nonflammable inhalation anesthetics such as enflurane and isoflurane, which are halogen derivatives of ethylmethyl ether.

Enflurane (2-chloro-1,1,2-trifluoroethyl difluoromethyl ether), a frequently used inhalation anesthetic, is an excellent relaxant of skeletal muscles that also provides a stable cardiac rhythm.

$$
\begin{array}{c}
\quad\; \text{Cl} \;\;\; \text{F} \qquad\quad \text{F} \\
\quad\;\; | \quad\;\; | \qquad\quad\; | \\
\text{H}-\text{C}-\text{C}-\text{O}-\text{C}-\text{H} \\
\quad\;\; | \quad\;\; | \qquad\quad\; | \\
\quad\;\; \text{F} \quad\;\; \text{F} \qquad\quad\; \text{F}
\end{array}
$$

enflurane

Another frequently used inhalation anesthetic is isoflurane (1-chloro-2,2,2-trifluoroethyl difluoromethyl ether), an isomer of enflurane. Isoflurane undergoes less metabolism than enflurane and does not stimulate the central nervous system.

$$
\begin{array}{c}
\quad\; \text{F} \;\;\; \text{Cl} \qquad\quad \text{F} \\
\quad\;\; | \quad\;\; | \qquad\quad\; | \\
\text{F}-\text{C}-\text{C}-\text{O}-\text{C}-\text{H} \\
\quad\;\; | \quad\;\; | \qquad\quad\; | \\
\quad\;\; \text{F} \quad\;\; \text{H} \qquad\quad\; \text{F}
\end{array}
$$

isoflurane

Summary

Alcohols are derivatives of hydrocarbons with one or more of the hydrogen atoms replaced by an —OH group. The —OH (hydroxyl) group is a functional group that imparts to alcohol its particular properties.

The general formula for an alcohol is ROH, where the R represents an alkyl group attached to the —OH group.

Alcohols do not ionize, they are not bases; their reactions are slower than those of inorganic hydroxides.

The simplest alcohol is methyl alcohol, CH_3OH, also known as methanol. This alcohol is poisonous and should never be used internally or externally.

Ethyl alcohol, C_2H_5OH, is also known as ethanol. It is used as a disinfectant because it has the property of coagulating protein. It can also be used for sponge baths to reduce body temperature. Solutions of medications in alcohol are called tinctures.

Isopropyl alcohol is used primarily as rubbing alcohol. It is toxic and should never be used internally.

Glycerol or glycerin is a trihydric alcohol; it has three —OH groups in its molecule. Glycerol is a constituent of fats.

Alcohols can be divided into three categories; primary where the —OH group is attached to a carbon atom having one or no carbon atoms attached to it; secondary, where the —OH group is attached to a carbon atom having two carbon atoms attached to it; and tertiary, where the —OH is attached to a carbon atom having three carbon atoms attached to it.

Thiols are sulfur analogs of alcohols and contain an — SH group. The simplest cyclic alcohol is phenol, a benzene ring with an —OH group attached.

Ethers are produced by the dehydration of an alcohol. Ethers are named according to the alcohol or alcohols from which they were produced.

Ethyl ether has been used as a general anesthetic. It is an excellent muscular relaxant and has little effect on the rate of respiration or pulse rate. However, ether is irritating to the membranes of the respiratory tract, it may cause nausea, and it is very flammable. It has been replaced by nonflammable, nonirritating anesthetics such as halothane and nitrous oxide.

Questions and Problems

1. What is the general formula for an alcohol? Indicate several general properties of alcohols.
2. Draw the structure and give the IUPAC name for
   a. ethyl alcohol
   b. ethylene glycol
   c. isopropyl alcohol
   d. methyl ether
   e. glycerol
3. Why should methyl alcohol never be used medicinally?
4. Why is ethylene glycol toxic if taken internally?
5. Explain why 70 percent alcohol is a better disinfectant than a 100 percent alcohol.
6. What is a tincture? Give two examples.
7. How can alcohol be prepared commercially? What is meant by the term *proof*?
8. Why is alcohol used for sponge baths? May water be substituted? Explain.
9. Why is glycerol important in the body?
10. How may nitroglycerin be prepared? Of what use is it medically?
11. List the three types of alcohols and give an example of each.
12. What is the general formula for an ether?

13. How can ethers be prepared?
14. Write the question for the formation of
    a. ethyl ether
    b. methyl ethyl ether
15. List the advantages and disadvantages of ether as a general anesthetic.
16. Name the following compounds according to both the IUPAC and common systems.
    a. $CH_2CH_2 — O — CH_2CH_2CH_3$
    b. $\begin{array}{cc} CH_2 — CH_2 \\ | \qquad | \\ OH \quad\ OH \end{array}$
    c. $CH_3CH_2CH_2SH$
    d.
    e.

    f.

    g. CH_3OCH_3
    h. $\begin{array}{c} CH_2CHCH_2 \\ | \quad | \quad | \\ OH\,OH\,OH \end{array}$
17. What is the antidote for etheylene glycol poisoning? Explain.

18. Draw structures for
    a. p-xylene
    b. methoxypropane
    c. dipropyl ether
    d. 3-methyl-1-butanol
    e. 1,2,3-trihydroxypropane
    f. phenol
    g. cresol
19. Identify a medical use for phenol; for cresol.
20. What is a functional group? Give an example.
21. Complete the following reactions:

$$CH_3CH_2CH_2OH \xrightarrow[\text{low temperature}]{H_2SO_4}$$

$$CH_3CH_2CH_2OH \xrightarrow[\text{high temperature}]{H_2SO_4}$$

22. Why does the dehydration of 2-octanol produce two different products?
23. How is methanol produced commercially? ethanol? What are these substances used for?
24. What is meant by the term *phenol coefficient*?
25. Draw and name some of the isomers of 2,5-dimethyl-hexanol.
26. Draw and name all of the isomers with the formula $C_5H_{12}O$.
27. Draw the structures for the following alcohols; decide whether each alcohol is a primary, secondary, or tertiary alcohol.
    a. 2-methyl-1-octanol
    b. 2,3-dimethyl-2-pentanol
    c. 3-cyclohexenol

Practice Test

1. The general formula for an ether is _____.
   a. RH   b. ROH   c. ROR   d. REt
2. The general formula for an alcohol is _____.
   a. RAl   b. ROH   c. ROR   d. Alc

3. $\overset{\displaystyle CH_2 - CH_2}{\underset{\displaystyle OH \quad\;\; OH}{|\qquad\; |}}$ is called _____.
   a. ethyl ether   b. ethyl alcohol
   c. ethylene glycol   d. glycerol
4. An example of a trihydric alcohol is _____.
   a. ethyl alcohol   b. glycerol
   c. glycol   d. isopropyl alcohol
5. An alcoholic solution of a medication is called a(n) _____.
   a. mercaptan   b. isotope
   c. emulsion   d. tincture
6. Eighty proof alcohol contains _____ percent alcohol?
   a. 20   b. 40   c. 80   d. 100
7. $CH_3OCH_2CH_3$ is called _____.
   a. methyl ether
   b. ethyl ether
   c. ethyl methyl ether
   d. methyl isopropyl ether

8. The antidote for ethylene glycol poisoning is _____.
   a. ethyl alcohol   b. methyl alcohol
   c. egg white   d. none of these

Use the following choices for questions 9 and 10.

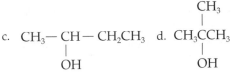

a. CH_3CH_2OH b. $CH_3OCH_2CH_3$

c. $CH_3 - \underset{\displaystyle OH}{\underset{\displaystyle |}{CH}} - CH_2CH_3$ d. $CH_3\underset{\displaystyle OH}{\underset{\displaystyle |}{\overset{\displaystyle CH_3}{\overset{\displaystyle |}{C}}}}CH_3$

9. Which is a primary alcohol?
10. Which is a tertiary alcohol?

11. 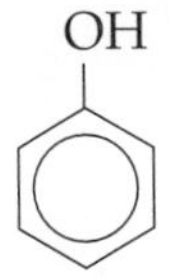 is called:
    a. hydroxy cyclohexene   b. cresol
    c. phenol   d. xylene

Aldehydes, Ketones, and Carboxylic Acids

Objectives

- To become familiar with the functional groups present in aldehydes, ketones, and carboxylic acids

- To write equations for the formation of aldehydes, ketones, and acids

- To distinguish between hemiacetals and acetals and between hemiketals and ketals

- To recognize the biologic importance of various aldehydes, ketones, acids

Organic acids are found in a wide variety of household items: Spinach and some cleansers contain oxalic acid; vinegar contains acetic acid; vitamin C tablets and lemons contain citric acid; and aspirin consists of acetylsalicylic acid.

The oxidation of primary and secondary alcohols gives different types of products. Tertiary alcohols are resistant to the usual oxidizing conditions in organic reactions. If we consider oxidation as the removal of hydrogens (see page 102), then the oxidation of an alcohol can be said to involve the removal of one hydrogen from the —OH group of the alcohol and of a second hydrogen from the carbon atom to which the —OH group is attached. The oxidation of a primary alcohol can be written as

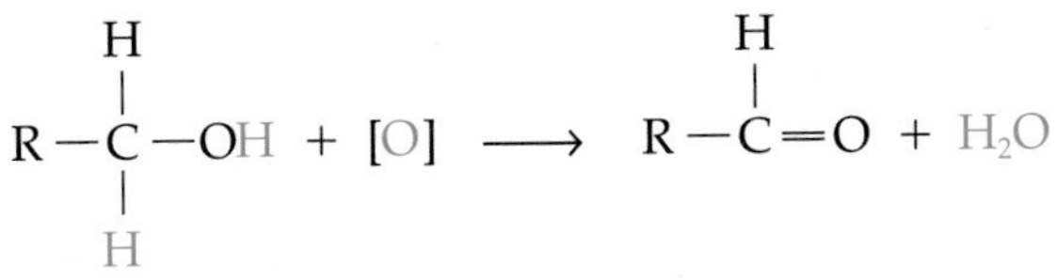

$$
\begin{array}{c}
\underset{\text{primary alcohol}}{
\overset{\displaystyle H}{\underset{\displaystyle H}{R-\overset{|}{\underset{|}{C}}-OH}}} + [O]
\quad \longrightarrow \quad
\underset{\text{aldehyde}}{
\overset{\displaystyle H}{R-\overset{|}{C}=O}} + H_2O
\end{array}
$$

The oxidation of an alcohol requires the use of some oxidizing agent such as $KMnO_4$, $K_2Cr_2O_7$, or CuO. However, for the sake of simplicity the oxidizing agent in the reactions shown in this chapter is simply listed as [O], which stands for any substance that will yield the oxygen needed for the reaction.

18-1 Aldehydes

Preparation by Oxidation of a Primary Alcohol

Recall that a primary alcohol has the —OH functional group bonded to a carbon with one or no other carbon atom attached to it. The following equation represents the oxidation of methyl alcohol (CH_3OH), a primary alcohol.

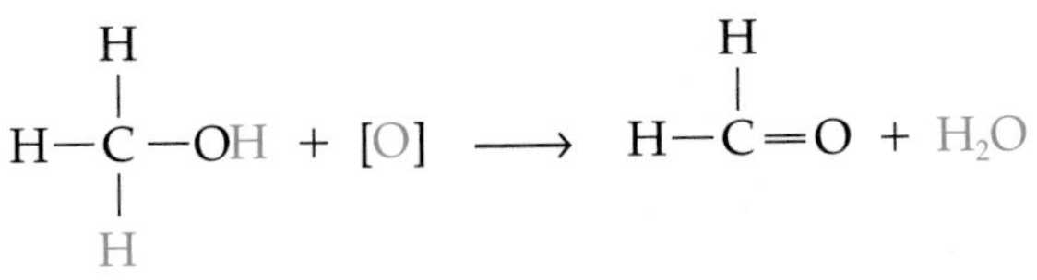

$$
\begin{array}{c}
\underset{\text{methyl alcohol}}{
\overset{\displaystyle H}{\underset{\displaystyle H}{H-\overset{|}{\underset{|}{C}}-OH}}} + [O]
\quad \longrightarrow \quad
\underset{\text{formaldehyde}}{
\overset{\displaystyle H}{H-\overset{|}{C}=O}} + H_2O
\end{array}
$$

Observe that during the oxidation one H was removed from the —OH group and another H from the carbon to which the —OH group was attached (the only carbon in this compound). Water is one product of this reaction; the other product is a new kind of compound called an aldehyde. In this example, the product is called formaldehyde. The formula for formaldehyde can also be written as HCHO.

The following reaction indicates the oxidation of ethyl alcohol (also a primary alcohol).

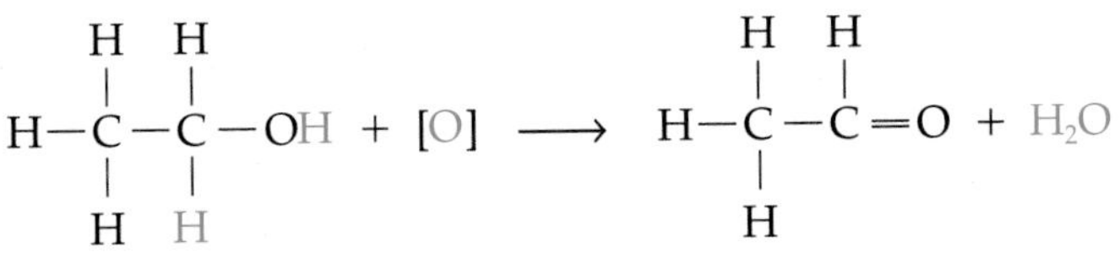

$$
\begin{array}{c}
\underset{\text{ethyl alcohol}}{
\overset{\displaystyle H \quad H}{\underset{\displaystyle H \quad H}{H-\overset{|}{\underset{|}{C}}-\overset{|}{\underset{|}{C}}-OH}}} + [O]
\quad \longrightarrow \quad
\underset{\text{acetaldehyde}}{
\overset{\displaystyle H \quad H}{\underset{\displaystyle H}{H-\overset{|}{\underset{|}{C}}-\overset{|}{C}=O}}} + H_2O
\end{array}
$$

The oxidation of ethyl alcohol, a primary alcohol, yields acetaldehyde, whose formula can also be written as CH_3CHO. (Acetaldehyde is partially responsible for the damage to the liver in cirrhosis.) In general,

$$\text{primary alcohol} \xrightarrow{\text{oxidation}} \text{aldehyde}$$

Aldehydes all have the —CHO (called a *carbonyl*) group at the end of the chain. The general formula for an aldehyde is RCHO, which indicates that some aliphatic alkyl group (R) is attached to a —CHO group at the end of the molecule.

Naming Aliphatic Aldehydes

As we have seen, the oxidation of methyl alcohol, a primary alcohol of one carbon atom, yields an aldehyde of one carbon atom, HCHO, formaldehyde. The oxidation of ethyl alcohol, a primary alcohol of two carbon atoms, yields an aldehyde of two carbon atoms, HC_3CHO, acetaldehyde. Note that the term *aldehyde* comes from the words *al*cohol *dehy*drogenation.

The IUPAC names for all aldehydes end in -al. To name an aldehyde according to the IUPAC system, take the name of the longest chain containing the aldehyde group, drop the ending -e, and replace it with the ending -al. Thus the following aldehyde, which contains four carbons, is called butanal.

$$CH_3—CH_2—CH_2—CHO$$

The aldehyde group is always at the end of the chain, at carbon 1, with that number being understood and not written.

Exercise 18-1

Name the following compound.

$$
\begin{array}{c}
\qquad\qquad\quad Br \\
\qquad\qquad\quad | \\
CH_3—CH—CH—CH—CH_2—CH_2—CH_2—CHO \\
\quad\;\; | \qquad\qquad\; | \\
\quad\;\; CH_3 \qquad\quad CH_3
\end{array}
$$

Solution

The largest chain containing the aldehyde group contains eight carbon atoms, so it is an octanal. The chain is numbered with the aldehyde group at carbon number 1.

$$
\begin{array}{c}
\qquad\qquad\qquad Br \\
\qquad\qquad\qquad | \\
\overset{8}{C}H_3—\overset{7}{C}H—\overset{6}{C}H—\overset{5}{C}H—\overset{4}{C}H_2—\overset{3}{C}H_2—\overset{2}{C}H_2—\overset{1}{C}HO \\
\qquad\quad\; | \qquad\qquad\;\; | \\
\qquad\quad\; CH_3 \qquad\quad CH_3
\end{array}
$$

There are methyl groups at carbon numbers 5 and 7 and a bromine at carbon number 6. Therefore, naming the groups in alphabetic order we have 6-bromo-5,7-dimethyloctanal.

Exercise 18-2

Draw the structure of 4,4-dichloro-2,3-dimethylpentanal.

Solution

Pentanal indicates a five-carbon aldehyde, with the —CHO group at carbon number 1, or

$$\overset{5}{C}-\overset{4}{C}-\overset{3}{C}-\overset{2}{C}-\overset{1}{C}HO$$

4,4-dichloro- indicates two chlorines, both at carbon number 4.
2,3-dimethyl indicates two methyls, one at carbon number 2 and one at carbon number 3, or

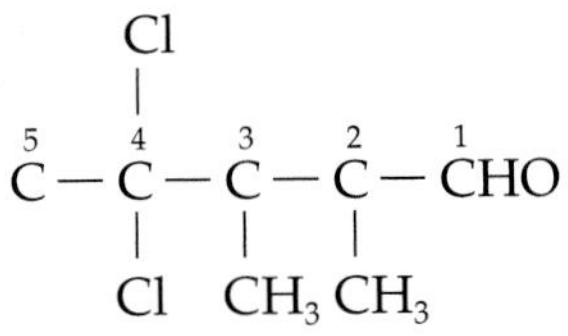

$$
\begin{array}{c}
\text{Cl} \\
| \\
\overset{5}{C}-\overset{4}{C}-\overset{3}{C}-\overset{2}{C}-\overset{1}{C}HO \\
| \quad | \quad | \\
\text{Cl} \quad \text{CH}_3 \; \text{CH}_3
\end{array}
$$

Adding H's to complete all the bonds, we have

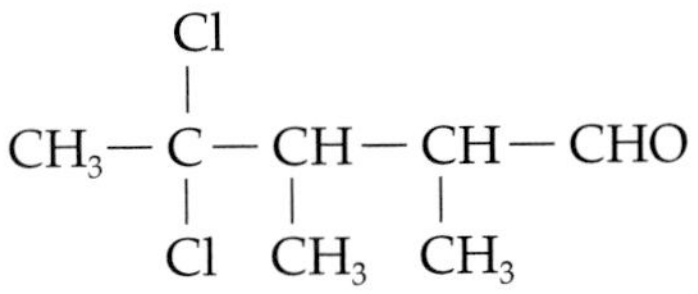

$$
\begin{array}{c}
\text{Cl} \\
| \\
\text{CH}_3-\text{C}-\text{CH}-\text{CH}-\text{CHO} \\
| \quad | \quad | \\
\text{Cl} \; \text{CH}_3 \; \text{CH}_3
\end{array}
$$

Self-test

Name

a. 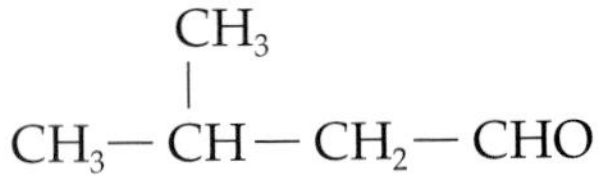
$$
\begin{array}{c}
\qquad \text{CH}_3 \\
\qquad | \\
\text{CH}_3-\text{CH}-\text{CH}_2-\text{CHO}
\end{array}
$$

Draw structures for
b. 2,3-dichloropropanal

c. 3,5-dimethyloctanal

Answers

a. 3-methylbutanal

b. 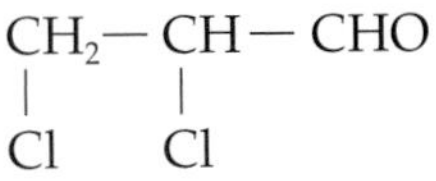
$$
\begin{array}{c}
\text{CH}_2-\text{CH}-\text{CHO} \\
| \quad | \\
\text{Cl} \quad \text{Cl}
\end{array}
$$

c.
$$
\begin{array}{c}
\text{CH}_3-\text{CH}_2-\text{CH}_2-\text{CH}-\text{CH}_2-\text{CH}-\text{CH}_2-\text{CHO} \\
\qquad\qquad\qquad\qquad | \qquad\qquad | \\
\qquad\qquad\qquad\qquad \text{CH}_3 \qquad\quad \text{CH}_3
\end{array}
$$

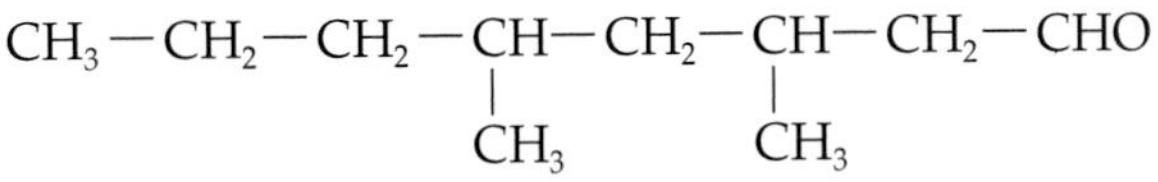

Aromatic Aldehydes

Aromatic aldehydes have the general formula **ArCHO**, where Ar stands for an aromatic ring. The simplest aromatic aldehyde is benzaldehyde, which consists of an aldehyde group attached to a benzene ring. Benzaldehyde is prepared by the mild oxidation of toluene, as shown in equation (18-1). Note that the side chain, the methyl group, is more susceptible to oxidation than the fairly stable benzene ring.

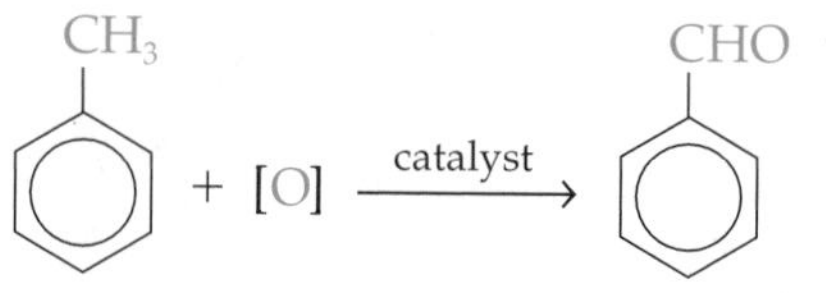

$$\underset{\text{toluene}}{CH_3\text{-}\bigcirc} + [O] \xrightarrow{\text{catalyst}} \underset{\text{benzaldehyde}}{CHO\text{-}\bigcirc} \qquad (18\text{-}1)$$

Benzaldehyde is a colorless, oily liquid with an cherry-almondlike odor. It is used in the preparation of flavoring agents, perfumes, drugs, and dyes.

Vanillin occurs in vanilla beans and gives the particular taste and odor to vanilla extract. It also has an aldehyde structure. Cinnamic aldehyde (cinnamalde-hyde) is present in oil of cinnamon, an oil found in cinnamon bark. Both vanillin and cinnamic aldehyde can be prepared synthetically, and both are used as flavoring agents.

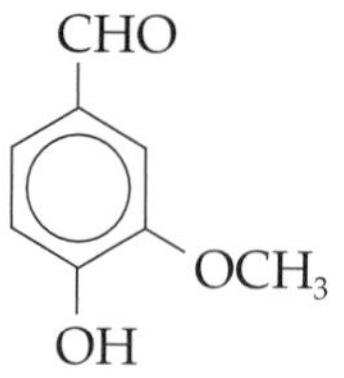

vanillin 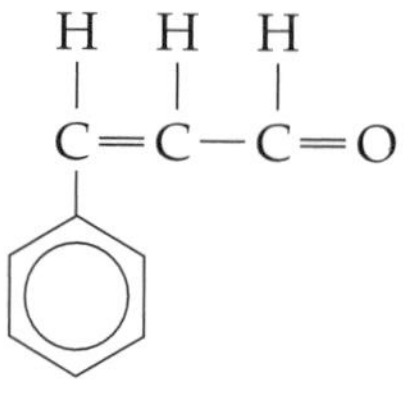   cinnamic aldehyde

Table 18-1 compares the IUPAC and common names for some simple aldehydes.

Uses of Aldehydes

Formaldehyde is a colorless gas with a very sharp odor. It is used in the laboratory as a water solution containing about 40 percent formaldehyde. The 40 percent solution commonly known as formalin is an effective germicide for the disinfection of excreta, rooms, and clothing. Formalin hardens protein, making it very insoluble in water. It is used in embalming fluids and also as a preservative for biologic specimens. Formaldehyde solutions should not be used directly on a patient or even in the room with a patient because of irritating fumes.

Formaldehyde and its oxidation product, formic acid, are primarily responsible for the systemic toxicity of methyl alcohol.

Glutaraldehyde is superior to formaldehyde as a sterilizing agent and therefore is replacing it in use. Glutaraldehyde is microcidal against all microorganisms, including spores and many viruses. Glutaraldehyde does not have the disagreeable odor that formaldehyde does, and it is less irritating to the eyes and skin.

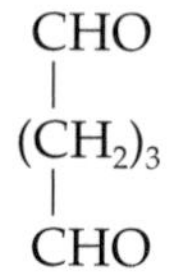

$$\begin{array}{c} CHO \\ | \\ (CH_2)_3 \\ | \\ CHO \end{array}$$

glutaraldehyde

Table 18-1 Comparison of the Names of Some Aldehydes		
Condensed Structural Formula	IUPAC Name	Common Name
HCHO	Methanal	Formaldehype
CH$_3$CHO	Ethanal	Acetaldehyde
CH$_3$—CH$_2$—CHO	Propanal	Propionaldehyde
CH$_3$—CH—CHO $\quad\quad$ | $\quad\quad$ CH$_3$	2-Methylpropanal	Isobutyraldehyde

Paraldehyde is formed by the polymerization (joining) of three molecules of acetaldehyde. Paraldehyde depresses the central nervous system. It is used as a hypnotic, a sleep producer. Paraldehyde is also used in the treatment of alcoholism (see following paragraph). In therapeutic dosages it is nontoxic; it does not depress heart action or respiration. Its disadvantages are its disagreeable taste and its unpleasant odor.

Acetaldehyde (ethanal) is responsible for many of the unpleasant side effects of ethyl alcohol consumption. The drug Antabuse, used to treat alcoholics, functions by increasing the concentration of acetaldehyde in the body.

Another aldehyde, glyceraldehyde, is an important component in the metabolism of carbohydrates (see page 333).

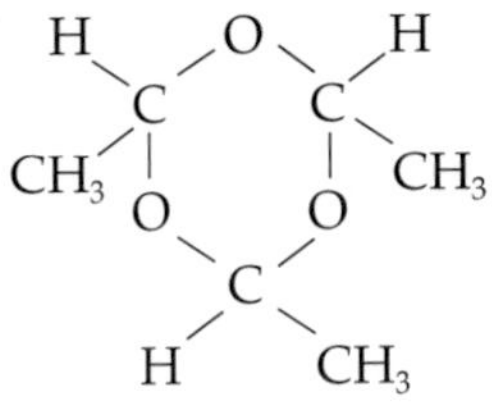

paraldehyde

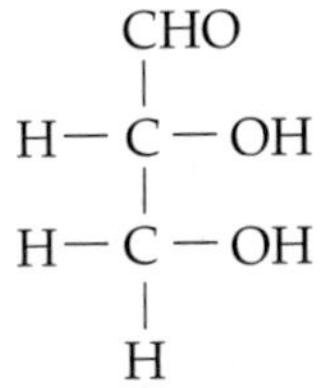

glyceraldehyde
2,3-dihydroxypropanal

Tests for Aldehydes

In general, aldehydes are good reducing agents. Laboratory tests for the presence of aldehydes are based on their ability to reduce copper(II) (cupric) ions to form copper(I) (cuprous) oxide. When an aldehyde is heated with Benedict's or Fehling's solution or treated with a Clini-test tablet (all of which contain Cu^{2+} complex ion), a red precipitate of copper(I) oxide (Cu_2O) is formed. This is actually the test for glucose (sugar) in urine, since glucose is an aldehyde (see page 344).

Another laboratory test for the presence of an aldehyde involves the use of Tollen's reagent, which contains an Ag^+ complex ion. In this test, the presence of an aldehyde causes the formation of a bright, shiny mirror on the inside of the test tube; hence the name "silver mirror test" (see page 344).

Reactions of Aldehydes

Oxidation   Aldehydes can be oxidized to form acids, a type of reaction that will be discussed later in this chapter.

$$RCHO + [O] \longrightarrow RCOOH$$
aldehyde   acid

Note that this type of oxidation involves the addition of an oxygen atom (see Section 6-2).

Reduction   Aldehydes can be reduced to the corresponding primary alcohols.

$$RCHO + [H] \longrightarrow RCH_2OH$$
aldehyde   primary alcohol

This is the reverse of the reaction whereby a primary alcohol was oxidized to yield an aldehyde.

aldehyde   primary alcohol
$$CH_3CHO + [H] \longrightarrow CH_3CH_2OH$$
ethanal   ethanol
(acetaldehyde)   (ethyl alcohol)

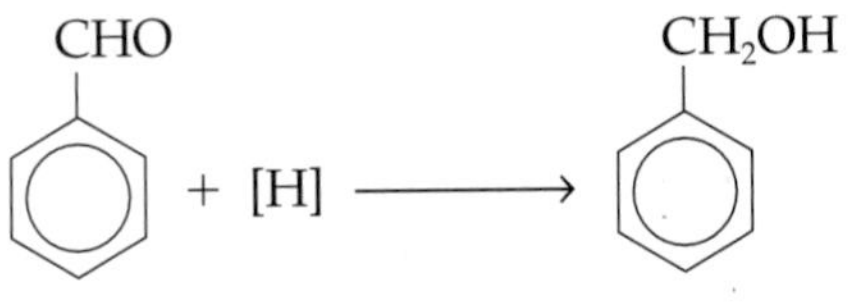

benzaldehyde   benzyl alcohol

Biologic oxidation-reduction in the body is carried out by substances called coenzymes (see Section 24-4). One coenzyme, nicotinamide adenine dinucleotide

(NAD^+), acts as an oxidizing agent and in turn is reduced to nicotinamide adenine dinucleotide hydride (NADH), as shown in the following reaction:

$$NAD^+ \ + \ CH_3CH_2OH \ \rightleftharpoons \ NADH \ + \ CH_3CHO + H^+$$

oxidized form   ethyl alcohol   reduced form   acetaldehyde
of coenzyme   of coenzyme

18-2 Ketones

Preparation by Oxidation of a Secondary Alcohol

CH₃—C—CH₃ with H on top and OH on bottom

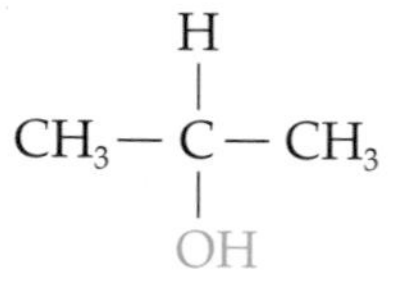

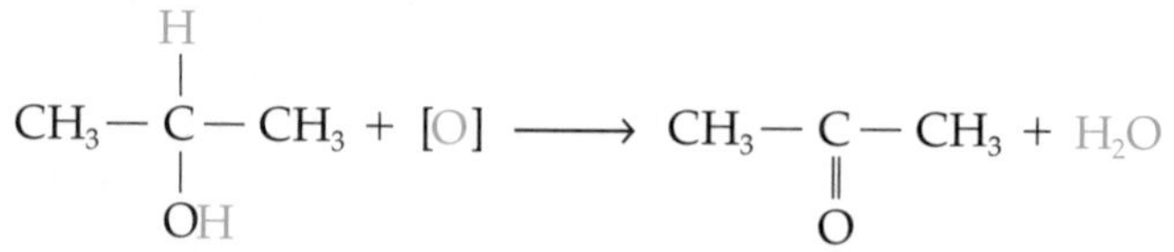

isopropyl alcohol
(2-propanol)

R—C=O with H on top

aldehyde

R—C—R with O below (double bond)

ketone

Recall that a secondary alcohol is one in which the —OH group is bonded to a carbon atom that is also bonded to two carbon atoms. Isopropyl alcohol is an example of a secondary alcohol. The oxidation of isopropyl alcohol is indicated by the equation

$$CH_3-\overset{\overset{\text{H}}{|}}{\underset{\underset{\text{OH}}{|}}{C}}- CH_3 + [O] \longrightarrow CH_3-\overset{\overset{}{}}{\underset{\underset{\text{O}}{||}}{C}}- CH_3 + H_2O$$

isopropyl alcohol   acetone
(2-propanol)   (propanone)

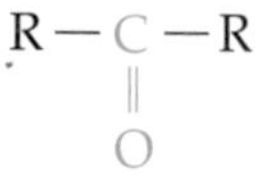

As before, the oxygen atom from the oxidizing agent reacts with the H from the —OH group and with the H attached to the same carbon as the —OH group, forming water and a new class of compounds called ketones.

The oxidation of a secondary alcohol yields a ketone, of the general formula RCOR. That is, a ketone has two alkyl groups attached to a $\diagup$C=O carbonyl, group. This carbonyl group is present in both aldehydes and ketones. However, the carbonyl group is at the end of the chain in an aldehyde and not at the end in a ketone.

Aromatic ketones have the general formula *ArCOAr'* or *ArCOR*. The simplest aromatic ketone is acetophenone.

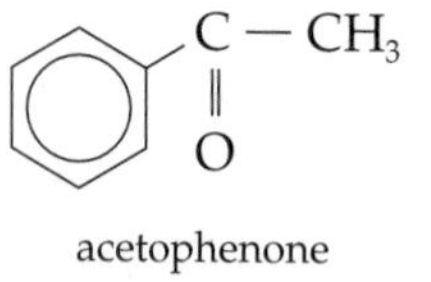

benzene ring—C—CH₃ with O below   benzene ring—C—CH₂Cl with O below

acetophenone 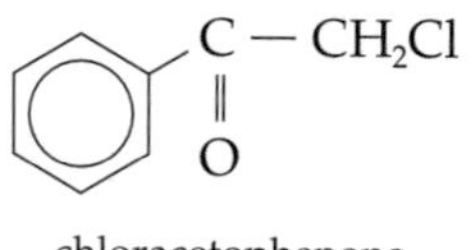   chloracetophenone

Acetophenone has been used as a hypnotic but has been supplanted for this purpose by newer and safer drugs.

Chloracetophenone is a *lacrimator and is used as a tear gas.

Among the aromatic ketones in the body are the sex hormones estrone, progesterone, testosterone, and androsterone (see Section 32-13).

Naming Aliphatic Ketones

CH₃—C—CH₂—CH₃ with O below

butanone

In the IUPAC system, the names of ketones end in -one. To name a ketone according to this system, take the name of the longest alkane containing the carbonyl group, drop the ending -e, and add -one. Thus, the four-carbon ketone is called butanone.

In the common system for naming ketones, each alkyl group attached to the carbonyl group is named and the word *ketone* is added afterward. Thus, the name of the preceding compound, according to the common system, is methyl ethyl ketone, since there is a methyl group attached to one end of the carbonyl group and an ethyl group attached to the other end.

Exercise 18-3

Name the following compound using both IUPAC and common systems.

$$CH_3 - CH_2 - \underset{\underset{O}{\|}}{C} - CH_2 - CH_2 - CH_3$$

Solution

The largest chain containing the ketone (carbonyl) group has six carbon atoms, so it is a hexanone. The carbonyl group should be given the lowest number, so it is on carbon number 3. Therefore, the IUPAC name is 3-hexanone. In the common system, there is an alkyl group on either side of the carbonyl group, or

$$CH_3 - CH_2 - \underset{\underset{O}{\|}}{C} - CH_2 - CH_2 - CH_3$$
ethyl   propyl

The common name for the compound is ethylpropylketone.

Exercise 18-4

Name the following compound according to the IUPAC system.

$$CH_3 - \underset{\underset{Cl}{|}}{\overset{\overset{Cl}{|}}{C}} - CH_2 - \underset{\underset{O}{\|}}{C} - CH_3$$

Solution

The largest chain containing the carbonyl group is five, so this is a pentanone. The carbonyl group is at carbon number 2, so it is a 2-pentanone.

$$\overset{5}{C}H_3 - \overset{4}{\underset{\underset{Cl}{|}}{\overset{\overset{Cl}{|}}{C}}} - \overset{3}{C}H_2 - \overset{2}{\underset{\underset{O}{\|}}{C}} - \overset{1}{C}H_3$$

There are two chlorine atoms, both at carbon number 4, so the name is 4,4-dichloro-2-pentanone.

Exercise 18-5

Draw the structure of 2,2-dimethyl-3-pentanone.

Solution

Pentanone indicates a five-carbon ketone with the carbonyl group at carbon number 3

$$C - C - C - C - C$$
$$\overset{\|}{O}$$

2,2-dimethyl indicates two methyl groups, both at carbon number 2

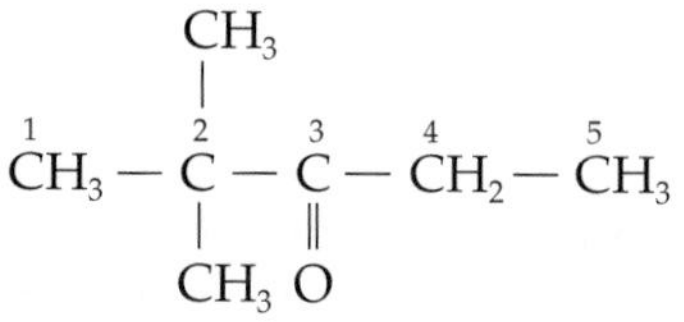

$$\underset{\overset{1}{CH_3}}{} - \underset{\underset{CH_3}{|}}{\overset{\overset{CH_3}{|}}{\underset{2}{C}}} - \underset{\overset{\|}{O}}{\overset{3}{C}} - \overset{4}{CH_2} - \overset{5}{CH_3}$$

Exercise 18-6

Draw the structure of methylethylketone.

Solution

Ketone indicates a carbonyl group attached to two alkyl groups. Methyl indicates an alkyl group with one carbon atom (CH_3—), ethyl indicates an alkyl group with two carbon atoms ($CH_3 CH_2$—), so the structure is

$$CH_3 - \underset{\overset{\|}{O}}{C} - CH_2 - CH_3$$

Self-test

Draw structures for
a. 3-chlorobutanone
b. hydroxyacetone

Answers

a. 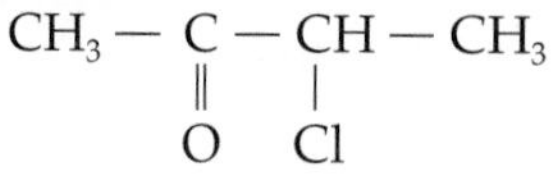
$$CH_3 - \underset{\overset{\|}{O}}{C} - \underset{\overset{|}{Cl}}{CH} - CH_3$$

b. 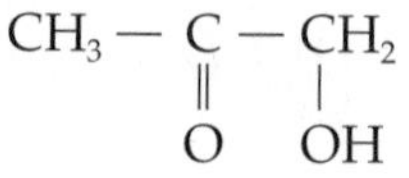
$$CH_3 - \underset{\overset{\|}{O}}{C} - \underset{\overset{|}{OH}}{CH_2}$$

Uses of Ketones

Acetone (propanone) is the simplest ketone. Acetone is a good solvent for fats and oils. It is also frequently used in fingernail polish and in polish remover. Acetone is normally present in small amounts in the blood and urine. In diabetes mellitus it is present in larger amounts in the blood and urine and even in the expired air (see page 483). Dihydroxyacetone is an intermediate in carbohydrate metabolism.

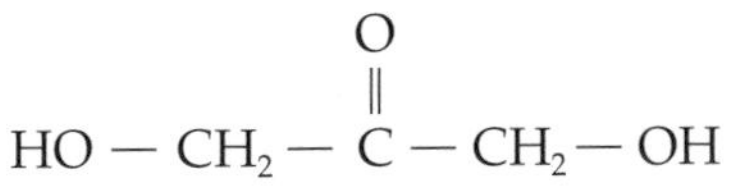

$$HO - CH_2 - \overset{\displaystyle O}{\overset{\|}{C}} - CH_2 - OH$$

dihydroxyacetone

Reactions of Ketones

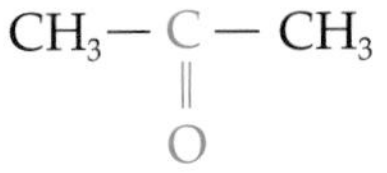

What happens when a ketone is oxidized? Consider the formula for the ketone acetone. There are no hydrogen atoms on the carbon atom of the carbonyl group.
Therefore, ketones are not easily oxidized. They are normally unreactive. Ketones
can be reduced, however, to the corresponding secondary alcohol. Ketones give a
negative test with such oxidizing agents as Benedict's solution or Clini-test tablets.
Recall that aldehydes give a positive test with these reagents.

$$CH_3 - \overset{\displaystyle }{\underset{\displaystyle O}{\overset{\|}{C}}} - CH_3$$

acetone

The test for acetone and ketone bodies makes use of the reaction between
sodium nitroprusside and ketones or ketone bodies to produce a lavender color.

18-3 Hemiacetals and Hemiketals

The reaction of an aldehyde or a ketone with an alcohol yields compounds known
as *hemiacetals or *hemiketals, respectively.

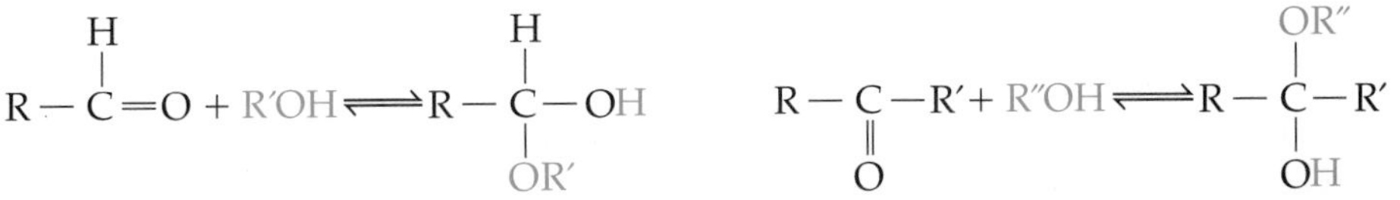

$$R - \overset{\displaystyle H}{\overset{\|}{C}}=O + R'OH \rightleftharpoons R - \overset{\displaystyle H}{\underset{\displaystyle OR'}{\overset{|}{C}}} - OH \qquad R - \overset{\displaystyle }{\underset{\displaystyle O}{\overset{\|}{C}}} - R' + R''OH \rightleftharpoons R - \overset{\displaystyle OR''}{\underset{\displaystyle OH}{\overset{|}{C}}} - R'$$

aldehyde   alcohol   hemiacetal   ketone   alcohol   hemiketal

These types of compounds are important in discussing the structures of monosaccharides (see page 339).

If a hemiacetal or hemiketal reacts with a second molecule of alcohol, an
*acetal or *ketal, respectively, is formed. These structures are important in disac-
charides and polysaccharides (Sections 20-5 and 20-7).

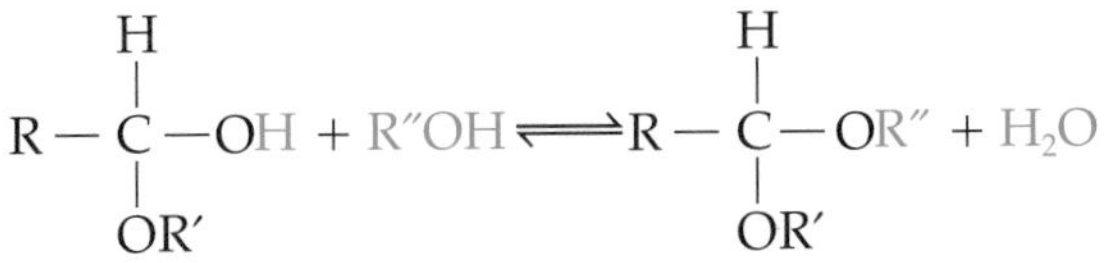

$$R - \overset{\displaystyle H}{\underset{\displaystyle OR'}{\overset{|}{C}}} - OH + R''OH \rightleftharpoons R - \overset{\displaystyle H}{\underset{\displaystyle OR'}{\overset{|}{C}}} - OR'' + H_2O$$

hemiacetal   acetal

$$R - \overset{\displaystyle OR''}{\underset{\displaystyle OH}{\overset{|}{C}}} - R' + R'''OH \rightleftharpoons R - \overset{\displaystyle OR''}{\underset{\displaystyle OR'''}{\overset{|}{C}}} - R' + H_2O$$

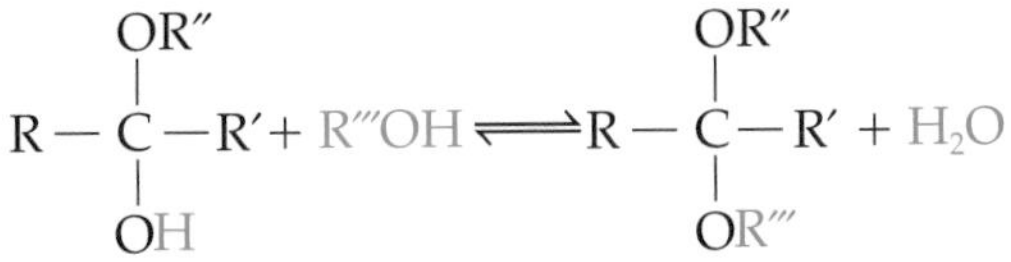

hemiketal   ketal

18-4 Organic Acids

Preparation by Oxidation of an Aldehyde

The oxidation of a primary alcohol yields an aldehyde. Aldehydes in turn can be
easily oxidized. When an aldehyde is oxidized, the reaction is

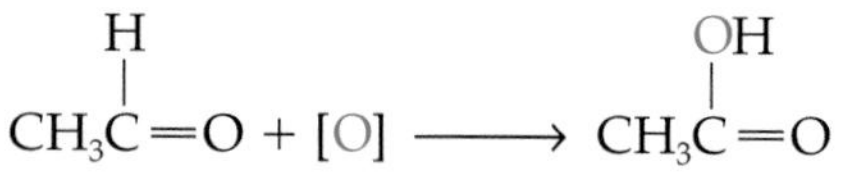

$$CH_3\overset{\overset{\displaystyle H}{|}}{C}{=}O + [O] \longrightarrow CH_3\overset{\overset{\displaystyle OH}{|}}{C}{=}O \quad \text{(also written as } CH_3COOH)$$

acetaldehyde   acetic acid

The resulting compound is acidic because it yields hydrogen ions in solution. (Note that this reaction involves oxidation because of a gain in oxygen; see Section 6-2.)

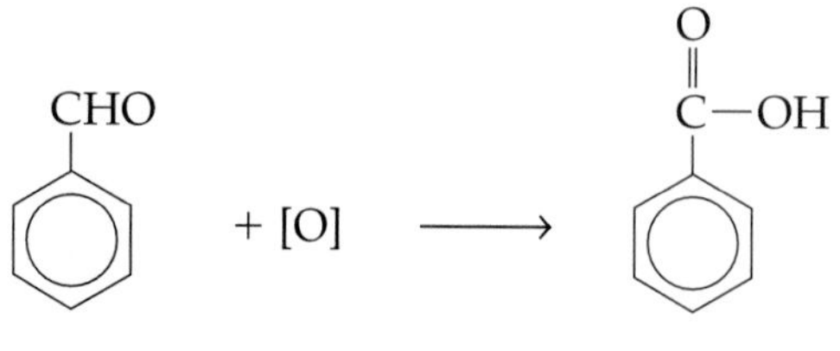

benzaldehyde   benzoic acid

The functional group of an organic acid is —COOH, so the oxidation of an aldehyde to an acid can be written functionally as

$$\underset{\text{aldehyde}}{R{-}CHO} \xrightarrow{[O]} \underset{\text{acid}}{R{-}COOH}$$

$$\underset{\text{aldehyde}}{ArCHO} \xrightarrow{[O]} \underset{\text{acid}}{R{-}ArCOOH}$$

The oxidation of methyl alcohol, a primary alcohol, to an aldehyde and then to an acid is illustrated in the following equation.

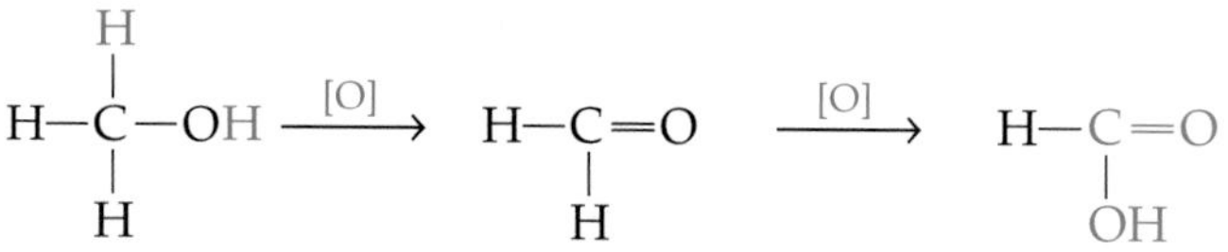

$$H\overset{\overset{\displaystyle H}{|}}{\underset{\underset{\displaystyle H}{|}}{C}}{-}OH \xrightarrow{[O]} H\overset{\overset{\displaystyle}{}}{\underset{\underset{\displaystyle H}{|}}{C}}{=}O \xrightarrow{[O]} H\overset{}{\underset{\underset{\displaystyle OH}{|}}{C}}{=}O \quad \text{(also written as HCOOH)}$$

methyl alcohol   formaldehyde   formic acid
(methanol)   (methanal)   (methanoic acid)

Primary alcohols can also be oxidized directly to acids.

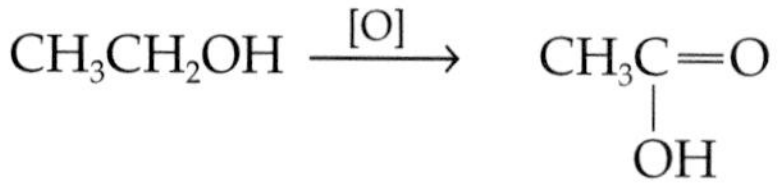

$$CH_3CH_2OH \xrightarrow{[O]} CH_3\overset{}{\underset{\underset{\displaystyle OH}{|}}{C}}{=}O$$

ethyl alcohol   acetic acid
(ethanol)   (ethanoic acid)

Naming Organic Acids

The IUPAC names for organic acids end in -*oic acid*. To name an acid according to the IUPAC system, take the longest alkane containing the acid group, drop the ending -*e* and add -*oic acid*. Thus, formula (18-2) shows ethanoic acid.

The common names of acids are derived from the names of the aldehyde from which they may be prepared. Thus, formula (18-2), derived from acetaldehyde, is called acetic acid.

The general formula for an acid is **RCOOH** or **ArCOOH**. All organic acids contain at least one —COOH group. This group is called the **carboxyl group,** and it is this group that yields hydrogen ions.

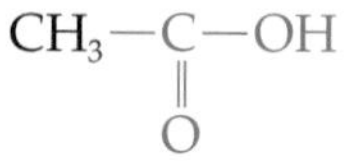

$$CH_3\overset{}{\underset{\underset{\displaystyle O}{\|}}{C}}{-}OH \qquad (18\text{-}2)$$

ethanoic acid

Table 18-2	Some Common Dicarboxylic Acids	
Structure	Common Name	IUPAC Name
COOH | COOH	Oxalic acid	Ethanedioic acid
COOH | CH$_2$ | COOH	Malonic acid	Propanedioic acid
COOH | CH$_2$ | CH$_2$ | COOH	Succinic acid	Butanedioic acid

Organic acids containing two carboxyl groups are called *dicarboxylic acids*. Those containing three carboxyl groups are called *tricarboxylic acids*. Table 18-2 gives the names and structures of some common dicarboxylic acids.

Exercise 18-7

Name the following compound.

$$CH_3—CH_2—CH_2—CH_2—COOH$$
$$\vert$$
$$Cl$$

Solution

The compound contains a COOH group, so it is a carboxylic acid. The longest chain contains five carbon atoms, so it is a pentanoic acid, numbered with the acid at carbon number 1.

$$\overset{5}{C}H_3—\overset{4}{C}H_2—\overset{3}{C}H_2—\overset{2}{C}H_2—\overset{1}{C}OOH$$
$$\vert$$
$$Cl$$

Attached to the chain is a chlorine at carbon number 2, so the name is 2-chloropentanoic acid.

Exercise 18-8

Draw the structure for 3-chlorobenzoic acid.

Solution

The structure for benzoic acid is

COOH

A chlorine is present at carbon number 3, numbered in either direction from the carboxyl group, which is at carbon number 1, or

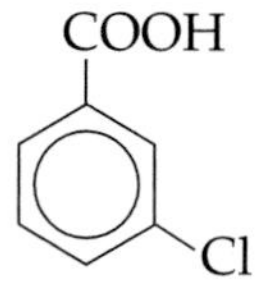

$$COOH$$

Properties and Reactions of Organic Acids

Most organic acids are relatively weak acids since they ionize only slightly in water.

$$CH_3COOH \rightleftharpoons H^+ + CH_3COO^-$$
acetic acid hydrogen ion acetate ion

Organic acids react with bases to form salts and water. The general reaction of an organic acid with a base to form a salt and water can be written as follows:

$$RCOOH + NaOH \rightleftharpoons RCOONa + H_2O$$
organic acid base organic salt water

$$CH_3COOH + NaOH \rightleftharpoons CH_3COONa + H_2O$$
acetic acid sodium hydroxide sodium acetate (a salt) water

Organic acids also react with bicarbonates and carbonates.

$$CH_3COOH + NaHCO_3 \longrightarrow CH_3COONa + CO_{2(g)} + H_2O$$
acetic acid sodium bicarbonate sodium acetate

$$2\,HCOOH + Na_2CO_3 \longrightarrow 2\,HCOONa + CO_{2(g)} + H_2O$$
formic acid sodium carbonate sodium formate

Organic acids containing few carbon atoms are soluble in water. As the length of the carbon chain increases, the solubility in water decreases.

Organic acids also react with alcohols to form a class of compounds called esters, which will be dealt with in the next chapter.

Medically Important Organic Acids

Formic acid ($HCOOH$) is a colorless liquid with a sharp, irritating odor. Formic acid is found in the sting of bees and ants and causes the characteristic pain and swelling when it is injected into the tissues. It is one of the strongest organic acids.

Acetic acid (CH_3COOH) is one of the components of vinegar, where it is usually found as a 4 to 5 percent solution. Acetic acid can be made by the oxidation of ethyl alcohol. The acetyl group,

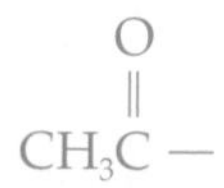

$$\overset{O}{\underset{\|}{CH_3C}} -$$

derived from acetic acid, is very important in metabolic reactions.

Citric acid is found in citrus fruits. Its formula indicates that it is an alcohol as well as an acid. Citric acid contains one alcohol (—OH) group and three acid (—COOH) groups. It is an example of a tricarboxylic acid.

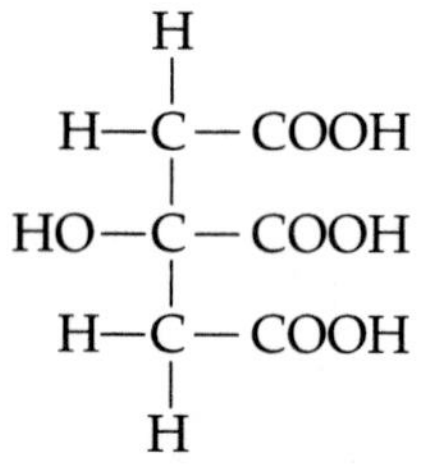

$$\begin{array}{c} H \\ | \\ H-C-COOH \\ | \\ HO-C-COOH \\ | \\ H-C-COOH \\ | \\ H \end{array}$$
citric acid

Magnesium citrate, a salt of citric acid, is used as a cathartic (a medication for stimulating the evacuation of the bowels). Sodium citrate, another salt of citric acid, is used as a blood anticoagulant. (It removes Ca^{2+} needed for coagulation from the blood.)

Lactic acid is found in sour milk. It is formed in the fermentation of milk sugar, lactose. Its formula is

$$\begin{array}{ccc} & H & H \\ & | & | \\ H\!-\!C\!-\!C\!-\!COOH \\ & | & | \\ & H & OH \end{array}$$

lactic acid

Lactic acid is also both an acid and an alcohol. It is formed whenever the body produces energy anaerobically (see Section 26-6).

Oxalic acid is another one of the strong, naturally occurring organic acids. Its formula is shown in Table 18-2. Oxalic acid is used to remove stains, particularly rust and potassium permanganate stains, from clothing. It is poisonous when taken internally. Oxalate salts also prevent clotting by *chelating Ca^{2+} from the blood. However, oxalate can be used only for blood samples that are to be analyzed in the laboratory because these salts are poisonous and cannot be added directly to the bloodstream.

Pyruvic acid is produced during the anaerobic phase of oxidation of glucose (see Section 26-6). It is a keto acid. In muscle, pyruvic acid is reduced to lactic acid during anaerobic exercise. In the tissues, pyruvic acid is changed to acetyl coenzyme A, which then enters the Krebs cycle (Section 26-8).

$$\begin{array}{c} CH_3\!-\!C\!-\!COOH \\ \| \\ O \end{array}$$

pyruvic acid

Tartaric acid is another organic acid that is both an acid and an alcohol. Tartaric acid is found in several fruits, particularly grapes. Potassium hydrogen tartrate, an acid salt called cream of tartar, is used in making baking powders. Rochelle salts, or potassium sodium tartrate, is used as a mild cathartic.

Stearic acid is a solid greaselike acid that is insoluble in water. It is an example of a fatty acid. Its formula is $C_{17}H_{35}COOH$. The sodium salt of stearic acid, sodium stearate, is a commonly used soap.

$$C_{17}H_{35}COOH + NaOH \longrightarrow C_{17}H_{35}COONa + H_2O$$

stearic acid   sodium stearate

Benzoic Acid Benzoic acid can be produced by the oxidation of benzaldehyde (see page 306) or by the oxidation of toluene.

toluene   benzoic acid

Benzoic acid is a white crystalline compound that is slightly soluble in cold water and more soluble in hot water. It is used medicinally as an antifungal agent. The sodium salt of benzoic acid, sodium benzoate, is used as a preservative.

Salicylic acid is both an alcohol and an acid, as can be seen from its structure. Salicylic acid is a white crystalline compound with properties similar to those of benzoic acid. It is used in the treatment of fungal infections and also for the removal of warts and corns.

Commonly used compounds of salicylic acid are the salt sodium salicylate and the ester methyl salicylate (see the following chapter).

salicylic acid

COONa
OH

sodium salicylate

COOCH₃
OH

methyl salicylate

COOH
OCOCH₃

acteylsalicylic acid
(aspirin)

OH

O
||
HN—C—CH₃
acetaminophen (Tylenol)

Sodium salicylate is used an an *antipyretic (to reduce fever) and also to relieve pain of arthritis, bursitis, and headache. *Methyl salicylate* is a liquid with a pleasant odor, that of wintergreen. It is used topically to relieve pain in muscles and joints.

The acetyl derivative of salicylic acid (the acetyl group is CH_3CO—) is *acetyl-salicylic acid*, more commonly known as aspirin. Aspirin is used as an *analgesic; as an antipyretic; for the treatment of colds, headaches, minor aches, and pains; and as a mild blood thinner, especially for individuals who have had a coronary artery bypass. Aspirin is also used in the treatment of rheumatic fever. More than 50 tons of aspirin tablets are used daily in the United States. Use of aspirin is contraindicated after surgery because it interferes with normal blood clotting and can induce hemorrhaging. Aspirin can also cause bleeding of the stomach and therefore should not be taken on an "empty" stomach. One person in 10,000 is allergic to aspirin.

The action of aspirin is related directly to that of the prostaglandins (see page 373). Aspirin stimulates respiration directly, and overdoses can cause serious acid-base balance disturbances.

Evidence appears to indicate that aspirin can prevent blood clots from forming by interfering with the action of the blood platelets. There is also evidence that one aspirin every other day helps prevent heart attacks.

Acetaminophen (Tylenol) has been used as a substitute for aspirin because it does not cause gastrointestinal bleeding and does not affect blood clotting. However, overdoses can lead to hepatic damage, as can be the case with aspirin itself. It is not effective against inflammation.

Summary

During the oxidation of an alcohol, the oxygen atom reacts with the H from the —OH group and with a hydrogen attached to the same carbon that has the —OH group.

The oxidation of a primary alcohol yields an aldehyde. Aldehydes contain the —CHO group and have the general formula RCHO. The aldehyde of one carbon atom is known as formaldehyde or methanal; that of two carbon atoms, as acetaldehyde or ethanal. A water solution of formaldehyde, known as formalin, is commonly used as a germicide. Paraldehyde, formed by polymerizing molecules of acetaldehyde, depresses the central nervous system.

Aldehydes are good reducing agents. When an aldehyde is heated with $Cu(OH)_2$, a red precipitate of Cu_2O is formed.

The oxidation of a secondary alcohol yields a ketone. A ketone cannot be further oxidized without decomposition. Likewise, tertiary alcohols are not easily oxidized.

The oxidation of an aldehyde yields an acid. The general formula for an acid is RCOOH. The —COOH group, called the carboxyl group, furnishes hydrogen ions and so causes the acidic properties. Organic acids react with bases to form organic salts.

Formic acid (HCOOH) is found in the sting of bees. Acetic acid (CH_3COOH) is one of the components of vinegar. Citric acid is found in citrus fruits, lactic acid in milk.

Questions and Problems

1. What is the general formula for an aldehyde? an acid? a ketone?
2. Write the equation for the formation of formaldehyde from methyl alcohol.
3. Write the equation for the oxidation of propyl alcohol to an aldehyde. What product is formed?
4. What does the suffix "al" indicate? "one"?
5. Describe a test for an aldehyde.
6. Indicate some uses for formaldehyde.
7. What is produced when isopropyl alcohol is oxidized?
8. The oxidation of a primary alcohol yields what type of product? a secondary alcohol? a tertiary alcohol?
9. The oxidation of an aldehyde yields what type of product? Illustrate by means of an equation.
10. What product is produced by the oxidation of formaldehyde? acetaldehyde? Name the product in each case.

11. Write the equation for the reaction of acetic acid with NAOH. Name the product.
12. What is a carbonyl group? In what type of compounds is it present?
13. Name the following compounds according to the IUPAC system.

    a. 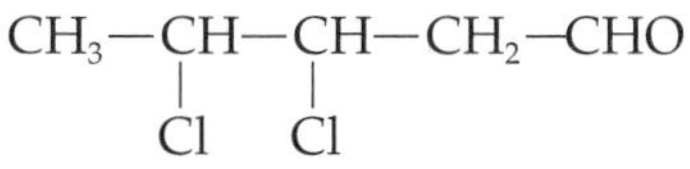

    b. 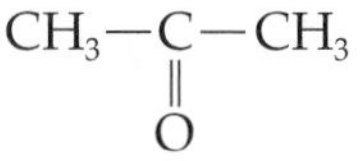

    c. 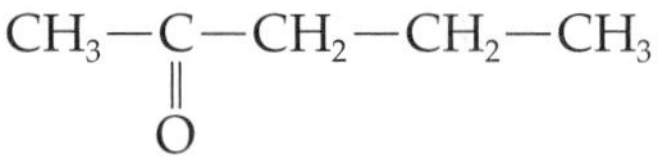

    d. 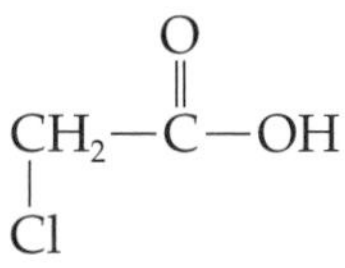

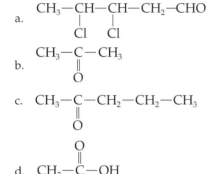

14. What is a hemiacetal? an acetal? Why are they important in the body?
15. Why do aldehydes give a positive test with Clini-test whereas ketones give a negative test?
16. Reduction of a ketone yields what type of product? reduction of an aldehyde?
17. What is a hemiketal? a ketal? Why are they important in the body?
18. Why does the removal of calcium ions from the blood prevent clotting?
19. Complete the following reactions and name the organic products.
    a. oxidation of 2-pentanol
    b. oxidation of 1-butanol
20. Write structures for and identify the major functional group for
    a. paraldehyde
    b. glyceraldehyde
    c. citric acid
    d. lactic acid
    e. acetone

Practice Test

1. The general formula for an aldehyde is ______.
   a. ROH b. ROR c. RCOR′ d. RCHO
2. The general formula for a ketone is ______.
   a. RCOR' b. RCHO c. RCOOH d. ROH
3. The general formula for an acid is ______.
   a. RCOR b. RCOOH c. RCHO d. RCOOR
4. The reaction of an aldehyde and an alcohol yields a(n) ______.
   a. hemiacetal b. hemiketal c. ester d. amino acid
5. The oxidation of a primary alcohol yields a(n) ___.
   a. ester b. ketone c. amide d. aldehyde
6. Which category of compounds yields a positive result with Clini-test?
   a. aldehydes b. ketones c. acids d. esters
7. The structure 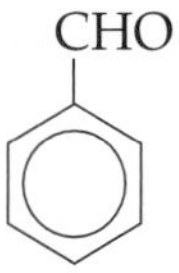 represents
   a. benzoic acid b. benzene oxide c. benzaldehyde d. benzyl alcohol
8. What is the name of the following compound?

   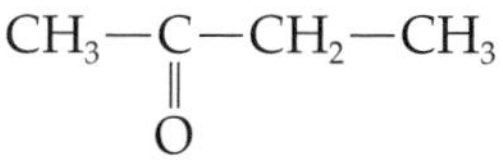

   a. 1-butanone
   b. 2-butanone
   c. 3-butanone
   d. 4-butanone
9. The common name for 2-pentanone is
   a. methyl propylketone
   b. ethyl methylketone
   c. methoxyketone
   d. ethyl propylketone
10. Which of the following is a carboxylic acid?
    a. lactic acid
    b. pyruvic acid
    c. stearic acid
    d. all of these

19

Esters, Amines, and Amides

- To become familiar with the functional groups of esters, amines, amides, and amino acids

- To write equations for the formation of esters, amines, and amides

- To recognize the biologic importance of various esters, amines, amides, and alkaloids

- To distinguish among primary, secondary, and tertiary amines

- To recognize the structure of amino acids

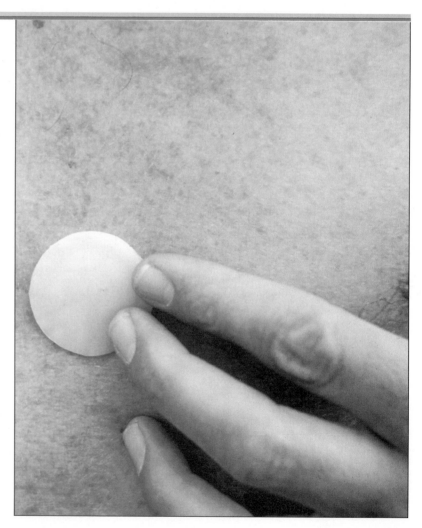

Nitroglycerin patch.

19-1 Esters

Esters are produced by the reaction of an organic acid with an alcohol and have the general formula

$$RCOOR' \quad \text{or} \quad RCOR'$$
$$ArCOOR \quad \text{or} \quad ArCOOAr'$$

The general reaction of an alcohol with an acid is illustrated by the following equation (where R and R' may be the same or different alkyl groups).

$$\underset{\text{acid}}{RCOOH} + \underset{\text{alcohol}}{R'OH} \rightleftharpoons \underset{\text{ester}}{RCOOR'} + \underset{\text{water}}{H_2O}$$

$$\underset{\text{acid}}{CH_3-\overset{\overset{\displaystyle O}{\|}}{C}-OH} + \underset{\text{alcohol}}{HOCH_3} \rightleftharpoons \underset{\text{ester}}{CH_3-\overset{\overset{\displaystyle O}{\|}}{C}-O-CH_3} + H_2O$$

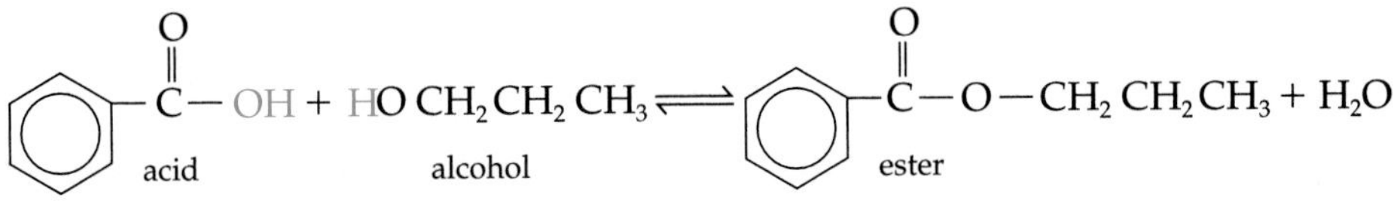

$$\underset{\text{acid}}{\text{C}_6H_5-\overset{\overset{\displaystyle O}{\|}}{C}-OH} + \underset{\text{alcohol}}{HO\,CH_2CH_2\,CH_3} \rightleftharpoons \underset{\text{ester}}{\text{C}_6H_5-\overset{\overset{\displaystyle O}{\|}}{C}-O-CH_2\,CH_2CH_3} + H_2O$$

Note that the reactions are written with a double arrow, indicating that these are equilibrium reactions; that is, the reverse reactions also take place. Thus, esters hydrolyze to form organic acids and alcohols. Esters do not readily ionize in water solution.

When a carboxylic acid reacts with an alcohol to form an ester, water is also produced. Esters have a carboxylic acid part and an alcohol part. The acid part is easily identified since it contains a carbonyl group (C=O).

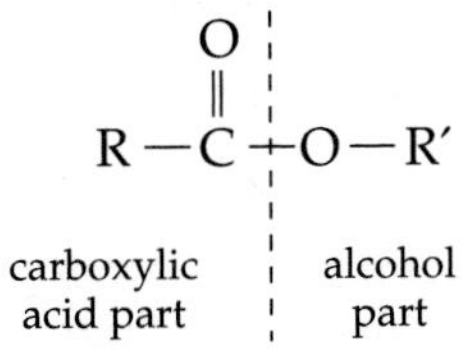

$$R-\overset{\overset{\displaystyle O}{\|}}{C}\,\vdots\,O-R'$$

$$\underset{\substack{\text{carboxylic} \\ \text{acid part}}}{} \qquad \underset{\substack{\text{alcohol} \\ \text{part}}}{}$$

If the formula of an ester is written with the formula HOH, water, placed below so that the OH part of the water molecule is below the carbonyl group and the H part of the water molecule is below the alcohol part, then the name of the ester can be easily determined by reading downward on each side of the dotted line.

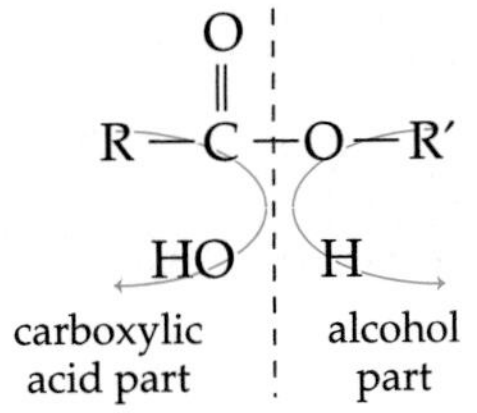

$$R-\overset{\overset{\displaystyle O}{\|}}{C}\,\vdots\,O-R'$$
$$HO\,\vdots\,H$$

$$\underset{\substack{\text{carboxylic} \\ \text{acid part}}}{} \qquad \underset{\substack{\text{alcohol} \\ \text{part}}}{}$$

Naming Esters

When a carboxylic acid reacts with an alcohol, the name of the ester is determined as follows:

IUPAC System

1. Write the IUPAC name of the alcohol minus the ending *-anol* and add the ending *-yl*
2. Write the IUPAC name of the acid minus the ending *-ic*
3. Add the ending *-ate* to the name of the acid

Common System

1. Write the common name of the alcohol
2. Write the name of the acid minus the ending *-ic*
3. Add the ending *-ate* to the name of the acid

Exercise 19-1

Give the IUPAC name and the common name for the following ester.

$$\overset{\displaystyle O}{\overset{\displaystyle \|}{H-C}}-O-CH_2CH_3$$

Solution

Writing water (HOH) below the formula of the ester as indicated previously, we have

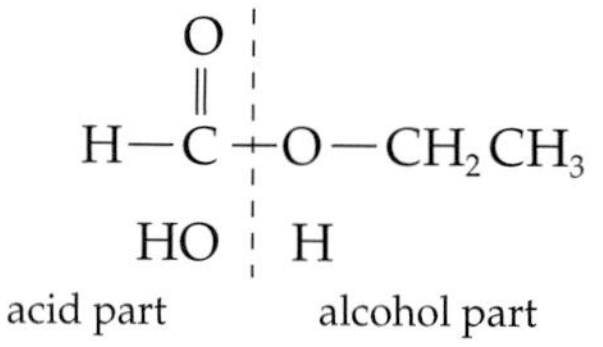

$$\overset{\displaystyle O}{\overset{\displaystyle \|}{H-C}}+O-CH_2CH_3$$

HO | H

acid part   alcohol part

IUPAC name

- **Step 1** The alcohol, on the right side above, is a two-carbon alcohol or ethanol.
- **Step 2** The acid part of the ester contains one carbon atom so, it is methanoic acid.
- **Step 3** Dropping the ending "ic" from the acid and adding the ending "ate," we have the name of the ester as ethyl methanoate.

Common name

- **Step 1** The name of the alcohol, as indicated above, is ethyl alcohol.
- **Step 2** The common name of the acid with one carbon atom is formic acid.
- **Step 3** Dropping the ending "ic" and adding the ending "ate," we have the common name of the ester as ethyl formate.

Self-test

Give the IUPAC and common name for the following ester.

$$CH_3-\overset{\displaystyle O}{\overset{\displaystyle \|}{C}}-O-CH_2-CH_2-CH_2-CH_3$$

Answer

IUPAC name: butyl ethanoate; common name: butyl acetate

Exercise 19-2

Give the IUPAC and common name for the following ester.

$$CH_3-\underset{\underset{CH_3}{\overset{|}{|}}}{\overset{\overset{H}{|}}{C}}-O-\overset{\overset{O}{\|}}{C}-CH_3$$

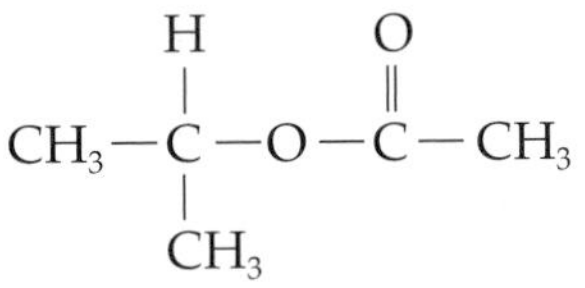

Solution

Writing HOH below the formula for the ester we have

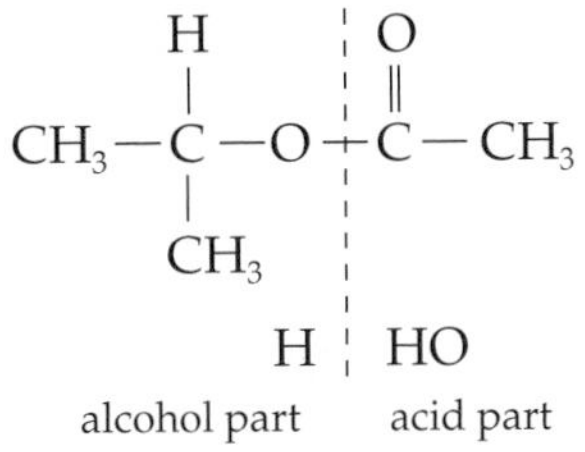

$$CH_3-\underset{\underset{CH_3}{\overset{|}{|}}}{\overset{\overset{H}{|}}{C}}-O-\overset{\overset{O}{\|}}{C}-CH_3$$

H   HO

alcohol part   acid part

IUPAC name

- **Step 1** The IUPAC name of the alcohol is 2-propanol. Dropping the "anol" and adding "yl" we have 2-propyl.
- **Step 2** The IUPAC name of the acid is ethanoic acid. Dropping the "ic" and adding "ate" we have 2-propyl ethanoate.

Common name

- **Step 1** The common name for the alcohol is isopropyl alcohol.
- **Step 2** The common name for the acid is acetic acid.
- **Step 3** Dropping the ending "ic" and adding "ate," the common name of the ester is isopropyl acetate.

Self-test

Give IUPAC name for

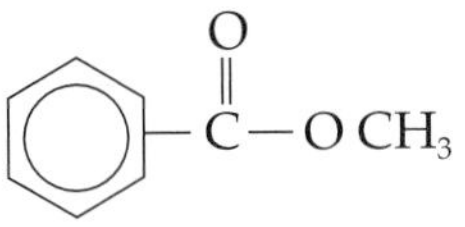

$$\bigcirc\!\!\!\!\!\!-\overset{\overset{O}{\|}}{C}-O\,CH_3$$

Answer

Methyl benzoate

Exercise 19-3

Give the IUPAC name for the following ester

$$\bigcirc\!\!\!\!\!\!-\overset{\overset{O}{\|}}{C}-O-CH_2-CH_2-CH_3$$

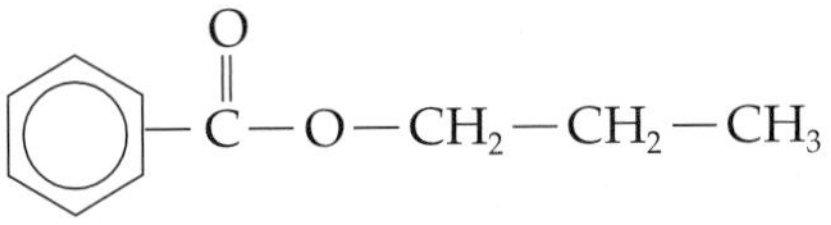

Table 19-1

Esters and Synthetic Flavors

Ester	Flavor
Amyl acetate	Banana
Ethyl butyrate	Pineapple
Amyl butyrate	Apricot
Isoamyl acetate	Pear
Octyl acetate	Orange

Solution

Writing HOH below the formula for the ester we have

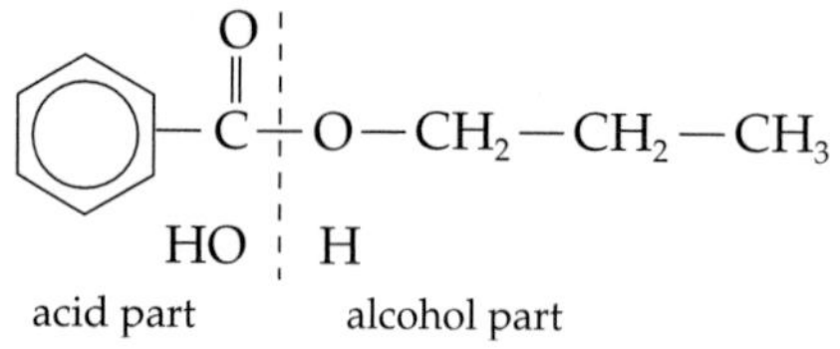

$$\underset{\text{acid part}}{\underset{\text{HO}}{\bigcirc}}-\overset{\overset{\text{O}}{\|}}{\text{C}}\mid\underset{\text{alcohol part}}{\text{O}-\text{CH}_2-\text{CH}_2-\text{CH}_3}$$

- **Step 1** The name of the alcohol is propanol. Dropping the ending "anol" and adding "yl" we have propyl.
- **Step 2** The name of the acid is benzoic acid. Dropping the ending "ic" and adding "ate" we have benzoate.
- **Step 3** The IUPAC name of the ester is propyl benzoate.

Esters are important as solvents, perfumes, and flavoring agents. Table 19-1 indicates some of the esters used in preparing synthetic flavors. Many esters are also used medicinally (see Table 19-2).

The hydrolysis of esters takes place in the body when fats and oil are digested. The reaction, in general, is

$$\underset{\text{ester}}{\text{RCOOR}'} + \text{H}_2\text{O} \xrightarrow{\text{enzymes}} \underset{\text{acid}}{\text{RCOOH}} + \underset{\text{alcohol}}{\text{R}'\text{OH}}$$

This type of reaction will be discussed in Chapter 21.

Esters also undergo saponification, a reaction with NaOH (see page 364).

$$\underset{\text{ester}}{\text{RCOOR}'} + \text{NaOH} \longrightarrow \underset{\text{salt of ester}}{\text{RCOONa}} + \underset{\text{alcohol}}{\text{R}'\text{OH}}$$

Thioesters

Thioesters have the general formula

$$\text{R}-\overset{\overset{\text{O}}{\|}}{\text{C}}-\text{S}-\text{R}'$$

where the prefix *thio-* indicates the presence of a sulfur atom (see Section 17-2). Many thioesters are found in biologic systems. The most important of these is acetyl coenzyme A, abbreviated acetyl CoA. Acetyl CoA is a key compound in metabolic reactions in the body. It is involved in the formation of fats (Section 27-7), ketone bodies (Section 27-4), and amino acids.

Table 19-2 Some Medical Uses of Esters

Esters	Use
Ethyl aminobenzoate (benzocaine)	Local anesthetic
Glyceryl trinitrate (nitroglycerin)	Used to dilate coronary arteries and lower blood pressure
Methyl salicylate (oil of wintergreen)	Used as a flavoring agent and also a counterirritant in many liniments
Phenyl mercuric acetate	Used to disinfect instruments and as an antiseptic on cutaneous and mucosal surfaces

Phosphate Esters

When an organic compound containing an —OH group reacts with phosphoric acid (H_3PO_4), phosphate esters are formed. Phosphoric acid is a triprotic acid, so it can yield three hydrogen ions,

$$HO-\overset{\overset{\displaystyle O}{\|}}{\underset{\underset{\displaystyle OH}{|}}{P}}-OH$$

and it can form various esters such as

$$HO-\overset{\overset{\displaystyle O}{\|}}{\underset{\underset{\displaystyle OH}{|}}{P}}-OCH_3 \qquad CH_3O-\overset{\overset{\displaystyle O}{\|}}{\underset{\underset{\displaystyle OH}{|}}{P}}-OCH_3 \qquad CH_3O-\overset{\overset{\displaystyle O}{\|}}{\underset{\underset{\displaystyle OCH_3}{|}}{P}}-OCH_3$$

methyl dihydrogen phosphate · dimethyl hydrogen phosphate · trimethyl phosphate

Before sugars can be metabolized, they must first be converted into phosphate esters (see Section 26-4). Phosphate esters serve many important functions in the body. They are present in

1. Adenosine triphosphate (ATP), adenosine diphosphate (ADP), and adenosine monophosphate (AMP), the body's energy compounds
2. Cyclic adenosine monophosphate (cAMP), a chemical messenger that regulates cellular enzymes
3. Deoxyribonucleic acid (DNA) and ribonucleic acid (RNA), the hereditary material of a cell
4. Phospholipids, which are found in cell membranes and also in all tissue, particulary brain, liver, and spine
5. Phosphorylases, enzymes involved in the adding of phosphate groups to carbohydrates

These uses of phosphate esters will be discussed later in appropriate chapters.

19-2 Aliphatic Amines

Amines are organic compounds derived from ammonia. There are three classes of amines: primary, secondary, and tertiary. Note that in relation to amines, the terms primary, secondary, and tertiary refer directly to the number of hydrogen atoms of ammonia that have been replaced by alkyl groups. In the case of alcohols, the terms primary, secondary, and tertiary refer to the number of carbon atoms attached to the carbon having the —OH group on it.

Primary amines are those in which one of the hydrogen atoms of ammonia (NH_3) has been replaced by an alkyl group. Primary amines have the general formula **RNH_2**.

Secondary amines are those in which two of the hydrogen atoms of the ammonia have been replaced by alkyl groups. Secondary amines have the general formula **R_2NH**.

Tertiary amines are those in which all three hydrogen atoms of the ammonia have been replaced by alkyl groups. Tertiary amines have the general formula **R_3N**.

Table 19-3 Ammonia and Simple Amines

Name	Formula	Example
Ammonia	NH_3 or 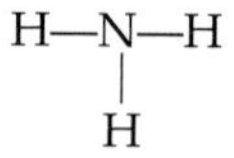	H—N—H with H below
Primary amine	RNH_2	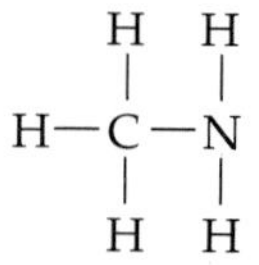 H—C—N (with H's) or CH_3NH_2
Secondary amine	R_2NH	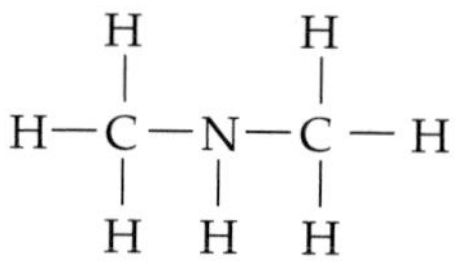 H—C—N—C—H (with H's) or $(CH_3)_2NH$
Tertiary amine	R_3N	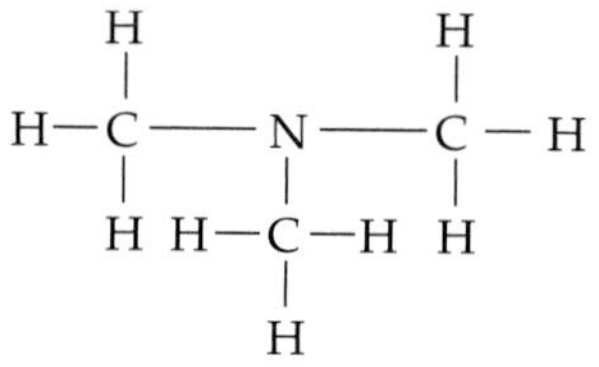 H—C—N—C—H (with H's) or $(CH_3)_3N$

Naming Aliphatic Amines

In the IUPAC system the —NH group is called an *amino* group and is indicated numerically according to its position on the carbon chain. If there is more than one carbon chain attached to the nitrogen, then the longest chain is the one used as the parent name. The other or others are indicated by the letter "N" to indicate that they are attached to the nitrogen.

In the common system, the word *amine* is added after the name of the alkyl group. The common system is used primarily for simple compounds. For more complex compounds, the IUPAC system is used. With this system the prefix "amino" is used to denote the presence of a —NH_2 group.

Exercise 19-4

Give the IUPAC and common name for the following.

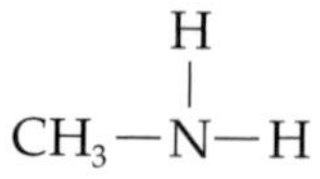

$$CH_3 - \overset{\overset{\displaystyle H}{|}}{N} - H$$

Solution

IUPAC name

There is only one carbon atom, so the compound is a derivative of methane. The —NH_2 group is called amino, so the compound is aminomethane. No number is needed since there is only one carbon atom.

Common name

The CH_3 group is the methyl group, so the name is methylamine.

Exercise 19-5

Give the IUPAC and common name for the following.

$$CH_3-CH-CH_3$$
$$|$$
$$NH_2$$

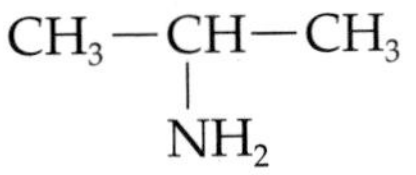

Solution

IUPAC name

The longer chain contains three carbons, so it is a propane derivative. Attached to the chain is an amino group at carbon number 2. Therefore, the IUPAC name is 2-aminopropane.

Common name

An alkyl group of three carbons is called a propyl group. If a substituent is attached to the central carbon, the name is changed to isopropyl. So, the common name is isopropylamine.

Exercise 19-6

Give the IUPAC name for the following (note that acids take precedence over alkanes and amines in the numbering system).

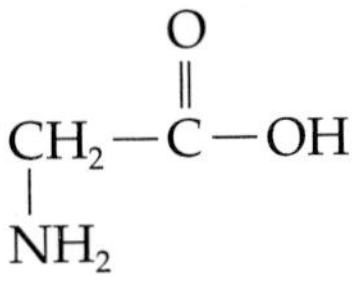

$$\overset{\displaystyle O}{\overset{\displaystyle \|}{CH_2-C-OH}}$$
$$|$$
$$NH_2$$

Solution

The chain consists of two carbon atoms and is a carboxylic acid. Its name is ethanoic acid. An amino group is attached to carbon 2, so the name is 2-aminoethanoic acid.

Self-test

Draw structures for:

a. 3-aminocyclohexanol   b. 4-amino-1-butene

Answers

a.

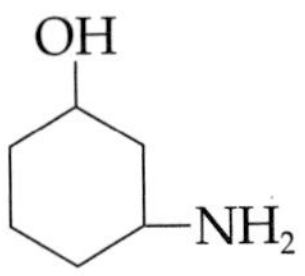

OH

(cyclohexane ring with OH at top and —NH₂ substituent)

b. $CH_2-CH_2-CH=CH_2$
    $|$
   NH_2

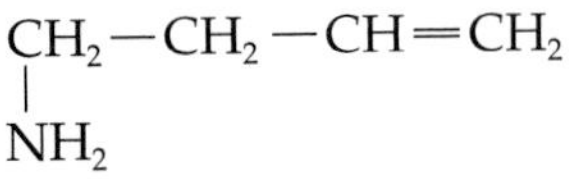

Exercise 19-7

Give the IUPAC and common name for the following.

$$CH_3$$
$$|$$
$$CH_3-N-CH_3$$

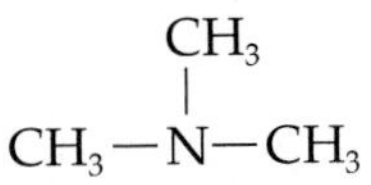

Solution

IUPAC name

The longer chain contains only one carbon, so this is a methane derivative. Attached to the methane is an amine group, so this is an aminomethane. There are two other methyl groups attached to the nitrogen. Each is indicated by the letter "N," so the name is N,N-dimethylaminomethane.

Common name

There are three methyl groups attached to the nitrogen of the amine, so this is a trimethyl compound. Its common name is trimethylamine.

Exercise 19-8

Give the IUPAC name for the following:

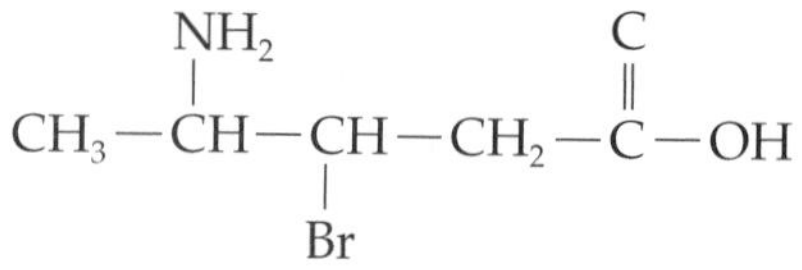

$$\begin{array}{ccccc}
& NH_2 & & & C \\
& | & & & \| \\
CH_3 & - CH - CH - CH_2 & - C - OH \\
& & | \\
& & Br
\end{array}$$

Solution

The longest chain is called pentanoic acid. There is an amino group at carbon number 4 and a bromo at carbon number 3. The name is 4-amino-3-bromopentanoic acid.

Among the amines found in the body or used medicinally are histamines (page 501), barbiturates, psychedelics, amphetamines, and nucleic acids (Section 23-1).

Amines produced during the decay of once-living matter include putrescine and cadaverine (Section 25-12).

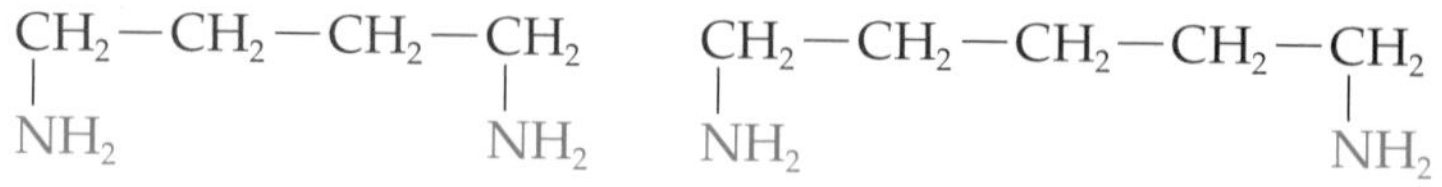

$$\begin{array}{cc}
CH_2 - CH_2 - CH_2 - CH_2 & \qquad CH_2 - CH_2 - CH_2 - CH_2 - CH_2 \\
| \qquad\qquad\qquad\qquad | & \qquad | \qquad\qquad\qquad\qquad\qquad\qquad\qquad | \\
NH_2 \qquad\qquad\qquad\quad NH_2 & \qquad NH_2 \qquad\qquad\qquad\qquad\qquad\qquad\quad NH_2
\end{array}$$

putrescine
(1,4-diaminobutane)

cadaverine
(1,5-diaminopentane)

Preparation of Amines

Amines can be produced by the reaction of alkyl halides with ammonia or with another amine. These reactions can be abbreviated, the inorganic by-products being ignored, as follows.

$$RCl \xrightarrow{NH_3} RNH_2$$

$$RCl \xrightarrow{R'NH_2} RNHR'$$

Examples of such reactions are

$$CH_3CH_2Cl \xrightarrow{NH_3} CH_3CH_2NH_2$$

$$CH_3CH_2Cl \xrightarrow{CH_3NH_2} CH_3CH_2NHCH_3$$

Reactions and Properties of Amines

When amines react with an acid such as hydrochloric acid, ammonium salts are produced. These ammonium salts are frequently named by placing the word "hydrochloride" after the name of the amine. Examples are thiamine hydrochloride (vitamin B$_1$) and procaine hydrochloride (Novocaine, a local anesthetic). The hydrochlorides are administered rather than the amines themselves because they are more soluble in water.

Amines in general are basic compounds. They react with inorganic acids to form salts. Recall that one definition of a base is "a substance that accepts protons (H$^+$)" (see Section 12-2). Compare the following reactions of ammonia and amines with HCl to the reaction of a base with an acid to form a salt.

$$NH_3 + HCl \longrightarrow NH_4^+ + Cl^-$$
$$RNH_2 + HCl \longrightarrow RNH_3^+ + Cl^-$$
$$R_2NH + HCl \longrightarrow R_2NH_2^+ + Cl^-$$
$$R_3N + HCl \longrightarrow R_3NH^+ + Cl^-$$

Quaternary ammonium salts are formed by the action of tertiary amines with organic halogen compounds. An example of this type of reaction is that between trimethylamine and methyl iodide. Tetramethylammonium iodide is formed.

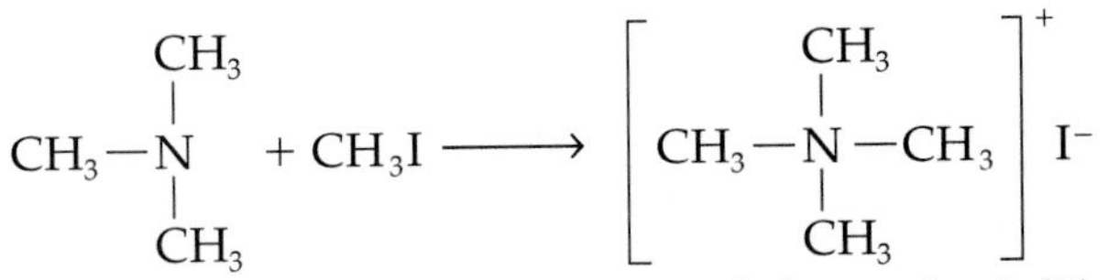

$$
\begin{array}{c}
\overset{\textstyle CH_3}{\underset{\textstyle CH_3}{\overset{|}{\underset{|}{CH_3-N}}}} + CH_3I \longrightarrow \left[\overset{\textstyle CH_3}{\underset{\textstyle CH_3}{\overset{|}{\underset{|}{CH_3-N-CH_3}}}}\right]^+ I^-
\end{array}
$$

trimethylamine   methyl iodide   tetramethylammonium iodide

Some quaternary ammonium salts have both a detergent action and antibacterial activity. They are used medicinally as antiseptics. Benzalkonium chloride (zephiran chloride) is used in a 0.1 percent solution for storage of sterilized instruments. A 0.01 to 0.02 percent solution is applied as a wet dressing to denuded areas. A 0.005 percent solution is used for irrigations of the bladder and the urethra. There are many others used as disinfectant soaps.

19-3 Amino Acids

Amino acids are organic acids that contain an amine group. Examples of amino acids are

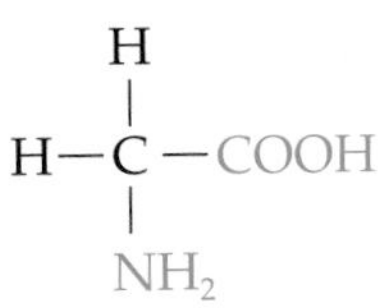

glycine, an α-amino acid 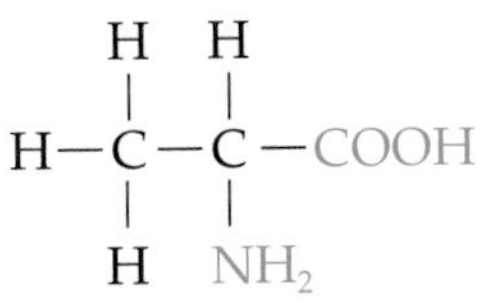   an α-amino acid 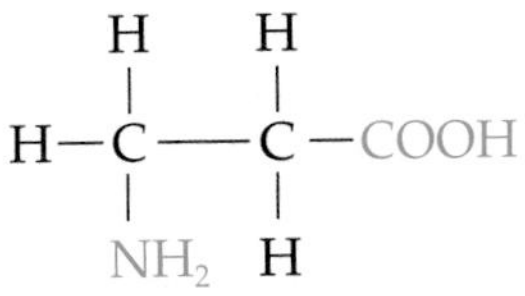   a β-amino acid

Note that Greek letters are used to designate the position of the amino group in the chain. The carbon atom next to the acid group, the —COOH group, is called the alpha (α) carbon. Then next in order come the beta (β), the gamma (γ), and the delta (δ) carbon atoms.

Amino acids contain an acid group, —COOH, which naturally is acidic, and also an —NH$_2$ group, which is basic (see above). Thus, amino acids exhibit both acidic and basic properties; amino acids can react with either acids or bases. Compounds that can act as, or react with, either acids or bases are called amphoteric compounds.

α-Amino acids are the building blocks of proteins. That is, proteins are polymers of α-amino acids (see Chapter 22).

Aromatic Amines

The simplest aromatic amine consists of an amino group attached to a benzene ring. Its IUPAC name is aniline. Aniline is prepared by the reduction of nitrobenzene.

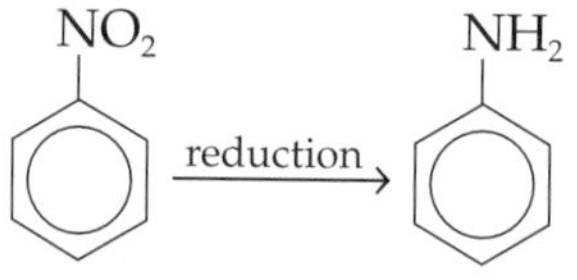

NO_2 $\xrightarrow{\text{reduction}}$ NH_2

nitrobenzene → aniline

Derivatives of aniline are named either by numbers or by prefixes, as shown on the following formulas:

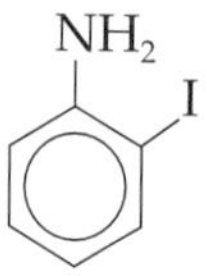

NH_2, I

2-iodoaniline
o-iodoaniline

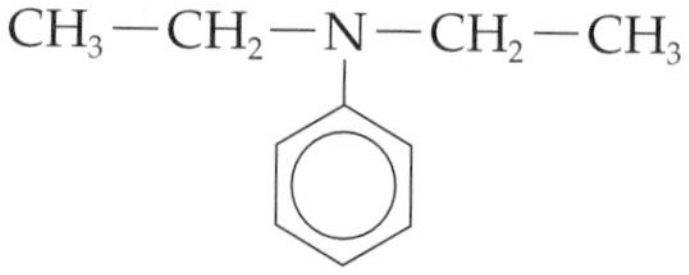

$CH_3-CH_2-N-CH_2-CH_3$

N,N-diethylaniline

Aniline is used commercially in the preparation of many dyes and drugs. When aniline is reacted with acetic acid, acetanilide is produced.

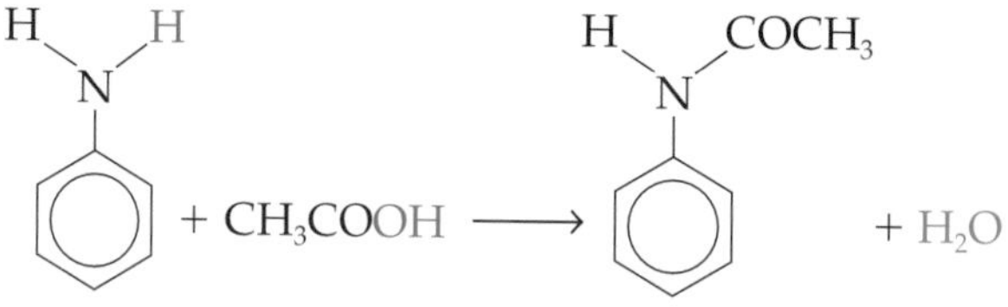

(H, H)N— + CH_3COOH ⟶ (H, $COCH_3$)N— + H_2O

analine   acetic acid   acetanilide

Acetanilide has been used as an antipyretic and as an analgesic. A related compound, phenacetin, has also been used for similar purposes, but its use has largely been discontinued because of liver and kidney toxicity.

$NHCOCH_3$

OC_2H_5
phenacetin

Alkaloids

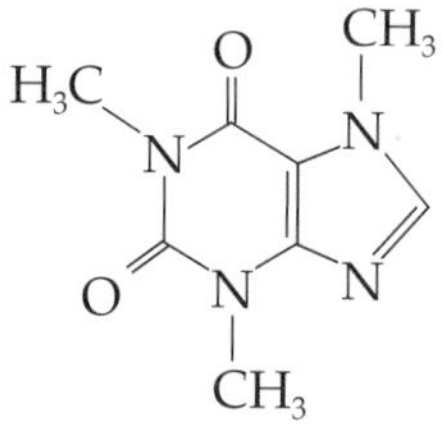

Alkaloids are nitrogen-containing compounds produced by plants. In general, they are heterocyclic (ring compounds with an element other than carbon in the ring). The term alkaloids (alkali-like) comes from the basic properties of amines.

The following are examples of alkaloids.

caffeine

Caffeine Caffeine occurs in coffee and tea. It stimulates the central nervous system and also acts as a *diuretic. Caffeine is used in various headache remedies in conjunction with analgesic drugs.

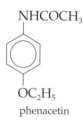

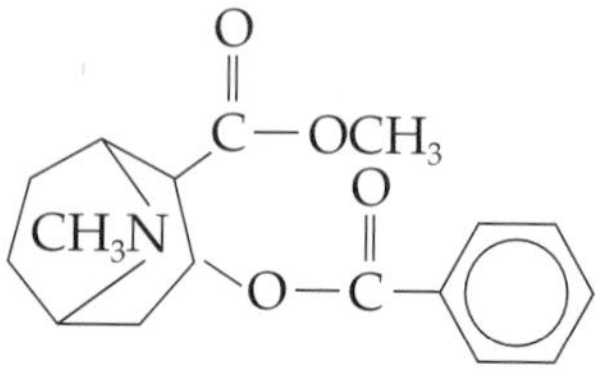

cocaine

Cocaine Obtained from the leaves of a coca shrub in South America, cocaine is administered medicinally in the form of a hydrochloride to enhance its solubility. Because it blocks nerve conduction, cocaine is widely used as a local anesthetic. However, it has largely been supplanted by synthetic drugs, which are less habit forming and less toxic. Pharmaceutical chemists have produced derivatives of cocaine that reduce the problem of habituation, such as procaine (Novocaine), lidocaine (Xylocaine), mepivacaine (Carbocaine), and bupivacaine (Marcaine).

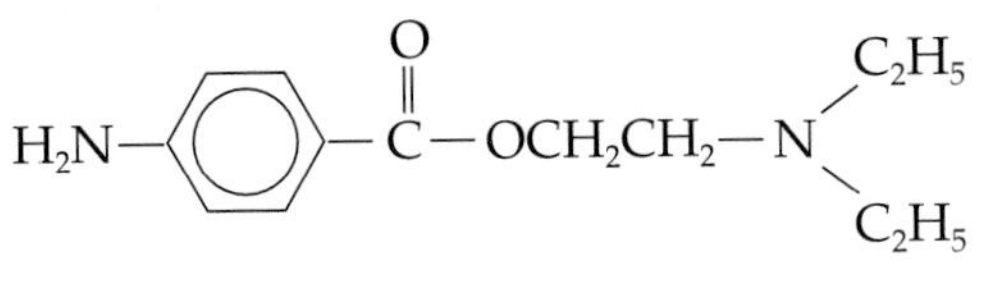

procaine

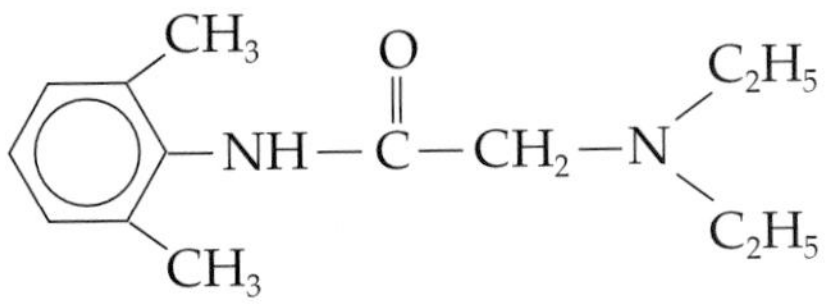

lidocaine

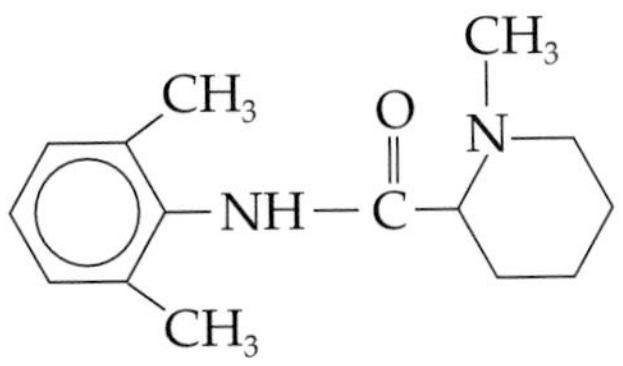

mepivacaine

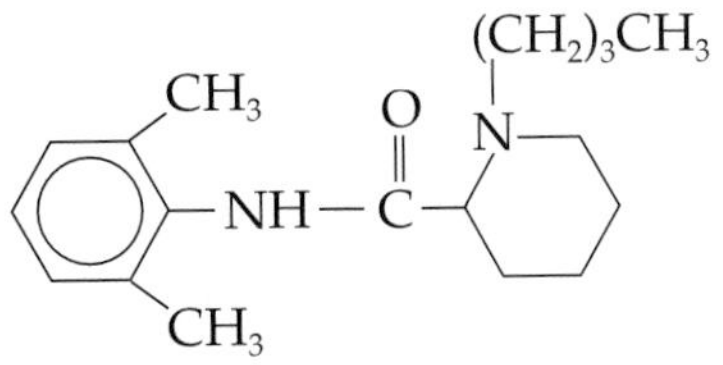

bupivacaine

Nicotine Found in tobacco leaves, nicotine is one of the few water-soluble alkaloids. It has no therapeutic use but is used as an insecticide. Nicotine is found in the body after cigarette smoking. It increases the blood pressure and pulse rate and constricts the blood vessels.

Ephedrine Obtained from twigs of a Chinese plant, ma-huang, ephedrine is used to elevate blood pressure during spinal anesthesia. It is also used to treat asthma, coughs, and colds and as a nasal decongestant.

Heterocyclic Amines Found in the Brain

The compounds serotonin, norepinephrine, and dopamine function as neurotransmitters in the brain.

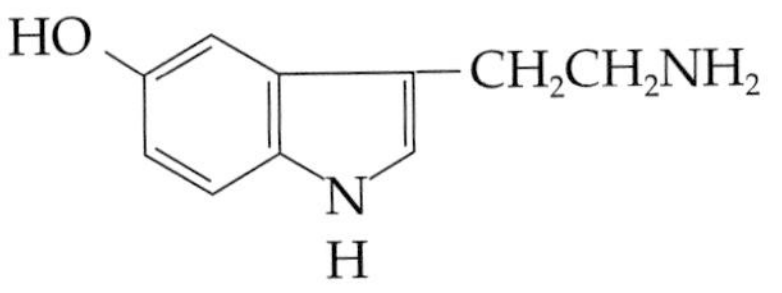

serotonin

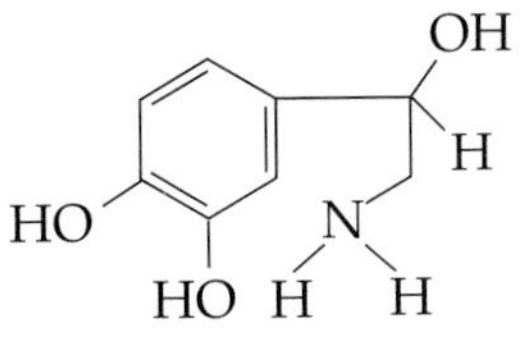

norepinephrine

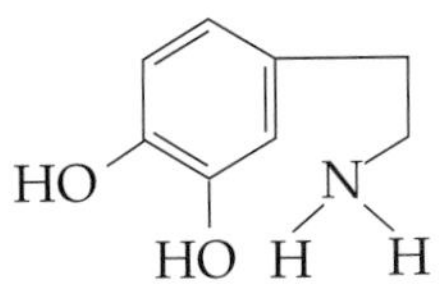

dopamine

Other examples of heterocyclic amines include hemoglobin, chlorophyll, vitamin B_2, vitamin B_{12}, nucleic acids, adenosine triphosphate (ATP), coenzyme A, barbiturates, amphetamines, and morphine.

19-4 Amides

Formation of Amides

Organic acids can react with ammonia or with amines to form a class of compounds called amides. Amides have a bond between a carbonyl group (—C═O) and a nitrogen.

$$\underset{\text{acid}}{R-\overset{\overset{\displaystyle O}{\|}}{C}-OH} + \underset{\text{ammonia}}{H-\overset{\overset{\displaystyle H}{|}}{N}-H} \xrightarrow{\text{heat}} \underset{\text{amide}}{R-\overset{\overset{\displaystyle O}{\|}}{C}-NH_2} + H_2O$$

Amides are named as follows.

IUPAC System

1. Drop ending -*oic* and the word acid from the name of the acid.
2. Add -*amide*.

Common System

1. Drop ending -*ic* and the word acid from the name of the acid.
2. Add -*amide*.

For example,

$$CH_3 - \overset{\overset{\textstyle O}{\|}}{C} - OH + H - \overset{\overset{\textstyle H}{|}}{N} - H \xrightarrow{\text{heat}} CH_3 - \overset{\overset{\textstyle O}{\|}}{C} - NH_2 + H_2O$$

| ethanoic acid (acetic acid) | ammonia (ammonia) | ethanamide (acetamide) |

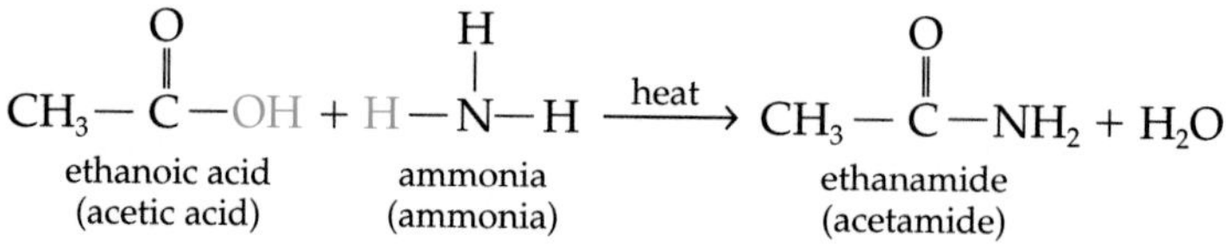

If an organic acid reacts with a primary or secondary amine, a substituted amide is formed.

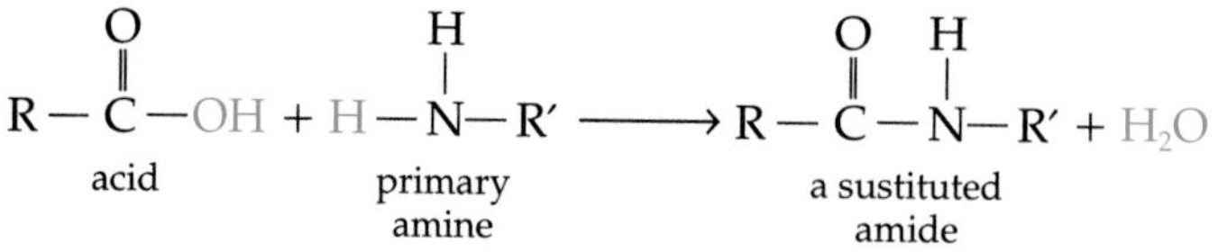

$$R - \overset{\overset{\textstyle O}{\|}}{C} - OH + H - \overset{\overset{\textstyle H}{|}}{N} - R' \longrightarrow R - \overset{\overset{\textstyle O}{\|}}{C} - \overset{\overset{\textstyle H}{|}}{N} - R' + H_2O$$

| acid | primary amine | a sustituted amide |

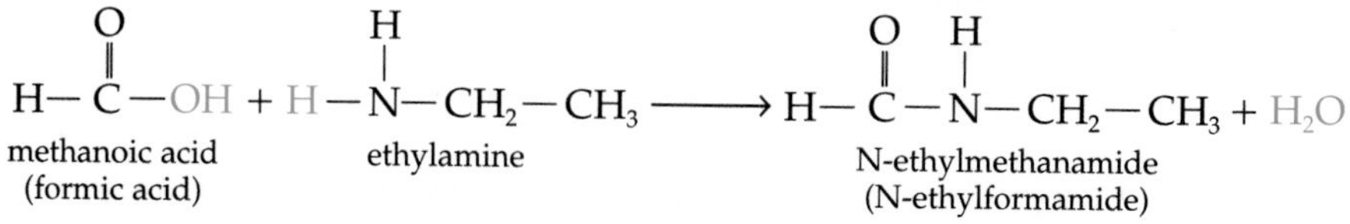

$$H - \overset{\overset{\textstyle O}{\|}}{C} - OH + H - \overset{\overset{\textstyle H}{|}}{N} - CH_2 - CH_3 \longrightarrow H - \overset{\overset{\textstyle O}{\|}}{C} - \overset{\overset{\textstyle H}{|}}{N} - CH_2 - CH_3 + H_2O$$

| methanoic acid (formic acid) | ethylamine | N-ethylmethanamide (N-ethylformamide) |

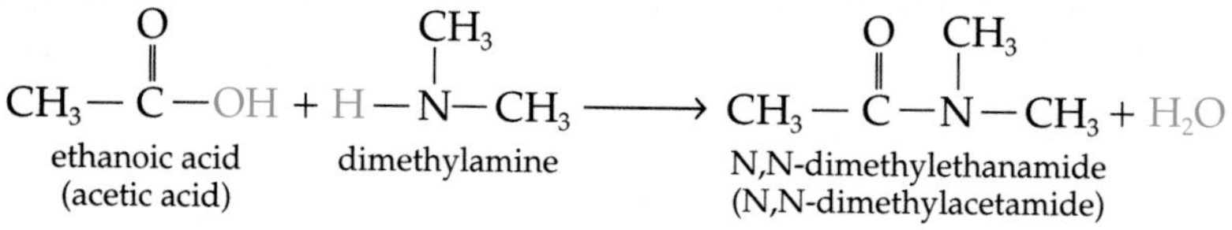

$$CH_3 - \overset{\overset{\textstyle O}{\|}}{C} - OH + H - \overset{\overset{\textstyle CH_3}{|}}{N} - CH_3 \longrightarrow CH_3 - \overset{\overset{\textstyle O}{\|}}{C} - \overset{\overset{\textstyle CH_3}{|}}{N} - CH_3 + H_2O$$

| ethanoic acid (acetic acid) | dimethylamine | N,N-dimethylethanamide (N,N-dimethylacetamide) |

Amides can also be prepared by reacting an acid with thionyl chloride ($SOCl_2$) to form a compound called an acyl chloride, which in turn reacts with ammonia to form an amide. The reactions are

$$RCOOH \quad + \quad SOCl_2 \quad \longrightarrow \quad RCOCl$$

| acid | thionyl chloride | acyl chloride |

$$RCOCl \quad + \quad NH_3 \quad \longrightarrow \quad RCONH_2$$

| acyl chloride | | amide |

Properties of Amides

Amides are neutral compounds compared with amines, which are basic. Acid hydrolysis of simple amides yields a carboxylic acid, whereas basic hydrolysis yields the salt of a carboxylic acid.

Acid hydrolysis $\qquad CH_3CONH_2 \xrightarrow[\text{H}_2\text{O}]{\text{HCl}} CH_3COOH$

Basic hydrolysis $\qquad CH_3CONH_2 \xrightarrow[\text{H}_2\text{O}]{\text{NaOH}} CH_3COONa$

If two amino acids combine, with the acid part of one reacting with the amine part of the other, the following type of reaction can occur.

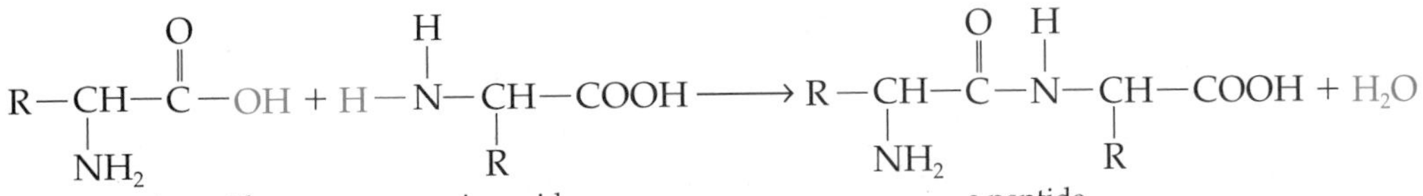

$$R-CH-\overset{\overset{\displaystyle O}{\|}}{C}-OH + H-\overset{\overset{\displaystyle H}{|}}{N}-CH-COOH \longrightarrow R-CH-\overset{\overset{\displaystyle O}{\|}}{C}-\overset{\overset{\displaystyle H}{|}}{N}-CH-COOH + H_2O$$

amino acid · · · · · · · amino acid · · · · · · · · · a peptide

These two amino acids are said to be linked by a peptide (amide) bond. Compounds of this type will be discussed in Chapter 22.

Niacin, one of the B vitamins, is administered as an amide, niacinamide. Urea, one of the metabolic products of protein metabolism, can be considered the diamide of carbonic acid.

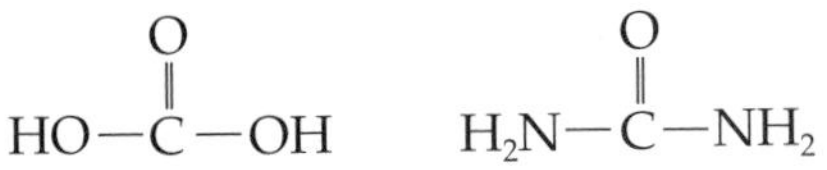

$$HO-\overset{\overset{\displaystyle O}{\|}}{C}-OH \qquad H_2N-\overset{\overset{\displaystyle O}{\|}}{C}-NH_2$$

carbonic acid · · · · · · · · urea

Aromatic Amides

The simplest aromatic amide is benzamide, which is formed by the reaction of benzoic acid with ammonia:

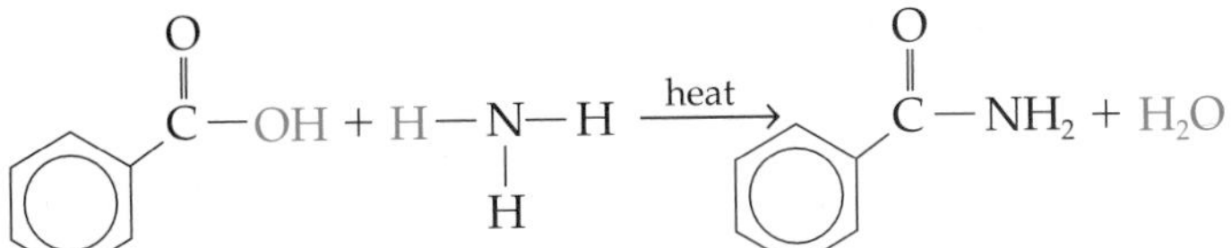

benzoic acid · · · · ammonia · · · · · · benzamide

Aromatic amides have the general formula $ArCONH_2$. The amide of niacin, niacinamide, is one of the B vitamins (see page 563).

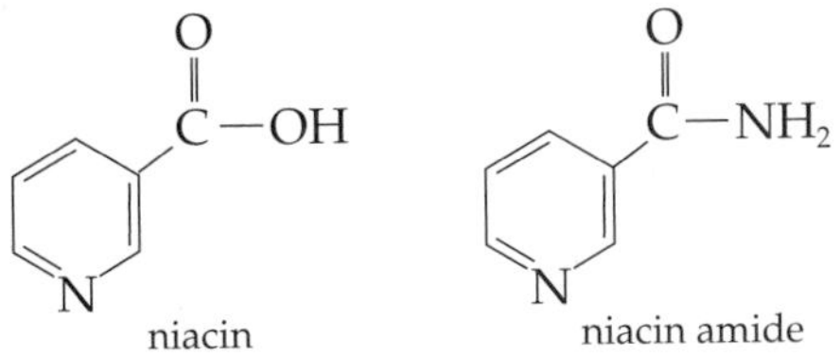

niacin · · · · · · · · · niacin amide

Summary

When an organic acid reacts with an alcohol, a compound called an ester is produced. Esters have the general formula RCOOR'. Esters are named according to the alcohol and acid from which they were made.

Amines are organic compounds in which an alkyl group has replaced one or more of the hydrogen atoms of ammonia, NH_3. Amines are basic compounds and readily form salts.

Amino acids are organic acids that contain an amine group. α-Amino acids are the building blocks of protein.

Amides are formed by the reaction of an organic acid with ammonia or with an amine. Amino acids are held together by peptide (amide) bonds.

Alkaloids are heterocyclic amines found chiefly in plants. They often have a marked physiologic effect on the human body by reacting with receptors for neurotransmitters.

Questions and Problems

1. What is the general formula for an amine? an ester? an amide?
2. The reaction of an organic acid and an alcohol yields what type of product?
3. Compare the structures of primary, secondary, and tertiary amines.
4. Compare the use of the terms primary, secondary, and tertiary when used for alcohols and for amines.
5. What is an amino acid? Give two examples. Where are amino acids used in the body?
6. Why are amino acids amphoteric?
7. What is an amide? a peptide bond?
8. Write the equation for the reaction that occurs when acetic acid and ammonia are heated.
9. What is the name of the product of the reaction described in question 8?
10. What are quaternary ammonium salts used for? Give an example.

11. Write an equation for the preparation of aniline.
12. Name two alkaloids and indicate their physiologic effects on humans.
13. Name three heterocyclic compounds other than alkaloids.
14. What heterocyclic compounds are found in the brain? What functions do they have?
15. Give the IUPAC name for:
    a. $CH_3CH_2NH_2$
    b. CH_3NHCH_3
    c. $(CH_3CH_2)_3N$
    d.

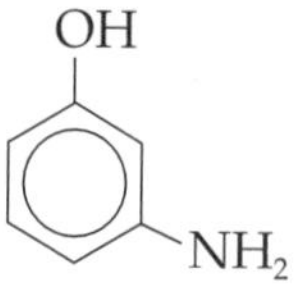

16. Draw structures for:
    a. 3-amino-2-hexene
    b. 3-chloroaniline
    c. N,N-diethyl aniline

17. List all of the functional groups found in the following compounds.

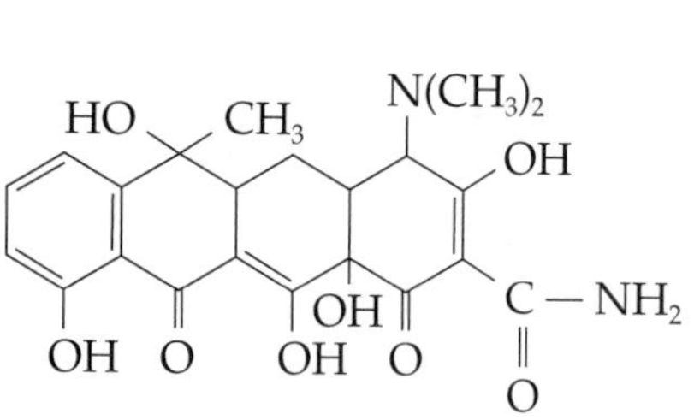

a. tetracycline

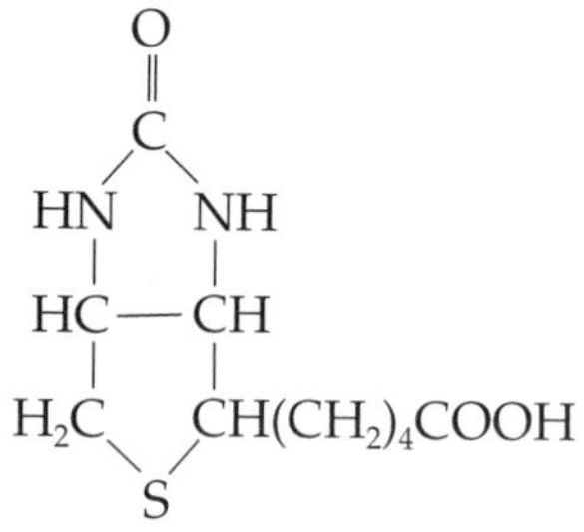

b. biotin

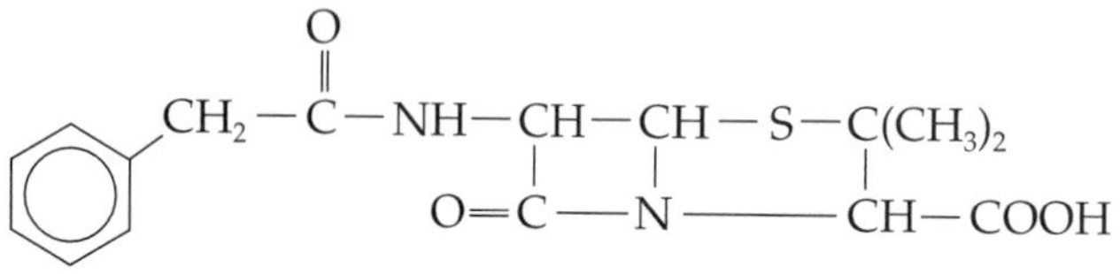

c. penicillin G

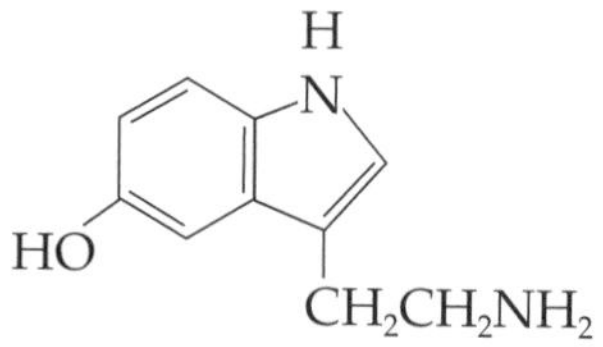

d. serotonin

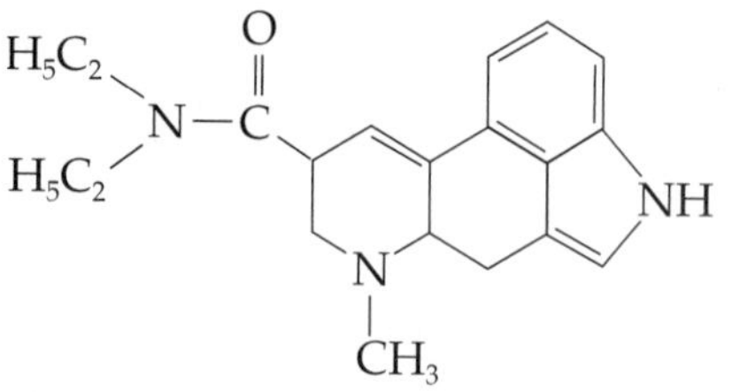

e. LSD

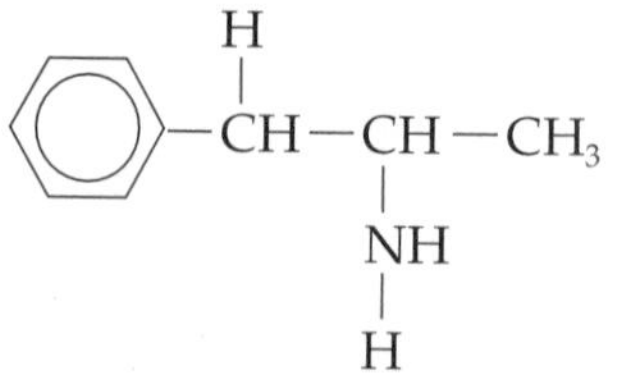

f. amphetamine (Benzedrine)

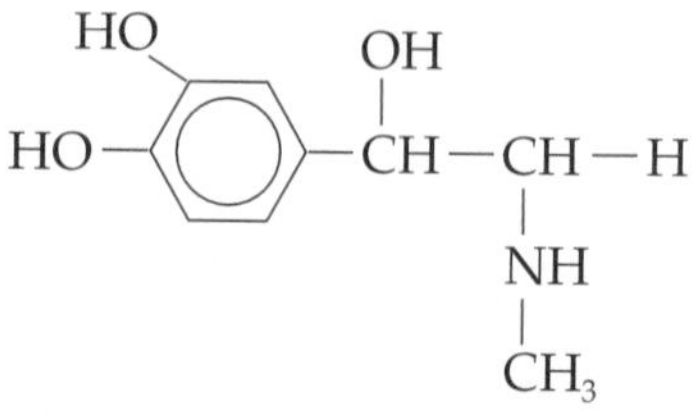

g. epinephrine (adrenaline)

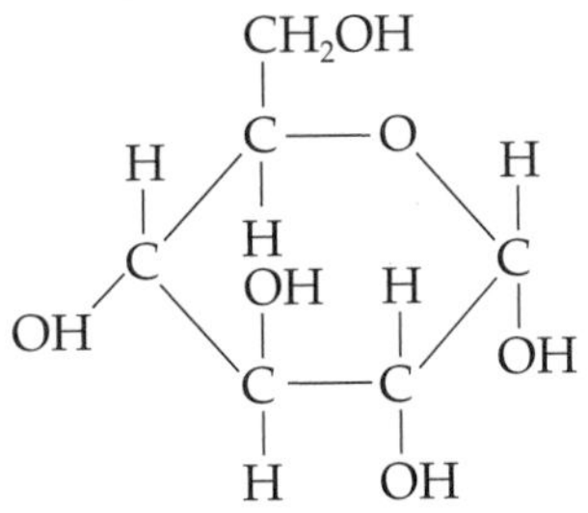

h. glucose

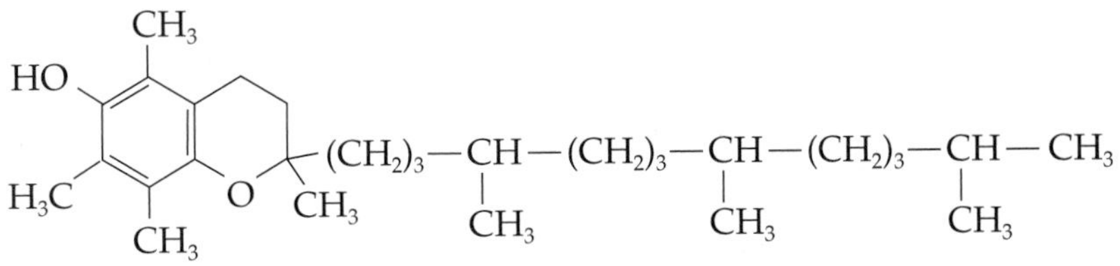

i. α-tocopherol (vitamin E)

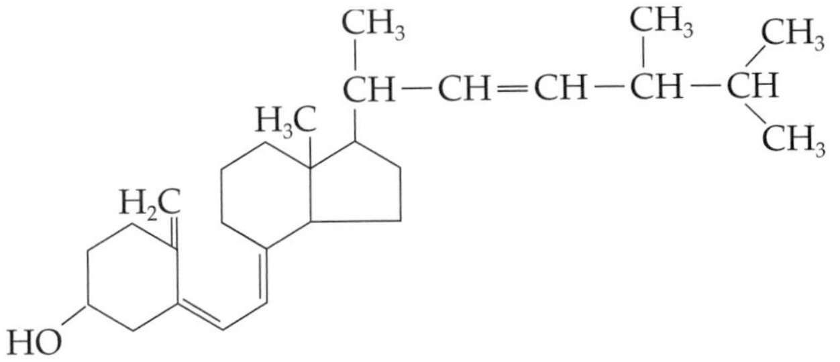

j. ergocalciferol (vitamin D$_2$)

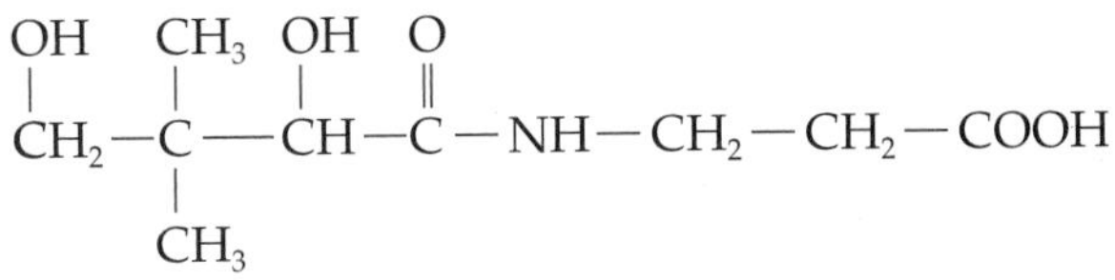

k. pantothenic acid

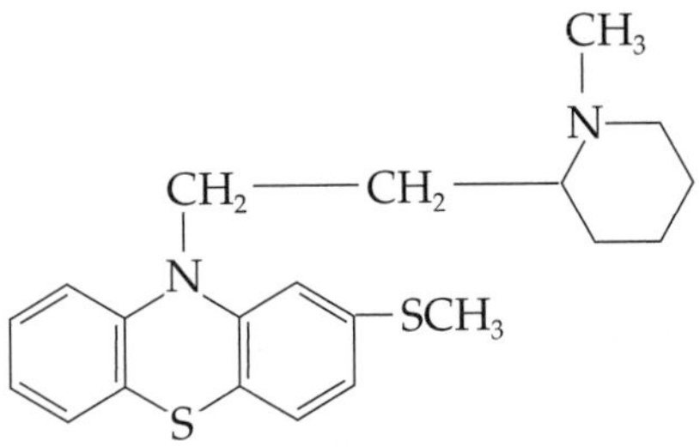

l. thioridazine (antipsychotic)

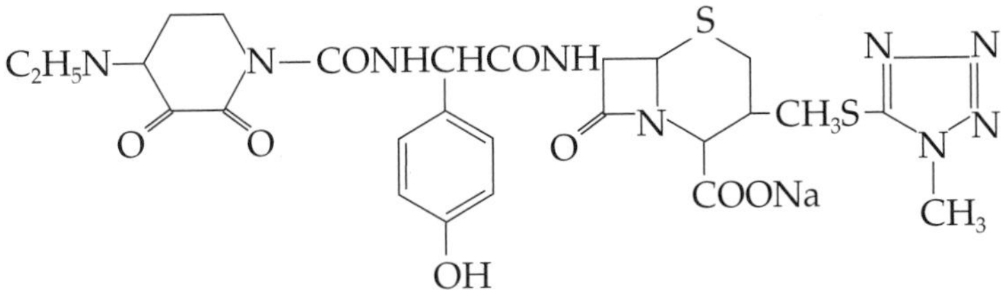

m. ceforanide (cephalosporin)

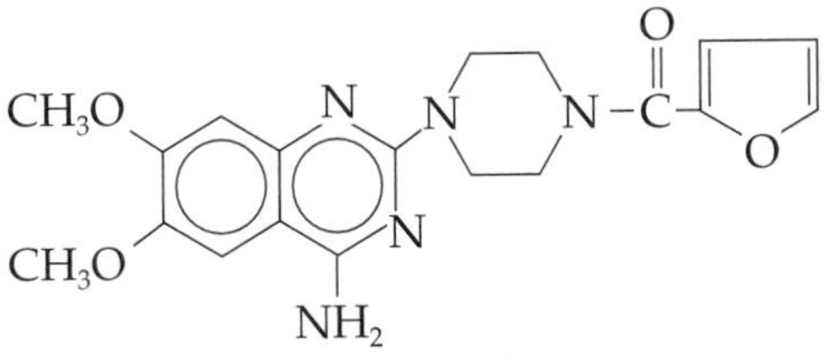

n. prazosin HCl (antihypertensive)

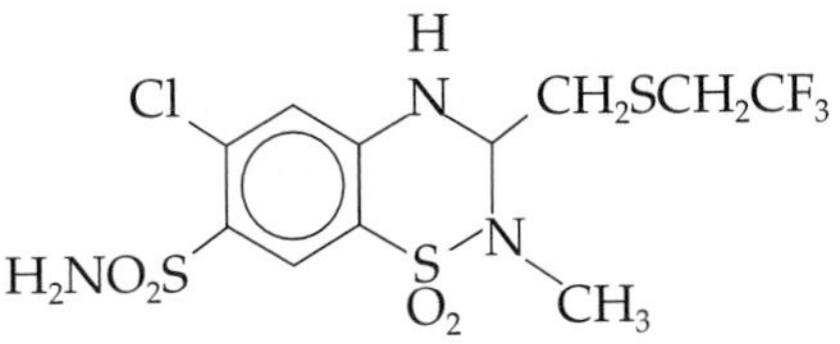

o. polythiazide (diuretic)

17. List all of the functional groups found in the following compounds—cont'd.

p. verapamil HCl

q. cromolyn sodium (allergic rhinitis inhalant)

18. Why are many drugs that contain the amine group given as the amine hydrochloride?

19. Give the IUPAC and the common name for:

$$CH_3-\overset{\overset{\displaystyle O}{\|}}{C}-O-CH_2-CH_3$$

20. Give the IUPAC name for:

$$CH_3CH_2\overset{\overset{\displaystyle }{\underset{\underset{\displaystyle O}{\|}}{C}}}{}-\overset{\overset{\displaystyle }{\underset{\underset{\displaystyle CH_3}{|}}{CH}}}{}-CH_3$$

21. Give one medical use for:
    a. nitroglycerin
    b. methyl salicylate
    c. benzocaine

22. Name a thioester; a phosphate ester. What are their biologic functions?

Practice Test

1. An ester is formed by the reaction between:
   a. a carboxylic acid and an alcohol
   b. a carboxylic acid and an amine
   c. an alkaloid and an amide
   d. an acetal and water

2. An example of an amphoteric compound is a(n):
   a. ester   b. primary amine
   c. amino acid   d. substituted amine

3. The name of the following compound is:

$$CH_3-\overset{\overset{\displaystyle }{\underset{\underset{\displaystyle CH_3}{|}}{N}}}{}-CH_3$$

   a. methylamine   b. trimethylamine
   c. triammonia   d. ethylmethylamine

4. Amides are found in what type of bond?
   a. peptide   b. hydrogen
   c. ionic   d. amino acid

5. Amino acids are components of:
   a. carbohydrates   b. fats
   c. proteins   d. alkaloids

6. Which of the following is not an alkaloid?
   a. nicotine   b. aniline
   c. ephedrine   d. caffeine

7. The name of $CH_3-\overset{\overset{\displaystyle }{\underset{\underset{\displaystyle }{|}}{N}}}{}-CH_3$ is:

   a. dimethylnitrobenzene
   b. N,N-dimethylnitrobenzene
   c. dimethylaniline
   d. N,N-dimethylaniline

8. The IUPAC name for $CH_3-\overset{\overset{\displaystyle O}{\|}}{C}-O-CH_3$ is:
   a. ethyl acetate   b. methyl acetate
   c. methyl methanoate d. methyl ethanoate

9. When a carboxylic acid reacts with ammonia, one of the products is an:
   a. amine   b. amide
   c. amino acid d. alkaloid

10. An example of a basic compound is a(n):
    a. ester   b. amine
    c. amide   d. hemiacetal

20

Carbohydrates

Objectives

- To understand the classification of carbohydrates

- To become familiar with stereoisomerism

- To draw the structures of and to indicate the reactions of several hexoses

- To compare the structures and reactions of disaccharides and polysaccharides

The breads and cereals we eat are composed almost entirely of carbohydrates.

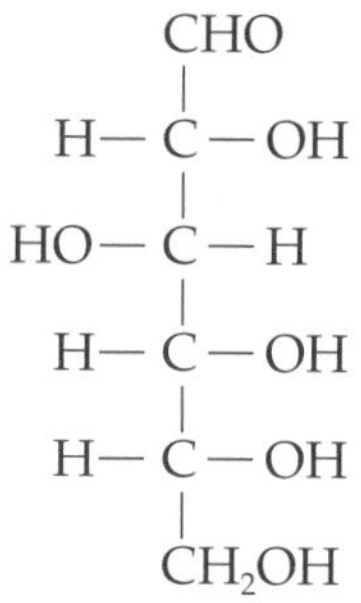

```
        CHO
         |
   H — C — OH
         |
  HO — C — H
         |
   H — C — OH
         |
   H — C — OH
         |
       CH₂OH
```

glucose, a
polyhydroxyaldehyde

Carbohydrates (which contain the elements carbon, hydrogen, and oxygen) form a class of organic compounds that includes sugars, starches, and cellulose. Originally all known carbohydrates were considered to be hydrates of carbon because they contain hydrogen and oxygen in the ratio of 2 to 1 just as in water. The formula for glucose, $C_6H_{12}O_6$, was written as $C_6(H_2O)_6$. Likewise, sucrose, $C_{12}H_{22}O_{11}$, was written as $C_{12}(H_2O)_{11}$. However, later investigation showed that rhamnose, another carbohydrate, has the formula $C_6H_{12}O_5$. It did not fit the general formula for a hydrate of carbon yet it was a carbohydrate. Also, such compounds as acetic acid ($C_2H_4O_2$) and pyrogallol ($C_6H_6O_3$) did fit such a system but were not carbohydrates.

Carbohydrates are now defined as polyhydroxyaldehydes or polyhodroxy ketones or substances that yield these compounds on hydrolysis. **Polyhydroxy** means "containing several alcohol groups." Thus, simple carbohydrates are alcohols and are also either aldehydes or ketones (they contain a carbonyl group, $\gtrdot C = O$; see Chapter 18).

20-1 Classification

Carbohydrates are divided into three major categories: monosaccharides, disaccharides, and polysaccharides.

Monosaccharides (*mono-* means one) are simple sugars. They cannot be changed into simpler sugars upon hydrolysis (reaction with water).

Disaccharides (*di-* means two) are double sugars. On hydrolysis, they yield two simple sugars.

$$\text{disaccharides} \xrightarrow{\text{hydrolysis}} \text{2 monosaccharides}$$

Polysaccharides (*poly-* means many) are complex sugars. On hydrolysis, they yield many simple sugars.

$$\text{polysaccharides} \xrightarrow{\text{hydrolysis}} \text{many simple sugars}$$

Monosaccharides, or simple sugars, are called either **aldoses** or **ketoses**, depending upon whether they contain an aldehyde (—CHO) or ketone ($\gtrdot C = O$) group. Aldoses and ketoses are further classified according to the number of carbon atoms they contain. An aldopentose is a five-carbon simple sugar containing an aldehyde group. A ketohexose is a six-carbon simple sugar containing a ketone group. Ribose, as shown below, is an aldose and fructose is a ketose.

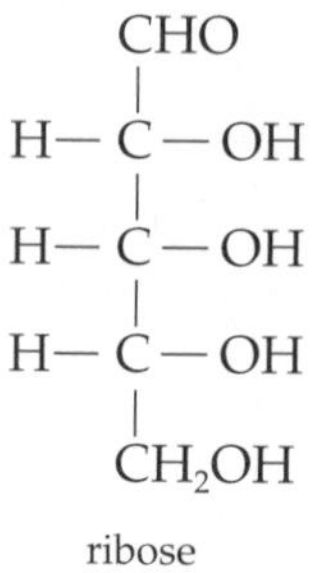

```
        CHO                      CH₂OH
         |                         |
   H — C — OH                    C = O
         |                         |
   H — C — OH               HO — C — H
         |                         |
   H — C — OH                H — C — OH
         |                         |
       CH₂OH                 H — C — OH
                                   |
                                 CH₂OH
```

ribose
an aldopentose

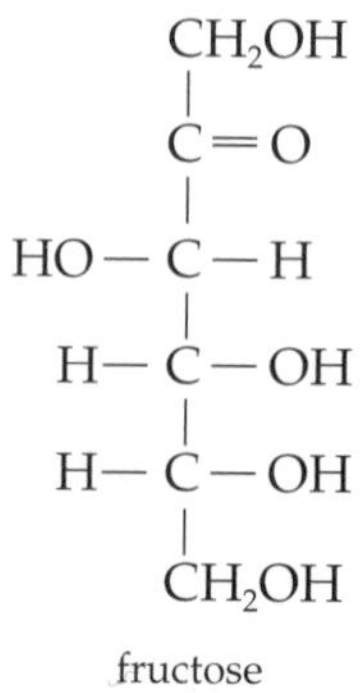

fructose
a ketohexose

Exercise 20-1

Classify the following monosaccharides by number of carbons and functional group.

```
                CHO
                 |
           H — C — OH
                 |
a.    H — C — OH
                 |
           H — C — OH
                 |
              CH₂OH
```

```
              CH₂OH
                 |
               C = O
                 |
        HO — C — H
                 |
b.    H — C — OH
                 |
        H — C — OH
                 |
              CH₂OH
```

Solution

a. There are five carbons, so it is a pentose; the function group is an aldehyde, so it is an aldose: aldopentose.
b. There are six carbons so it is a hexose; the functional group is a ketone, so it is a ketose: ketohexose.

Self-test

Classify the following monosaccharide.

```
             CHO
              |
      H — C — OH
              |
           CH₂OH
```

Answer

Aldotriose

Although there are simple sugars with three carbon atoms (trioses), four carbon atoms (tetroses), and five carbon atoms (pentoses), the hexoses (six-carbon simple sugars) are the most common in terms of the human body because they are the body's main energy-producing compounds.

```
      CHO
       |
H — C — OH
       |
    CH₂OH
```
glyceraldehyde
(an aldotriose)

```
    CH₂OH
      |
    C = O
      |
H — C — OH
      |
    CH₂OH
```
erythrulose
(a ketotetrose)

```
       CH₂OH
         |
       C = O
         |
HO — C — H
         |
HO — C — H
         |
       CH₂OH
```
xylulose
(a ketopentose)

```
        CHO
         |
  H — C — H
         |
HO — C — H
         |
HO — C — H
         |
  H — C — OH
         |
       CH₂OH
```
galactose
(an aldohexose)

Polysaccharides are sometimes called hexosans or pentosans, depending on the type of monosaccharide they yield on hydrolysis. That is, hexosans on hydrolysis yield hexoses, and pentosans yield pentoses.

20-2 Origin

Plants pick up carbon dioxide from the air and water from the soil and combine them to form carbohydrates in a process called photosynthesis. Enzymes, chlorophyll, and sunlight are necessary. The overall reaction is represented in equation (20-1).

$$6\ CO_2 + 6\ H_2O \xrightarrow[\substack{\text{chlorophyll,}\\\text{enzymes}}]{\text{sunlight}} \underset{\substack{\text{glucose and}\\\text{other carbohydrates}}}{C_6H_{12}O_6} + 6\ O_{2(g)} \tag{20-1}$$

However, it should be understood that even though reaction (20-1) appears simple, it is very complex, with many intermediate steps between the original reactants and the final products. During photosynthesis, oxygen is given off into the air, thus renewing our vital supply of this element.

The carbohydrate produced in reaction (20-1), $C_6H_{12}O_6$, is a monosaccharide. Plant cells also have the ability to combine two molecules of a monosaccharide into one of a disaccharide, as indicated in (20-2).

$$2\ \underset{\text{monosaccharide}}{C_6H_{12}O_6} \longrightarrow \underset{\text{disaccharide}}{C_{12}H_{22}O_{11}} + H_2O \tag{20-2}$$

Reaction (20-2) is the reverse of hydrolysis—water is removed when two molecules of a monosaccharide combine.

Plant (and animal) cells can also combine many molecules of monosaccharide into large polysaccharide molecules. The n in equation (20-3) represents a number larger than 2.

$$n\ \underset{\text{monosaccharide}}{C_6H_{12}O_6} \longrightarrow \underset{\text{polysaccharide}}{(C_6H_{10}O_5)_n} + n\ H_2O \tag{20-3}$$

This is an example of a condensation polymerization reaction (see Section 16-1).

Polysaccharides occur in plants as cellulose in the stalks and stems and as starches in the roots and seeds. Monosaccharides and disaccharides are generally found in plants in their fruits. Plants as well as animals are able to convert carbohydrates into fats and proteins.

Oxygen–Carbon Dioxide Cycle in Nature

Although plants have the ability to pick up carbon dioxide from the air and water from the ground to form carbohydrates, animals are unable to do this and must rely on plants for their carbohydrates.

Animals oxidize carbohydrates in their bodies to yield carbon dioxide, water, and energy:

$$C_6H_{12}O_6 + 6\ O_2 \longrightarrow 6\ CO_2 + 6\ H_2O + \text{energy}$$

Again, this reaction is not as simple as it appears; many steps are involved between the reactants and the products, and many different enzymes are required.

It should be noted that this overall reaction during metabolism is the reverse of the one taking place during photosynthesis. Both reactions can be summarized by equation (20-4).

$$\text{energy} + 6\ CO_2 + 6\ H_2O \underset{\substack{\text{animal}\\\text{metabolism}}}{\overset{\substack{\text{plant}\\\text{photosynthesis}}}{\rightleftharpoons}} C_6H_{12}O_6 + 6\ O_2 \tag{20-4}$$

Thus, there is a cycle in nature. During photosynthesis, plants pick up carbon dioxide from the air and give off oxygen; during metabolism both plants and animals pick up oxygen from the air and give off carbon dioxide.

During photosynthesis, the energy from the sun is needed for the aerobic reaction (that is, the reaction is *endothermic). During catabolism of these carbohydrates in animals, this same amount of energy is liberated (the reaction is *exothermic). Thus, energy from the metabolism of carbohydrates by animals comes primarily from the sun. Plants store solar energy in carbohydrates, and this energy is utilized by living organisms during the catabolic process. It has been estimated that only about 1 percent of the total solar energy falling on plants is converted into useful stored energy.

20-3 Stereoisomerism

Stereoisomers are compounds with the same molecular formula but different structures that are mirror images of one another.

The mirror image of an object is the reflection of that object in a plane mirror. If you hold your left hand in front of a plane mirror, you will see its mirror image, one that is identical to your right hand (Figure 20-1). Your left hand is not the same as your right hand, nor is your left foot the same as your right foot. A glove for your left hand will not fit on your right hand nor will a "left handed" substrate fit a "right-handed" enzyme.

If you place your left hand on top of your right hand, you can see that they are *not* superimposable. Neither are your left foot and right foot superimposable. These objects and their mirror images are not superimposable.

Now consider an Erlenmeyer flask and its mirror image. They are identical and are superimposable.

Similarly, some molecules are superimposable with their mirror images and others are not.

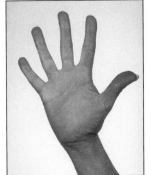

a

glyceraldehyde

mirror image of glyceraldehyde

Glyceraldehyde and its mirror image are *not* superimposable; they are said to be *chiral*. An object that is superimposable on its mirror image is said to be *achiral*. The methane molecule is identical with its mirror image; it is achiral.

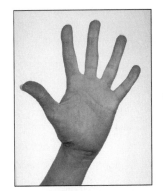

methane molecule

mirror image

b

FIGURE 20-1

(a) Left hand. (b) Mirror image left hand.

The glyceraldehyde molecule shown above is chiral. It is not superimposable on its mirror image.

In general, any object or molecule with a plane of symmetry is achiral. An Erlenmeyer flask has a plane of symmetry; it is achiral. Likewise, the methane molecule has a plane of symmetry; it is achiral.

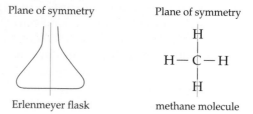

Conversely, if an object or molecule does not have a plane of symmetry, it is chiral.

Another method for determining whether a molecule is chiral is to see if there are four different groups attached to a central carbon atom.

The simplest carbohydrate is glyceraldehyde (Figure 20-2). Glyceraldehyde has an aldehyde group at one end of the molecule, a primary alcohol at the other end, and a secondary alcohol in the middle. The central carbon atom in glyceraldehyde is chiral—it has four different groups attached to it. They are

$$-\text{CHO} \qquad -\text{OH} \qquad -\text{CH}_2\text{OH} \qquad -\text{H}$$

Stereoisomers, which are chiral, are called enantiomers and are said to be optically active. They rotate in the plane of polarized light equally but in opposite directions. The enantiomers of glyceraldehyde (see Figure 20-2) are not superimposable. They are chiral. This compound can exist in two optically active forms.

Polarized light vibrates in one plane only, in contrast to ordinary light, which vibrates in all planes. When polarized light is passed through a solution of an optically active substance, the plane of polarized light is rotated (Figure 20-3). What causes such a rotation of the plane of polarized light? According to the van't Hoff theory, such an effect on the plane of polarized light is due to the presence of one or more chiral carbon atoms.

Enantiomers have a three-dimensional structure and can be represented as follows: bonds that extend toward the front are shown as solid wedges and those projecting toward the back as dotted wedges. In such a representation, the structure of glyceraldehyde is shown as

$$
\begin{array}{cc}
\text{CHO} & \text{CHO} \\
\text{H}\blacktriangleright\text{C}\blacktriangleleft\text{OH} & \text{HO}\blacktriangleright\text{C}\blacktriangleleft\text{H} \\
\text{CH}_2\text{OH} & \text{CH}_2\text{OH}
\end{array}
$$

structures of glyceraldehyde

A two-dimensional method of indicating the structure of an enantiomer is called a Fischer projection. In a Fischer projection, the horizontal lines indicate bonds extending forward from the paper, and the vertical lines indicate bonds extending backward from the paper. The Fischer projection formulas are always written with the aldehyde (or ketone) group—the most highly oxidized—at the top.

$$
\begin{array}{cc}
\text{CHO} & \text{CHO} \\
\text{H}-\text{C}-\text{OH} & \text{HO}-\text{C}-\text{H} \\
\text{CH}_2\text{OH} & \text{CH}_2\text{OH}
\end{array}
$$

Fischer structures of glyceraldehyde

CHO
|
H—C—OH
|
CH₂OH

glyceraldehyde

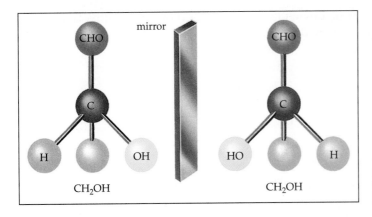

FIGURE 20-2
Ball-and-stick models of the two forms of glyceraldehyde.

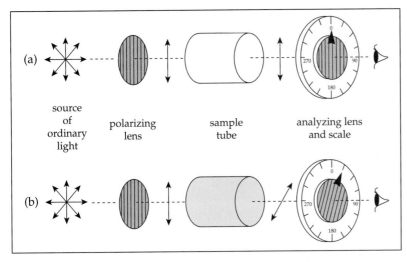

FIGURE 20-3
(a) Schematic diagram of polarimeter showing production of polarized light. (b) Rotation of polarized light by an optically active substance.

In these Fischer formulas, the —H and —OH groups project forward from the paper and the —CHO and —CH$_2$OH groups project backward from the paper.

D and L Enantiomers

The parent compound for enantiomers is glyceraldehyde, which can be written in two ways, according to the Fischer projection, as shown. Both structures have a chiral carbon atom to which are attached four different groups.

Fischer called the enantiomer with the OH group on the left side of the chiral carbon the L (for *levo*, meaning left) compound and the one with the OH group on the right side of the chiral carbon atom the D (for *dextro*, meaning right) compound.

$$
\begin{array}{c}
\text{CHO} \\
\mid \\
\text{H}-\text{C}-\text{OH} \\
\mid \\
\text{CH}_2\text{OH}
\end{array}
\qquad \text{and} \qquad
\begin{array}{c}
\text{CHO} \\
\mid \\
\text{HO}-\text{C}-\text{H} \\
\mid \\
\text{CH}_2\text{OH}
\end{array}
$$

D-glyceraldehyde   L-glyceraldehyde

Most carbohydrates have longer carbon chains than glyceraldehyde does and contain more than one chiral carbon atom. In these cases, the carbonyl group is again written at the top of the structure and the CH$_2$OH group at the bottom. *The position of the OH group on the chiral carbon atom farthest from the carbonyl group determines whether the compound will be of the L or D type.*

In the following structure of glucose, the OH group on the fifth carbon atom (the aldehyde group is assigned position number 1) is indicated on the right side, so that this structure represents D-glucose.

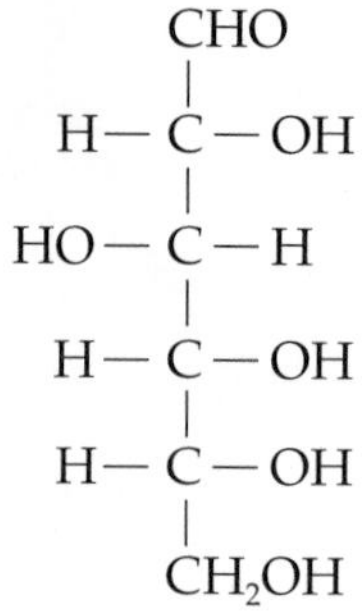

$$
\begin{array}{c}
\text{CHO} \\
|\\
\text{H}-\text{C}-\text{OH} \\
|\\
\text{HO}-\text{C}-\text{H} \\
|\\
\text{H}-\text{C}-\text{OH} \\
|\\
\text{H}-\text{C}-\text{OH} \\
|\\
\text{CH}_2\text{OH}
\end{array}
$$

D-glucose

In the L series, the OH group on carbon number 5 of glucose would be on the left side and would be called L-glucose.

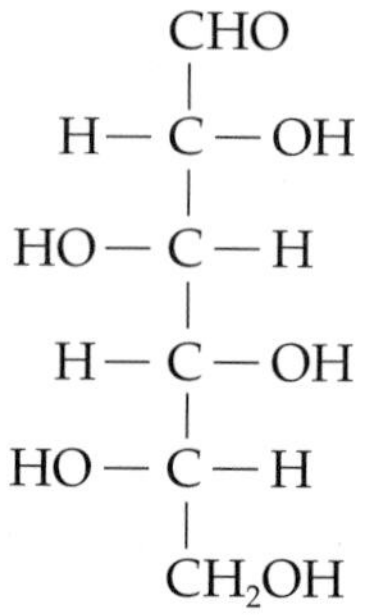

$$
\begin{array}{c}
\text{CHO} \\
|\\
\text{H}-\text{C}-\text{OH} \\
|\\
\text{HO}-\text{C}-\text{H} \\
|\\
\text{H}-\text{C}-\text{OH} \\
|\\
\text{HO}-\text{C}-\text{H} \\
|\\
\text{CH}_2\text{OH}
\end{array}
$$

L-glucose

Likewise, the following structure represents xylulose, a ketopentose. Note that again the carbonyl group is written at the top of the structure. The chiral carbon atom farthest from the carbonyl group is carbon number 4.

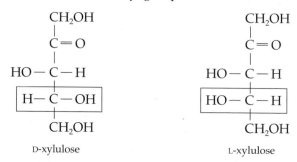

D-xylulose   L-xylulose

Figure 20-4 illustrates the D family of aldoses. The colored areas indicate the D conformation. A similar L family also exists.

The simplest ketose is dihydroxyacetone. This compound is produced during the metabolism of glucose. Dihydroxyacetone does not contain a chiral carbon atom.

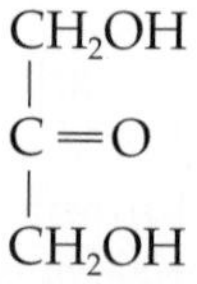

$$
\begin{array}{c}
\text{CH}_2\text{OH} \\
|\\
\text{C}=\text{O} \\
|\\
\text{CH}_2\text{OH}
\end{array}
$$

dihydroxyacetone

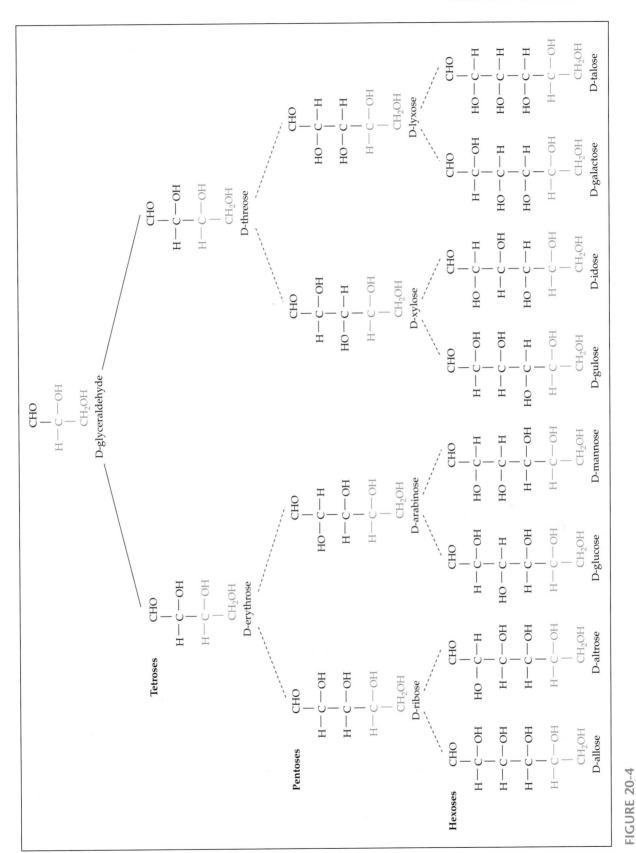

FIGURE 20-4

The D family of aldoses.

Some D-isomers of ketoses are shown below:

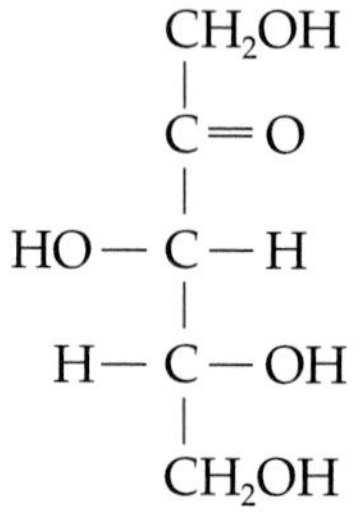

$$
\begin{array}{ccc}
& & \text{CH}_2\text{OH} \\
& \text{CH}_2\text{OH} & | \\
\text{CH}_2\text{OH} & | & \text{C}=\text{O} \\
| & \text{C}=\text{O} & | \\
\text{C}=\text{O} & | & \text{HO}-\text{C}-\text{H} \\
| & \text{HO}-\text{C}-\text{H} & | \\
\text{HO}-\text{C}-\text{H} & | & \text{H}-\text{C}-\text{OH} \\
| & \text{H}-\text{C}-\text{OH} & | \\
\text{H}-\text{C}-\text{OH} & | & \text{H}-\text{C}-\text{OH} \\
| & \text{H}-\text{C}-\text{OH} & | \\
\text{CH}_2\text{OH} & | & \text{H}-\text{C}-\text{OH} \\
& \text{CH}_2\text{OH} & | \\
& & \text{CH}_2\text{OH}
\end{array}
$$

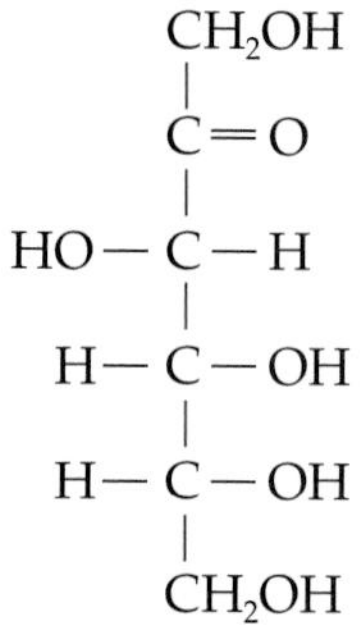

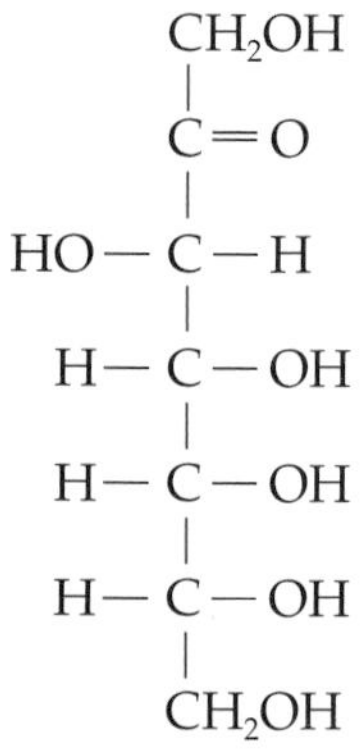

D-xylulose	D-fructose	D-sedoheptulose
(5 carbons)	(6 carbons)	(7 carbons)

L-isomers of ketoses also exist.

The number of optical isomers depends on the number of chiral carbon atoms present in a compound and can be calculated by using the formula 2^n, where n is the number of chiral carbons. Thus, glyceraldehyde, which has one chiral carbon, has 2^1, or 2, optical isomers, as was shown. Glucose, which will be discussed later in this chapter, has four chiral carbons and so has 2^4, or 16, optical isomers. Of these 16 isomers, eight belong to the D series and 8 to the L series (one set of eight is the mirror image of the other set).

Stereoisomerism is of great importance in the body because many enzymes will interact with only one particular enantiomer. In the human body, the D series is the primary configuration for carbohydrates, whereas the L series is the primary one for proteins.

Stereoisomers that are mirror images are called *enantiomers*. They are not superimposable.

Stereoisomers that are not mirror images are called *diastereomers*. They are not superimposable.

Consider the following structures:

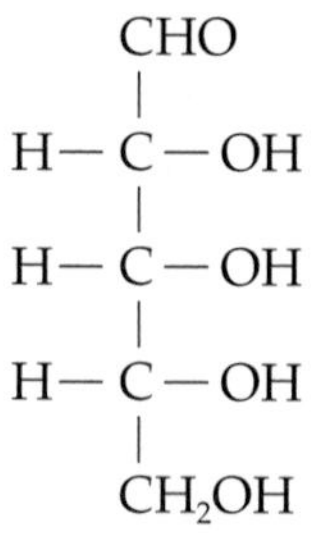

$$
\begin{array}{cccc}
\text{CHO} & \text{CHO} & \text{CHO} & \text{CHO} \\
| & | & | & | \\
\text{H}-\text{C}-\text{OH} & \text{H}-\text{C}-\text{OH} & \text{HO}-\text{C}-\text{H} & \text{HO}-\text{C}-\text{H} \\
| & | & | & | \\
\text{H}-\text{C}-\text{OH} & \text{H}-\text{C}-\text{OH} & \text{HO}-\text{C}-\text{H} & \text{HO}-\text{C}-\text{H} \\
| & | & | & | \\
\text{H}-\text{C}-\text{OH} & \text{HO}-\text{C}-\text{H} & \text{H}-\text{C}-\text{OH} & \text{HO}-\text{C}-\text{H} \\
| & | & | & | \\
\text{CH}_2\text{OH} & \text{CH}_2\text{OH} & \text{CH}_2\text{OH} & \text{CH}_2\text{OH} \\
\text{D-ribose} & \text{L-ribose} & \text{D-lyxose} & \text{L-lyxose}
\end{array}
$$

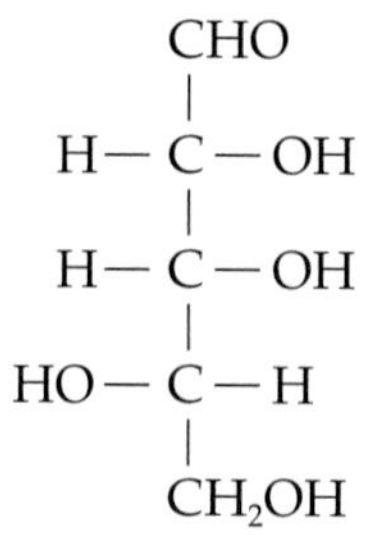

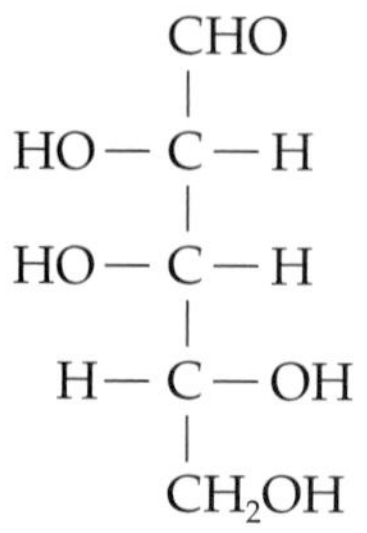

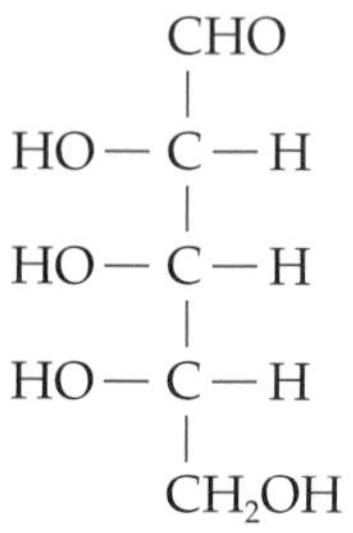

D-ribose and L-lyxose are enantiomers. They are mirror images of each other. Also, L-ribose and D-lyxose are enantiomers. D-ribose and L-ribose are diastereomers. They are not mirror images of each other. Also diastereomers:

D-ribose and D-lyxose
L-ribose and L-lyxose
D-lyxose and L-lyxose

Table 20-1	Single Enantiomers and Their Uses
Enantiomer	**Use**
Dexfenfluramine	Obesity
Ibuprofen	Pain
Indinavir	AIDS
Levofloxacin	Antibiotic
Levomoprolol	Hypertension
Lisinopril	Hypertension
Paclixatel	Ovarian cancer
Paroxetine	Psychiatric depression

Chiral Drugs

The number of single enantiomers of chiral drugs has expanded considerably in the past few years. Table 20-1 indicates some of these single enantiomers and their uses.

20-4 Monosaccharides

Monosaccharides are simple sugars; they cannot be broken down into other sugars. They are categorized according to the number of carbons they contain.

Trioses

A triose is a three-carbon simple sugar. Trioses are formed during the metabolic breakdown of hexoses in muscle metabolism. An example of a triose is glyceraldehyde (glycerose), whose optical isomers are shown on page 335.

Tetroses

Tetroses are four-carbon sugars. One tetrose, erythrose, is an intermediate in the hexose monophosphate shunt for the oxidation of glucose (see Section 26-7).

Pentoses

Pentoses are five-carbon sugar molecules. The most important of these are ribose and deoxyribose, which are found in nucleic acids. Ribose forms part of ribonucleic acid (RNA), and deoxyribose forms part of deoxyribonucleic acid (DNA). Both DNA and RNA are components of every cell nucleus and cytoplasm. The prefix de- means without, so *deoxy-* means without oxygen. Note that deoxyribose has one less oxygen atom than does ribose.

Other pentoses of physiologic importance include D-ribulose, which is an intermediary in the pentose phosphate shunt; D-lyxose, which is found in heart muscle; and D-xylose and D-arabinose, which are components of glycoproteins.

Structures (20-5), the Fischer projection representations for the pentoses, are called open-chain structures. However, the predominant form for pentoses is a ring structure. *Recall that aldehydes react with alcohols to form hemiacetals* (Section 18-3). In the case of ribose, the aldehyde can react with the alcohol at carbon number 4 to form two different compounds.

$$
\begin{array}{c}
CHO \\
| \\
H-C-OH \\
| \\
H-C-OH \\
| \\
H-C-OH \\
| \\
CH_2OH
\end{array}
$$

D-ribose

(20-5)

$$
\begin{array}{c}
CHO \\
| \\
H-C-H \\
| \\
H-C-OH \\
| \\
H-C-OH \\
| \\
CH_2OH
\end{array}
$$

D-deoxyribose

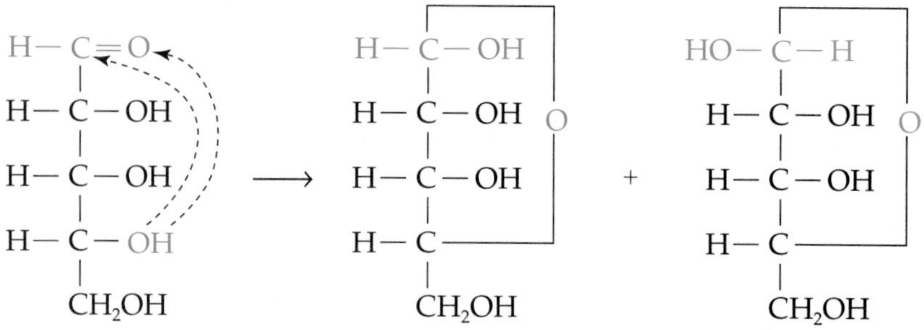

ribose   α-ribose   β-ribose

Another representation for the hemiacetal (ring) structure of ribose, and other monosaccharides, is called the Haworth projection (shown below).

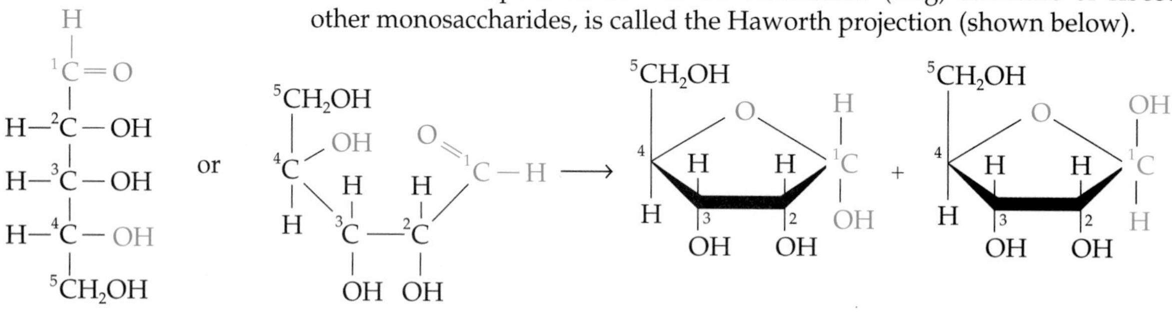

D-ribose   α-D-ribose   β-D-ribose   (20-6)

In this projection, the ring, consisting of carbons at each corner and an oxygen where indicated, is considered to be in a plane perpendicular to the paper. The heavy line in the ring indicates the section closest to you; the lighter line, that part of the ring farther from you. The —H and —OH groups are above and below that plane. The Fischer and Haworth projections are related as follows.

1. The groups on the right side of the Fischer projection are written below the plane in the Haworth projection. Those on the left side are written above the plane.

2. One exception to rule 1 occurs at carbon 4 in pentoses and at carbon 5 in hexoses because of the nature of the reaction occurring there. At these carbons, rule 1 is reversed.

3. At carbon 1 the α form is indicated by the —OH being written below the plane; the β form has the —OH above the plane.

4. In both projections the CH$_2$OH group, which has no chiral carbon, is written as a unit.

A simplified Haworth projection shows only the positions of the OH groups. For ribose, the simplified structures are

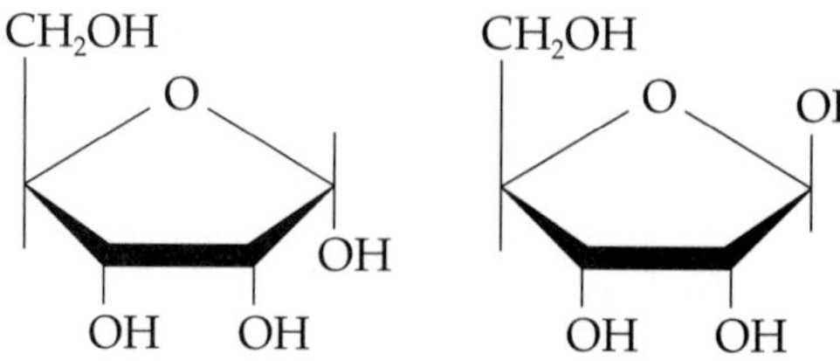

α-D-ribose   β-D-ribose

Hexoses

The hexoses, the six-carbon sugars, are the most common of all the carbohydrates. Of the several hexoses, the most important as far as the human body is concerned are *glucose, galactose,* and *fructose*. All three of these hexoses have the same molecular formula, $C_6H_{12}O_6$, but have different structural formulas; they are isomers.

Glucose   Glucose ($C_6H_{12}O_6$) is an aldohexose (Section 20-1) and can be represented structurally as

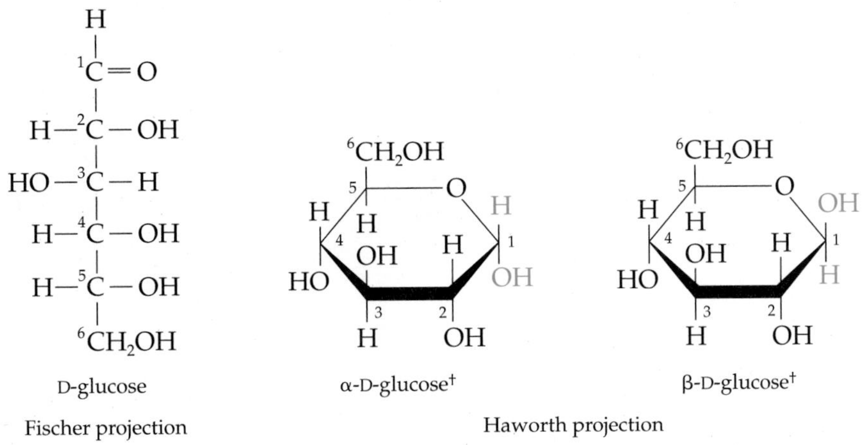

| | D-glucose | α-D-glucose[†] | β-D-glucose[†] |

Fischer projection   Haworth projection

Note that glucose contains four chiral centers (numbers 2, 3, 4, 5) and so has 2^4, or 16, optical isomers.

Medically, when the word glucose is mentioned, the D-isomer is meant because that is the biologically active isomer. Likewise, for the other hexoses, the D-isomer is commonly called by name only, without the prefix D.

Glucose is known commonly as dextrose, or grape sugar. It is a white crystalline solid that is soluble in water and insoluble in most organic liquids. It is found, along with fructose, in many fruit juices. It can be prepared by the hydrolysis of sucrose, a disaccharide, or by the hydrolysis of starch, a polysaccharide.

Glucose is a most important monosaccharide. It is normally found in the bloodstream and in the tissue fluids. As will be discussed in Chapter 26, "Metabolism of Carbohydrates," glucose requires no digestion and can be given intravenously to patients who are unable to take food by mouth (see Figure 20-5).

Glucose is found in the urine of patients suffering from diabetes mellitus and is an indication of this disease. The presence of glucose in the urine is called glycosuria. Glucose may also show up in the urine during extreme excitement (emotional glycosuria), after ingestion of large amounts of sugar (alimentary glucosuria), or because of other factors that will be discussed in the chapter on the metabolism of carbohydrates.

[†]The chemical names for these ring compounds are α-D-glucopyranose and β-D-glucopyranose, respectively, but this system of naming is of interest primarily to organic chemists.

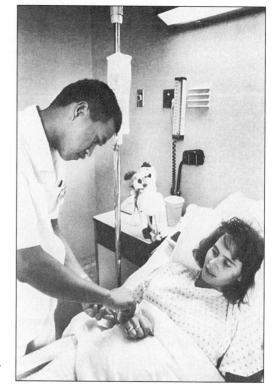

FIGURE 20-5
Intravenous infusion of glucose.

Galactose   Galactose, an isomer of glucose, is also an aldohexose. The structures of galactose are

D-galactose   α-D-galactose   β-D-galcatose

Glucose and galactose differ from each other only in the configuration of the H and OH about a single carbon atom. Two sugars that differ only in the configuration about a single carbon atom are called **epimers**. D-Galactose is converted to D-glucose in the liver by a specific enzyme called an epimerase. Galactose is present in some glycoproteins and glycolipids (see page 372).

Galactosemia, a severe inherited disease, results in the inability of infants to metabolize galactose because of a deficiency of either the enzyme galactose 1-phosphate uridyl transferase or the enzyme galactokinase. The galactose concentration increases in the blood and urine (galactosuria).

Fructose   Fructose is a ketohexose. Its molecular formula, like that of glucose and galactose, is $C_6H_{12}O_6$. It, too can be represented as a straight-chain or as a ring com-

■ **PLATE 1**

A nurse looks after a premature infant in a pediatric intensive care unit. The plexiglass incubator provides the infant with the warm, stable environment it needs.

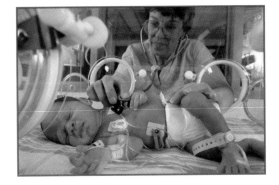

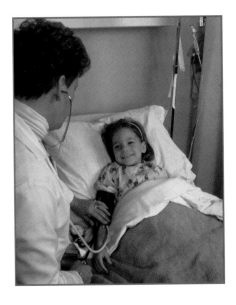

■ **PLATE 2**

A young girl has her blood pressure measured with a sphygmomanometer. This device is relatively simple to use, yet provides vital information about the cardiovascular system.

■ **PLATE 3**

The use of high technology in health care is becoming increasingly evident. A far cry from the traditional paper charts, this bedside computer is used to display and update patient records.

■ **PLATE 4**

A histologist examines a thin slice of tissue for the presence of cancer. Modern pathology laboratories are well-equipped to diagnose disorders found in biopsied tissue and rapidly report the results to the attending physician by telephone or fax.

■ PLATE 5

A laboratory technician prepares a culture of a suspected pathogen (e.g., bacteria).

■ PLATE 6

Laboratory technicians use a pipette to deliver specified volumes of suspended cells to an incubation heater. Such pipettes are ideal for experimental protocols that demand high accuracy and precision: pipettes can reliably dispense volumes as small as a microliter (1/1,000,000th of a liter)!

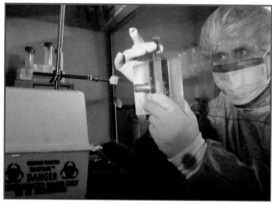

■ PLATE 7

A scientist at work in a maximum containment laboratory at the Centers for Disease Control where little-known contagious viruses are studied. Because the hazards of infection are great, laboratory workers handling virulent pathogens must wear containment suits.

■ PLATE 8

Pulmonary function is measured as an individual is subjected to physical stress. Such tests quantify the performance of the lungs and readily discriminate between normal and substandard function.

■ PLATE 9

A patient is weighed while submerged. This weight can be compared to his weight in the air to determine his density and his percentage of total body fat.

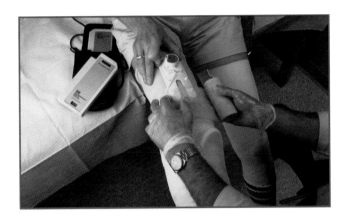

■ PLATE 10

This orthopedic electrical device, used to speed healing of bone fractures, delivers a low-intensity current to bone tissue, promoting rapid bone growth and fusion of the fractured ends.

■ PLATE 11

A Geiger counter is used to evaluate thyroid functioning. It detects radiation emitted by a previously injected radiolabel. The distribution of this substance within the thyroid reflects the overall condition of the gland.

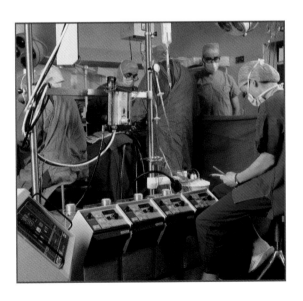

■ PLATE 12

The heart and lung machine is used often in thoracic surgery. The patient's blood is shunted to the machine (bypassing the heart and lungs), which oxygenates it and pumps it through the patient, hence its name.

■ PLATE 13

A unit of blood is removed from cold storage prior to use in surgery. Each unit is marked according to blood type for ease of identification. Freshly collected units of blood are tested and then stored until needed for a future operation. The need for blood is great; donated units are frequently used within a matter of days.

■ PLATE 14

An anesthesiologist carefully monitors the flow of a general anesthetic to a patient about to undergo surgery.

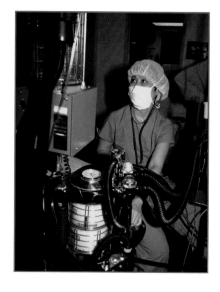

■ PLATE 15

The anesthesiologist checks the patient's vital signs during surgery.

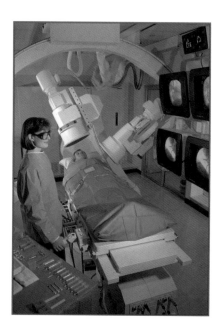

■ PLATE 16

Angiography is a diagnostic test used to spot potential structural problems in the blood vessels. In this procedure, an x-ray—opaque dye is injected into the circulatory system and its flow is monitored by continuous x-ray examination. The dye outlines the blood vessels and reveals structural defects that may be present.

■ **PLATE 17**

A patient receives a chest x ray. This relatively inexpensive procedure is often among the first of a series of diagnostic tests.

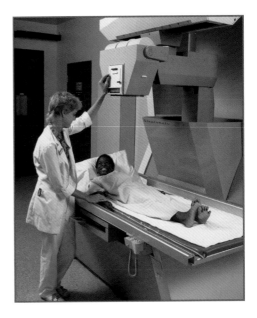

■ **PLATE 18**

A woman receives a sonogram in the third trimester of her pregnancy. Also known as ultrasound, sonograms provide a convenient method of monitoring fetal health in utero.

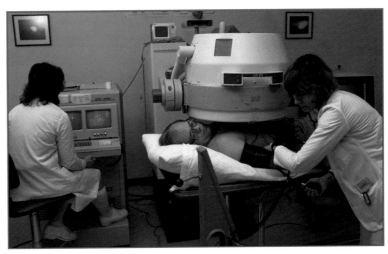

■ **PLATE 19**

The patient is undergoing a heart stress test.

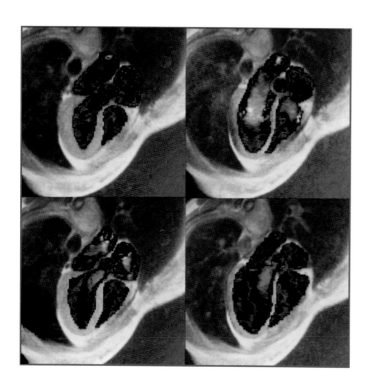

Four slices of a cine display of a cardiac cycle, color enhanced to show direction and velocity of blood flow: red indicates upward flow and blue downward. Green indicates a lack of distinct directional flow. In addition to quantitative flow management, the MRI technique used in this study helps assess wall motion and valve function. The blue patches show blood flow from the left atrium to the left ventricle; red patches show blood flow from the left ventricle to the aorta.

■ PLATE 21

Through the use of computer analysis, modern scanning images can be produced with color enhancement. Values are assigned to each color to show the density of organs and tissues.

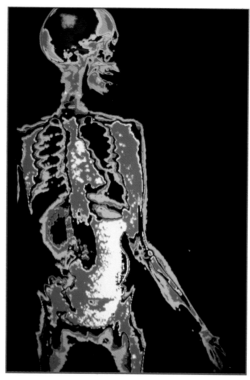

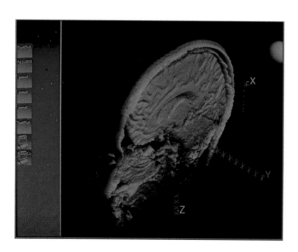

■ PLATE 22

SQUID information is superimposed onto a CAT scan. Fourteen sites were chosen to measure the brain's electromagnetic response to stimuli as well as record and measure the direction and intensity of brain signals.

■ PLATE 23

These PET scans indicate regions of brain activity during a series of sensory, motor, and cognitive tasks. This procedure is not only useful in locating normal areas of activity, but is very helpful in the identification of many cognitive disorders.

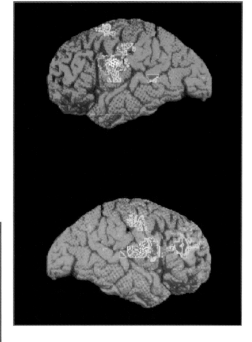

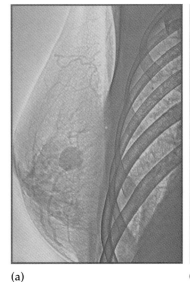 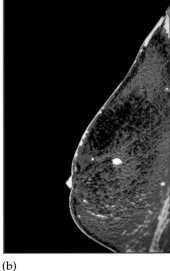

■ PLATE 24

Two mammograms showing cancer of the breast. (a) Xeroradiogram reveals a well-defined tumor (dark circle at center). (b) This false-color xeroradiogram betrays the presence of another tumor, visible as the white object at center.

(a)   (b)

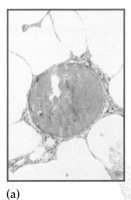 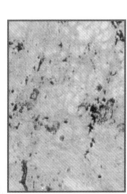 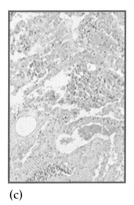

(a)   (b)   (c)

■ PLATE 25

Light micrographs of normal and damaged lung tissue. (a) Tissue from a normal lung. Healthy alveoli surround a capillary (orange object at center). (b) Section of lung from a heavy smoker. Black tar deposits are evident. (c) Tissue from lung of patient with pneumonia. The large-scale destruction of alveoli has resulted in the formation of large pockets of air (irregularly shaped white patches). Pink areas are connective tissue and reddish-brown objects are clumps of blood cells.

■ PLATE 26

The patient is undergoing radiotherapy for Hodgkin's disease.

■ PLATE 27

The lithotripter eliminates kidney stones with the use of shock waves.

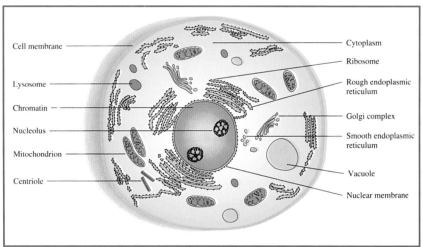

■ PLATE 28

A typical cell.

Cell membrane

Lysosome

Chromatin

Nucleolus

Mitochondrion

Centriole

Cytoplasm

Ribosome

Rough endoplasmic reticulum

Golgi complex

Smooth endoplasmic reticulum

Vacuole

Nuclear membrane

■ PLATE 29

CT scans reveal intracranial anomalies. (a) This two-dimensional horizontal section through the head reveals a lesion in the right optic nerve. (The intact left optic nerve is visible as the green tract running from the left eye to the brain.) (b) Two tumors (in red) are visible in this three-dimensional section of the head. Orientation: The posterior aspect of the skull faces the viewer; the anterior portion is toward the top of the picture.

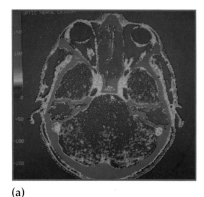

(a)

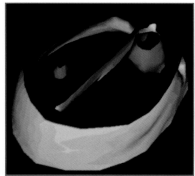

(b)

pound (20-7). The ring structure is predominant. Note that the ring structure represents a hemiketal (see Section 18-3).

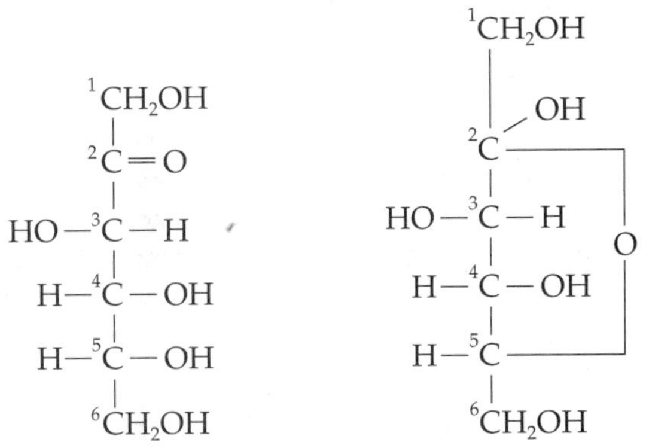

D-fructose

D-fructose

(20-7)

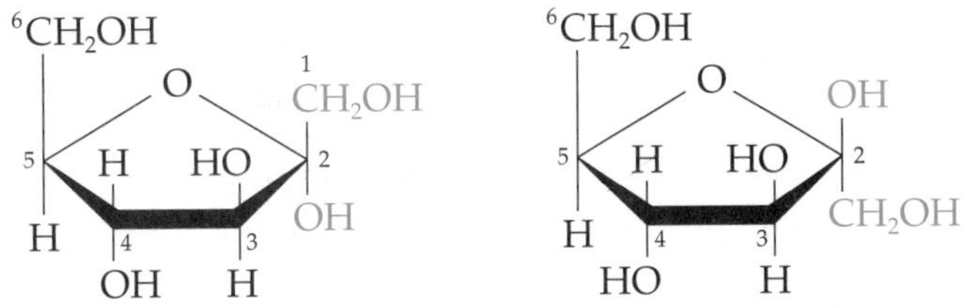

α-D-fructose

β-D-fructose

Fructose is often called levulose, or fruit sugar. It occurs naturally in fruit juices and honey. It can be prepared by the hydrolysis of sucrose, a disaccharide, and also by the hydrolysis of inulin, a polysaccharide found in Jerusalem artichokes. Fructose is the most soluble and also the sweetest of all sugars, being 75 percent sweeter than glucose.

Fructosemia, fructose intolerance, is an inherited disease due to a deficiency of the enzyme fructose 1-phosphate aldolase. An infant suffering from this disease experiences hypoglycemia, vomiting, and severe malnutrition. Such a condition is treated by placing the infant on a low-fructose diet.

Reactions of the Hexoses

Hexoses, which are either aldoses or ketoses, show reducing properties. This reducing property is the basis of the test for sugar in the urine and in the blood. When a reducing agent is treated with an oxidizing agent such as Cu^{2+} complex ion,[†] a red-orange precipitate of copper(I) oxide (Cu_2O) is formed. The unbalanced equation for the reaction of an aldehyde with copper(II) complex ion can be written as follows:

$$\text{aldehyde} + \underset{\substack{\text{deep blue} \\ \text{solution}}}{Cu^{2+}} \xrightarrow[\text{NaOH}]{\text{heat}} \text{acid} + \underset{\substack{\text{red-orange} \\ \text{precipitate}}}{Cu_2O_{(s)}} + \text{water}$$

In this reaction the aldehyde is oxidized to the corresponding acid. When glucose is treated with Cu^{2+} complex ion and the mixture is heated, the reaction is as follows:

[†]Fehling's solution is an alkaline solution of Cu^{2+} ions complexed with tartrate; Benedict's solution is an alkaline solution of Cu^{2+} ions complexed with citrate.

$$
\begin{array}{c}
\text{H} \\
| \\
{}^{1}\text{C}=\text{O} \\
| \\
\text{H}-{}^{2}\text{C}-\text{OH} \\
| \\
\text{HO}-{}^{3}\text{C}-\text{H} \\
| \\
\text{H}-{}^{4}\text{C}-\text{OH} \quad + \quad \text{Cu}^{2+} \quad \xrightarrow[\text{NaOH}]{\text{heat}} \quad \text{Cu}_2\text{O}_{(s)} \quad + \\
| \\
\text{H}-{}^{5}\text{C}-\text{OH} \\
| \\
{}^{6}\text{CH}_2\text{OH}
\end{array}
\qquad
\begin{array}{c}
{}^{1}\text{COOH} \\
| \\
\text{H}-{}^{2}\text{C}-\text{OH} \\
| \\
\text{HO}-{}^{3}\text{C}-\text{H} \\
| \\
\text{H}-{}^{4}\text{C}-\text{OH} \quad + \quad \text{H}_2\text{O} \\
| \\
\text{H}-{}^{5}\text{C}-\text{OH} \\
| \\
{}^{6}\text{CH}_2\text{OH}
\end{array}
$$

D-glucose   copper(II)   copper(I) oxide   D-gluconic acid
                 complex ion   (red-orange
                 (deep blue color) precipitate)

Laboratory tests for the presence of glucose in urine use Benedict's solution of Fehling's solution, both of which contain copper(II) complex ion. Clini-test tablets, which also contain a copper(II) complex, give a rapid quantitative measurement of the concentration of glucose present. If the blue liquid turns green, a trace of sugar (glucose) is present. This is frequently recorded as +. A yellow color, indicated by ++, indicates up to 0.5 percent sugar; an orange color, +++, 0.5 to 1.5 percent; and a red color, ++++, over 1.5 percent sugar.

Glucose does not normally appear in the urine for any extended period of time. Its persistent presence usually indicates that something is wrong with the metabolism of carbohydrates, such as diabetes mellitus.

Another laboratory test for the presence of a reducing sugar uses Tollens' reagent, which contains Ag^+ ion in the form of $Ag(NH_3)_2^+$. In this reaction, glucose is oxidized to gluconic acid as before and the silver complex ion is reduced to free silver, which appears as a bright shiny mirror on the inside of the test tube.

$$
\text{glucose} + \text{Ag}^+ \xrightarrow[\text{NH}_4\text{OH}]{\text{heat}} \text{gluconic acid} + \text{Ag}_{(s)} + \text{water}
$$

Tollens'   silver mirror
reagent

Oxidation   An aldose contains an aldehyde group, as well as several —OH groups. If the aldehyde end of the molecule is oxidized, the product is named an *-onic acid*. When the aldehyde end of glucose is oxidized, the product is called gluconic acid (see above formula). When the aldehyde end of galactose is oxidized, the product is called galactonic acid. If the alcohol at the end opposite the aldehyde (the other end of the molecule) is oxidized, the product is called a *-uronic acid*. The oxidation of the alcohol end of glucose yields glucuronic acid. Glucuronic acid is a minor product of glucose metabolism. It is also part of the heparin molecule (see page 353).

$$
\begin{array}{c}
{}^{1}\text{CHO} \\
| \\
\text{H}-{}^{2}\text{C}-\text{OH} \\
| \\
\text{HO}-{}^{3}\text{C}-\text{H} \\
| \\
\text{H}-{}^{4}\text{C}-\text{OH} \\
| \\
\text{H}-{}^{5}\text{C}-\text{OH} \\
| \\
{}^{6}\text{CH}_2\text{OH}
\end{array}
\xrightarrow[\substack{\text{(in several} \\ \text{steps)}}]{\text{oxidation}}
\begin{array}{c}
{}^{1}\text{CHO} \\
| \\
\text{H}-{}^{2}\text{C}-\text{OH} \\
| \\
\text{HO}-{}^{3}\text{C}-\text{H} \\
| \\
\text{H}-{}^{4}\text{C}-\text{OH} \\
| \\
\text{H}-{}^{5}\text{C}-\text{OH} \\
| \\
{}^{6}\text{COOH}
\end{array}
$$

glucose   glucuronic acid

Likewise, if the alcohol end of galactose is oxidized, galacturonic acid is formed. If both ends of the glucose molecule are oxidized at the same time, the product is called saccharic acid.

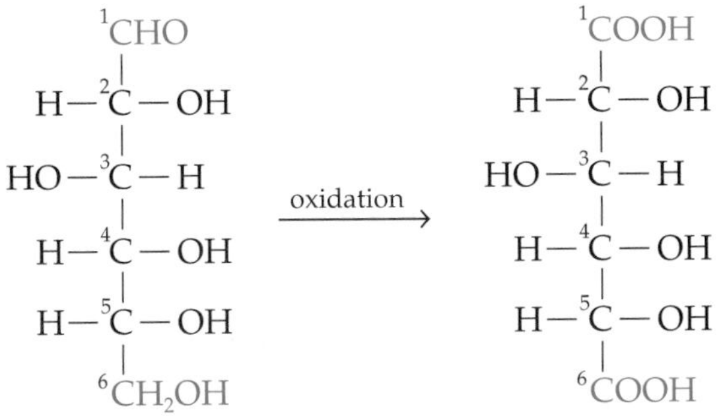

$$
\begin{array}{ccc}
^1CHO & & ^1COOH \\
H-^2C-OH & & H-^2C-OH \\
HO-^3C-H & \xrightarrow{\text{oxidation}} & HO-^3C-H \\
H-^4C-OH & & H-^4C-OH \\
H-^5C-OH & & H-^5C-OH \\
^6CH_2OH & & ^6COOH \\
\text{glucose} & & \text{saccharic acid}
\end{array}
$$

Reduction   The aldohexoses can be reduced to alcohols. When glucose is reduced, sorbitol is formed. Sorbitol accumulation in the eye is a major factor in the formation of cataracts due to diabetes.

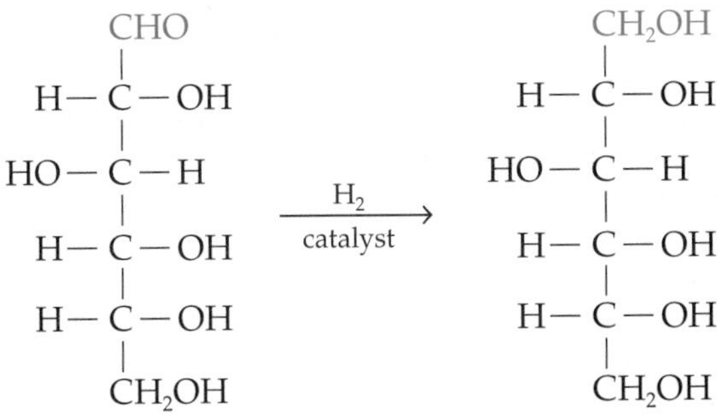

$$
\begin{array}{ccc}
CHO & & CH_2OH \\
H-C-OH & & H-C-OH \\
HO-C-H & \xrightarrow[\text{catalyst}]{H_2} & HO-C-H \\
H-C-OH & & H-C-OH \\
H-C-OH & & H-C-OH \\
CH_2OH & & CH_2OH \\
\text{glucose} & & \text{sorbitol}
\end{array}
$$

Reduction of galactose yields dulcitol, and reduction of fructose yields a mixture of mannitol and sorbitol.

Mannitol is now used in the treatment of malignant brain tumors. Chemotherapeutic agents are unable to cross the blood-brain barrier, which is the brain's natural defense mechanism. The blood-brain barrier allows substances such as glucose, alcohol, and hallucinogenic agents to pass through, but it inhibits the entry of bacteria and toxic substances, including chemotherapeutic drugs. However, a high concentration of mannitol can be injected directly into the brain's main arteries, temporarily shrinking the cells that line those blood vessels. This shrinkage lowers the blood-brain barrier for approximately 30 minutes and allows the chemotherapeutic agents to be administered intravenously.

Fermentation   Glucose ferments in the presence of yeast, forming ethyl alcohol and carbon dioxide. This reaction will not readily occur in the absence of yeast. Yeast contains certain enzymes that catalyze this particular reaction.

$$
\underset{\text{glucose}}{C_6H_{12}O_6} \xrightarrow{\text{enzymes}} \underset{\text{ethyl alcohol}}{2\,C_2H_5OH} + 2\,CO_2
$$

Fructose will also ferment; galactose will not readily ferment. Pentoses do not ferment in the presence of yeast.

Formation of Phosphate Esters   Phosphate esters such as D-glyceraldehyde 3-phosphate and dihydroxyacetone phosphate are triose phosphate esters involved in glycolysis (see Figure 26-6). Glucose forms phosphate esters at carbons 1 and 6, as shown,

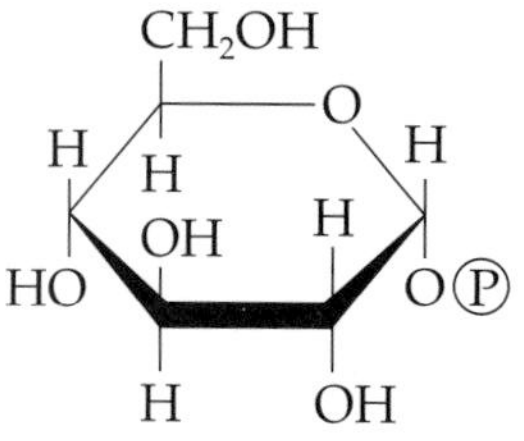

glucose 1 phosphate 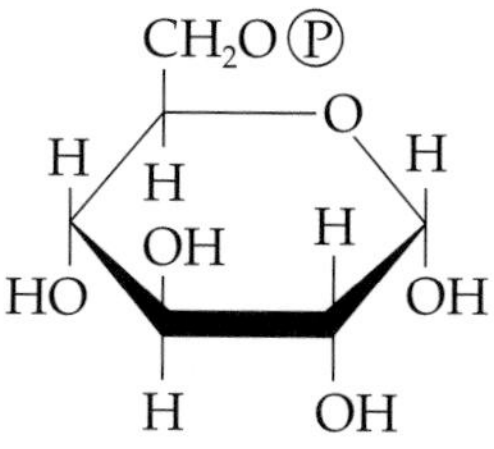   glucose 6-phosphate

and ribose forms a 1,5-diphosphate ester.

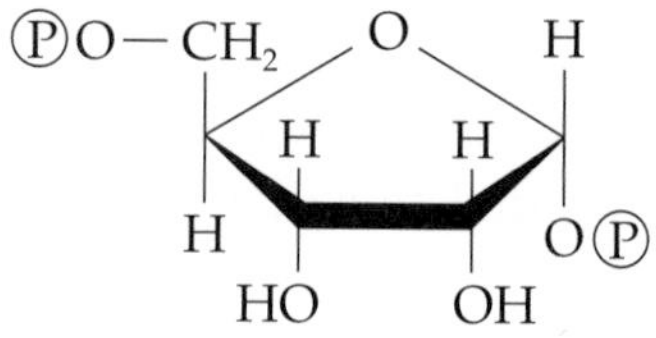

D-ribose 1,5-diphosphate

The phosphate group is abbreviated as Ⓟ, and the compound names are sometimes shortened, for example, glucose 1-P.

Amino Sugars

Amino sugars (hexosamines) contain an amino group in place of an —OH group. Three amino sugars have been found in nature. They are

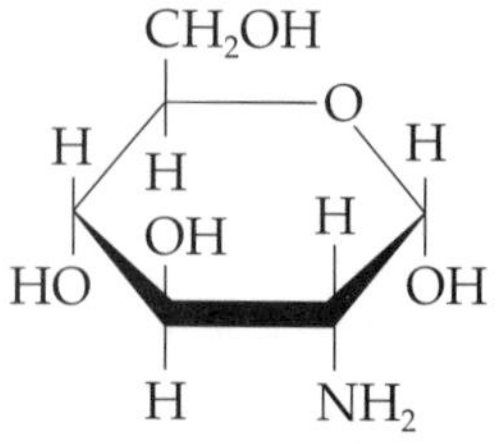

D-glucosamine 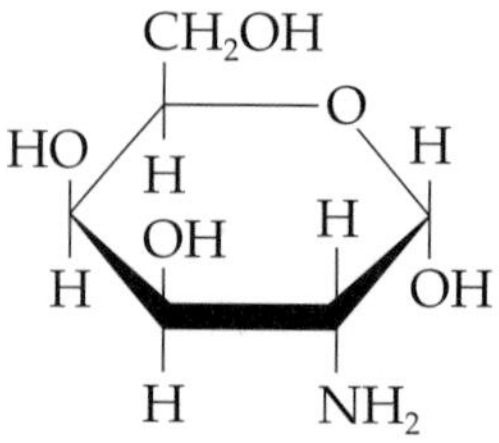   D-galactosamine 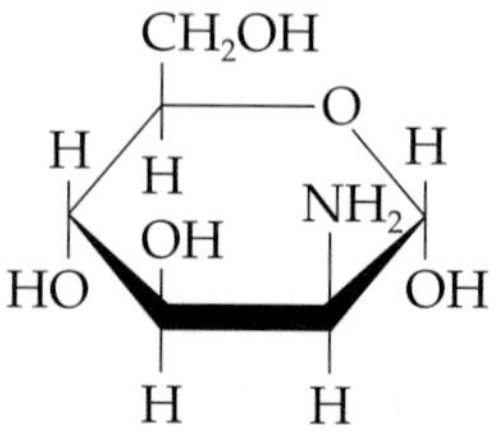   D-mannosamine

Erythromycin and carbomycin are examples of antibiotics that contain amino sugars.

Protein-Sugar Interactions

When plasma glucose concentration is elevated over a period of time, normal hemoglobin covalently binds to glucose. The amount of glucosylated hemoglobin (HbA_{ic}) in the blood is used as a measure of the effectiveness of blood glucose control in a diabetic patient because the concentration of HbA_{ic} directly reflects the elevation of blood glucose over the preceding several days. (Blood glucose levels directly reflect the instantaneous glucose level.) About 7 to 11 percent of a diabetic patient's hemoglobin is HbA_{ic}, compared to 4 to 6 percent for a nondiabetic person. The question arises whether high levels of glucose might combine with other proteins to produce some of the other complications associated with diabetes.

20-5   Disaccharides

There are three common disaccharides—sucrose, maltose, and lactose—all of which are isomers with the molecular formula $C_{12}H_{22}O_{11}$. On hydrolysis these disaccharides yield two monosaccharides. The general reaction can be written as follows:

$$C_{12}H_{22}O_{11} + H_2O \xrightarrow{\text{hydrolysis}} C_6H_{12}O_6 \quad + \quad C_6H_{12}O_6$$

a disaccharide	a monosaccharide	a monosaccharide
sucrose	glucose	fructose
maltose	glucose	glucose
lactose	glucose	galactose

The disaccharides, just like the monosaccharides, are white, crystalline, sweet solids. Sucrose is very soluble in water; maltose is fairly soluble; and lactose is only slightly soluble. The disaccharides are also optically active; they rotate the plane of polarized light. However, even though they are soluble in water, they are too large to pass through cell membranes.

Structure

Disaccharides are formed by the combination of two monosaccharides. In the last section we noted that monosaccharides were either hemiacetals or hemiketals. If a hemiacetal or hemiketal (a monosaccharide) combines with an alcohol (another monosaccharide), an acetal or a ketal will be formed (see Section 18-3). Such a bond between the two monosaccharides is called a *glycosidic linkage.

Consider the combination of a molecule of α-glucose with a molecule of β-glucose, as shown in equation (20-8). The products of such a reaction are β-maltose and water.

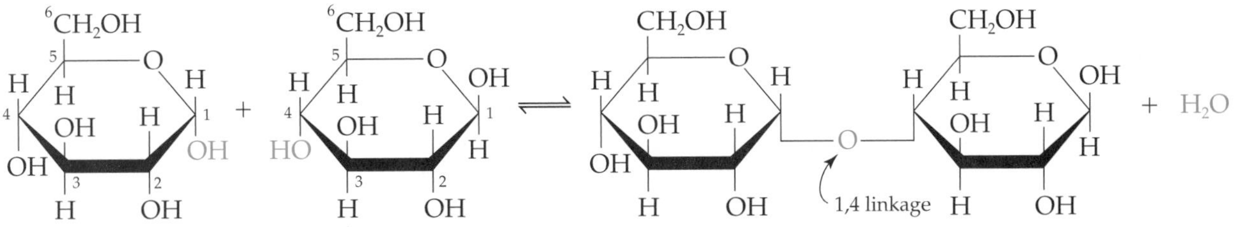

α-glucose   β-glucose   β-maltose   (20-8)

The linkage in β-maltose is between carbon 1 of one glucose and carbon 4 of the other glucose. Such a linkage is called an α-1,4 linkage; α-maltose has an —OH on carbon 1 of the second glucose molecule below the plane rather than above. Both α- and β-maltose exist, but the predominant form is the β.

If a molecule of β-glucose combines with another molecule of β-glucose, cellobiose, a compound with a β-1,4 glycosidic linkage, is formed.

Enzymes are specific in the type of glycosidic linkage (α or β) whose hydrolysis they can catalyze. For example, maltase (page 428) catalyzes the hydrolysis of maltose, which contains an α glycosidic linkage. It does not catalyze the hydrolysis of cellobiose, which contains a β glycosidic linkage.

Conversely, an enzyme that catalyzes the hydrolysis of β glycosidic linkages catalyzes the hydrolysis of cellobiose. Such β glycosidic linkage hydrolysis enzymes are not found in the human digestive system. That is why humans cannot digest cellulose and cellobiose, both of which have β glycosidic linkages.

Consider the reaction of a molecule of α-glucose with a molecule of β-fructose. The products of such a reaction are sucrose and water.

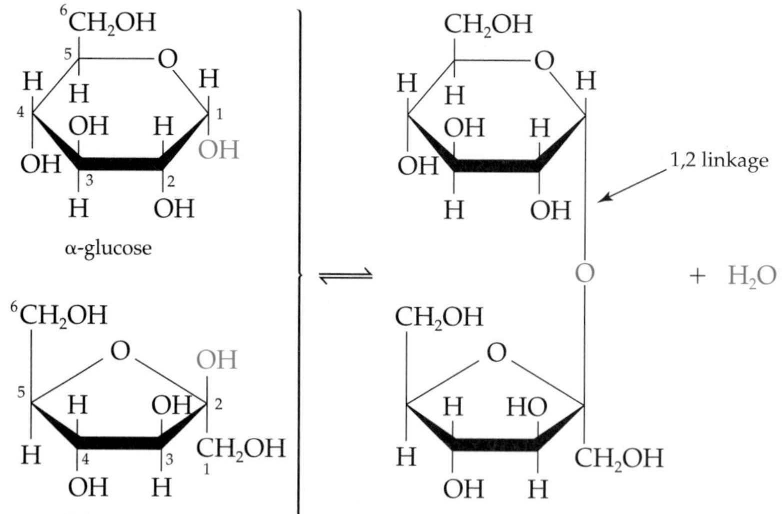

The linkage in sucrose is an α-1,2 glycosidic linkage because it occurs between carbon 1 of the glucose molecule and carbon 2 of the fructose. There is only one form of sucrose; α and β forms do not exist.

If a molecule of β-galactose reacts with a molecule of glucose (α or β), the products are water and lactose (α or β). The linkage is β-1,4.

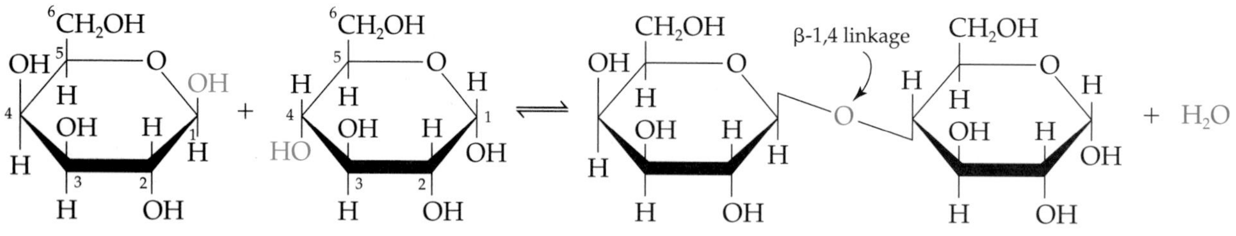

Reducing Properties

In maltose the aldehyde groups are at carbon 1 in each of the original glucose molecules. Since the linkage is 1,4, one free aldehyde group remains. Therefore, maltose acts as a reducing sugar (see pages 301 and 343).

In sucrose, the glucose part had the aldehyde at carbon 1, and the fructose part had the ketone group at carbon 2. Since the linkage is 1,2, neither group is free. Therefore, sucrose is not a reducing sugar.

Lactose, which has a 1,4 linkage, acts as a reducing sugar because both of the original aldehyde groups were on carbon 1, and one of them is free to react.

Fermentation

Sucrose and maltose will ferment when yeast is added because yeast contains the enzymes sucrase and maltase; lactose will not ferment when yeast is added because yeast does not contain lactase. The identity of a disaccharide can be deduced on the basis of its fermentation reaction and its reducing properties.

Suppose that a test tube contains a disaccharide, $C_{12}H_{22}O_{11}$. Is it sucrose, lactose, or maltose? The identity can be determined by the following method:

1. Mix the unknown disaccharide with alkaline Cu^{2+} complex and warm gently. If there is no reaction, the disaccharide must be sucrose. In this case, no further test is necessary to prove the identity of the disaccharide.
2. If the unknown disaccharide gives a positive test with alkaline Cu^{2+} complex, it must be either maltose or lactose. In this case, another sample of the disaccharide is mixed with yeast and allowed to stand to observe whether or not fermentation takes place. If the disaccharide does ferment, it must be maltose; if it does not ferment, it must be lactose.

The same two laboratory tests can be performed in reverse order with the same results.

1. Mix the unknown disaccharide with yeast, allow the mixture to stand, and observe whether fermentation takes place. If no fermentation is observed, the disaccharide is lactose, and no further test is necessary.
2. If fermentation does occur, the disaccharide is either sucrose or maltose. In this case take another sample of the unknown disaccharide, mix it with alkaline Cu^{2+} complex, and warm gently. If the color remains blue, the unknown disaccharide must be sucrose. If the unknown gives a positive test with alkaline Cu^{2+} complex, it must be maltose.

Sucrose

Sucrose is the sugar used ordinarily in the home. It is also known as cane sugar. Sucrose is produced commercially from sugar cane and sugar beets. It also occurs in sorghum, pineapple, and carrot roots.

When sucrose is hydrolyzed, it forms a mixture of glucose and fructose. This 50:50 mixture of glucose and fructose is called **invert sugar** because it reverses the rotation of polarized light. Honey contains a high percentage of invert sugar.

Maltose

Maltose, commonly known as malt sugar, is present in germinating grain. It is produced commercially by the hydrolysis of starch.

Lactose

Lactose, commonly known as milk sugar, is present in milk. It differs from the preceding sugars in that it has an animal origin. Certain bacteria cause lactose to ferment, forming lactic acid. When this reaction occurs, the milk is said to be sour. Lactose is used in high-calcium diets and in infant foods. Lactose can be used for increasing calorie intake without adding much sweetness. Lactose is found in the urine of pregnant women and, since it is a reducing sugar, it gives a positive test with Cu^{2+} complex ions.

Table 20-2	
Relative Sweetness of Sugars and Other Compounds	
Fructose	175
Galactose	32
Glucose	75
Lactose	16
Maltose	32
Sucrose	100
Aspartame	15,000
Cyclamate	3,000
Saccharin	35,000

20-6 Sweetness and Sugar Substitutes

When we speak of sugar, we think of a substance with a sweet taste. However, sweetness is not a specific property of carbohydrates. Many sugars have some degree of sweetness, but several synthetic compounds have much more sweetness without any calories. Table 20-2 shows the relative sweetness of a variety of sugars and synthetic compounds. Because of the *greater* sweetness of fructose, this substance is used in some low-calorie foods because less is needed to provide the same level of sweetness.

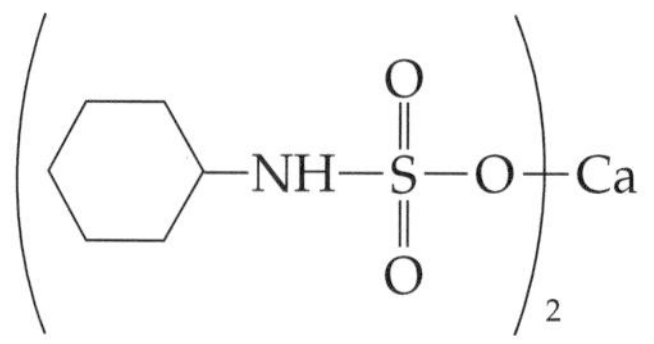

calcium cyclamate 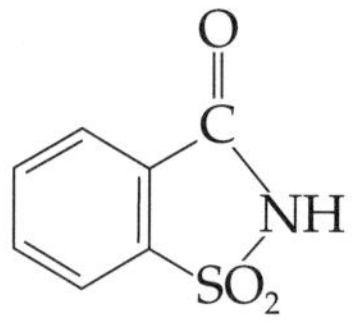 saccharin

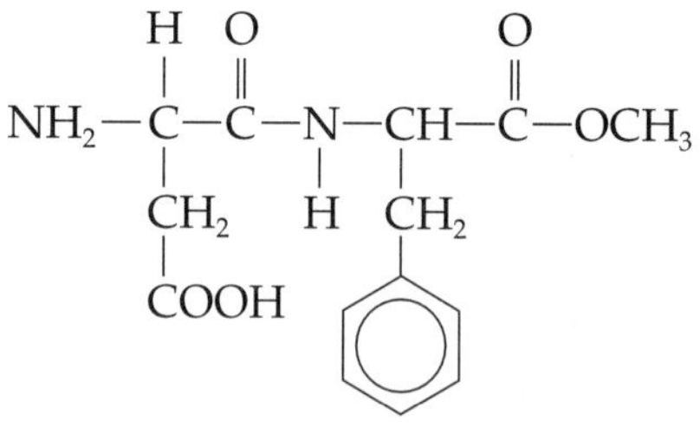

aspartame

Cyclamates were first marketed in 1950, but after 1969 findings that large doses caused cancer in rats, they were banned in the United States, although they are still in use in Canada. There has never been a connection between cyclamates and cancer or any other disease in humans.

The wide acceptance of aspartame is partially attributable to its lack of a bitter aftertaste, as is associated with saccharin. Aspartame is hydrolyzed to aspartic acid, phenylalanine, and methanol. Studies have indicated that aspartame is safe, and only those persons on a low phenylalanine diet (see page 411) need to avoid it.

20-7 Polysaccharides

Polysaccharides are polymers of monosaccharides. Complete hydrolysis of polysaccharides produces many molecules of monosaccharides. The polysaccharides differ from monosaccharides and disaccharides in many ways, as is indicated in Table 20-3.

Polysaccharides can be formed from pentoses (five-carbon sugars) or from hexoses (six-carbon sugars). Polysaccharides formed from pentoses are called

Table 20-3	Comparison of Polysaccharides with Monosaccharides and Disaccharides	
Property	Monosaccharides and Disaccharides	Polysaccharides
Molecular mass	Low	Very high
Taste	Sweet	Tasteless
Solubility in water	Soluble	Insoluble
Size of particles	Pass through a membrane[†]	Do not pass through a membrane
Test with Cu^{2+} complex ions (an oxidizing agent)	Positive (except for sucrose)	Negative

[†]*Only monosaccharides.*

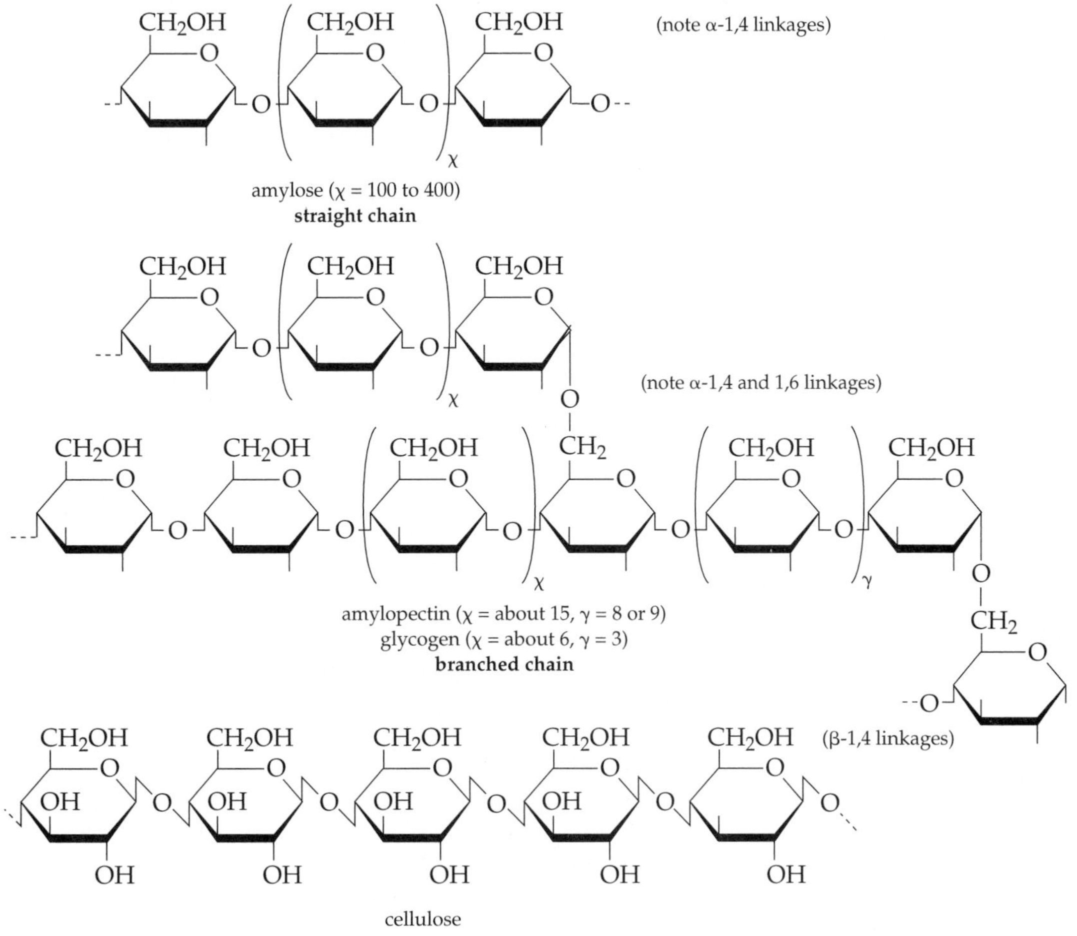

FIGURE 20-6
Structures of some polysaccharides—glucose polymers in amylose, amylopectin, and cellulose.

pentosans. Those formed from hexoses are called hexosans (or sometimes glucosans).

The hexosans (or glucosans) are the most important in terms of physiology.
The hexosans have the general formula $(C_6H_{10}O_5)_x$, where x is some large number.
Some of the common hexosans are starch, cellulose, glycogen, and dextrin. All are
made up of only glucose molecules, as shown in Figure 20-6.

Starch

Plants store energy primarily in the form of starch granules. Starch is actually a
mixture of the polysaccharides amylopectin and amylose (see Figure 20-6). Amylopectin is a branched polysaccharide present in starch to a large extent (80 to 85
percent). It is usually present in the covering of the starch granules. Amylose is a
nonbranched polysaccharide present in starch to an extent of 15 to 20 percent.

Starch is insoluble in water. When starch is placed in boiling water, the granules
rupture, forming a paste that gels on cooling. When a small amount of starch is added
to a large amount of boiling water, a colloidal dispersion of starch in water is formed.

Starch gives a characteristic deep blue color with iodine. This test is used to
detect the presence of starch because it is conclusive even when only a small

amount of starch is present. That is, if iodine is added to an unknown and a blue color is produced, starch is present. This test can also be used to check for the presence of iodine. If starch is added to an unknown and a blue color is produced, iodine must be present.

When starch is hydrolyzed, it forms dextrins (amylodextrin, erythrodextrin, achroodextrin), then maltose, and finally glucose. Erythrodextrins turn red in the presence of iodine. Both maltose and glucose produce no color in the presence of iodine. Thus, it is possible to follow the hydrolysis of the starch by observing the changing colors when iodine is added.

$$\underset{\text{blue}}{\text{starch}} \longrightarrow \underset{\text{red}}{\text{erythrodextrins}} \longrightarrow \underset{\text{colorless}}{\text{maltose}} \longrightarrow \underset{\text{colorless}}{\text{glucose}}$$

Cellulose

Wood, cotton, and paper are composed primarily of cellulose. Cellulose is the supporting and structural substance of plants. Like starch, cellulose is a polysaccharide composed of many glucose units. It is not affected by any of the enzymes present in the human digestive system and so cannot be digested. However, it does serve a purpose when eaten with other foods; it gives bulk to the feces and prevents constipation.

Cellulose does not dissolve in water or in most ordinary solvents. It gives no color test with iodine and gives a negative test with Cu^{2+} complex ions.

Cotton is nearly pure cellulose. When cotton fibers are treated with a strong solution of sodium hydroxide and then stretched and dried, the fibers take on a high luster. Such cotton is called *mercerized* cotton.

Cellulose is also used to make rayon. In this process, purified wood pulp (nearly pure cellulose) is converted into a viscous liquid called viscose by treatment with sodium hydroxide and carbon disulfide. When the viscose is forced through small openings in a block suspended in an acid solution, the cellulose is regenerated into fibers that can then be formed into threads.

Glycogen

Glycogen is present in the body and is stored in the liver and the muscles, where it serves as a reserve supply of glucose. Glycogen has an animal origin, as opposed to the plant origin of starch. The structure of glycogen is similar to that of amylopectin (see Figure 20-6) but not as branched.

Glycogen forms a colloidal dispersion in water and gives a red color with iodine. It gives no test with alkaline Cu^{2+} complex. Glycogen is formed in the body cells from molecules of glucose. This process is called **glycogenesis** (see Section 26-4). When glycogen is hydrolyzed into glucose, the process is called **glycogenolysis**.

$$\text{glucose} \underset{\text{glycogenolysis}}{\overset{\text{glycogenesis}}{\rightleftharpoons}} \text{glycogen}$$

Exercise 20-2

What is the difference between amylose and cellulose?

Solution

Amylose has α-1,4 linkages between glucose units, whereas cellulose has β-1,4 linkages between glucose units.

Self-test

What is the difference between amylopectin and glycogen?

Answer

Glycogen is more highly branched.

Dextrin

Dextrin is produced during the hydrolysis of starch. Dextrin is an intermediate between starch and maltose. It forms sticky colloidal suspensions with water and is used in the preparation of adhesives. The glue on the back of postage stamps is a dextrin. Dextrin is also used when digestion of starch might be a problem, as with infants and elderly persons.

Heparin

Heparin is a polysaccharide used as a blood anticoagulant. It accelerates the inactivation of thrombin and other blood-clotting agents (see Section 30-11). The structure of heparin consists of repeating units of glucuronic acid and glucosamine with some sulfate groups on the amino and hydroxyl groups. Heparin is the strongest organic acid present in the body.

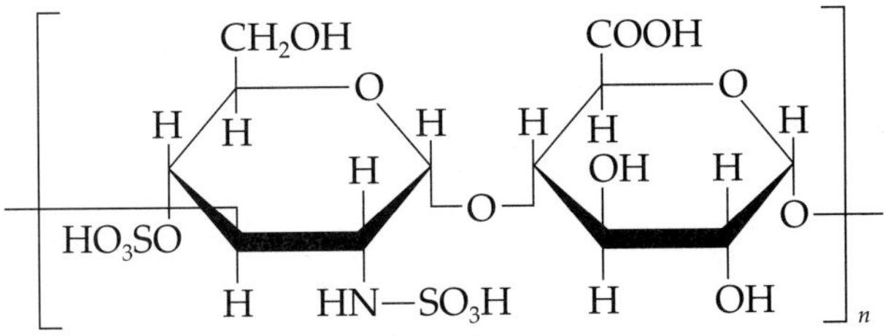

heparin

Dextran

Dextran (not the same as dextrin) is a polysaccharide produced by certain bacteria when they are grown on sucrose. There are various types of dextrans, differing in chain length and degree of branching. Medically, dextrans are used as blood extenders to hold water in the bloodstream and help prevent drops in blood volume and blood pressure. Dextrans growing on the surfaces of teeth are an important component of dental plaque.

Summary

Carbohydrates are polyhydroxyaldehydes, or polyhydroxyketones, or substances that yield these compounds on hydrolysis. Carbohydrates are divided into three categories: monosaccharides, disaccharides, and polysacchrides—based on hydrolytic possibilities.

Stereoisomers are compounds that have the same molecular formula but different structures that are mirror images of one another. Stereoisomers that are chiral are called enantiomers and are said to be optically active.

The parent compound of enantiomers is glyceraldehyde. Stereoisomers that are mirror images are called enantiomers; stereoisomers that are not mirror images are called diastereomers.

Carbohydrates are formed in plants by a process called photosynthesis. Plant cells take carbon dioxide from the air and water from the ground and combine them in the presence of sunlight and chlorophyll to produce monosaccharides, at the same time giving off oxygen into the air. Plant cells also have the ability to convert the monosaccharides thus formed into disaccharides and polysaccharides. When carbohydrates are catabolized in the body, carbon dioxide and water are formed, thus returning these substances for reuse by plants.

Monosaccharides, or simple sugars, are either aldoses or ketoses, depending upon whether they contain an aldehyde or a ketone group. The six-carbon monosaccharides, the hexoses, are the most common in terms of the human body.

The most important hexoses in the human body are glucose, fructose, and galactose. These compounds are isomers with the molecular formula $C_6H_{12}O_6$.

Glucose is an aldohexose whose structure may be represented as a linear or ring-shaped molecule. Glucose is commonly known as dextrose, or grape sugar. It is the most important of all monosaccharides and is normally found in the bloodstream and in the tissue fluids.

Galactose is also an aldohexose. It occurs in nature as one of the constituents of lactose.

Fructose, a ketohexose, is commonly known as levulose, or fruit sugar. Fructose is the sweetest of all sugars. It occurs in nature as one of the constituents of sucrose and is found free in fruit juices and in honey.

The hexoses are either aldehydes or ketones and can act as reducing agents. When a hexose is treated with alkaline Cu^{2+} complex (Fehling's solution, Benedict's solution, or Clini-test), a red-orange precipitate of Cu_2O is formed. This reaction is the basis for the test for sugar (hexoses) in the urine. Hexoses will also reduce Tollens' reagent (Ag^+ complex) to free silver.

Hexoses will ferment in the presence of enzymes found in yeast.

When the aldehyde end of a monosaccharide is oxidized, an *-onic* acid is formed.

When the alcohol end of a monosaccharide is oxidized, a *-uronic* acid is formed.

The three common disaccharides—sucrose, maltose, and lactose—are isomers with the molecular formula $C_{12}H_{22}O_{11}$. On hydrolysis a disaccharide yields two monosaccharides.

Of the three disaccharides, only maltose and lactose show reducing properties with alkaline Cu^{2+} complex ions. Sucrose is not a reducing sugar.

Sucrose and maltose will ferment with yeast, owing to the presence of the enzymes sucrase and maltase. Lactose will not ferment with yeast because of the absence of the enzyme lactase.

Polysaccharides are polymers of monosaccharides and yield monosaccharides when hydrolyzed.

Polysaccharides have a high molecular mass, are insoluble in water, are tasteless, and give negative tests for reducing sugars. These properties are the opposite of those for monosaccharides and disaccharides.

Three common polysaccharides are starch, cellulose, and glycogen. Plants store their food as starch; plants use cellulose as supporting and structural parts; animals use glycogen as a reserve supply of carbohydrate.

Questions and Problems

1. What are carbohydrates? Where did the word come from?

2. What is a stereoisomer? an enantiomer? a diastereomer?

3. What are aldoses? ketoses? How is the number of carbon atoms indicated in these compounds?

4. What is the difference between ordinary light and polarized light?

5. What effect does solution of an optical isomer have on polarized light?

6. If a compound has four chiral carbon atoms, how many optical isomers are possible?

7. Are all carbon atoms chiral? Explain.

8. Draw the structure for
   a. *cis*-1,2-dichloro-2-butene
   b. *trans*-1,2-dimethylcyclopropane

9. What do the letters D and L refer to in terms of optical isomers? Which is the predominant one in terms of the carbohydrates used by the body?

10. What is the reference compound for the configuration of carbohydrates? Draw the D and L structures of this compound.

11. What is a hexosan? a pentosan? epimers?

12. How are carbohydrates formed in nature? Write a reaction for this process. Are animals able to synthesize carbohydrates?

13. What are the three types of carbohydrates? Give an example of each.

14. What are trioses? tetroses? pentoses? hexoses? Give an example of each.

15. Draw the linear and ring structures for glucose, galactose, fructose. What are their molecular formulas?

16. Which carbohydrate can be given intravenously? Why not others?

17. Where is glucose found in nature? fructose? galactose? maltose? sucrose? lactose?

18. What product is formed when the aldehyde end of glucose is oxidized? the alcohol end? both ends?

19. What product is formed when glucose is reduced?

20. Where is glucose found in the body? What function does it serve?

21. How do hexoses affect alkaline Cu^{2+} complex ions, and what use is made of this reaction?

22. Name some agents containing alkaline Cu^{2+} complex ions.

23. Do all three disaccharides act as reducing agents? Why or why not?

24. Do all hexoses ferment in the presence of yeast? Why or why not? all three disaccharides?

25. What tests could be performed to identify a disaccharide?

26. Compare the properties of monosaccharides, disaccharides, and polysaccharides.

27. Compare the ring structures of the three disaccharides and use them to explain their reaction to alkaline Cu^{2+} complex ions.

28. What is starch? What type of mixture does it form in boiling water?

29. What products are formed when starch is slowly hydrolyzed? How can the presence of these products be identified?

30. What is glycogenesis? glycogenolysis?

31. What is a Fischer projection? a Haworth projection? How can it indicate three-dimensional structures?

32. Draw the Fischer and Haworth projections for D-glucose.

33. What is a hemiacetal? a hemiketal? Give one structure for each.

34. What is a glycosidic linkage? What type of such a linkage is present in sucrose? maltose? lactose?

35. Why are alpha and beta forms of maltose possible?

36. The ATP system is important for energy storage. List as many other functions of ATP as you can.

37. Classify the following (for example, as an aldohexose)

a.
$$HO-CH_2-\overset{\overset{\displaystyle O}{\|}}{C}-CH_2-OH$$

b.
$$
\begin{array}{c}
CHO \\
| \\
HO-C-H \\
| \\
HO-C-H \\
| \\
H-C-OH \\
| \\
H-C-OH \\
| \\
CH_2OH
\end{array}
$$

c.
$$
\begin{array}{c}
CHO \\
| \\
HO-C-H \\
| \\
H-C-OH \\
| \\
CH_2OH
\end{array}
$$

Practice Test

1. A carbohydrate that can be given intravenously is _____.
   a. cellulose
   b. sucrose
   c. lactose
   d. glucose

2. If a compound contains three chiral carbon atoms, how many optical isomers are possible?
   a. 2   b. 4   c. 8   d. 16

3. A chiral carbon atom has _____ different groups attached to it.
   a. 1   b. 2   c. 3   d. 4

4. An example of a disaccharide is _____.
   a. glucose
   b. maltose
   c. cellulose
   d. starch

5. An example of a hexose is _____.
   a. sucrose
   b. lactose
   c. galactose
   d. ribose

6. Animals store carbohydrates in the form of _____.
   a. cellulose
   b. glucose
   c. starch
   d. glycogen

7. An example of a polysaccharide is _____.
   a. starch
   b. glycogen
   c. cellulose
   d. all of these

8. When sucrose is hydrolyzed, it yields _____.
   a. glucose only
   b. glucose and fructose
   c. glucose and galactose
   d. ribose and galactose

9. Alkaline Cu^{2+} complex ions are used to test for the presence of _____.
   a. aldehydes
   b. ketones
   c. acids
   d. oxidizing sugars

10. An example of a pentose is _____.
    a. maltose
    b. talose
    c. glyceraldehyde
    d. ribose

21

Lipids

Objectives

- To become acquainted with the general properties of lipids and fatty acids

- To understand the classification of lipids

- To distinguish between fats and oils

- To distinguish between soaps and detergents

- To understand the role of phospholipids in cell membranes

- To understand the role of the eicosanoids

Lipids are the major constituent of the cell's external and internal membranes, a role for which their unique properties equips them well. Soap solution, chemically similar to a lipid, forms bubbles reminiscent of the lipid membranes of cells.

21-1 General Properties

A second group of organic compounds that serve as food for the body is the lipids. Lipids are organic compounds of biologic origin and in general

1. Are insoluble in water
2. Are soluble in nonpolar organic solvents such as ether, acetone, and carbon tetrachloride
3. Contain carbon, hydrogen, and oxygen; sometimes contain nitrogen and phosphorus
4. In most cases yield fatty acids on hydrolysis or combine with fatty acids to form esters
5. Take part in plant and animal metabolism

21-2 Fatty Acids

Both simple and compound lipids yield fatty acids on hydrolysis. Fatty acids are straight-chain organic acids. The fatty acids found in natural fats usually contain an even number of carbon atoms. Fatty acids can be either saturated or unsaturated. Saturated fatty acids contain only single bonds between carbon atoms. Unsaturated fatty acids contain a few double bonds between carbon atoms. Polyunsaturated fatty acids contain many double bonds.

Table 21-1 lists some common fatty acids and where they are found in nature. Note that they all contain an even number of carbon atoms.

Unsaturated fatty acids have lower melting points than the corresponding saturated fatty acids, and the greater the degree of unsaturation, the lower the melting point. The 18-carbon saturated fatty acid, stearic acid, melts at 70° C. The 18-carbon fatty acid with one double bond, oleic acid, melts at 13° C. The 18-carbon fatty acid with two double bonds, linoleic acid, melts at −5° C; and the 18-carbon fatty acid with three double bonds, linolenic acid, melts at −10° C.

Table 21-1 Common Fatty Acids

Name	Formula	Source
Saturated fatty acids		
Butyric	C_3H_7COOH	Butter fat
Caproic	$C_5H_{11}COOH$	Butter fat
Caprylic	$C_7H_{15}COOH$	Coconut oil
Capric	$C_9H_{19}COOH$	Palm oil
Lauric	$C_{11}H_{23}COOH$	Laurel
Myristic	$C_{13}H_{27}COOH$	Nutmeg oil, coconut oil
Palmitic	$C_{15}H_{31}COOH$	Palm oil, lard, cottenseed oil
Stearic	$C_{17}H_{35}COOH$	Plant and animal fats such as lard, peanut oil
Arachidic	$C_{19}H_{39}COOH$	Peanut oil
Unsaturated fatty acids		
Oleic	$C_{17}H_{33}COOH$ (contains 1 double bond)	Olive oil
Linoleic	$C_{17}H_{31}COOH$ (contains 2 double bonds)	Linseed oil
Linolenic	$C_{17}H_{29}COOH$ (contains 3 double bonds)	Linseed oil
Arachidonic	$C_{19}H_{31}COOH$ (contains 4 double bonds)	Animal tissues, corn oil, linseed oil

Unsaturated fatty acids can be subdivided into the following categories.

1. Monounsaturated, those that contain only one double bond
2. Polyunsaturated, those that contain many double bonds
3. Eicosanoids, which include the prostaglandins, leukotrienes, prostacyclins, and thromboxanes (see Section 21-9)

Linoleic acid is called the nutritionally essential fatty acid—it is essential for the complete nutrition of the human body. It cannot be synthesized in the body and must be supplied from food we eat. Arachidonic and linolenic acids, which were formerly also designated as essential fatty acids, can be synthesized in the body from linoleic acid. Linoleic acid is found in large concentrations in corn, cottonseed, peanut, and soybean oils but *not* in coconut or olive oils. One of the functions of this essential fatty acid is in the synthesis of prostaglandins (see page 373).

The absence of the essential fatty acid from the diet of an infant causes loss of weight and also *eczema. These conditions can be cured by administering corn oil or linseed oil. Commercial boiled linseed oil should never be used for this purpose because it can contain litharge, a poisonous lead compound.

The percentages of fatty acids in corn oil, linseed oil, butter, and lard are listed in Table 21-2. The percentages are given as averages because the percent composition of a fat or oil can vary considerably because of weather conditions or the type of food eaten by the animal or both.

Oleic acid ($C_{17}H_{33}COOH$) occurs in nature as the *cis* configuration (see page 270), as do most naturally occurring unsaturated fatty acids. The *trans* form is called elaidic acid.

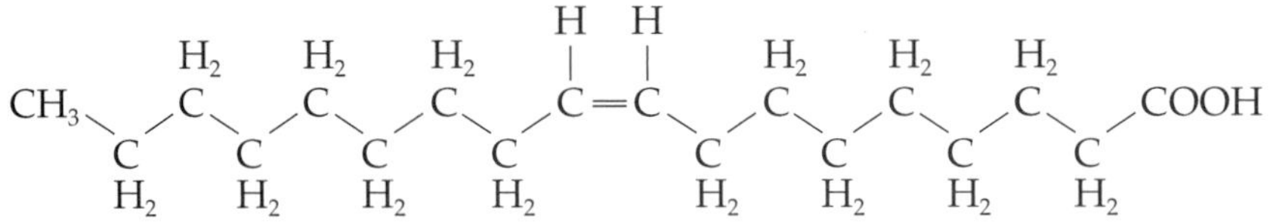

oleic acid (*cis* form)

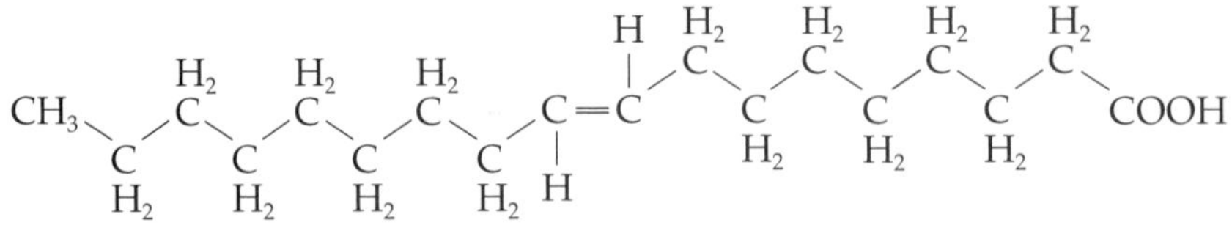

elaidic acid (*trans* form)

Table 21-2 Average Percentage of Fatty Acids in Fats and Oils

	Saturated				Unsaturated			
	Myristic Acid	Palmitic Acid	Stearic Acid	Other Acids	Oleic Acid	Linoleic Acid	Other Acids	Iodine Number
Vegetable oils								
Cottonseed oil	0–3	17–23	1–3		23–44	34–55	0–1	103–115
Corn oil	0–2	8–10	1–4		36–50	34–56	0–3	116–130
Animal fats								
Butter	8–13	25–32	8–13	4–11	22–29	3	3–9	26–45
Lard	1	25–30	12–16		41–51	3–8	5–8	46–66

The need to lower the amount of saturated fat in the diet has been well publicized, and many individuals as well as commercial establishments have switched to vegetable oils for food preparation. Saturated fats are found in meat and dairy products and oils such as palm oil. Dietary saturated fats increase the blood levels of low-density lipoproteins (LDL), which aid in the deposition of cholesterol on artery walls.

Partially hydrogenated vegetable oils have been substituted for saturated fats as a way to lower both cholesterol and LDL levels in the blood. However, partially hydrogenated oils such as those found in solid vegetable shortening and margarines have an effect opposite to that which was desired.

Natural vegetable oils contain primarily *cis* isomers, but partial hydrogenation produces a mixture of *cis* and *trans* isomers. It is the *trans* isomers that cause many undesirable effects such as lowering HDL (the good cholesterol) levels, raising LDL (the bad cholesterol) levels, and raising total cholesterol levels.

21-3 Classification of Lipids

Lipids are divided into three main categories: simple, complex, and precursor and derived.

Simple Lipids

Simple lipids are esters of fatty acids. The hydrolysis of a simple lipid may be expressed as

$$\text{simple lipid} + H_2O \xrightarrow{\text{hydrolysis}} \text{fatty acid(s)} + \text{alcohol}$$

If the hydrolysis of a simple lipid yields three fatty acids and *glycerol*, the simple lipid is called a *fat* or an *oil*. If the hydrolysis of a simple lipid yields a fatty acid and a high molecular mass monohydric alcohol, the simple lipid is called a *wax*.

Complex Lipids

Complex lipids on hydrolysis yield one or more fatty acids, an alcohol, and some other type of compound. In this category are phospholipids and glycolipids (also called cerebrosides because they are found in the cerebrum of the brain).

Phospholipids undergo hydrolysis as follows:

$$\text{phospholipid} + H_2O \xrightarrow{\text{hydrolysis}} \text{fatty acid} + \text{alcohol} + \text{phosphoric acid} + \text{a nitrogen compound}$$

Phospholipids are further subdivided into (1) phosphogylicerides, in which the alcohol is glycerol, and (2) phosphosphingosides, in which the alcohol is sphingosine.

Glycolipids (glycosphingolipids) undergo hydrolysis as follows.

$$\text{glycolipid} + H_2O \xrightarrow{\text{hydrolysis}} \text{fatty acid} + \text{a carbohydrate} + \underset{\substack{\text{a nitrogen-containing} \\ \text{alcohol}}}{\text{sphingosine}}$$

Other complex lipids include the sulfolipids and aminolipids as, well as the lipoproteins.

Precursor and Derived Lipids

Precursor lipids are compounds produced when simple and complex lipids undergo hydrolysis. They include such substances as fatty acids, glycerol, sphingosine, and other alcohols (see page 371). Derived lipids are formed by metabolic transformation of fatty acids. They include ketone bodies, steroids, fatty aldehydes, prostaglandins, and lipid-soluble vitamins.

21-4 Fats and Oils

Structure

Fats are esters formed by the combination of a fatty acid with one particular alcohol, glycerol. If one molecule of glycerol reacts with one molecule of stearic acid (a fatty acid), glyceryl monostearate is formed.

$$
\begin{array}{ccccc}
 & & H & & H \\
 & & | & & | \\
C_{17}H_{35}COOH + & HO-C-H & & C_{17}H_{35}COO-C-H & \\
 & | & & | & \\
 & HO-C-H & \longrightarrow & HO-C-H + H_2O & \\
 & | & & | & \\
 & HO-C-H & & HO-C-H & \\
 & | & & | & \\
 & H & & H & \\
\end{array}
$$

stearic acid   glycerol   glycerol monostearate

The product of this reaction can react with a second molecule and then with a third molecule of stearic acid.

Glyceryl tristearate (also called tristearin) is formed by the reaction of one molecule of glycerol with three molecules of stearic acid. Since stearic acid is a saturated fatty acid, the product is a fat. As the degree of unsaturation of the fatty acid increases, the melting point decreases (See Section 21-2). Fats with a melting point below room temperature are called oils.

$$
\begin{array}{ccccc}
 & & H & & H \\
 & & | & & | \\
 & C_{17}H_{35}COO-C-H & & C_{17}H_{35}COO-C-H & \\
 & | & & | & \\
C_{17}H_{35}COOH + & HO-C-H & \longrightarrow & C_{17}H_{35}COO-C-H + H_2O & \\
 & | & & | & \\
 & HO-C-H & & HO-C-H & \\
 & | & & | & \\
 & H & & H & \\
\end{array}
$$

glycerol distearate

$$
\begin{array}{ccccc}
 & & H & & H \\
 & & | & & | \\
 & C_{17}H_{35}COO-C-H & & C_{17}H_{35}COO-C-H & \\
 & | & & | & \\
C_{17}H_{35}COOH + C_{17}H_{35}COO-C-H & & \longrightarrow & C_{17}H_{35}COO-C-H + H_2O & \\
 & | & & | & \\
 & HO-C-H & & C_{17}H_{35}COO-C-H & \\
 & | & & | & \\
 & H & & H & \\
\end{array}
$$

glycerol tristearate,
a fat

The glycerol molecule contains three—OH groups and so combines with three fatty acid molecules. However, these fatty acids molecules do not have to be the same. Fats and oils can contain three different fatty acid molecules, which can be saturated, unsaturated, or some combination of these.

Exercise 21-1

Write out the structure for a tryglyceride formed by glycerol and three palmitic acids.

Solution

The formula for palmitic acid is $C_{15}H_{31}COOH$. Attaching three palmitic acids to glycerol will produce the following:

$$\begin{array}{l} H_2COOCC_{15}H_{31} \\ \;\;| \\ HCOOC_{15}H_{31} \\ \;\;| \\ H_2COOC_{15}H_{31} \end{array}$$

Self-test

Write out the structural formula for the triglyceride containing a stearic acid, a myristic acid, and a capric acid.

Answer

$$\begin{array}{l} H_2COOC_{17}H_{31} \\ \;\;| \\ HCOOC_{13}H_{27} \\ \;\;| \\ H_2COOC_9H_{19} \end{array}$$

An example of a mixed trigylceride[†] formed from the reaction of glycerol with three different fatty acid molecules follows. The fatty acids are oleic, stearic, and linoleic.

$$\begin{array}{rl} & H \\ & | \\ CH_3CH_2CH_2CH_2CH_2CH_2CH_2CH_2CH{=}CHCH_2CH_2CH_2CH_2CH_2CH_2CH_2COO- & C-H \\ \text{(from oleic acid—one double bond)} & | \\ CH_3CH_2CH_2CH_2CH_2CH_2CH_2CH_2CH_2CH_2CH_2CH_2CH_2CH_2CH_2CH_2CH_2COO- & C-H \\ \text{(from stearic acid—saturated—no double bonds)} & | \\ CH_3CH_2CH_2CH_2CH_2CH{=}CHCH_2CH{=}CHCH_2CH_2CH_2CH_2CH_2CH_2CH_2COO- & C-H \\ \text{(from linoleic acid—two double bonds)} & | \\ \text{a mixed triglyceride} & H \end{array}$$

Oleic acid has a cis configuration around its double bond; linoleic acid has a *cis-cis* configuration.

The preceding formula for a mixed triglyceride can be written in condensed form as shown in structure (21-1).

$$\begin{array}{rl} & H \\ & | \\ CH_3(CH_2)_7CH{=}CH(CH_2)_7COO- & C-H \\ & | \\ CH_3(CH_2)_{16}COO- & C-H \\ & | \\ CH_3(CH_2)_4CH{=}CHCH_2CH{=}CH(CH_2)_7COO- & C-H \\ & | \\ & H \end{array} \qquad (21\text{-}1)$$

[†]According to the International Union of Pure and Applied Chemistry (IUPAC) and the International Union of Biochemistry (IUB), monoglycerides are to be designated as monoacylglycerols, diglycerides as diacylglycerols, and triglycerides as triacylglycerols. However, the glyceride name continues in general use.

or simple as

$$
\begin{array}{c}
\text{H} \\
| \\
\text{C}_{17}\text{H}_{33}\text{COO}-\text{C}-\text{H} \\
| \\
\text{C}_{17}\text{H}_{35}\text{COO}-\text{C}-\text{H} \\
| \\
\text{C}_{17}\text{H}_{31}\text{COO}-\text{C}-\text{H} \\
| \\
\text{H}
\end{array}
\tag{21-2}
$$

A more correct representation of the triglyceride structure (21-2) is

$$
\begin{array}{c}
\text{H} \\
| \\
\text{H}-\text{C}-\text{OOCC}_{17}\text{H}_{33} \\
| \\
\text{C}_{17}\text{H}_{35}\text{COO}-\text{C}-\text{H} \\
| \\
\text{H}-\text{C}-\text{OOCC}_{17}\text{H}_{31} \\
| \\
\text{H}
\end{array}
$$

indicating the L configuration of most naturally occurring triglycerides, but for simplicity we will use structure (21-2).

Iodine Number

Unsaturated fats and oils will readily combine with iodine, whereas saturated fats and oils will not do so readily. The more unsaturated the fat or oil, the more iodine it will react with.

The iodine number of a fat or oil is the number of grams of iodine that will react with the double bonds present in 100 g of that fat or oil. The higher the iodine number, the greater the degree of unsaturation of the fat or oil. The iodine number of some fats and oils are listed in Table 21-2.

In general, animal fats have a lower iodine number than vegetable oils. This indicates that vegetable oils are more unsaturated. This increasing unsaturation is also accompanied by a change of state: animal fats are solid, and vegetable oils are liquid. Fats have iodine numbers below 70, oils above 70.

Animal and vegetable oils should not be confused with mineral oil, which is a mixture of saturated hydrocarbons, or with essential oils, which are volatile aromatic liquids used as flavors and perfumes.

Exercise 21-2

Using Table 21-1, decide which fat or oil is the most unsaturated and the least unsaturated.

Solution

Using the iodide number in the table, we find that corn oil is the most unsaturated, with a number of 116; butter has an iodine number of 26, making it the least unsaturated.

Self-test

Which fat or oil in Table 21-1 has the most linoleic acid? the most myristic acid?

Answer

Cottonseed or corn oil; butter

Use of Fats in the Body

Fats serve as a fuel in the body, producing more energy per gram than either carbohydrate or protein. Metabolism of fat produces 9 kcal/g, whereas the metabolism of either carbohydrate or protein produces 4 kcal/g.

Fats also serve as a reserve supply of food and energy for the body. If a 70 kg person stored energy in the form of carbohydrate rather than fat, that person would weigh an additional 55 kg. Fat is stored in the adipose tissue and serves as a protector for the vital organs; that is, fats surround the vital organs to keep them in place and also act as shock absorbers. Fats in the outer layers of the body act as heat insulators, helping to keep the body warm in cold weather. Fats act as electrical insulators and allow rapid propagation of nerve impulses. The fat content of nerve tissue is particularly high. Fats are a constituent of lipoproteins, which are found in cell membranes and in the mitochondria and also serve as a means of transporting lipids in the bloodstream.

Physical Properties

Pure fats and oils are generally white or yellow solids and liquids, respectively. Pure fats and oils are also odorless and tasteless. However, over a period of time fats become rancid; they develop an unpleasant odor and taste.

Fats and oils are insoluble in water but are soluble in such organic liquids as benzene, acetone, and ether. Fats do not diffuse through a membrane. Fats are lighter than water and have a greasy feeling. Fats and oils form a temporary emulsion when shaken with water. The emulsion can be made permanent by the addition of an emulsifying agent such as soap. Fats and oils must be emulsified by bile in the body before they can be digested.

Chemical Reactions

Hydrolysis   When fats are treated with enzymes, acids, or bases, they hydrolyze to form fatty acids and glycerol. When tripalmitin (glyceryl tripalmitate) is hydrolyzed, it forms palmitic acid and glycerol and requires three molecules of water. Recall that in the formation of a fat, water is a product.

When fats are hydrolyzed to fatty acids and glycerol, the glycerol separates from the fatty acids and can be drawn off and purified. Glycerol is used both medicinally and industrially.

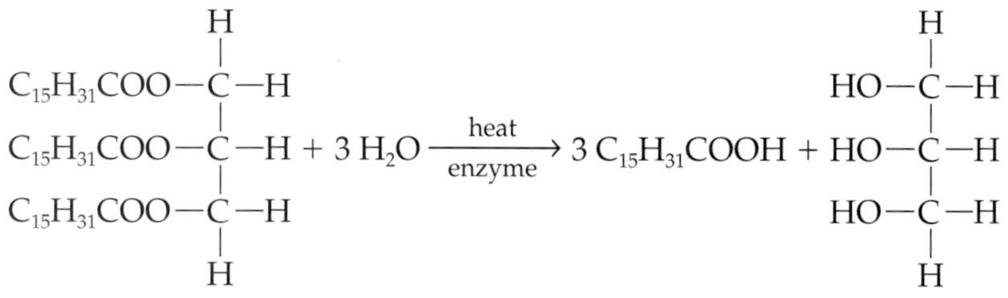

$$C_{15}H_{31}COO-CH_2$$
$$C_{15}H_{31}COO-CH + 3\,H_2O \xrightarrow[\text{enzyme}]{\text{heat}} 3\,C_{15}H_{31}COOH + \text{glycerol}$$
$$C_{15}H_{31}COO-CH_2$$

tripalmitin   palmitic acid   glycerol

Saponification   Saponification is the heating of a fat with a strong base such as sodium hydroxide to produce glycerol and the salt of a fatty acid.

$$
\begin{array}{c}
\text{H} \\
|\\
C_{17}H_{35}COO\text{-}\overset{\displaystyle H}{\underset{\displaystyle H}{C}}\text{—H} \\
C_{17}H_{35}COO\text{-}C\text{—H} + 3\ \text{NaOH} \xrightarrow{\ \text{heat}\ } 3\ C_{17}H_{35}COONa + \\
C_{17}H_{35}COO\text{-}C\text{—H} \\
|\\
\text{H}
\end{array}
\qquad
\begin{array}{c}
\text{H}\\
|\\
\text{HO—C—H}\\
\text{HO—C—H}\\
\text{HO—C—H}\\
|\\
\text{H}
\end{array}
$$

tristearin   sodium stearate,   glycerol
                                a soap

The sodium (or potassium) salt of a fatty acid is called a soap. Reactions and properties of soaps are discussed later in this chapter.

Hydrogenation   Fats and oils are similar compounds except that oils are more unsaturated; that is, oils contain many double bonds. These double bonds can change to single bonds upon the addition of hydrogen. Vegetable oils can be converted to fats by the addition of hydrogen in the presence of a catalyst. This process is called hydrogenation. Hydrogenation is used to produce the so-called vegetable shortenings used in the home. Oleomargarine is prepared by hydrogenating certain fats and oils, then adding flavoring and coloring agents, plus vitamins A and D. Compounds that give butter its characteristic flavor are sometimes added.

$$
\begin{array}{c}
\text{H}\\
|\\
C_{17}H_{33}COO\text{-}C\text{—H}\\
C_{17}H_{33}COO\text{-}C\text{—H} + 3\ H_2 \xrightarrow{\ \text{catalyst}\ }\\
C_{17}H_{33}COO\text{-}C\text{—H}\\
|\\
\text{H}
\end{array}
\qquad
\begin{array}{c}
\text{H}\\
|\\
C_{17}H_{35}COO\text{-}C\text{—H}\\
C_{17}H_{35}COO\text{-}C\text{—H}\\
C_{17}H_{35}COO\text{-}C\text{—H}\\
|\\
\text{H}
\end{array}
$$

triolein, an oil   tristearin, a fat
(contains double bonds)   (contains single bonds)

In actual practice, vegetable oils are not completely hydrogenated. Enough hydrogen is added to produce a solid at room temperature. If the oil were completely hydrogenated, the solid fat would be hard and brittle and unsuitable for cooking purposes.

As should be expected, hydrogenation lowers the iodine number to a value within the range of fats.

Acrolein Test   The acrolein test, which is a test for the presence of glycerol, is sometimes used as a test for fats and oils, since all fats and oils contain glycerol.

When glycerol is heated to a high temperature, especially in the presence of a dehydrating agent such as potassium bisulfate ($KHSO_4$), a product called acrolein results.

$$
\begin{array}{c}
\text{H}\\
|\\
\text{H—C—OH}\\
\text{H—C—OH}\\
\text{H—C—OH}\\
|\\
\text{H}
\end{array}
\xrightarrow[\ KHSO_4\ ]{\ \text{heat}\ }
\begin{array}{c}
\text{H—C}=\text{O}\\
\text{H—C}\\
\ \ \|\\
\text{H—C}\\
|\\
\text{H}
\end{array}
+\ 2\ H_2O
$$

glycerol   acrolein

This substance is easily recognized by its strong, pungent odor. When fats or oils are heated to a high temperature or are burned, the disagreeable odor is that of acrolein.

Rancidity   Fats develop an unpleasant odor and taste when allowed to stand at room temperature for a short period of time. That is, they become rancid. Rancidity is due to two types of reactions—hydrolysis and oxidation.

Oxygen present in the air can oxidize some unsaturated parts of fats and oils. If this oxidation reaction produces short-chain acids or aldehydes, the fat turns rancid, as evidenced by a disagreeable odor and taste. Since oxidation, as well as hydrolysis, takes place more rapidly at higher temperatures, fats and foods containing a high percentage of fats should be stored in a cool place. Oxidation of fats, especially in hydrogenated vegetable compounds, can be inhibited by the addition of antioxidants, substances that prevent oxidation. Two naturally occurring antioxidants are vitamic C and vitamin E.

When butter is allowed to stand at room temperature, hydrolysis takes place between the fats and the water present in the butter. The products of this hydrolysis are fatty acids and glycerol. One of the fatty acids produced, butyric acid, has the disagreeable odor that causes one to say that the butter is rancid. The catalysts necessary for the hydrolysis reaction are produced by the action of microorganisms present in the air acting on the butter. At room temperature this reaction proceeds rapidly so that the butter soon turns rancid. This effect can be overcome by keeping the butter refrigerated and covered.

21-5   Soaps

Soaps are produced by the saponification of fats. Soaps are salts of fatty acids. When the saponifying agent used is sodium hydroxide, a sodium soap is produced. Sodium soaps are bar soaps. When the saponifying agent used is potassium hydroxide, a potassium soap is produced. Potassium soaps are soft or liquid soaps.

Soaps can also be produced by the reaction of a fatty acid with an inorganic base, although this method is much too expensive to be of commercial value.

$$C_{17}H_{35}COOH + NaOH \longrightarrow C_{17}H_{35}COONa + H_2O$$

stearic acid   sodium   sodium stearate,
                    hydroxide   a soap

Various substances can be added to soaps to give them a pleasant color and odor. Floating soaps contain air bubbles. Germicidal soaps contain a germicide. Scouring soaps contain some abrasive. Tincture of green soap is a solution of a potassium soap in alcohol.

Calcium and magnesium ions present in hard water react with soap to form insoluble calcium and magnesium soaps.

$$2\,Na\,soap + Ca^{2+} \longrightarrow Ca\,soap_{(s)} + 2\,Na^+$$

$$2\,Na\,soap + Mg^{2+} \longrightarrow Mg\,soap_{(s)} + 2\,Na^+$$

The soap "precipitate" is mostly organic and floats to the top rather than sinking to the bottom as most precipitates do. This precipitated soap is seen as "the ring around the bathtub." More soap is required to produce a lather in hard water than in soft water (see Section 9-11).

Zinc stearate is an insoluble soap used as a dusting powder for infants. It has antiseptic properties but is irritating to mucous membranes. Zinc undecylenate is used in the treatment of athlete's foot.

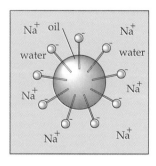

FIGURE 21-1
Soap in an oil-water mixture

Children who suffer from celiac disease cannot absorb fatty acids from the small intestine. The unabsorbed fatty acids combine with calcium ions to form insoluble calcium compounds, or soaps. These calcium compounds are eliminated from the body, and the body will become deficient in calcium unless additional amounts of this necessary element are given to the child.

Cleansing Action

Soaps are cleansing agents. Consider a soap molecule such as sodium stearate.

$$CH_3—(CH_2)_{16}—COONa$$

The long-chain aliphatic part is nonpolar, whereas the carboxylate part is polar. A simplified representation of soap molecule is ———◯, the line representing the nonpolar part and the circle the polar part.

In general, nonpolar compounds dissolve in nonpolar liquids, and polar compounds dissolve in polar liquids (see page 163). If soap is added to a mixture of water and oil and then shaken rapidly, the nonpolar end of the soap molecule will dissolve in the oil, a nonpolar liquid. At the same time, the polar end of the soap molecule will dissolve in the water, a polar liquid. The nonpolar end of the soap molecule is said to be *hydrophobic (water-repelling). The polar end is *hydrophilic (water-loving). The carboxylate end of the soap molecule, which is in the water, yields sodium ions, which are free to move about. Structures such as this are called micelles (Figure 21-1).

Note that the oil drop has a negative charge because of the negative ends of the soap molecules sticking out into the water. This negatively charged oil drop will repel all other oil drops, which will have acquired a like charge. That is, the oil will have become emulsified, with the soap acting as the emulsifying agent.

This is the manner in which soap cleanses, since most dirt is held on skin and clothing by a thin layer of grease or oil. Mechanical washing causes the oil or grease to break up into small drops. the soap then emulsifies that oil or grease, which can then be easily washed away. Soap also acts as a surfactant (page 169); it lowers the surface tension of the water, making emulsification easier.

Soap has little effect as an antibacterial agent. Nurses and surgeons in the operating room scrub for at least 10 minutes to remove most of the debris, such as keratin and natural fats, from the skin. A germicidal soap, one that contains a germ-killing compound, usually is used.

21-6 Detergents

Detergents (syndets) are synthetic compounds used as cleansing agents. They work like soaps but are free of several of the disadvantages that soaps have.

Detergents work as well in hard water as they do in soft water. That is, calcium and magnesium salts of detergents are soluble and do not precipitate out of solution (as do calcium and magnesium soaps). Recall that soaps do not work as well in hard water because insoluble calcium and magnesium salts precipitate out of solution. Detergents are generally neutral compounds compared with soaps, which are usually alkaline or basic substances. Therefore, detergents can be used on silks and woolens but soaps cannot. Detergents are used for washing clothes and also as cleansing agents in toothpastes and toothpowders.

Detergents are sodium salts of long-chain alcohol sulfates. For example, sodium lauryl sulfate can be prepared by treating lauryl alcohol, a 12-carbon alcohol, with sulfuric acid and then neutralizing with sodium hydroxide. The reactions are

$$C_{11}H_{23}CH_2OH + H_2SO_4 \longrightarrow C_{11}H_{23}CH_2OSO_3H + H_2O$$

lauryl alcohol   lauryl hydrogen sulfate

$$C_{11}H_{23}CH_2OSO_3H + NaOH \longrightarrow C_{11}H_{23}CH_2OSO_3Na + H_2O$$

lauryl hydrogen   sodium lauryl sulfate,
sulfate   a detergent

Note that the detergent, like a soap, has a nonpolar part and a polar part.

Detergents containing straight chains are biodegradable and do not cause water pollution, whereas those containing branched chains are nonbiodegradable and cause pollution.

21-7 Waxes

A wax is a compound produced by the reaction of a fatty acid with a high molecular mass monohydric alcohol such a myricyl alcohol ($C_{30}H_{61}OH$) or ceryl alcohol ($C_{26}H_{53}OH$). Carnauba wax is largely $C_{25}H_{31}COOC_{30}H_{61}$, an ester of myricyl alcohol. Beeswax is largely $C_{15}H_{31}COOC_{30}H_{61}$, also an ester of myricyl alcohol.

Note that waxes are primarily esters of long-chain fatty acids with an even number of carbon atoms and long-chain alcohols, also with an even number of carbon atoms. The number of carbon atoms is usually 26 to 34. The alcohol may also be a steroid such as lanosterol. The wax thus produced, lanolin, is widely used in cosmetics and ointments.

Waxes are insoluble in water, nonreactive, and flexible; hence, waxes make excellent protective coatings. Excessive loss of water through the feathers of birds, through the fur of animals, and through the leaves of plants is prevented by the presence of waxes.

Some of the common waxes are listed in Table 21-3.

Paraffin wax is different from these waxes because it is merely a mixture of hydrocarbons and is not an ester.

21-8 Complex Lipids

Phospholipids

Phospholipids are phosphate esters and can be divided into two categories—phosphoglycerides and phosphosphingosides—depending on whether the alcohol is glycerol or sphingosine. As indicated in Section 21-3, phospholipids also contain a nitrogen compound. Phospholipids are found in all tissues in the human body, particularly in brain, liver, spinal tissue, and cell membranes.

Phospholipids also occur in the membranes of all cells. Their peculiar properties are responsible for passage of various substances into and out of the cells.

Table 21-3	Common Waxes	
Name	Source	Use
Beeswax	Honeycomb of bee	Polishes and pharmaceutical products
Spermaceti	Sperm whale	Cosmetics and candles
Carnauba	Carnauba palm	Floor waxes and polishes
Lanolin	Wool	Skin ointments

Consider a phosphoglyceride whose structure can be represented as

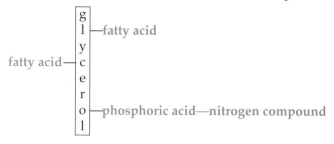

At carbons 1 and 2 of the glycerol there are esters of fatty acids. At carbon 3 there is a phosphate group, which in turn is bonded to a nitrogen compound. There are many different phosphoglycerides, depending on the types of fatty acids bonded to the glycerol and also on the identity of the nitrogen compound bonded to the phosphate group. Most phosphoglycerides have a saturated fatty acid connected at carbon 1 and an unsaturated fatty acid at carbon 2.

The phosphate group and the nitrogen compound are polar substances, whereas the fatty acid molecules are nonpolar.

The fatty acid chains, are *hydrophobic—they point away from water. The other end of the molecule, the one containing the nitrogen compound and phosphoric, acid is *hydrophilic and dissolves in water. Molecules of this type, with a hydrophobic (nonpolar) and a hydrophilic (polar) end, are said to be *amphipathic. Soaps (Section 21-5) are amphipathic, as are phospholipids.

Cell Membranes Cell membranes serve two important functions. (1) They act as a mechanical support to separate the contents of a cell from its external environment, and (2) a structural support for certain proteins that serve to transport ions and polar molecules across the membrane. Some of these proteins act as "gates" and "pumps" to move certain materials through the membrane but exclude others. Other membrane proteins act as "receptor sites" by which molecules outside the cell can send messages inside the cell. An example of such a protein is the hormone insulin, which regulates the metabolism of glucose in certain cells but does not cross membranes itself. Instead, the insulin reacts with the specific receptor protein on the outer surface of the membrane, and that receptor protein in turn communicates the specific message to the inside of the cell.

Cell membranes are composed, on the average, of 40 to 50 percent lipids and 50 to 60 percent protein. However, these percentages vary considerably even among the cells of the same individual, as shown in Table 21-4. In addition, membranes also contain cholesterol and a small amount of carbohydrate. The cell membrane is relatively fluid, as it must be to account for the flexibility of the cell and its deformation without disruption of structure.

How do lipids form a membrane, and where do the proteins fit into that membrane? Recall that phospholipids have a polar end and a nonpolar end. In the marginal diagram, the long lines represent the nonpolar (hydrophobic) end, the fatty acid chains, whereas the circles represent the polar (hydrophilic) end, the end that dissolves in water. Such amphipathic molecules can form a bilayer, as shown at the left.

Such a simple model is not satisfactory for cell membranes because this type of bilayer would be highly impermeable to ions and most polar molecules, which cannot pass through the center of such a system. The current theory of cell membranes involves a phospholipid bilayer in which are embedded proteins, as shown in Figure 21-2.

Table 21-4

Percentages of Lipids in Certain Membranes

Myelin	82
Red blood cells	48
Mitochondria	
Outer membrane	48
Inner membrane	24

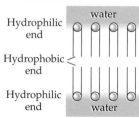

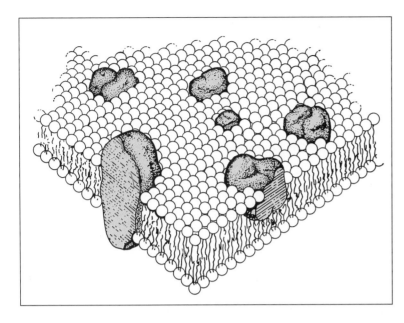

FIGURE 21-2
Fluid-mosaic model of plasma membrane structure.

In this type of membrane, called the fluid mosaic model, some proteins are embedded in the surface of the membrane, others are embedded in the center, and still others provide channels all the way through to transport polar molecules and ions through the membrane. The proteins are believed to be mobile and to move through the bilayer, as well as across its surface.

Phosphoglycerides

One category of phospholipid is the phosphoglycerides, whose general structure is shown below. Phosphoglycerides can in turn be subdivided into several types, depending on the nitrogen compound present. Among these are the lecithins and the cephalins.

glycerol	
g—fatty acid	g—fatty acid
l	l
y	y
fatty acid—c	fatty acid—c
e	e
r	r
o—fatty acid	o—phosphoric acid—N compound
l	l

a fat   a phosphoglyceride

Lecithins Lecithins, now called phosphatidylcholines, are compounds that are particularly important in the metabolism of fats by the liver. In lecithins, the nitrogen compound is choline, an alcohol. Choline is an example of a quaternary ammonium compound (see page 321).

$$HOCH_2—CH_2—\overset{+}{N}—(CH_3)_3\ OH^-$$

choline

A typical formula for a lecithin (phosphatidylcholine) is shown as structure (21-3). In most lecithins, the fatty acid at carbon 1 of the glycerol is saturated, whereas the fatty acid at carbon 2 is unsaturated. The carbon marked with an asterisk is chiral, indicating optical activity. Naturally occurring lecithins have the L form (see page 335).

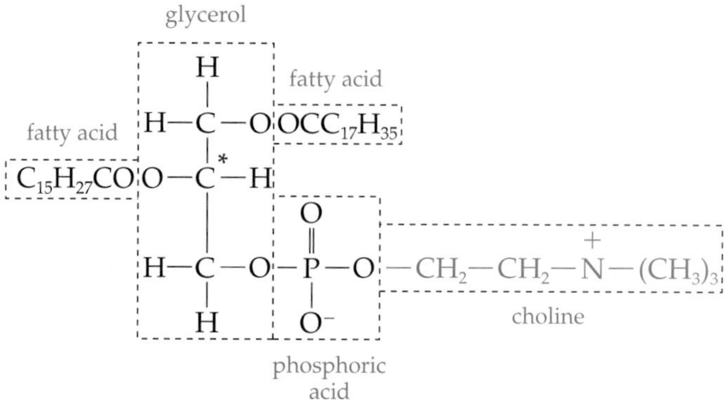

$$
\begin{array}{ll}
\text{glycerol} & \\
\text{fatty acid} \quad H-C-O-OCC_{17}H_{35} \quad \text{fatty acid} & \\
C_{15}H_{27}CO-O-\overset{*}{C}-H & \\
H-C-O-\overset{O}{\underset{O^-}{\overset{\|}{P}}}-O-CH_2-CH_2-\overset{+}{N}-(CH_3)_3 & (21\text{-}3) \\
\quad\quad H & \text{choline} \\
\text{phosphoric acid} &
\end{array}
$$

phosphatidycholine,
a lecithin

Lecithins are insoluble in water but are good emulsifying agents. They are also good sources of phosphoric acid, which is needed for the synthesis of new tissue. Lecithin is abundant in egg yolk and soybeans. It is used commercially as an emulsifying agent in dairy products and in the manufacture of mayonnaise.

Fats are partly converted to lecithins in the body and are transported as lecithins from one part of the body to another. Lecithins are widely distributed in all cells and have both metabolic and structural functions in membranes.

Dipalmitoyl lecithin (lecithin in which the two fatty acids are palmitic acid) is a very good surface active agent (see page 169). It prevents adherence of the inner surfaces of the lungs. The absence of dipalmitoyl lecithin from the lungs of premature infants causes *respiratory distress syndrome*.

Removal of one molecule of fatty acid from lecithin produces lysolecithin. The removal of this molecule of fatty acid is catalyzed by the enzyme lecithinase A, which is found in the venom of poisonous snakes. The venom is poisonous because it produces lysolecithin, which in turn causes hemolysis—the destruction of the red blood cells.

Cephalins Cephalins (21-4) are similar to lecithins except that another nitrogen compound, ethanolamine, is present instead of choline. The newer name for this compound is phosphatidylethanolamine.

$$HO-CH_2-CH_2-NH_2$$
ethanolamine

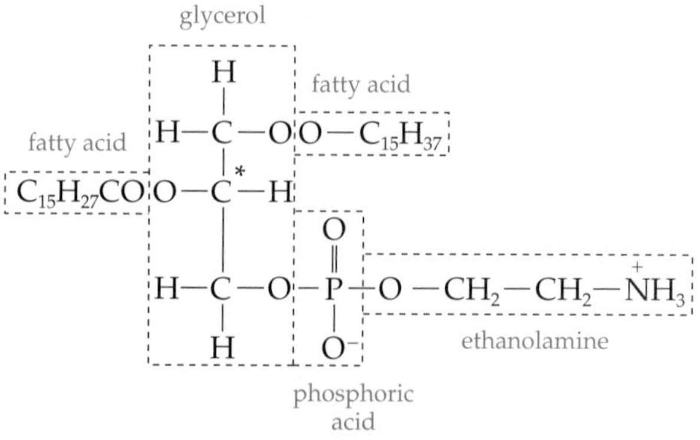

$$
\begin{array}{ll}
\text{glycerol} & \\
\text{fatty acid} \quad H-C-O-O-C_{15}H_{37} \quad \text{fatty acid} & \\
C_{15}H_{27}CO-O-\overset{*}{C}-H & \\
H-C-O-\overset{O}{\underset{O^-}{\overset{\|}{P}}}-O-CH_2-CH_2-\overset{+}{N}H_3 & (21\text{-}4) \\
\quad\quad H & \text{ethanolamine} \\
\text{phosphoric acid} &
\end{array}
$$

a cephalin

Cepahlins are important in the clotting of the blood and also are sources of phosphoric acid for the formation of new tissue.

Similar compounds, phosphatidylserine and phosphatidylinositol, are abundant in brain tissue. In these compounds, serine and inositol, respectively, occur in place of the choline or ethanolamine of (21-3) and (21-4).

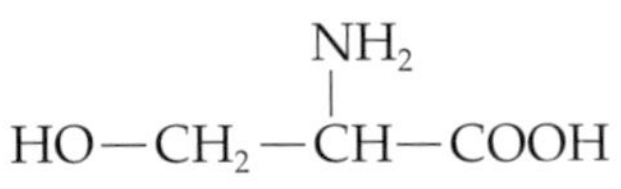

serine 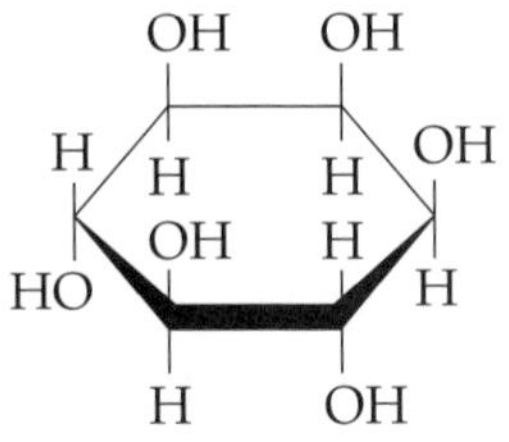   inositol

Plasmalogens   Plasmalogens structurally resemble lecithin and cephalin but have an unsaturated ether at carbon 1 instead of an ester. The fatty acid at carbon 2 of the glycerol is usually unsaturated. Plasmalogens constitute up to 10 percent of the phospholipids found in membranes of brain and muscle cells. The structure of a plasmalogen is

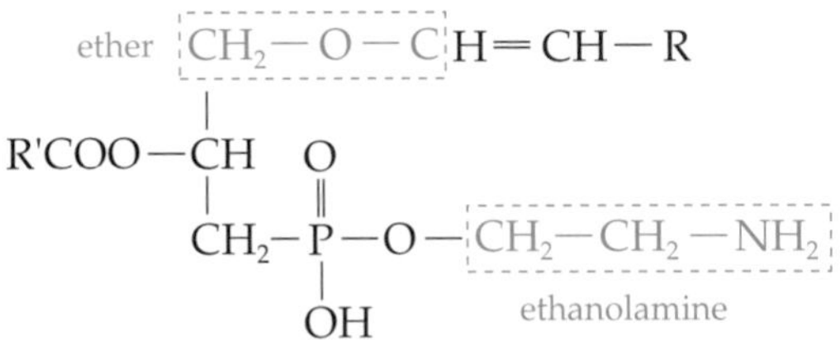

Phosphosphingosides   Phosphosphingosides, also called sphingolipids, differ from phosphoglycerides in that they contain the alcohol sphingosine in place of glycerol. One particular type of sphingolipid, called sphingomyelin, is present in large amounts in brain and nerve tissue. The general formula for a sphingolipid, is given below and the structural formula for sphingomyelin is structure (21-5).

In Niemann-Pick disease, a disease of infancy or early childhood, sphingomyelins accumulate in the brain, liver, and spleen. Accumulation of the sphingomyelins results in mental retardation and early death. It is caused by the lack of a specific enzyme, sphingomyelinase.

a sphingolipid

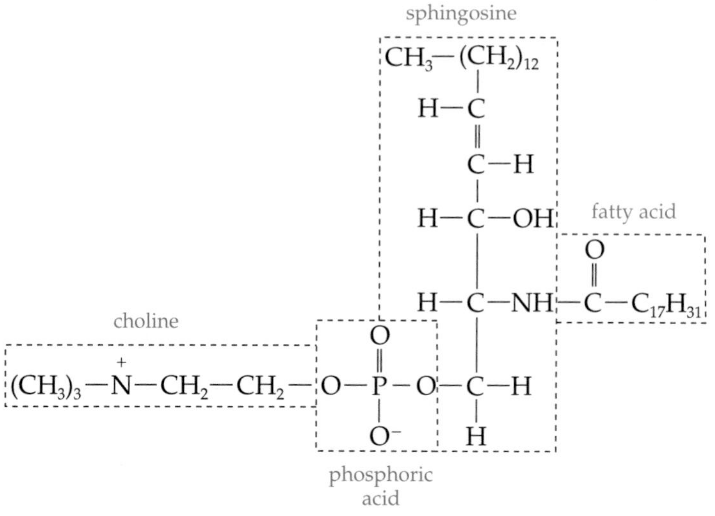

sphingosine

$$CH_3-(CH_2)_{12}$$
$$|$$
$$H-C$$
$$||$$
$$C-H$$
$$|$$
$$H-C-OH \quad \text{fatty acid}$$
$$O$$
$$||$$
$$H-C-NH-C-C_{17}H_{31}$$

choline
$$O$$
$$||$$
$$(CH_3)_3-\overset{+}{N}-CH_2-CH_2-O-P-O-C-H$$
$$|$$
$$O^- \quad H$$

phosphoric
acid

sphingomyelin

(21-5)

Note that the fatty acid in sphingomyelin is bonded to an —NH₂ group rather than to an —OH group as in phosphoglycerides.

In multiple sclerosis, a demyelinating process, there is a loss of phospholipids and sphingolipids. An increased phosphate level occurs in the cerebrospinal fluid.

Glycolipids

Glycolipids are similar to sphingomyelins except that they contain a carbohydrate, often galactose, in place of the choline and phosphoric acid. The general structure for a glycolipid is

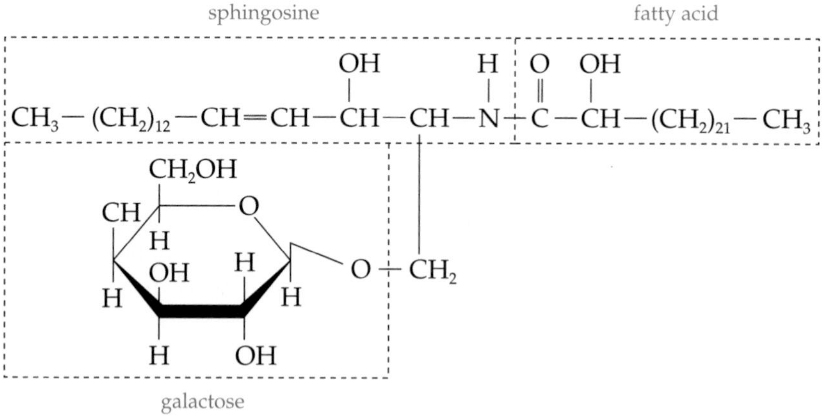

sphingosine   fatty acid

$$\text{OH} \qquad \text{H} \quad \text{O} \quad \text{OH}$$
$$| \qquad | \quad || \quad |$$
$$CH_3-(CH_2)_{12}-CH=CH-CH-CH-N-C-CH-(CH_2)_{21}-CH_3$$

$$CH_2OH$$

galactose

As in sphingomyelins, the fatty acid is bonded to an —NH₂ group on the sphingosine molecule.

Glycolipids produce no phosphoric acid on hydrolysis because they do not contain this compound. Glycolipids are also called cerebrosides because they are found in large amounts in the brain tissue.

Among the glycolipids are kerasin, cerebron, nervon, oxynervon, and the gangliosides. These compounds differ primarily in the identity of the fatty acid attached to the sphingosine.

In Gaucher's disease glycolipids accumulate in the brain and cause severe mental retardation and death by age 3. Juvenile and adult forms of this disease are characterized by enlarged spleen and kidneys, hemorrhaging, mild anemia, and fragile bones. This disease is caused by the lack of a specific enzyme, β-glucosidase.

In the absence of a particular enzyme, hexosaminidase A, glycolipids accumulate in the tissues of the brain and eyes. This effect, called Tay-Sachs disease, is usually fatal to infants before they reach age 2.

21-9 Derived Lipids

Eicosanoids

The eicosanoids are a biologically active group of compounds derived from arachidonic acid. They are extremely potent compounds with a variety of actions, as will be discussed in the following paragraphs. Among the eicosanoids are the prostaglandins, the thromboxanes, prostacyclin, and the leukotrienes.

The Prostaglandins   The prostaglandins consist of 20-carbon unsaturated fatty acids containing a five-membered ring and two side chains. One side chain has seven carbon atoms and ends with an acid group (COOH). The other chain contains eight carbon atoms with an —OH group on the third carbon from the ring (see structure given). The E series of prostaglandins has, in addition to four chiral carbon atoms, a *trans-* configuration.

Prostaglandins are derived from arachidonic acid, which is formed from the nutritionally essential fatty acid linoleic acid. The structures of arachidonic acid and prostaglandin E_1 (PGE_1) are

arachidonic acid   prostaglandin E_1 (PGE_1)

The abbreviation PGE_1 refers to prostaglandin E with one double bond. Likewise, PGE_2 refers to prostaglandin E with two double bonds.

Prostaglandins have been isolated from most mammalian tissues, including the male and female reproductive systems, liver, kidneys, pancreas, heart, lungs, brain, and intestines. The richest source of prostaglandins is human seminal fluid.

Prostaglandins have a wide range of physiologic effects. They seem to be involved in the body's natural defenses against all forms of change including those induced by chemical, mechanical, physiologic, and pathologic stimuli. Aspirin and other anti-inflammatory drugs appear to partially operate by inhibiting prostaglandin synthesis. Prostaglandins are involved at the cellular level in regulating many body functions, including gastric acid secretion, contraction and relaxation of smooth muscles, inflammation and vascular permeability, body temperature, and blood platelet aggregation. Prostaglandins stimulate steroid production by the adrenal glands and also stimulate the release of insulin from the pancreas. Prostaglandins markedly stimulate the movement of calcium ions from bone. Excessive production of prostaglandins by malignant tissue may provide a partial answer for the *hypercalcemia and *osteolysis observed in patients afflicted with such a condition.

Prostaglandins have also been used clinically to induce abortion or to induce labor in a term prognancy, to treat hypertension, to relieve bronchial asthma, and to heal peptic ulcers.

Prostaglandins increase cyclic adenosine monophosphate (cAMP) (see page 576) in blood platelets, thyroid, corpus luteum, adenohypophysis, and lungs but decrease cAMP in adipose tissue.

Prostaglandin E_1 (PGE$_1$) is now used to strengthen babies born with cyanotic congenital heart disease ("blue babies") to prepare them for corrective surgery.

Prostacyclin and Thromboxanes   Prostacyclin is so called because it contains a second five-membered ring in addition to the ring found in prostaglandins. Thromboxanes have a cyclopentane ring interrupted by an oxygen atom. The structures of prostacyclin and thromboxane B$_2$ (TXB$_2$) are shown below.

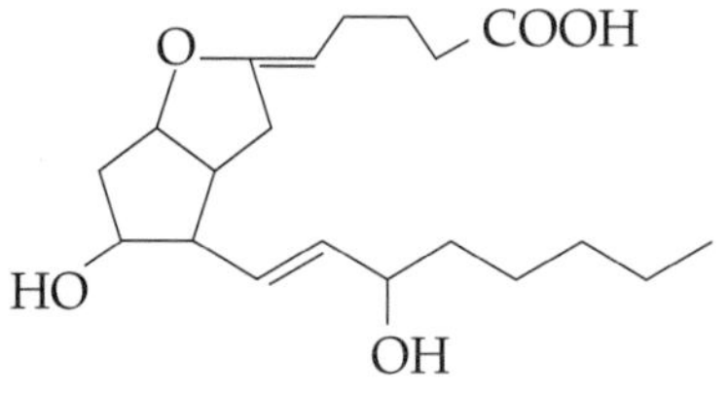

prostacyclin 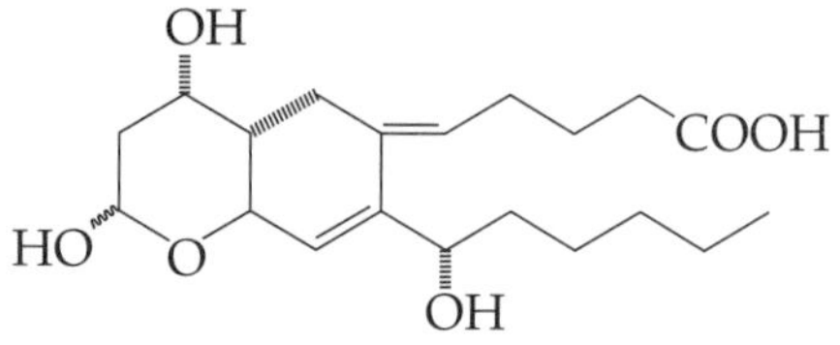   TXB$_2$

Prostacyclin is a potent inhibitor of platelet aggregation and is a powerful vasodilator.

Thromboxanes have an effect opposite to that of prostacyclin. They are potent aggregators of blood platelets and have a profound contractive effect on a variety of smooth muscles.

Thromboxanes function in conjunction with prostacyclin in maintaining a healthy vascular system. Both substances exert their influence by regulating the production of cAMP, with the thromboxanes acting as inhibitors and prostacyclin as a stimulator of cAMP production.

Leukotrienes   Leukotrienes are another group of eicosanoids derived from arachidonic acid. The *tri* refers to three alternate sets of double bonds in the molecule. One of this group, leukotriene C, is involved in the body's allergic responses. It constricts air passages to the bronchi during an asthma attack. The structure of leukotriene C is

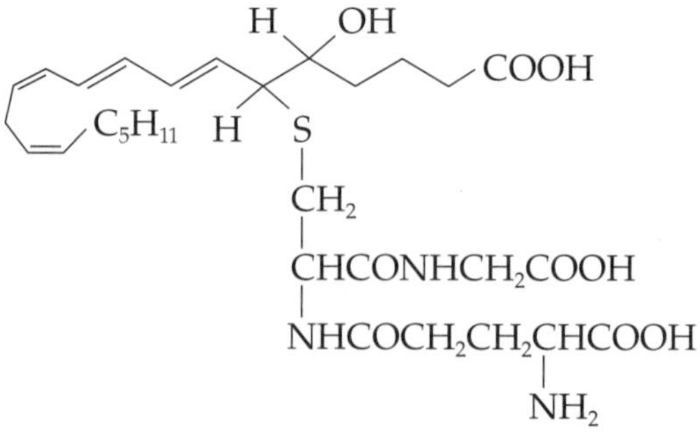

leukotriene C

Steroids

Steroids are high molecular mass tetracyclic (four-ring) compounds. Those containing one or more —OH groups and no C=O groups are called sterols. The most common sterol is cholesterol, which is found in animal fats but not in plant fats. Cholesterol is found in all animal tissues, particularly in brain and nervous tissue, in the bloodstream, and as gallstones. Cholesterol aids in the absorption of fatty acids from the small intestine.

Most of the body's cholesterol is derived or synthesized from other substances such as carbohydrates and proteins, as well as from fats. The rest comes from the diet (see Table 21-5).

Table 21-5 Cholesterol Values of Some Foods		
Food	Amount/Serving	Cholesterol/mg
Butter	1 tbsp	30
Margarine	1 tbsp	0
Skim milk	8 oz	5
Whole milk	8 oz	35
Chicken (no skin)	3 oz	75
Chicken (fried)	3 oz	111
Hamburger	3 oz	85
Liver (beef)	3 oz	375

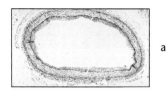

a

b

FIGURE 21-3
(a) Normal artery; (b) blood vessel with atherosclerosis.

Atherosclerosis, a form of arteriosclerosis, results from the deposition of excess lipids, primarily triglycerides and cholesterol, from the bloodstream (see Figure 21-3). Of these two, cholesterol poses a greater threat to the well-being of a person, although excess triglycerides also present a significant risk. One way of combating heart disease and atherosclerosis is to reduce the concentration of lipids in the bloodstream—either by reducing lipid intake or by the use of antihyperlipidemic drugs, those that tend to reduce blood lipid levels. It has been found that certain unsaturated fish and vegetable oils, when substituted for saturated fats, lower the serum cholesterol level.

Cholesterol levels in humans should be in the range of 200 to 220 mg/dL, with slightly higher levels being normal for older individuals. Elevated cholesterol levels should be controlled, usually by diet. In extreme cases, cholesterol-lowering drugs such as pravastatin or lovastatin may be prescribed.

Cholesterol does not occur in plants.

Ergosterol is a sterol similar to cholesterol. When ergosterol is irradiated (exposed to radiation) with ultraviolet light, one of the products formed is calciferol (vitamin D_2, see page 556).

Other steroids include bile salts, the sex hormones, and the hormones of the adrenal cortex. The similarities of the structure of some of these steroids are indicated in the following structures:

cholesterol

estradiol

testosterone

$$CH_3$$
$$|$$
$$HC-CH=CH-CH-CH$$

ergosterol

21-10 Anabolic Steroids

Anabolic steroids are hormones that control the synthesis of larger molecules from smaller ones. Athletes have used these substances (even though they are illegal) to increase muscle mass and, hence, body strength. An example of such an anabolic steroid is the male hormone testosterone. While it does increase muscle mass, it has several undesirable side effects. In men these side effects include testicular atrophy, impotence, hypercholesterolemia, breast growth, and liver cancer. Women using anabolic steroids will develop larger muscles and greater strength at the expense of increased masculinity, formation of a greater amount of body hair, deepening of the voice, and menstrual irregularities. Another drawback of the use of such anabolic steroids is that they cannot be taken orally; they must be injected. It is standard practice today during athletic competitions to routinely test an athlete's urine for the presence of these illegal substances.

Summary

Lipids yield fatty acids on hydrolysis or combine with fatty acids to form esters. Lipids are insoluble in water but are soluble in organic solvents such as ether, acetone, and carbon tetrachloride.

Lipids can be classified into three types: simple, complex, and derived. Simple lipids are esters of fatty acids. A simple lipid that yields fatty acids and glycerol upon hydrolysis is called a fat or oil. A simple lipid that, on hydrolysis, yields fatty acids and a high molecular mass alcohol is called a wax.

Complex lipids yield fatty acids, alcohol, and some other type of compound on hydrolysis.

Precursor lipids are compounds produced when simple or compound lipids undergo hydrolysis. Derived lipids are formed by metabolic transformation of fatty acids. Fatty acids are straight-chain organic acids. Those found in nature usually contain an even number of carbon atoms. Saturated fatty acids have only single bonds between carbon atoms. Unsaturated fatty acids have one or more double bonds in the molecule and occur in nature in the *cis* form.

The nutritionally essential fatty acid, so called because it is necessary in the diet, is linoleic acid.

Unsaturated fats and oils react with iodine, whereas saturated ones do not. The iodine number of a fat or oil is the number of grams of iodine that will react with (the double bonds present in) 100 g of that fat or oil.

Fats serve as fuel for the body—1 g of fat produces 9 kcal, compared with only 4 kcal/g of carbohydrate. Fats protect nerve endings and also act as insulators to keep the body warm in cold weather.

Fats and oils are odorless and tasteless when pure. They are insoluble in water but are soluble in organic solvents. Fats and oils must be emulsified before being digested.

When fats are hydrolyzed, they form fatty acids and glycerol. When fats are saponified, they form salts of fatty acids (soaps) and glycerol.

When oils are hydrogenated, the double bonds are changed to single bonds, and the (liquid) oil becomes a (solid) fat.

When fats or oils are heated to a high temperature, especially in the presence of a dehydrating agent, a product known as acrolein is produced. The odor of burning fat or oil is due to the presence of acrolein.

When fats are allowed to stand at room temperature, they become rancid because of hydrolysis and oxidation. Keeping fats cool prevents their becoming rancid.

Soaps are salts of fatty acids—sodium soaps are solid or bar soaps and potassium soaps are soft or liquid soaps.

Calcium and magnesium soaps are insoluble in water and are formed when sodium or potassium soaps are used in hard water. The precipitated calcium and magnesium soaps are seen as the "ring around the bathtub."

Detergents are similar to soaps in their cleansing properties. However, detergents do not precipitate in hard water because their calcium and magnesium compounds are soluble.

A wax is a compound produced by the reaction of a fatty acid with a high molecular mass monohydric alcohol.

Phospholipids contain fatty acids, an alcohol, a nitrogen compound, and phosphoric acid. Two types of phospholipids are phosphoglycerides and phosphosphingosides.

Phosphoglycerides include phosphatidylcholines (lecithins) and phosphatidylethanolamines (cephalins). Lecithins are important in the metabolism of fats by the liver and also are a source of phosphoric acid, which is needed for the formation of new tissue. Cephalins are important in clotting of the blood and also are a source of phosphoric acid for the formation of new tissue.

Sphingomyelins are an example of phosphosphingosides and are present in large amounts in brain and nerve tissue.

Glycolipids, also called cerebrosides, are found in large amounts in brain tissue.

Phospholipids are found in all cell membranes as bilayers and are responsible for the passage of various substances into and out of the cells.

The eicosanoids are a biologically active group of compounds derived from arachidonic acid. Among the eicosanoids are the prostaglandins, the thromboxanes, prostacyclin, and leukotrienes.

Steroids are high molecular mass, four-ring compounds. Those containing —OH groups are called sterols. The most common sterol in the body is cholesterol. Other steroids are vitamin D, bile salts, sex hormones, and hormones of the adrenal cortex.

Questions and Problems

1. Compare the properties of lipids with those of carbohydrates.

2. What are simple lipids? complex lipids? precursor lipids? Give an example of each.

3. What is a fatty acid? What is the difference between saturated and unsaturated fatty acids?

4. What is the difference between a fat and an oil? between a fat and a wax?

5. Do fatty acids in the body contain (usually) an even or an odd number of carbon atoms? Why?

6. Which is the nutritionally essential fatty acid? Why is it important?

7. What determines the iodine number of a fat or oil? Which have higher iodine numbers, fats or oils?

8. Draw the structure of the compound formed by the reaction of one molecule of glycerol with three molecules of palmitic acid; one molecule each of stearic, palmitic, and oleic acid.

9. What products are formed when a fat is hydrolyzed? saponified?

10. What is the difference between an unsaturated fatty acid and a polyunsaturated one?

11. How can an unsaturated fatty acid be changed to a saturated one? Of what use is this process?

12. Describe the test for the presence of a fat or oil.

13. Compare soaps and detergents on the basis of structure; on their reactions with hard water.

14. What causes rancidity in fat or oil? How can it be prevented?

15. Which types of soap are germicidal or have germicidal properties? Explain.

16. List the types of phospholipids and give an example of each.
17. What causes celiac disease?
18. Compare the structure of a phospholipid with that of a glycolipid.
19. List three phosphoglycerides and give a function of each.
20. What is meant by the term *amphipathic*?
21. How do phospholipids help control the passage of materials into or out of the cell?
22. What causes Gaucher's disease? Neiman-Pick disease? respiratory distress syndrome? Tay-Sachs disease?
23. What causes atherosclerosis and what might be done to prevent it?

24. Name several steroids in the body and indicate the function of each.
25. What purpose does cholesterol serve in the body? What can be done to reduce serum cholesterol?
26. Compare the terms *hydrophobic* and *hydrophilic*.
27. Are waxes important biologically? Why or why not?
28. How does the fluid mosaic model account for the movement of polar substances through a membrane?
29. What is prostacyclin? PGE$_1$? thromboxane? What function does each serve?
30. From what compound are the eicosanoids derived? List several eicosanoids.

Practice Test

1. Which of the following is *not* a complex lipid?
   a. lecithin
   b. sphingomyelin
   c. cephalin
   d. prostaglandin
2. Which substance can be used to convert a fatty acid into a soap?
   a. CaCl$_2$
   b. O$_2$
   c. H$_2$
   d. NaOH
3. An example of a steroid is _____.
   a. arachidonic acid
   b. sphingomyelin
   c. cholesterol
   d. thromboxane
4. Which of the following is an unsaturated fatty acid?
   a. stearic
   b. lauric
   c. linoleic
   d. palmitic
5. Molecules with polar and nonpolar ends are called _____.
   a. isomers
   b. dipolar
   c. amphoteric
   d. amphipathic
6. A lipid necessary for blood clotting is _____.
   a. cephalin
   b. lecithin
   c. sphingomyelin
   d. leukotriene

7. Fats and oils contain which of the following?
   a. a phosphate group
   b. lecithin
   c. glycerol
   d. cholesterol
8. Oils have more _____ than fats.
   a. O$_2$
   b. CO$_2$
   c. double bonds
   d. hydrogen bonds
9. The higher the iodine number of a fat or oil, the greater the degree of _____.
   a. rancidity
   b. acidity
   c. unsaturation
   d. hydrolysis
10. Lipids _____.
    a. yield fatty acids upon hydrolysis
    b. are insoluble in water
    c. are soluble in organic solvents
    d. all of these

22

Proteins

Objectives

- To understand the sources and the needs for protein

- To become familiar with the structures and reactions of amino acids

- To indicate the classification of proteins according to solubility, composition function, and shape

- To understand the significance of protein denaturation

- To draw the structures of dipeptides

- To describe the primary, secondary, tertiary, and quaternary structure of a protein

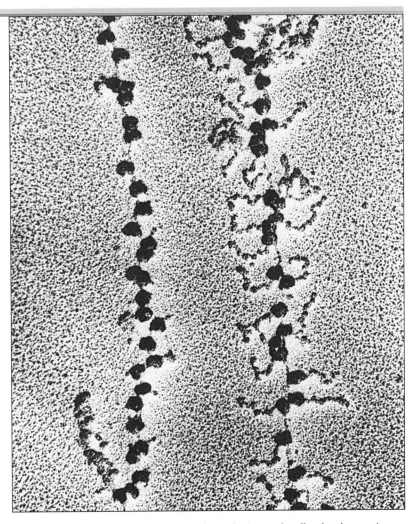

Proteins make major structural and functional contributions to the cell and to the organism as a whole. Because of the continual use and degradation of existing protein, new proteins must be synthesized almost constantly. In this photograph, proteins (squiggly lines) are being produced by ribosomes (large black dots), which assemble them from precursors according to specific instructions supplied by the messenger RNA molecules (straight vertical lines) along which they move.

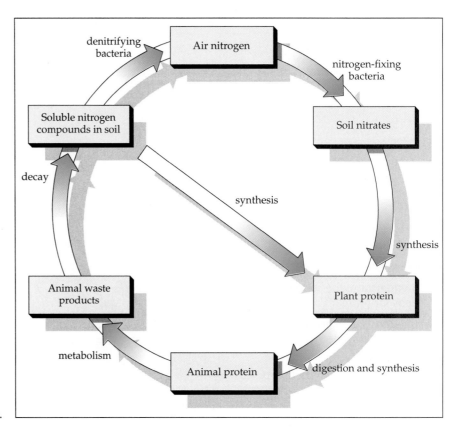

FIGURE 22-1
The nitrogen cycle.

Other than water, proteins are the chief constituents of all cells of the body. Proteins are much more complex than either carbohydrates or fats. All proteins contain the elements carbon, hydrogen, oxygen, and nitrogen. Most proteins also contain sulfur, some contain phosphorus, and a few, such as hemoglobin, contain some other element.

22-1 Sources

Plants synthesize proteins from inorganic substances present in the air and in the soil. Animals cannot synthesize proteins from such materials. Animals must obtain proteins from plants or from other animals who in turn have obtained them from plants.

Animals excrete waste materials containing many nitrogen compounds. These nitrogen compounds along with decaying animal and plant matter are converted into soluble nitrogen compounds by soil bacteria. Plants in turn use these soluble nitrogen compounds to manufacture more protein, thus completing a cycle. A simplified version of the nitrogen cycle is shown in Figure 22-1. As the figure shows, another part of the cycle involves bacteria and gaseous nitrogen in the air.

22-2 Functions

The word protein is derived from the Greek word *proteios*, which means "of first importance." Proteins function in the body in the building of new cells, the maintenance of existing cells, and the replacement of old cells. Thus, proteins are the most important type of compound in the body. Proteins are also a valuable source of energy in the body. The oxidation of 1 g of protein yields 4 kcal—just as does the oxidation of 1 g of carbohydrate.

Proteins are involved in the regulation of metabolic processes (hormones), in the catalysis of biochemical reactions (enzymes), in the transportation of oxygen (hemoglobin), in the body's defense against infection (antibodies), in the transmission of impulses (nerves), in the transmission of hereditary characteristics (nucleoprotein), and in muscular activity (contraction). Proteins are components of skin, hair, and nails, as well as connecting and supporting tissue.

22-3 Molecular Masses

Proteins have very high molecular masses. A comparison of the molecular masses of proteins with those of carbohydrates and fats can be seen in Table 22-1.

22-4 Amino Acids

Proteins are polymers built up from simple units called amino acids. Hydrolysis of proteins yields amino acids. There are 20 known amino acids that can be produced by the hydrolysis of protein.[†] All these amino acids, except glycine, which has no chiral carbon, have the L configuration. Compare structures of L-amino acid and L-glyceraldehyde in structure (22-1). Certain microorganisms can prepare D-amino acids, which are used as antibiotics.

Composition

An amino acid is an organic acid that has an amine ($—NH_2$) group attached to a chain containing an acid group. Although the amine group can be anywhere on the chain, amino acids found in nature usually have the amine group on the alpha (α)

[†]γ-Aminobutyric acid (GABA) is an amino acid that is not incorporated into protein. It is manufactured almost exclusively in the brain and spinal cord and functions as an inhibitory transmitter. A specific deficiency of GABA in the brain occurs in Huntington's *chorea, an inherited neurologic *syndrome characterized by uncontrollable body movements.

Table 22-1 Molecular Masses of Various Proteins, Carbohydrates, and Fats

Type of Compounds	Molecular Mass	Type of Compounds	Molecular Mass
Inorganic compounds		Proteins	
Water	18	Insulin	12,000–48,000
Sodium chloride	58.5	Lactalbumin	17,500
Plaster of paris	290	Hemoglobin	68,000
Organic compounds		Serum globulin	180,000
Benzene	78	Fibrinogen	450,000
Ethyl alcohol	46	Thyroglobin	630,000
Carbohydrates		Hemocyanin	9,000,000
Glucose	180	Tobacco mosaic virus	59,000,000
Sucrose	342		
Lipids			
Tristearin	891		
Cholesterol	384		

carbon—that is, the carbon atom next to the acid group. (The second carbon from the acid group is the [β] carbon; then come the gamma [γ] and delta [δ] carbons.) α-Amino acids can be represented by the general formula in (22-1), where R can be many different groups.

$$
\begin{array}{cc}
\text{COOH} & \text{CHO} \\
| & | \\
\text{H}_2\text{N}-\text{C}-\text{H} \qquad & \text{HO}-\text{C}-\text{H} \\
| & | \\
\text{R} & \text{CH}_2\text{OH} \\
\text{\small L-amino acid} & \text{\small L-glyceraldehyde}
\end{array}
\qquad (22\text{-}1)
$$

Amino acids, can be divided into two groups, polar and nonpolar, depending on the polarity of the R group attached to the α carbon. If the R group is nonpolar, then the amino acid will be less soluble in water than one containing a polar group.

An R group that is polar, such as —OH, —SH, —NH$_2$, or —COOH, produces an amino acid that is polar. Such amino acids are soluble in water. Table 22-2 lists the polar and nonpolar amino acids.

The body can synthesize some, but not all, of the amino acids it needs. Those that it cannot synthesize must be supplied from food consumed. These are called the nutritionally essential amino acids; they are listed in Table 22-3 along with the daily requirements per kilogram of body weight.

Amphoteric Nature

Amino acids contain the —COOH group, which is acidic, and the —NH$_2$ group, which is basic (see Section 19-3). In solution, the carboxyl group can donate a hydrogen ion to the amino group, forming a dipolar ion, called a zwitterion.

$$
\begin{array}{ccc}
\text{R}-\text{CH}-\text{COOH} & \longrightarrow & \text{R}-\text{CH}-\text{COO}^- \\
| & & | \\
\text{NH}_2 & & \text{NH}_3{}^+ \\
\text{\small amino acid} & & \text{\small zwitterion form of an amino acid}
\end{array}
$$

Amino acids are amphoteric compounds; that is, they can react with either acids or bases. When an amino acid is placed in a basic solution, it forms a negatively charged ion that will be attracted toward a positively charged electrode. In an acid solution, the amino acid forms a positively charged ion that will be attracted toward a negatively charged electrode.

$$
\begin{array}{ccccc}
\text{R}-\text{CH}-\text{COOH} & \underset{\text{H}^+}{\rightleftharpoons} & \text{R}-\text{CH}-\text{COO}^- & \overset{\text{OH}^-}{\rightleftharpoons} & \text{R}-\text{CH}-\text{COO}^- \\
| & & | & & | \\
\text{NH}_3{}^+ & & \text{NH}_3{}^+ & & \text{NH}_2 \\
\text{\small positively charged ion} & & \text{\small zwitterion} & & \text{\small negatively charged ion} \\
\text{\small (in acid solution)} & & & & \text{\small (in basic solution)}
\end{array}
$$

Since amino acids are *amphoteric, proteins, which are made up of amino acids, are also amphoteric. This amphoteric nature of proteins accounts for their ability to act as buffers in the blood; they can react with either acids or bases to prevent an excess of either.

At a certain pH (that is, a certain hydrogen ion concentration) amino acids will not migrate toward either the positive or the negative electrode. At this pH, amino acids will be neutral; there will be an equal number of positive and negative ions. This point is called the *isoelectric point (see Table 22-2).

Proteins, which are composed of amino acids, also have an isoelectric point, which is different for each protein. At its isoelectric point, a protein has a minimum solubility, a minimum viscosity, and also a minimum osmotic pressure. At a pH

Text continued on p. 386.

Table 22-2 Nonpolar and Polar Amino Acids

Structure	Name	IUPAC Abbreviation	Isoelectric Point
Nonpolar			
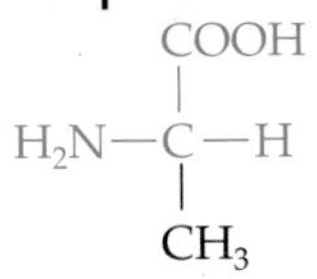	Alanine	Ala	6.00
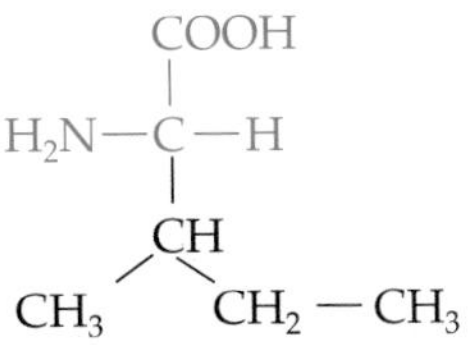	Isoleucine	Ile	6.02
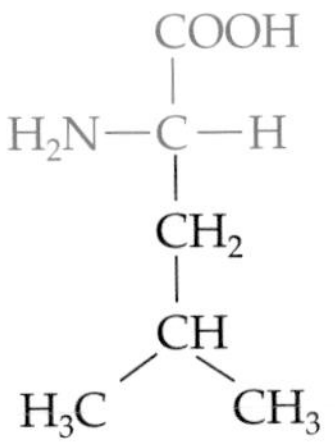	Leucine	Leu	5.98
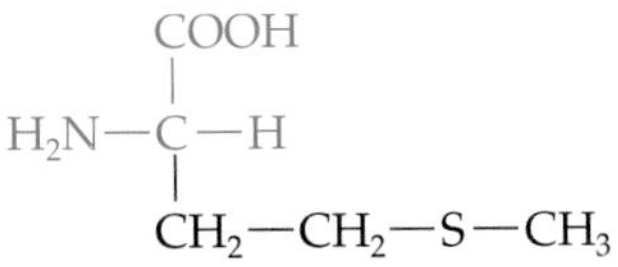	Methionine	Met	5.74
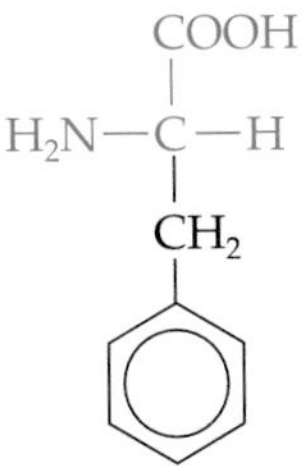	Phenylalanine	Phe	5.48
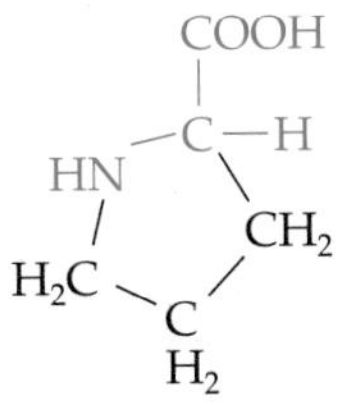	Proline	Pro	6.30
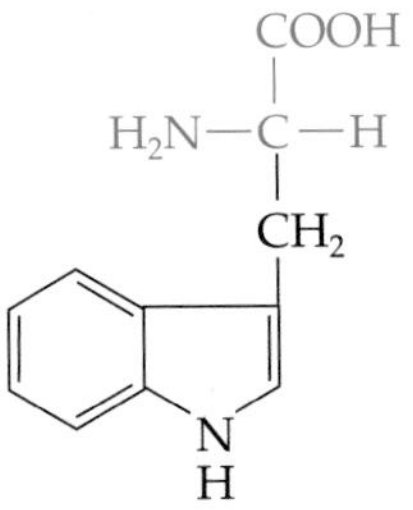	Tryptophan	Trp	5.89

Continued.

Table 22-2 Nonpolar and Polar Amino Acids—cont'd

Structure	Name	IUPAC Abbreviation	Isoelectric Point
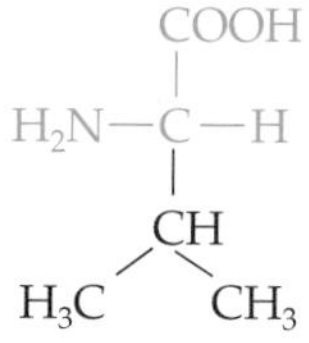 COOH H₂N—C—H CH H₃C CH₃	Valine	Val	5.96

Polar

Structure	Name	IUPAC Abbreviation	Isoelectric Point
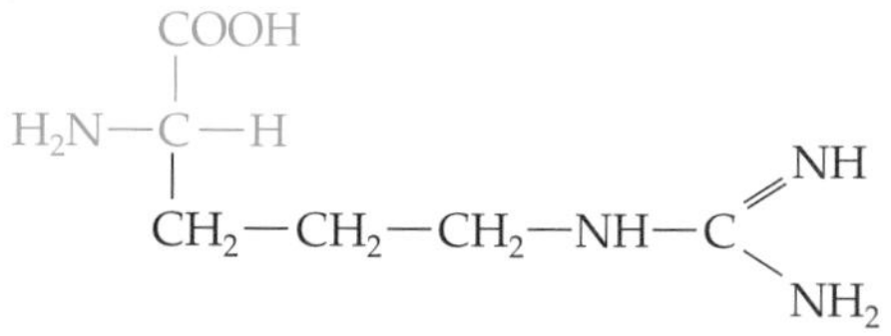 COOH H₂N—C—H CH₂—CH₂—CH₂—NH—C(=NH)(NH₂)	Arginine	Arg	10.76
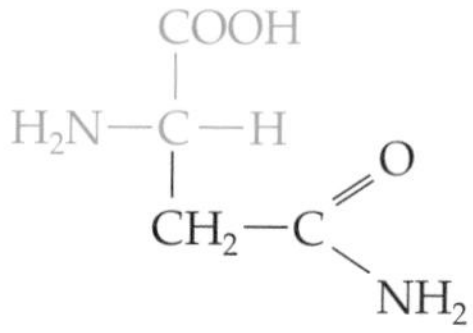 COOH H₂N—C—H CH₂—C(=O)(NH₂)	Asparagine	Asn	5.41
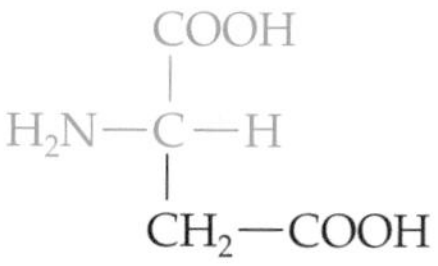 COOH H₂N—C—H CH₂—COOH	Aspartic acid	Asp	2.97
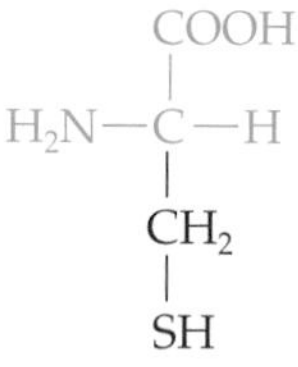 COOH H₂N—C—H CH₂ SH	Cysteine	Cys	5.07
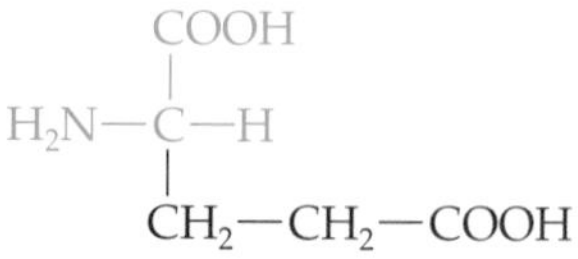 COOH H₂N—C—H CH₂—CH₂—COOH	Glutamic acid	Glu	3.22
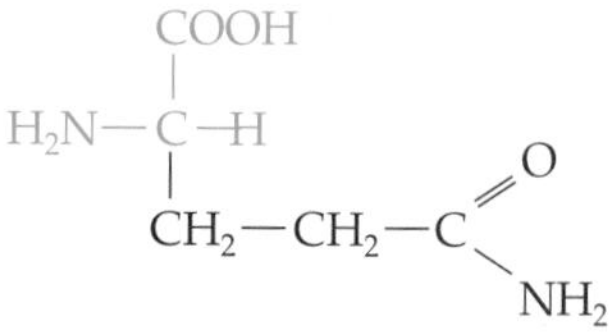 COOH H₂N—C—H CH₂—CH₂—C(=O)(NH₂)	Glutamine	Gln	5.65
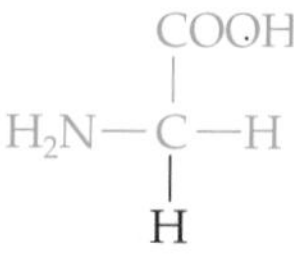 COOH H₂N—C—H H	Glycine	Gly	5.97

Table 22-2 Nonpolar and Polar Amino Acids—cont'd

Structure	Name	IUPAC Abbreviation	Isoelectric Point
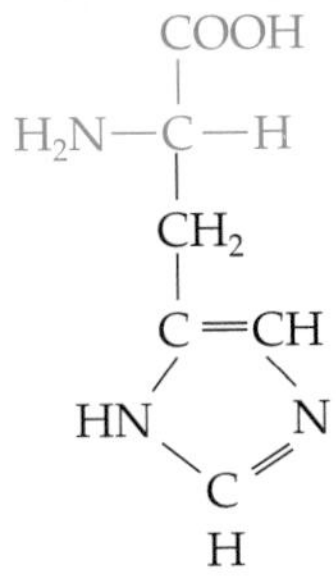 (histidine structure)	Histidine	His	7.59
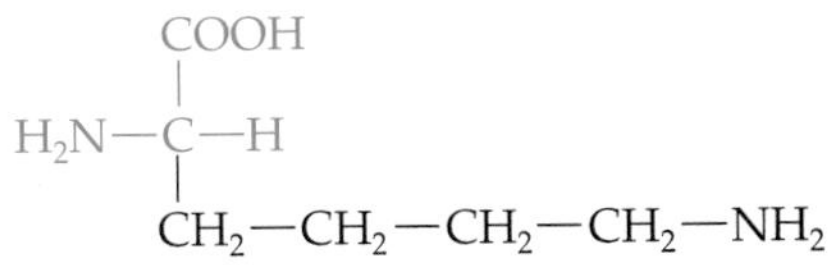 (lysine structure)	Lysine	Lys	9.74
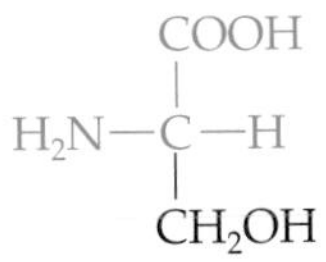 (serine structure)	Serine	Ser	5.68
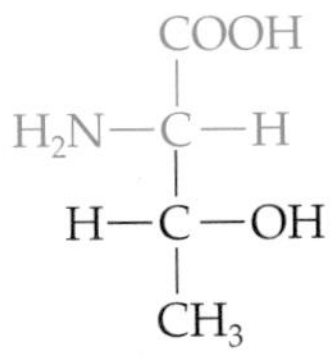 (threonine structure)	Threonine	Thr	5.60
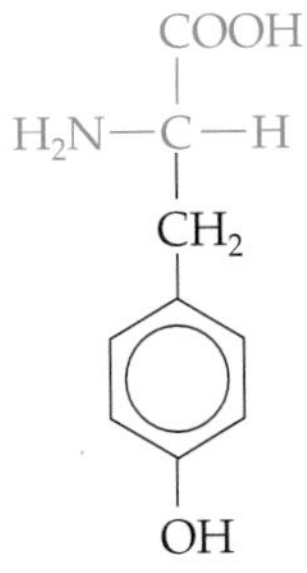 (tyrosine structure)	Tyrosine	Tyr	5.66

Table 22-3 Essential Amino Acids and Their Daily Requirements in Milligrams per Kilogram of Body Weight

	Adult	Infant		Adult	Infant
Isoleucine	28	70	Threonine	28	87
Leucine	42	161	Tryptophan	33	12
Lysine	44	103	Valine	35	93
Methionine	22	58	Histidine		28
Phenylalanine	22	135			

Table 22-4

Isoelectric Points of
Some Proteins

Protein	Isoelectric Point (pH)
Egg albumin	4.7
Casein	4.6
Hemoglobin	6.7
Insulin	5.3
Serum globulin in blood	5.4
Fibrinogen in blood	5.6

above the isoelectric point, a protein has more negative than positive charges. At a pH below the isoelectric point, a protein has more positive than negative charges. The isoelectric points of a few proteins are listed in Table 22-4.

Exercise 22-1

Draw the dipolar structure for serine.

Solution

The structure of serine is

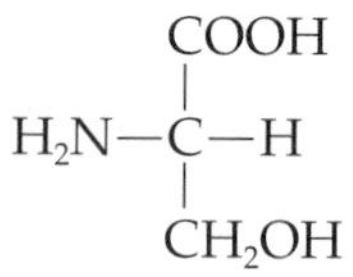

$$
\begin{array}{c}
COOH \\
| \\
H_2N-C-H \\
| \\
CH_2OH
\end{array}
$$

The dipolar structure for serine would be drawn as
$$
\begin{array}{c}
COO^- \\
| \\
H_3\overset{+}{N}-C-H \\
| \\
CH_2OH
\end{array}
$$

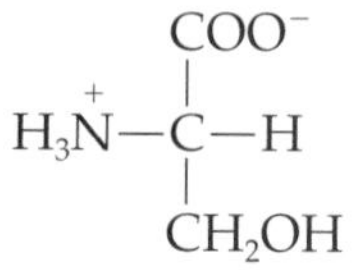

Self-test

Draw the dipolar structure for proline.

Answer

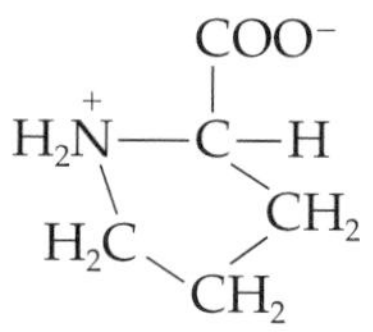

$$
\begin{array}{c}
COO^- \\
| \\
H_2\overset{+}{N}-C-H \\
\diagdown \quad | \\
\quad CH_2 \\
H_2C \quad | \\
\diagdown CH_2
\end{array}
$$

22-5 Dipeptides

Proteins consist of many amino acids joined together by what is called a **peptide linkage** or a **peptide bond** (see page 325). Suppose that a glycine molecule reacts with an alanine molecule. This reaction can occur in two different ways. The amine part of the glycine may react with the acid part of the alanine (equation [22-2]) or the acid part of the glycine may react with the amine part of the alanine (equation [22-3]).

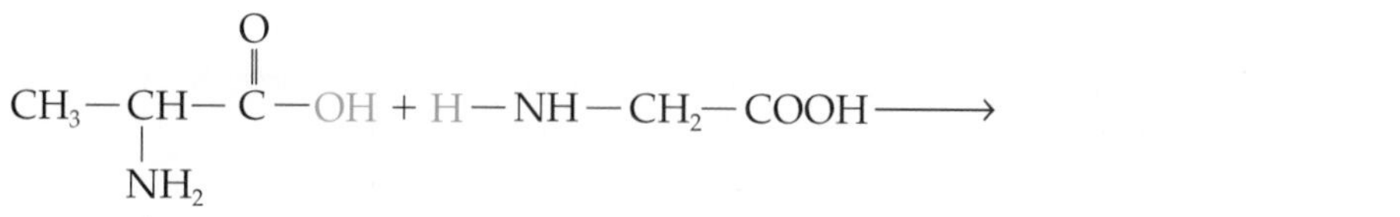

$$
\underset{\text{alanine}}{CH_3-\underset{\underset{NH_2}{|}}{CH}-\overset{\overset{O}{\|}}{C}-OH} + \underset{\text{glycine}}{H-NH-CH_2-COOH} \longrightarrow
$$

(22-2)

$$
\underset{\text{alanylglycine (Ala-Gly)}}{CH_3-\underset{\underset{NH_2}{|}}{CH}-\overset{\overset{O}{\|}}{C}\underset{\text{peptide bond}}{\overset{}{-}}NH-CH_2-COOH} + H_2O
$$

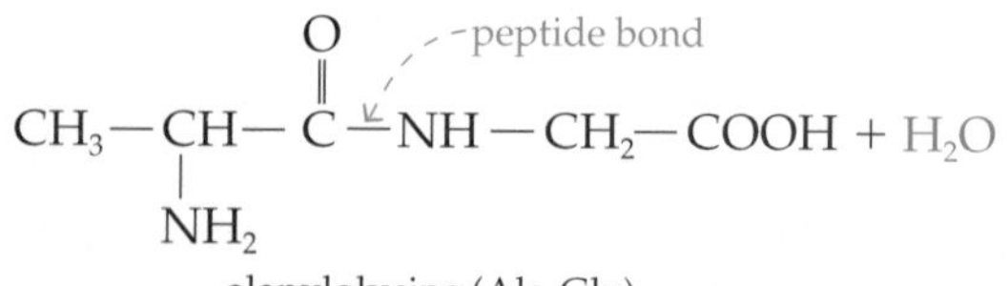

$$NH_2-CH_2-\overset{\overset{\displaystyle O}{\|}}{C}-OH + HNH-\underset{\underset{\displaystyle CH_3}{|}}{CH}-COOH \longrightarrow$$

glycine   alanine   (22-3)

- - peptide bond

$$NH_2-CH_2-\overset{\overset{\displaystyle O}{\|}}{C}-NH-\underset{\underset{\displaystyle CH_3}{|}}{CH}-COOH + H_2O$$

glycylalanine (Gly-Ala)

When two amino acids combine, the product is called a **dipeptide**. When three amino acids combine, the product is called a tripeptide. Whey many amino acids join together, the product is called a **polypeptide**. For just two amino acids, glycine and alanine, two different combinations have already been indicated—glycylalanine and alanylglycine, in which the first member of each group acts as the one furnishing the —OH from the acid group. For three different amino acids—such as glycine, alanine, and valine—there are six possible combinations (or tripeptide linkages).

1. Glycylalanylvaline (Gly-Ala-Val)
2. Glycylvalylalanine (Gly-Val-Ala)
3. Alanylglycylvaline (Ala-Gly-Val)
4. Alanylvalylglycine (Ala-Val-Gly)
5. Valylglycylalanine (Val-Gly-Ala)
6. Valylalanylglycine (Val-Ala-Gly)

By convention, peptides are written with the —NH$_2$ end (the N terminal) at the left and the —COOH (the C terminal) at the right.

Proteins contain a large number of peptide linkages, and the number of possible combinations of the many amino acids in the formation of a protein is beyond all comprehension (for 20 amino acids it is $20 \times 19 \times 18 \times 17 \times 16 \times 15 \times 14 \times 13 \times 12 \times 11 \times 10 \times 9 \times 8 \times 7 \times 6 \times 5 \times 4 \times 3 \times 2$, an unimaginable number). Insulin (see Figure 22-2) illustrates the peptide linkages. It has an A chain, containing 21 amino acids, and a B chain, which contains 30 amino acids. Note that the two chains are connected by two disulfide bridges (see pages 289 and 390).

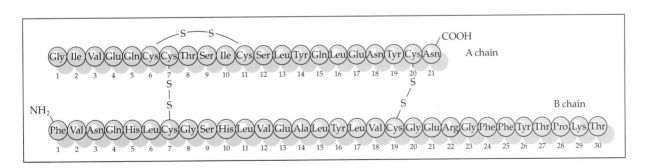

FIGURE 22-2

Structure of human insulin. Note that insulin contains disulfide bridges between Cys groups. Breaking these disulfide bridges inactivates insulin.

Exercise 22-2

Write all the possible combinations for a tripeptide formed from two alanines and one glycine.

Solution

Consider each amino acid as a letter and make as many different words as possible. Let A represent alanine and G represent glycine. The possible combinations are AAG; AGA; GAA.

Self-test

Write all possible tripeptides formed from an alanine (A), glycine (G), and leucine (L).

Answer

AGL; ALG; GAL; GLA; LAG; LGA

22-6 Structure

When a protein is hydrolyzed (by acids, bases, or certain enzymes), it breaks down into smaller and smaller units, eventually forming amino acids. Likewise, when amino acids combine (under the influence of certain enzymes), they first form dipeptides, then tripeptides, then polypeptides, and so on, until they eventually form a protein.

$$\text{protein} \xrightleftharpoons{H_2O} \text{proteoses} \xrightleftharpoons{H_2O} \text{peptones} \xrightleftharpoons{H_2O} \text{polypeptides} \xrightleftharpoons{H_2O}$$

$$\text{tripeptides} \xrightleftharpoons{H_2O} \text{dipeptides} \xrightleftharpoons{H_2O} \text{amino acids}$$

Proteins have a three-dimensional structure that can be considered as being composed of simpler structures.

The primary structure of a protein refers to the number and sequence of the amino acids in the protein. These amino acids are held together by peptide bonds. The primary structure of human insulin is indicated in Figure 22-2.

A slight change in the amino acid sequence, such as the replacement of a single amino acid with another, can change the entire protein (see normal hemoglobin and sickle cell hemoglobin, page 410).

The secondary structure of a protein refers to the regular recurring arrangement of the amino acid chain. One such arrangement, called the α helix, occurs when the amino acids form a coil, or spiral. The coil consists of loops of amino acids held together by hydrogen bonds (between the —H of the —NH_2 of one amino acid and the O of the C=O of the acid part of another amino acid [see Figure 22-3]). Each turn of the helix contains an average of 3.6 amino acids. Such a structure is both flexible and elastic. Hair and wool are examples of protein in such a helical structure.

When the amino acids are coiled, they can form either a right- or left-handed spiral. However, since α-amino acids in protein are all of the L configuration, the coils always are right-handed.

A second type of secondary structure, the β pleated sheet (also called the pleated sheet) consists of parallel strands of polypeptides held together by

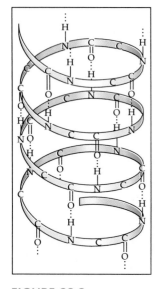

FIGURE 22-3

The α-helical secondary structure of a protein.

hydrogen bonds. Such a structure (Figure 22-4) is flexible but not elastic. Silk has such a pleated sheet structure. It is strong, but resistant to stretching. This type of structure is less common than the α helix.

The tertiary structure of a protein refers to the specific folding and bending of the coils into specific layers or fibers (see Figure 22-5). It is the tertiary structure that gives proteins their specific biologic activity (see enzymes, page 434). Tertiary structures are stabilized by several types of bonds, as indicated in Figure 22-6. Salt bridges (a) are formed between positively and negatively charged groups within the protein molecule. Examples of such groups are the carboxyl and amino side chains found in glutamic acid, lysine, arginine, and aspartic acid. Hydrogen bonds (b) can form between different segments of the coil. Disulfide bonds (c) can form between cysteine groups in different parts of the coil. Hydrophobic bonds (d) can be formed. In general, nonpolar amino acids are folded on the "inside" of the protein, and polar amino acids are on the "outside" (e), where they can react with water molecules to form polar group interactions (also hydrogen bonds).

Some proteins have a quaternary structure, which occurs when two or more protein units, each with its own primary, secondary, and tertiary structure, combine to form a more complex unit. An example of a protein with a quaternary structure is hemoglobin (Figure 22-7). It consists of two identical α chains (light-colored) and two identical β chains (dark-colored). Each chain enfolds a heme (iron-containing) group.

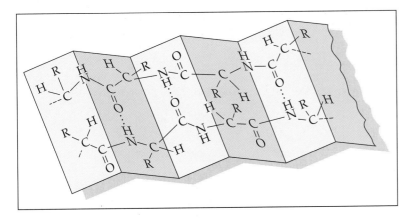

FIGURE 22-4

A β-pleated sheet structure formed by hydrogen bonds between two polypeptide chains.

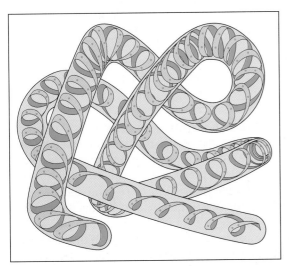

FIGURE 22-5

Tertiary structure of a protein. Note that the coiled helix represents the secondary structure, which in turn is made up of the various amino acids in the sequence specified by mRNA (see page 404).

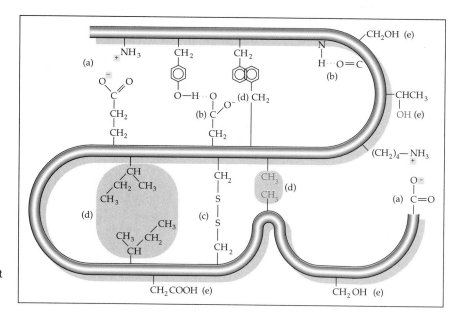

FIGURE 22-6
Various types of bonds that stabilize the tertiary structure of a protein.

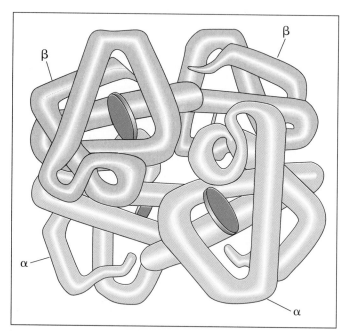

FIGURE 22-7
Quaternary structure of hemoglobin.

22-7 Percent Composition

The average percentage of nitrogen present in protein is 16 percent; that is, about one sixth of protein is nitrogen. Because protein is the major food that contains nitrogen, the chemist can determine the amount of protein present in a food substance by determining the amount of nitrogen present. This amount is about one sixth of the amount of protein present. Therefore, the amount of protein in the food can be calculated by multiplying the weight of nitrogen by 6 and converting this to a percentage of the total. For example, suppose that a 100 g sample of food yielded 4 g of nitrogen on chemical analysis. Since the amount of nitrogen in protein is one sixth of the total amount of protein present, the amount of protein present is 6 × 4 g, or 24 g. Then the percentage of protein present in the original 100 g is 24 percent.

22-8 Classification

Proteins are divided into three categories—simple, conjugated, and derived. On hydrolysis, simple proteins yield only amino acids or derivatives of amino acids. On hydrolysis, conjugated proteins yield amino acids plus some other type of compound. Conjugated protein consists of a simple protein combined with a nonprotein compound. Derived proteins are produced by the action of chemical, enzymatic, and physical forces on the other two classes of protein. Derived proteins include proteoses, peptones, polypeptides, tripeptides, and dipeptides. They also can be hydrolyzed to amino acids.

Proteins are classified according to their solubility, composition, function, or shape.

Classification According to Solubility

Simple proteins are classified according to their solubility in various solvents and also as to whether they are coagulated by heat (see Table 22-5).

Classification According to Composition

Conjugated proteins are classified according to the nature of the nonprotein portion of the molecule (see Table 22-6).

Table 22-5 Properties of Simple Proteins

Type of Protein	Solubility	Coagulated by Heat	Examples
Albumins	Soluble in water and salt solutions	Yes	Egg albumin; serum albumin; lactalbumin
Globulins	Slightly soluble in water; soluble in salt solutions	Yes	Serum globulin; lactoglobulin; vegetable globulin
Albuminoids	Insoluble in all neutral solvents and in dilute acid and alkali	No	Keratin in hair, nails, feathers; collagen
Histones	Soluble in salt solutions; insoluble in very dilute NH_4OH	No	Nucleohistone in thymus gland; globin in hemoglobin

Table 22-6 Conjugated Proteins

Type	Prosthetic Group (Nonprotein Portion of the Combination)	Examples
Nucleoproteins	Nucleic acid	Chromosomes
Glycoproteins	Carbohydrates	Mucin in saliva
Phosphoproteins	Phosphate	Casein in milk
Chromoproteins	Chromophore group (color-producing group)	Hemoglobin, hemocyanin, flavoproteins, cytochrome
Lipoproteins	Lipids	Fibrin in blood
Metalloproteins	Metals	Ceruloplasmin (containing Cu) and siderophilin (containing Fe) in blood plasma

Nucleoproteins   Nucleoproteins are proteins combined with nucleic acids (see Chapter 23).

Glycoproteins   Glycoproteins are proteins containing carbohydrates in varying amounts. Glycoproteins have molar masses from 15,000 to more than 1 million. The carbohydrates present in glycoproteins include the hexoses mannose and galactose, the pentoses arabinose and xylose, and sialic acids, which are derivatives of neuraminic acid.

neuraminic acid
(sialic acids have an R—C = O group
in place of the H marked with the asterisk)

Glucose is not found in glycoproteins, except for collagen.

Glycoproteins are present in most organisms, including animals, plants, bacteria, viruses, and fungi. Human cell membranes are about 5 percent carbohydrate, which is present as glycoproteins and glycolipids. Glycophorin is a glycoprotein found in the membranes of human *erythrocytes. Heparin, which inhibits clotting of blood, is also a glycoprotein.

In addition to their functions in membranes, glycoproteins serve in the following ways: as structural proteins (collagen); as lubricants (mucin and mucous secretions); as transportation molecules for vitamins, lipids, minerals, and trace elements; as immunoglobulins such as interferon; as hormones such as thyrotropin (TSH); as enzymes such as the hydrolases and nucleases; as hormone receptor sites; and for the specification of human blood types.

Lipoproteins   Lipoproteins, which are proteins containing lipids, are part of cell membranes (see Section 21-8).

Lipids such as cholesterol and triglycerides are not soluble in water and thus must be complexed to a water-soluble carrier protein (lipoprotein). **Plasma lipoproteins** consist of a neutral lipid core of triglyceride and cholesterol ester that is surrounded and stabilized by free cholesterol, protein, and phospholipid. The relative proportions of nonpolar lipid, protein, and polar lipid determine the density, size, and charge of the resulting lipoproteins. The density of lipoproteins has been used to classify them, as indicated in Table 22-7.

Chylomicrons are produced in the intestinal mucosa and are used to transport dietary lipids into the blood plasma via the thoracic lymph duct. They are removed from the plasma with a half-life of 5 to 15 minutes. They are responsible for the creamed-tomato-soup appearance of blood following a meal containing fats.

Very low density lipoproteins (VLDL) transport triglycerides synthesized by the liver to the other parts of the body. Their breakdown leads to the production of the transient intermediate density lipoproteins (IDL) and the end product low density lipoprotein (LDL). LDL provides cholesterol for cellular needs. LDL is thought to promote coronary heart disease by first penetrating the coronary artery wall and then depositing cholesterol to form atherosclerotic plaque.

Table 22-7 Lipoproteins

Type	Density (g/mL)	% Protein	Triglycerides	Cholesterol Free	Cholesterol Ester	Phospholipids	Fatty Acids
Chylomicrons	Less than 0.9	1–2	88	1	3	8	
Very low density lipoproteins (VLDL)	0.9–1.006	7–10	56	8	15	20	1
Intermediate density lipoproteins (IDL)	1.006–1.019	11	29	9	34	26	1
Low density lipoproteins (LDL)	1.019–1.063	21	13	10	4	28	1
High density lipoproteins (HGL$_2$)	1.063–1.125	33	16	31	10	43	
High density lipoproteins (HDL$_3$)	1.125–1.20	57	13	6	29	46	6

Percentage of Total Lipids

High density lipoproteins (HDL) are involved in the catabolism of other lipoproteins. They incorporate the cholesterol and phospholipid released by a lipoprotein. HDLs may also remove excess cholesterol from peripheral tissue.

A new lipoprotein, lipoprotein (a), which is similar to LDL, has been detected recently. The striking structural similarity of lipoprotein (a) to human plasminogen has stimulated intense studies as to a possible link between atherosclerosis and thrombosis. Primarily, results have indicated that lipoprotein (a) is an independent risk factor (similar to total cholesterol) for coronary heart disease.

Exercise 22-3

Using Table 22-7, determine which lipoprotein has the highest percentage of triglycerides.

Solution

The table shows that chylomicrons have the highest level at 88 percent.

Self-test

Which lipoprotein has the highest combined (free + ester) percentage of cholesterol?

Answer

LDL

Elevated LDL levels have been associated with an increased risk of developing coronary artery disease, whereas elevated HDL levels appear to reduce the risk. Women have higher HDL levels than men (55 vs. 45 mg/100 mL), and this

Table 22-8 Proteins Classified According to Function		
Type of Protein	Example	Use
Structural	Collagen	In structure of connective tissue
	Keratin	In structure of hair and nails
Contractile	Myosin, actin	In muscle contraction
Storage	Ferritin	In storage of iron needed to make hemoglobin
Transport	Hemoglobin	In carrying oxygen
	Serum albumin	In carrying fatty acids
Hormones	Insulin	In metabolism of carbohydrates
Enzymes	Pepsin	In digestion of protein
Protective	Gamma globulin	In antibody formation
	Fibrinogen	In blood clotting
Toxins	Venoms	Poisons

may account for women's lower rate of heart disease. Aerobic exercise increases HDL levels (marathon runners average 65 mg/100 mL).

Classification According to Function

Proteins can also be classified according to their biologic function. Table 22-8 lists the various categories of proteins classified by this method.

Classification According to Shape

Proteins can also be classified according to their shape and dimensions. Globular proteins consist of polypeptides folded into the shape of a "ball." They have a length-to-width ratio of less than 10. Globular proteins are soluble in water or form colloidal dispersions and have an active function. Proteins classified as globular are hemoglobin, albumin, and the globulins.

Fibrous proteins consist of parallel polypeptide chains that are coiled and stretch out. They have a length-to-width ratio greater than 10. Fibrous proteins are insoluble in water. Examples include collagen, fibrin, and myosin.

22-9   Properties

Colloidal Nature

Proteins form colloidal dispersions in water. Being colloidal, protein will pass through a filter paper but not through a membrane. The inability of protein to pass through a membrane is of great importance in the body. Proteins present in the bloodstream cannot pass through the capillaries and should remain in the bloodstream. Since proteins cannot pass through membranes, there should be no protein material present in the urine. The presence of protein in the urine indicates damage to the membranes in the kidneys—possibly nephritis.

Denaturation

Denaturation of a protein refers to the unfolding and rearrangement of the secondary and tertiary structures of a protein without breaking the peptide bonds (see Figure 22-8). A protein that is denatured loses its biologic activity. When the con-

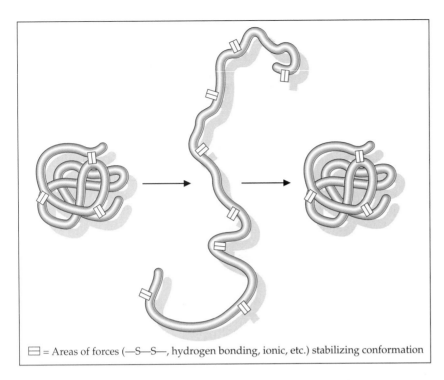

□ = Areas of forces (—S—S—, hydrogen bonding, ionic, etc.) stabilizing conformation

FIGURE 22-8
Reversible denaturation.

ditions for denaturation are mild, the protein can be restored to its original conformation by carefully reversing the conditions that caused the denaturation. This is called reversible denaturation. If the conditions that caused the denaturation are drastic, the process is irreversible; the protein will coagulate or precipitate from solution. Proteins can be denatured by a variety of agents, as indicated in the following section.

Reagents or Conditions That Cause Denaturation

Alcohol   Alcohol coagulates (precipitates) all types of protein except prolamines. Alcohol (70 percent) is used as a disinfectant because of its ability to coagulate the protein present in bacteria (see page 284). Alcohol denatures protein by forming hydrogen bonds that compete with the naturally occurring hydrogen bonds in the protein. Such a process is not reversible.

Salts of Heavy Metals   Heavy metal salts, such as mercuric chloride (bichloride of mercury) or silver nitrate (lunar caustic), precipitate protein. These denature protein irreversibly by disrupting the salt bridges and the disulfide bonds present in the protein. They are very poisonous if taken internally because they coagulate and destroy protein present in the body. The antidote for mercuric chloride or silver nitrate when these poisons are taken internally is egg white. The heavy metal salts react with the egg white and precipitate out. (The egg white colloid has a charge opposite to that of the heavy metal ion and so attracts it.) The precipitate thus formed must be removed from the stomach by an emetic or the stomach will digest the egg white and return the poisonous material to the system.

Dilute silver nitrate solution is used as a disinfectant in the eyes of newborn infants. Stronger solutions of silver nitrate are used to cauterize fissures and destroy excessive granulation tissues.

Heat   Gentle heating causes reversible denaturation of protein, whereas vigorous heating denatures protein irreversibly by disrupting several types of bonds. Egg white, a substance containing a high percentage of protein, coagulates on heating. Heat coagulates and and destroys protein present in bacteria. Hence, sterilization of instruments and clothing for use in operating rooms requires the use of high temperatures. The presence of protein in the urine can be determined by heating a sample of urine, which will cause the coagulation of any protein material that is present.

Alkaloidal Reagents   Alkaloidal reagents, such as tannic acid and picric acid, form insoluble compounds with proteins. Alkaloidal reagents denature protein irreversibly by disrupting salt bridges and hydrogen bonds.

Tannic acid has been used extensively in the treatment of burns. When this substance is applied to a burn area, it causes the protein to precipitate as a tough covering, thus reducing the amount of water loss from the area. It also reduces exposure to air. Newer drugs have taken the place of tannic acid for burns, but an old-fashioned remedy still in use for emergencies involves the use of wet tea bags (which contain tannic acid).

Radiation   Ultraviolet or x rays can cause protein to coagulate. The radiation denatures irreversibly by disrupting the hydrogen bonds and the hydrophobic bonds present in the protein. In the human body the skin absorbs and stops ultraviolet rays from the sun so they do not reach the inner cells. Proteins in cancer cells (rapidly dividing cells) are more susceptible to radiation than those present in normal cells, so x irradiation is used to destroy cancerous tissue (see page 227).

pH   Changes in pH can disrupt hydrogen bonds and salt bridges, causing irreversible denaturation. Proteins are coagulated by such strong acids as concentrated hydrochloric, sulfuric, and nitric acids. Casein is precipitated from milk as a curd when it comes into contact with the hydrochloric acid of the stomach. Heller's ring test is used to detect the presence of albumin in urine. A layer of concentrated nitric acid is carefully placed under a sample of urine in a test tube. If albumin is present, it will precipitate out as a white ring at the interface of the two liquids. If acid or base remains in contact with protein for a long period of time, the peptide bonds will break.

Oxidizing and Reducing Agents   Oxidizing agents such as bleach and nitric acid and reducing agents such as sulfites and oxalates denature protein irreversibly by disrupting disulfide bonds (see page 390).

Salting Out   Most proteins are insoluble in saturated salt solutions and precipitate out unchanged. To separate a protein from a mixture of other substances, the mixture is placed in a saturated salt solution (such as NaCl, Na_2SO_4, or $[NH_4]_2SO_4$). The protein precipitates out and is removed by filtration. The protein can then be purified from the remaining salt by the process of dialysis (see page 179).

Summary

Proteins are high molecular mass polymers containing the elements carbon, hydrogen, oxygen, and nitrogen. Some proteins also contain other elements. Proteins are the chief constituents of all cells of the body.

Animals cannot synthesize protein from raw materials. They must obtain their protein from plants or from other animals, which in turn have obtained the protein from plants.

Protein serves to build new cells, to maintain existing cells, and to replace old cells in the body. Protein is necessary for the formation of the enzymes and hormones in the body. The oxidation of protein yields energy in the amount of 4 kcal/g.

Proteins are polymers of amino acids. Hydrolysis of protein yields amino acids with an amine group attached to the α carbon. The body can synthesize very few amino acids. Those that it needs and cannot synthesize are called essential amino acids. All the amino acids in the body, except glycine, have the L configuration. Amino acids in which the R group is nonpolar are generally insoluble in water. Amino acids with polar R groups are polar compounds and are soluble in water.

Amino acids are amphoteric—they react with either acids or bases because they contain an acid group (—COOH) and a basic group (—NH₂).

When an amino acid is placed in an acid solution, it forms a positive ion and migrates toward the negative electrode. When an amino acid is placed in a basic solution it forms a negative ion and migrates toward the positive electrode. At a certain pH, the isoelectric point, the amino acid will be neutral; it will not migrate toward either electrode.

When two amino acids combine, a dipeptide is formed. When four or more amino acids combine, a polypeptide is formed. Polypeptides in turn form peptones, then proteoses, and finally protein. The hydrolysis of protein proceeds through the same types of compounds in reverse order, forming proteoses, peptones, polypeptides, dipeptides, and amino acids.

The primary structure of a protein refers to the number and sequence of the amino acids in the protein chain. The secondary structure refers to the regular recurring arrangement of the amino acid chain into a coil or pleated sheet. The tertiary structure refers to the specific folding and bending of the coils into specific layers or fibers. The quaternary structure of a protein occurs when several protein units combine to form a more complex unit.

In general, the nitrogen content of protein is 16 percent.

Proteins can be classified according to solubility, composition, function, or shape.

Denaturation of a protein refers to the unfolding and rearrangement of the secondary and tertiary structures of a protein.

Proteins do not dissolve in water; rather they form colloidal dispersions. Proteins, being colloids, cannot pass through membranes and should not normally be present in the urine.

Proteins can be coagulated (precipitated) by means of alcohol, concentrated salt solutions, salts of heavy metals, heating, the use of alkaloidal reagents, concentrated inorganic acids, and x rays.

Questions and Problems

1. All proteins contain which elements? Which additional elements may be present?
2. Where do plants obtain their protein? animals?
3. Describe the nitrogen cycle in nature.
4. How does the energy value of protein compare with that of carbohydrates? lipids?
5. What are amino acids? Give the names of three amino acids and draw their structures.
6. Name the essential amino acids.
7. What is meant by the term *isoelectric point* of a protein? What happens to the solubility of a protein at its isoelectric point?
8. Why are many amino acids amphoteric? optically active?
9. What is a peptide linkage? What is a polypeptide?
10. Write the reaction of leucine with serine in two different ways.
11. What amino acids are present in insulin?
12. Describe the types of protein structures.
13. When proteins are slowly hydrolyzed what products are formed?

14. What percentage of protein is nitrogen? What use is made of this fact?
15. List several types of simple proteins; conjugated proteins.
16. Is protein normally found in urine? Why or why not?
17. Why is 100 percent alcohol not used as a disinfectant?
18. What happens when protein is heated? What use is made of this effect?
19. Of what importance are proteins to the body?
20. What types of bonding stabilize the tertiary structure of a protein?
21. Compare the polar and nonpolar amino acids in terms of solubility in water; in terms of structure.
22. Where are D-amino acids formed? What function do they have?
23. Why are coils of protein right-handed?
24. Compare reversible and irreversible denaturation.
25. What is a globular protein? fibrous protein? How do they differ?

26. What is an α-helix? What holds it together? Compare the α-helical structure with a β-helical structure.
27. Give one example of a protein or polypeptide used for transportation; for storage; for hormones; for enzymes.
28. Explain why pyruvate is central to the metabolic pathways of glucose, fatty acids, and amino acids.
29. What is the origin of the nitrogen in urea?
30. Which amino acids are synthesized from aromatic amino acids?
31. What route of administration would be most effective for Bacitram, a peptide antibiotic? Why?

32. Which amino acids have more than one pair of enantiomers?
33. Complete hydrolysis of an octapeptide yields the following amino acids: threonine, 2 alanines, histidine, leucine, lysine, proline, and tyrosine. Partial hydrolysis yields the following tripeptides: leu-ala-tyr; thr-pro-leu; lys-ala-his; and his-thr-pro. Draw the structure of this octapeptide.
34. Describe the method used to separate lipoproteins.

Practice Test

1. An example of an amino acid is _____.
   a. benzene
   b. toluene
   c. glycine
   d. histamine
2. Which of the following will not denature protein?
   a. acid   b. water   c. heat   d. alcohol
3. Which structure refers to the number and sequence of amino acids in a protein?
   a. primary
   b. secondary
   c. tertiary
   d. quaternary
4. Amino acids exhibit which property?
   a. allotropism
   b. amphipathism
   c. amphoterism
   d. all of these
5. How many different peptide configurations are possible for three different amino acids?
   a. 2   b. 4   c. 6   d. 8

6. Proteins form which type of mixture in water?
   a. solution
   b. suspension
   c. colloid
   d. emulsion
7. Proteins are classified according to:
   a. shape
   b. solubility
   c. function
   d. all of these
8. All proteins contain the elements C, H, O, and _____.
   a. N   b. S   c. P   d. Fe
9. An example of a lipoprotein is(are):
   a. heparin
   b. chylomicrons
   c. cholesterol
   d. casein
10. Protein can be denatured by:
    a. heat
    b. alkaloidal reagents
    c. radiation
    d. all of these

23

Nucleic Acids

Objectives

- To recognize the structure of nucleic acids
- To become familiar with the metabolism of nucleoproteins
- To understand the roles of DNA and RNA
- To become familiar with the DNA code
- To understand the causes of genetic mutations
- To become aware of various genetic diseases and their causes
- To become familiar with the classifications of genetic diseases
- To understand recombinant DNA techniques

The physical characteristics of an individual are determined largely by the genetic makeup inherited from parents. The child in this multiracial family has inherited features from each of his parents.

23-1 Nucleic Acids

Nucleic acids were first isolated from the cellular nucleus, hence the name. Nucleic acids are macromolecules, huge polymers with molecular masses of over 100 million.

There are two main types of nucleic acids, **deoxyribonucleic acid (DNA)** and **ribonucleic acid (RNA).** DNA is primarily responsible for the transfer of genetic information, whereas RNA is primarily concerned with the synthesis of protein.

Hydrolysis of nucleic acids gives nucleotides, which can be considered the units that make up the polymer. A nucleotide consists of three parts, a heterocyclic base, a sugar, and phosphoric acid.

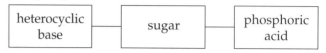

a nucleotide

The sugar in nucleotides, and so in nucleic acids, is a pentose. In RNA and its nucleotides the sugar is *ribose*, whereas in DNA and its nucleotides it is *deoxyribose*. The prefix *deoxy-* means "without oxygen." Note in the following structural diagrams that deoxyribose contains one less oxygen atom on carbon number 2 than ribose.

ribose deoxyribose

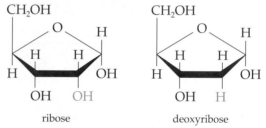

Nucleotides can be hydrolyzed to yield nucleosides and phosphoric acid. A nucleoside consists of two parts—a heterocyclic base and a sugar (a pentose).

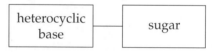

a nucleoside

These hydrolysis reactions can be summarized as follows:

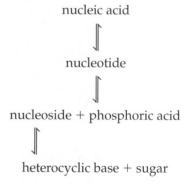

nucleic acid

⇅

nucleotide

⇅

nucleoside + phosphoric acid

⇅

heterocyclic base + sugar

Heterocyclic Bases

The heterocyclic bases present in nucleic acids are divided into two types—purines and pyrimidines. The two purines present in both DNA and RNA are **adenine** and **guanine.** The pyrimidine **cytosine** is present in both DNA and RNA, whereas thymine is found in DNA only and uracil is present in RNA only. The structures of these heterocyclic bases are indicated in the following sections.

The Pyrimidines   Pyrimidine is a six-membered heterocyclic ring containing two nitrogen atoms. Three important derivatives of pyrimidine found in nucleic acids are thymine (2,4-dioxy-5-methylpyrimidine), cytosine (2-oxy-4-aminopyrimidine), and uracil (2,4-dioxypyrimidine). Their structures are as follows:

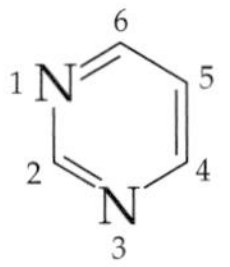

pyrimidine

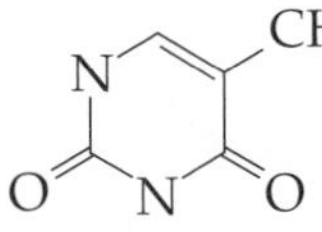

thymine 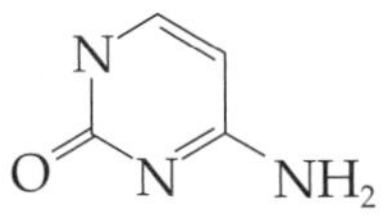   cytosine 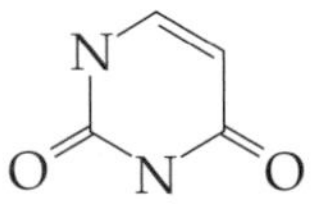   uracil

Other important compounds containing pyrimidines are thiamin (vitamin B_1) and the barbiturates.

The Purines   The purines found in nucleic acids are derivatives of a substance, purine, that does not occur naturally. As indicated by their structures, adenine is 6-amino-purine and guanine is 2-amino-6-oxypurine.

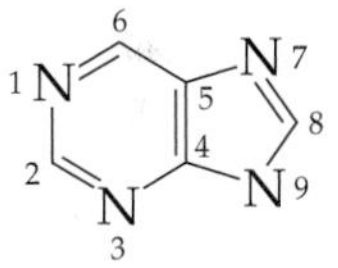

purine

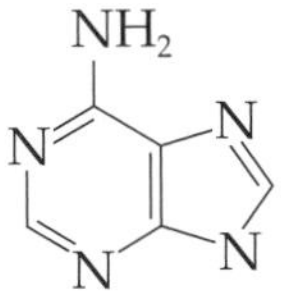

adenine 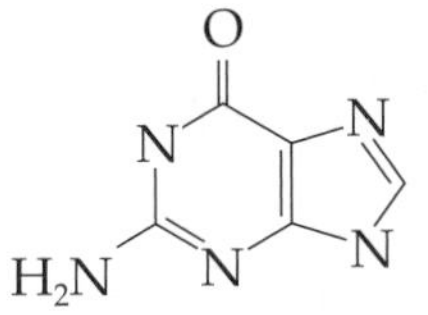   guanine

Other purines include caffeine and theophylline. Caffeine is a stimulant for the central nervous system and also a diuretic. Caffeine is found in coffee and tea. Its chemical name is 1,3,7-trimethyl-2,6-dioxypurine. Theophylline, 1,3-dimethyl-2,6-dioxypurine, is found in tea and is used medicinally as a diuretic and for bronchial asthma.

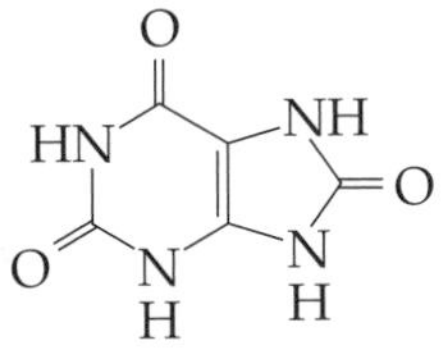

uric acid 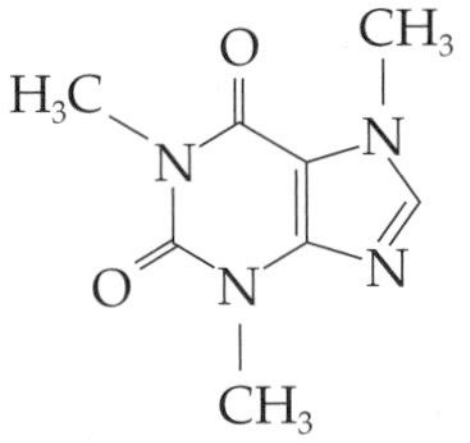   caffeine 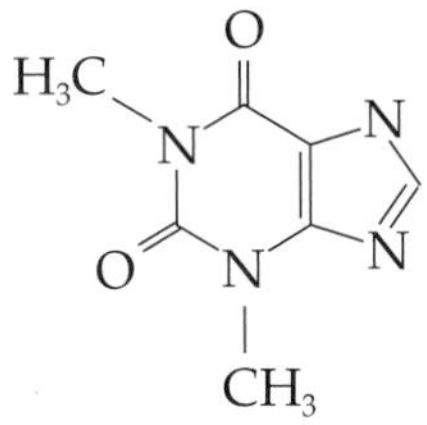   theophylline

Uric acid is the end product of purine metabolism.

The Adenosine Phosphates

One important nucelotide is adenosine monophosphate (AMP). It is formed by the reaction of adenosine (a nucleotide) with one molecule of phosphoric acid. If two phosphate groups react with adenosine, adenosine disphosphate (ADP) is formed; ATP, adenosine triphosphate, is formed when three phosphate groups react.

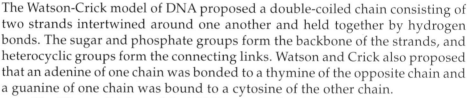

AMP (adenosine monophosphate)
ADP (adenosine diphosphate)
ATP (adenosine triphosphate)

ATP, ADP, and AMP are involved in various metabolic processes involving the storage and release of energy from their phosphate bonds.

23-2 The Watson-Crick Model of DNA

FIGURE 23-1

Dr. J. D. Watson, Harvard University, who was awarded the Nobel Prize for Medicine and Physiology in 1962 along with Dr. F. H. C. Crick and Dr. M. H. F. Wilkins, for the discovery of the molecular structure of deoxyribonucleic acid (DNA).

The Watson-Crick model of DNA proposed a double-coiled chain consisting of two strands intertwined around one another and held together by hydrogen bonds. The sugar and phosphate groups form the backbone of the strands, and heterocyclic groups form the connecting links. Watson and Crick also proposed that an adenine of one chain was bonded to a thymine of the opposite chain and a guanine of one chain was bound to a cytosine of the other chain.

The Watson-Crick model of DNA can also be shown as indicated in Figure 23-2, as a spiral ladder with the two sides held together by rungs consisting of heterocyclic nitrogen compounds (adenine, thymine, guanine, and cytosine).

If the chain is untwisted and straightened out, it can be represented as follows:

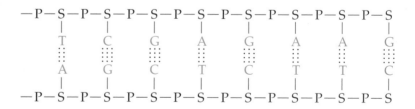

The solid lines indicate ordinary chemical bonding, and the dotted lines indicate hydrogen bonding. Recall that hydrogen bonds are much weaker than ordinary chemical bonds. Note that there are two hydrogen bonds between adenine and thymine and three hydrogen bonds between guanine and cytosine.

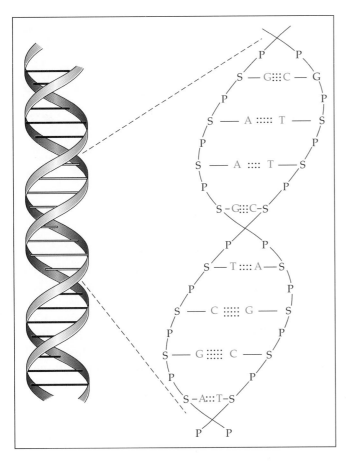

FIGURE 23-2

Structure of DNA. P is phosphate; S, sugar; G, guanine; C, cytosine; A, adenine; and T, thymine.

23-3 The DNA Code

All DNA molecules have the same sequence of deoxyribose and phosphates in the ladder part of the chain. The difference lies in the order of the adenine, thymine, cytosine, and guanine parts of the chain. This difference in sequence of the heterocyclic compounds constitutes the genetic code.

This code consists of only four letters—A, T, G, and C—representing adenine, thymine, guanine, and cytosine, respectively. In the DNA molecule these letters of the code are grouped in threes (which we shall discuss later). Thus the code might be ATC, TAC, AAA, CCT, and so on.

Simple bacterial viruses contain about 5500 nucleotides in their DNA molecules; 5500 nucleotides grouped in threes can produce 5500/3 or approximately 1800 coded pieces of information. These 1800 coded pieces of information are sufficient to describe that particular virus. If each piece of information corresponded to one letter of our alphabet, it would take 1800 letters to describe, genetically, that virus; 1800 letters corresponds roughly to the number of letters on one page of this book.

One DNA molecule in a human contains approximately 5,000,000,000 nucleotides, or enough to form 1,700,000,000 coded pieces of information (the three nucleotides to a group). If each of these coded pieces corresponded to a letter of our alphabet, it would take approximately 2500 volumes, each the size of this book, to describe a human being genetically.

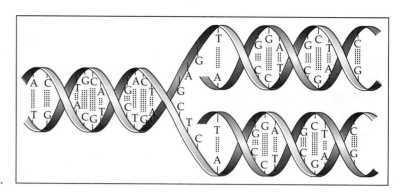

FIGURE 23-3
Unwinding of DNA.

23-4 Replication of DNA

When a cell divides, it produces two new cells with identical characteristics. This means that the DNA originally present must duplicate (or replicate) itself. How does the DNA molecule direct the synthesis of another identical DNA molecule?

Watson and Crick theorized that the DNA molecule unwinds, and each half acts as a template for nucleotide units to collect on and form a new chain. Recall that if adenine (A) is present on one chain, it can attract and hold only thymine (T), and cytosine (C) can attract and hold only guanine (G). Thus, each half of the chain is highly specific in what it attracts. It forms complementary chains that coil up again and form two new DNA molecules.

This can be represented diagramatically, as shown in Figure 23-3. This representation is necessarily quite general and approximate. The actual mechanism of DNA unwinding, untwisting, and joining for replication is quite complex, involving several enzymes, such as DNA polymerase and DNA ligase, and is beyond the scope of this book.

23-5 Transfer of Information

The preceding paragraph indicates how the DNA molecule duplicates itself, but how does the DNA molecule transfer its coded information to the cell's ribosomes where the actual production of the protein called for by the code takes place?

The information contained in the DNA molecule is carried by another molecule called messenger RNA (mRNA). During replication, DNA "unzips" and synthesizes two new chains to produce two identical molecules. When the DNA molecule synthesizes mRNA, the DNA again "unzips"; however, only one part of one strand of DNA acts as a template for the formation of mRNA (see Figure 23-4). The enzyme required is called RNA polymerase. The process is known as transcription.

When mRNA is produced, there is one major difference in the attraction of nucleotides between it and DNA. In RNA, the partner of adenine (A) is uracil (U) and not thymine (T). Another difference is that RNA contains the pentose ribose, whereas DNA contains the pentose deoxyribose.

There are at least three types of RNA: (1) messenger RNA, (2) ribosomal RNA, and (3) transfer RNA. Messenger RNA (mRNA) is concerned with the transmission of genetic information from DNA to the site of protein synthesis, the ribosomes; it is found in the nucleus and the cytoplasm of the cell. Ribosomal RNA (rRNA), which is the major faction of the total RNA, combines with protein to form the ribosomes; therefore, ribosomes are an example of nucleoprotein. It is in the

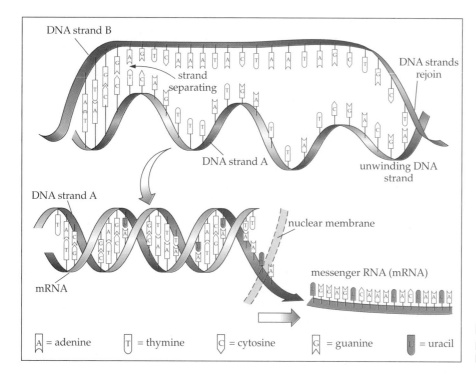

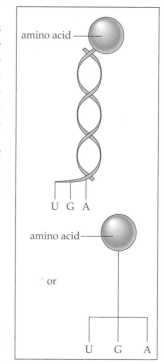

DNA strand B

strand
separating

DNA strands
rejoin

DNA strand A

unwinding DNA
strand

DNA strand A

nuclear membrane

mRNA

messenger RNA (mRNA)

| ⬡ = adenine | ⬡ = thymine | ⬡ = cytosine | ⬡ = guanine | ⬡ = uracil |

FIGURE 23-4
Formation of messenger
RNA from DNA.

ribosomes that the "translation" of the DNA message takes place. **Transfer RNA (tRNA)** holds a specific amino acid for incorporation into a protein molecule, as indicated in the following paragraphs. A fourth type of RNA, heterogeneous RNA (hRNA), may be a precursor for mRNA. hRNA is found in the nucleus of the cell.

Let us see how mRNA and tRNA function in the ribosomes in the synthesis of a protein. The mRNA moves from the nucelus to the cytoplasm and then to the ribosomes, where it acts as a template for the formation of protein. In the cytoplasm is tRNA, which is a cloverleaf form of RNA containing relatively few nucleotides. At one end of the tRNA are three nucleotides that are specific for a certain code on the mRNA. Amino acids in the cytoplasm are activated by ATP and are coupled to the other end of tRNA. This can be illustrated as shown in Figure 23-5.

If mRNA contains the coded units ACU, the tRNA that would attach itself must have the code UGA. (Recall that in RNA the adenine [A] is bonded to uracil [U] and cytosine [C] to guanine [G].) If the coded message in mRNA is UGC, the corresponding code in tRNA must be ACG:

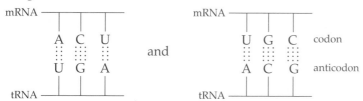

As the mRNA travels across the ribosome, the first coded group (of three letters) picks up and holds a corresponding tRNA. As the second coded group of the mRNA passes, it too picks up a corresponding tRNA. Attached to the opposite end of each tRNA is an amino acid. The amino acid attached to the end of the second tRNA becomes bonded to the amino acid at the end of the first tRNA. The amino acid at the end of the third tRNA in turn becomes bonded to the second amino acid.

FIGURE 23-5
Diagrammatic forms of
transfer RNA (tRNA).

Thus, as each tRNA attaches itself to the mRNA, the tRNA gives up its amino acid to form a chain of amino acids, or a protein. After giving up its amino acid, the tRNA leaves the mRNA and goes in search of another amino acid that it can pick up and use to repeat the sequence.

Thus, as the mRNA passes through the ribosome, it directs the gathering, in specified sequence, of the tRNAs, which in turn bond together their amino acids to form a protein.

This effect can be illustrated diagrammatically as shown in Figure 23-6. In these diagrams, the first coded group in the mRNA—group CAG— attracts tRNA coded GUC. Attached to tRNA coded GUC is an amino acid that we have simply labeled 1. As the mRNA passes further along into the ribosome, the next coded group, code UUA, attracts tRNA coded AAU. The amino acid attached to the end of this tRNA is labeled 2. Amino acid 2 bonds itself to amino acid 1 as illustrated. At the same time, the tRNA coded GUC that held amino acid 1 goes off in search of another amino acid labeled 1 so it can be used again whenever the code calls for

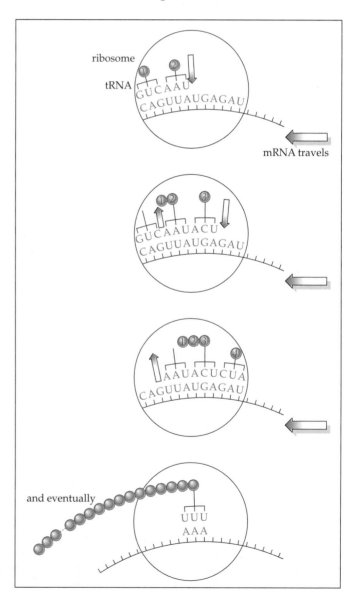

FIGURE 23-6

mRNA directs the proper sequence of tRNAs to form a protein.

it. The third coded group in the mRNA, code UGA, attracts tRNA coded AGU, with its attached amino acid 3. Then amino acid 3 bonds to amino acid 2, which is already bonded to amino acid 1. So the chain of amino acids begins and continues as illustrated in the fourth figure. This process continues until the end of the coded groups of the mRNA. At this time, the protein is complete and moves out of the ribosome into the cytoplasm.

23-6 The Triplet Code

The DNA code consists of four letters, A, T, C, and G, grouped in threes. It is believed that the code in the DNA molecule specifies individual amino acids. Since there are 20 primary amino acids that occur in nature, there must be at least 20 different coded groups in the DNA molecule.

The 20 primary amino acids and their abbreviations are as follows:

alanine	Ala	glycine	Gly	proline	Pro
arginine	Arg	histidine	His	serine	Ser
asparagine	Asn	isoleucine	Ile	threonine	Thr
aspartic acid	Asp	leucine	Leu	tryptophan	Trp
cysteine	Cys	lysine	Lys	tyrosine	Tyr
glutamic acid	Glu	methionine	Met	valine	Val
glutamine	Gln	phenylalinine	Phe		

Decoding the Code

How can the DNA code be decoded? That is, how can we tell which amino acid is specified by a certain coded group?

In 1961 it was found that if a synthetic RNA composed only of uracil nucleotides was substituted for mRNA in a protein-synthesis system, a polypeptide was formed that contained only the amino acid phenylalanine. Since the synthetic mRNA contained only uracil, it must have the code group UUU. The corresponding group in the DNA molecule must be AAA. Thus, we can say that the coded group AAA in the DNA molecule corresponds to the code UUU in mRNA, which in turn specifies the amino acid phenylalanine. Another synthetic mRNA, consisting solely of adenine nucleotides, produced a polypeptide containing only lysine. Thus the mRNA code for lysine is AAA and the corresponding DNA code must be TTT. Table 23-1 indicates the amino acids and their coded groups.

Some experiments indicated that there may be more than one code group for a specific amino acid. That is, alanine is indicated by the mRNA codes GCU, GCC, GCA, and also GCG. Note that all of these code groups begin with the letters "GC," and thus the code is called degenerate. In general, the first two "letters" of the code are more important than the third.

The genetic code is universal. That is, all known organisms use this same code for the synthesis of their proteins.

Figure 23-7 illustrates the synthesis of a polypeptide chain from mRNA, indicating the various amino acids specified by that mRNA.

Regulation of Protein Synthesis

When mRNA acts as a template for the synthesis of protein, only a small amount of the total information in DNA is used at one time to produce a certain type of protein. What determines whether mRNA is formed from a certain segment of the DNA? What "turns on" the DNA and what "turns it off'?

Table 23-1	Genetic Code in mRNA				
First Letter	**Second Letter**				**Third Letter**
	A	C	G	U	
A	Lys	Thr	Arg	lle	A
	Asn	Thr	Ser	lle	C
	Lys	Thr	Arg	Met‡	G
	Asn	Thr	Ser	lle	U
C	Gln	Pro	Arg	Leu	A
	His	Pro	Arg	Leu	C
	Gln	Pro	Arg	Leu	G
	His	Pro	Arg	Leu	U
G	Glu	Ala	Gly	Val	A
	Asp	Ala	Gly	Val	C
	Glu	Ala	Gly	Val	G
	Asp	Ala	Gly	Val	U
U	†	Ser	†	Leu	A
	Tyr	Ser	Cys	Phe	C
	†	Ser	Trp	Leu	G
	Tyr	Ser	Cys	Phe	U

†Chain terminator.
‡Chain initiator.

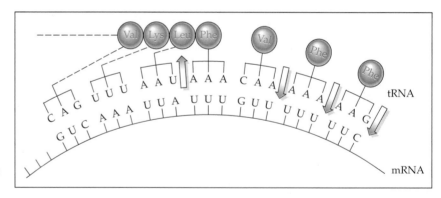

FIGURE 23-7
Synthesis of a polypeptide from mRNA.

Since protein is not synthesized continuously but only as needed, DNA must normally be a "repressed state." A **repressor,** which is a polypeptide, binds to a small segment of the DNA. This segment is called the **operator site.** As long as the repressor is bonded to the DNA, no mRNA is produced and so production of protein is inhibited. When a particular protein is needed, an **inducer** is formed. The inducer combines with the repressor, changing its shape so that it can no longer bind to the DNA. Once the repressor is removed from the DNA, synthesis of mRNA and hence protein can begin. When sufficient protein has been produced, the inducer is removed and the repressor once again binds to the DNA, stopping protein synthesis.

Certain drugs, such as tetracyclines and streptomycin, bind to the ribosomes of bacteria and prevent synthesis of protein. Hence these drugs are effective antibiotics.

Mutations

Suppose that one of the letters (nucleotides) in a DNA code group was substituted by another letter. Then the message would be miscopied and a mutation would occur. Such a change in the sequence of nucleotides may (1) give no detectable effect, particularly if the change is in the third letter of the code (see page 408); (2) cause a different amino acid to be incorporated into the chain (the results may be acceptable, partially acceptable, or totally unacceptable to the function of that protein); or (3) cause the termination of the chain prematurely so that the protein cannot function normally.

A mutation will also occur if one of the code letters is omitted or if one is added or if the order of the code letters is rearranged. Mutations occur because of exposure to radiation (in industry, medically, because of naturally occurring radiation and cosmic rays, or because of fallout) and possibly because of exposure to certain chemicals.

23-7 The Triple Helix

Collagen, the primary macromolecule of connective tissue, exists as a triple helix, a coiled structure of three polypeptide subunits. One distinct characteristic of collagen is that the amino acid glycine constitutes one third of one of the coils.

Triplex DNA is triple-stranded DNA. Among its uses are as a molecular scissors for cutting strands of DNA and as an agent for treating viral diseases. The latter is due to its ability to disrupt the copying of genetic material onto mRNA.

23-8 The Human Genome Project

The Human Genome Project is an international effort with the goal of sequencing the entire human *genome as well as the genomes of other organisms (such as *E. coli*) that have been used in the study of genetics. Once the human genome is completely sequenced, it may be possible to determine the molecular basis of genetic diseases and perhaps develop the technology to diagnose and treat these diseases.

23-9 Genetic Diseases

Most genetic diseases are caused by a defective gene, which results in a loss of activity of some enzyme. Even though the body has thousands of enzymes, the loss of only one may be disastrous.

Consequences of Inherited Enzyme Deficiency

Consider a simple enzymatic reaction in which compound X is changed to compound Y under the influence of enzyme x_1, and then compound Y is changed to compound Z under the influence of enzyme y_1.

$$X \longrightarrow Y \longrightarrow Z$$
$$\quad x_1 \qquad \quad y_1$$

If enzyme y_1 is deficient, one of the following consequences might occur:

$$X \xrightarrow{x_1} Y \not\rightarrow_{y_1} Z$$

a. The body would be lacking in compound Z, with the resulting effect of that lack

$$X \xrightarrow{x_1} Y \not\rightarrow_{y_1} Z$$
$$\searrow$$
$$R$$

b. Compound Y might be changed to a harmful by-product, R.

$$\begin{array}{c} Y \\ X \xrightarrow{x_1} Y \not\rightarrow_{y_1} Z \\ Y \end{array}$$

c. Compound Y might accumulate and could be harmful if present in large amounts

In these cases, the genetic disease is caused by a lack of enzyme y_1, which in turn is due to a defective gene—one that might have been copied incorrectly, or changed, or even be lacking altogether. We shall see in the following paragraphs how a single simple change in the sequence of a code group in mRNA can lead to a genetic defect.

Common Genetic Diseases

The most common genetic disease in the United States is cystic fibrosis, which affects 1 in 2000 births, although approximately 1 in 20 persons may carry the genetic trait. Although lung function appears normal at birth, early in childhood a thick mucus develops and results in bronchiolar obstruction. Infection occurs and becomes difficult to eradicate, even with antibiotics.

When such an infection develops in the lungs, the subsequent inflammatory reaction results in the destruction of bacteria, leukocytes, and tissue. The process of cellular destruction releases DNA, which in turn substantially increases the viscosity of the mucus. An experimental treatment for cystic fibrosis involves the use of DNase, an enzyme that degrades extracellular DNA but has no effect on the DNA within intact cells. DNase is produced by recombinant DNA technology (see page 414) and is composed of a single chain of 260 amino acids.

Another genetic disease, phenylketonuria, affects 1 in 10,000; most states now require routine screening of newborn infants for this disease. Tay-Sachs disease affects 1 in 1000 among persons of eastern European Jewish origin. One out of every 10 black persons in the United States carries the gene for sickle cell disease.

In almost every ethnic, demographic, or racial group certain genetic diseases occur at much higher frequencies among their members than in the general population. Paget's disease, a disease of the bones, occurs frequently among people of English descent. Polynesians are prone to have clubfoot, and Finns tend to have kidney disease. There is a higher than average frequency of deafness in Eskimos and of glaucoma in Icelanders. Native Americans and Mexicans are very susceptible to diabetes and gallbladder disease. Impacted wisdom teeth occur frequently among Europeans and Asians but rarely among Africans. However, impacted wisdom teeth may also be dietary in origin. Today there are at least 1200 distinct inherited genetic diseases that have been identified.

Sickle Cell Anemia

Notice the arrangement for the 146 amino acids in the β chain of normal hemoglobin, shown in Figure 23-8. Each of these 146 amino acids was designated by a certain arrangement of three nucleotides in mRNA. Altogether, these must have been 438 (3×146) nucleotides, arranged in the proper sequence in order to form this molecule of the β chain of hemoglobin.

```
 1          5                    10                    15
Val-is-Leu-Thr-Pro-Glu-Glu-Lys-Ser-Ala-Val-Thr-Ala-Leu-Trp-Gly-Lys-Val-
   20            25                30                    35
Asn-Val-Asp-Glu-Val-Gly-Gly-Glu-Ala-Leu-Gly-Arg-Leu-Leu-Val-Val-Tyr-Pro-
       40                45                50
Trp-Thr-Gln-Arg-Phe-Phe-Glu-Ser-Phe-Gly-Asp-Leu-Ser-Thr-Pro-Asp-Ala-Val-
   55            60                65                    70
Met-Gly-Asn-Pro-Lys-Val-Lys-Ala-His-Gly-Lys-Lys-Val-Leu-Gly-Ala-Phe-Ser-
         75                80                85                    90
Asp-Gly-Leu-Ala-His-Leu-Asp-Asn-Leu-Lys-Gly-Thr-Phe-Ala-Thr-Leu-Ser-Glu-
              95               100               105
Leu-His-Cys-Asp-Lys-Leu-His-Val-Asp-Pro-Glu-Asn-Phe-Arg-Leu-Leu-Gly-Asn-
   110           115               120               125
Val-Leu-Val-Cys-Val-Leu-Ala-His-His-Phe-Gly-Lys-Glu-Phe-Thr-Pro-Pro-Val-Gln-
         130           135           140                       146
Ala-Ala-Tyr-Gln-Lys-Val-Val-Ala-Gly-Val-Ala-Asn-Ala-Leu-Ala-His-Lys-Tyr-His
```

FIGURE 23-8

Sequence of amino acids in normal hemoglobin β chain.

Look at amino acid 6, the one in color. This amino acid, Glu, is glutamic acid. The mRNA code group for glutamic acid is either GAA or GAG. If the middle codon of this group is changed from A to U, the sequence becomes either GUA or GUG, both of which designate the amino acid valine (Val). That is, if there is a change in only one of the nucleotides, from A to U, on the sixth codon of the 146 amino acid chain of the β chain of hemoglobin, a different type of molecule is produced. This type of hemoglobin is called hemoglobin S and causes the genetic disease sickle cell anemia.

The red blood cells normally have a concave shape when deoxygenated (see Figure 23-9a). Red blood cells containing hemoglobin S look like a sickle (see Figure 23-9b). These sickle cells are more fragile than normal red blood cells, leading to anemia. They can also occlude capillaries, leading to thrombosis. The points and abnormal shapes of the sickle cells cause slowing and sludging of the red blood cells in the capillaries, with resulting hypoxia of the tissues. This produces such symptoms as fever, swelling, and pain in various parts of the body. Eventually the spleen is affected. Many victims of severe sickle cell anemia die in childhood.

Sickle cell anemia is a hereditary condition found primarily among Hispanics and blacks. Many of these people have the sickle cell trait but are relatively unaffected by it until there is a sharp drop in blood oxygen level, such as might be caused by strenuous exercise at high altitudes, underwater swimming, and alcohol intoxication.

Phenylketonuria

Phenylketonuria (PKU) results when the enzyme phenylalanine hydroxylase is absent. A person with PKU cannot convert phenylalanine to tyrosine, and so the phenylalanine accumulates in the body, resulting in injury to the nervous system. In infants and in children up to age 6, an accumulation of phenylalanine leads to retarded mental development.

This disease can be readily diagnosed from a sample of blood or urine. Treatment consists of giving the affected person a diet low in phenylalanine and adding tyrosine to the diet. Tyrosine is essential in the absence of phenylalanine.

Galactosemia

Galactosemia results from the lack of the enzyme uridyl transferase, which catalyzes the formation of glucose from galactose. This disease may result in an increased concentration of galactose in the blood. Galactose in the blood is reduced

a

b

FIGURE 23-9

(a) Normal red blood cells.
(b) A sickled cell.

in the eye to galacticol, which accumulates and causes a cataract. Ultimately, if galactose continues to accumulate, liver failure and mental retardation will occur. This disease can be controlled by the administration of a diet free of galactose.

Wilson's Disease

Wilson's disease is caused by the body's failure to eliminate excess Cu^{2+} ions because of a lack of ceruloplasmin (see page 547) or a failure in the bonding of copper ions to the copper-bonding globulin, or both factors. In this disease, copper accumulates in the liver, kidneys, and brain. There is also an excess of copper in the urine. If deposition of copper in the liver becomes excessive, cirrhosis may develop. In addition, accumulation of copper in the kidneys may lead to damage of the renal tubules, leading to increased urinary output of amino acids and peptides.

Albinism

Albinism is caused by the lack of the enzyme tyrosinase, which is necessary for the formation of melanin, the pigment of the hair, skin, and eyes. Consequently albinos have very white skin and hair. Although this disease is not serious, persons affected by it are very sensitive to sunburn.

Hemophilia

Hemophilia is caused by a missing protein, an antihemophilic globulin, which is important in the normal clotting process of the blood. Consequently, any cut may be life threatening to hemophiliacs, but the primary damage is the crippling effect of repeated episodes of internal bleeding into body joints.

Muscular Dystrophy

One form of muscular dystrophy, Duchenne's muscular dystrophy, is caused by the lack of a protein called dystrophin. This disease primarily affects boys and causes progressive weakness and wasting of muscles. Victims are usually confined to a wheelchair by age 12 and die before age 30 because of respiratory failure. Currently, there is no treatment for this disease.

Dystrophin is a large protein with a molecular mass of 400,000. It is normally found in the muscle cell membrane, on the side facing the cell interior.

Other Genetic Diseases

Niemann-Pick disease is caused by a lack of the enzyme sphingomyelinase, which causes an accumulation of sphingomyelin in the liver, spleen, bone marrow, and lymph nodes. This disease affects the brain and causes mental retardation and early death.

Gaucher's disease is caused by a lack of the enzyme glucocerebrosidase, which is necessary for the cleavage of glucocerebrosides into glucose and ceramide. This disease is characterized by the accumulation of glycolipids in the spleen and liver. In children, Gaucher's disease causes severe mental retardation and early death (see page 372). In adults, the spleen and liver enlarge progressively but the disease is compatible with long life.

Tay-Sachs disease is due to a lack of the enzyme hexosaminidase A, leading to the accumulation of glycolipids in the brain and the eyes. Red spots show up in the retina, and there is also muscular weakness. This disease is fatal to infants before the age of 4.

Farber's disease, due to a lack of the enzyme ceramidase, is characterized by hoarseness, dermatitis, skeletal degeneration, and mental retardation.

Krabbe's disease, due to a lack of the enzyme β-galactosidase, is characterized by mental retardation and the absence of myelin.

Lou Gehrig's Disease

Amyotrophic lateral sclerosis, ALS, also known as Lou Gehrig's disease, is characterized by the degeneration of the nerves that control motor activity in the brain and spinal chord. This degeneration results in a progressive wasting of the muscles. Individuals with such a disease lose their ability to walk, talk, and swallow, Their intellect, however, is unimpaired. Most patients die within 2 to 5 years after the first symptoms, and death usually comes from suffocation

Recent evidence has shown that patients with ALS have an accumulation of glutamate in the fluid between the cells in the brain. As a result of this accumulation, excess calcium ions flow into the nerve cells. The excess calcium ions, in turn, have a deadly effect on the neurons that control motor activity, and those cells soon die.

23-10 Classification of Genetic Diseases

Genetic diseases may be classified into three major categories:

1. *Chromosomal disorders*, in which there is an excess or loss of chromosomes, deletion of part of a chromosome, or translocation of a chromosome. Examples include Down's syndrome and chronic myelogenous leukemia.
2. *Monogenic disorders*, which involve one mutant gene. This category may be further subdivided into
   (a) Autosomal dominant (example, Huntington's chorea)
   (b) Autosomal recessive (example, cystic fibrosis)
   (c) X-linked (example, hemophilia)
3. *Multifunctional disorders* involve the action of a number of genes. One example is essential hypertension.

Treatment of Genetic Diseases

One of the following methods may be used in the treatment of genetic diseases.

1. Correct the metabolic consequences of the disease by supplying the missing product. For example, a patient with familial goiter can be treated with the missing hormone L-thyroxine.
2. Replace the missing enzyme or hormone. Examples: injection of β-glucosidase in the treatment of Gaucher's disease and injection of factor VIII in the treatment of hemophilia.
3. Remove excess stored substance. Example: administration of vitamin B_{12} in the treatment of methylmelanic aciduria.
4. Correct the major genetic abnormality. One example is the use of a liver transplant in a patient with galactosemia (primarily used on patients with advanced cases of this disease).

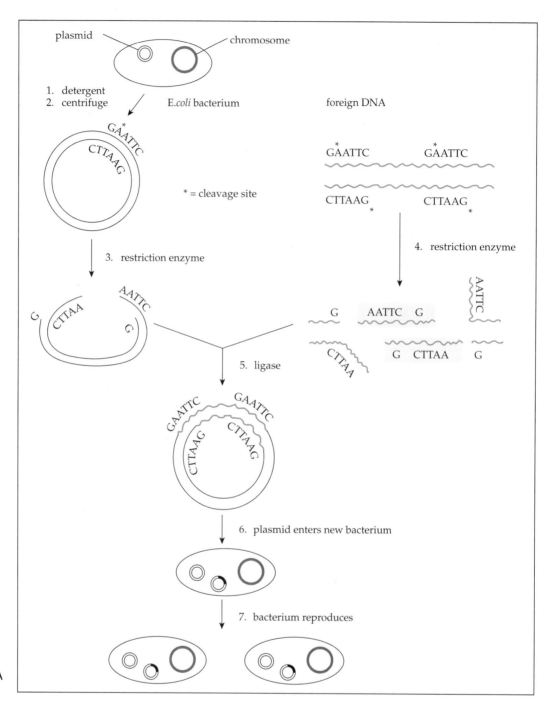

FIGURE 23-10
Recombinant DNA
in *Escherichia coli*.

23-11 Recombinant DNA (Gene Splicing)

The term recombinant DNA refers to DNA molecules that have been artificially created by splicing segments of DNA from one organism into the DNA of a completely different organism. Almost any organism (animal, plant, bacteria, virus) can serve as a DNA donor. Theoretically, any organism also can serve as a DNA acceptor, but to date the bacterium *Escherichia coli* (E. coli) has been used almost exclusively as a DNA acceptor. E. coli was selected because its genetic makeup has been thoroughly studied. Most of *E. coli's* 3000 to 4000 genes are contained within a large, single, ringed chromosome. In addition to this large chromosome, there are plasmids, smaller closed loops of DNA consisting of only a few genes. It is these plasmids that act as DNA acceptors for *E. coli*, since the plasmids are known to multiply independently within the host cell as the cell itself replicates.

How does gene splicing work? First, the *E. coli* is placed in a detergent solution to break open the cells. The plasmids are treated with *restriction enzymes*, which cut the plasmid DNA at specific points. DNA from another organism (also cleaved by a similar method) is inserted into the *E. coli* plasmid. The plasmid ring is then closed, and the plasmid is put back into *E. coli*. The new cells are isolated and allowed to replicate (*E. coli* duplicates itself every 20 to 30 min). As the cells replicate, they synthesize the proteins coded for by their DNA, including those coded for by the inserted DNA (see Figure 23-10).

Recombinant DNA research has produced human insulin, interferon, growth hormone, somatostatin, and some vaccines. Recombinant DNA research has led to two widely used drugs to treat lowered red blood cell levels in dialysis patients and neutropenia in chemotherapy patients. Recently, recombinant DNA products have been tried in patients with cystic fibrosis and emphysema. Theoretically, gene splicing could be used to find cures for genetic diseases and for cancer and to manufacture various enzymes, hormones, antibodies, and other substances that are difficult to isolate and produce pharmaceutically.

Opponents of recombinant DNA research point out the considerable risks involved. They fear that a synthetic strain of *E. coli* might be accidentally released and cause a worldwide epidemic and the death of large populations. They also believe that this type of research might be used politically to control human behavior.

23-12 Genetic Markers

One of the newer topics of cancer research is genetic markers, defects in one part of a chromosome that it is believed can cause a certain disease. A genetic marker for Alzheimer's disease has been found on chromosome 21, the same chromosome that is associated with Down's syndrome. Since a chromosome can include many, many genes, much work is yet to be done in this area.

23-13 Oncogenes

Oncogenes are genes that appear to trigger uncontrolled, or cancerous, growth. Cancerous cells exhibit three general characteristics:

1. Uncontrolled growth
2. Invasion of body tissues
3. Spread to other body parts (metastasis)

In cancer, the genes controlling growth are abnormal, but little is known about how the cell growth is controlled.

Cancer is the second most common cause of death in the United States, after cardiovascular disease.

The incidence of cancer increases with age, and a wide variety of body organs can be affected. The abnormal production of enzymes, hormones, and proteins is frequently associated with cancer. These products are called *tumor markers*. Measurement of such markers is helpful in diagnosing and treating cancer. For example, the detection of the enzyme prostatic acid phosphatase may be associ-

Table 23-2 Some Oncogenic Viruses	
Class	Members
DNA virus	
Papovavirus	Polyomavirus, SV 40 virus
Herpesvirus	Epstein-Barr virus, herpes simplex type 2 virus
Hepadnavirus	Hepatitis B virus
RNA virus	
Retrovirus type C	Leukemia virus

ated with cancer of the prostate gland, and abnormal amounts of calcitonin may indicate carcinoma of the thyroid gland.

What causes cancer? It is believed that all cancer-causing agents can be grouped into three main categories: radiant energy, chemical compounds, and viruses.

Radiant energy such as x rays and γ rays are carcinogenic because of the formation of *free radicals in the tissues. Radiation can also damage DNA and so is *mutagenic.

Many *chemical compounds* in common use are known to be carcinogenic. As much as 75 percent of human cancers are caused by chemicals in the environment. Among these carcinogenic chemicals are benzene, asbestos, and fused-ring compounds such as the benzpyrene found in cigarette smoke (see page 279).

Oncogenic viruses contain either DNA or RNA. Under certain circumstances, infection of appropriate cells with polyoma virus or SV 40 virus can result in a malignant transformation. Table 23-2 lists some oncogenic viruses.

24-14   DNA Fingerprinting

Fingerprint analysis was and still is an effective method for identifying individuals since no two persons have the same fingerprints. However, modern genetic "fingerprinting" can accomplish the same thing since no two individuals have the same DNA unless they are identical twins. Samples of DNA can be taken from saliva, skin, hair, semen, or bone. With current laboratory procedures only a small part of the DNA sequence is examined. The primary method used in DNA fingerprinting is called *restriction fragment length polymorphism* (RFLP). In this method, special enzymes are used to "cut" samples of DNA into sections called restriction fragments, which are then separated by gel electrophoresis according to size. The electrophoresis patterns are quite distinctive and when exposed to photographic film produce a "fingerprint." A match can have an accuracy of 99%.

Samples too small to analyze by the RFLP method can be identified by the *polymerase chain reaction* (PCR) method, whereby the DNA is recopied many times until enough is present for a proper analysis. This method can be used to identify the DNA present in one tiny drop of blood or in one strand of hair.

Summary

The two types of nucleic acids are deoxyribonucleic acid (DNA), which is primarily responsible for the transfer of genetic information, and ribonucleic acid (RNA), which is primarily responsible for protein synthesis.

The Watson-Crick model of DNA proposed a double-coiled chain consisting of two strands intertwined around one another. The chains are composed of alternating sugar (deoxyribose) and phosphates, with the sugar being bonded to one of four different heterocyclic compounds—adenine, guanine, cytosine, and thymine (A, G, C, T). The model suggests that the adenine of one chain is always bonded to a thymine of the opposite chain and that cytosine is always bonded to guanine.

All DNA molecules have the same sequence of deoxyribose and phosphates in the ladder part of the chain. The difference lies in the order of adenine, thymine, guanine, and cytosine.

When a cell divides, the DNA molecule uncoils and each half acts as a template for the formation of a new chain.

During replication, the DNA molecule uncoils and both halves produce molecules identical to the original. However, when the DNA molecule synthesizes RNA, part of the coil again unwinds, but only one half acts as the template to produce RNA. There are four kinds of RNA: ribosomal RNA (rRNA), messenger RNA (mRNA), transfer RNA (tRNA), and heterogeneous RNA (hRNA).

When mRNA is formed from a gene in DNA, guanine is still bonded to cytosine but adenine is bonded to uracil (U) rather than to thymine.

Messenger RNA moves through the cytoplasm to the ribosomes, where it acts as a template for the formation of protein. The ribosomes contain tRNA; tRNA carries a specific amino acid to the mRNA. The identity of this amino acid is determined by the code on the end of the tRNA, and this code is specific for another code on the mRNA.

As the mRNA moves through the ribosome, the first of its coded groups picks up and holds a corresponding coded group on tRNA. The second, third, and succeeding groups on mRNA do likewise. The specific amino acids carried by the tRNA become bonded to one another to form the designated protein, and the tRNAs are released to find more of the specific amino acids and begin the procedure again.

The DNA code consists of four letters (A, T, C, and G) arranged in groups of three. The three-letter coded groups specify the 20 primary amino acids occurring in nature.

If one of the code letters is changed or is missing, the information copied will be incorrect and a mutation will occur. Mutations occur because of exposure to radiation, both naturally and in industry and medicine, and because of certain chemicals.

DNA directs the synthesis of enzymes (which are also proteins). If the message is transferred incorrectly, the proper enzyme will not be synthesized and so will be lacking in the body. This lack of a specific enzyme may lead to a genetic disease such as phenylketonuria, galactosemia, albinism, pentosuria, or hemophilia.

Genetic diseases may be classified as chromosomal disorders, as monogenic disorders, or as multifunctional disorders.

Oncogenes are genes that appear to trigger cancerous growth. Cancerous cells exhibit uncontrolled growth, invasion of body tissues, and metastasis.

Questions and Problems

1. Diagram the Watson-Crick model of DNA. What types of bonds are present in DNA? Are they all of equal strength?

2. Describe DNA replication.

3. What are the types of RNA? What is the function of each?

4. List the difference(s) in nucleotides in mRNA and DNA?

5. Describe the relationship between chromosomes, genes, and DNA?

6. How do mRNA and tRNA function in the formation of protein?

7. List several common genetic diseases.

8. How many primary amino acids occur in nature? Name them and give their abbreviations.

9. How was the DNA code decoded?

10. Which amino acid does each of the following coded groups of mRNA represent?
    a. UUU   b. CCC
    c. ACG   d. GCA
    e. CGA

11. What happens to DNA to cause a mutation?

12. What is sickle cell disease? What causes it? What is the difference in the shapes of the red blood cells? What effect does this have on their ability to carry oxygen?

13. What causes the following genetic diseases, and what are the symptoms of each?
    a. Tay-Sachs disease   b. albinism
    c. hemophilia   d. PKU
    e. galactosemia   f. Wilson's disease

14. Explain genetic diseases in terms of enzyme deficiency.

15. Why is cytosine more basic than uracil or thymine?

Practice Test

1. The insertion of the wrong nucleotide into a DNA code group could result in _____.
   a. mutation
   b. splicing
   c. recombination
   d. cloning

2. The most common genetic disease in the United States is _____.
   a. muscular dystrophy
   b. galactosemia
   c. hemophilia
   d. cystic fibrosis

3. The end product of purine metabolism is _____.
   a. ammonia
   b. creatinine
   c. urea
   d. uric acid

4. The RNA concerned with the transcription of genetic information is _____.
   a. homogeneous
   b. messenger
   c. transfer
   d. ribosomal

5. The number of nucleotides necessary to code for an amino acid is _____.
   a. 1
   b. 2
   c. 3
   d. 4

6. How many different amino acids are involved in the genetic code?
   a. 5
   b. 10
   c. 20
   d. an infinite number

7. Which of the following is part of a nucleotide?
   a. phosphoric acid
   b. purine or pyrimidine base
   c. five-carbon sugar
   d. all of these

8. The two chains of DNA are held together by _____ bonds.
   a. disulfide
   b. hydrogen
   c. ionic
   d. covalent

9. Which of the following is *not* a genetic disease?
   a. phenylketonuria
   b. goiter
   c. albinism
   d. sickle cell anemia

10. Guanosine pairs with _____.
    a. uracil
    b. cytosine
    c. thymine
    d. adenosine

Enzymes

Objectives

- To understand the functions of enzymes
- To distinguish between activators and inhibitors
- To understand the mode of enzyme activity
- To distinguish between apoenzymes and coenzymes
- To become familiar with the classification of enzymes
- To become aware of the clinical significance of plasma enzyme concentrations

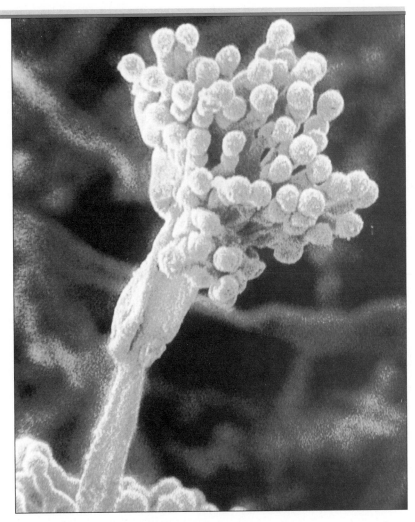

Enzymes catalyze the chemical reactions necessary for growth, repair, and maintenance of a living organism: Without them, it could not survive. The fungus Penicillium, pictured here, produces the potent antibiotic penicillin. Penicillin is lethal to bacteria because it incapacitates an enzyme instrumental in the construction of the bacterial cell wall. Because humans lack similar enzymes, we are immune to penicillin's effects.

Enzymes are biologic catalysts. Catalysts are substances that increase the speed of a chemical reaction by lowering the energy requirement. Although a catalyst influences a chemical reaction, it is not itself permanently changed, nor does it cause the reaction to occur; that is, a catalyst can increase the speed of a reaction but cannot cause that reaction if it would not occur in the absence of that catalyst. Since catalysts are not used up, they can be used over and over again.

Enzymes are organic catalysts produced by living organisms. The reactant in an enzyme-catalyzed reaction is called the *substrate*. The active site of an enzyme is that small portion of the molecule that is responsible for the catalytic action of the enzyme. Enzymes provide a chemical pathway that has a lower *activation energy than the same reaction uncatalyzed. How do enzymes differ from nonbiologic catalysts, and why are there so many in the body? Enzymes are superior to other catalysts in several ways.

1. They have a much greater catalytic power. Consider the reaction

$$CO_2 + H_2O \xrightleftharpoons{\text{carbonic anhydrase}} H_2CO_3$$

which takes place in red blood cells (see page 534). The enzyme carbonic anhydrase increases the reaction rate more than 10 million times that of the same reaction without the presence of the enzyme. In general, enzymes increase the rate of reaction from more than 1 million to more than 1 trillion times faster than a corresponding reaction without an enzyme. Few catalysts can cause such an increase in reaction rate.

2. Enzymes are highly specific with varying degrees of specificity. Some enzymes exhibit *absolute specificity*. They act on one substrate and only on that substrate. For example, the enzyme succinic dehydrogenase (see page 466) catalyzes the conversion of succinic acid to fumaric acid. Not only does succinic dehydrogenase act specifically on succinic acid, it produces only a *trans* isomer (fumaric acid) and never the *cis* isomer (malic acid).

fumaric acid (a trans isomer)   malic acid (a cis isomer)

A second example of such an enzyme is nitric oxide synthase (NO synthase), which catalyzes the production of NO. While there is a plentiful supply of oxygen in all cells, there is only one source of nitrogen for this particular reaction. It is the amino acid arginine:

$$\text{arginine} \xrightarrow{\text{NO synthase}} \text{citrulline} + NO$$

Some enzymes exhibit *stereospecificity*. Such enzymes can detect the difference between optical isomers (mirror images) and select only one of such isomers. For example, the enzyme arginase catalyzes the hydrolysis of L-arginine but has no effect on the hydrolysis of D-arginine.

Some enzymes are *linkage specific*. That is, they catalyze the reaction that breaks the bonds only between specific groups. For example, the enzyme thrombin will break the bonds between the amino acids arginine and glycine and does not affect bonds between other amino acids.

Other enzymes exhibit *reaction specificity*; they catalyze certain types of reactions. Esterases (see page 429) catalyze the hydrolysis of esters in general, and carbohydrases catalyze the hydrolysis of carbohydrates.

Still other enzymes exhibit *group specificity*. Chymotrypsin catalyzes the hydrolysis of only those proteins that contain phenylalanine, tryptophan, or tyrosine.

3. The activity of enzymes is closely regulated, whereas that of catalysts is difficult to control. There are substances in cells that can increase or decrease the activity of an enzyme and thus control the rate of the particular reaction that a cell requires.

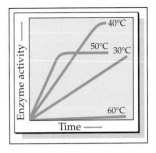

FIGURE 24-1
Effect of temperature on enzyme activity.

24-1 Enzyme Reactions

Enzymes are proteins[†] and therefore undergo all the reactions that proteins do. That is, enzymes can be coagulated by heat, alcohol, strong acids, and alkaloidal reagents. Many enzymes have now been prepared in crystalline form.

Temperature Requirement

The speed of all chemical reactions is affected by temperature: the higher the temperature, the faster the rate of the reaction (Figure 24-1). This is also true for reactions involving enzymes. However, if the temperature is raised too much, the enzyme (protein) will be inactivated by denaturation and will be unable to function.

The best temperature for enzyme function—the temperature at which the rate of a reaction involving an enzyme is the greatest— is called the optimum temperature for that particular enzyme. At higher temperatures, the enzyme will coagulate and will be unable to function. At temperatures below the optimum temperature, the rate of reaction will be slower than the maximum rate.

Because many enzymes have an optimum temperature near 40° C, or close to that of body temperature, they function at maximum efficiency in the body.

Role of pH

Each enzyme has a pH range within which it can best function (Figure 24-2). This is called the optimum pH range for that particular enzyme. For example, the optimum pH of pepsin, an enzyme found in gastric juice, is approximately 2.0, whereas the optimum pH of trypsin, an enzyme found in pancreatic juice, is near 8.2. If the pH of a substrate is too far from the optimum pH required by the enzyme, that enzyme cannot function at all. However, since body fluids contain buffers, the pH usually does not vary too far from the optimum values.

Effects of Concentrations

As with all chemical reactions, the speed is increased with an increase in concentration of reactants. With an increased concentration of substrate, the rate of the reaction will increase until the available enzyme becomes saturated with substrate. Also, with an increase in the amount of enzyme, the rate of reaction will increase, assuming an unlimited supply of substrate.

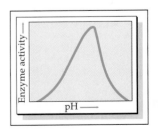

FIGURE 24-2
Effect of pH on enzyme activity.

[†]Recent discoveries show that ribonucleic acid (RNA) (see Section 23-1) can act as an enzyme, but such enzymatic reactions are beyond the scope of this book.

24-2   Activators and Inhibitors

Inorganic substances that tend to increase the activity of an enzyme are called activators. For example, the magnesium ion (Mg^{2+}) is an inorganic activator of the enzyme phosphatase, and the zinc ion (Zn^{2+}) is an activator for the enzyme carbonic anhydrase (see page 524).

An enzyme inhibitor is any substance that will make an enzyme less active or render it inactive. Enzyme inhibitors that bind reversibly to the active site and so block access by the substrate (see Figure 24-6) are called competitive inhibitors. Their effect can be overcome by increasing the concentration of the substrate. Other inhibitors that bind to another site on the enzyme to render it less active or inactive are called noncompetitive inhibitors. They act by changing the conformation of the enzyme, thereby reducing or stopping its activity. *Irreversible inhibitors* form strong covalent bonds with the enzyme, rendering it inactive. This effect cannot be overcome by increasing the concentration of the substrate.

Zalcitabin (dideoxycytidine) is used in patients with advanced HIV infection. Inside the cell, zalcitabin is converted into its active metabolite dideoxycytidine 5'-triphosphate (ddCTP). ddCTP in turn acts as a competitive inhibitor of the natural substrate deoxycytidine triphosphate (dCTP) for the active site of viral reverse transcriptase and serves as an alternate substrate to dCTP for HIV reverse transcriptase. As a result of this competitive inhibition, replication of HIV-1 is inhibited.

Heat, changes in pH, strong acids, alcohol, and alkaloidal reagents can all denature protein. These are examples of nonspecific inhibitors. They affect all enzymes in the same manner. Specific inhibitors affect one single enzyme or group of enzymes. In this category are most poisonous substances, such as cyanide ion (CN^-), which inhibits the activity of the enzyme cytochrome oxidase (see Section 26-10).

Poisons   Many enzyme inhibitors are poisonous because of their effect on enzyme activity. Mercury and lead compounds are poisonous because they react with sulfhydryl (—SH) groups of an enzyme and so change its conformation (Figure 24-3). The subsequent loss of enzyme activity leads to the various symptoms of lead and mercury poisoning, such as loss of equilibrium, hearing, sight, and touch, which are generally irreversible.

Organic phosphorus compounds are frequently poisonous because they act as irreversible inhibitors for cholinesterase, an enzyme that breaks down acetylcholine in the synaptic gap of nerve cells in the brain. Such organic phosphorus compounds are called nerve poisons. Among them are the insecticides malathion and parathion and various war gases.

Once a neurotransmitter is blocked, the receptor sites in the brain "fire" repeatedly. This overstimulates the muscles. The heart beats rapidly and irregu-

FIGURE 24-3

Mercury and lead compounds are noncompetitive inhibitors, reacting with an enzyme to change its conformation.

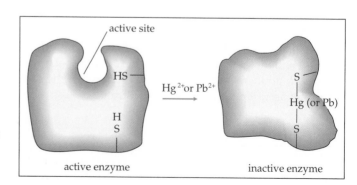

larly; convulsions occur, and death ensues rapidly. However, there antidotes for several organic phosphorus poisons. Some function by displacing the poisonous substance from the enzyme. Others, such as atropine, function by blocking the effects of excess acetylcholine.

Drugs While some enzyme inhibitors are poisonous, others are beneficial to life. Penicillin acts as an enzyme inhibitor for transpeptidase, a substance that bacteria need to build their cell walls (see page 430). If the cell wall is lacking, osmotic pressure causes the bacterial cell to burst and die. However, new strains of bacteria have developed an enzyme, penicillinase, that inactivates penicillin. To destroy these new strains, synthetically modified penicillins have been prepared so that this antibiotic remains effective.

Cyclic AMP (cAMP) acts as a chemical messenger to regulate enzyme activity within the cells that store carbohydrate and fat. Without cAMP the activity of all the enzymes working at maximum speed within the cells would soon create chaos. It also appears that an inadequate supply of cAMP can lead to one type of the uncontrolled cell growth that we call cancer.

24-3 Mode of Enzyme Activity

How do enzymes act? Why are they so specific toward certain substrates?

Each enzyme contains an "active site"—that section of the molecule at which combination with the substrate takes place. The active site consists of different parts of the protein chain (the enzyme). These parts are brought close together by the folding and bending of the protein chain (the secondary and tertiary structures), so that the active site occupies a relatively small area. The fact that enzymes (proteins) can be denatured by heat (changed in three-dimensional configuration) indicates the importance of structural arrangement (Figure 24-4).

It is believed that enzyme activity occurs in two steps. First, the active site of the enzyme combines with the substrate to form an enzyme-substrate complex (see Figure 24-5b). This enzyme-substrate complex then breaks up to form the products and the free enzyme, which can react again (Figure 24-5c). According to this theory

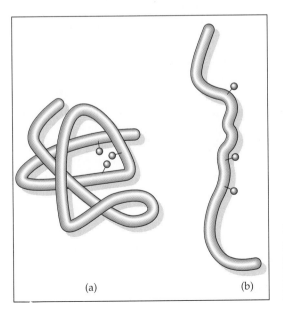

(a)   (b)

FIGURE 24-4

(a) Representation of an active site in an enzyme. (b) Denatured enzyme— parts of the active site are no longer in close proximityl.

FIGURE 24-5

Schematic representation of the interaction of enzyme and substrate. (a) The active site on the enzyme and the substrate have complementary structures and hence fit together as a key fits a lock. (b) While they are bonded together in the enzyme–substrate complex, the catalytic reaction occurs. (c) The products of the reaction leave the surface of the enzyme, freeing the enzyme to combine with another molecule of substrate.

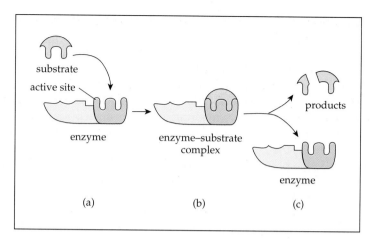

FIGURE 24-6
Enzyme–inhibitor complex.

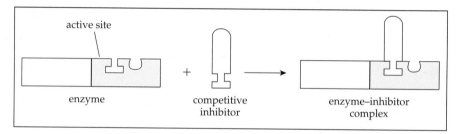

(the lock-and-key model), the substrate must "fit" into the active site of the enzyme—hence the specificity of that enzyme. For substrates that have an optically active site, it is believed that there must be three points of attachment between the substrate and the active site of the enzyme so that only one of the two optical isomers "fits." In the body, enzymes are specific for the L-amino acids and the D-carbohydrates.

A more recent version of the activity of an enzyme, the induced-fit model, suggests that the active site is not rigid, as in the lock-and-key model, but flexible. That is, the site changes in conformation upon binding to a substrate in order to yield an enzyme-substrate fit.

24-4 Apoenzymes and Coenzymes

As was mentioned previously, most enzymes are proteins. However, some enzymes are simple proteins—they yield only amino acids on hydrolysis. Examples of such enzymes are pepsin and trypsin (Section 25-6). Other enzymes are conjugated proteins—they contain a protein and nonprotein part. Both parts must be present before the enzyme can function. The protein part is called the *apoenzyme and the nonprotein (organic) part is called the *coenzyme.

Coenzymes

Coenzymes are not proteins and so are not inactivated by heat. Examples of coenzymes are the vitamins or compounds derived from vitamins.

The reaction involving a coenzyme can be written as follows:

$$\text{coenzyme} + \text{apoenzyme} \longrightarrow \text{enzyme}$$

Coenzyme A (CoA, structure [24-1]) is essential in the metabolism of carbohydrates, lipids, and proteins in the body. It also functions in certain acetylation reactions. Hydrolysis of CoA yields pantothenic acid (a B vitamin), adenine (a purine), ribose (a sugar), phosphoric acid, and mercaptoethanolamine (a sulfur compound).

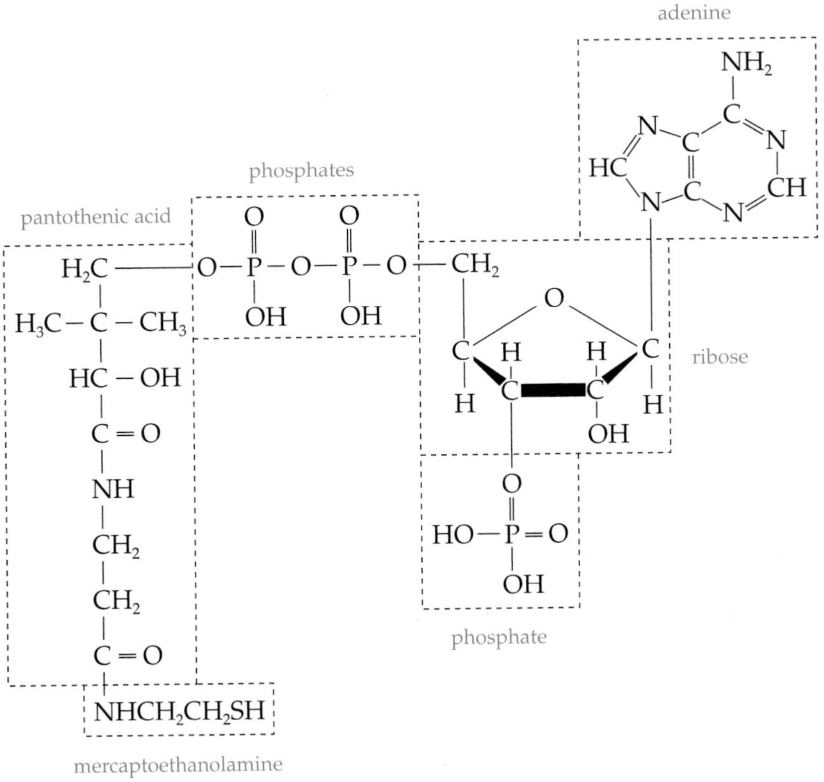

coenzyme A

Acetyl coenzyme A (acetyl CoA) has an acetyl group attached to the sulfur atom. Acetyl CoA functions in the oxidation of food in the Krebs (critic acid) cycle (see Chapter 26).

Nicotinamide is a very important constituent of two coenzymes—nicotinamide adenine dinucleotide (NAD^+) and nicotinamide adenine dinucleotide phosphate ($NADP^+$). These coenzymes are involved in most oxidation–reduction reactions in the mitochondria. They also take part in the Krebs cycle. The structure of NAD^+ is indicated in formula (24-2).

When NAD^+ is reduced to NADH, another hydrogen atom is attached to the carbon atom indicated by asterisk in structure (24-2). The charge on the nitrogen atom is eliminated, and the double bond between that nitrogen and the marked carbon is changed to a single bond. Recent evidence indicates that a defect in the enzyme responsible for the conversion of NADH to NAD^+, NADH dehydrogenase, may be involved in cystic fibrosis.

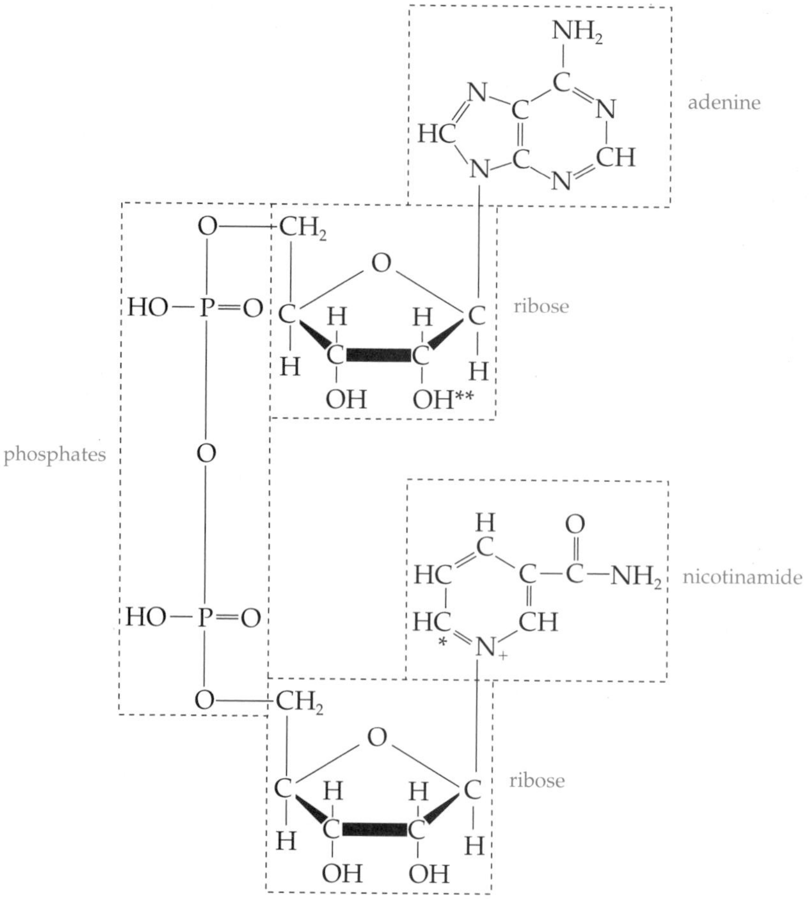

$$(24-2)$$

nicotinamide adenine dinucleotide (NAD^+)

In $NADP^+$, another phosphate group has been added in the position marked with a double asterisk in NAD^+ (structure [24-2]), so that $NADP^+$ contains three phosphate groups.

Coenzyme Q is found in the *mitochondria and has a structure similar to those of vitamin K and vitamin E (see pages 559 and 560). The structure of coenzyme Q is

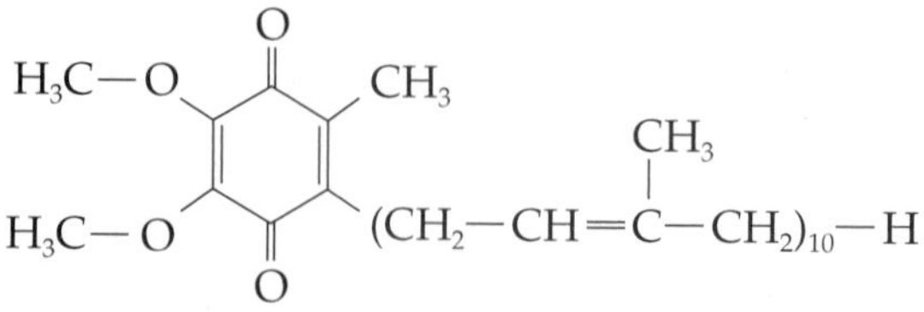

coenzyme Q

Coenzyme Q is also called coenzyme Q_{10} because in humans it contains ten side chain units. Coenzyme Q functions in electron transport and in oxidative phosphorylation.

Coenzymes frequently contain B vitamins as part of their structure. Many coenzymes involved in the metabolism of amino acids contain pyridoxine, vitamin B_6. Other B vitamins, such as riboflavin, pantothenic acid, nicotinamide, thiamine, and lipoic acid, are found in coenzymes involved in oxidation-reduction reactions.

B vitamins such as folic acid and cobalamin (vitamin B_{12}) are part of different coenzymes.

Reactions involving enzymes and coenzymes usually are written as a series of interconnected equations. An example of such a series of reactions is diagram (24-3). In step 1 a substrate, S, is oxidized by NAD^+, which in turn is reduced to NADH. However, the NAD^+ must be regenerated so that it can act again. Therefore, in step 2, NADH is oxidized to NAD^+ by a flavoprotein, which in turn is reduced to flavoprotein H_2. In step 3, the flavoprotein is regenerated by the cytochrome in which Fe^{2+} is oxidized to Fe^{3+}. Finally, in step 4, the cytochrome is regenerated by oxygen, with the final product being water. Such a series of reactions is called a chain. Series (24-3) illustrates the steps in the respiratory chain (see Figure 26-1).

(24-3)

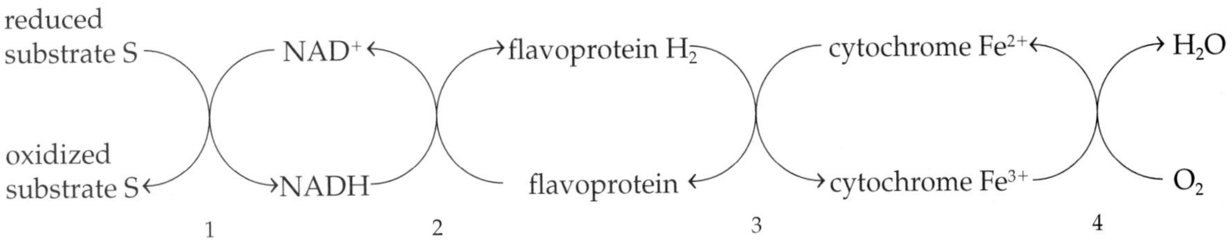

24-5 Nomenclature

Formerly enzymes were given names ending in *-in*, with no relation being indicated between the enzyme and the substance it affects—the substrate. Some of the enzymes named under this system are listed in Table 24-1.

The current system for naming enzymes uses the name of the substrate or the type of reaction involved, with the ending *-ase*. Table 24-2 lists some enzymes and substrates named under the preferred system.

24-6 Classification

The Commission on Enzymes of the International Union of Biochemistry has classified enzymes into six divisions. Each of these divisions can be further subdivided into several classes. The following paragraphs indicate these divisions and some of the classes. The older names of the enzymes are still in common use and are given throughout this text merely for simplicity. It is much easier to write *sucrase* than α-*glucopyrano-β-fructofuranohydrolase*.

Table 24-1

Enzymes Named Under the Older System

Enzyme	Substrate
Rennin	Casein
Pepsin	Protein
Trypsin	Protein
Ptyalin	Carbohydrate

Table 24-2	Enzymes and Substrates or Reaction Types
Enzyme	**Substrate or Reaction Type**
Maltase	Maltose
Urease	Urea
Proteases	Proteins
Carbohydrases	Carbohydrates
Lipases	Lipids
Hydrolases	Hydrolysis reactions
Deaminases	Removing amines
Dehydrogenases	Removing hydrogens

Oxidoreductases

Oxidoreductases are enzymes that catalyze oxidation-reduction reactions between two substrates. The enzymes that catalyze oxidation-reduction reactions in the body are important because these reactions are responsible for the production of heat and energy. Recall that oxidation-reduction requires a transfer of electrons. Many of these enzymes are present in the mitochondria, which are the most prominent structural and functional units in animal cells, except for the nucleus (see Plate 28). Mitochondria are sausage-shaped objects ranging in size from 0.2 to 5 μm (1 μm is one millionth of a meter). The cytoplasm of a typical cell contains from 50 to 50,000 mitochondria.

Inside the mitochondrion is a double-layered membrane that is folded back and forth to form sacs called *cristae*. These cristae have a tremendous amount of surface area. On the surfaces of these membranes are enzyme-containing particles. Coenzymes are present in the fluid between the membranes. The mitochondria are highly organized and hold their enzymes in a definite spatial arrangement for the most efficient control of various cellular processes.

The mitochondria play an important part in oxidative phosphorylation in the Krebs cycle, which relates to carbohydrate metabolism (Chapter 26). Mitochondria contain enzymes needed for the oxidation of fatty acids. They also are involved in the metabolism of amino acids.

In addition to these functions, the mitochondria manufacture adenosine triphosphate (ATP), the main energy-supplying substance in the cell.

Dehydrogenases catalyze the removal of hydrogen from a substrate. Oxidases activate oxygen so that it will readily combine with a substrate. Catalases catalyze the decomposition of hydrogen peroxide to water and oxygen. Peroxidases catalyze the decomposition of organic peroxides to hydrogen peroxide and water.

Transferases

Transferases are enzymes that catalyze the transfer of a functional group between two substrates.

Hydrolases—The Hydrolytic Enzymes

The hydrolytic enzymes—hydrolases—catalyze the hydrolysis of carbohydrates, esters, and proteins. They are named for the substrate upon which they act. Some hydrolases are present in the cytoplasm in organelles called lysosomes.

Carbohydrases are enzymes that catalyze the hydrolysis of carbohydrates into simple sugars. The various carbohydrases are as follows:

1. *Ptyalin*, or salivary amylase, for the hydrolysis of starch to dextrins and maltose
2. *Sucrase*, for the hydrolysis of sucrose to glucose and fructose; sucrase occurs in the intestinal juice
3. *Maltase*, for the hydrolysis of maltose to glucose; maltase occurs in the intestinal juice
4. *Lactase*, for the hydrolysis of lactose to glucose and galactose; lactase occurs in the intestinal juice
5. *Amylopsin*, or pancreatic amylase, for the hydrolysis of starch to dextrins and maltose; pancreatic amylase occurs in the pancreatic juice

Esterases are enzymes that catalyze the hydrolysis of esters into acids and alcohols. The various types of esterases are as follows:

1. *Gastric lipase*, for the hydrolysis of fats to fatty acids and glycerol
2. *Steapsin*, or pancreatic lipase, for the hydrolysis of fats to fatty acids and glycerol
3. *Phosphatase*, for the hydrolysis of phosphoric acid esters to phosphoric acid

Proteases are enzymes that catalyze the hydrolysis of protein to derived protein and amino acids. These are of two types: proteinases and peptidases.

The proteinases, for the hydrolysis of proteins to peptides, are as follows:

1. *Pepsin*, found in the gastric juice, for the hydrolysis of protein to polypeptides
2. *Trypsin*, found in the pancreatic juice, for the hydrolysis of protein to polypeptides
3. *Chymotrypsin*, found in the pancreatic juice, for the hydrolysis of protein to polypeptides

The **peptidases**, for the hydrolysis of polypeptides to amino acids, are as follows:

1. *Aminopeptidases*, from the intestinal juices
2. *Carboxypeptidases*, from the pancreatic juice

Nucleases are enzymes that catalyze the hydrolysis of nucleic acids. Examples are ribonuclease and deoxyribonuclease.

Lyases

Lyases are enzymes that catalyze the removal of groups from substrates by means other than hydrolysis, usually with the formation of double bonds. An example is fumarase, which catalyzes the change of fumaric acid to L-malic acid in the Krebs cycle (see Figure 26-12).

Isomerases

Isomerases are enzymes that catalyze the interconversion of cis-trans isomers. One example is retinal isomerase, which catalyzes the conversion of 11-*trans*-retinal to 11-*cis*-retinal (see page 555). Another example is alanine racemase, which catalyzes the conversion of L-alanine to D-alanine, which is the form the body can use.

Ligases

Ligases, or synthetases, are enzymes that catalyze the coupling of two compounds with the breaking of pyrophosphate bonds (see page 457). One example is the enzyme that catalyzes the formation of malonyl CoA during lipogenesis (see Section 27-7).

24-7 Enzymes of the Kidneys

If an individual's blood pressure drops, as in the case of hemorrhaging or in hypokalemia (page 544), the kidneys secrete the enzyme renin (sometimes considered a hormone) into the bloodstream. The following reactions occur.

$$\text{angiotensinogen} \xrightarrow{\text{renin}} \text{angiotensin I} \xrightarrow{\text{converting enzyme}} \text{angiontensin II}$$

Angiotensin II increases the force of the heartbeat and constricts the arterioles, thus causing an increase in blood pressure. Angiotensin II brings about the contraction of smooth muscle and also triggers the release of the hormone aldosterone (page 585), which aids in the retention of water. Actually, angiotensin II is the most powerful *vasoconstrictor known. It is an octapeptide; angiotensin I is a decapeptide.

Angiotensinogen is an α-globulin produced in the liver.

Various conditions can cause the kidneys to release renin. Among these are a decrease in blood pressure and increased amounts of prostaglandins and β-adrenergic agents. On the other hand, factors that tend to decrease the release of renin include vasopressin, angiotensin II, β-adrenergic antagonists, prostaglandin inhibitors, and an increase in blood pressure.

Angiotensins are deactivated by enzymes called *angiotensinases*.

Other kidney enzymes include glucose-6-phosphatase, which is involved in the removal of the phosphate group from glucose-6-phosphate, thereby enabling glucose to diffuse from the cell into the bloodstream; glutaminase, which is involved in the conversion of glutamine into glutamic acid and NH_4^+; and a hydroxylase, which is involved in the synthesis of calcitriol (see page 595).

24-8 Chemotherapy

Chemotherapy is the use of chemicals to destroy infectious microorganisms and cancerous cells without damaging the host's cells. These chemicals function by inhibiting certain cellular enzyme reactions. Among the chemotherapeutic agents are the antibiotics and the *antimetabolites.

Antibiotics

Antibiotics are compounds produced by one microorganism that are toxic to another microorganism. They function by inhibiting enzymes that are essential to bacterial growth. Among the most commonly used antibiotics are penicillin and tetracycline.

penicillin G

tetracycline

Various strains of penicillin contain different groups attached to the cysteine-valine combination. Recall that both of these substances are amino acids.

The antibacterial action of penicillins is a result of their inhibition of a final stage in bacterial cell wall synthesis. In these final stages long polysaccharide chains are cross-linked together by short peptide chains. The last step in this process is the formation of a peptide bond between alanine on one peptide chain and glycine on another (Figure 24-7). Penicillin inhibits this step, probably by irreversibly bonding to the active site of the enzyme. The similarity between alanylalanine (Ala-Ala) from the peptide chain and penicillin is shown in Figure 24-8. There is no human enzyme that uses alanylalanine (Ala-Ala) as a substrate, and penicillin therefore does not interfere with our own enzymatic machinery. Allergic sensitivity to penicillin is due to the breakdown products of penicillin, especially to 6-aminopenicillinic acid.

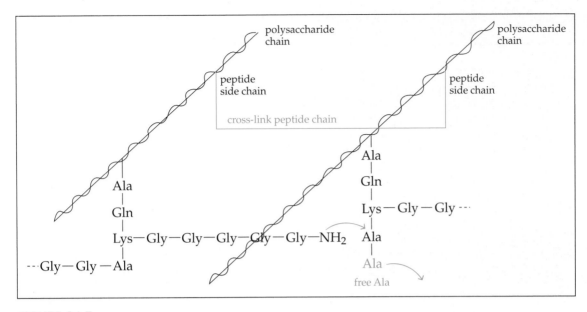

FIGURE 24-7

The final cross-linkages of polysaccharide chains in the bacterial cell wall.

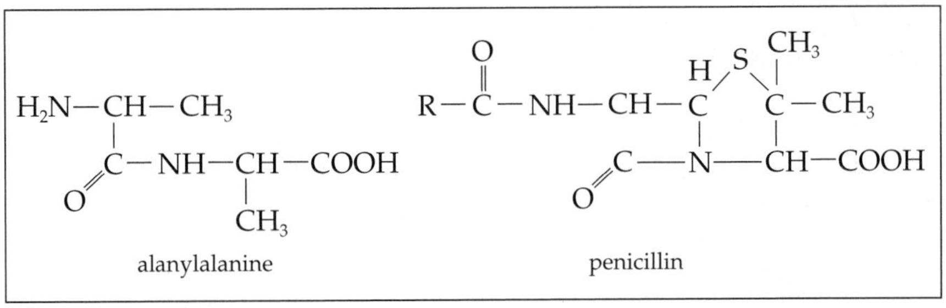

FIGURE 24-8

Comparison of the structures of alanylalanine (D-alanyl-D-alanine) and penicillin.

Bacterial resistance to penicillin is caused by the secretion of an enzyme, penicillinase, which cleaves the ring amide bond.

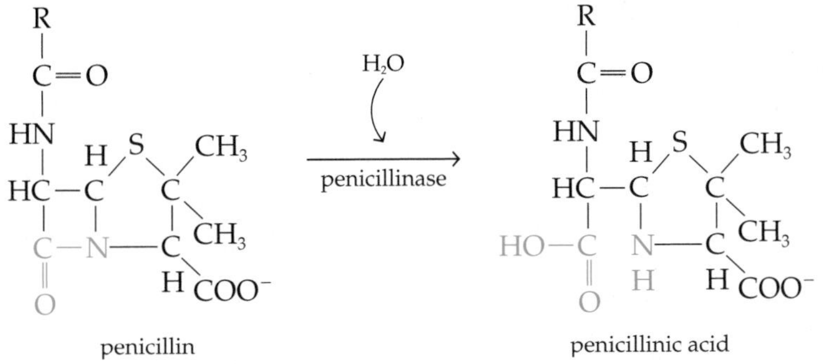

This enzyme probably evolved as a detoxification mechanism since its absence does not have any effect on the bacterial cell.

Antimetabolites

Antimetabolites are chemicals that have structures closely related to those of the substrates enzymes act on, thus inhibiting enzyme activity.

Case Study 24-1   Enzyme Inhibition by Antibiotics

A 12-year-old boy is brought to the rural pediatric clinic with an infected ankle. The mother reports that her son had developed a blister on his heel 2 days after wearing his new shoes. Three days later the ankle appears swollen, red, and was warm to the touch. The boy is limping and complains of pain in his ankle. The nurse practitioner gives the mother a prescription for penicillin G and requests a return visit in 7 days. At the return visit there was no improvement, so the nurse practitioner prescribes oxacillin, and another appointment is made for 5 days later. By the third visit the boy is walking without a limp and the ankle is free of redness and swelling.

Questions

1. How is penicillin inactivated?
2. Bacteria that are sensitive to penicillin live when grown in an isotonic medium. What does this suggest about a possible mechanism of action of penicillin?
3. Some penicillins can be excreted in the urine. What does this signify in terms of metabolism?

Some *chemotherapeutic agents used in the treatment of cancer *antineoplastic agents) are antimetabolites (mercaptopurine used in the treatment of leukemias); some are antibiotics (adriamycin used in the treatment of Hodgkin's disease); and others are alkylating agents, hormones, or natural products.

One of the most promising new chemotherapeutic agent in decades is *taxol*, a natural product obtained from the bark of Pacific yew trees. Taxol acts by interfering with cellular growth and function and is very effective in shrinking a variety of tumors, particularly in advanced cases of ovarian and breast cancers.

24-9   Clinical Significance of Plasma Enzyme Concentrations

The measurement of plasma enzyme levels can be of great diagnostic value. For example, the levels of glutamic pyruvic transaminase increase with infectious hepatitis; the levels of trypsin increase during acute disease of the pancreas; ceruloplasmin levels decrease during Wilson's disease (see page 547); and glutamic oxaloacetic transaminase levels rise rapidly after myocardial infarction (see page 496). Many other plasma enzymes are useful in the diagnosis of various diseases (Table 24-3).

24-10   Isozymes

Isozymes, or isoenzymes, are enzymes with the same function but slightly different structural features. The reason for their existence is not known, but they are made use of clinically. Lactate dehydrogenase (LDH) (see page 462), creatine kinase, and alkaline phosphatase all occur in isoenzyme form and are of diagnostic value. LDH has five forms, and Table 24-4 lists some clinically useful diagnoses

Case Study 24-2 Myocardial Infarction and Creatine Kinase

An 80-year-old man is brought by ambulance to the hospital's emergency department after his car hit a light pole. The patient remembers feeling chest pain and severe dizziness. He thinks he blacked out before the crash. The physician suspects either a myocardial infarction or cerebrovascular accident and orders the patient admitted to the intensive care unit for observation. Blood samples are drawn for creatine kinase (CK) assay, and other enzyme assays are sent to the laboratory.

Questions

1. In which tissues is CK found? From which tissues can CK be released into the bloodstream?
2. Would you expect to see an elevated CK level following a stroke?
3. What other enzyme tests might this physician have ordered?
4. Is it possible to correlate the CK level to the extent of tissue damage?

Table 24-3 Enzymes of Diagnostic Value

Serum Enzyme	Major Diagnostic Use
Glutamic oxaloacetic transaminase (SGOT)	Myocardial infarction
Glutamic pyruvic transaminase (SGPT)	Infectious hepatitis
Trypsin	Acute pancreatic disease
Ceruloplasmin	Wilson's disease
Amylase	Liver and pancreatic disease
Acid phosphatase	Prostate cancer
Alkaline phosphatase	Liver or bone disease
Creatine phosphokinase (CPK)	Myocardial infarction, muscle disorders
Lactate dehydrogenase (LDH)	Myocardial infarction, leukemia, anemia
Renin	Hypertension

Table 24-4 Clinical Significance of Relative Amounts of LDH

Condition	Isoenzyme Pattern
Myocardial infarction	Moderate elevation of LDH_1; slight elevation of LDH_2
Acute hepatitis	Large elevation of LDH_5; moderate elevation of LDH_4
Muscular dystrophy	Elevation of LDH_1, LDH_2, LDH_3
Megaloblastic anemia	Large elevation of LDH_1
Sickle-cell anemia	Moderate elevation of LDH_1, LDH_2
Arthritis with joint effusions	Elevation of LDH_5

based on these isoenzymes. Different tissues will contain different relative amounts of each of these isoenzymes. This difference allows the determination of what tissues have been damaged.

24-11 Allosteric Regulation

Most of the body's biochemical processes take place in several steps, each one catalyzed by a particular enzyme, such as

$$A \xrightarrow{a_1} B \xrightarrow{b_1} C \xrightarrow{c_1} D \xrightarrow{d_1} E$$

where a_1, b_1, c_1, d_1 are the enzymes needed to react with the given substrates. Thus B, the product of the first reaction is in turn the substrate for the second enzyme reaction, and so on. The final product, E, may inhibit the activity of enzyme d_1. Such an inhibitor may be either competitive or noncompetitive (see page 422). Therefore, as the concentration of E increases, the activity of enzyme d_1 decreases and soon stops. Conversely, when the concentration of E is low, all the reactions proceed rapidly. Such a mechanism for regulating enzyme activity is called *feedback control*.

Enzymes whose activity can be changed by molecules other than those of the substrate are called **allosteric enzymes.** It is thought that the other molecules bind to the enzyme at sites other than the active site. This binding changes the three-dimensional structures of the enzymes. Some molecules increase the catalytic rate (effectors), and some decrease it (noncompetitive inhibitors). This control of key enzymes is of utmost importance to ensure that biologic processes remain coordinated at all times to meet the immediate metabolic needs of the cells. The key glycolytic enzyme phosphofructokinase (see page 461) is one such allosteric enzyme.

When an effector binds to a site on an enzyme, it changes the conformation of that enzyme not only at the active site but also at neighboring sections. Therefore, the substrate can bind more readily, and so the reaction rate is increased, as shown in Figure 24-9. In Figure 24-9a the active sites X_1 and X_2 are not readily available for substrate S. In Figure 24-9b an effector has changed the shape of the enzyme so that when one molecule of the substrate reacts, a second molecule can also react more easily.

A noncompetitive inhibitor binds to an enzyme, changing its shape so that the active site is less available to the substrate, thus decreasing enzyme activity. The inhibitor binds at another part of the enzyme, sometimes but not always near the active site, deforming the enzyme so that it cannot form an enzyme-substrate

FIGURE 24-9

An effector changes the shape of an enzyme so the substrate binds more readily.

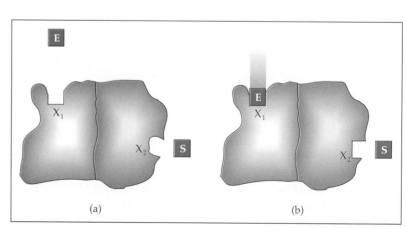

(a)   (b)

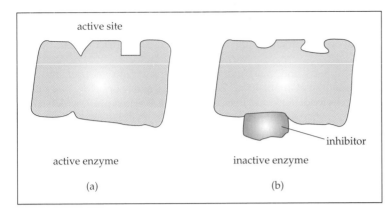

active site

active enzyme

inactive enzyme

inhibitor

(a)

(b)

FIGURE 24-10

A noncompetitive inhibitor changes the shape of the enzyme so the substrate cannot react.

complex. Noncompetitive inhibitors do not need to have a structure similar to that of the substrate because they are not competing for the active site (Figure 24-10).

24-12 Zymogens

Zymogens are inactive precursors of enzymes. Most digestive and blood-clotting enzymes exist in the zymogen form until activated. In the case of digestive enzymes, this is necessary to prevent digestion of pancreatic and gastric tissue. For blood clotting, it is to avoid premature formation of blood clots.

Zymogen		Active Form of Enzyme
pepsinogen	$\longrightarrow$	pepsin
trypsinogen	$\longrightarrow$	trypsin
prothrombin	$\longrightarrow$	thrombin

Lactose Intolerance

Individuals who cannot eat food containing lactose are said to be *lactose intolerant*. They lack the enzyme lactase, which is required for the hydrolysis of lactose. As a result, lactose accumulates in the intestinal tract and pulls water out of the tissues by osmosis. This in turn causes abdominal cramps, distention, and diarrhea.

Normally, individuals have enough of the enzyme lactase, especially in their early years when cow's milk is a major food source. As an individual ages, the levels of this enzyme decrease and lactose intolerance may result.

The hydrogen breath test measures the amount of hydrogen in the breath of a person, an amount that is normally very low. Undigested lactose in the large intestine can be fermented by bacteria, thereby producing among other gases, hydrogen. The hydrogen gas is absorbed into the blood and carried to the lungs where it is exhaled. In the hydrogen breath test, a patient drinks a lactose-rich beverage and his or her breath is analyzed at regular intervals. Increased or high levels of hydrogen indicate lactose intolerance.

Some population groups such as the Chinese have a high incidence of lactose intolerance when compared with groups of persons of European ancestry. This genetic difference was probably caused by an absence of or deficiency of cow's milk in the Chinese diet.

To overcome such an effect today, an individual may take Lactaid orally to supply this missing enzyme.

Summary

Enzymes are biologic catalysts that increase the speed of a chemical reaction but do not themselves change.

Most enzymes are proteins and will undergo all the reactions of proteins. The enzymes in the body function best at about 40° C. Temperatures above or below body temperature will decrease the activity of enzymes. Each enzyme has a certain pH at which it can function best.

An increase in the amount of enzyme will increase the rate of reaction. An increase in the amount of substrate will increase the rate of the reactions.

Compounds that increase the activity of an enzyme are called *activators*. Compounds that interfere with the activity of an enzyme are called *inhibitors*.

Enzymes contain an "active site" that binds to the substrate to form an enzyme-substrate complex. This complex yields the products and regenerates the enzyme.

Many enzymes contain two parts—a protein and a nonprotein part. The protein part of an enzyme is called the *apoenzyme*.

Some enzymes require the presence of a substance called a *coenzyme* before they can act effectively. Coenzymes frequently contain the B vitamins or compounds derived from the B vitamins.

Under the older system of naming enzymes the substrate was not mentioned; the newer system indicates the substrate being acted on. The names of enzymes under this system end in *-ase*.

Enzymes can be classified as oxidoreductases (enzymes that catalyze oxidation-reduction reactions between two substrates), transferases (which catalyze the transfer of a functional group between two substrates), hydrolases (which catalyze hydrolysis reactions), lyases (which catalyze the removal of groups from substrates by means other than hydrolysis), isomerases (which catalyze the interconversion of cis-trans isomers), and ligases (which catalyze the coupling of two compounds with the breaking of pyrophosphate bonds).

Some hydrolytic enzymes are found in the lysosomes of the cytoplasm. The cytoplasm also contains mitochondria. These structural and functional units contain most of the oxidative enzymes and are deeply involved in the electron transport system of oxidation-reduction. The mitochondria also produce ATP, the cells' chief source of energy.

Chemotherapy is the use of chemicals to destroy infectious microorganisms and cancerous cells. Among chemotherapeutic agents are antibiotics and antimetabolites.

Abnormal plasma enzyme concentrations are of clinical significance in the diagnosis of certain diseases.

Isozymes are enzymes with the same function but slightly different structural features.

Allosteric enzymes are key metabolic enzymes whose activity can be changed by molecules other than the substrate.

Zymogens are the precursors of enzymes.

Questions and Problems

1. What effect does temperature have on the activity of an enzyme? What effect does pH have?
2. What effect do enzymes have on the speed of a reaction? What is the effect of enzyme concentration on the speed of a reaction? the effect of substrate concentration?
3. Enzymes are what type of compounds?
4. What is an apoenzyme? coenzyme? optimum temperature? optimum pH?
5. How are enzymes classified?
6. How do enzymes function?

7. What are the various degrees of specificity of enzymes?
8. What are carbohydrases? esterases? proteases? Name two of each, indicating where they are found in the body and the substrate they act on.
9. What is a nuclease? transferase? an isomerase? Give an example of each.
10. Name three oxidoreductases and indicate where they function.
11. What is coenzyme A? coenzyme Q? What are their functions?
12. What functions do the mitochondria serve?

13. Distinguish between specific and nonspecific inhibitors and between competitive and noncompetitive inhibitors.
14. What effect does denaturing have on the active site of an enzyme? Why?
15. What is chemotherapy? Name two chemotherapeutic agents.
16. How can plasma enzyme concentration be used in the diagnosis of disease?
17. What are zymogens? isozymes? allosteric enzymes?

18. Discuss enzymes in terms of activation energy.
19. Compare the lock-and-key theory of enzymes with the induced-fit theory.
20. What is renin? Where is it found? Of what diagnostic value is it?
21. Describe an experiment that would allow you to decide whether an enzyme inhibitor was competitive or noncompetitive.
22. Why do we need acid as well as alkaline phosphatases?

Practice Test

1. Which enzyme would be useful in determining whether a patient has a myocardial infarction?
   a. SGOT
   b. renin
   c. ceruloplasmin
   d. aldosterone
2. Most enzymes are _____.
   a. carbohydrates
   b. lipids
   c. proteins
   d. nucleic acids
3. The protein part of an enzyme is called the _____.
   a. inhibitor
   b. coenzyme
   c. apoenzyme
   d. activator
4. Organic compounds of which element inhibit the enzyme cholinesterase?
   a. sulfur
   b. lead
   c. phosphorus
   d. mercury
5. Inactive precursors of enzymes are called _____.
   a. antibiotics
   b. zymogens
   c. antimetabolites
   d. allosterases

6. Penicillin is used to treat _____.
   a. viruses
   b. cancer
   c. bacterial infection
   d. all of these
7. Many _____ act as coenzymes.
   a. vitamins
   b. hormones
   c. phospholipids
   d. carbohydrates
8. An example of a carbohydrase is _____.
   a. lipase
   b. ptyalin
   c. pepsin
   d. sucrose
9. The optimum pH of gastric juice is _____.
   a. 9
   b. 7
   c. 5
   d. 2
10. Many oxidoreductases are found in the _____.
    a. cytoplasm
    b. mitochondria
    c. cell membranes
    d. cell nuclei

25

Digestion

Objectives

- To follow the path of salivary, gastric, and intestinal digestion

- To describe the absorption of carbohydrates, lipids, proteins, vitamins, and minerals

- To understand the formation of feces

- To recognize the effects of improper digestion and absorption of carbohydrates

Through the process of digestion, the complex foods we eat are broken down into simpler substances that our cells can readily absorb.

Most foods (carbohydrates, fats, and proteins) are composed of large molecules that are usually not soluble in water. Before these foods can be absorbed through the alimentary canal, they must be broken down into smaller soluble molecules. Digestion is the process by which food molecules are broken down into simpler molecules that can be absorbed into the blood through the intestinal walls.

Digestion involves the use of hydrolases—the hydrolytic enzymes (Chapter 24). The hydrolases catalyze the hydrolysis of carbohydrates to monosaccharides, fats to fatty acids and glycerol, and proteins to amino acids.

However, not all foods require digestion. Monosaccharides are already in their simplest form and do not require digestion. Inorganic salts and vitamins also do not require digestion.

Digestion of food takes place in the mouth, the stomach, and the small intestine, each area having its own particular enzyme or enzymes that catalyze the hydrolytic reactions.

25-1 Salivary Digestion

During chewing, the food is mixed with saliva. The saliva moistens the food so that swallowing is easier. Saliva is approximately 99.5 percent water. The remaining 0.5 percent consists of mucin, a glycoprotein that acts as a lubricant; several inorganic salts that act as buffers; salivary amylase, an enzyme that catalyzes the hydrolysis of starches; and a lingual lipase that catalyzes the hydrolysis of triglycerides. Saliva also acts as an excretory fluid for certain drugs, such as morphine and alcohol, and for certain inorganic ions, such as K^+, Ca^{2+}, HCO_3^-, and SCN^-. Saliva has a pH range of 5.75 to 7.0, with an optimum pH of 6.6.

25-2 Gastric Digestion

When food is swallowed, it passes down the esophagus into the stomach, where it is mixed with the gastric juice. The gastric juice is secreted by glands in the walls of the stomach. When food enters the stomach, it causes the production of the hormone gastrin. Gastrin diffuses into the blood stream, which carries it back to the stomach, where it then stimulates the flow of gastric juice.

Approximately 2 to 3 L of gastric juice is secreted daily. Gastric juice is normally a clear, pale yellow liquid with a pH of 1.0 to 2.0. It is 97 to 99 percent water and up to 0.5 percent free hydrochloric acid. It is the presence of this hydrochloric acid that causes the gastric juice to have such a low pH. In certain pathologic conditions, the acidity of the stomach may be less than normal. Such a condition is known as *hypoacidity and is commonly associated with stomach cancer and pernicious anemia. Hyperacidity is a condition in which the stomach has too high an acid concentration. It is indicative of gastric ulcers, hypertension, or gastritis (inflammation of the stomach walls).

Hydrochloric acid in the stomach denatures protein so that the tertiary structure is lost. Thus the polypeptide chains unfold, making it easier for the proteases to act. The low pH of the stomach also destroys most of the microorganisms that enter the body through the mouth.

The gastric juice contains the zymogen pepsinogen and the enzyme (gastric) lipase. In addition to these enzymes, a substance known as the intrinsic factor is secreted by the parietal cells in the walls of the stomach. Vitamin B_{12} must undergo a reaction with this intrinsic factor before it can be absorbed into the bloodstream. A lack of the intrinsic factor is associated with pernicious anemia.

25-3 Intestinal Digestion

The food in the stomach is very acidic. When this acid material enters the small intestine, it stimulates the mucosa to release the hormone secretin. Secretin, in turn, stimulates the pancreas to release the pancreatic juice into the small intestine.

Three different digestive juices enter the small intestine. These digestive juices are alkaline and neutralize the acid contents coming from the stomach. The three digestive juices entering the small intestine are (1) pancreatic juice, (2) intestinal juice, and (3) bile.

Pancreatic Juice

Pancreatic juice contains several enzymatic substances. Among these are trypsinogen, chymotrypsinogen, carboxypeptidase, pancreatic lipase, proelastase, and pancreatic amylase.

The pancreatic juice also contains cholesterol ester hydrolase, which acts on cholesterol esters and converts them into free cholesterol plus fatty acids; ribonuclease, which converts ribonucleic acids into nucleotides; deoxyribonuclease, which converts deoxyribonucleic acids into nucleotides; and phospholipase A_2, which acts on phospholipids and converts them into fatty acids and lysophospholipids.

Intestinal Juice

Intestinal juice contains several enzymes. Among these are aminopeptidase and dipeptidase. The intestinal juice also contains enzymes that catalyze the hydrolysis of phosphoglycerides, nucleoproteins, and organic phosphates. The intestinal mucosal cells contain the enzymes sucrase, maltase, and lactase.

Bile

Bile is produced in the liver and stored in the gallbladder. The gallbladder absorbs some of the water and other substances from the liver bile and so changes its composition slightly. When meat or fats enter the small intestine, they cause it to secrete a hormone, cholecystokinin (CCK). This hormone enters the bloodstream and is carried to the gallbladder, where it causes that organ to contract and empty into the duodenum through the bile duct.

Bile is a yellowish brown to green viscous liquid with a pH of 7.8 to 8.6. Because it is alkaline, it serves to neutralize the acid entering from the stomach. Primarily, bile contains bile salts, bile pigments, and cholesterol. Bile contains no digestive enzymes. Bile is the end product of cholesterol metabolism and is a primary determinant in cholesterol synthesis. Figure 25-1 illustrates the interrelation of cholesterol and bile. Bile acts in the removal of many drugs and poisons from the body, in addition to removing such inorganic ions as Ca^{2+}, Zn^{2+}, and Hg^{2+}.

Bile Salts Sodium glycocholate and sodium taurocholate are the two most important bile salts. They are both derived from cholic acid, a steroid similar to cholesterol in structure (see page 375).

cholic acid

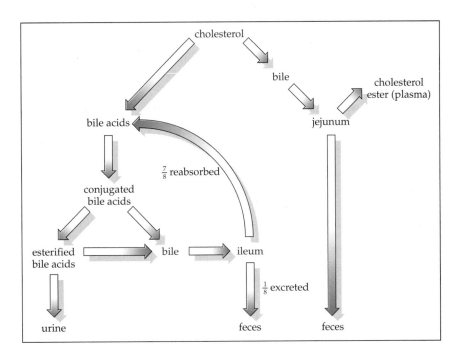

FIGURE 25-1
Metabolism of cholesterol and bile acids in the liver.

Bile salts have the ability to lower surface tension and increase surface area, thus aiding in the emulsification of fats. They also increase the effectiveness of pancreatic lipase (steapsin) in its digestive action on emulsified fats.

In addition, bile salts aid the absorption of fatty acids through the walls of the intestine. After absorption of these fatty acids, the bile salts are removed and carried back by portal circulation to the liver, where they are again returned to the bile. Bile salts also help stimulate intestinal motility.

Bile Pigments The average red blood cell lasts about 120 days and is then destroyed. On average 100 million to 200 million red blood cells are destroyed per hour, a figure that roughly corresponds to $\frac{1}{5}$ ounce of hemoglobin daily. The hemoglobin is broken down into globin and heme. The body removes the iron from the heme and reuses it. The heme, with the iron removed, becomes biliverdin. Biliverdin is reduced in the reticuloendothelial cells of the liver, spleen, and bone marrow to form bilirubin, the main bile pigment excreted into the bile by the liver. In the intestines some bilirubin is converted to stercobilinogen and stercobilin, a pigment that gives the feces its characteristic yellow-brown color. Some bilirubin is absorbed into the bloodstream and comes to the liver where it is converted to urobilinogen and then to urobilin, which appears in the urine, giving that fluid its characteristic color. These reactions can be written as follows:

$$
\text{hemoglobin} \longrightarrow
\begin{array}{c} \text{heme} \\ + \\ \text{globin} \end{array}
\longrightarrow
\begin{array}{c} \text{biliverdin} \\ + \\ \text{iron} \end{array}
\longrightarrow \text{bilirubin}
\begin{array}{l} \nearrow \text{stercobilinogen} \longrightarrow \text{stercobilin} \\ \searrow \text{urobilinogen} \longrightarrow \text{urobilin} \end{array}
$$

If the bile duct is blocked, the bile pigments remain in the bloodstream, producing jaundice. This disorder is recognizable by the yellow pigmentation of the skin. If the bile duct is blocked, no bile pigments can enter the intestine and the feces will appear clay-colored or nearly colorless.

Cholesterol   The body's excess cholesterol is excreted by the liver and carried to the small intestine in the bile. Sometimes the cholesterol precipitates in the gallbladder, producing gallstones. Figure 25-2 shows gallstones present in a gallbladder.

The body excretes about 1 g of cholesterol daily. Half is eliminated in the feces after conversion to bile acids; the other half is excreted as neutral steroids. However, much of the cholesterol secreted in the bile is reabsorbed in the jejunum.

25-4 Digestion of Carbohydrates

Digestion begins in the mouth, as chewing action reduces the size of the food particles. Thus they will have more surface area in contact with the digestive enzymes.

In the Mouth

Saliva contains salivary amylase (ptyalin), which catalyzes the hydrolysis of starch into maltose. However, this enzyme becomes inactive at a pH below 4.0 so that its activity ceases when it is mixed with the contents of the stomach, where the pH falls to about 1.5. Salivary amylase does not serve a very important function in digestion because the food does not remain in the mouth long enough for any appreciable hydrolysis to take place. Some hydrolysis of carbohydrates catalyzed by salivary amylase may take place in the stomach before the food is thoroughly mixed with the contents of the stomach, but this is of little importance because there are intestinal enzymes capable of hydrolyzing starch and maltose. The principal function of saliva is to lubricate and moisten the food so it can be easily swallowed.

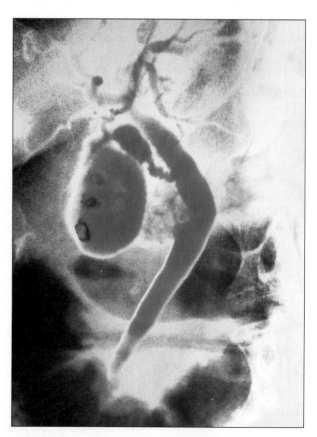

FIGURE 25-2
Granulated gallstones.

In the Stomach

The stomach contains no carbohydrases, so no digestion of carbohydrates occurs there except for that catalyzed by salivary amylase. The activity of salivary amylase ceases as soon as it becomes mixed with the acid contents of the stomach.

In the Small Intestine

The major digestion of carbohydrates takes place in the small intestine through the action of enzymes in the pancreatic and intestinal juices.

The pancreatic juice contains the enzyme pancreatic amylase, which catalyzes the hydrolysis of starch and dextrins into maltose. The maltose thus produced is hydrolyzed to glucose through the activity of the enzyme maltase from the intestinal mucosal cells. The optimum pH of pancreatic amylase is 7.1. The intestinal mucosal cells also contain the enzymes sucrase and lactase, which catalyze the hydrolysis of sucrose and lactose, respectively.

$$\underset{\text{starch}}{(C_6H_{10}O_5)_n} + H_2O \xrightarrow{\overset{\text{pancreatic}}{\text{amylase}}} \underset{\text{maltose}}{C_{12}H_{22}O_{11}}$$

$$\underset{\text{maltose}}{C_{12}H_{22}O_{11}} + H_2O \xrightarrow{\text{maltase}} \underset{\text{glucose}}{2\ C_6H_{12}O_6}$$

If a monosaccharide such as glucose is eaten, digestion is not necessary because the monosaccharide is already in its simplest form and can easily undergo absorption into the bloodstream.

25-5 Digestion of Fats

In the Mouth

A lingual lipase secreted by the dorsal surface of the tongue acts on triglycerides, particularly of the type found in milk. Lingual lipase has an optimum pH of 4.0 to 4.5 and pH range of activity of 2.0 to 7.5, so it can continue its activity even at the low pH of the stomach.

In the Stomach

Although gastric lipase is present in the stomach, very little digestion of fats takes place because the pH of the stomach (1.0 to 2.0) is far below the optimum pH of that enzyme (7.0 to 8.0). Also, fats must be emulsified before they can be digested by lipase, and there is no mechanism for emulsification of fats in the stomach. However, if emulsified fats are eaten, a small amount of hydrolysis may take place in the stomach. In infants, whose stomach pH is higher, fat hydrolysis of milk may be of some importance.

In the Small Intestine

In the small intestine, the pancreatic lipase catalyzes the hydrolysis of fats into fatty acids and glycerol. This action is aided by the bile, which emulsifies the fats so that they can be acted upon readily by pancreatic lipase.

$$\text{fat} + \text{water} \xrightarrow{\overset{\text{pancreatic}}{\text{lipase}}} \text{fatty acids} + \text{glycerol}$$

25-6   Digestion of Proteins

In the Mouth

As the saliva contains no enzymes for the hydrolysis of protein, there is no digestion of protein in the mouth.

In the Stomach

The precursor enzyme pepsinogen is converted to pepsin when it is mixed with the hydrochloric acid of the stomach. Pepsin catalyzes the hydrolysis of protein to polypeptides.

$$\text{protein} + \text{water} \xrightarrow{\text{pepsin}} \text{polypeptides}$$

Rennin, an enzyme present in the gastric juice of calves, coagulates casein to form paracasein, which is precipitated by the calcium ions present in milk. Coagulated milk remains in the stomach longer than uncoagulated milk and so is more readily digested there. According to modern theories, rennin is not present in the gastric juice of humans.

In the Small Intestine

In the small intestine, the zymogen trypsinogen from the pancreatic juice is changed into trypsin by the intestinal enzyme enterokinase. Trypsin in turn changes chymotrypsinogen, another pancreatic zymogen, into chymotrypsin. Both trypsin and chymotrypsin catalyze the hydrolysis of protein, proteases, and peptones to polypeptides. The optimum pH of trypsin and chymotrypsin is 8.0 or 9.0.

The intestinal enzymes aminopeptidase and dipeptidase catalyze the hydrolysis of polypeptides and dipeptides into amino acids. Carboxypeptidase, an enzyme of the pancreatic juice, also catalyzes the hydrolysis of polypeptides to amino acids. Carboxypeptidase contains the element zinc.

Proelastase from the pancreatic juice is converted to elastase by trypsin. Elastase acts on protein and polypeptides to convert them into polypeptides and dipeptides, respectively.

Table 25-1 summarizes the steps in digestion.

25-7   Absorption of Carbohydrates

As we have seen, the principal digestion of carbohydrates takes place in the small intestine, where the polysaccharides and disaccharides are hydrolyzed into monosaccharides—glucose, fructose, and galactose. The monosaccharides are transported through the walls of the small intestine directly into the bloodstream by diffusion and by active transport of galactose and glucose. The blood carries the monosaccharides to the liver and then into the general circulation to all parts of the body. The monosaccharides can be oxidized to furnish heat and energy. Some of the monosaccharides are converted to glycogen, a polysaccharide, and stored in the liver or the muscles, and the rest are converted to fat and stored in the adipose tissue.

25-8   Absorption of Fats

The digestion of fats takes place primarily in the small intestine. The end products of digestion of fats—monoglycerides and diglycerides, fatty acids, and glycerol—pass through the intestinal mucosa, where they are reconverted into triglycerides

Table 25-1	Summary of Digestion		
Location of Digestion	Digestive Juice and Enzymes	Substrate	Product
Mouth	Saliva		
	Salivary amylase (ptyalin)	Starch, glycogen	Maltose and dextrins
	Lingual lipase	Milk	Fatty acids + 1,2-diglycerides
Stomach	Gastric juice		
	Hydrochloric acid	Pepsinogen	Pepsin
	Pepsin	Protein	Polypeptides
	Lipase	Fats	Fatty acids + glycerol
Small intestine	Intestinal juice		
	Enterokinase	Trypsinogen	Trypsin
	Aminopeptidase	Polypeptides	Amino acids
	Dipeptidase	Peptides	Amino acids
	Maltase	Maltose	Glucose
	Sucrase	Sucrose	Glucose + fructose
	Lactase	Lactose	Glucose + galactose
	Pancreatic juice		
	Trypsin	Protein	Polypeptides
	Chymotrypsin	Protein	Polypeptides
	Pancreatic amylase	Starch + glycogen	Maltose
	Pancreatic lipase	Fats	Fatty acids + glycerol
	Carboxypeptidase	Polypeptides	Amino acids
	Phosphatase	Organic phosphates	Free phosphates
	Polynucleotidases	Nucleic acids	Nucleotides
	Nucleosidases	Nucleosides	Purines, pyrimidines, phosphates, pentoses
	Elastase	Protein, polypeptides	Polypeptides, dipeptides
	Deoxyribonuclease	Deoxyribonucleic acid	Nucleotides
	Cholesterol ester hydrolases	Cholesterol esters	Cholesterol, fatty acids
	Phospholipase A_2	Phospholipids	Fatty acids, phosphates

and phosphoglycerides, which then enter the lacteals, the lymph vessels in the villi in the walls of the small intestine. From the lacteals these products pass into the thoracic duct (a main lymph vessel) and then into the bloodstream.

Bile salts are necessary for this absorption process. After the absorption of the fatty acids the bile salts are returned to the liver to be excreted again into the bile. Fatty acids of less than 10 to 12 carbon atoms are transported through the intestinal walls directly into the bloodstream, as is any free glycerol present.

25-9 Absorption of Proteins

The end products of the hydrolysis of proteins are amino acids. Absorption of amino acids occurs chiefly in the small intestine and is an active, enzyme-requiring process resembling the active transport of glucose. There are six or more specific transport systems for amino acids being carried into the bloodstream.

Case Study 21-1 Acute Pancreatitis

An obese 42-year-old white woman is brought to the hospital by her husband at 6:00 PM. She is complaining of an acute onset of severe, sharp pain in the upper right quadrant. The pain worsens with movement, and she appears pale and sweaty. Her respiratory rate is 34 breaths/min and very shallow. She states that it hurts to breathe deeply. Her blood pressure is 108/58 mm Hg, pulse is 110 beats/min and regular, and temperature is 37.7° C. The physician notes a rigid abdomen with tenderness on palpation.

The patient is admitted with orders that nothing be given by mouth. A nasogastric tube is placed for continuous aspiration of stomach contents. An intravenous catheter is placed, and the patient is started on a glucose and saline solution.

An intravenous cholangiography is performed because of the possiblity of gallbladder disease. The common bile duct appears normal, and the results are equivocal. The laboratory results show a significant elevation of serum amylase, with a lesser increase in lipase. The absence of a previous history of peptic ulcer leads to a diagnosis of acute pancreatitis. Two weeks later the patient is asymptomatic, and most laboratory values have returned to normal. However, the fasting blood glucose level continues to be elevated for 3 weeks, and the postprandial blood glucose level continues to be high for an additional month. Three months later, the patient's gallbladder is surgically removed.

Questions

1. The pancreas produces which digestive enzymes? Can the laboratory evaluate these enzymes?
2. The pancreas produces which hormones? In what form are they stored in the pancreas?
3. How would digestion and absorption be affected if the pancreatic enzymes were lacking?
4. What are the major ionic components of the pancreatic secretions? Is the concentration of these components constant?
5. What electrolyte problems are seen with lengthy nasogastric suction?
6. Would the intravenous administration of glucose and saline solution correct the electrolyte problems?
7. What controls regulate the production of pancreatic secretions?

1. A system for small, netural amino acids such as glycine
2. A system for large, neutral amino acids such as phenylalanine
3. A system for basic amino acids such as lysine
4. A system for acidic amino acids such as aspartic acid
5. A system for proline
6. A system for very small peptides

Amino acids compete with one another for absorption via a particular pathway. Thus, high levels of leucine lower the absorption of isoleucine and valine.

Occasionally, proteins also escape digestion and are absorbed directly into the blood. This occurs more often in the very young since the permeability of their intestinal mucosa is greater, allowing the passage of antibodies of colostral milk.

This passage of protein into the blood may be sufficient to cause immunologic sensitization and related food allergies.

The L-amino acids are absorbed more rapidly than the D isomers and pass through the capillaries of the villi directly into the bloodstream, which carries them to the tissues to be used to build or replace tissue. The amino acids can also be oxidized to furnish energy. Although the body can store carbohydrate and fat, it cannot store protein.

25-10 Absorption of Iron

Iron differs from practically all other inorganic materials in that the amount in the body is controlled by its absorption, not its excretion. Body stores of iron are conserved very efficiently except during menstrual flow.

Normally, only 5 to 10 percent of orally ingested iron is absorbed. Most of it is absorbed in the upper duodenum region of the small intestine. The stomach acidity along with reducing agents such as vitamin C (ascorbic acid) and proteins help reduce dietary iron to stable, absorbable, water-soluble Fe^{2+} complexes. Several factors make absorption of iron difficult.

1. Most iron has to be reduced from Fe^{3+} to Fe^{2+}.
2. The relatively high pH of the small intestine increases the formation of insoluble iron compounds.
3. Iron forms insoluble compounds with bile salts.
4. Phosphates form insoluble iron compounds.
5. Iron absorption requires stomach acidity.

A lack of HCl in the stomach (achlorhydria) prevents the absorption of most of the iron, and stomach antacids may also reduce iron absorption.

25-11 Absorption of Vitamins

Most vitamins are absorbed in the upper small intestine. The fat-soluble vitamins (A, D, E, K) need fat and bile salts to be absorbed. Taking a multivitamin capsule with water does not provide the fat necessary for the absorption of the fat-soluble vitamins. Vitamin B_{12} absorption depends on its binding to an intrinsic factor produced in the stomach (see page 439). This complex, along with calcium ions, finds

Case Study 25-2 Iron Overload

A 65-year-old man is admitted to the hospital's emergency department with thrombophlebitis in his leg. The physical examination reveals an enlargement of the liver. His laboratory tests reveal a high serum iron concentration. A significant deposition of hemosiderin in the hepatocytes is noted.

Questions

1. What is the mechanism of iron absorption?
2. What are the different forms of iron, and which foods are high in iron?
3. How is excess iron removed from the body?
4. What are the causes of excess iron concentration in the body?

Table 25-2 Sites of Absorption of Body Nutrients	
Small Intestine	**Large Intestine**
Glucose, other monosaccharides	Bile salts
Monoglycerides	Some electrolytes
Glycerol	Water
Cholesterol	
Amino acids	
Peptides	
Vitamins	
Some electrolytes, e.g., Fe^{2+}, Ca^{2+}, PO_4^{3-}	

acceptor sites in the lower small intestine. The sites of absorption of body nutrients are indicated in Table 25-2.

25-12 Formation of Feces

After the absorption of monosaccharides, glycerol, fatty acids, and amino acids, the remaining contents of the small intestine pass into the large intestine. The large intestine contains undigestible material (such as cellulose), undigested food particles, unused digestive juices, epithelial tissues from the walls of the digestive system, bile pigments, bile salts, and inorganic salts. The material passing into the large intestine is semifluid and contains much water. Most of this water and some salts are reabsorbed through the walls of the large intestine, leaving behind a residue called *feces*. Little or no digestion takes place in the large intestine.

The conditions in the large intestine are ideal for the growth of bacteria; usually one fourth to one half of the feces consists of bacteria. The bacteria cause the fermentation of carbohydrates to produce hydrogen, carbon dioxide, and methane gases, as well as acetic, butyric, and lactic acids. The gases can cause distention and swelling of the intestinal tract, producing a feeling of discomfort. The acids may be irritating to the intestinal mucosa and cause diarrhea. Particularly in infants, the acids can cause *excoriated buttocks. Infants who have this condition are usually given a diet that is high in protein and low in carbohydrate because the fermentation bacteria act on carbohydrates and not on protein. The bacteria also produce vitamin K.

Some of the amino acids undergo decarboxylation because of the action of intestinal bacterial to produce toxic amines called ptomaines (see page 501). For example, decarboxylation produces cadaverine from the amino acid lysine, putrescine from ornithine, and histamine from histidine. These toxic substances are reabsorbed from the large intestine, carried to the liver where they are detoxified, and then excreted in the urine.

The amino acid tryptophan undergoes a series of reactions to form the compounds indole and skatole, which are primarily responsible for the odor of the feces.

tryptophan   indole   skatole (methylindole)

Table 25-3 Effects of Malabsorption of Nutrients	
Malabsorbed Substance	**Symptoms**
Lactose	Milk intolerance
Vitamin K	Bleeding, bruising
Iron, vitamin B_{12}, folates	Anemia
Protein products	Edema
Calcium, magnesium, vitamin D	Tetany
Calcium, protein products	Osteoporosis
Neutral amino acids	Hartnup disease

25-13 Dietary Fiber

Dietary fiber denotes substances that cannot be digested by the enzymes in the human digestive tract. It includes cellulose, lignin, pectins, gums, and pentosans. Even though such substances cannot be digested in the human body, they exhibit a beneficial effect by aiding water retention during passage of food through the gastrointestinal system, thereby producing feces that are softer and larger. A diet high in fiber is associated with reduced incidences of cardiovascular disease, diabetes mellitus, cancer of the colon, and diverticulosis. The more insoluble fibers such as those found in brans are beneficial to colon function, whereas the more soluble fibers aid in lowering serum cholesterol.

25-14 Defects of Carbohydrate Digestion and Absorption

Some individuals have a deficiency of lactase that causes an intolerance to milk. Many newborns also have a transient lactose intolerance after a bout of diarrhea has temporarily decreased the level of lactase enzymes in the intestinal tract. Treatment consists of withholding lactose-containing milk for several days until the intestinal tract has repaired itself. Adults who lack lactase enzymes must avoid milk and milk products.

A few individuals have an inherited sucrase deficiency. Symptons, which occur in early childhood, are similar to those of lactase deficiency.

Some persons have a disaccharidase deficiency that causes disacchariduria, an increase in disaccharide excretion.

Malabsorption of glucose and galactose is a congenital condition resulting from a defect in a carrier mechanism. However, fructose absorption is normal because absorption of fructose does not depend on that carrier system

If absorption of nutrients is disturbed, various types of symptoms may occur, as illustrated in Table 25-3.

Summary

Digestion is the process by which foods are broken down (hydrolyzed) into simple molecules that can then be absorbed through the intestinal walls. Digestion takes place in the mouth, stomach, and small intestine.

Digestion begins in the mouth. Food is mixed with saliva, which moistens the food and makes swallowing easier. Saliva contains an enzyme that begins the hydrolysis of carbohydrates (starch). Saliva has a pH of 5.75 to 7.0 and is approximately 99.5 percent in water.

In the stomach the food is mixed with the gastric juices. The gastric juices contain hydrochloric acid and have a pH of 1 to 2. The gastric juice contains the zymogen pepsinogen and the enzyme gastric lipase. Upon contact with hydrochloric acid, pepsinogen is converted into pepsin, which then catalyzes the hydrolysis of protein to polypeptides. Gastric lipase is not an important enzyme because it can act only on emulsified fats and there is practically no emulsified fat present in the stomach.

The food leaving the stomach enters the small intestine, where it is mixed with three digestive juices—those from the pancreas, those from the intestinal walls, and bile from the gallbladder.

Bile is produced in the liver and stored in the gallbladder. Bile is alkaline and neutralizes the acid entering the small intestine from the stomach. Bile salts lower surface tension and help emulsify fats; they aid in the absorption of fatty acids; they help stimulate intestinal motility.

Bile also contains pigments that come from the breakdown of hemoglobin. These bile pigments give the feces and the urine their characteristic color. If cholesterol crystallizes in the gallbladder, gallstones are produced.

Digestion of carbohydrates begins in the mouth. No digestion of carbohydrate takes place in the stomach except that catalyzed by salivary amylase from the saliva. Even this activity ceases when the food from the mouth reaches the low pH of the stomach. The major digestion of carbohydrates takes place in the small intestine with the aid of the enzymes pancreatic amylase, maltase, lactase, and sucrase.

No digestion of fats takes place in the mouth. No digestion of fats takes place in the stomach either unless the fats are already emulsified. In that case their hydrolysis is catalyzed by gastric lipase, although to a very limited extent. Fats are emulsified in the small intestine by the action of bile and then acted on by the enzyme pancreatic lipase.

Digestion of protein begins in the stomach with the aid of pepsin. The digestion of protein continues in the small intestine with the aid of the enzymes trypsin, chymotrypsin, carboxypeptidase, aminopeptidase, and dipeptidase.

The monosaccharides produced by the digestion of carbohydrates pass through the villi of the small intestine and enter the bloodstream.

Fatty acids and glycerol, products of the digestion of fats, are converted into glycerides in the intestinal mucosa by the action of bile and pass through the lacteals into the thoracic duct and then into the bloodstream. Fatty acids of less than 10 to 12 carbon atoms pass directly into the bloodstream through the villi.

Amino acids, from the digestion of protein, pass through the villi of the small intestine into the bloodstream.

Undigested food, undigestible foods, unused digestive juices, epithelial tissues from the walls of the digestive system, bile salts, inorganic salts, and water pass from the small intestine into the large intestine. Most of the water and some salts are reabsorbed. The remaining material is excreted as feces.

Questions and Answers

1. Do all foods require digestion? Why or why not?
2. Where does digestion take place?
3. What does saliva contain? What is its pH? What does it do?
4. What is hyperacidity? hypoacidity? What may cause them?
5. What is the pH of the gastric juice? What causes this pH?
6. What is the difference between pepsin and pepsinogen? How may one be converted to the other?
7. Why is gastric lipase considered an unimportant enzyme?
8. What is the function of secretin?
9. What enzymes are present in the gastric juice? in the intestinal juice? What function does each serve?
10. Where is bile produced? stored? What is its pH? What does it contain?
11. What are the functions of the bile salts?
12. Where do the bile pigments come from? How are they excreted?
13. What causes gallstones?
14. What is urobilin? How is it formed?
15. What happens when the bile duct is blocked?

16. What is the intrinsic factor? What disease is caused by a lack of this substance?
17. What is enterokinase? Where does it function?
18. Describe the formation of feces.
19. Describe the digestion of carbohydrates; of lipids; of proteins. Where are the end products absorbed?
20. What are the functions of the large intestine?
21. What causes the characteristic odor of feces?
22. Why is dietary fiber important?

This scenario pertains to questions 23 to 26:

A 44-year-old woman has been complaining of severe intermittent abdominal pain. The pain seems to follow meals, which many times are high in fat. She has not had any jaundice or gastrointestinal bleeding. Her physician prescribes antacids and a bland diet. The patient continues to have abdominal pain and returns to her physician's office 4 weeks later. The physician orders a cholecys-togram, which reveals the presence of gallstones. Surgery is performed and her gallbaldder removed. Laboratory analysis of the gallstones shows them to be composed of cholesterol.

23. Explain the function of bile in digestion.
24. What is the relationship between cholesterol and bile acids?
25. What factors cause cholesterol gallstones to form?
26. Explain how cholesterol is kept soluble in normal human bile.
27. What effect on digestion would the removal of the gallbladder have?
28. A patient with serum hypercholesterolemia (elevated plasma cholesterol) had not responded to drug therapy, so the patient is to undergo surgical removal of part of the intestinal tract. What part of the intestinal tract should be removed?

Practice Test

1. The function of HCl in the stomach is to _____.
   a. convert pepsinogen into pepsin
   b. denature protein
   c. destroy microorganisms
   d. all of these

2. Biles salts _____.
   a. stimulate intestinal motility
   b. aid in the absorption of fatty acids
   c. lower surface tension
   d. all of these

3. Most of the digestion of food takes place in the _____.
   a. stomach
   b. large intestine
   c. mouth
   d. small intestine

4. Adults who cannot tolerate milk or milk products lack the enzyme _____.
   a. lactase
   b. sucrase
   c. amylase
   d. enterokinase

5. Dietary fiber is associated with reduced incidence of _____.
   a. gallstones
   b. colon cancer
   c. Graves' disease
   d. Alzheimer's disease

6. Most of the absorption of food takes place in the _____.
   a. mouth
   b. stomach
   c. small intestine
   d. large intestine

7. The process of digestion involves _____.
   a. hydrolysis
   b. hemolysis
   c. oxidation
   d. isomerization

8. Bile is produced in the _____.
   a. stomach
   b. liver
   c. small intestine
   d. pancrease

9. The end product of starch digestion is _____.
   a. galactose
   b. lactose
   c. maltose
   d. glucose

10. The part of the gastrointestinal system that has the lowest pH is the _____.
    a. mouth
    b. stomach
    c. small intestine
    d. large intestine

Metabolism of Carbohydrates

Objectives

- To recognize the significance of normal and abnormal blood sugar levels

- To describe the role of ATP in metabolism

- To understand glycogenesis and glycogenolysis

- To understand glycolysis and the hexose monophosphate shunt

- To describe the role of oxidation in the citric acid cycle

- To understand the electron transport system

- To understand the role of gluconeogenesis

- To discuss the overall scheme of carbohydrate metabolism

- To become familiar with the hormones involved in the regulation of blood sugar level

Much of our "quick energy" comes from the metabolization of carbohydrates—glycogen stored in muscle tissue. Most people have enough glycogen to provide a full day's worth of energy; however, these reserves are used up to four times as quickly during intensive athletic competition, as in this rowing regatta.

26-1   General Principles of Metabolism

There are two major questions in biochemistry:

1. How do cells obtain energy from their environment?
2. How do cells synthesize the building blocks of their macromolecules?

The answers lie with the chemical reactions collectively known as metabolism. Although there are hundreds of reactions in metabolism, the number of kinds of reactions is quite small. The metabolic pathways are regulated in simple, common ways.

Pathways that produce ATP (adenosine triphosphate), the currency of energy in biologic systems, pertain to the first question. ATP is used for three major purposes: muscle contraction and movement, active transport of molecules and ions, and synthesis of biologic molecules.

Metabolism occurs in stages. In the first stage (biodegradation), large molecules in food are broken down into small units. This stage involves digestion and absorption. In the second stage (biosynthesis), these numerous small molecules are converted into a few very simple units that play the central role in metabolism. The third stage consists of the energy production mainly associated with the Krebs cycle and oxidative phosphorylation (Figure 26-1).

An important general principle of metabolism is that the biosynthetic and biodegradative pathways are almost always separated. This separation is necessary both for energetic reasons and for metabolic control. Frequently this separation is enhanced by compartmentalization of the pathways. For example, fatty acid oxidation occurs in the mitochondria, whereas fatty acid synthesis takes place in the cytoplasm.

An overview of the principal pathways to be studied in this and the following chapters is given in Figure 26-2.

26-2   Concentration of Sugar in the Blood (Blood Sugar Level)

The end products of carbohydrate digestion are the monosaccharides glucose, fructose, and galactose. Both fructose and galactose are converted to glucose in the liver so that the major monosaccharide remaining in the bloodstream is glucose.

The amount of glucose present in the blood will vary considerably, depending on whether the measurements were taken $1/2$ hour after eating, 1 hour after eating, or during a period of fasting. The normal quantity of glucose present in 100 mL of blood taken after a period of fasting is 70 to 100 mg. This value is called the normal fasting blood sugar. Soon after a meal, the blood sugar level may rise to 120 to 130 mg per 100 mL of blood or even higher. However, the level soon drops, so that after $1^1/2$ to 2 hours it again returns to its normal fasting value.

Blood sugar levels below 70 produce hypoglycemia, and levels over 100 produce hyperglycemia (Figure 26-3). The brain metabolizes approximately 120 g of glucose daily, and hypoglycemia can reduce the brain's energy supply, causing dizziness and loss of consciousness.

During the time of fasting, even though the body is continuously using glucose for the production of heat and energy, the amount of glucose present in the blood remains fairly constant, usually in the range of 70 to 100 mg per 100 mL. How does the body regulate the amount of glucose present in the blood, and what happens when these control mechanisms do not function property?

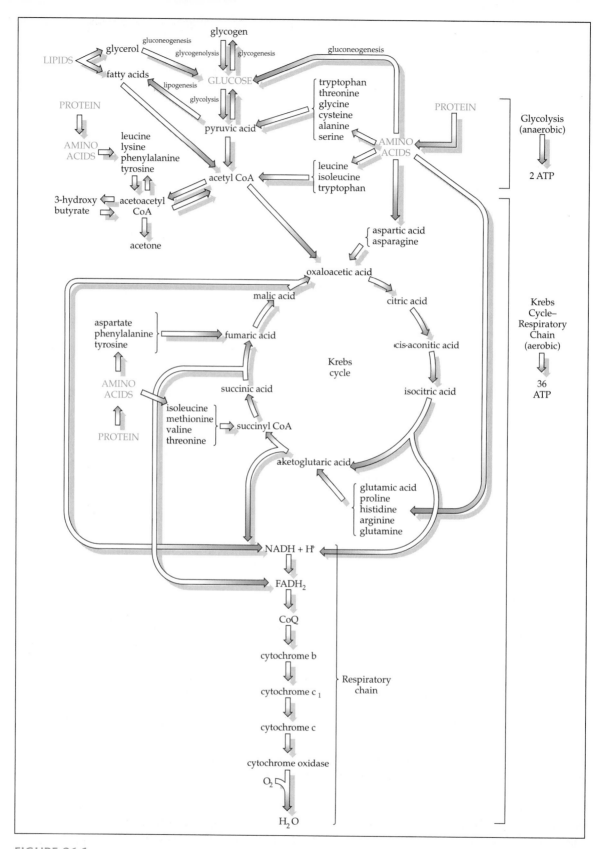

FIGURE 26-1

The interrelationships in the metabolism of carbohydrates, lipids, and proteins.

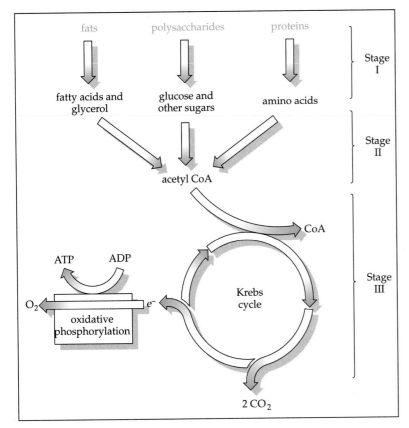

FIGURE 26-2
Three stages of metabolism.

When the glucose level in the blood rises, as happens after the digestion of a meal, the liver removes the excess glucose and converts it to glycogen, a polysaccharide. This process is called glycogenesis. This glycogen may be stored in the liver and in muscle tissue. However, only a certain amount of glycogen can be stored in the liver and muscle; the rest is changed to fat and stored as such.

Cellular uptake of glucose is greatly increased by the release of insulin (see page 472). Glucose is also removed from the blood by the normal oxidative reactions that take place continuously throughout the body.

Glucose does not normally appear in the urine except in amounts too small to be detected by Benedict's solution. However, if the blood sugar level rises above 170 to 180 mg per 100 mL of blood, the sugar "spills over" into the urine. The point at which the sugar spills over into the urine is called the renal threshold. The presence of glucose in the urine is called glycosuria.

Thus, the factors that remove excess glucose from the blood are (1) insulin secretion, (2) glycogenesis and storage as glycogen, (3) conversion to fat, (4) normal oxidation reactions in the body, and (5) excretion through the kidneys when the renal threshold is exceeded.

Two to 3 hours after the ingestion of food, the liver converts glycogen back to glucose. This process is called glycogenolysis. Liver glycogen will be exhausted after 10 hours. The body then switches to making glucose from amino acids; this process is called gluconeogenesis (see Figure 26-14).

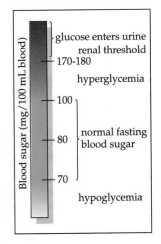

FIGURE 26-3

Blood sugar levels in milligrams per 100 mL of blood.

26-3 Source of Energy

Where does a muscle get its energy to contract? Where does the body get the energy necessary to synthesize protein, to send nerve impulses, to perform countless other functions?

The energy necessary for the body functions comes from certain high-energy compounds, compounds that yield a large amount of energy on hydrolysis. The key compound of this type is adenosine triphosphate (ATP). Hydrolysis of ATP to adenosine diphosphate (ADP) and inorganic phosphate liberates about 7600 cal/mol. This hydrolysis breaks one of the high-energy phosphate bonds, designated by ~ in structures (26-1) and (26-2).

(26-1)

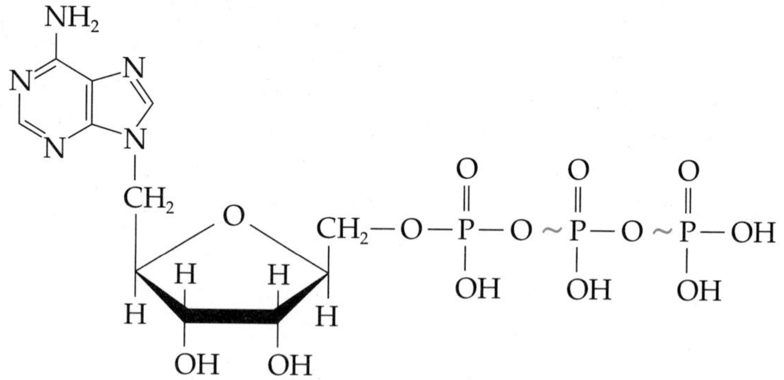

adenosine triphosphate (ATP)

This formula for ATP can be abbreviated as

$$\boxed{\text{adenosine}}-O-\overset{\overset{O}{\|}}{\underset{\underset{OH}{|}}{P}}-O\sim\overset{\overset{O}{\|}}{\underset{\underset{OH}{|}}{P}}-O\sim\overset{\overset{O}{\|}}{\underset{\underset{OH}{|}}{P}}-OH \quad \text{or} \quad A-P\sim P\sim P \qquad (26\text{-}1)$$

(26-2)

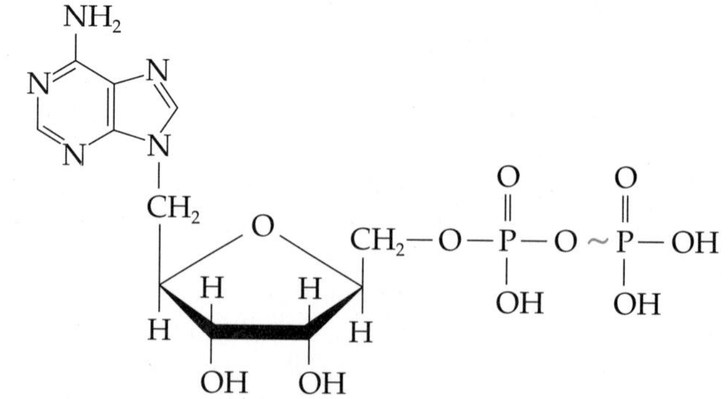

adenosine diphosphate (ATP)

This formula for ADP can be abbreviated as

$$\boxed{\text{adenosine}}-O-\overset{\overset{O}{\|}}{\underset{\underset{OH}{|}}{P}}-O\sim\overset{\overset{O}{\|}}{\underset{\underset{OH}{|}}{P}}-OH \quad \text{or} \quad A-P\sim P \qquad (26\text{-}2a)$$

However, the supply of ATP in the body is limited. There must be some mechanism for regenerating this high-energy compound so it will be available for continued use. How does the body change ADP back to ATP? To accomplish this, a high-energy phosphate group must be added to the ADP. This process is called *phosphorylation* and can be represented by the equation

$$\text{ADP} + \text{phosphate ion} + \text{fuel} \longrightarrow \text{ATP} + \text{fuel residue}$$

One of the fuels used in this rection is glucose. The oxidation of glucose, the steps and enzymes required, the products produced, and the involvement of ADP and ATP in these processes will be discussed later in this chapter.

The metabolism of carbohydrates in humans is categorized as follows:

1. Glycogenesis, the synthesis of glycogen from glucose
2. Glycogenolysis, the breakdown of glycogen to glucose
3. Glycolysis, the oxidation of glucose or glycogen to pyruvic or lactic acid
4. Hexose monophosphate shunt, an alternative oxidative path for glucose
5. Krebs (citric acid) cycle and electron transport chain, the final oxidative paths to carbon dioxide and water
6. Gluconeogenesis, the formation of glucose from noncarbohydrate sources

26-4 Glycogenesis

Glycogenesis is the formation of glycogen from glucose. This process occurs primarily in the liver and the muscles. The liver may contain up to 5 percent glycogen after a high-carbohydrate meal but may contain almost no glycogen after 12 hours of fasting.

The overall conversion of glucose to glycogen and vice versa can be written as

$$n\,\underset{\text{glucose}}{C_6H_{12}O_6} \underset{\text{glycogenolysis}}{\overset{\text{glycogenesis}}{\rightleftharpoons}} \underset{\text{glycogen}}{(C_6H_{10}O_5)_n} + n\,H_2O$$

However, the reaction is by no means as simple as this equation indicates. There are several steps involved, each one catalyzed by a particular enzyme.

The first step in glycogenesis involves the conversion of glucose to glucose 6-phosphate (abbreviated as glucose 6-P), a phosphate ester of glucose. ATP from the liver cells serves as a source of the phosphate group. After the loss of the phosphate group, ADP is left. The enzyme glucokinase is necessary to catalyze this reaction. Insulin is involved in the phosphorylation of glucose by glucokinase.

$$\text{glucose} + \text{ATP} \xrightarrow[\text{insulin}]{\text{glucokinase}} \text{glucose 6-P} + \text{ADP}$$

Glucose 6-phosphate is then rearranged so that the phosphate group moves from the number 6 position to the number 1 position, producing glucose 1-phosphate. The enzyme required for this reaction is phosphoglucomutase.

$$\text{glucose 6-P} \xrightarrow{\text{phosphoglucomutase}} \text{glucose 1-P}$$

Glucose 1-phosphate then reacts with uridine triphosphate (UTP) to form uridine diphosphate glucose (UDPG). (Uridine triphosphate is similar to adenosine triphosphate except that uracil takes the place of adenine.) The enzyme is UDPG pyrophosphorylase.

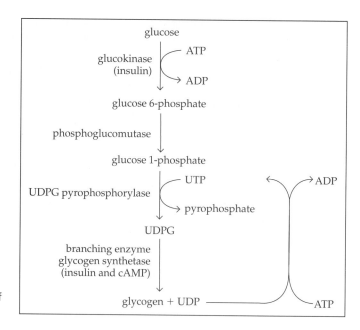

FIGURE 26-4
Overall reaction of glycogenesis.

$$\text{glucose 1-P} + \text{UTP} \xrightarrow{\text{UDPG pyrophosphorylase}} \text{UDPG} + \text{pyrophosphate}$$

Then the glucose molecules in UDPG (activated glucose molecules) are joined together to form glycogen. The enzymes necessary here are glycogen synthetase and a branching enzyme. The former enzyme is regulated by both insulin and cyclic adenosine monophosphate (cAMP). The latter enzyme aids in the formation of 1,4- and 1,6-glycosidic linkages.

$$\text{UDPG} \xrightarrow[\text{branching enzyme}]{\text{glycogen synthetase}} \text{glycogen} + \text{UDP}$$

The UDP formed in this reaction then reacts with ATP to regenerate UTP.

$$\text{UDP} + \text{ATP} \longrightarrow \text{UTP} + \text{ADP}$$

A diagram of the overall reaction is shown in Figure 26-4.

26-5 Glycogenolysis

In glycogenolysis we might expect the reverse of all the reactions of glycogenesis, but it should be noted that the first reaction, the one involving glucokinase, is not a reversible reaction.

In glycogenolysis, glycogen is converted to glucose 1-phosphate by the enzyme phosphorylase a and then to glucose 6-phosphate by the enzyme phosphoglucomutase.

Glucose 6-phosphate is then converted to glucose by the enzyme glucose 6-phosphatase, an enzyme found in the liver but not in the muscle. Therefore, muscle glycogen cannot serve as a source of blood glucose.

cAMP (see page 577) is involved in the conversion of glycogen into glucose 6-phosphate in both the liver and the muscles. When the body is under stress, it produces hormones that are carried by the bloodstream to the liver cells, where they activate the enzyme adenyl cyclase, which in turn causes the production of cAMP from ATP. cAMP then activates a protein kinase, which in turn activates a phosphorylase b kinase. This phosphorylase b kinase then activates phosphorylase a, which triggers the conversion of glycogen into glucose.

The same type of reaction is involved in the conversion of muscle glycogen into glucose 6-phosphate. cAMP also deactivates the enzyme glycogen synthetase, thereby stopping glycogenesis.

Glycogenesis and glycogenolysis can be summarized as shown in Figure 26-5.

Glycogen Storage Diseases

Glycogen is normally stored in the liver and the muscles. In several inherited diseases, glycogen cannot be reconverted to glucose, so it begins to accumulate.

In von Gierke's disease the enzyme glucose 6-phosphatase is lacking in the liver, and glycogen accumulates in that organ. As this occurs, hypoglycemia, ketosis, and hyperlipemia also occur, and the liver enlarges because of the increase in stored glycogen.

In Pompe's disease glycogen accumulates in the lysosomes because of a lack of a lysosomal enzyme that acts in the breakdown of glycogen.

In Forbes's disease or Cori's limit dextrinosis a debranching enzyme is absent so, as in von Gierke's disease, glycogen accumulates. However, the symptoms of Forbes' disease are not as severe as those of von Gierke's disease.

Other glycogen storage diseases include Andersen's disease (in which death occurs in the first year of life because of liver and cardiac failure), McArdle's syndrome (in which inviduals have a greatly diminished tolerance to exercise because of a lack of a muscle enzyme involved in glycogenolysis), and Tarui's disease (which is caused by phosphorylase deficiency in the liver).

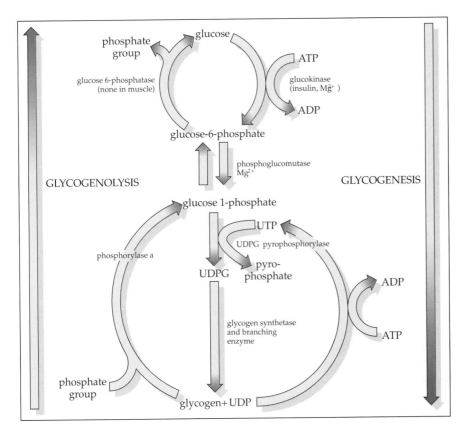

FIGURE 26-5

Combined reactions of glycogenolysis and glycogenesis.

Case Study 26-1 Von Gierke's Disease

An 11-year-old girl is admitted to the hospital with a history of frequent bouts of weakness and sweating, which are relieved by eating. Her abdomen is grossly distended and she is pale. A developmental history shows that she was able to sit unsupported at 1 year of age and did not walk until 2 years of age. Currently her school performance is poor.

Physical examination findings include blood pressure, 112/60 mm Hg; temperature, 38.1° C; weight, 22.5 kg (low); and height, 130 cm (low). The liver is enlarged and firm.

A fasting blood sample reveals the following:

	Patient	Reference values
Glucose (mmol/L)	2.8	3.9-5.6
Lactate (mmol/L)	6.5	0.56-2.0
Pyruvate (mmol/L)	0.44	0.05-0.10
Free fatty acids (mmol/L)	1.8	0.3-0.8
Triclycerides (g/L)	3.2	1.5
Total ketone bodies	390	30
pH	7.24	7.35-7.44
Total CO_2 (mmol/L)	14	24-30

A liver biopsy is performed; it reveals the hepatic cells to be bulging and dilated. The enlarged liver is firm but not cirrhotic and it is buff colored. The portal areas are shrunken and compressed. Staining for carbohydrates shows a positive result for large amounts of carbohydrates in the parenchymal cells, which were removed by digestion with salivary amylase.

Glycogen levels and lipid content are both elevated. However, the hepatic glycogen structure is within normal limits. The following are enzyme assay results from the liver tissue biopsy:

Glucose-6-phosphatase	20	214 ±4
Glucose-6-phosphate dehydrogenase	0.06	0.05-0.13
Phosphoglucomutase	28	25 ±4
Phosphorylase	23	22
Fructose-1,6-bisphosphatase	8.5	10 ±6

Questions

1. What is the normal structure for hepatic glycogen?
2. What change would you see with a deficiency in branching enzyme?
3. What other tissues would accumulate excessive amounts of glycogen?
4. Explain the reasons for the elevated free fatty acid, the ketonemia, and the metabolic acidoses.

26-6 Embden-Meyerhof Pathway: Glycolysis

The breakdown of glycogen resupplies the energy used up during muscle contraction. This breakdown involves a series of steps, each catalyzed by a particular enzyme.

The breakdown of glycogen to pyruvate (pyruvic acid) and lactate (lactic acid), called the Embden-Meyerhof pathway, is the first phase of muscle contraction. This process is also called *glycolysis* and is an *anaerobic process—that is, each step takes place without oxygen. Glycolysis supplies the ATP needed for muscle contraction. More than 90 percent of the energy in the red blood cells is produced by glycolysis.

The overall reaction of glycolysis can be summarized as

$$\text{glucose 6-P} \quad + \text{2 ADP} \longrightarrow \text{2 pyruvic acid (or pyruvate)} + \text{2 ATP}$$
glucose 6-phosphate
(from glycogen or glucose)

The ATP formed is available for muscular work. As ATP is used, it is changed to ADP and must then be regenerated. This regeneration can be accomplished through the above pathway or by the use of another anaerobic sequence,

$$\text{creatine phosphate} + \text{ADP} \longrightarrow \text{creatine} + \text{ATP}$$

The steps in glycolysis are given below and, with the structural formulas of the intermediary products, in Figure 26-6.

Glycogen is changed to glucose 1-phosphate by the catalytic action of the enzyme phosphorylase a. Glucose 1-phosphate is then changed to glucose 6-phosphate by the enzyme phosphoglucomutase. Glucose 6-phosphate could also be formed directly from glucose by the action of ATP and the enzyme glucokinase.

Step A Glucose 6-phosphate is changed to fructose 6-phosphate by the action of the enzyme phosphoglucose isomerase.

Step B Fructose 6-phosphate is changed to fructose 1,6-diphosphate by the action of the enzyme phosphofructokinase. Note that during this reaction, ATP is converted to ADP.

Step C Fructose 1,6-diphosphate is changed to the three-carbon compounds glyceraldehyde 3-phosphate and dihydroxyacetone phosphate. The enzyme involved is aldolase. Dihydroxyacetone phosphate is converted to glyceraldehyde 3-phosphate by the action of the enzyme phosphotriose isomerase.

Step D Glyceraldehyde 3-phosphate is changed to 1,3-diphosphoglycerate by the action of the enzyme glyceraldehyde 3-phosphate dehydrogenase. During this reaction, nicotinamide adenine dinucleotide (NAD^+) is reduced to $NADH + H^+$.

Step E 1,3-Diphosphoglycerate is changed to 3-phosphoglycerate by the action of the enzyme phosphyglycerokinase. In this reaction, two ADPs (One for each three-carbon compound) are changed to two ATPs.

Step F 3-Phosphoglycerate is changed to 2-phosphoglycerate by the action of the enzyme phosphyglyceromutase.

Step G 2-Phosphoglycerate is changed to phosphoenolpyruvate by the action of the enzyme enolase.

Step H Phosphoenolpyruvate is changed to pyruvate by the action of the enzyme pyruvic kinase. During this reaction, two ADPs are changed to two ATPs (one for each three-carbon compound).

Step I Pyruvate is changed to lactate through the enzyme lactic dehydrogenase. At the same time, NADH and H^+ are changed to NAD^+.

The sequence of reactions involved in glycolysis is summarized in Figure 26-7. Three of these reactions are irreversible—the conversions of glucose to glucose 6-phosphate, fructose 6-phosphate to fructose 1,6-diphosphate, and phosphoenolpyruvate to pyruvate. The rate of glycolysis is primarily controlled by the allosteric enzyme phosphofructokinase. This enzyme is inhibited by high levels of

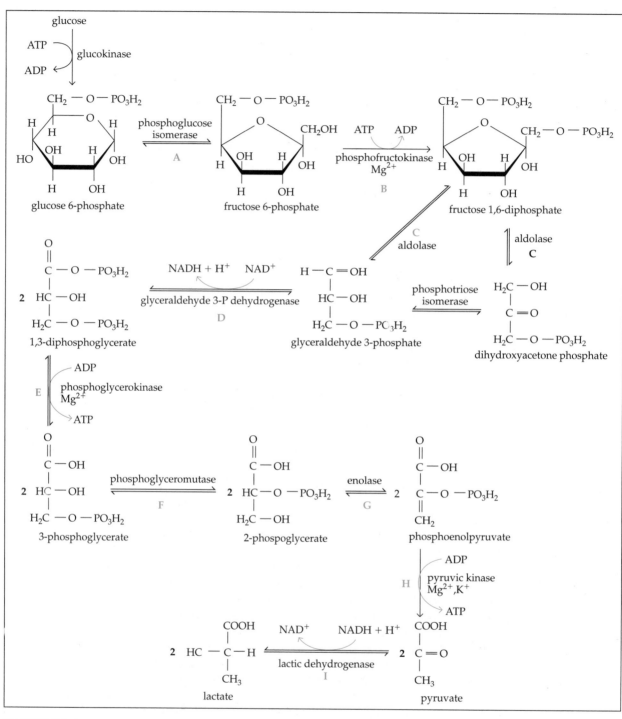

FIGURE 26-6

Glycolysis—the Embden-Meyerhof pathway.

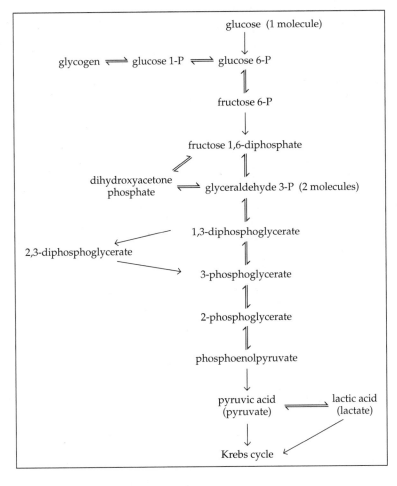

FIGURE 26-7

The sequence of events in glycolysis.

ATP and, conversely, is stimulated by low ATP levels. Glycolysis also provides compounds for biosynthesis, and thus phosphofructokinase is regulated by other sources for these compounds, such as fatty acids and citrate as well.

In addition, red blood cells have two enzymes that allow for the formation of 2,3-diphosphoglycerate (2,3-DPG) from 1,3-diphosphoglycerate. 2,3-DPG is a regulator of oxygen transport in red blood cells. When the O_2 level in the blood drops, 2,3-DPG increases and more O_2 is delivered to the cells.

According to these reactions, muscle glycogen is changed into pyruvic acid and then to lactic acid. However, only about one fifth of the lactic acid thus formed is oxidized to carbon dioxide and water, resupplying the energy used up during muscle contraction. The other four fifths of the lactic acid is changed back to glycogen, reversing the previously described reactions. Part of the lactic acid is changed back to glycogen in the muscle. The rest of the lactic acid is carried to the liver by the bloodstream, where it is converted to liver glycogen. In addition, for lactic acid to be utilized for energy, it must first be converted to pyruvic acid. Lactic acid is a metabolic dead end.

The oxidation of some of the lactic acid, the aerobic sequence, produces a large amount of energy. The oxidation of one molecule of lactic acid converts 18 molecules of ADP to ATP.

After a muscle contracts and relaxes, the net change is a partial loss of glycogen. Glycogen can be replenished by the conversion of blood glucose to muscle glycogen (muscle glycogenesis).

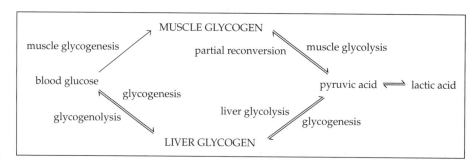

FIGURE 26-8
The lactic acid or Cori cycle.

This conversion of glycogen to lactic acid (lactate) and partial reconversion to glycogen, the lactic acid or Cori cycle, is shown in Figure 26-8.

Lactacidosis

Arsenite ions and mercuric ions react with the —SH groups in lipoic acid and inhibit the formation of the enzyme lactic dehydrogenase (see previously described reactions, step I). A dietary deficiency of thiamin (vitamin B_1) has the same effect. A lack of or a deficiency of lactic dehydrogenase allows lactic acid to accumulate, leading to lactacidosis, which, if untreated, can be fatal.

26-7 Another Oxidative Pathway: Hexose Monophosphate Shunt

The oxidation of glucose to lactic acid can also proceed through a series of reactions called the **hexose monophosphate shunt,** or the **pentose shunt.** This sequence is important because it provides the five-carbon sugars needed for the synthesis of nucleic acids and nucelotides and because it makes available NADPH, the reduced form of nicotinamide adenine dinucleotide phosphate ($NADP^+$), a coenzyme necessary for the synthesis of fatty acids and steroids. This pathway is much nore active in adipose tissue than in muscle. Note that ATP is not generated in this sequence. The ribose can be further metabolized to glyceraldehyde 3-phosphate.

A genetic deficiency of glucose 6-phosphate dehydrogenase (see Figure 26-9) is a major cause of hemolytic anemia. A lack of glucose 6-phosphate dehydrogenase also provides resistance to malaria. However, at the same time a lack of this enzyme reduces the level of NADPH in red blood cells. The major role of NADPH in red blood cells is the maintenance of proper levels of the antioxidase glutathione. Without adequate glutathione, red blood cells are easily oxidized by a variety of drugs.

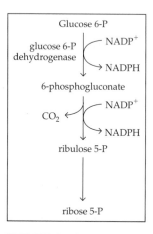

FIGURE 26-9
The hexose monophosphate or pentose shunt.

26-8 The Aerobic Sequence: The Citric Acid Cycle

The aerobic sequence converts lactic and pyruvic acids (from anaerobic glycolysis) through a series of steps to carbon dioxide and water. This series of reactions is called the **citric acid cycle** (Krebs cycle) (see Figure 26-10). This cycle uses oxygen transported to the cells by hemoglobin, hence the term *aerobic*. This cycle takes place in the mitochondria (see Section 24-6 and Figure 26-13).

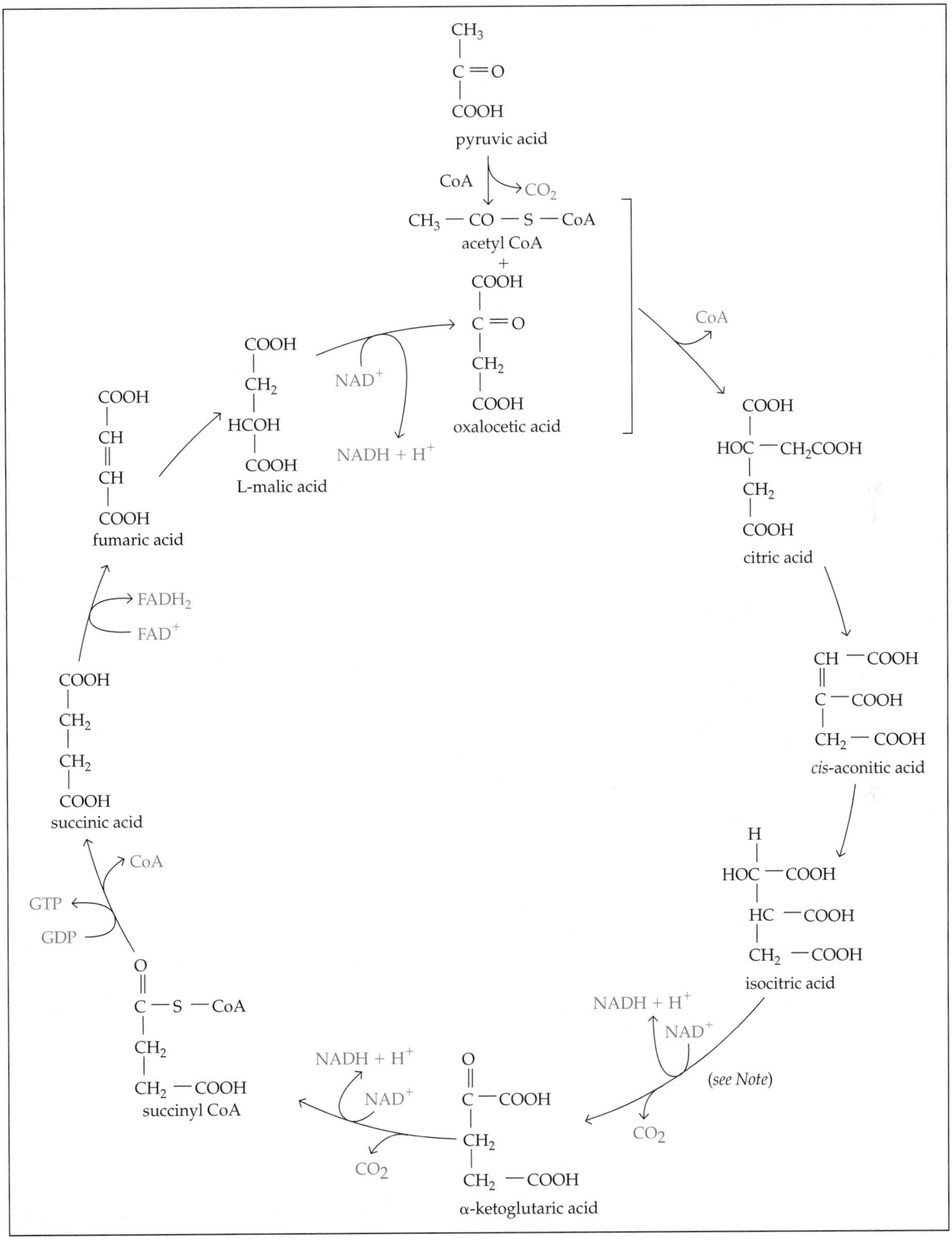

FIGURE 26-10

The citric acid cycle.
NOTE: An enzyme-bound complex of oxalocuccinic acid occurs here as an intermediary.

	Table 26-1 Production of ATP From the Oxidation of One Molecule of Glucose		
Source of ATP	**Number of Molecules of ATP Formed From One Molecule of Glucose**	**Other Compounds Formed**	
Glycolysis	2	2 NADH	
Pyruvic acid $\longrightarrow$ acetyl CoA	0	2NADH	
Krebs cycle	2	$2\,GTP + 6\,NADH + 2\,FADH_2$	
Oxidative phosphorylation of 2 NADH from glycolysis	6		
Transportation across mitochondrial membrane	−2		
2 NADH from pyruvic acid $\longrightarrow$ acetyl CoA reaction	6		
6 NADH from Krebs cycle	18		
2 FADH$_2$ from Krebs cycle	4		
	36		

ATP = adenosine triphosphate; FADH$_2$ = reduced form of flavin, adenine dinucleotide; GTP = guanosine triphosphate; NADH = reduced form of nicotinamide adenine dinucleotide.

During the complete oxidation of one molecule of glucose, 36 molecules of ATP are produced, as indicated in Table 26-1. Note that most of the ATP formed comes from oxidative phosphorylation.

Assuming 7.6 kcal per high-energy phosphate bond, the overall sum is 36×7.6 kcal, or 274 kcal. Theoretically, 686 kcal should be produced from 1 mol of glucose; thus, the efficiency of conversion is approximately 40 percent.

The first step in the aerobic process is the formation of active acetate from pyruvic acid. This active acetate is the acetyl derivative of coenzyme A, or acetyl CoA. Acetyl CoA is the converting substance in the metabolism of carbohydrates, fats, and proteins. Acetyl CoA becomes the "fuel" for the Krebs cycle. As will be noted, acetyl CoA reacts with oxaloacetic acid and goes through the cycle. At the end of the cycle, oxaloacetic acid is regenerated and picks up another molecule of acetyl CoA to carry it through the sequence. During the cycle, acetyl CoA is oxidized to carbon dioxide and at the same time NADH and FADH$_2$ are produced.

These enter into the electron transport chain that functions on the inner membranes of the mitochondria. The overall reaction for the citric acid cycle can be summarized by the following equation.[†]

$$\text{acetyl CoA} + 3\,NAD^+ + FAD + GDP + P_i + 2\,H_2O \longrightarrow$$
$$2\,CO_2 + CoA + 3\,NADH + 2\,H^+ + FADH_2 + GTP$$

26-9 The Role of B Vitamins in the Citric Acid Cycle

Four B vitamins are necessary for the proper functioning of the citric acid cycle:

Riboflavin in the form of flavin adenine dinucleotide (FAD^+), a cofactor in the α-ketoglutarate dehydrogenase complex and also in succinate dehydrogenase

---

[†]FAD = flavin adenine dinucleotide; GDP = guanosine diphosphate.

Thiamine, vitamin B_1, which is the coenzyme for the decarboxylation in the α-ketoglutarate dehydrogenase reaction; lipoic acid also is involved in the decarboxylation of α-keto acids

Niacin in the form of nicotinamide adenine dinucleotide (NAD^+), the coenzyme for the dehydrogenases in the citric acid cycle

Pantothenic acid, which is part of coenzyme A

26-10 Oxidative Phosphorylation: Electron Transport System

In the electron transport chain, or electron transport system, electrons are transferred from NADH through a series of steps to oxygen, with the regeneration of NAD^+ and FAD, and the formation of H_2O. The energy yielded is used for the formation of ATP (Figure 26-11). The overall reaction in the electron transport chain is also called oxidative phosphorylation and proceeds because large amounts of NADH and $FADH_2$ are produced from the Krebs cycle (Figure 26-12) and also because there is a plentiful supply of oxygen in the tissues. The overall reaction is

$$NADH + H^+ \, 3\,ADP + P_i + \tfrac{1}{2}O_2 \longrightarrow NAD^+ + 3\,ATP + H_2O$$

Involved in this electron transport chain are NADH, $FADH_2$, coenzyme Q, and several cytochromes, which are complexes containing heme (recall that heme is part of the hemoglobin molecule).

One of the cytochromes, cytochrome oxidase, is a complex that binds oxygen, reduces it with electrons received from other cytochromes in the electron transport system, and finally converts that oxygen to water. Note that the oxygen in this sequence (which is therefore an aerobic sequence) reacts only at the last step.

One unusual fact about cytochrome oxidase is that it contains two metals—iron and copper. This particular enzyme is mainly responsible for the introduction of oxygen into the metabolic processes of the Krebs cycle and so is considered absolutely vital to life. It is believed that the cytochrome oxidase functions by transferring electrons from copper to iron to oxygen.

Note that another high-energy compound produced by the Krebs cycle is guanosine triphosphate (GTP). GTP is considered to be interchangeable with ATP.

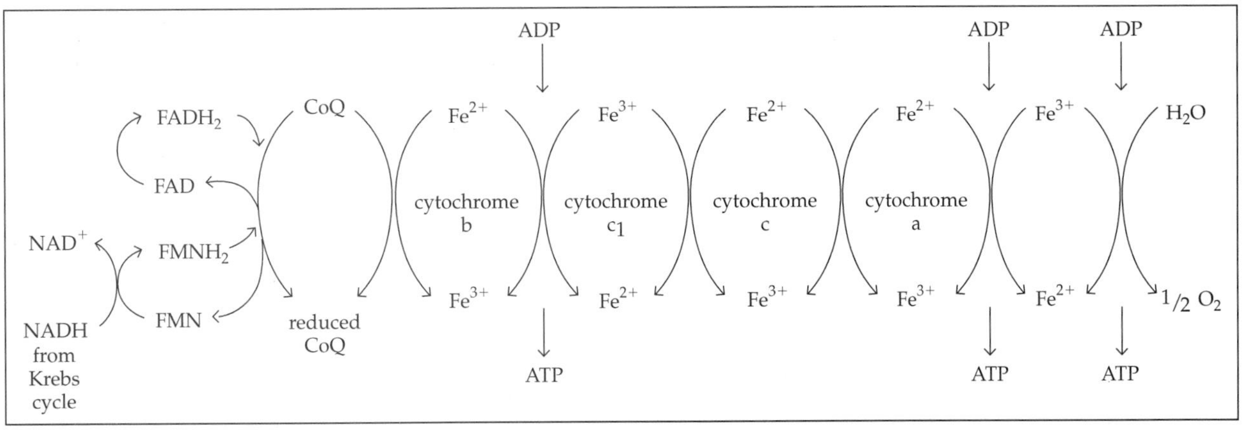

FIGURE 26-11
The electron transport system.

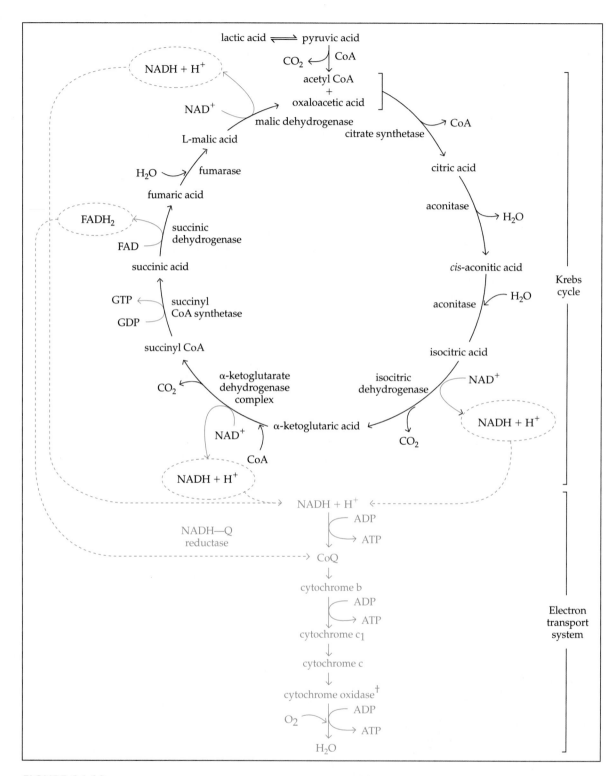

FIGURE 26-12
Oxidative phosphorylation.
†Cytochrome oxidase, also designated as cytochrome aa₃, was formerly believed to be two sep-
arate substances, cytochromes a and a₃.

Poisons That Inhibit the Respiratory Chain

Inhibitors to the respiratory chain act at three different sites in that chain, frequently with fatal results.

1. **Poisons that inhibit the formation of coenzyme Q**   Among such poisons are barbiturates such as amobarbital, antibiotics such as piericidin A, and insecticides such as rotenone. At high doses these substances can prove fatal.
2. **Poisons that inhibit the conversion of cytochrome b to cytochrome c**   Among such poisons are hydrogen sulfide (H_2S), carbon monoxide (CO), and cyanides such as HCN.
3. **Poisons that act as "uncouplers" to oxidative phosphorylation**   Recall that the action of the respiratory chain is also called oxidative phosphorylation, so that any substance that uncouples oxidation in the respiratory chain causes uncontrolled reactions in that system. Among such poisonous uncouplers are 2,4-dinitrophenol, dinitrocresol, m-chlorocarbonyl cyanide phenylhydrazone, and snake venoms.

The Chemiosmotic Theory

The chemiosmotic theory, proposed by Peter Mitchell in England in 1961 endeavors to explain oxidative phosphorylation in terms of the movement of protons (H^+) through the mitochondrial membrane (Figure 26-13). The main principles of this theory are

1. The mitochondrial membrane is impermeable to ions, particularly to H^+, which accumulates outside the membrane, causing an electrochemical potential difference across the membrane.
2. The synthesis of ATP occurs under the influence of an enzyme on the inside of the inner mitochondrial membrane.
3. ATP synthesis occurs because of the movement of protons (H^+) through special ports in the membrane (not through the membrane itself) from the outside to the inside of the inner mitochondrial membrane.
4. The potential difference drives membrane-located ATP synthase.
5. The respiratory chain is folded into three oxidation-reduction loops in the membrane, each loop corresponding to a part of the respiratory chain.

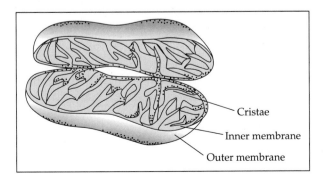

Cristae

Inner membrane

Outer membrane

FIGURE 26-13

Three-dimensional representation of a mitochondrion.

26-11 Gluconeogenesis

Gluconeogenesis is the formation of glucose from noncarbohydrate substances such as amino acids and glycerol. This process takes place primarily in the liver, although it also occurs to a small extent in the kidneys.

A continuous supply of glucose is necessary for normal body functions. If blood glucose levels fall too low (severe hypoglycemia), brain dysfunction may occur, which can lead to coma and eventually to death. Also, glucose is the only fuel that supplies energy to the skeletal muscles under anaerobic conditions.

Glucose is the precursor of lactose in the mammary glands and also is actively used by the fetus.

Thus gluconeogenesis meets the body's needs for glucose when sufficient carbohydrate is unavailable.

Gluconeogenesis is increased by high-protein diets and decreased by high-carbohydrate diets. During starvation, gluconeogenesis supplies glucose from the amino acids of the tissue protein. In severe diabetes, gluconeogenesis not only from food protein but also from tissue protein may lead to emaciation. Since several reactions in glycolysis are not reversible, gluconeogenesis includes some additional reactions (Figure 26-14).

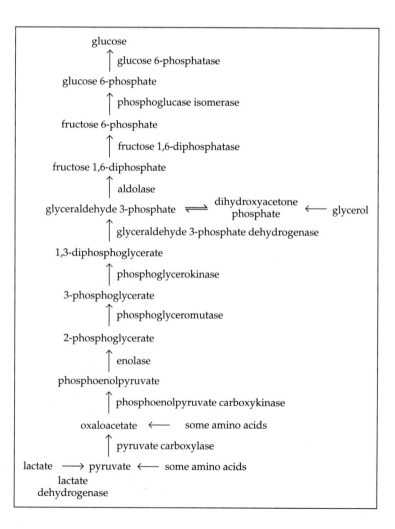

FIGURE 26-14

Gluconeogenesis.

Case Study 26-2 Galactosemia

Three days after delivery a male infant develops increasing jaundice and becomes difficult to feed. His delivery was normal, and birth weight was 3.7 kg. There was no blood group incompatibility. However, at 6 days of age his serum bilirubin is found to be elevated to 504 µmol/L, with a resultant weight loss of 15 percent. He is readmitted to the hospital with increased muscular tonus and has begun to have seizure activity. Exchange blood transfusions have been performed, but the bilirubin level remains high. Three days later it is noted that the infant's liver is enlarged and he begins vomiting.

Six days after birth, a positive test result for reducing sugars in his urine is noted. On day 7 the test is repeated and is found to be positive; however, a Clinistix test done at the same time is negative for D-glucose. Hereditary galactosemia is confirmed by the following results:

Hemoglobin	12.6 mmol/L (200 g/L)
Bilirubin (max)	550 µmol/L (at day 7)
Galactose 1-P uridyl transferase (in erythrocytes)	0†

Formula feeding is stopped on day 9 and replaced by intraveneous glucose infusion. A galactose-free diet is started and the patient's condition improves.

Questions

1. Explain the biochemical effects of galactosemia.
2. If a mother is homozygous for galactosemia, would she produce lactose in her milk?
3. Is there another source of tissue galactose for a patient on a galactose-free diet?

†See back of book for normal laboratory blood values.

26-12 Interconversion of Hexoses

A typical meal containing starch, sucrose, and lactose loads the liver with galactose and fructose. These sugars must be converted into glucose. Figure 26-15 summarizes the reactions involved in these conversions. These reactions indicated that there is no essential carbohydrate. Adequate levels of any hexose can be used to produce the others.

Galactosemia is a disease associated with the inability to convert galactose into glucose. The reverse reaction, the conversion of glucose into galactose, is not affected, so galactose accumulates. High levels of galactose produce mental retardation and cataracts and stunt growth. The removal of all galactose-containing foods reverses these symptoms except mental retardation, which may or may not be reversible. Galactose-deprived patients are able to provide the galactose necessary for biosynthetic purposes by utilizing the glucose-to-galactose conversion.

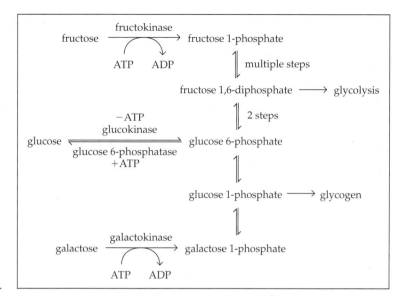

FIGURE 26-15
Interconversion of hexoses.

26-13 Overall Scheme of Metabolism

The interrelationship of glycogenesis, glycogenolysis, glycolysis, gluconeogenesis, and the Krebs (citric acid) cycle, along with the metabolsim of fats and proteins, is indicated in Figure 26-1.

26-14 Hormones Involved in Regulating Blood Sugar

The liver plays a vital function in controlling the normal blood sugar level by removing sugar from and adding sugar to the blood. The acitivity of the liver in maintaining the normal blood sugar level is in turn controlled by several different hormones. Among these are insulin, epinephrine, and glucagon. The hormones of the anterior pituitary, the adrenal cortex, and the thyroid also have a definite effect upon carbohydrate metabolism.

Insulin

Insulin is a hormone produced by the β cells of the islets of Langerhans in the pancreas. Insulin performs the following functions:

1. It aids in the transportation of glucose across cell membranes.
2. It accelerates the oxidation of glucose in the cells.
3. It increases the transformation of glucose to glycogen (glycogenesis) in the muscle and also in the liver. (Insulin controls the phosphorylation of glucose to glucose 6-phosphate by means of the enzyme glucokinase; see page 457.)
4. It depresses the production of glucose (glycogenolysis) in the liver.
5. It promotes the formation of fat from glucose.

Thus, the principal function of insulin may be said to be the removal of glucose from the bloodstream and a consequent lowering of the blood sugar level.

Diabetes Mellitus

If the amount of insulin is decreased or eliminated (either because of decreased activity of the islets of Langerhans or by the degeneration of these cells), the blood sugar level will rise. Increased blood sugar level (hyperglycemia) leads to glycosuria (glucose in the urine) because the renal threshold is exceeded.

Also, the lack of insulin in diabetes leads to an increased oxidation of fatty acids as a source of ATP. Increased oxidation of fatty acids leads to an accumulation of acetoacetic acid, β-hydroxybutyric acid, and acetone. These substances, commonly known as ketone bodies, form faster than they can be oxidized and removed and so accumulate in the blood (and urine). A higher-than-normal concentration of these substances in the blood is known as ketosis.

The presence of ketone bodies affects the pH of the blood, since two of the three compounds are acids. If the ketone bodies accumulate and lower the pH of the blood, a condition known as acidosis exists. A decreased pH reduces the ability of hemoglobin to carry oxygen; therefore acidosis exists. A decreased pH reduces the ability of hemoglobin to carry oxygen; therefore acidosis can be very serious. Prolonged acidosis first causes nausea, then depression of the central nervous system, severe dehydration, deep coma (known as diabetic coma), and finally death.

Prompt injection of insulin will alleviate the symptoms accompanying high blood sugar. Persons suffering from diabetes mellitus can lead normal lives provided that they receive insulin as needed. Since insulin is a protein, it cannot be taken orally (it would be digested as are all proteins) and so must be administered by injection.

Epinephrine

Epinephrine is a hormone secreted by the *medulla of the adrenal glands. It stimulates the formation of glucose from glycogen in the liver (glycogenolysis) (see page 458) and so has an action opposite to that of insulin. Insulin removes glucose from the bloodstream, whereas epinephrine increases the amount of glucose present in the blood.

During periods of strong emotional stress, such as anger or fright, epinephrine is secreted into the bloodstream, where it promotes glycogenolysis in the liver. This increases the amount of glucose in the blood, making that glucose readily available as the body needs it to meet the emergency situation. The amount of glucose may then exceed the renal threshold (hyperglycemia) and sugar will appear in the urine. This is one example of how the presence of sugar in the urine may be the result of a condition other than diabetes.

Glucagon

Glucagon is a hormone produced by the α cells of the pancreas. Its effects are opposite to those of insulin. Glucagon raises blood sugar levels by stimulating the activity of the enzyme phosphorylase in the liver, which changes liver glycogen to glucose. The activity of phosphorylase depends upon cAMP. Glucagon also increases gluconeogenesis from amino acids and from lactic acid. Glucagon has no effect on phosphorylation in the muscles.

26-15 Glucose Tolerance Test

A positive Benedict's test on a urine specimen indicates that the patient *may* have diabetes mellitus. However, this test is by no means conclusive proof because the presence of sugar in the urine may be due to other conditions, such as pregnancy,

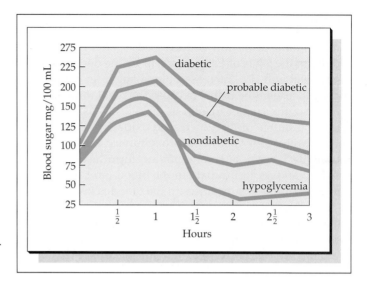

FIGURE 26-16

Criteria used for interpretation of glucose tolerance tests.

emotional disturbances, large intake of fruit or fruit juices, and genetic disorders (idiopathic pentosuria, fructosuria, or galactosuria).

A patient with a suspected problem with glucose metabolism is given a glucose tolerance test. For 3 days prior to the test, the patient must be placed on a diet containing adequate protein, calories, and 150 g of carbohydrate per day. This stimulates the production of the liver enzymes necessary for the conversion of glucose into glycogen.

Then, after 12 hours of fasting, the patient is given approximately 1 g of glucose for each kilogram of body weight. The patient's blood sugar level is checked by withdrawing blood samples at regular intervals over several hours. The samples are chemically analyzed, and the concentration of sugar in the blood is plotted against time.

In a normal person the blood sugar level rises from about 80 mg per 100 mL of blood to 130 mg per 100 mL of blood in about 1 hour. Then the blood sugar level gradually returns to normal after about $2\frac{1}{2}$ hours. In a diabetic patient, because no insulin is being secreted, the blood sugar level rises to an even higher level than in the normal individual and remains there for a much longer period of time. Then it slowly begins to return toward its normal value.

These results are illustrated in the graphs in Figure 26-16.

Also in the diabetic patient, large amounts of glucose appear in the urine because of the spillover when the renal threshold is exceeded.

The glucose tolerance test is a valuable diagnostic tool because it indicates the ability of the body to utilize carbohydrate. A decreased utilization may indicate diabetes, whereas an increased utilization may indicated Addison's disease, hypopituitarism, or hyperinsulinism.

The American Diabetes Association recommends the criteria given in Table 26-2 for establishing a diagnosis of diabetes mellitus. A similar test, a galactose tolerance test, helps in the diagnosis of hepatic function.

Table 26-2	Wilkerson Point System for Interpretation Of Oral Glucose Tolerance Test	
Time (hr after glucose)	Plasma Glucose Concentration (mg/dL)	Points[a]
0	130	1
1	195	0.5
2	140	0.5
3	140	1

*From H. L. C. Wilkerson. H. L. C. Diagnosis: Oral Glucose Tolerance Tests. In Danowski, T. S. (Ed. Diabetes Mellitus.) New York: American Diabetes Association, 1964, p. 31. *A total of 2 points establishes diabetes.*

Summary

The end products of carbohydrate digestion are the monosaccharides. The major monosaccharide in the bloodstream is glucose. Some of the blood glucose is converted to glycogen in the liver and in the muscle. This process is called glycogenesis. Other glucose is constantly being oxidized to furnish energy for the body. If the blood sugar level rises too much, the excess spills over into the urine. The presence of glucose in the urine is known as glycosuria.

The oxidation of glucose produces energy, which is stored in high-energy compounds, especially adenosine triphosphate (ATP). Hydrolysis of ATP to adenosine diphosphate (ADP) and inorganic phosphate liberates about 7.6 kcal.

Glycogenesis—the formation of glycogen from glucose—takes place in the liver and in muscle tissue. There are many intermediate steps involved in this conversion, each being catalyzed by a specific enzyme.

Glycogenolysis—the conversion of glycogen to glucose—takes place primarily in the liver.

The breakdown of muscle glycogen to pyruvic acid (pyruvate) and lactic acid (lactate), a process that requires no oxygen (anaerobic), is called glycolysis. Glycolysis supplies most of the ATP needed for muscle contraction. Glycolysis proceeds through a series of steps, each catalyzed by a specific enzyme. In this process three molecules of ADP are converted to ATP, which is then available for muscular work.

About one fifth of the lactic acid formed in the glycolysis is oxidized to carbon dioxide and water. The other four fifths is converted back to liver glycogen. The cycle of glucose-glycogen-lactic acid-glycogen is known as the lactic acid cycle.

The oxidation of glucose to lactic acid can also proceed through a series of reactions called the hexose monophosphate shunt or the pentose shunt. This sequence is important because it provides five-carbon sugars needed for the synthesis of nucleic acids and nucleotides and also because it makes available the reduced form of $NADP^+$, a coenzyme necessary for the synthesis of fatty acids.

The aerobic sequence for the oxidation of lactic and pyruvic acids is called the Krebs or citric acid cycle. This cycle takes place in the mitochondria.

The first step in the Krebs cycle is the formation of acetyl CoA from pyruvic acid. This acetyl CoA, also called active acetate, is the fuel for the Krebs cycle. Acetyl CoA reacts with oxaloacetic acid and then goes through a series of steps, each catalyzed by a particular enzyme. At the completion of the cycle, oxaloacetic acid is regenerated and then picks up another molecule of acetyl CoA to carry through the same cycle again. The NADH and $FADH_2$ produced in the Krebs cycle enter the electron transport chain in the mitochondria. The overall reaction, called oxidative phosphorylation, involves oxygen and produces ATP. Several coenzymes and cytochromes are involved.

Glucose can also be formed from noncarbohydrate substances such as amino acids, fatty acids, and glycerol. Such a process is called *gluconeogenesis*.

Thus carbohydrate metabolism interrelates the processes of glycogenesis, glycogenolysis, glycolysis, gluconeogenesis, and the Krebs cycle.

The liver controls the blood sugar level. This activity is governed by several hormones. Among these are insulin, ephinephrine, and glucagon.

Insulin, a hormone secreted by the pancreas, accelerates oxidation of glucose in the cells, increases glycogenesis, decreases glycogenolysis, and promotes the formation of fat from glucose. Thus insulin removes glucose from the bloodstream.

Epinephrine, a hormone of the adrenal medulla, changes liver glycogen to glucose and muscle glycogen to lactic acid. Epinephrine is secreted into the bloodstream during periods of emotional stress.

Glucagon, a hormone secreted by the pancreas, has an effect opposite to that of insulin. It raises blood sugar levels.

The glucose tolerance test is performed when a patient is suspected of having diabetes. He or she is fed glucose and the blood sugar level is checked over several hours.

Questions and Problems

1. What is normal fasting blood sugar? renal threshold?
2. What happens to carbohydrates that cannot be immediately metabolized or converted to glycogen?
3. Which monosaccharide is the principal one remaining in the bloodstream after passing through the liver?
4. Describe the process of glycogenesis, indicating the intermediate products and the enzymes involved.
5. What is glycosuria? What causes it?
6. How can glucose be removed from the bloodstream?
7. What types of compounds does the body use to store energy?
8. What is glycogenolysis? gluconeogenesis?
9. Name two ways in which ADP can be converted to ATP.
10. Indicate the hydrolysis reaction of ATP. How much energy is produced in this reaction?
11. What is glycolysis? Is it an aerobic or anaerobic sequence? Does it supply most of the body's energy?
12. Diagram and label the lactic acid cycle.
13. After a muscle contracts and relaxes, how does it replenish its glycogen?
14. Is all of the lactic acid formed in glycolysis oxidized to carbon dioxide and water? Explain.
15. What is the hexose monophosphate shunt? Why is it important?
16. Describe briefly what happens in the citric acid cycle.
17. What is the function of acetyl CoA in the citric acid cycle?
18. How does the energy produced during the citric acid cycle compare with that produced during glycolysis?
19. Diagram and label the Krebs cycle.
20. Show the interrelationship between glycolysis, gluconeogenesis, the citric acid cycle, glycogenesis, and glycogenolysis.
21. What are the functions of insulin? Where is it produced?
22. What is the function of glucagon? Where is it formed?
23. Describe the steps in the electron transport system.
24. What is cytochrome oxidase? What does it do? Why is it essential to life?
25. What is the function of epinephrine? Where is it secreted?
26. The ATP system is important for energy storage. List as many other functions of ATP as you can.

The following scenario pertains to questions 27 to 30:

A 55-year-old man undergoes cardiac bypass surgery during which an oxygenator is used. The extracorporeal circuit was primed with 2-day-old acid citrate dextrose blood diluted with bicarbonate-buffered saline solution to an 8 percent hemoglobin content. The volume of the priming blood was 40 to 50 percent of the patient's blood volume, and during the bypass the mixed blood averaged 10 to 11 percent hemoglobin. The 2,3-bisphospho-D-glycerate level is expressed as millimoles of phosphorus per liter of packed RBCs. The $pO_{2_{50}}$ is the oxygen tension in kilopascals at which the hemoglobin was 50 percent saturated with oxygen. During the determination of the pO_2 it was shown that the 2,3-bisphospho-D-glycerate and pH_{RBC} was measured in packed, hemolyzed cells. The pH of whole blood was consistently 7.4 throughout the surgery.

	2,3-Bisphospho-D-glycerate	pO$_{2_{50}}$	pH RBC
Priming blood	3.8	1.86	
Patient control	4.8	3.46	7.2
Duration of bypass (min)			
10		2.4	7.0
20	2.7		
40	2.5		
60	2.88	2.66	6.9
After bypass (hr)			
1	2.4	3.1	7.2
24	4.45	3.6	7.2

Priming blood contained 3 percent carbon monoxide compared with 1 percent in the patient's blood.

27. Can the change in the 2,3-bisphospho-D-glycerate content of the red blood cells be explained by admixture of the priming blood with the patient's blood?
28. Explain the effect of 2,3-bisphospho-D-glycerate and pH on the oxygen saturation curve of hemoglobin.
29. What will be the effect on the oxygen transport to tissues which results from the combined decrease of 2,3-bisphospho-D-glycerate and pH$_{RBC}$ in the patient?

30. Do you believe that the priming blood donor was a smoker and that the bypass patient smoked less? What effect does carbon monoxide have on oxygen transport?

Practice Test

1. The chief end product of glycolysis is _____.
   a. carbon dioxide
   b. pyruvic acid
   c. acetyl CoA
   d. glucose
2. Gluconeogenesis takes place primarily in the _____.
   a. pancreas
   b. small intestine
   c. liver
   d. stomach
3. Which of the following is not produced directly in the citric acid cycle?
   a. ATP
   b. FADH$_2$
   c. GTP
   d. NADH
4. ATP contains how many high-energy bonds?
   a. 1
   b. 2
   c. 3
   d. 4
5. Glycogenesis takes place primarily in the _____.
   a. kidneys
   b. liver
   c. stomach
   d. brain
6. The hexose monophosphate shunt is important because it produces _____.
   a. GTP
   b. ATP
   c. NADPH
   d. NADH
7. The citric acid cycle converts _____ to CO$_2$.
   a. glucose
   b. acetyl CoA
   c. lactate
   d. pyruvate

8. Which of the following functions is (are) performed by insulin?
   a. increase cellular oxidation of glucose
   b. decrease the breakdown of glycogen into glucose
   c. increase glucose transportation into cells
   d. all of these
9. Normal fasting blood sugar levels are approximately _____ mg/100 mL.
   a. 50-70
   b. 70-100
   c. 100-150
   d. 150-180
10. 2,3-DPG, a regulator of oxygen transport in red blood cells, is produced during _____.
    a. glycogenesis
    b. the citric acid cycle
    c. glycolysis
    d. gluconeogenesis

27

Metabolism of Fats

Objectives

- To describe the mechanism of fat oxidation

- To become familiar with the amount of energy produced during the oxidation of a fat

- To become aware of the significance of the role of ketone bodies

- To become acquainted with some metabolic diseases caused by impaired oxidation of fatty acids

- To follow the path of lipogenesis and of the formation of cholesterol

The bulk of an individual's energy stores resides in fat deposits. These portly sumo wrestlers have mastered the art of energy storage.

27-1 Plasma Lipid Levels

During digestion, fats and phospholipids are emulsified and then hydrolyzed into fatty acids and glycerol. The products are synthesized into triglycerides in the intestinal mucosa and flow into the thoracic duct and then into the bloodstream. However, such substances are insoluble in water and likewise in the blood. To be transported by the blood, fats and phospholipids form a complex with plasma (water-soluble) protein. Such complexes are called lipoproteins (see page 392). The table at the back of the book indicates the normal range for lipid levels in the blood.

Abnormalities of lipid metabolism lead to various types of hypolipoproteinemia or hyperlipoproteinemai. The most common type of abnormality is diabetes, where a deficiency of insulin leads to hypertriacylglycerolemia.

27-2 Absorption of Fat

The digestion of fats takes place primarily in the small intestine, with hydrolysis yielding fatty acids and glycerol. Before they are digested, the fats have been emulsified by the bile salts. The products of fat digestion pass through the *lacteals of the *villi into the *lymphatics, where they appear as resynthesized fats. From the lymphatics, the fats flow through the thoracic duct into the bloodstream and then to the liver. After a meal the fat content of the blood rises and remains at a high level for several hours, then gradually decreases to the fasting level.

In the liver some of the fats are changed to phospholipids, so the blood leaving the liver contains both fats and phospholipids. These phospholipids, such as sphingomyelin and lecithin (see page 372), are necessary for the formation of nerve and brain tissue. Lecithins (phosphatidyl cholines) are also involved in the transportation of fat to the tissues. Cephalin, another phospholipid, is involved in the normal clotting of the blood. From the liver some fat goes to the cells, where it is oxidized to furnish heat and energy. The fat in excess of what the cells need is stored as adipose tissue.

Lipolysis, the hydrolysis of triacylglycerols (triglycerides) to fatty acids and glycerol is primarily controlled by the amount of cAMP present in the tissues. Hormones that stimulate the production of cAMP and so increase lipolysis include epinephrine, norepinephrine, glucagon, adrenocorticotropic hormone (ACTH), α and β melanocyte–stimulating hormones (MSH), thyroid-stimulating hormone (TSH), growth hormone (GH), and vasopressin. However, insulin and the prostaglandins (see Section 21-9) depress the levels of cAMP and so decrease the rate of lipolysis.

The enzyme that degrades cAMP is inhibited by methyl xanthines such as caffeine and theophylline. One unusual effect of this inhibition is the presence of a definite and prolonged elevation of free fatty acid levels in the body in those who drink large amounts of coffee (which contains caffeine).

27-3 Oxidation of Fat

The oxidation of fat (triglyceride) actually involves the oxidation of the two hydrolysis products—glycerol and fatty acids. General aspects of the controls on triglyceride breakdown into glycerol and fatty acids are summarized in Figure 27-1. This sequence is blocked by insulin or high levels of glucose.

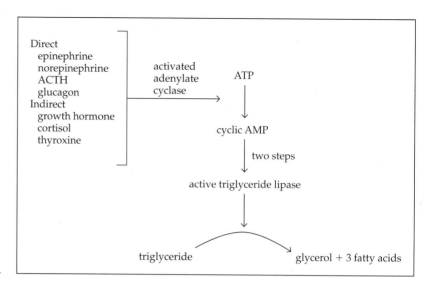

FIGURE 27-1

Breakdown of triglyceride in adipose tissue.

Oxidation of Glycerol

The glycerol part of a fat is oxidized to dihydroxyacetone phosphate, as is indicated in the following sequence. Recall that dihydroxyacetone phosphate is part of the glycolysis sequence (see Section 26-6). This compound can be converted into glycogen in the liver or muscle tissue or into pyruvic acid, which enters the Krebs cycle. Thus, the glycerol part of a fat is metabolized through the carbohydrate sequence.

$$\begin{array}{c} \text{CH}_2\text{OH} \\ | \\ \text{CHOH} \\ | \\ \text{CH}_2\text{OH} \\ \text{glycerol} \end{array} \quad \xrightarrow[\text{phosphatase}]{\overset{\text{ATP} \quad \text{ADP}}{\text{glycero-kinase}}} \quad \begin{array}{c} \text{CH}_2\text{OH} \\ | \\ \text{HO}-\text{C}-\text{H} \\ | \\ \text{CH}_2-\text{O}-\text{PO}_3\text{H}_2 \\ \alpha\text{-glycerophosphate} \end{array} \quad \xrightarrow[\text{dehydrogenase}]{\overset{\text{NAD}^+ \quad \text{NADH} + \text{H}^+}{\text{glycerophosphate}}} \quad \begin{array}{c} \text{CH}_2\text{OH} \\ | \\ \text{C}=\text{O} \\ | \\ \text{CH}_2-\text{O}-\text{PO}_3\text{H}_2 \\ \text{dihydroxyacetone} \\ \text{phosphate} \end{array}$$

Oxidation of Fatty Acids

There are several theories about the oxidation of fatty acids. The original one, proposed by Knoop in 1905 and still preferred today, is called the β-oxidation theory. This theory involves the oxidation of the second carbon atom from the acid end of the fatty acid molecule—the β carbon atom. In this process β oxidation removes two carbon atoms at a time from the fatty acid chain. That is, an 18-carbon fatty acid is oxidized to a 16-carbon fatty acid, then to a 14-carbon fatty acid, and so on, until the oxidation process is complete. A simplified version of such an oxidation is shown in Figure 27-2.

The acetyl CoA thus produced enters the Krebs cycle and the new molecule of active fatty acid goes through the same sequence again, each time losing two carbon atoms until the entire fatty acid molecule has been oxidized. This sequence presupposes the presence of fatty acids containing an even number of carbon atoms, a condition usually found in nature. The FADH_2 and the $\text{NADH} + \text{H}^+$ enter the respiratory chain.

If fatty acids containing an odd number of carbon atoms are oxidized, they follow the same steps except that the final products are acetyl CoA and propionyl CoA. The propionyl CoA is changed in a series of steps to succinyl CoA, which

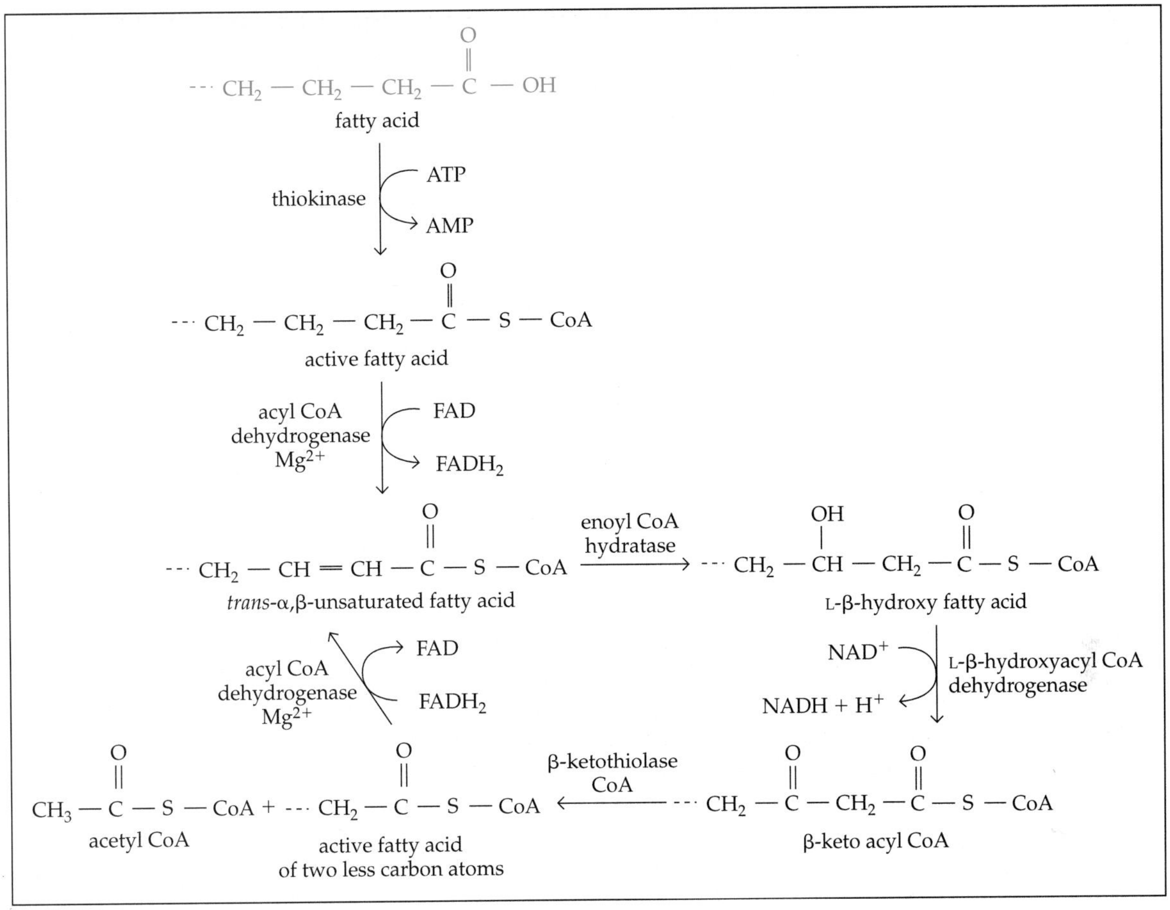

FIGURE 27-2
Oxidation of a fatty acid.

then enters the Krebs cycle, as does the acetyl CoA. These reactions require the presence of vitamin B_{12} and also biotin.

The unsaturated fatty acids are metabolized slowly. They must first be reduced by some of the dehydrogenases found in the cells. Then they can follow the fatty acid cycle for oxidation.

Energy Produced by Oxidation of Fatty Acids

The oxidation of 1 g of fat produces more than twice as much energy as the oxidation of 1 g of carbohydrate. Let us see why.

The oxidation of acetyl CoA through the Krebs cycle yields 12 high-energy phosphate bonds (ATP) per molecule of acetyl CoA. If we consider the oxidation of palmitic acid, a 16-carbon fatty acid, eight two-carbon units will be formed during the β oxidation cycle. These eight two-carbon units will yield $8 \times 12 = 96$ ATP. However, 2 ATP are used up in the initial activation of the fatty acid. In addition, it has been calculated that palmitic acid will produce 35 ATP as it goes through the fatty acid cycle (7 $FADH_2$, each equivalent to 2 ATP, and 7 NADH, each equivalent to 3 ATP). That is, the net number of ATP molecules produced will be $96 - 2 + 35 = 129$ (see Table 27-1).

Table 27-1	ATP Formed From the Oxidation of A 16-Carbon Fatty Acid
Source	Number of ATP Molecules Formed per C_{16} Molecule
7 $FADH_2$	14
7 NADH	21
Initial activation of fatty acid	−2
8 acetyl CoA	96
	$\overline{129}$

Considering each mole of ATP as requiring 7.6 kcal for formation, 129 × 7.6 kcal, or 980 kcal is needed. The theoretic yield from 1 mol of palmitic acid is 2340 kcal, so that the efficiency of conversion is 980/2340, or 42 percent, with the remainder of the energy being produced as heat. (Other fatty acids and glycerol are also oxidized, so the net result is that fats produce much more energy than do carbohydrates.)

27-4 Ketone (Acetone) Bodies

In a diabetic patient, or in any other situation in which carbohydrate metabolism is restricted, the body uses oxaloacetate to produce glucose for the brain and muscles. This reduces the amount of oxaloacetate available for the Krebs cycle, and acetyl CoA cannot be properly metabolized. When this occurs, the acetyl CoA is changed to acetoacetyl CoA, which is converted into acetoacetic acid in the liver by the enzyme deacylase. Acetoacetic acid may be changed into acetone and β-hydroxybutyric acid, as shown in Figure 27-3.

These three substances—acetoacetic acid, β-hydroxybutyric acid, and acetone—are commonly called acetone bodies, or ketone bodies. They are carried by the blood to the muscles and tissues, where they are converted back to acetoacetyl CoA and then oxidized normally. However, in patients with diabetes the production of these ketone bodies by the liver exceeds the ability of the muscles and tissues to oxidize them, so they accumulate in the blood.

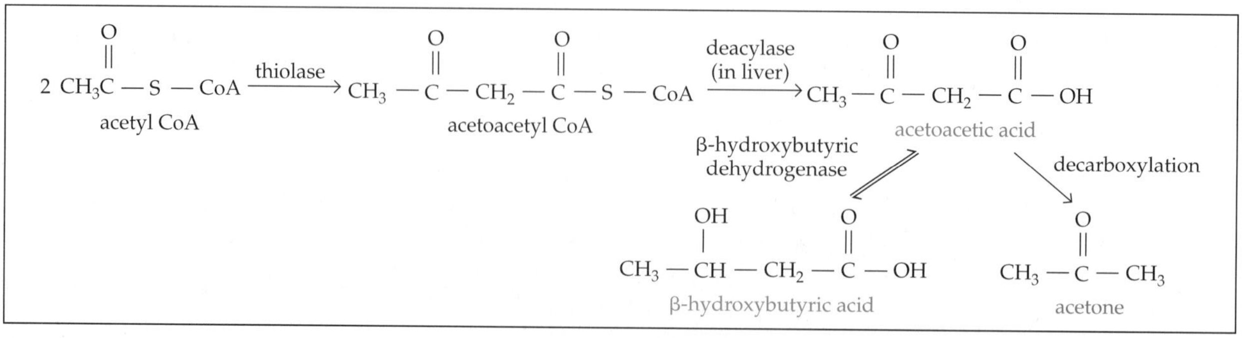

FIGURE 27-3
Formation of ketone (acetone) bodies.

Ketosis

The excess accumulation of ketone bodies in the blood is called ketonemia. The excess accumulation of ketone bodies in the urine is called ketonuria. The overall accumulation of ketone bodies in the blood and the urine is called *ketosis*. During ketosis, acetone can be detected on the patient's breath because it is a volatile compound—tending to evaporate rapidly—and is easily excreted through the lungs. Ketosis may occur with diabetes mellitus, in starvation or severe liver damage, or on a diet high in fats and low in carbohydrates.

In diabetes mellitus the body is unable to oxidize carbohydrates and instead oxidizes fats, leading to an accumulation of ketone bodies in the blood and the urine. These ketone bodies are acidic and tend to decrease the pH of the blood. The lowering of the pH of the blood is called *acidosis* and may lead to a fatal coma. During acidosis an increased amount of water intake is needed to eliminate the products of metabolism. Unless the water intake of a diabetic patient is increased, dehydration will occur. Dehydration in patients with diabetes can also be caused by polyuria due to an increased amount of glucose in the urine.

Likewise, during prolonged starvation or on a high-fat, low-carbohydrate diet, the body tends to burn fat instead of carbohydrate, leading to ketosis and acidosis.

Case Study 27-1   **Diabetic Ketoacidosis**

A 20-year-old woman with a history of juvenile-onset diabetes from the age of 14 is admitted to the hospital with a recent onset of fever, vomiting, and diarrhea. She had not taken her insulin since 24 hours before admission because she was unable to eat. On admission she was semiconscious and had an odor of acetone on her breath. She was moderately dehydrated. Her urine was 4+ for glucose and positive for ketones. The blood serum was also positive for ketones. She was given 100 U of insulin, and an intravenous drip was started of hypotonic saline solution with one amp of sodium bicarbonate. Her first laboratory results were as follows[†]:

Blood sugar	53 mmol/L	Sodium	140 mmol/L
BUN	6.5 mmol/L	Potassium	6.0 mmol/L
creatinine	250 µmol/L	Chloride	95 mmol/L
CO_2	3 mmol/L	pCO_2	19 mm Hg
Arterial blood gases	pH 7.08		

The intravenous drip has been continued and she is given an additional 200 U of insulin because of her high blood sugar level and low CO_2. Two hours later she is alert and oriented, with a blood sugar level of 14 mmol/L; potassium, 4.1 mmol/L; total CO_2, 16 mmol/L; and pH, 7.35. Plasma ketones are now trace positive.

Questions

1. Why did this patient develop hyperglycemia?
2. How did ketosis develop along with an elevated blood glucose?
3. How did the insulin work to correct the metabolic problem?
4. Describe the mechanism of action in acidosis. Explain the rationale for replacement of fluid and electrolytes.

[†]*See back of book for normal laboratory values.*

In severe liver damage, the liver cannot store glycogen in the required amounts. The resulting shortage of carbohydrates needed for the normal oxidation of fats leads to ketosis.

27-5 Metabolic Diseases Caused by Impaired Oxidation of Fatty Acids

Jamaican vomiting disease is caused by the ingestion of the unripe fruit of the akee tree. This unripe fruit contains hypoglycin, a substance that inactivates acyl-CoA-dehydrogenase, thereby inhibiting β-oxidation and so causing hypoglycemia.

Refsum's disease, a rare genetic disease, is caused by the accumulation of phytanic acid. Phytanic acid, in turn, blocks β-oxidation.

Hepatic carnitine palmitoyltransferase deficiency produces hypoglycemia and a low level of plasma ketone bodies.

Muscle carnitine palmitoyltransferase deficiency leads to impaired oxidation of fatty acids, causing muscular weakness.

27-6 Storage of Fat

Fat in excess of that required for the normal oxidative processes of the body is stored as adipose tissue under the skin and around the internal organs. This stored fat serves several important purposes:

1. Reserve supply of food
2. Support for the internal organs
3. Shock absorber for the internal organs
4. Insulation of interior of the body against sudden external changes in temperature

The fat stored in the body is in equilibrium with that in the bloodstream; that is, the fats stored in the adipose tissue do not merely remain there as inert compounds until they are needed. They are continuously being used and replaced, and there is always a dynamic transfer of fats between the bloodstream and the storage tissues.

Obesity (20 percent or more over normal weight) is a condition in which excess fat is deposited as adipose tissue. An obese person eats more food than his or her body can burn up, and the excess is converted to fat and stored as adipose tissue. For every 9 kcal of food eaten in excess of the body's requirements, 1 g of fat is deposited.

Most people have a tendency to become overweight as they grow older because they require less food for the maintenance of their bodies and exercise less than younger people.

In general, obesity leads to a shortened life expectancy, as indicated in Figure 27-4. An overweight person runs a higher than normal risk of developing cardiovascular disease, diabetes, or liver disease. A weight greater than 10 percent above that considered normal for a person's age and height can cause medical problems. The answer to obesity lies in proper dieting under a physician's supervision because the metabolism of the body is a highly intricate mechanism that can very easily be disturbed.

Excessive accumulation of triglycerides in the liver leads to cirrhosis and impaired liver function. Such an accumulation may be caused either by increased levels of free plasma fatty acids or by blockage in the production of plasma lipoproteins from the free fatty acids.

Chronic alcoholism also leads to hyperlipidemia and eventually to cirrhosis.

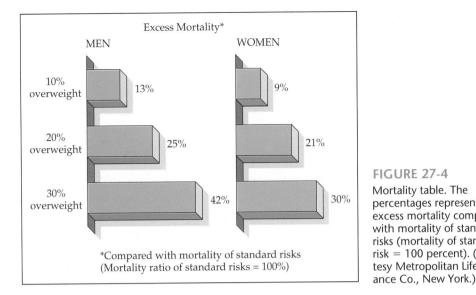

FIGURE 27-4

Mortality table. The percentages represent excess mortality compared with mortality of standard risks (mortality of standard risk = 100 percent). (Courtesy Metropolitan Life Insurance Co., New York.)

27-7 Lipogenesis

Lipogenesis—the conversion of glucose to fats—takes place in the liver and adipose tissue, with the latter place predominating. Insulin is necessary for lipogenesis both in the liver and in the adipose tissue. The main factor controlling the rate of lipogenesis is the nutritional state of the organism. If a person is on a high-carbohydrate diet, the rate of lipogenesis increases. If a person is on a calorie-restricted or high-fat diet or when there is a deficiency of insulin, as in diabetes mellitus, the rate of lipogenesis decreases. There is also an increase in the concentration of plasma free fatty acids associated with a decrease in the rate of lipogenesis.

The synthesis of fatty acids occurs in the mitochondria and in the cell cytoplasm, especially the latter. The process in the mitochondria involves the lengthening of fatty acid chains of moderate length, whereas the cytoplasmic processes involve the synthesis of fatty acids from acetyl CoA.

Steps in the synthesis of fatty acids from acetyl CoA are

Step 1 Acetyl CoA is changed to malonyl CoA.

$$CH_3-\overset{\overset{\displaystyle O}{\|}}{C}-S-CoA \xrightarrow[\text{ATP, biotin}]{Mn^{2+}} \underset{\underset{\displaystyle COOH}{|}}{CH_2}-\overset{\overset{\displaystyle O}{\|}}{C}-S-CoA$$

acetyl CoA   malonyl CoA

Note that one carbon atom has been added to the chain.

Step 2 Malonyl CoA reacts with another molecular of acetyl CoA to form acetoacetyl complex.

$$\underset{\underset{\displaystyle COOH}{|}}{CH_2}-\overset{\overset{\displaystyle O}{\|}}{C}-S-CoA + CH_3-\overset{\overset{\displaystyle O}{\|}}{C}-S-CoA \longrightarrow CH_3-\overset{\overset{\displaystyle O}{\|}}{C}-CH_2-\overset{\overset{\displaystyle O}{\|}}{C}-complex + CO_2 + H_2O$$

malonyl CoA   acetyl CoA   acetoacetyl complex

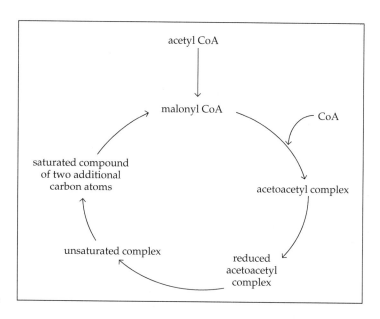

FIGURE 27-5
Synthesis of a fatty acid.

Note that the additional carbon atom from the previous equation has been removed. It was used mainly to activate the α carbon in the acetyl CoA so that the condensation reaction could take place. The enzyme complex required here is called fatty acid synthetase and consists of seven enzymes.

Step 3 The keto group of the acetoacetyl complex is reduced to the corresponding alcohol by NADPH.

$$CH_3-\overset{\overset{O}{\|}}{C}-CH_2-\overset{\overset{O}{\|}}{C}-complex \xrightarrow{NADPH} CH_3-\overset{\overset{OH}{|}}{CH}-CH_2-\overset{\overset{O}{\|}}{C}-complex$$

Step 4 The alcohol produced in step 3 is dehydrated to form an unsaturated compound.

$$CH_3-\overset{\overset{OH}{|}}{CH}-CH_2-\overset{\overset{O}{\|}}{C}-complex \longrightarrow CH_3-CH=CH-\overset{\overset{O}{\|}}{C}-complex + H_2O$$

Step 5 The unsaturated compound in step 4 is reduced by NADPH to the corresponding saturated compound.

$$CH_3-CH=CH-\overset{\overset{O}{\|}}{C}-complex \xrightarrow{NADPH} CH_3-CH_2-CH_2-\overset{\overset{O}{\|}}{C}-complex$$

The product of step 5 goes through the cycle again and again, each time adding two carbon atoms until the fatty acid the body requires has been synthesized. The process is diagrammed in Figure 27-5.

27-8   Synthesis of Phospholipids

The phospholipids are very important because they form (along with protein) the framework of most of the cell membrane system (see Section 21-8). Figure 27-6 shows the formation of a diglyceride as an intermediate step in the formation of a fat (a triglyceride).

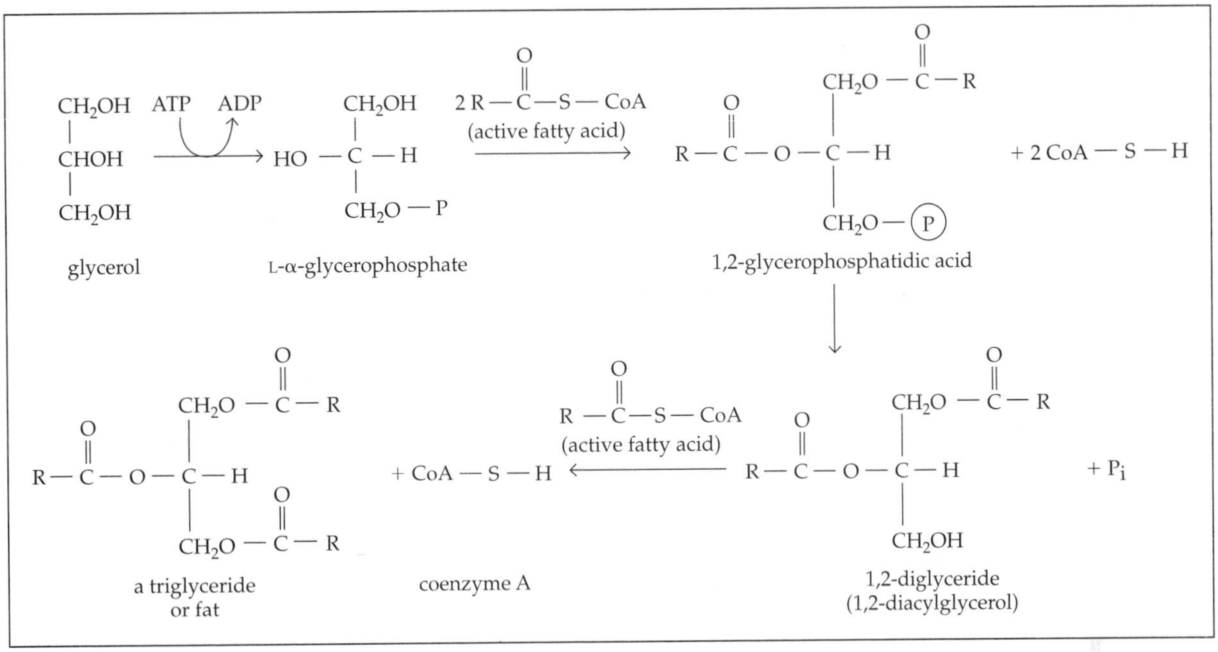

FIGURE 27-6
Synthesis of a triglyceride.

The 1,2-diglyceride may also be converted into a phospholipid such as lecithin as follows:

$$CH_2-CH_2-\overset{+}{N}(CH_3)_3 \xrightarrow{\text{ATP}} \textcircled{P}-CH_2-CH_2-\overset{+}{N}(CH_3)_3$$
$$|$$
$$OH$$

choline   choline phosphate

$$\text{choline}-\textcircled{P} + \underset{\substack{\text{cytidine}\\\text{triphosphate}}}{\text{CTP}} \longrightarrow \text{CDP choline}$$

In this reaction the CTP acts as an activator for the choline phosphate.

The CDP choline reacts with a diglyceride to form phosphatidylcholine (lecithin).

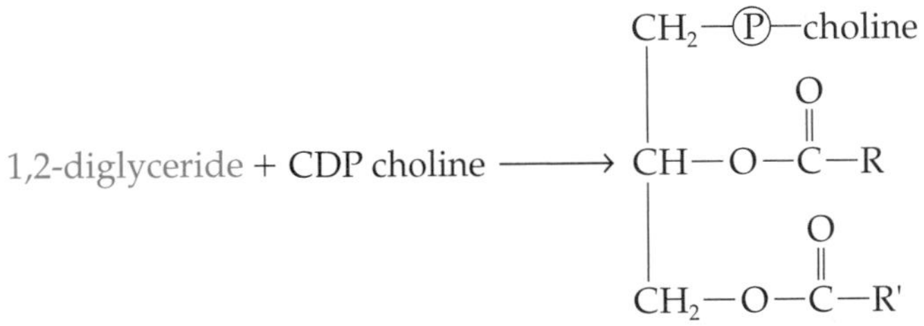

$$1,2\text{-diglyceride} + \text{CDP choline} \longrightarrow$$

$$CH_2-\textcircled{P}-\text{choline}$$
$$|\qquad\qquad O$$
$$|\qquad\qquad ||$$
$$CH-O-C-R$$
$$|\qquad\qquad O$$
$$|\qquad\qquad ||$$
$$CH_2-O-C-R'$$

phosphatidylcholine
(lecithin)

If ethanolamine, inositol, or serine is used in place of choline, the corresponding phospholipid—phosphatidylethanolamine, phosphatidylinositol, or phosphatidylserine—is formed.

27-9 Cholesterol

Cholesterol is found in all cells of the body, but particularly in brain and nerve tissue. Cholesterol occurs in animal fat but not in plant fat. An adult normally ingests about 500 mg cholesterol daily from such foods as egg yolk, meat fats, liver, and liver oils. In addition, the body manufactures 500 mg of cholesterol daily. Of the cholesterol manufactured by the body, 50 percent is produced by the liver, 15 percent by the gut, and the skin produces most of the balance. However, nearly all tissues containing nucleated cells are capable of synthesizing cholesterol.

Cholesterol normally is eliminated in the bile. However, sometimes it settles out in the gallbladder as gallstones. If cholesterol deposits in the walls of the larger arteries, the condition is known as atherosclerosis, a type of hardening of the arteries (Figure 27-7). When this occurs, there is a decrease in the usable diameter of the blood vessels. The elasticity of the arterial walls decreases. There is an interference with the rate of blood flow as a result of greater friction caused by the irregular lining of the blood vessels. This irregular lining may also cause clots as the blood flows over that type of surface and lead to myocardial infarction.

Cholesterol is important as a precursor of several important steroids such as vitamin D, the sex hormones, and the adrenocortical hormones (the hormones of the cortex of the adrenal glands).

Synthesis

Cholesterol is synthesized primarily in the liver, but the adrenal cortex, skin, testes, aorta, and intestines are also able to synthesize it. This synthesis takes place in the microsomal and cytosomol fraction of the cell. Acetyl CoA is the starting material and is also the source of all the carbon atoms in cholesterol (Figure 27-8).

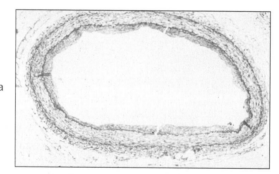

a

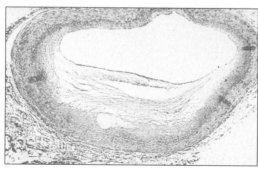

b

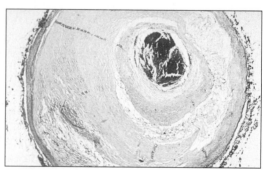

c

FIGURE 27-7
Comparison of normal artery wall (a) with those containing fatty deposits (b and c).

Lowering the Serum Cholesterol

If a low cholesterol diet does not bring serum cholesterol levels down to their proper value, the use of certain hypolipidemic drugs may be indicated. Among such drugs are the statins (lovastatin, pravastatin, simvastatin, and fluvastatin), which inhibit cholesterol synthesis in step one of the chain; sitosterol (Cytellin),

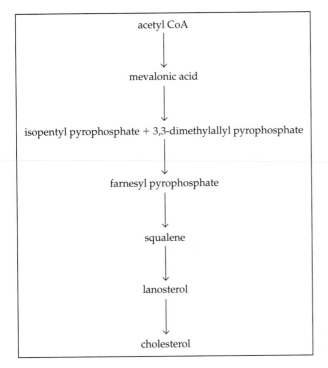

FIGURE 27-8
Biosynthesis of cholesterol.

Case Study 27-2 Hypercholesterolemia

A 45-year-old man has been diagnosed with hypercholesterolemia. A second set of laboratory results[†] shows a consistent blood plasma cholesterol concentration of 8.8 mmol/L. His LDL values are also elevated. His internist put him on a very low cholesterol diet after learning that the patient routinely consumes 1.6 mmol, or 600 mg, of cholesterol per day. At his return visit 3 months later, the patient's cholesterol continues to be high at 7.7 mmol/L. His physician prescribes colestipol hydrochloride, a bile acid–binding resin. This resin binds bile acids in the intestines, which are then excreted in the feces. Two months later the man's cholesterol has declined to 5.8 mmol/L, a level that is acceptable to his physician.

Questions

1. How does the body use dietary cholesterol?
2. Since this man was very compliant with his low cholesterol diet for 3 months, why did he continue to have an elevated cholesterol level?
3. How does an increased excretion of bile acids reduce cholesterol?

[†]*See back of book for normal laboratory values.*

which blocks the absorption of cholesterol from the gastrointestinal tract; cholestyramine (Questran), which acts by preventing the reabsorption of bile salts; gemfibrozol (Lopid), which increases the oxidation of fatty acids and decreases secretion of triacylglycerols: probucol (Lorelco), which increases LDL catabolism; and nicotinic acid, which inhibits VLDL production in the liver.

27-10 Lipid Storage Diseases

Glycolipids (see page 372) are components of nerve and brain tissue. If these compounds accumulate because of a breakdown in the enzyme system, genetic abnormalities such as Tay-Sachs disease, Niemann-Pick disease, and Gaucher's disease occur (see page 412).

Summary

Emulsified fats pass through the lacteals of the villi into the lymphatics through the thoracic duct to the liver. In the liver some of the fats are changed to phospholipids, which are necessary for the formation of nerve and brain tissue. Some fat is stored in the adipose tissue; some is oxidized to furnish energy.

Glycerol from a fat is oxidized to dihydroxyacetone phosphate, which is part of the glycolysis sequence. That is, the glycerol part of a fat is metabolized through the carbohydrate sequence.

Fatty acids pass through a β-oxidation cycle in which two carbon atoms are removed at a time and converted to acetyl CoA, which then enters the Krebs cycle.

The oxidation of one molecule of a 16-carbon fatty acid yields a total of 129 ATP molecules, with an efficiency of about 42 percent.

Ketone (acetone) bodies are normally produced in the β-oxidation process. However, they are produced only in small amounts and do not normally accumulate. If excess ketone bodies accumulate in the blood, a condition known as ketonemia exists. Accumulation of ketone bodies in the urine is termed ketonuria. The overall accumulatlion of ketone bodies is called ketosis; it occurs in the abnormal metabolism of carbohydrates.

Fat in excess of the body's needs is stored as adipose tissue. Stored fat serves as a reserve supply of food, as a support for the internal organs, as a shock absorber for the internal organs, and as an insulator for the body.

When more than a normal amount of fat is deposited in the adipose tissue, the resulting condition is called obesity. Obesity may be the result of glandular disorder or simply be caused by overeating.

The conversion of glucose to fats, lipogenesis, takes place in the liver and in the adipose tissue.

The synthesis of fatty acids occurs both inside and outside the mitochondria, with the latter being the predominant site.

Phospholipids can be synthesized from glycerol. If the 1,2-diglyceride formed from glycerol reacts with cytidine diphosphate choline, then the phospholipid lecithin is formed. If instead the 1,2-diglyceride combines with cytidine diphosphate ethanolamine, then the phospholipid cephalin is formed.

Cholesterol is found in all cells of the body but particularly in the nerve tissue. Cholesterol is normally eliminated in the bile, but if it settles out in the gallbladder, gallstones are formed. If cholesterol deposits on the walls of the arteries, atherosclerosis occurs.

Cholesterol is an important precursor for vitamin D, for the sex hormones, and for the adrenocortical hormones.

Cholesterol is synthesized in the liver, beginning with acetyl CoA and proceeding through a series of steps.

Questions and Problems

1. Where are lipids absorbed? How do they travel to the liver? What happens to them in the liver?

2. What types of products are formed during the oxidation of fatty acids with an even number of carbon atoms? an odd number of carbon atoms?

3. What is the β-oxidation theory of fatty acids? What are the end products of this cycle? What happens to these products?

4. Describe the oxidation of the glycerol part of a fat.

5. Compare the energy produced by the oxidation of 1 g carbohydrate with that from 1 g fat.

6. In the oxidation of a fat, in which part of the sequence is most of the energy produced?

7. What happens to unsaturated fatty acids before they can be oxidized?

8. The oxidation of one molecule of a 16-carbon fatty acid produces how many ATP molecules?

9. What are ketone bodies? How are they interrelated structurally? Where are they formed?

10. Where is excess fat stored? What functions does it have?

11. What is ketonemia? ketonuria? ketosis?

12. Describe the synthesis of lecithin. How does this synthesis compare with that of cephalin?

13. Where is cholesterol found in the body? Where is it produced? Why is it important?

14. If cholesterol is deposited in the walls of the large arteries, what is the condition termed? If the deposit occurs in the gallbladder, what is the condition termed?

15. What is lipogenesis? Where does it occur?

16. How does lipogenesis in the mitochondria compare with the oxidation of a fat?

17. Why are phospholipids important?

18. If a person is on a totally cholesterol-free diet, will he or she be completely lacking in that substance? Explain.

19. What hormones increase lipolysis? decrease it?

20. Name two metabolic diseases caused by impaired oxidation of fat.

Practice Test

1. The glycerol portion of a triglyceride is metabolized through the _____ sequence.
   a. nucleic acid
   b. fatty acid
   c. amino acid
   d. carbohydrate

2. Fat serves as a(n) _____.
   a. insulator for the body
   b. support for the internal organs
   c. reserve supply of fuel
   d. all of these

3. Restricted carbohydrate metabolism causes _____.
   a. ketosis
   b. acidosis
   c. accumulation of acetone
   d. all of these

4. Which of the following hormones does not cause triglyceride breakdown?
   a. cortisol
   b. insulin
   c. glucagon
   d. epinephrine

5. In the oxidation of a fatty acid, how many carbon atoms are removed at a time?
   a. 1   b. 2   c. 3   d. 4

6. The end product of fatty acid oxidation is _____.
   a. acetyl CoA
   b. acetone
   c. pyruvate
   d. lactate

7. Lipogenesis takes place primarily in the _____.
   a. adipose tissue
   b. liver
   c. both of these
   d. neither of these

8. Cholesterol is involved in _____.
   a. atherosclerosis
   b. gallstone formation
   c. formation of sex hormones
   d. all of these

9. Fat is transported through the bloodstream as _____.
   a. lipoproteins
   b. glycoproteins
   c. phospholipids
   d. all of these

10. The production of ketone bodies occurs whenever the body has limited _____ to metabolize.
    a. nucleic acids
    b. carbohydrates
    c. lipids
    d. amino acids

Metabolism of Proteins

Objectives

- To understand the significance of nitrogen balance
- To understand the synthesis of protein
- To understand transamination, deamination, and decarboxylation
- To follow the path of urea formation
- To become familiar with some metabolic disorders associated with the urea cycle
- To understand the metabolism of hemoglobin

The liver enzyme catalase speeds up the decomposition of hydrogen peroxide into water and oxygen. Catalase protects liver cells from the destructive effects of hydrogen peroxide, a product of other enzymatic activity.

28-1 Functions of Protein in the Body

During digestion, proteins are hydrolyzed into amino acids, which are then absorbed into the bloodstream through the villi of the small intestine. These amino acids enter the amino acid pool of the body (Figure 28-1).

The amino acids from the amino acid pool serve many functions. For example, they

1. Convert to tissue protein to build new tissue
2. Convert to tissue protein to replace old tissue
3. Aid in the formation of hemoglobin
4. Aid in the formation of some hormones
5. Aid in the formation of enzymes
6. Are used in synthesis of other amino acids that the body needs
7. Serve as a source of energy when they are catabolized
8. Are used to form nucleic acids, neurotransmitters, and other substances needed for body functions

28-2 Nitrogen Balance

The body can store carbohydrates (as glycogen in the liver and the muscles) and fats (as adipose tissue and in tissues around the internal organs). However, the body cannot store protein. The amino acids that result from the digestion of protein are either used for the synthesis of new tissues, the replacement of old tissues, and the formation of various required body substances such as hormones and enzymes, or they are converted to fat or oxidized to furnish energy.

All protein in the body is constantly being degraded and then resynthesized. The time required to degrade a protein and to resynthesize it is the *protein turnover rate*. The turnover rate varies with the type of protein involved. It ranges from a few minutes for proteins such as simple hormones and enzymes to about 3 years for structural protein such as collagen.

Individual proteins are degraded at different rates. The susceptibility of an enzyme (a protein) to degradation is expressed in terms of its half-life ($t_{1/2}$), the time required for its concentration to be reduced to 50 percent of its original value. Enzymes with short half-lives include many regulatory substances such as tryptophan oxygenase and tyrosine transaminase. These have a $t_{1/2}$ of about 2 hours.

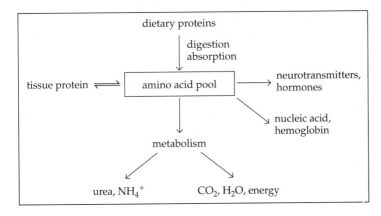

FIGURE 28-1
Protein metabolism.

Enzymes such as aldolase and the cytochromes have a $t_{1/2}$ of more than 100 hours. The rate of degradation of these regulatory enzymes may be increased or decreased, depending on physiologic demand, thereby altering enzyme levels.

Because the body cannot store protein and because protein contains nitrogen, the amount of nitrogen taken in each food should usually equal the amount of nitrogen excreted per day (for a normal adult whose weight remains constant). This takes into consideration the normal replacement of worn-out tissue where the reaction is merely one of exchange between one amino acid and another.

A person whose body excretes as much nitrogen per day as is taken in with food is said to be in nitrogen balance. Children exhibit a positive nitrogen balance because they take in more nitrogen in their food than they excrete. This is because children need amino acids to build growing body tissues. Any body condition marked by the growth of new tissues will exhibit a positive nitrogen balance. An example of this is seen in persons recovering from a wasting illness where the body needs to rebuild tissues and so does not excrete as much nitrogen (amino acids) as it takes in.

Conversely, if the body excretes more nitrogen than it acquires from food, a negative nitrogen balance exists. Conditions that can produce a negative nitrogen balance are starvation, malnutrition, prolonged fever, and various wasting illnesses. However, a person can survive for a reasonable period of time in negative nitrogen balance, as when dieting, since body protein will be used for essential purposes.

However, when inadequate protein intake leads to a negative nitrogen balance, the individual will eventually develop a protein deficiency disease called kwashiorkor. This disease is characterized by a wasting away of fat and muscle and a degeneration of many of the internal organs (see page 530).

Protein malnutrition also develops in seriously ill patients who cannot eat by mouth. Such patients include those who have had major surgery or trauma. If protein is not provided along with other foods and minerals, the person will have a negative nitrogen balance and will have a much slower rate of recovery. In such cases, a procedure called *hyperalimentation* is used whereby glucose and hydrolyzed protein are administered intravenously.

A person consuming about 300 g of carbohydrate, 100 g of fat, and 100 g of protein (2500 kcal) per day will excrete about 16.5 g of nitrogen per day. About 95 percent of this nitrogen is eliminated through the kidneys and the remaining 5 percent in the stool. Most of the nitrogen is excreted in the form of urea (see page 498).

28-3 Synthesis of Protein

As discussed in Chapter 22, proteins are synthesized from amino acids through the various intermediaries such as peptides and polypeptides. The body takes the amino acids produced by the digestion of protein and recombines them into protein that it needs in the various parts of the body. (See Chapter 23 for the mechanism of protein synthesis.)

Some amino acids that the body needs can by synthesized from other amino acids. However, there are certain amino acids that the body needs but cannot synthesize. These amino acids must be supplied in the food if the body is to function normally.

The eight nutritionally essential (ten for infants) and the nonessential amino acids (with their precursors) are listed in Table 28-1. Recall that the body requires the essential amino acids in the diet but can manufacture the nonessential amino acids that it requires from other amino acids (see Section 22-4).

Table 28-1 Amino Acids in the Body

Nutritionally Essential Amino Acids	Nutritionally Nonessential	
	Amino Acid	Precursor
Isoleucine	Alanine	Pyruvate
Leucine	Arginine	Glutamate
Lysine	Asparagine	Aspartate
Methionine	Aspartate	Oxaloacetate
Phenylalanine	Cystein	Serine, homocysteine
Threonine	Glutamate	α-Ketoglutarate
Tryptophan	Glutamine	Glutamate
Valine	Glycine	Serine
For infants:	Hydroxylysine	
Arginine*	Proline	Arginine
Histidine*	Serine	3-Phosphoglycerate
	Tyrosine	Phenylalanine

*Infants cannot manufacture arginine or histidine and must obtain them in their diet.

Some proteins contain all of the nutritionally essential amino acids; those that do not are called incomplete proteins. Two common incomplete proteins are gelatin, which is lacking in tryptophan, and zein (from corn), which is lacking in both tryptophan and lysine.

28-4 Biosynthesis of Nonessential Amino Acids

Most of the time the mixture of amino acids provided by the protein in our diet is not in the proportions required by our bodies. Consequently, it is necessary to rearrange the amino acid pool metabolically. Table 28-1 shows the sources of nonesssential amino acids that can be synthesized by the body provided that there are adequate supplies of amine nitrogen.

Many of these syntheses involve transamination (see following section). One particular biosynthetic pathway deserves a closer look. This pathway involves the methyl cycle, which is illustrated in Figure 28-2. Note that the methyl cycle consumes three high-energy phosphate bonds. Several important methylated compounds are produced as by-products. The methyl carrier is tetrahydrofolate (THF), which is the active form of the B vitamin folic acid (see page 566). Several anticancer drugs, such as methotrexate, block the synthesis of methylated DNA by preventing the transfer of methyl groups into the cycle by THF.

Transamination

Transamination is a reaction in which one or more amino acids are converted into other amino acids. When an α-amino acid and an α-keto acid react, they interconvert to form another α-amino acid and another α-keto acid. All other amino acids can be converted to glutamic acid by transamination. In this way the body can manufacture the amino acids it needs.

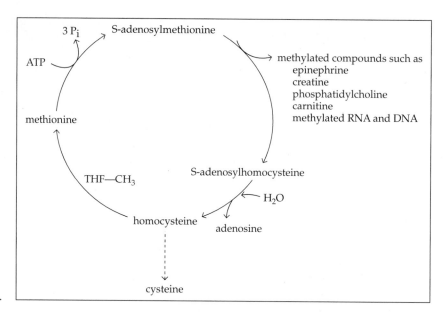

FIGURE 28-2
The methyl cycle.

Transamination is catalyzed by enzymes called transaminases or aminotransferases. An essential part of the active site of a transaminase is pyridoxal phosphate, the coenzyme form of vitamin B_6. Transaminases are used in the diagnosis of a variety of disorders. For example, serum glutamic oxaloacetic transaminase (SGOT) levels are increased after myocardial infarction and with cirrhosis of the liver, and serum glutamic pyruvic transaminase (SGPT) levels are increased during infectious hepatitis. Decreased serum transaminase levels occur during pregnancy and with vitamin B_6 deficiency.

An example of transamination is the reaction of glutamic acid (an α-amino acid) and oxaloacetic acid (an α-keto acid) to form α-ketoglutaric acid (another α-keto acid) and aspartic acid (another α-amino acid).

$$
\begin{array}{cccc}
\text{COOH} & \text{COOH} & \text{COOH} & \text{COOH} \\
| & | & | & | \\
\text{H}-\text{C}-\text{NH}_2 & \text{C}=\text{O} & \text{C}=\text{O} & \text{H}-\text{C}-\text{NH}_2 \\
| & | & | & | \\
\text{CH}_2 \quad + & \text{CH}_2 \xrightarrow{\text{GOT}} & \text{CH}_2 \quad + & \text{CH}_2 \\
| & | & | & | \\
\text{CH}_2 & \text{COOH} & \text{CH}_2 & \text{COOH} \\
| & & | & \\
\text{COOH} & & \text{COOH} & \\
\text{glutamic} & \text{oxaloacetic} & \text{α-ketoglutaric} & \text{aspartic} \\
\text{acid} & \text{acid} & \text{acid} & \text{acid}
\end{array}
$$

Note that this transamination reaction is catalyzed by the enzyme glutamic oxaloacetic transaminase (GOT). This enzyme occurs in high concentration in heart muscle. Increased levels of GOT in the bloodstream (called serum GOT or SGOT) indicate myocardial infarction, which results from the reduction in blood flow to the heart muscle caused by a clot in the coronary artery.

In addition to transamination, the body has other processes for the synthesis of several nonessential amino acids.

Some genetic diseases (see Section 23-8) are caused by a deficit or lack of enzymes that catalyze the synthesis of certain amino acids. One such disease is PKU (phenylketonuria). Normally the body converts phenylalanine (an α-amino acid) to tyrosine (another α-amino acid). The required enzyme is phenylalanine hydroxylase.

Case Study 28-1 Hepatic Coma in Cirrhosis

A patient with advanced liver cirrhosis was admitted to the intensive care unit of a university hospital. His condition has deteriorated from a disoriented state to a coma. He was initially treated with fluids. His BUN rose, and on the fourth day the BUN was 21 mmol/L (120 mg/dL)[†]. His blood ammonia level rose to 110 µmol/L (190 µg/dL)[†]; α-ketoglutarate was added to the intravenous infusion in an attempt to lower the blood ammonia concentration, but the patient died during the infusion.

Questions

1. Is the conversion of α-ketoglutarate to glutamate readily reversible?
2. Transamination occurs in which cellular compartment?
3. Is the hepatic coma caused solely by the increase in blood ammonia levels?

[†]*See back of book for normal laboratory values.*

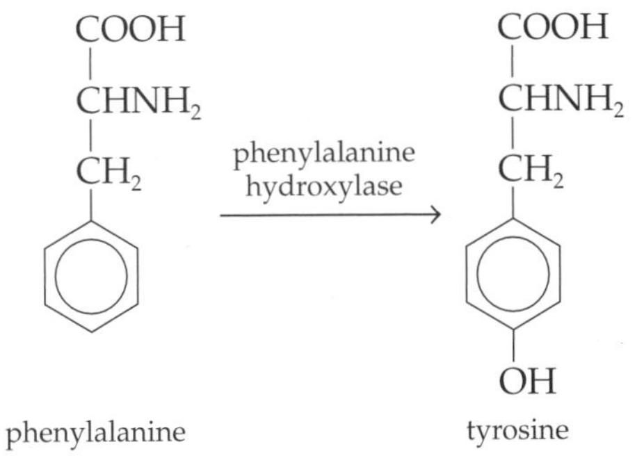

$$
\underset{\text{phenylalanine}}{\begin{array}{c} \text{COOH} \\ | \\ \text{CHNH}_2 \\ | \\ \text{CH}_2 \\ | \\ \bigcirc \end{array}} \xrightarrow[\text{hydroxylase}]{\text{phenylalanine}} \underset{\text{tyrosine}}{\begin{array}{c} \text{COOH} \\ | \\ \text{CHNH}_2 \\ | \\ \text{CH}_2 \\ | \\ \bigcirc \\ | \\ \text{OH} \end{array}}
$$

If this enzyme is lacking (as a result of genetic deficiency), tyrosine cannot be produced. Instead, phenylalanine is converted into phenylpyruvic acid (a transamination reaction).

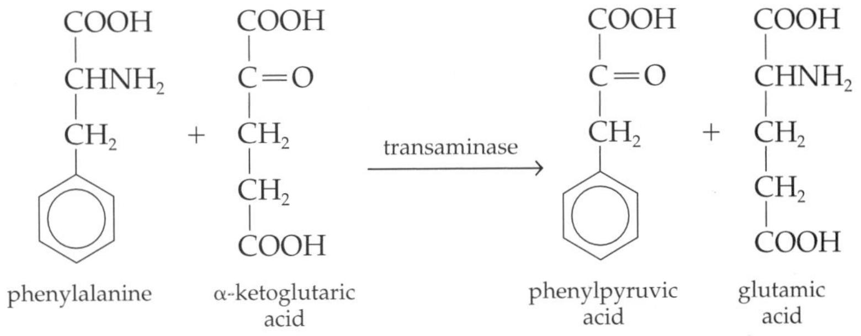

$$
\underset{\text{phenylalanine}}{\begin{array}{c} \text{COOH} \\ | \\ \text{CHNH}_2 \\ | \\ \text{CH}_2 \\ | \\ \bigcirc \end{array}} + \underset{\substack{\text{α-ketoglutaric}\\ \text{acid}}}{\begin{array}{c} \text{COOH} \\ | \\ \text{C}=\text{O} \\ | \\ \text{CH}_2 \\ | \\ \text{CH}_2 \\ | \\ \text{COOH} \end{array}} \xrightarrow{\text{transaminase}} \underset{\substack{\text{phenylpyruvic}\\ \text{acid}}}{\begin{array}{c} \text{COOH} \\ | \\ \text{C}=\text{O} \\ | \\ \text{CH}_2 \\ | \\ \bigcirc \end{array}} + \underset{\substack{\text{glutamic}\\ \text{acid}}}{\begin{array}{c} \text{COOH} \\ | \\ \text{CHNH}_2 \\ | \\ \text{CH}_2 \\ | \\ \text{CH}_2 \\ | \\ \text{COOH} \end{array}}
$$

Phenylpyruvic acid accumulates in the bloodstream and is eliminated in the urine. PKU refers to the presence of phenyl ketones in the urine. PKU is characterized by severe mental retardation.

28-5 The Body's Requirements of Protein

A certain minimum daily amount of protein is required for the normal replacement of body tissues. This amount, however, may be greatly increased during periods of increased metabolism, such as during high fevers. However, for the normal adult keeping a constant weight, the recommended daily intake of protein is approximately 0.8 g per kilogram of body weight. This amounts to approximately 46 g of protein per day for the adult female and 56 g per day for the adult male.

28-6 Catabolism of Amino Acids

The amino acids that the body does not need for tissue building or that are not of the correct type for this purpose are broken down to ammonia, carbon dioxide, and water, at the same time producing heat and energy. Such a process is called *catabolism*, which is defined as the breakdown or oxidation of large molecules into smaller molecules with the release of energy. *Anabolism* is the buildup of large molecules necessary for life and is a process requiring energy. The total of all anabolic and catabolic reactions is called *metabolism*.

Deamination

Deamination (also called oxidative deamination) is a catabolism reaction whereby the α-amino group of an amino acid is removed, forming an α-keto acid and ammonia. Deamination occurs primarily in the liver and the kidneys under the catalysis of the enzyme amino acid oxidase.

$$CH_3-\underset{\underset{NH_2}{|}}{CH}-COOH \xrightarrow{\text{amino acid oxidase}} CH_3-\underset{\underset{O}{||}}{C}-COOH + NH_3$$

alanine   pyruvic acid   ammonia
(an α-amino acid)   (an α-keto acid)

The α-keto acid produced by this process can undergo several types of reactions:

1. It can be catabolized to carbon dioxide, water, and energy in the citric acid cycle.
2. It can be converted to carbohydrates (glycogen) or to fat.
3. It can be reconverted to a different amino acid by transamination.

Formation of Urea

The ammonia formed from the deamination of amino acids combines with carbon dioxide to form urea and water. This process takes place in the liver. The overall reaction can be written as

$$2\,NH_3 + CO_2 \xrightarrow{\text{enzymes}} NH_2CONH_2 + H_2O \qquad (28\text{-}1)$$

ammonia   carbon dioxide   urea   water

but the actual process is certainly not as simple as that. It consists of a series of steps, each catalyzed by an appropriate enzyme.

Ammonia is a toxic by-product of the deamination of amino acids and must be removed from the body, predominantly in the form of the compound urea. Three amino acids are involved in the conversion of ammonia to urea—arginine, citrulline, and ornithine. The pathway for the conversion of ammonia to urea is called the urea cycle (see Figure 28-3).

The first step in this cycle is the reaction of ammonia and carbon dioxide to form carbamoyl phosphate. In this reaction, ATP is converted to ADP. This reaction is catalyzed by *N*-acetylglutamic acid and the enzyme carbamoyl phosphate synthetase in the presence of magnesium ions. Lack of this enzyme produces the very serious disorder hyperammonemia. The body has only one principal way to dispose of excess nitrogen and this is by the excretion of urea.

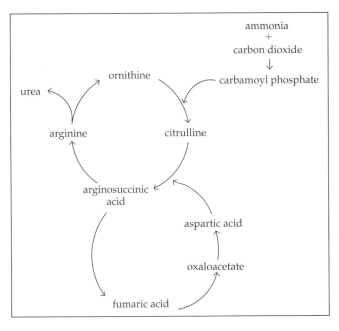

FIGURE 28-3
The urea cycle.

$$NH_3 + CO_2 + H_2O + 2\,ATP \xrightarrow[\substack{N\text{-acetylglutamic}\\ \text{acid,}\\ \text{carbamoyl phosphate}\\ \text{synthetase}}]{Mg^{2+}} \underset{\substack{\text{carbamoyl}\\ \text{phosphate}}}{H_2N-\overset{\displaystyle O}{\overset{\|}{C}}-O-\textcircled{P}} + 2\,ADP + P_i$$

In the second step, carbamoyl phosphate combines with ornithine to form citrulline. This reaction is catalyzed by the liver enzyme ornithine *trans*-carbamoyl transferase. Lack of this enzyme will produce a different hyperammonemia.

$$\begin{array}{c}
CH_2-NH_2 \\
| \\
(CH_2)_2 \\
| \\
HC-NH_2 \\
| \\
COOH \\
\text{ornithine}
\end{array}
\quad + \quad
\begin{array}{c}
O \\
\| \\
H_2N-C-O-\textcircled{P} \\
\text{carbamoyl} \\
\text{phosphate}
\end{array}
\xrightarrow[\text{transferase}]{\substack{\text{ornithine}\\ \textit{trans}\text{-carbamoyl}}}
\begin{array}{c}
O \\
\| \\
CH_2-NH-C-NH_2 \\
| \\
(CH_2)_2 \\
| \\
HC-NH_2 \\
| \\
COOH \\
\text{citrulline}
\end{array}$$

Next citrulline reacts with aspartic acid (derived from transamination of oxaloacetate) to form arginosuccinic acid. This reaction takes place in the presence of ATP, magnesium ions, and the enzyme arginosuccinate synthetase.

$$\begin{array}{c}
O \\
\| \\
CH_2-NH-C-NH_2 \\
| \\
(CH_2)_2 \\
| \\
HC-NH_2 \\
| \\
COOH \\
\text{citrulline}
\end{array}
\quad + \quad
\begin{array}{c}
COOH \\
| \\
H_2N-CH \\
| \\
CH_2 \\
| \\
COOH \\
\text{aspartic acid}
\end{array}
\xrightleftharpoons[\substack{Mg^{2+}\\ \text{arginosuccinate}\\ \text{synthetase}}]{ATP \quad AMP}
\begin{array}{c}
NH \qquad COOH \\
\| \qquad\quad | \\
CH_2-NH-C-NH-CH \\
| \qquad\qquad\quad | \\
(CH_2)_2 \qquad\quad CH_2 \\
| \qquad\qquad\quad | \\
HC-NH_2 \qquad COOH \\
| \\
COOH \\
\text{arginosuccinin acid}
\end{array}$$

Arginosuccinic acid is cleaved (split) hydrolytically into arginine and fumaric acid. Some fumaric acid may be converted back to aspartic acid, and some enters the Krebs cycle (Chapter 26).

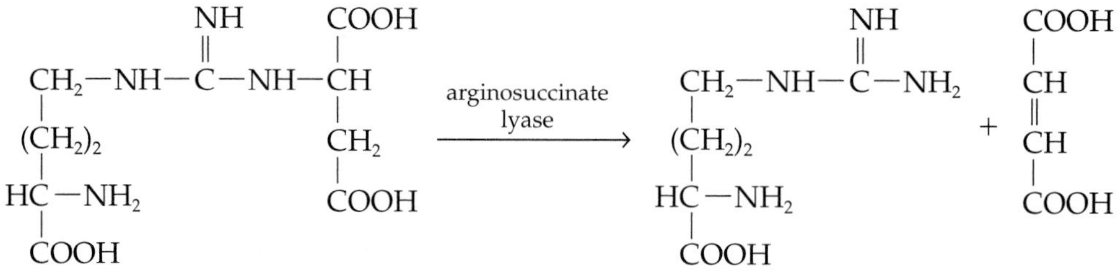

arginosuccinic acid   arginine   fumaric acid

Finally, arginine is split hydrolytically by the liver enzyme arginase into ornithine and urea. The ornithine can then go through the cycle again, and the urea is excreted.

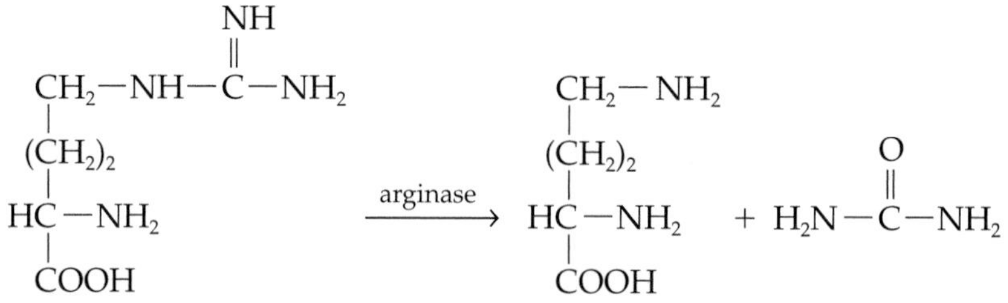

arginine   ornithine   urea

To summarize, the urea cycle is shown diagrammatically in Figure 28-3.

The blood picks up the urea from the liver and carries it to the kidneys, where it is excreted in the urine. Urea is the principal nitrogen end product of protein metabolism and contains a large percentage of the total nitrogen excreted by the body. (This is one means the body has of removing ammonia.)

28-7 Metabolic Disorders Associated With the Urea Cycle

A lack or deficiency of enzymes necessary for the urea cycle, usually associated with a genetic disease (see Section 23-8), produces different metabolic disorders, all of which are quite similar.

Correction of these disorders may be brought about by a low-protein diet and/or by the frequent intake of small amounts of food so as not to cause rapid increases in blood amino acid levels.

Among these metabolic disorders are:

Hyperammonemia, which is caused by a lack of the enzyme carbamoyl phosphate synthetase. It is characterized by increased levels of NH_4^+ in the blood. There are also increased levels of glutamine in the cerebrospinal fluid, the blood, and the urine. One of the effects of hyperammonemia is mental retardation.

Citrullinemia is caused by an absence of or decreased levels of the enzyme arginine succinate synthetase. It is characterized by the excretion of large amounts of citrulline into the urine. In addition, large amounts of citrulline are also found in the blood plasma and in the cerebrospinal fluid.

Arginosuccinase aciduria is caused by a lack of the enzyme arginosuccinase and is characterized by elevated levels of arginosuccinic acid in the blood plasma. This disorder is usually fatal to children at about age 2.

Case Study 28-2   Hereditary Hyperammonemia

An 8-month-old infant is admitted to the hospital with a diagnosis of failure to thrive. Twenty-four hours after admission the nurse notes that the baby is very difficult to arouse and has an elevated temperature of 39.6° C and a heart rate of 140 beats/min. The nurse contacts the resident physician on call, who notes a hepatomegaly. The resident orders an immediate electroencephalogram, which yields abnormal findings. An intravenous solution of glucose is given and the baby is alert within 24 hours. A paper chromatogram of the urine shows high amounts of glutamine and uracil. Laboratory testing shows a high blood ammonium ion concentration.

Questions

1. Which two enzymatic defects can cause hereditary hyperammonemia?
2. Based on the above case study information, which enzyme appears to be defective in this case?
3. Explain the high urine glutamine.
4. How does treatment with arginine lower ammonium levels?
5. This disease is often fatal in males but not in females. Explain.
6. Children with this disorder are alkalotic. Balance the equation for urea formation from carbon dioxide and ammonium ions. With the balanced equation, explain the alkalosis.

Decarboxylation

The decarboxylation (removal of —COOH group) of an amino acid yields a primary amine. The carboxyl group that is removed is converted to carbon dioxide. The enzyme involved in a decarboxylation reaction requires pyridoxal phosphate as a coenzyme. The decarboxylation reaction can be summarized as follows:

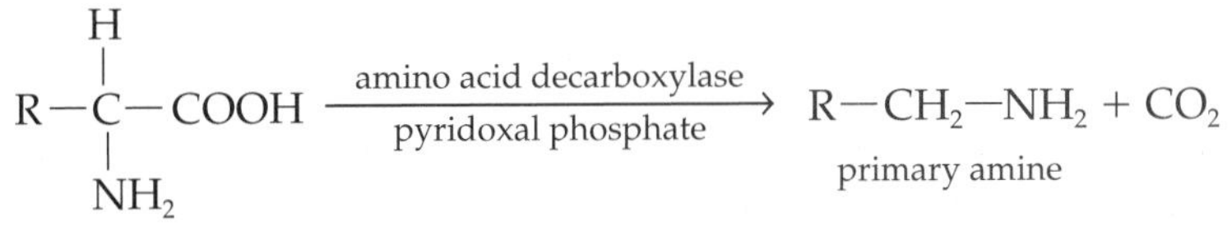

$$
\underset{\alpha\text{-amino acid}}{R-\overset{\displaystyle H}{\underset{\displaystyle NH_2}{\overset{|}{\underset{|}{C}}}}-COOH} \quad\xrightarrow[\text{pyridoxal phosphate}]{\text{amino acid decarboxylase}}\quad \underset{\text{primary amine}}{R-CH_2-NH_2} + CO_2
$$

Several naturally occurring amines are formed by the decarboxylation of amino acids, for example,

$$\textit{amino acid} \xrightarrow{\text{decarboxylation}} \textit{primary amine}$$

amino acid		primary amine
histidine	$\longrightarrow$	histamine
lysine	$\longrightarrow$	cadaverine
ornithine	$\longrightarrow$	putrescine
tyrosine	$\longrightarrow$	tyramine

Some decarboxylation reactions are brought about by intestinal bacteria that attack amino acids, producing toxic amines called ptomaines (see Section 25-12). This process is common in the spoilage of food protein.

The amino acid tryptophan undergoes a series of reactions to form the compounds indole and methylindole (skatole). These two compounds produce the characteristic odor of feces.

trytophan   indole   skatole

Decarboxylation of certain amino acids sometimes followed by other metabolic changes leads to the formation of several physiologically active compounds. For example, histidine forms histamine, which plays an important role in allergic reactions; tryptophan can be converted to serotonin, a neurotransmitter, melatonin, a pineal hormone, and melanin, which plays a role in skin pigmentation; glutamic acid is converted to γ-aminobutyric acid, a neurotransmitter; tyrosine can be converted into dopamine, norepinephrine, or epinephrine, all of which are neurotransmitters; glycine is involved in the synthesis of heme, the purines, and creatine; and arginine produces polyamines as well as nitric oxide (see page 420).

28-8 Metabolism of the Carbon Portion of Amino Acids

Once the nitrogen has been removed from an amino acid, the carbon portion can be used as an energy source. The carbon skeletons can be converted into a variety of compounds (Figure 28-7). The ones converted to acetyl CoA and acetoacetyl CoA are ketogenic, whereas those converted to pyruvate of Krebs cycle intermediates are glucogenic, since the body can synthesize glucose from them. Only leucine is purely ketogenic. The metabolism of lysine is poorly understood, and although Figure 28-7 shows lysine to be ketogenic, feeding experiments have shown that it is only slightly involved in producing ketone bodies. Amino acid catabolism thus provides energy and a wide variety of precursors and intermediates.

28-9 Metabolism of Hemoglobin

A red blood cell has a life span of about 120 days. After that period of time, the hemoglobin is catabolized. The globin (protein) part is metabolized as is any other protein. The heme is metabolized and excreted as waste products, but the iron is reused. The normal diet supplies about 12 to 15 mg of iron per day, but of this amount only about 1 mg per day may be absorbed. When hemoglobin is metabolized, 20 to 25 mg of iron is released per day. This amount must be reused or else the body will suffer a serious loss of iron. Stomach acidity is needed to aid in the absorption of iron.

The body synthesizes hemoglobin at the same rate as it metabolizes it. There are three important component parts of hemoglobin: Fe^{2+}, globin (a protein), and the porphyrin ring. Figure 28-4 summarizes the biosynthesis of hemoglobin.

When an *erythrocyte ruptures, the hemoglobin ring is broken and the products formed are globin, iron ions, and biliverdin, a blue-green pigment. This process takes place in the reticuloendothelial cells of the liver, spleen, and bone marrow. Biliverdin is rapidly reduced to bilirubin, an orange-yellow pigment, by the enzyme bilirubin reductase, also in the reticuloendothelial cells. From there the bilirubin is transported to the liver as a bilirubin-albumin complex with the aid of serum albumin. In the liver, bilirubin is converted to bilirubin diglucuronide, which is then excreted into the bile. The bile flows into the small intestine. In the

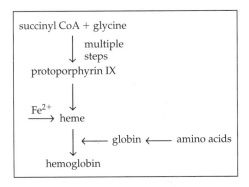

FIGURE 28-4
Biosynthesis of hemoglobin.

small intestine the bilirubin diglucuronide is changed to stercobilinogen and then to stercobilin for excretion into the stool and also into urobilinogen and then to urobilin for excretion into the urine. These reactions are shown in Figure 28-5. *Jaundice is the condition in which abnormal amounts of bilirubin accumulate in the blood. Patients with jaundice exhibit a characteristically yellow skin as a result of the presence of bilirubin.

If hemolysis takes place at an abnormally high rate so that bilirubin accumulates in the blood, the condition is hemolytic jaundice. If the bile duct is obstructed so that bile cannot enter the intestinal tract, bilirubin again accumulates in the blood. This condition is obstructive jaundice and is characterized by white or clay-colored stools because decomposition products of bilirubin are not present. If the liver is damaged in such diseases as infectious hepatitis or cirrhosis, bilirubin cannot be removed and a jaundiced condition results.

There are two types of bilirubin. Direct bilirubin is bilirubin diglucuronide (a conjugated form), which is quite water soluble; indirect bilirubin (nonconjugated) is not very water soluble.

28-10 Metabolism of Nucleoproteins

Nucleoproteins are composed of proteins conjugated with nucleic acids. DNA and RNA, two nucleic acids, are essential constituents of chromosomes, viruses, and cell nuclei. Figure 28-6 illustrates the hydrolysis products of nucleoprotein.

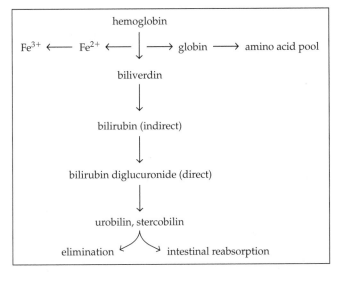

FIGURE 28-5
Metabolism of hemoglobin.

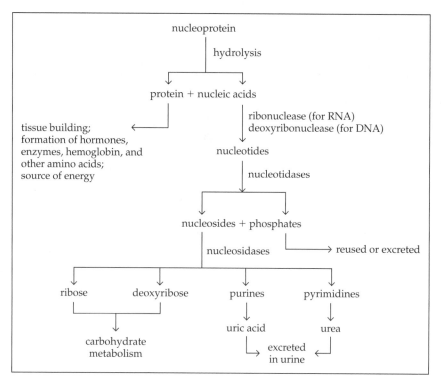

FIGURE 28-6

Metabolism of nucleoproteins.

Ribose and deoxyribose enter the normal carbohydrate metabolic pathway. The phosphates either are used to prepare new phosphate compounds or are excreted in the urine. The purines are converted through a series of reactions to uric acid, which is eliminated in the urine. The reactions are as follows:

adenine (a purine) → hypoxanthine → xanthine → uric acid

The uric acid concentration in the blood is increased during the disease called gout (see page 513 and Table 30-2). In this disease, uric acid crystals along with other substances are deposited in and around the joints and in the cartilage, chiefly in the large toe and in the ear lobe. Uric acid crystals may also be deposited in the form of gallstones or kidney stones.

The pyrimidines are eliminated in the urine in the form of urea:

thymine (a pyrimidine) → dihydrothymine → urea $H_2N-C-NH_2$ ($\underset{\parallel}{O}$)

28-11   Overview of Protein Metabolism

The metabolism of protein produces compounds such as pyruvic acid that can enter the Krebs cycle, other products that can enter the glycogen-forming cycle, and still others that can enter the lipogenesis cycle, as indicated in Figure 28-7.

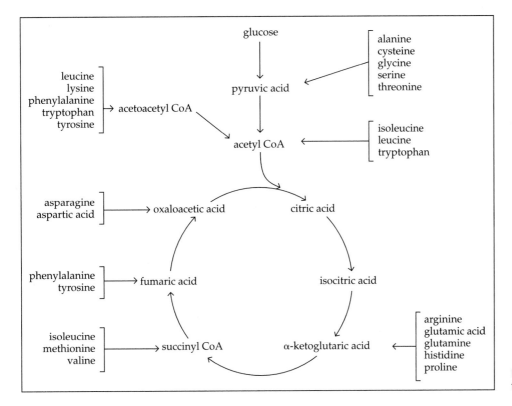

FIGURE 28-7

Metabolism of amino acids from proteins.

Case Study 28-3   Lesch-Nyhan Syndrome

On a routine visit to her child's pediatrician, the mother of an 8-month-old infant complains that the boy seems to be developing slower than her neighbor's child. She also has noticed orange crystals in the baby's diaper. She reports that the baby has recently begun biting his fingers and lips. The pediatrician suspects Lesch-Nyhan syndrome and orders a urinary uric acid test. It is found to be almost three times the normal value. The serum urate level is also markedly elevated at 0.60 mmol/L (10 mg/dL). The pediatrician refers the child to the city's medical center. Through additional testing doctors find that the boy also has megaloblastic anemia and that his urine has high amounts of 5-amino imidazole.

Questions

1. What were the orange crystals in the diaper?
2. Patients with Lesch-Nyhan syndrome have an enzymatic defect in hypoxanthine-guanine phosphoribosyltransferase (HGPRT). Explain how this enzymatic defect results in excess urate excretion.

Summary

During digestion, proteins are hydrolyzed into amino acids. These amino acids may be used for the synthesis of new tissue, for the replacement of old tissue, and for the formation of enzymes and hormones, or they may be oxidized to furnish energy. The body cannot store protein (amino acids).

A person who excretes as much nitrogen daily as he or she takes in is said to be in nitrogen balance. Children have a positive nitrogen balance because they need extra proteins for growth and so excrete less. Malnutrition and prolonged fever may lead to a negative nitrogen balance.

Amino acids that the body cannot synthesize and that must be supplied in the food are called essential amino acids. A protein that does not contain all the essential amino acids is called an incomplete protein.

Normal adults require 46 to 56 g of protein per day.

Amino acids that the body does not need are catabolized into carbon dioxide and water. This catabolism may be oxidative deamination (deamination) whereby the α-amino group is removed, forming an α-keto acid and ammonia. The α-keto acid may be catabolized to carbon dioxide and water and energy through the Krebs cycle; it may be converted to glycogen or to fat; or it may undergo transamination, whereby a different amino acid is formed.

The ammonia formed from the deamination of an amino acid unites with carbon dioxide and water to form urea. This process, which takes place in the liver, proceeds through a cycle called the urea cycle.

Amino acids may also be decarboxylated to form primary amines.

When hemoglobin is metabolized, the protein part is metabolized as usual. The iron is used over again, and the remaining heme part goes through a series of steps, eventually ending up as urobilinogen and urobilin in the urine and stercobilinogen and stercobilin in the feces.

When excessive amounts of bilirubin, one of the intermediate products of the metabolism of hemoglobin, accumulate in the blood, the condition is known as jaundice.

Nucleoproteins are composed of proteins conjugated with nucleic acids.

Questions and Problems

1. Can the body store carbohydrates? fat? protein? If so, in what form?

2. What is meant by the term *nitrogen balance*? What might cause a positive nitrogen balance? a negative nitrogen balance?

3. What are the principal functions of protein in the body?

4. What is an essential amino acid? Name five.

5. What is the normal daily requirement of protein? What can affect this amount?

6. What is deamination? Where does it occur in the body? What products are produced during this process?

7. What is transamination? What is its function in the body?

8. What is decarboxylation? What type of product is produced by the decarboxylation of an amino acid? Give two examples of a decarboxylation reaction.

9. GOT is found in high concentrations in which part of the body? What use is made of this fact?

10. What various processes can an α-keto acid undergo?

11. What are the end products of the metabolism of hemoglobin? Which are found in the urine? in the feces?

12. What is biliverdin? Where is it formed?

13. What is jaundice? What might cause it?

14. What is the difference between obstructive jaundice and hemolytic jaundice?

15. What are ptomaines? How are they produced?

16. Indicate the relationship between the metabolism of carbohydrates, lipids, and proteins.

17. What is hyperalimentation and where is it used?

18. What is protein turnover rate? Is it constant for all protein?

19. Describe two metabolic disorders associated with the urea cycle.

20. Which amino acid is associated with the formation of dopamine? histamine? NO?

21. Explain why pyruvate is central to the metabolic pathways of glucose, fatty acids, and amino acids.

22. What is the origin of the nitrogen in urea?

23. What amino acids are synthesized from aromatic amino acids?

24. What route of administration, oral or intravenous, would be most effective for a peptide antibiotic? Why?

25. Compare anabolism and catabolism.

Questions 26 and 27 pertain to the following scenario:

A fair, blue-eyed male infant of Norwegian descent has been referred to the pediatric clinic of the university hospital. The mother states that the child has been irritable and has a musty odor. The child's skin has an eczematous, dry appearance. Tests indicate that the child is mentally retarded. The urine sample tested with ferric chloride solution gives a blue-green color. The blood concentration of phenylalanine is 0.5 mmol/L (10 mg/dL). The infant is diagnosed as having phenylketonuria (PKU), and a diet low in phenylalanine is prescribed.

26. What is the connection between phenylpyruvate and pyruvate?

27. What is the metabolic connection between phenylalanine and phenylpyruvate?

Questions 28 to 30 pertain to the following scenario:

A morbidly obese man volunteers to begin a starvation diet study at the local medical center. This study is to investigate the amino acid metabolism that occurs during a prolonged fast. The patient has period blood samples drawn during the 6-week fast. During the first week his levels of valine, leucine, isoleucine, methionine, and γ-aminobutyrate are increased, but later drop below the initial levels. His glycine, threonine, and serine levels increase more slowly, but eventually the 13 amino acids decrease. Alanine has the largest decrease of 70 percent in the first week of the fast. Total plasma amino nitrogen decreases by only 10 percent.

28. How is the reduction in plasma alanine concentration caused by gluconeogenesis?

29. Explain starvation and the resultant ketosis and acidosis.

30. What changes would you expect in metabolism of carbohydrates and lipids at the beginning of a fast?

Questions 31 to 38 pertain to the following scenario:

A 54-year-old man was admitted to the hospital. He had experienced a bout of hematemesis (vomiting of blood) after lifting a heavy box. The emergency department resident physician's notes are as follows:

The patient appears very thin, weighing 10 kg less than normal for his height. He appears dehydrated and confused. The patient has a fetid breath and several carious teeth. The liver is enlarged and firm to palpation, with a distended abdomen and pedal edema. The temperature is 41.5° C; the pulse is 105 and irregular. Admission labs are: hemoglobin (10.3 g/dL) with a hematocrit level of 39%. The urine specimen was strongly positive for protein and was tea colored. A stool specimen was positive for blood. The AST was elevated as was the ALP. The total serum protein was 0.7 mmol/L (50 g/L); the serum bilirubin was 84 μmol/L (5.0 mg/dL). Serum electrolytes were K^+ 3.2 mmol/L, Na^+ 150 mmol/L, Cl^- 90 mmol/L, and total CO_2 30 mmol/L. The BUN was 2.9 mmol/L (8.0 mg/dL). This patient has a history of alcohol dependency.

31. What is the significance of the increased AST? Would you also expect an elevated ALT?

32. Why is the serum bilirubin increased?

33. Would the increase be in the direct or indirect bilirubin fraction?

34. What metabolite is responsible for the different urine color?

35. Why is the hemoglobin level so low?

36. How does a damaged liver relate to the clotting properties of the blood?

37. Alcohol is metabolized to acetyl CoA. Which coenzymes are involved in this process?

38. How is alcohol distributed in the tissues?

Questions 39 to 44 pertain to the following scenario:

A 60-year-old man had just attended his youngest daughter's wedding reception the night before. The next day he felt a dull pain in his upper left flank, which progressively got worse. He arrives in the emergency department with a temperature of 39.4° C, a heart rate of 100, and continued pain. His history is unremarkable except for a metatarsal arthralgia, which he attributed to his daily tennis game. He acknowledges that he had several alcoholic drinks during the reception. His urine has a pH of 4.7 and is positive for protein. Urine examination shows fine crystalline material and casts. A 24-hour urine sample is collected; it reveals 110 mg of protein and 1.52 g (9 mmol) of uric acid. His serum uric acid level is 0.70 mmol/L (11.8 mg/dL).

39. Would a positive family history of gout be useful? Explain the biochemical defect in gout.

40. List foods that are high in purines and pyrimidines.

41. Are dietary purines and pyrimidines required in the diet?

42. Would a high-protein diet be advised for this patient?

43. What is the mechanism of action of drugs used to treat gout?

44. How does alcohol metabolism relate to uric acid excretion? Explain whether the alcohol consumption at his daughter's reception contributed to this attack.

Questions 45 to 48 pertain to the following scenario:

On his recent annual physical examination, a 40-year-old man had a fasting plasma cholesterol of 6.5 mmol/L (normal = 3.1-5.7) and a plasma triglyceride concentration of 5.5 mol/L (normal = 1.1-1.7). His physician initially recommended a daily diet based on 15 percent protein, 15 percent fat, and 70 percent carbohydrate, with 300 mg cholesterol. Although the patient's compliance to this diet was excellent, his repeat laboratory test results showed a worsening of the plasma lipid levels. The physician now ordered a diet to reduce carbohydrates; the new diet called for 20 percent protein, 55 percent fat, and 25 percent carbohydrates, with 400 mg of cholesterol. Within 72 hours a marked reduction in plasma lipids was seen. The plasma cholesterol continued at 5.6 mmol/L and triglycerides were 2.2 mmol/L. The physician considered these values acceptable, and no medications were prescribed.

45. Explain hyperlipoproteinemia.

46. What are the long-term dangers of a very low fat diet?

47. Explain the type of hyperlipoproteinemia in this case. Explain how the plasma lipid levels were improved on a low-carbohydrate diet.

48. How can the plasma triglyceride concentration continue to be elevated despite the very low fat diet?

Practice Test

1. The metabolism of hemoglobin produces _____.
   a. aspartic acid
   b. bilirubin
   c. uric acid
   d. creatine

2. A positive nitrogen balance occurs _____.
   a. in young children
   b. during a prolonged fever
   c. during malnutrition
   d. all of these

3. The principal end product of protein metabolism is _____.
   a. amino acids
   b. phosphate
   c. creatine
   d. urea

4. The basic way the body has to eliminate nitrogen is to convert it to _____.
   a. urea
   b. uric acid
   c. NO
   d. NH_3

5. The decarboxylation of an amino acid yields _____.
   a. an aldehyde
   b. uric acid
   c. an α-keto acid
   d. a primary amine

6. Deamination takes place in the _____.
   a. pancreas
   b. liver
   c. stomach
   d. small intestine

7. The cause of PKU is the inability to convert phenylalanine to _____.
   a. acetyl CoA
   b. alanine
   c. tyrosine
   d. dopamine

8. An example of an essential amino acid is _____.
   a. serine
   b. dopamine
   c. proline
   d. lysine

9. Protein is used in the formation of _____.
   a. hemoglobin
   b. enzymes
   c. some hormones
   d. all of these

10. Transamination of glutamic acid will produce the amino acid _____.
    a. aspartic acid
    b. alanine
    c. histamine
    d. glutaric acid

Body Fluids: Urine

Objectives

- To follow the path of urine formation
- To become familiar wilth the general properties of urine
- To become aware of the significance of abnormal urine constituents

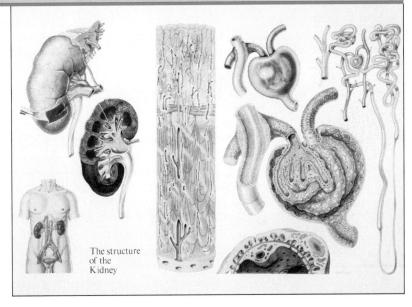

The structure
of the
Kidney

The kidney is an elaborate and elegant filtration system that removes wastes from the blood through dialysis, concentrating them as urine. This illustration depicts various levels of structure in the anatomy of a kidney.

29-1   Excretion of Waste Material

The waste products of the body are excreted through the lungs, the skin, the intestines, and the kidneys. The liver also excretes waste products—the bile, pigments and cholesterol.

The lungs eliminate water and carbon dioxide through the expired air. The skin eliminates water in the form of perspiration. Included in the perspiration are small amounts of inorganic and organic salts. The feces, excreted from the large intestine, contain undigested and undigestible material plus the excretory products from the liver—the bile pigments and cholesterol—some water, and some organic and inorganic salts. The primary excretory organs of the body, however, are the kidneys, which excrete water and water-soluble compounds including nitrogen compounds from the catabolism of amino acids.

The kidneys are important not only for their excretory function but also because of their role in the control and regulation of water, electrolyte, and acid-base balances in the body.

29-2   Formation of Urine

Blood flows to the kidneys through the renal arteries. From the renal arteries the blood passes into the arterioles and then into the capillaries of the kidneys. These capillaries coil up to form a glomerulus, a rounded ball of capillaries. Around the glomerulus is a structure called Bowman's capsule (Figure 29-1). Each Bowman's capsule is connected by a tubule to a larger tube, which in turn carries the urine to the bladder, where it is stored until it is excreted.

As blood flows into the kidney, the various soluble components diffuse into the glomeruli (there are over a million glomeruli in each kidney). The protein material in the blood cannot pass through the membrane. (Recall that proteins are colloids, and colloids do not pass through membranes.) The driving force for this diffusion of fluid through the walls of the glomerulus is the blood pressure. The liquid in the glomerulus thus has approximately the same composition as blood plasma except for the protein material.

As the fluid in the glomerulus passes down the tubule, a large proportion of the water is reabsorbed into the bloodstream. Also reabsorbed are the glucose,

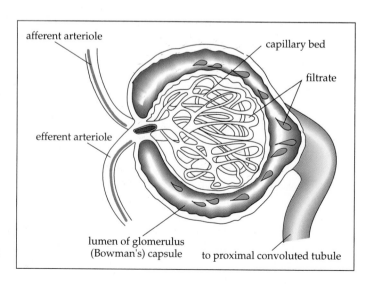

afferent arteriole

capillary bed

filtrate

efferent arteriole

lumen of glomerulus (Bowman's) capsule

to proximal convoluted tubule

FIGURE 29-1
Glomerulus and Bowman's capsule in the kidney.

Table 29-1	Comparison of Composition Of Blood Plasma And Urine	
Constituent	Percentage in Blood Plasma	Percentage in Urine
Water	90-93	95
Protein	7	0
Sodium	0.3	0.35
Glucose	0.09	0
Ammonia	0.004	0.05
Phosphate	0.009	0.5
Urea	0.03	2.0
Sulfate	0.002	0.18

amino acids, and most inorganic ions. The remaining liquid containing urea and other waste products, flows to the collecting tubules and then to the bladder.

Thus, the kidneys act as a very efficient filter, removing the waste materials but not the needed nutrients from the blood. Approximately 1 L of blood is filtered through the kidneys every minute. Of this amount, most of the water is reabsorbed, so the amount of urine excreted per day is less than 1 percent of the total amount of liquid filtered. If the kidneys are not functioning normally, an artificial kidney machine may be used (see page 180).

Table 29-1 compares the composition of blood plasma and urine.

29-3 General Properties of Urine

Volume

A normal adult excretes 600 to 2500 mL of urine per day. The amount depends on the liquid intake and also on the weather conditions. In hot weather, more water is lost through perspiration; therefore, the amount of urine formed is less. Conversely, a greater amount of urine is formed during cold weather or when the humidity is high, when little evaporation of perspiration takes place. Drugs such as caffeine (in coffee or tea) and also alcoholic beverages have a *diuretic effect; that is, they increase the flow of urine.

A decreased flow of urine is called oliguria. Such a condition may occur during a high fever when most of the water lost by the body is in the form of perspiration. Certain kidney diseases may also cause oliguria.

Anuria means a total lack of urine excretion. Anuria indicates extensive kidney damage such as may be caused by a blood transfusion of the wrong type. In this condition the blood cells disintegrate, releasing hemoglobin, which clogs the glomeruli and does not allow any excretion of urine. Bichloride of mercury also affects the kidneys and may cause oliguria or anuria.

Polyuria is a condition in which the amount of urine excreted is much greater than normal. It may be due to excessive intake of water or to certain pathologic conditions. Polyuria may be caused by such diuretics as alcohol or caffeine. Urea, a normal constituent of urine, is also a diuretic. A person on a high-protein diet will excrete more urea, which in turn causes the formation of more urine.

During diabetes insipidus the hormone vasopressin, which controls the reabsorption of water in the kidneys, is lacking or deficient. In this case the amount of urine is greatly increased, sometimes as high as 30 L per day.

Patients with diabetes mellitus show a definite polyuria because glucose is a diuretic. Excessive water loss during diabetes mellitus can lead to dehydration.

Density

The density of the urine depends on the concentration of the solutes. The greater the concentration of the solutes, the greater the density. A normal range of the density of the urine is 1.003 to 1.030 g/mL. In cases of diabetes mellitus, the density will be higher because of a high concentration of sugar in the urine. In cases of diabetes insipidus, the density of the urine will be very low (close to 1.000 g/mL) because of the large amounts of water being excreted.

pH

Urine is normally slightly acidic, with a pH range of 4.6 to 8.0 and an average value of about 6.3. However, the pH of urine varies with the diet. Protein foods, such as meats, increase the acidity of the urine (lower the pH) because of the formation of phosphates and sulfates. The acidity of the urine is also increased during acidosis and with fever. Conversely, the urine may tend to become alkaline on a diet high in vegetables and fruits or because of alkalosis, a condition that may be produced by excess vomiting.

Color

Normal urine is pale yellow or amber. The color, however, varies with the amount of urine produced and also with the concentration of the solutes in the urine. The larger the volume of urine excreted, the lighter the color. The greater the concentration of solutes, the darker the color. The color of the urine is caused by urobilin and urobilinogen (see page 502). Various other components of the urine may cause it to have different colors. The presence of blood in the urine gives it a reddish color. A reddish color may also result from eating beets or rhubarb, or taking cascara, a carthartic. Homogentisic acid (an intermediary in the metabolism of phenylalanine and tyrosine) colors the urine brown. The drug methylene blue colors the urine green.

Freshly voided urine is clear and usually contains no sediment. However, when it stands for a while it may become cloudy and develop sediment because of the precipitation of calcium phosphate.

Odor

Fresh urine has a distinctive odor, but this odor may be modified by the presence of other substances. In patients with ketosis the odor of acetone may be detected. Diet can also modify the odor of urine. For example, when asparagus is eaten, the urine may have a sulfurlike odor.

29-4 Normal Constituents

Approximately 50 to 60 g of dissolved solid material is excreted daily in the urine of the average person. This solid material has both inorganic and organic constituents (Table 29-2). The inorganic constituents of urine make up approxi-

Table 29-2 Constituents of Urine (Amount Excreted per Day)

Constituent	Amount (g)	Approximate Percent
Organic		
Urea	25-30	40-50
Uric acid	0.35-0.6	1
Creatinine	1.4	2.5
Creatine	0.06-0.1	0.1-0.2
Others	0.1-1	0.1-1
Inorganic		
Chloride ion	9-16	15-25
Sodium ion	4	6
Phosphates	2	3
Sulfates	2.5	4
Ammonium ions	0.7	1
Other ions and inorganic constituents	2.5	4

mately 45 percent of the total solids; the organic constituents compose the other 55 percent.

Organic Constituents

Urea The principal end product of the metabolism of protein, urea, comprises about one half of the total solids in the urine.

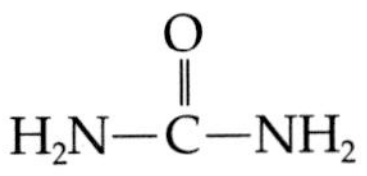

$$H_2N-\overset{\overset{\displaystyle O}{\|}}{C}-NH_2$$

urea

Uric Acid Uric acid, a product of the metabolism of purines (see page 504) from nucleoprotein, is only slightly soluble in water and is excreted primarily as urate salts. The structure of uric acid is

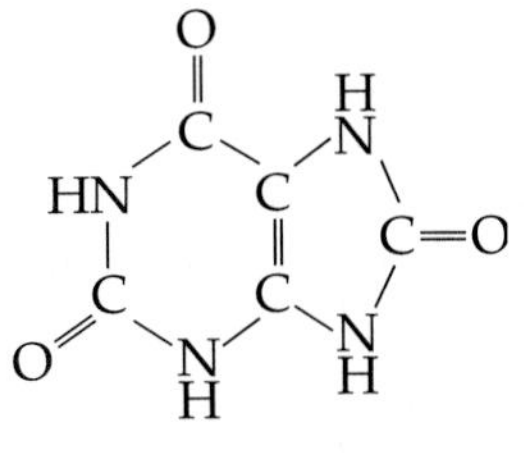

uric acid

When urine is allowed to stand, uric acid may crystallize and settle out since it is only slightly soluble in water or acid solution.

The average daily excretion of uric acid is about 0.7 g, but an increase in nucleoproteins in the diet will cause an increased excretion of uric acid. The output of uric acid is also increased in leukemia, in severe liver disease, and in various stages of gout. Deposits of urates and uric acid in the joints and tissues are also characteristic of gout, so this disease appears to be a form of arthritis (Figure 29-2).

Under certain conditions uric acid or urates crystallize in the kidneys and are called kidney stones, or *calculi (Figure 29-3). Kidney stones can also be formed from calcium oxalate.

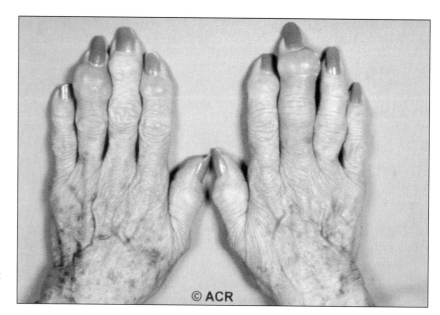

FIGURE 29-2

The ravaging effects of gout are manifest in the deformed hands of a patient afflicted with this crippling disease.

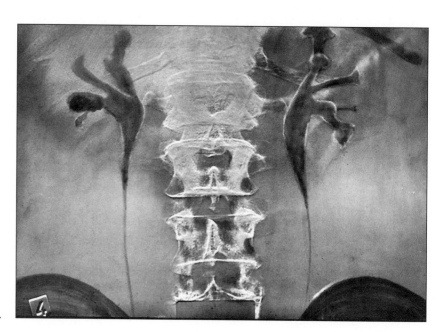

FIGURE 29-3

Stone in right kidney.

Creatinine   Creatinine is a product of the breakdown of creatine. The amount of creatinine excreted per day is fairly constant regardless of the protein intake. The number of milligrams of creatinine excreted in the urine within a 24-hr period per kilogram of body weight is the creatinine coefficient of the individual. The creatinine coefficient of the normal male is 20 to 26; that of females is 14 to 22.

The average adult excretes 1.4 g of creatinine daily and about 0.06 to 0.1 g of creatine in the same period of time. Formation of creatinine appears necesssary for the excretion of most of the creatine.

$$
\begin{array}{c}
\overset{\displaystyle NH}{\underset{\displaystyle \parallel}{}} \\
C-NH_2 \\
| \\
N-CH_2-COOH \\
| \\
CH_2
\end{array}
\longrightarrow
\begin{array}{c}
\overset{\displaystyle NH}{\underset{\displaystyle \parallel}{}} \\
C\text{————}NH \\
| \qquad\quad | \\
N-CH_2-\ C=O + H_2O \\
| \\
CH_3
\end{array}
$$

creatine   creatinine

Creatine   Creatine is produced in the body from three amino acids: arginine, methionine, and glycine. Creatine is normally present in muscle, brain, and blood, both as free creatine and as creatine phosphate. Recall the reaction

creatine phosphate + ADP $\longrightarrow$ creatine + ATP

Creatinuria is a condition in which abnormal amounts of creatine occur in the urine. It may occur during starvation, diabetes mellitus, prolonged fevers, wasting diseases, and hyperthyroidism. Creatinuria may also occur in pregnancy.

Other Organic Constituents   Also present in small amounts are amino acids, allantoin (from partial oxidation of uric acid), hippuric acid, urobilin, urobilinogen, and biliverdin. The presence of cAMP helps diagnose parathyroid function. Patients with hyperparathyroidism excrete significantly more cAMP and those with hypoparathyroidism excrete significantly less cAMP than persons with normal parathyroid glands (see page 584).

Urobilinogen excretion is increased in hemolytic anemias and in liver disease. Biliverdin excretion is increased in certain liver and biliary diseases.

Tests for pregnancy are based on the fact that the implantation of a fertilized ovum in the placental tissue produces the hormone chorionic gonadotropin, which is excreted in the urine. Tests for the presence of chorionic gonadotropin (for pregnancy) are based on the fact that this hormone will react with specific *antibodies and can yield a result in the physician's office or at home in a very short time.

In addition to these constituents, the urine also normally contains very small amounts of vitamins, other hormones, and enzymes. Urinalysis for these substances is of diagnostic value.

Inorganic Constituents

The inorganic constituents of urine are the various positive and negative ions that make up the inorganic compounds being excreted. Among these are the following.

Chloride Ions   Between 9 and 16 g of chloride ion is excreted daily, mostly as sodium chloride. The amount of chloride ion varies with the intake, which is primarily sodium chloride. The excretion of sodium chloride is decreased in fevers and in some stages of *nephritis.

Sodium Ions   The amount of sodium ion excreted varies with the intake and the body's requirement. However, it is usually about 4 g per day.

Phosphates   The amount of phosphates present in the urine also depends upon the diet; the amount is higher when the diet contains foods high in phosphorus (nucleoproteins and phospholipids). An increase in excreted phosphates is found in certain bone diseases and in hyperparathyroidism. A decrease in phosphates is found in hypoparathyroidism, in renal diseases, and during pregnancy.

Sulfates   The sulfates in the urine are derived from the metabolism of sulfur-containing proteins, so the amount of sulfur is influenced by the diet. Sulfates are found as both organic and inorganic salts.

Ammonium Ions   The hydrolysis of urea produces such ammonium compounds as chlorides, sulfates, and phosphates in the urine.

Other Ions   In addition to the sodium and ammonium ions, other positive ions present in the urine are calcium, potassium, and magnesium.

The amount of calcium ions in the urine is increased in hyperthyroidism, hyperparathyroidism, and osteoporosis and decreased in hypoparathyroidism and vitamin D deficiency.

Nitrite ions in the urine indicate the presence of reducing bacteria in urinary tract.

The normal ratio of sodium to potassium in the urine is 2 parts sodium to 1 part potassium. The ratio of sodium to potassium is increased in Addison's disease.

Magnesium concentration in the urine is decreased in chronic alcoholism.

29-5 Abnormal Constituents

Protein

Because proteins are colloids and because colloids cannot pass through membranes, urine should not normally contain protein. Proteinuria denotes the presence of protein in the urine. Sometimes it is called albuminuria because albumin is the smallest plasma protein and is the protein most frequently found in the urine. In cases of kidney disease, such as nephritis and nephrosis, and in severe heart disease, protein appears in the urine. The presence of protein resulting from such disorders is frequently called renal proteinuria or renal albuminuria to distinguish it from false albuminuria, which is a temporary harmless condition. False albuminuria, often called *orthostatic albuminuria*, is found in certain patients who stand for a long period of time. It is caused by the constriction of the kidneys' blood vessels and disappears when the patient lies down. Small amounts of protein may also be found in the urine after severe muscular exercise, but they soon disappear.

The tests for the presence of protein in the urine are based on the fact that protein coagulates when heated. When a sample of urine is heated, any protein (albumin) present will precipitate out as a white cloud. However, phosphates may also precipitate when the urine is heated. To prove that the cloudy substance is albumin, the urine, after heating, is acidified with dilute acetic acid. The acid will dissolve the phosphates but not the protein, so a cloudy precipitate in the urine after heating and acidification is a verification of the presence of protein.

Glucose

The presence of glucose in the urine is called glycosuria. Normally there is always a very small amount of glucose present in the urine, but this amount is too small to give a positive result when the urine is tested with Benedict's solution. Glucose may be found in the urine after severe muscular exercise, but this condition clears up when the body returns to normal. Glucose may also be found in the urine after a meal high in carbohydrates. Glycosuria may be caused by such diseases as diabetes mellitus or renal diabetes or to liver damage.

Other Sugars

Lactose and galactose may occur in the urine during pregnancy and lactation. Both of these sugars give a positive result to Benedict's test. Pentoses may occur in the urine after consumption of foods such as plums, grapes, and cherries, which contain large amounts of these carbohydrates.

Ketone Bodies

Ketone (acetone) bodies are present in the urine during diabetes mellitus, in starvation, or in other circumstances with inadequate carbohydrate intake. Such a condition is termed ketonuria. The excretion of the ketone bodies, which are acidic compounds, requires alkaline compounds. This results in a depletion of the alkaline reserve of the blood and leads to acidosis. The kidneys produce more ammonia to neutralize these ketone bodies.

The test for the presence of ketone bodies in the urine is performed by adding sodium nitroprusside to a sample of urine and then making the mixture alkaline with ammonium hydroxide. The presence of ketone bodies is indicated by a pink-red color. Normal urine gives no color with this test.

Blood

The presence of blood in the urine is called hematuria. It may result from lesions or stones in the kidneys or urinary tract. The presence of free hemoglobin in the urine, hemoglobinuria, results from hemolysis of the red blood cells caused by an injection of hypotonic solution, severe burns, or blackwater fever.

Large amounts of blood in the urine can be detected by the reddish color. Small amounts do not color the urine enough to show any color change but can be detected by the use of a test dipstick.

Bile

Normally, bile is excreted by the liver into the small intestine and eventually ends up in the feces. The presence of bile in the urine indicates obstruction to the flow of bile to the intestines. Bile in the urine is indicated by a greenish brown color. Bile in the urine is also indicated by the presence of a yellow foam when the urine is shaken.

Phenylpyruvic Acid

Phenylpyruvic acid, an intermediate product in the metabolism of the essential amino acid phenylalanine, is not normally present in urine. However, its presence can be detected in the urine of a person with phenylketonuria (PKU) (see page 411). If this disease is not detected and treated early, mental retardation results. For this reason, some states require screening of newborns for this disease.

Bilirubin

Bilirubin is present in the urine of patients with conjugated hyperbilirubinemia. Such patients are said to have choluric jaundice. Such a condition can be caused by (a) Dubin-Johnson syndrome, which is caused by a genetic defect in the hepatic secretion of bilirubin into the bile, or (b) biliary tree obstruction because of a blocking of the hepatic or common bile ducts.

Urobilinogen

Increased levels of urobilinogen and the absence of biliverdin in the urine is an indication of hemolytic jaundice. In addition, increased levels of urobilinogen can indicate pernicious anemia.

Creatine

Normally, small amounts of creatine are present in the urine (see page 515). Elevated levels of this compound can indicate muscular dystrophy, myesthenia gravis, muscle wasting poliomyelitis, or hyperthyroidism.

Uric Acid

Elevated urine levels of uric acid can indicate gout, Lesch-Nyhan syndrome, and some cases of glycogen storage disease. Decreased levels can indicate renal insufficiency.

29-6 Diuretics

Diuretics are drugs that promote loss of water and salts through the urine. Examples of diuretics are ethyl alcohol, caffeine, mannitol, and thiazides.

Kidney-function tests are based on the amount of urea eliminated in the urine compared to the amount present in the blood (urea clearance test) or the change in the density of the urine after the patient's fluid intake is restricted (concentration test).

29-7 Reagent Tablets, Papers, and Dipsticks

To help nonlaboratory personnel perform a variety of urine (or blood) tests easily, quickly, and economically, many tablets, papers, and dipsticks have been developed. In all cases, it is essential that the manufacturer's directions be followed.

Reagent tablets and some papers and dipsticks are used to test for one substance only. That is, a tablet or paper or dipstick can be used to test for glucose, but another kind of tablet, paper, or dipstick must be used to test for ketone bodies or any other substance.

Some papers and dipsticks contain several reagent strips. These can be used to test for several substances at the same time.

In all cases, whether using tablets, papers, or dipsticks, the color produced is compared to a color chart prepared by the manufacturer in order to read the results.

Summary

The principal excretory organs of the body are the kidneys, which also control and regulate the water balance, electrolyte balance, and pH of body fluids.

The waste materials in the blood are picked up by the kidneys and are excreted in the urine.

Approximately 600 to 2500 mL of urine is excreted daily, the amount depending on fluid intake, weather conditions, humidity, and certain diuretic substances.

A decreased flow of urine is called oliguria; anuria is a total lack of urine; polyuria is excess urine formation.

The density of urine varies between 1.003 and 1.03 g/mL, and the pH ranges from 4.6 to 8.0, with an average value of 6.3. Urine is normally pale yellow or amber, but certain components may cause another color.

Approximately 50 to 60 g of solid material, both organic and inorganic, is excreted daily. The principal organic constituent of urine is area, the end product of protein metabolism. Another important constituent is uric acid, a product of the metabolism of purines. Gout is characterized by an increase of uric acid in the urine and blood and the deposition of uric acid or urate salts in the joints and tissues. Uric acid and urates may also crystallize in the kidneys as kidney stones.

The average adult excretes 0.06 to 0.1 g of creatine and 1.4 g of creatinine daily. Creatine is produced from three amino acids: arginine, methionine, and glycine. Creatinine is produced from creatine by a dehydration reaction. Creatinuria is the condition in which abnormal amounts of creatine occur in the urine.

Inorganic constituents of urine are the following ions: chloride, sodium, phosphate, sulfate, ammonium, and some small amounts of calcium, potassium, and magnesium.

Abnormal constituents in the urine are protein (proteinuria, or albuminuria) resulting from kidney disease; glucose (glycosuria) from diabetes mellitus or liver damage; ketone bodies from diabetes mellitus or starvation (ketonuria); blood (hematuria) from lesions in the kidneys or urinary tract; and bile from an obstruction of the flow of bile to the intestines.

Diuretics increase the output of water and salts in the urine.

Questions and Problems

1. What are the principal excretory organs of the body? What additional functions do they have?

2. What is a glomerulus? Bowman's capsule? What function do they have in the formation of urine? Where is urine stored?

3. What forces fluids through the membranes in the kidneys?

4. What volume of urine does an adult excrete daily? What might affect this amount?

5. What is a diuretic? Name two.

6. What is oliguria? anuria? polyuria? What might cause each of these?

7. What does the hormone vasopressin do? If this hormone is lacking, what will be the effect on the body?

8. What is the pH range of urine?

9. What might affect the color of urine? the pH? What might make urine cloudy on standing?

10. What is the principal organic constituent of urine? Where does it come from? Give its structural formula.

11. Where does uric acid in the urine come from? Draw the structure of uric acid.

12. What might affect the amount of uric acid excreted daily?

13. What is gout?

14. What is the creatinine coefficient?

15. How does the body produce creatine? How does the structure of creatine compare with that of creatinine?

16. What might cause kidney stones?

17. What is creatinuria? proteinuria? What might cause these conditions?

18. What is albuminuria? false albuminuria?

19. How is the amount of urinary phosphate affected by disease?

20. Describe the test for the presence of protein in the urine.

21. What is glycosuria? What causes it?

22. Could more than one sugar be present in urine? Explain.

23. What causes the presence of acetone (ketone bodies) in urine?

24. What causes hematuria?

25. The presence of urinary bile might indicate what condition?

Practice Test

1. An example of a diuretic is _____.
   a. creatinine
   b. caffeine
   c. morphine
   d. glucose

2. Polyuria occurs in _____.
   a. high fever
   b. kidney damage
   c. diabetes mellitus
   d. all of these

3. Urine is normally _____.
   a. slightly acidic
   b. slightly alkaline
   c. neutral
   d. variable over a large range in pH

4. Kidney stones may contain _____.
   a. uric acid
   b. creatinine
   c. urea
   d. bilirubin

5. The principal organic constituent of urine is _____.
   a. uric acid
   b. creatinine
   c. urea
   d. glucose

6. The color of urine is caused by the presence of _____.
   a. stercobilin
   b. urobilin
   c. hemoglobin
   d. bilirubin

7. Glycosuria indicates the presence of urinary _____.
   a. glycose
   b. glucose
   c. glycogen
   d. all of these

8. Urine normally does not contain _____.
   a. protein
   b. ketone bodies
   c. uric acid
   d. urea

9. The principal anion in urine is _____.
   a. Cl^- b. SO_4^{2-} c. PO_4^{3-} d. HCO_3^-

10. A decreased output of urine is called _____.
    a. anuria
    b. oliguria
    c. polyuria
    d. hematuria

Body Fluids: Blood

Objectives

- To understand the composition of blood
- To become aware of the clinical significance of abnormal values of blood components
- To become familiar with the structure of hemoglobin
- To become aware of the significance of plasma lipid levels
- To understand the role of plasma proteins in water balance
- To understand blood clotting
- To become familiar with the transportation of oxygen to and carbon dioxide from the cells
- To understand the acid-base balance, including metabolic and respiratory acidosis and alkalosis
- To understand how the body regulates water balance as well as electrolyte balance
- To become aware of the clinical importance of mineral anions and cations

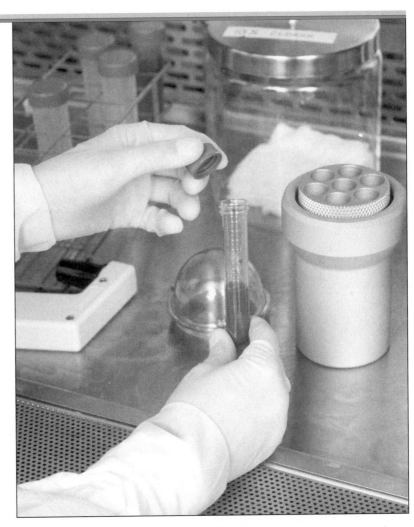

Blood is the fluid that supplies cells with nutrients and oxygen, removes their wastes, and provides transport for hormones produced by the endocrine system and the cells of the immune system. In this photograph, a sample of blood is tested for AIDS at the National Institutes of Health.

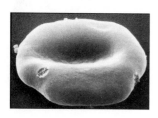

FIGURE 30-1

Scanning electron micrograph of a human red blood cell showing the typical biconcave disc shape.

30-1 Functions

Blood has been called a circulating tissue. It carries oxygen, minerals, and food to the cells and carries carbon dioxide and other waste products away from the cells. It also transports hormones, enzymes, and blood cells. Blood regulates body temperature by transferring heat from the interior to the surface capillaries. The blood buffers maintain the pH of the body at its optimum value. Blood contains a clotting system that protects the body against hemorrhage and defense mechanisms against infection.

30-2 Composition

Blood consists of two parts—the suspended particles and the suspending liquid, the plasma. The suspended particles in the blood are the red blood cells (see Figure 30-1), white blood cells, and platelets.

Red Blood Cells (Erythrocytes)

Normally there are 4.5 million to 5.0 million red blood cells per cubic millimeter of blood. Since there are about 6 quarts of blood in the human body, the total number of red blood cells is approximately 30 trillion (30,000,000,000,000). An excess of red blood cells is called **polycythemia**; a shortage is called **anemia.** Each day 200 billion new red blood cells are formed in the bone marrow.

Erythropoietin, a hormone that stimulates red blood cell formation, is a glycoprotein with a molecular mass of approximately 35,000. Erythropoietin is formed by the action of a substance produced by the kidneys (renal erythropoietic factor) on a globulin in the blood plasma. The production of this hormone is increased by hypoxia in the kidneys, by cobalt salts, and by androgens. If the kidneys do not function properly, the patient may become anemic (see Section 30-7). Table 30-1 lists some of the many disorders affecting red blood cells.

White Blood Cells (Leukocytes)

The number of white blood cells normally present in the blood ranges from 5000 to 10,000 per cubic millimeter. White blood cells are larger than red blood cells. White blood cells have a nucleus; red blood cells do not. There are several types of white blood cells; among these are the basophils, eosinophils, lymphocytes, monocytes, and neutrophils.

Table 30-1 Disorders Affecting Red Blood Cells	
Disorder	**Cause**
Iron-deficiency anemia	Inadequate intake of iron, excess loss of iron
Sickle-cell anemia	Genetic disorder (see page 410)
Methemoglobinemia	Intake of excessive oxidants
α- and β-Thalessemias	Genetic mutation
Megaloblastic anemia	Decreased absorption of vitamin B_{12}
	Decreased intake of folic acid

The white blood cells attack and destroy harmful microorganisms and thus serve as one of the body's defenses against infection. A white blood cell count above normal usually indicates an infection. For diagnostic purposes, a **differential count** is sometimes ordered. This count gives the percentages of each of the various types of leukocytes present.

Platelets (Thrombocytes)

The number of thrombocytes or platelets ranges from 250,000 to 400,000 per cubic millimeter of blood. Platelets are smaller than red blood cells and do not have a nucleus. Platelets contain cephalin (phosphatidylethanolamine), a phospholipid that is involved in the clotting of the blood.

Blood Plasma

Approximately 92 percent of the plasma is water. The solids dissolved or colloidally dispersed in the blood make up the other 8 percent. The most important of all plasma solids are the blood proteins. These include serum albumin, the globulins, and fibrinogen.

The plasma proteins (primarily albumin) maintain the osmotic pressure of the blood, thus regulating the water and acid-base balance in the body.

The globulins are a mixture of protein molecules labeled α, β, and γ. The γ-globulins or immunoglobulins (antibodies) function against infection and disease. The α- and β-globulins include glycoproteins and lipoproteins, which function in the transportation of oligosaccharides, lipids, steroids, and hormones. Two particular globulins, transferrin and ceruloplasmin, transport iron and copper, respectively, in the plasma. Fibrinogen and prothrombin, two other globulins, function in the clotting of the blood.

Blood plasma also contains small amounts of lipids and carbohydrates (glucose), inorganic salts, waste products (such as urea, uric acid, carbon dioxide, creatinine, and ammonia), enzymes, vitamins, hormones, and antibodies. The inorganic ions present in the blood plasma serve to regulate the acid-base balance of the body.

On standing, freshly drawn blood soon forms a clot. When the clot settles, a yellowish liquid remains. This liquid is called **blood serum**. Blood serum is blood without the blood cells and without the fibrinogen necessary for the clotting of the blood. Blood plasma can be separated from the solid parts of the blood by centrifuging, during which the cells settle at the bottom of the test tube and the plasma remains above them. To keep the blood from clotting, an anticoagulant must be added before centrifuging.

30-3 Plasma Substitutes

Plasma volume can be greatly reduced in cases of severe burns, when water intake is too low, or following severe diarrhea, excessive vomiting, or polyuria. In such cases, plasma volume can be increased by intravenous infusion of an isotonic saline solution (0.9 percent NaCl).

However, if there has been a loss of blood volume as a result of hemorrhaging, intravenous infusion of saline solution is not adequate because such solutions lack plasma protein, salts, and blood cells. In such cases, transfusions of whole blood, plasma, or plasma substitutes may be used. If whole blood is to be used for a transfusion, careful matching of blood types is essential.

Blood volume extenders, such as dextrans, have been used. Although such substances are able to maintain the osmotic pressure of the blood, they have an adverse effect on the blood clotting process.

One group of plasma substitutes includes the perfluorocarbons, which are clear, colorless emulsions fortified with inorganic salts and other ingredients. These compounds have the ability to carry oxygen to the tissues and carbon dioxide from the tissues. However, they cannot assume the functions of white blood cells, platelets, clotting factors, and other blood components.

Most transfusions are given to replace lost blood volume and red blood cells, and plasma substitutes will work well in such cases. Also, plasma substitutes can be given to all patients regardless of blood type. Plasma substitutes are particularly useful in transplantation surgery, in which large volumes of blood are required for perfusion of organs outside the body. Use of plasma substitutes in such cases saves a large volume of whole blood for other needs.

Plasma substitutes are also useful in certain types of anemias in which the ability of the blood to carry oxygen is greatly reduced.

Patients whose religious beliefs prevent them from receiving blood transfusions can be given blood substitutes, such as perfluorocarbons, which have a shelf life measured in years compared with just a few weeks for whole blood. But the emulsion must be kept frozen until use or it begins to break down.

30-4 General Properties of Blood

Oxygenated blood has a characteristic bright-red color; deoxygenated blood has a dark purplish color.

The density of whole blood ranges from 1.054 to 1.060 g/mL, whereas that of blood plasma ranges from 1.024 to 1.028 g/mL.

Blood is normally slightly alkaline, with a pH range of 7.35 to 7.45. If the pH of the blood falls slightly below 7.35, the condition is called **acidosis.** If the pH of the blood rises slightly above 7.45, the condition is called **alkalosis.** If the pH of the blood changes more than a few tenths from the normal values, the results are usually fatal.

The viscosity of the blood is approximately 4.5 times that of water and varies according to the number of cells, the quantity of protein, the temperature, and the amount of water present in the body (see page 169).

30-5 Blood Analysis

For most laboratory tests, 5 mL of blood is collected from a vein in the arm before the patient is given breakfast. If blood plasma is to be used for the test, an anticlotting agent such as potassium oxalate is added to the blood sample. If blood serum is to be tested, the blood is allowed to clot and the serum is poured off.

Blood chemistry tests and the clinical significance of these tests are indicated in Table 30-2.

30-6 Blood Volume

Approximately 8 to 9 percent of the total body weight is blood. The volume of the blood in the body amounts to 5 to 7 L in the adult. The volume increases in fever and pregnancy and decreases during diarrhea and hemorrhaging. Since blood volume can be rapidly replaced, a small loss of blood from bleeding or from donating blood has no deleterious effect on the body.

Table 30-2 Clinical Significance of Substances in Blood

Determination	Clinical significance	
	Increased in	Decreased in
Calcium	Hyperparathyroidism, Addison's disease, malignant bone tumor, hypervitaminosis D	Hypoparathyroidism, rickets, malnutrition, diarrhea, chronic kidney disease, celiac disease
Cholesterol	Diabetes mellitus, obstructive jaundice, hypothyroidism, pregnancy	Pernicious anemia, hemolytic jaundice, hyperthyroidism, tuberculosis
Uric acid	Gout, leukemia, pneumonia, liver and kidney disease	
Urea nitrogen	Mercury poisoning, acute glomerulonephritis, kidney disease	Pregnancy, low-protein diet, severe hepatic failure
Nonprotein nitrogen	Kidney disease, pregnancy, intestinal obstruction, congestive heart failure	Low-protein diet
Creatine	Nephritis, renal destruction, biliary obstruction, pregnancy	
Creatinine	Nephritis, chronic renal disease	
Glucose	Diabetes mellitus, hyperthyroidism, infections, pregnancy, emotional stress, after meals	Starvation, hyperinsulinism, Addison's disease, hypothyroidism, extensive hepatic damage
Chlorides	Nephritis, anemia urinary obstruction	Diabetes mellitus, diarrhea, pneumonia, vomiting, burns
Phosphate, inorganic	Hypoparathyroidism, Addison's disease, chronic nephritis	Hyperparathyroidism, diabetes mellitus
Sodium	Kidney disease, heart disease, pyloric obstruction	Vomiting, diarrhea, Addison's disease, myxedema, pneumonia, diabetes mellitus
Potassium	Addison's disease, oliguria, anuria, tissue breakdown	Vomiting, diarrhea
Carbon dioxide	Tetany, vomiting, intestinal obstruction, respiratory disease	Acidosis, diarrhea, anesthesia, nephritis
Hemoglobin	Polycythemia	Anemia

Effects of Meals and Position on Concentrations of Blood Substances

Normally, blood samples are taken after a period of fasting, usually 8 to 12 hours. Other than water, no foodstuff should be taken. Why? When blood is drawn 3 to 4 hours after breakfast, very little of the various blood substance levels are altered. If blood is drawn 3 to 4 hours after lunch, there may be a great variation between those values and those taken after a fasting period. For example, the SGOT level can vary as much as 30 percent, and the lactase dehydrogenase levels by 5 percent. Other levels are less affected.

Position may also affect concentrations of blood substances. A person who has been resting horizontally for several hours will have a 12 to 15 percent greater blood volume than a person who is standing. This greater blood volume, in turn, causes a corresponding decrease in concentrations of blood substances. Results for a person who has been sitting will be intermediate between those standing and those lying down.

30-7 Hemoglobin

pyrrole

Hemoblobin is a conjugated protein made up of a protein part, globin, and an iron-containing part, heme. Heme contains four pyrrole groups joined together with an iron ion in the center.

The structure of heme is indicated in Figure 30-2. Various hydrocarbon side chains are attached to the pyrrole rings in this compound. Hemoglobin consists of 2 α and 2 β chains with each chain containing a heme group. Each chain is roughly comparable to the myoglobin chain (see page 528).

The structures of cytochrome c, part of the oxidative phosphorylation sequence of the Krebs cycle (see Section 26-8), and of chlorophyll, a plant pigment, are shown in Figure 30-2 for comparison with that of heme. Note that whereas both heme and cytochrome c have an iron ion (Fe^{2+}) in the center of four pyrrole rings, chlorophyll has a magnesium ion (Mg^{2+}).

Fetal Hemoglobin

Fetal hemoglobin binds oxygen more tightly than hemoglobin so the fetus can draw oxygen from its mother's blood. The structure of fetal hemoglobin is similar to that of hemoglobin, and soon after birth fetal hemoglobin transforms to normal hemoglobin.

Anemia

If the hemoglobin content of the blood falls below normal, the condition is called anemia. Anemia can result from a decreased rate of production of red blood cells, from an increased destruction of red blood cells, or from an increased loss of red blood cells.

A decreased rate of production of red blood cells may be due to various diseases that destroy or suppress the activity of the blood-forming tissues. Among these diseases are leukemia, *multiple myeloma, and Hodgkin's disease. Radiation and certain drugs such as benzene and gold salts also decrease the activity of the blood-forming tissues. Another frequent cause of decreased red blood cell production is a diet lacking in iron and protein, particularly in infancy and childhood and during pregnancy. Anemia may also be related to a genetic defect that affects the production of hemoglobin (see Chapter 23). Pernicious anemia, a failure of red blood cell production, is due to a lack of vitamin B_{12} or of the intrinsic factor (see Section 25-2).

A decrease in red blood cells may be caused by several poisons or infections that can cause hemolysis. Carbon monoxide (CO) is a poisonous gas because hemoglobin combines with it approximately 210 times as fast as it does with oxygen. The compound formed between hemoglobin and carbon

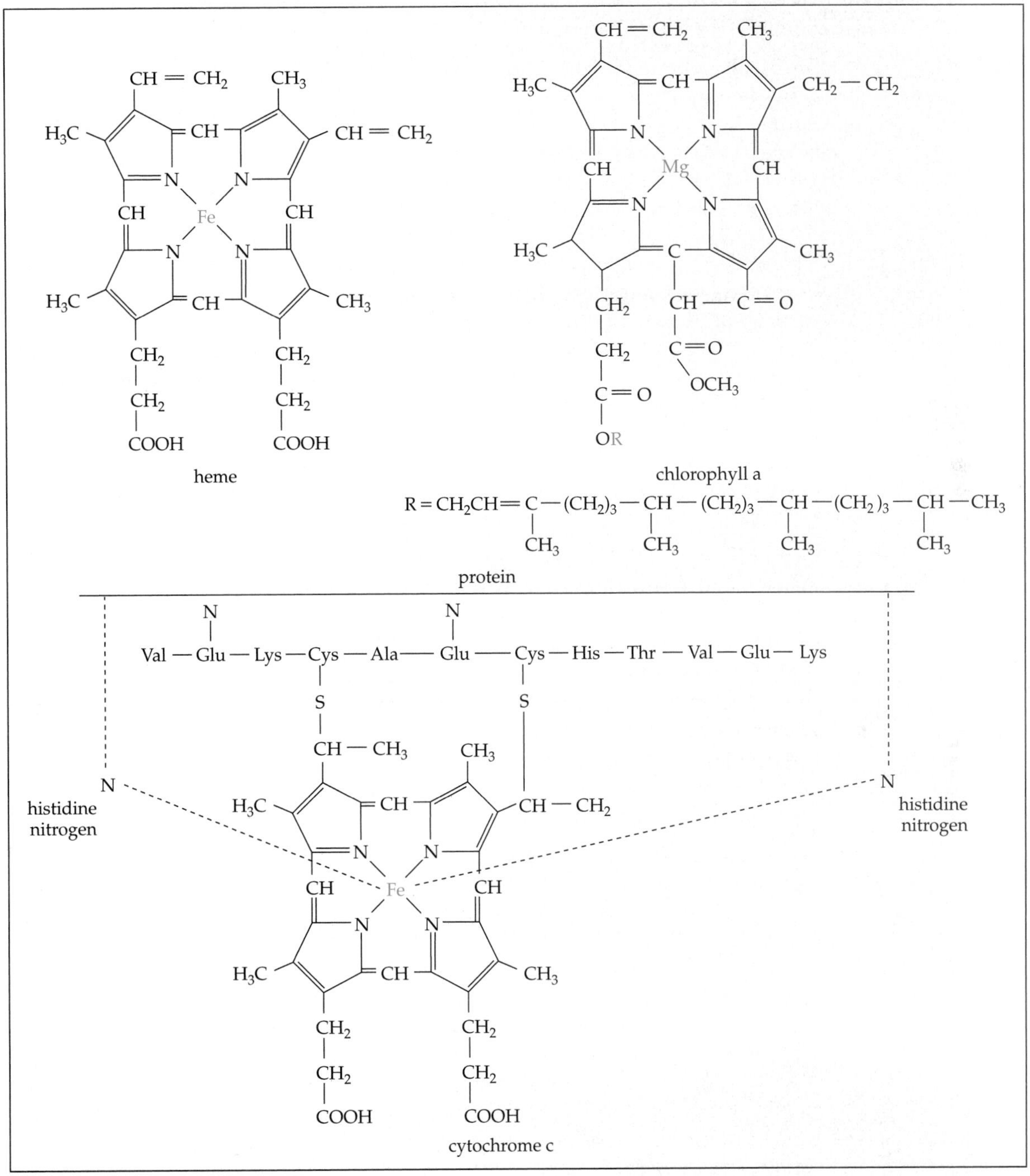

FIGURE 30-2
Structural formulas of heme, cytochrome c, and chlorphyll a.

Case Study 30-1   Carbon Monoxide Poisoning

A 73-year-old woman who lives alone has come to the emergency department complaining of chest pain, vertigo, nausea, chills, headache, and dizziness. She is immediately admitted to the cardiac intensive care unit. Her initial vital signs were BP, 110/90 mm Hg; heart rate, 112 beats/min; and respirations, 30 breaths/min. Arterial blood gases assessed while the patient was receiving 4 L/min of O_2 per nasal cannula reveal an oxygen saturation of 80%, pO_2, 101 mm Hg; pCO_2, 32 mm Hg; pH, 7.45; HCO_3^-, 22.2 mmol/L; and carboxyhemoglobin, 15.2%. Later investigation reveals that the woman's gas furnace has a rusted flue.

Questions

1. What is the struction of carboxyhemoglobin?
2. How is carboxyhemoglobin determined in a blood test?
3. What effect does carbon monoxide have on the oxygen-carrying capacity of the blood?

monoxide—carboxyhemoglobin—is very stable, so only a small amount of hemoglobin is left to carry oxygen. If the CO content of the air is 0.02 percent, nausea and headache occur; if the CO content of the air rises to 0.1 percent, unconsciousness will occur within 1 hour and death within 4 hours. Other poisonous gases such as hydrogen sulfide (H_2S) and hydrocyanic acid (HCN) have similar effects on hemoglobin.

An excessive loss of hemoglobin may be caused by hemorrhaging.

30-8 Myoglobin

Myoglobin is a globular protein similar to hemoglobin in that it also contains a heme group. However, myoglobin differs in that it has a single chain of 153 amino acids attached to the heme group.

Hemoglobin is involved in the transportation of oxygen to the tissues. Myoglobin stores oxygen in red muscle tissue such as heart muscle. Actually, the red color of such muscle tissues is due to the presence of myoglobin. Like fetal hemoglobin, myoglobin binds oxygen to itself more strongly than normal hemoglobin does and so stores it for the needs of the heart.

Under conditions of oxygen deprivation (such as severe exercise), myoglobin releases oxygen to the heart muscle for the synthesis of ATP. The oxygen must then be replaced from the hemoglobin.

30-9 Plasma Lipids

Plasma lipids can be divided into several classes, as shown in Table 30-3.

Metabolism of lipids produces much of the body's energy. In order to be metabolized, lipids must first be transported by the bloodstream. But lipids are hydrophobic and are not soluble in blood (mostly water). The transportation of lipids in the bloodstream is accomplished by associating insoluble lipids with

Table 30-3	Lipids In Human Blood Plasma
Lipid	Range (mmol/L)
Triglycerides	0.9-2.0
Phospholipids	1.8-5.9
Cholesterol (total)	1.8-8.3
Free fatty acids	0.2-0.6

phospholipids (which are polar) and then combining them with cholesterol and protein to form a *hydrophilic lipoprotein complex*. In this manner triglycerides, which are derived from intestinal absorption of fat or from the liver, are transported as chylomicrons and very low density lipoproteins (VLDL) (see Section 22-8). Free fatty acids, which are released from the adipose tissue, are carried in the blood stream in an unesterified form as an albumin-fatty acid complex. That is, many types of lipids are transported in the bloodstream as plasma lipoproteins.

30-10 Plasma Proteins

The plasma proteins constitute about 7 percent of the plasma and are usually divided into three groups: albumin, globulins, and fibrinogen. Approximately 55 percent of the plasma protein is albumin, 38.5 percent globulins and 6.5 percent fibrinogen.

Albumin

Albumin in the blood functions in the regulation of osmotic pressure. The control of osmotic pressure in turn affects the water balance in the body.

Albumin, like other plasma proteins, cannot pass through the walls of the blood vessels (because they are colloids, and colloids cannot pass through membranes). Since albumin is the principal plasma protein and the smallest plasma protein both in size and mass (it consists of a single chain of 610 amino acids), it accounts for most of the colloid osmotic pressure of the blood.

The effect of albumin (and other plasma proteins) on water balance (both filtration and reabsorption) has been hypothesized by Starling, as indicated in the following paragraphs and also in Figure 30-3.

When blood enters the arterial end of a capillary, it exerts a *hydrostatic (blood) pressure of 35 mm Hg, forcing fluid outward from the blood vessel. At the same time, the colloid osmotic pressure of the plasma, 25 mm Hg, pulls fluid back into the blood vessel. The *interstitial fluid exerts a hydrostatic pressure of 2 mm Hg, which forces fluid out of the tissues back into the blood. Since there is almost no protein in the tissue fluids at the end of the capillary, the colloid osmotic pressure of the interstitial fluid is 0 mm Hg. The net result of the pressure acting outward from the blood (35 and 0 mm Hg) and the pressure acting inward toward the blood (25 and 2 mm Hg) causes a net pressure of 8 mm Hg, forcing fluid out of the blood at the arterial end of the capillary. Thus, there is a net outward filtration from the capillary.

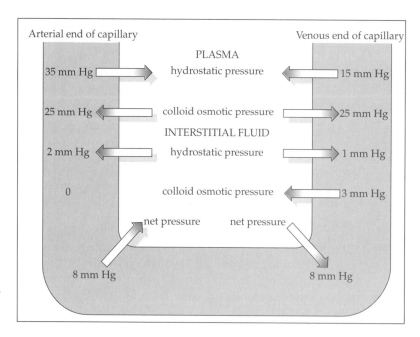

FIGURE 30-3

Effect of albumin (and other plasma proteins) on water balance.

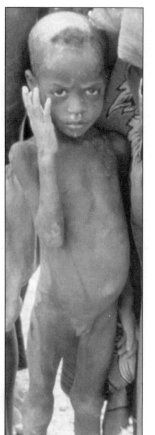

FIGURE 30-4

The swollen belly of this undernourished child is a sign of malnutrition.

At the venous end of the capillary, the following conditions exist: The hydrostatic pressure of the blood is 15 mm Hg, forcing fluid outward. (Note that the hydrostatic pressure of the blood at the venous end is less than at the arterial end of the capillary.) The colloid osmotic pressure of the plasma remains at 25 mm Hg, pulling fluid inward. The interstitial fluid hydrostatic pressure of 1 mm Hg forces fluid out of the tissues back into the blood, whereas the interstitial fluid colloid osmotic pressure of 3 mm Hg causes fluid to flow back to the tissues.[†] The net result of two pressures acting outward at the venous end of the capillary (15 and 3 mm Hg) and two pressures acting inward (25 and 1 mm Hg) gives a net pressure of 8 mm Hg inward, causing reabsorption of materials. This reabsorption is aided by the *lymphatics.

If the plasma proteins (primarily albumin) are present in decreased amounts (as in nephritis or during a low-protein diet), the osmotic pressure of the plasma decreases. This decreased osmotic pressure of the blood causes a greater net pressure outward at the arterial end of the capillary and a lower net inward venous pressure at the venous end of the capillary. When this occurs, water (fluid) accumulates in the tissues. Such a condition is known as *edema.

Kwashiorkor (Figure 30-4), a protein deficiency disease, is characterized by edema of the abdomen and extremities. In children, a swollen belly is characteristic of the disease. Kwashiorkor is caused by a drop in plasma protein, particularly albumin. Under these circumstances, water moves from the bloodstream into the tissues, causing swelling.

Edema can also occur because of heart disease, whereby there is an increase in venous hydrostatic pressure. Many terminal illnesses are accompanied by edema. This becomes a serious problem, and tapping and draining may be necessary. Concentrated albumin infusions (25 g in 100 mL diluent) are helpful in the treatment of shock, to increase blood volume, and to remove fluid from the tissues.

[†]This colloid osmotic pressure is caused by small amounts of plasma protein that pass through the capillary membranes and tend to accumulate at the venous end of the capillaries.

The amount of albumin present in the blood is lowered in liver disease because albumin is formed in the liver.

Another function of albumin in the blood is to act as a carrier for fatty acids, trace elements, and many drugs.

Globulins

The globulins present in the plasma can be separated into different groups by a process known as *electrophoresis (see page 178), whereby the charged protein particles migrate at varying rates to electrodes of opposite charge, with albumin migrating the fastest. The distribution of the plasma proteins is shown in Figure 30-5. As can be seen in the illustration, the globulins are subdivided into alpha (α), beta (β), and gamma (γ). The globulins form complexes (loose combinations) with such substances as carbohydrates (mucoprotein and glycoprotein), lipids (lipoprotein), and metal ions (transferrin for iron and ceruloplasmin for copper). The amount of transferrin is decreased in such diseases as pernicious anemia and liver disease. The amount of ceruloplasmin is decreased in Wilson's disease (see page 547). These complexes can be transported to all parts of the body.

The γ-globulins (immunoglobulins) include the antibodies with which the body fights infectious diseases. γ-Globulin has been found to contain as many as 20 different antibodies for immunity against such diseases as measles, infectious hepatitis, poliomyelitis, mumps, and influenza. The most important use of serum electrophoresis is as an aid in the diagnosis of diseases in which abnormal proteins appear in the blood (*multiple myeloma and *macroglobulinemia), or when protein component is either present in decreased amounts or lacking altogether, as in agammaglobulinemia.

Some people lack the ability to make γ-globulin. These people are quite susceptible to infections because they have no antibodies to counteract such diseases. The lack of γ-globulin is called **agammaglobulinemia** and can be counteracted by the administration of γ-globulin.

Fibrinogen

Fibrinogen is the plasma protein involved in the clotting of the blood. It is manufactured in the liver, so any disease that destroys liver tissue causes a decrease in the amount of fibrinogen in the plasma. Fibrinogen is a soluble plasma glycoprotein with a molecular mass of 340,000.

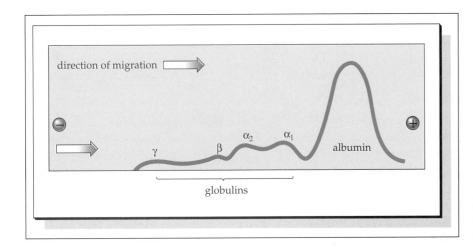

FIGURE 30-5

Distribution of plasma proteins during electrophoresis.

30-11   Blood Clotting

When the skin is ruptured, blood flows out and soon forms a clot. When blood is taken from a vein and placed in a test tube, it soon forms a clot. Why does blood clot when it is removed from its normal place in the circulatory system? Why doesn't it clot in the blood vessels themselves?

When blood clots, a series of reactions occurs in which the soluble plasma protein fibrinogen is converted into insoluble fibrin. Fibrin precipitates in the form of long threads that cling together to form a spongy mass, which entraps and holds the blood cells, forming a clot.

When a blood vessel is cut, the blood comes into contact with the exposed tissue. This contact activates two separate systems of coagulation—the intrinsic system, in which all components necessary to form a clot are found in the blood, and the extrinsic system, which needs a component released by the tissues. Figure 30-6 illustrates the three mechanisms of *hemostasis—vasoconstriction, platelet aggregation, and fibrin clot. A vasoconstriction occurs promptly at the site of the injury, thus reducing blood flow. Also, the platelets form a temporary plug in the injured capillaries. Prostaglandins and thromboxanes are required for this latter action. Since aspirin interferes with the action of prostaglandins, aspirin interferes with the clotting process.

Calcium ions are necessary for the clotting of blood. The coagulation of freshly drawn blood samples can be prevented by adding a substance (such as potassium oxalate) that removes calcium ions from solution. However, since oxalates are poisonous, this method can be used only if the blood is to be analyzed in a laboratory.

To prevent clotting in blood used for transfusions, sodium citrate is added. This substance removes the calcium ions from the blood by forming calcium citrate, a compound that is almost completely nondissociated. A deficiency of vitamin K reduces the production of prothrombin, without which the blood cannot clot.

In one form of hemophilia, factor X_a is missing, leading to an increase in blood clotting time.

Anticoagulant drugs such as bishydroxycoumarin (Dicumarol) reduce the conversion of prothrombin to thrombin and so keep blood from clotting rapidly. Heparin speeds up the removal of thrombin from several minutes to a few seconds.

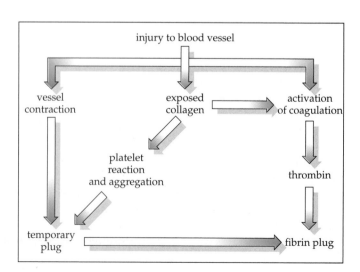

FIGURE 30-6

Reactions involved in hemostasis.

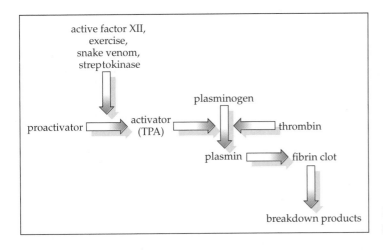

FIGURE 30-7
Fibrinolytic system.

Anticoagulant drugs may be used after surgery to prevent clots from forming in the cut blood vessels. A clot formed in a blood vessel is called a **thrombus.** A thrombus in a blood vessel does no harm if it remains where it was formed because it is slowly reabsorbed. However, if the clot breaks loose and travels through the blood vessels, it may lodge in and obstruct a blood vessel leading to the heart or brain, causing paralysis or death.

Breakdown of fibrin in blood clots (fibrinolysis) occurs within a few days after formation and is the result of the action of the enzyme plasmin. Plasmin is activated by a variety of circumstances, including stress, exercise, snake venom, and streptokinase from streptococci. Two drugs for dissolving and preventing blood clots, streptokinase and the genetically engineered TPA (tissue plasmin activator), both activate plasmin. Figure 30-7 shows a condensed version of the fibrinolytic system.

30-12 Respiration

The tissues require oxygen for their normal metabolic processes; they must also eliminate carbon dioxide. Oxygen is carried from the lungs to the tissues by the hemoglobin of the blood. In the tissues the oxygen is given up by the hemoglobin and the waste carbon dioxide from the tissues is picked up and carried to the lungs.

The inspired air has a higher concentration of oxygen than does the blood in the alveoli of the lungs. Gases always diffuse from an area of high concentration to one of lower concentration, so the oxygen diffuses from the lungs (high concentration) into the blood (lower concentration). In the blood the oxygen combines with the hemoglobin. Very little oxygen is actually dissolved (uncombined) in the blood. When the oxygen-rich blood reaches the tissues, it gives up its oxygen to the cells because those cells are using up oxygen and have a lower concentration of that gas than the blood. At the same time, the cells have a higher concentration of carbon dioxide than the blood; therefore, that gas diffuses from the cells into the bloodstream. The blood carries the carbon dioxide to the lungs. There it is in contact with air, which has a lower carbon dioxide concentration, and so it passes from the blood into the lungs where it is exhaled.

The transportation of gases in respiration includes the nine steps described in the next section.

Transportation of Oxygen and Carbon Dioxide

The transportation of oxygen and carbon dioxide is summarized in Figure 30-8, where the interrelationships of the following steps are shown.

1. As oxygen passes from the alveoli of the lungs into the bloodstream, some of it dissolves in the blood plasma. However, most of it reacts with hemoglobin (here represented by the formula HHb) to form oxyhemoglobin, HbO_2^-.

$$HHb + O_2 \longrightarrow HbO_2^- + H^+$$

Oxygen is carried to the cells in this form.

2. Carbon dioxide, which is produced during metabolic processes in the cells, diffuses from the tissues into the blood. A small amount dissolves directly in the plasma, but most of the carbon dioxide diffuses into the red blood cells, where it reacts with water in the cells to form carbonic acid. This reaction takes place rapidly under the influence of the enzyme carbonic anhydrase.

$$CO_2 + H_2O \xrightarrow{\text{carbonic anhydrase}} H_2CO_3$$

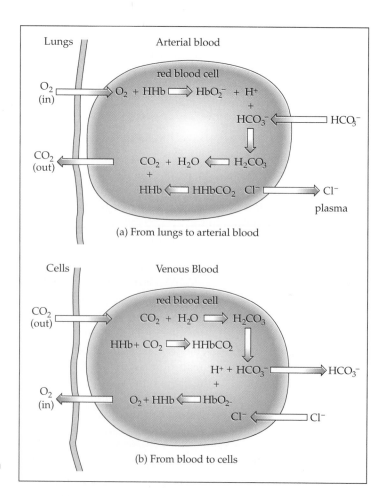

(a) From lungs to arterial blood

(b) From blood to cells

FIGURE 30-8

Transportation of oxygen and carbon dioxide.

3. The carbonic acid thus formed ionizes to yield hydrogen and bicarbonate ions.

$$H_2CO_3 \longrightarrow H^+ + HCO_3^-$$

4. In the tissues, oxyhemoglobin reacts with the hydrogen ions to yield oxygen and hemoglobin (HHb).

$$HbO_2^- + H^+ \longrightarrow HHb + O_2$$

The release of oxygen to the tissues is enhanced by a decrease in the pH and by the presence of 2,3-diphosphoglyceric acid. Most of the hemoglobin travels back to the lungs to pick up more oxygen, but some of the hemoglobin reacts with carbon dioxide to form carbaminohemoglobin (here represented as $HHbCO_2$).

$$HHb + CO_2 \longrightarrow HHbCO_2$$

5. The bicarbonate ions from step 3 do not remain in the red blood cells because those cells can hold only a small amount of that ion. Therefore, the excess bicarbonate ions diffuse outward into the blood plasma. Red blood cells cannot stand a loss of negative ions, so to counteract this outflow of bicarbonate ions, chloride ions from the blood plasma flow into the red blood cells. This process is called the chloride shift.

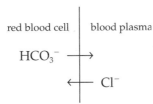

The bicarbonate ions can then act as buffers in the plasma.

6. In the lungs, the bicarbonate ions react with the hydrogen ions produced in step 1 to form carbonic acid.

$$HCO_3^- + H^+ \longrightarrow H_2CO_3$$

7. The carbonic acid thus formed rapidly decomposes into carbon dioxide and water under the influence of the enzyme carbonic anhydrase.

$$H_2CO_3 \xrightarrow{\text{carbonic anhydrase}} H_2O + CO_2$$

8. As the bicarbonate ions are used up in steps 6 and 7, more bicarbonate ions from the plasma flow into the red blood cells. At the same time, chloride ions migrate outward from the red blood cells. This is the reverse chloride shift.

red blood cell | blood plasma

$$Cl^- \longrightarrow$$
$$\longleftarrow HCO_3^-$$

9. At the same time, the carbaminohemoglobin formed in step 4 decomposes in the lungs to yield hemoglobin and carbon dioxide.

$$HHbCO_2 \longrightarrow HHb + CO_2$$

30-13   Acid-Base Balance

The normal pH range of the blood is 7.35 to 7.45. When the pH falls below this range, the condition is called *acidosis. *Alkalosis occurs when the pH rises above its normal value. Acidosis is more common than alkalosis because many of the metabolic products produced during digestion are acidic. The ability of the blood buffers to neutralize acid is called the alkaline reserve of the blood. In acidosis, the alkaline reserve decreases; during alkalosis, it increases.

How does the blood maintain the pH when acidic or basic substances are continuously being added to it?

The blood retains its fairly constant pH because of the presence of buffers. These buffers are present both in the blood plasma and in the red blood cells. Those in the plasma are primarily sodium buffers; those in the blood cells are mainly potassium buffers. Recall that buffers are substances (usually a weak acid and a salt of a weak acid) that resist change in pH (see Section 12-9). The blood buffers consist of

1. Bicarbonate buffers
2. Phosphate buffers
3. Protein buffers (including hemoglobin and oxyhemoglobin)

Bicarbonate Buffers

The bicarbonate buffer system in the red blood cells consists of carbonic acid (H_2CO_3) and potassium bicarbonate ($KHCO_3$). The bicarbonate buffer system in the blood plasma consists of carbonic acid and sodium bicarbonate ($NaHCO_3$). Actually, the blood bicarbonate buffers consist of H_2CO_3 and HCO_3^-, with the Na^+ and K^+ ions acting merely as spectator ions. However, for simplicity we will discuss the buffer systems as $H_2CO_3/NaHCO_3$ and $H_2CO_3/KHCO_3$. If a strong acid (such as HCl) is added to a sample of blood, it will react with the salt part of the buffer and undergo the following reactions:

$$HCl + KHCO_3 \longrightarrow H_2CO_3 + KCl \qquad \text{(in blood cells)}$$

$$HCl + NaHCO_3 \longrightarrow H_2CO_3 + NaCl \qquad \text{(in blood plasma)}$$

The carbonic acid (H_2CO_3) produced is part of the original buffer. Note that the strong acid, HCl, has been replaced by a very weak one, H_2CO_3. The other products, KCl and NaCl, are neutral salts and will not affect the pH of the system.

If a strong base such as KOH or NaOH is added to a sample of blood, the following reactions will occur with the bicarbonate buffer systems:

$$KOH + H_2CO_3 \longrightarrow KHCO_3 + H_2O \qquad \text{(in blood cells)}$$

$$NaOH + H_2CO_3 \longrightarrow NaHCO_3 + H_2O \qquad \text{(in blood plasma)}$$

The salts $KHCO_3$ and $NaHCO_3$ are part of the original buffer systems and the water produced is neutral, so the pH again is unaffected.

In both cases (reaction with a strong acid or a strong base), more of the buffer is produced plus a neutral compound.

The bicarbonate buffers and the blood protein buffers play a major part in the control of the pH; the phosphate buffers have an important role inside the cell and in the urine.

Phosphate Buffers

The phosphate buffers consist of mixtures of K_2HPO_4 and KH_2PO_4 (also Na_2HPO_4 and NaH_2PO_4), which function similarly to the bicarbonate buffers in neutralizing excess acid and base.

$$HCl + K_2HPO_4 \longrightarrow KH_2PO_4 + KCl$$

$$KOH + KH_2PO_4 \longrightarrow \underset{\text{more buffer}}{K_2HPO_4} + \underset{\text{neutral compound}}{H_2O}$$

Hemoglobin Buffers

The hemoglobin buffers account for more than half of the total buffering action in the blood. There are hemoglobin buffers and oxyhemoglobin buffers.

    These buffers, as well as other proteins that act as buffers in the bloodstream, pick up excess acid or base to help keep the pH of the blood within its normal range.

Hemoglobin buffer
HHb
KHb

Oxyhemoglobin buffer
$HHbO_2$
$KHbO_2$

Function of the Kidneys

The kidneys help maintain the acid-base balance of the blood by excreting or absorbing phosphates and also by forming ammonia. The acid substances in the blood combine with ammonia and are excreted as ammonium salts through the kidneys, saving sodium and potassium ions for the buffer systems.

Metabolic Acidosis

Metabolic acidosis is characterized by a drop in plasma pH as a result of a decrease in the HCO_3^- concentration and compensatory drop in the partial pressure of carbon dioxide pCO_2. Metabolic acidosis may be caused by either the addition of H^+ or the loss of HCO_3^-.

    The most common causes of metabolic acidosis are uncontrolled diabetes mellitus with ketosis (see Section 27-4), renal failure, poisoning with acid substances, severe diarrhea with loss of HCO_3^-, lactic acidosis, and severe dehydration.

    As the pH decreases, more H_2CO_3 is produced from the HCO_3^-. The H_2CO_3 in turn decomposes to CO_2 and H_2O.

$$HCO_3^- + H^+ \rightleftharpoons H_2CO_3 \rightleftharpoons H_2O + CO_2$$

If the blood buffers cannot control the pH, then the excess carbon dioxide produced stimulates the respiratory center of the brain, making the person breathe faster, thus removing more carbon dioxide from the blood. The increased rate of respiration (hyperventilation) continues until the amount of carbon dioxide is too small to further stimulate the respiratory center of the brain, and breathing returns to its normal rate. In addition to hyperventilation, the kidneys respond by excreting H^+, primarily as NH_4^+.

    Treatment, if the kidneys are functioning normally, consists of intravenous administration of HCO_3^-. This lowers the H^+ concentration and so increases pH to normal. Fluids are administered to replace lost water. A diabetic is given insulin therapy. Hemodialysis may be required if the kidneys are not functioning normally.

Metabolic Alkalosis

Metabolic alkalosis is characterized by an increase in plasma pH, an increase in HCO_3^- concentration, and a compensatory increase in pCO_2. Metabolic alkalosis is caused by a loss of H^+ or a retention of HCO_3^-.

Common causes of metabolic alkalosis are prolonged vomiting, gastric suction (both of which remove acid contents from the stomach), overdose of HCO_3^- (in the treatment of gastric ulcer), loss of K^+ along with Cl^- in severe exercise, renal disease, massive blood transfusions, and diuretic therapy.

Normally the kidneys are able to correct this situation by excreting large amounts of HCO_3^-. Therefore, persistence of metabolic alkalosis indicates impaired renal excretion.

Symptoms include slow respiration (hypoventilation) as the lungs try to conserve CO_2 (which comes from the H_2CO_3 needed to neutralize the excess alkalinity), numbness, convulsions, weakness, and muscle cramps.

The usual treatment involves the administration of NaCl, with KCl if the K^+ concentration is low. The increased Na^+ will enhance the excretion of HCO_3^-, thereby correcting the situation. Patients with metabolic alkalosis resulting from diuretic therapy can be given acetazolamide (Diamox) to increase HCO_3^- excretion.

Respiratory Acidosis

Respiratory acidosis is characterized by a decrease in plasma pH, an increase in pCO_2, and a variable compensatory increase in HCO_3^- concentration. Respiratory acidosis is caused by impaired ventilation, either from inhibition of the respiratory center in the brain (sedatives, anesthetics, opiates, cardiac arrest) or impaired gas exchange across the lungs to the blood capillaries (poliomyelitis, emphysema, severe asthma, acute pulmonary edema, pneumonia).

The response to respiratory acidosis takes time. Buffers are ineffective, because HCO_3^- cannot buffer H_2CO_3. The kidneys respond by increasing H^+ excretion, but this occurs over hours to days. Once renal response has developed, it is usually capable of handling excessive acidity. However, the kidneys will not be able to correct the problem. Chronic respiratory acidosis is a relatively common problem most often seen in smokers with pulmonary disease.

Case Study 30-2   Emphysema

A 65-year-old man has a history of shortness of breath that has become progressively worse over the past 5 years. His laboratory values are as follows[†]:

pO_2	44 mm Hg (6.00 kPa)
pCO_2	63 mm Hg (8.66 kPa)
pH	7.40
BUN	8.0 mmol/L
Total CO_2	42.5 mmol/L
Cl^-	88.0 mmol/L
K^+	3.6 mmol/L
Na^+	134.0 mmol/L

Questions

1. Explain the acid-base imbalance.
2. What is the probable disease that is apparent in this case?
3. If a urinalysis were done, what unusual findings might you expect?
4. Explain the conclusions from the pCO_2 value.

[†]*See back of book for normal laboratory values.*

Table 30-4	Typical Changes in Plasma pH, pCO_2, and HCO_3^- Concentration for Various Acid-Base Disorders		
	pH	$[HCO_3^-]$	pCO_2
Metabolic acidosis	↓	↓	↓
Metabolic alkalosis	↑	↑	↑
Respiratory acidosis	↓	↑	↑
Respiratory alkalosis	↑	↓	↓

Symptoms of respiratory acidosis include headache, blurred vision, fatigue, and weakness. Treatment of acute respiratory acidosis includes returning the breathing to normal. This may include mechanical assistance, use of bronchodilators, and small infusions of $NaHCO_3$. Chronic respiratory acidosis is usually not treated. Care must be taken with chronic patients since opiates and anesthetics will act as further respiratory depressants and induce further hypoventilation.

Respiratory Alkalosis

Respiratory alkalosis is characterized by an increase in plasma pH, a drop in pCO_2, and a variable decrease in HCO_3^- concentration. Such a condition may be caused by hyperventilation (hysteria), residence at high altitudes, congestive heart failure, pulmonary disease, aspirin overdose, excessive exercise, and cirrhosis. The problem originates with excessive excretion of CO_2. For instance, overstimulation of the respiratory control center in the brain by chemicals (aspirin) or the need for more O_2 (congestive heart failure, residence at high altitudes) leads to excessive loss of CO_2. The kidneys will attempt to compensate by increasing HCO_3^- excretion.

Symptoms of respiratory alkalosis are hyperventilation that cannot be controlled, possible convulsions, and lightheadedness. Treatment involves removal of the cause of the problem. Breathing into and out of a paper bag (to inhale CO_2 that has been exhaled) is also helpful.

Table 30-4 compares the changes in pH, pCO_2, and HCO_3^- concentration for respiratory and metabolic acidosis and alkalosis. The normal pH of the blood is 7.35 to 7.45; normal pCO_2 is 35 to 40 mm Hg; normal HCO_3^- is 25 to 30 mEq/L.

30-14 Fluid-Electrolyte Balance

Normally in humans water intake is balanced by water output. If the intake of water exceeds the output, **edema** results. If the output of water exceeds the input, **dehydration** may occur (Figure 30-9).

Water Balance

The body replenishes its water supply in three ways:

1. By the ingestion of liquids (which are primarily, if not wholly, water)
2. By the ingestion of foods such as meats, vegetables, and fruits, all of which contain a very high percentage of water
3. By metabolic processes taking place normally in the body. When carbohydrates, fats, and proteins are metabolized, water is produced. Approxi-

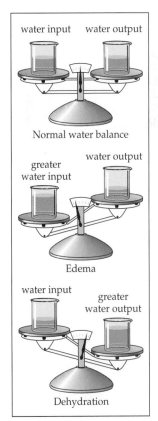

FIGURE 30-9

Water balance in the body.

Case Study 30-3   Cholera

A 50-year-old man was vacationing in southeast Asia, then trekked in the Himalayas in Nepal. During the trek he developed malaise, watery diarrhea, abdominal pain, vomiting, and anorexia. The symptoms continued for 5 days; he became progressively weak, and the watery diarrhea continued. The diarrhea and vomitus were copious and clear. Since he was in a remote village without medical facilities he decided to call his wife in the United States. She arranged for an emergency medical transport to Katmandu so that he could be treated. When he arrived at the medical facility in Katmandu, his laboratory values were as follows[†]:

Na^+	140 mmol/L
K^+	44 mmol/L
Cl^-	105 mmol/L
pCO_2	28 mm Hg (3.73 kPa)
pH	7.20
HCO_3^-	10 mmol/L
Glucose	12 mmol/L (180 mg/dl)
Total protein	100 g/L
Plasma specific gravity	1.039
Plasma osmolarity	325 mOsm/L
Hematocrit	46%

The patient improved with fluid replacement and oral administration of tetracycline. In 14 days he returned home to the United States. Toxigenic *Vibrio cholerae* had been isolated from his stool.

Questions

1. When the patient was admitted to the hospital in Katmandu, what was his acid-base status?
2. List the signs of dehydration.
3. Explain how cholera is produced by *V. cholerae*.
4. Explain why glucose should be present in fluid and electrolyte replacements.

[†]*See back of book for normal laboratory values.*

mately 14 mL of water is formed for every 100 kcal of energy released by the oxidation of foods. The oxidation of 100 g of carbohydrate produces 55 g of H_2O; the oxidation of 100 g of fat produces 107 g of H_2O; the oxidation of 100 g of protein produces 41 g of H_2O.

The total normal input of water in the body is approximately 2500 mL per day. The body loses water in several ways.

1. Through the kidneys, as urine
2. Through the skin, as perspiration
3. Through the lungs, as exhaled moisture
4. Through the feces

The total of these water losses should approximate that of the water intake, 2500 mL per day. However, the amounts of water lost by these methods may vary considerably. For example, the amount of moisture lost through the skin (by evaporation of perspiration) and through the lungs increases in the following circumstances.

1. During vigorous muscular exercise
2. With an increased respiratory rate
3. In a hot, dry environment
4. During a fever
5. When the skin receives severe burns

Conversely, the amount of water lost through the kidneys increases following the ingestion of large amounts of water within a short period of time. It also increases because of the presence of large amounts of waste products in the bloodstream. However, when the body loses excess water through vomiting and/or diarrhea, the output of water through the kidneys is immediately lessened.

Distribution of Water in the Body

The water of the body is considered to be distributed in two major areas: intracellular (within the cells), 55 percent, and extracellular (outside the cells), 45 percent. The extracellular water in turn can be further divided into four areas:

1. Intravascular (plasma): fluid within the heart and blood vessels (7.5 percent)
2. Interstitial and lymph: fluids outside the cells (20 percent)
3. Dense connective tissue, cartilage, and bone (15 percent)
4. Transcellular fluids: extracellular fluid collections, including the salivary glands, thyroid gland, gonads, mucous membranes of the respiratory and gastrointestinal tracts, kidneys, liver, pancreas, cerebrospinal fluid, and fluid in the spaces within the eyes (2.5 percent)

Approximately 55 percent of the weight of an adult male and 50 percent of the weight of an adult female is water.

The principal difference between the blood plasma (the intravascular fluid) and the interstitial fluid is in the protein content. Proteins cannot pass through membranes and so they remain in the blood vessels. Thus, the protein content of the interstitial fluid is very low compared to that of the intravascular fluid, although the soluble electrolytes in each are approximately the same.

There are distinct differences in salt concentration in the intracellular and extracellular fluids.

The extracellular water in the dense connective tissue, cartilage, and bone and the extracellular water in the transcellular fluids do not readily interchange fluids and the electrolytes with the rest of the body water. The rest of the body water—intracellular, intravascular, interstitial, and lymph—freely moves from one area to another within the body.

Electrolyte Balance

Table 30-5 indicates the electrolyte concentrations in the extracellular and intracellular fluids. The ions are divided into two groups—cations (positively charged ions) and anions (negatively charged ions). Concentrations of anions and cations are expressed in units of mmol/L, or using the former unit of milliequivalents per liter (mEq/L). The unit milliequivalent measures the chemical and physiologic activity of an ionized substance, the electrolyte. Since milliequivalents are based on ions, the term milliequivalents represents the number of charged particles or the

Table 30-5	Electrolyte Concentrations of Body Fluids			
	Extracellular		Intracellular	
	mmol/L	mEq/L	mmol/L	mEq/L
Cations				
Na^+	140	140	10	10
K^+	4	4	140	140
Mg^{2+}	1.5	3.0	30	60
Ca^{2+}	2.5	5.0	—	—
Anions				
Cl^-	100	100	4	4
HCO_3^-	27	27	10	10
PO_4^{3-}	2	6	60	180
Others		13		16

number of both positive and negative charges present in a solution of an electrolyte. Recall that the number of positive charges must always equal the number of negative charges.

Table 30-5 shows that potassium is the principal cation in the intracellular fluid and sodium is the principal cation in the extracellular fluid.

Osmotic activity depends on the number of particles present in a solution regardless of whether they carry a charge or not. Thus, sodium ions, potassium ions, and chloride ions cause osmotic pressure, but so does glucose, which is a nonelectrolyte; that is, it carries no charge.

Osmotic activity is expressed by the unit milliosmol, which is a measure of the amount of work that dissolved particles can do in drawing a fluid through a semipermeable membrane. Osmotic activity is measured by means of an instrument called an osmometer.

30-15 Clinical Importance of Mineral Cations and Anions

Minerals required for the body can be divided into two main groups: *macrominerals*, which are required in amounts greater than 100 mg per day, and *microminerals* (trace elements), which are needed in amounts less than 100 mg per day.

Macrominerals

Sodium Ions Sodium ions are the primary cations of the extracellular fluids. The principal functions of sodium ions are

1. To maintain the osmotic pressure of the extracellular fluid
2. To control water retention in tissue spaces
3. To help maintain blood pressure
4. To maintain the body's acid-base balance by means of the bicarbonate buffer system.
5. To regulate the irritability of the nerve and muscle tissue and of the heart

The average daily adult intake of sodium, as NaCl, is 5 to 15 g. About 95 percent of the sodium lost by the body passes through the kidneys. The body's sodium ion concentration is influenced by aldosterone, a hormone of the adrenal cortex. This hormone promotes the reabsorption of sodium ions in the kidney tubules. The antidiuretic hormone (ADH) promotes water absorption in the kidneys and so has a definite effect on extracellular sodium ion concentration.

Hyponatremia, a lower-than-normal serum sodium ion concentration, may be due to such causes as vomiting, diarrhea, hormone disorders, starvation, extensive skin burns, loss of sodium ions because of kidney damage, or use of diuretics. The clinical symptoms of hyponatremia are cold, clammy extremities, lowered blood pressure, weak and rapid pulse, oliguria, muscular weakness, and cyanosis (a dark purplish discoloration of the skin and mucous membranes due to decreased oxygenation of the blood). In addition, because of an increased plasticity of the tissues, hyponatremia frequently shows as fingerprinting over the sternum. In hyponatremia the density of the urine is less than 1.010 g/mL.

Hypernatremia, a higher-than-normal serum sodium ion concentration, may be due to such causes as deficient water intake, excessive sweating, excessive water output (such as caused by diabetes insipidus), poor kidney excretion, rapid administration of sodium salts, hyperactivity of the adrenal cortex (as in Cushing's disease), and some causes of cerebral disease. The clinical symptoms of hypernatremia are dry, itchy mucous membranes, intense thirst, oliguria or anuria, rough dry tongue, and elevation of temperature. The density of the urine rises above 1.030 g/mL. In an extreme case of hypernatremia the symptoms include tachycardia (rapidly beating heart), edema, and cerebral disturbances.

Potassium Ions   Potassium ions are the principal cations of the intracellular fluid. Since the kidneys do not conserve potassium ions as well as they preserve sodium ions and since the body cannot store potassium ions, depletion of this substance occurs readily in a patient whose diet is low in potassium or who is excreting more potassium than he or she takes in.

The principal functions of potassium ions in the body are to maintain

1. The osmotic pressure of the cells
2. The electric potential of the cells
3. The size of the cells
4. Proper contraction of the heart
5. Proper transmission of nerve impulses

Potassium ions move into the cells during anabolic activity and move out of the cells during catabolic activity. The concentration of potassium ion is usually measured in terms of serum potassium because this is a much easier quantity to measure than cellular potassium concentration.

Hypokalemia, a lower-than-normal serum potassium ion concentration, can occur under the following conditions:

1. Too low an intake of potassium ions:
   During starvation or malnutrition
   In a diet deficient in potassium
   During intravenous infusions of fluids low or lacking in potassium ions
2. Too great an output of potassium ions because of:
   The use of diuretics
   The use of corticosteroids (these hormones promote retention of sodium ions at the expense of potassium ions)

Prolonged vomiting

Gastric suction and intestinal drainage

Diarrhea

Polyuria

In addition, hypokalemia may be caused by a sudden shift of potassium ions from the extracellular fluid to the intracellular fluid. This could occur, for example, in the treatment of diabetic acidosis with insulin and glucose.

In general, hypokalemia occurs most frequently in conjunction with some other pathologic condition. The general symptoms of hypokalemia are a general feeling of being ill, lack of energy, muscular weakness, numbness of fingers and toes, apathy, dizziness on rising, and cramps, particularly in the calf muscles.

As hypokalemia develops to a greater extent, symptoms relating to the heart become evident. Among these are weak pulse, falling blood pressure, faint heart sounds, and changes in the ECG—first a flattening of the T wave, then inverted T waves with a sagging ST segment and AV block, and finally cardiac arrest.

Hypokalemia can be treated or prevented by giving the patient potassium intravenously (in the form of a potassium salt) or orally by the use of high-potassium foods such as veal, chicken, beef, pork, bananas, orange and pineapple juices, broccoli, and potatoes. Note that although there are many other foods high in potassium, they are usually also high in sodium. A patient on a high-potassium diet usually also has a low-sodium requirement; therefore, these other types of foods are not recommended.

Hyperkalemia, an increased serum potassium ion level, occurs

1. If the intake of potassium ions is too great because of too rapid an infusion of potassium ions or the administration of excess potassium ions intravenously

2. If the output of potassium ions is too low because of renal failure or acute dehydration

3. If there is a sudden shift of potassium ions from the intracellular fluid to the extracellular fluid because of severe burns, crush injuries (both of these could release potassium ions from the cells into the blood stream), or during acidosis

The symptoms of hyperkalemia are a general feeling of illness, muscular weakness, listlessness, mental confusion, slower heart beat, poor heart sounds, *bradycardia, and eventually, cardiac arrest. Characteristic changes in the ECG are elevated T waves, widening of the QRS complex, gradual lengthening of the PR interval, and final disappearance of the P wave.

Excess potassium ions can be removed either by dialysis or by the administration of glucose and insulin.

Calcium Ions Most of the body's calcium is found in the bones and the teeth in the form of calcium carbonate and calcium phosphate. If the blood calcium ion concentration falls, it can readily be replenished from the bone. Conversely, if the blood calcium ion concentration rises, the amount replenished from the bones decreases (see pages 582 and 584).

The daily intake (adult) for calcium varies from 200 to 1500 mg and comes primarily from milk and milk products. Ionized calcium is present in body fluids and is important in blood coagulation (see Section 30-11), regulation of membrane permeability, and normal functioning of nerve, heart, and muscle tissue.

Because of the great amount of calcium present in the bones, administration of supplemental calcium is not required during intravenous therapy. In addition, calcium-containing solutions are not suitable for infusions because, if mixed with citrated blood, they may cause a clot in the drip tube.

Calmodulin is a calcium-modulating protein found in all cells. It has a molecular mass of 16,700 and consists of a chain of 148 amino acids. Calmodulin affects the synthesis and action of the prostaglandins and enhances blood platelet aggregation.

Phenothiazines, which are major tranquilizers used to treat psychoses, and phenytoin, which is used in the treatment of leprosy, may inactivate calmodulin, indicating a link between these conditions and calcium metabolism.

Hypocalcemia, a low serum calcium concentration, may be due to a hypoactive parathyroid gland (see page 584), the surgical removal of the parathyroid glands, or a large infusion of citrated blood. The symptoms of hypocalcemia include tingling of the fingertips, abdominal and muscle cramps, and tetany.

Hypercalcemia, an increased serum calcium concentration, may be caused by an overactive parathyroid or by a tumor of that gland. It may also be caused by the administration of excess vitamin D. The symptoms of hypercalcemia include hypotonicity of muscles, kidney stones, deep bone pain, and bone cavitation.

The serum calcium ion concentration decreases during hypoparathyroidism and rises during hyperparathyroidism (see page 585).

Magnesium Ions   Magnesium ion, like potassium ion, is found primarily in the intracellular fluid. Magnesium ions are essential for the proper functioning of the neuromuscular system.

Magnesium acts as an activator for more enzymes than any other metal ion in the body; it is necessary for over 100 metabolic reactions. The normal intake of magnesium is 7.5 to 15 mmol (15 to 30 mEq) per day. Another unusual fact about magnesium is that it is the only positively charged ion that has a higher concentration in the cerebrospinal fluid than in the blood serum.

A deficiency of serum magnesium ions, hypomagnesemia, because of dietary intake is unusual because magnesium is a necessary element for chlorophyll, which is found in all green plant foods. A lower-than-normal magnesium ion concentration may be caused by such factors as

1. Chronic alcoholism
2. Diabetic acidosis
3. Prolonged intravenous infusion without magnesium ions
4. Hypoparathyroidism
5. Prolonged nasogastric suction
6. Acute pancreatitis
7. Severe malabsorption

The clinical symptoms of a deficiency of magnesium ions are

1. Muscular tremors
2. Convulsions
3. Delirium
4. Delusions
5. Disorientation
6. Hyperirritability
7. Elevated blood pressure

An excess of serum magnesium ions can be caused by severe dehydration or renal insufficiency. Excess magnesium ions act as a sedative. Extreme excesses may cause coma, respiratory paralysis, or cardiac arrest.

Chloride Ions   The chloride ion is the primary anion of the extracellular fluid. The body's intake of chloride ion is closely related to that of the sodium ion (see page 543).

One of the principal functions of the chloride ion is as a component of gastric hydrochloric acid. The chloride ion also serves an important function in the transportation of oxygen and carbon dioxide in the blood (see Section 30-12).

Hypochloremia, a lower-than-normal serum chloride ion concentration, can be brought about by prolonged vomiting, profuse sweating, and diarrhea. This condition causes an alkalosis because of an increased concentration of bicarbonate ions. Hypochloremia can also occur when there is a marked loss of potassium ions.

Phosphate Ions   The phosphate ion is the primary anion of the intracellular fluid. Diets that are adequate in calcium usually contain more than enough phosphorus for the body's needs.

Most of the body's phosphate is present in the bones as calcium phosphate, a substance that gives the bones their rigidity, but phosphate is found in every cell of the body. Phosphate ions are important in the acid-base balance of the body. They constitute one of the body's buffer systems. Phosphates are also of great importance in the production of ATP, the body's principal energy compound. Serum phosphate levels are low in hyperparathyroidism and high in hypoparathyroidism and celiac disease.

Microminerals

Iron Ions   Iron ions are involved almost exclusively in cellular respiration. Iron is part of hemoglobin, myoglobin, and cytochromes as well as several oxidative enzymes. The formation of hemoglobin requires the presence of traces of copper. The best dietary sources of iron are the "organ meats," such as liver, heart, and kidneys. Other sources are egg yolk, fish, beans, and spinach.

Most of the iron present in the food we eat is in the iron (III) ion (Fe^{3+}) form. In the digestive system, iron (III) ions are reduced to iron (II) ions (Fe^{2+}), which are then absorbed into the bloodstream from the stomach and duodenum. In the blood plasma, iron (II) ions are oxidized to iron (III) ions, which then become part of a specific protein—transferrin, a β-globulin. The conversion of iron (II) ions to iron (III) ions in the blood plasma is catalyzed by the enzyme ceruloplasmin, a copper-containing compound (see following paragraphs).

The liver, spleen, and bone marrow are able to extract the iron from transferrin and to store that iron in the form of two proteins: ferritin and hemosiderin. The bone marrow is also able to extract the iron from transferrin and to use that iron for the production of hemoglobin.

A deficiency of iron (iron-deficiency anemia) can result from a low intake of iron because of a diet high in cereal and low in meat, because of poor absorption of iron due to gastrointestinal disturbances or diarrhea, or because of excessive loss of blood. This type of anemia can be treated with a daily dose of ferrous sulfate in the diet, if absorption is normal.

Copper Ions   In addition to being required for the synthesis of hemoglobin, copper ions are necessary for certain oxidative enzymes, such as cytochrome oxidase (part of the oxidative phosphorylation sequence; see Section 26-10) and uricase (which catalyzes the oxidation of uric acid to allantoin). Copper ions act as a cofactor for the enzyme superoxide dismutase, which removes superoxide ions from the body. Elevated superoxide levels have been implicated in arthritis, aging, and radiation sickness.

Copper is found in the brain in the form of cerebrocuprein, in the blood cells as erythrocuprein, and in the blood plasma as ceruloplasmin, an α-globulin. In **Wilson's disease**, there is a decreased concentration of ceruloplasmin in the blood. This disease is characterized by the presence of large amounts of copper in the brain along with an excessive urinary output of copper.

The average daily diet contains about 2.5 to 5 mg of copper, an amount that is considered to be adequate for the normal adult. The richest sources of copper are liver, nuts, kidney, raisins, and dried legumes.

Menkes' syndrome is caused by an inherited abnormality in copper metabolism. It affects male infants and can be fatal. Menkes' syndrome is also known as "kinky hair" or "steely hair" syndrome.

Zinc Ions Zinc is essential for normal growth and reproduction. It has a beneficial effect on wound healing and tissue repair. Zinc complexes with insulin are present in the β cells of the pancreas. Zinc is an essential component of several enzymes, such as alcohol dehydrogenase, alkaline phosphatase, carbonic anhydrase, and retinene reductase (found in the retina). Acetazolamide (Diamox) inhibits carbonic anhydrase activity by binding to the zinc atom present in the enzyme molecule. The absorption of zinc by the intestines involves pyridoxine, a B vitamin.

A deficiency of zinc can stunt growth and cause impaired wound healing, decreased sense of smell and taste, and hypogonadism.

An excess of zinc can cause gastrointestinal disturbances and vomiting.

Food sources of zinc include seafood (particularly oysters and clams), meat, liver, eggs, milk, and whole grain cereals. The recommended daily allowance for zinc is 15 mg per day for adults with an additional 15 mg per day during pregnancy and 10 mg per day during lactation. For children, the recommended allowance is 6 to 10 mg per day.

Other Ions Cobalt is a constituent of vitamin B_{12} that is necessary for the formation of red blood cells. Cobalt is present in almost all foods, so cobalt deficiency is quite rare.

Manganese is essential for normal bone structure, reproduction, and normal functioning of the central nervous system. Manganese is a constituent of several mitochondrial enzyme systems. A deficiency of manganese is not known in humans, indicating that the average daily diet supplies a sufficient amount of this substance. Good sources of manganese are nuts, whole-grain cereals, vegetables, and fruits. Poor sources are meats, poultry, and fish.

Chromium is involved in the metabolism of glucose and in the proper activity of insulin. A deficiency of chromium leads to impaired glucose tolerance.

Iodine is necessary for the production of thyroid hormones (see Section 32-7). A deficiency in children causes cretinism; in adults it causes goiter and myxedema. An excess of iodine produces psychotic symptoms and parkinsonism.

Molybdenum is involved in the metabolism of nucleic acids.

Lithium salts are being used to treat manic-depressive psychoses and as antidepressants for some psychiatric patients. At one time, lithium chloride was employed as a sodium substitute, with dangerous and frequently fatal results. Patients on a therapeutic dosage of lithium salts may complain of fatigue, muscular weakness, nausea, and diarrhea. Slurred speech and hand tremors are noticeable. In larger doses, the central nervous system is affected and the patient may become unconscious or even go into a coma. Abnormalities in the *electroencephalogram are also common.

Table 30-6	Recommended Daily Intake of Minerals												
Mineral (unit)	Na (g)	K (g)	Cl (g)	Ca (g)	P (g)	Mg (mg)	Fe (mg)	Zn (mg)	Mn (mg)	Cu (mg)	I (µg)	Mo (µg)	Se (µg)
Female	1.1-3.3	2-6	1.7-5.1	0.8-1.2	0.8-1.2	280	15	12	2-5	1-3.0	150	75-250	55
Male	1.1-3.3	2-6	1.7-5.1	0.8-1.2	0.8-1.2	350	10	15	2-5	1-3.0	150	75-250	70

Table 30-6 indicates the recommended daily intake of many of the aforementioned minerals.

Summary

Blood carries oxygen, minerals, and food to the cells and carries carbon dioxide and other waste products from the cells. Blood also carries hormones, enzymes, antibodies, and blood cells. Blood regulates body temperature and maintains the pH of the body fluids.

Blood consists of two parts: the suspended particles, red blood cells, white blood cells, and platelets; and the suspending liquid, the plasma. When freshly drawn blood is allowed to clot and settle, the yellow liquid remaining is called blood serum.

Oxygenated blood has a bright red color, whereas deoxygenated blood has a dark purplish color. Blood has a pH range of 7.35 to 7.45. If the pH falls below 7.35, the condition is called acidosis. If the pH rises above 7.45, the condition is called alkalosis. The volume of blood in the body is 5 to 7 L.

The chemical analysis of blood samples is of great clinical significance. Increased or decreased amounts of some substances may indicate certain diseases.

Blood contains hemoglobin, a conjugated protein containing iron. Hemoglobin is composed of heme and globin. Cytochrome c, part of the oxidative phosphorylation sequence in the Krebs cycle, has a structure similar to that of hemoglobin. Chlorophyll a has a structure similar to heme but with a magnesium ion at the center instead of the iron.

If the hemoglobin content of the blood falls below normal, the condition is called anemia. Anemia may result from a decreased rate of production of red blood cells, an increased destruction of red blood cells, or an increased loss of red blood cells.

Plasma proteins are divided into three groups: albumin, globulins, and fibrinogen. Albumin regulates the osmotic pressure of the blood and controls the water balance of the body. α-Globulins and β-globulins form loose combinations with carbohydrates, metal ions, and lipids so that these substances can be transported to all parts of the body. γ-Globulins contain the antibodies with which the body fights infectious diseases. Fibrinogen is the plasma protein involved in the clotting of the blood.

When a blood vessel is cut, blood comes into contact with the exposed tissues. This activates both the extrinsic and intrinsic coagulation systems, producing a clot.

In respiration, the hemoglobin picks up oxygen to form oxyhemoglobin in the lungs. The oxyhemoglobin is carried to the tissues, where it gives up its oxygen. At the same time the blood picks up carbon dioxide and other waste products from the cells. Bicarbonate ions are involved in the hemoglobin-oxyhemoglobin cycle and also in the carbon dioxide removal cycle. Excess bicarbonate ion in the blood cells is shifted to the plasma in a process called the chloride shift, whereby chloride ions take the place of the bicarbonate ions. In the lungs, as carbon dioxide is removed from the blood, a reverse chloride shift takes place.

The blood maintains a constant pH by the use of buffers, which react with acidic (or basic) substances entering the blood and form new substances that are either neutral compounds or more of buffer salts.

Buffer systems in the blood are bicarbonate buffers, phosphate buffers, hemoglobin buffers, and protein.

Changes in CO_2 level in the blood and corresponding changes in HCO_3^- and pH may be caused by metabolic acidosis, metabolic alkalosis, respiratory acidosis, or respiratory alkalosis. Each condition involves different changes in these three factors.

Water intake must be balanced by water output. If water intake is greater than water output, edema occurs. If water output is greater than water intake, dehydration may occur.

Water in the body is considered to be distributed in two major areas: intracellular and extracellular. Extracellular water is further subdivided into interstitial water; intravascular and lymph water; water in dense connective tissues, cartilage, and bone; and water in transcellular fluids.

Each of the body's water compartments has its own concentration of electrolytes, all concentrations being expressed as mmol/L or milliequivalents per liter. An increase or decrease in the concentration of any one of the ions will have some effect on the body.

Serum osmotic pressure is affected primarily by the concentrations of sodium ions, bicarbonate ions, and chloride ions. Osmotic activity is measured in the unit milliosmol.

Questions and Problems

1. What is the difference between blood plasma and blood serum?

2. What are the functions of blood?

3. What is the normal concentration of red blood cells in the body? white blood cells? What is the function of each?

4. What is polycythemia? anemia? What might cause them?

5. Name several types of leukocytes.

6. What are the functions of the inorganic ions in the plasma?

7. What is the normal pH range of blood? the viscosity? the density?

8. Under what conditions might blood volume increase? decrease?

9. What type of compound is hemoglobin? What might cause an increased loss of hemoglobin?

10. How does the structure of heme compare with that of chlorophyll a? with cytochrome c?

11. What is pernicious anemia? What causes it?

12. What conditions might cause a decreased production of erythrocytes?

13. What are the three groups of plasma proteins? What is the function of each?

14. Describe the effects of albumin on osmotic pressure.

15. What is edema? What could cause it?

16. What are the different types of globulins? What are their functions? What is agammaglobulinemia?

17. Describe the clotting mechanism of the blood.

18. Why should oxalates never be used to prevent blood from clotting when that blood is to be used for transfusion?

19. What is the effect of vitamin K deficiency on blood clotting?

20. What is a thrombus? What might result from the presence of a thrombus in the brain? in a major blood vessel?

21. Describe the process whereby blood carries oxygen to the cells; carbon dioxide from the cells.

22. What is the function of carbonic anhydrase? What metallic element does it contain?

23. What is the alkaline reserve of the blood? What might affect it?

24. What is the chloride shift? the reverse chloride shift? Where do they occur?

25. What types of buffers are present in the blood? Where are the sodium buffers located? the potassium buffers?

26. Describe the reaction (in equation form) of a bicarbonate buffer with an acid and with a base.

27. How does the body replenish its supply of water? How does it lose its water?

28. Into what areas can the body's water be considered as being subdivided?

29. What may cause hyponatremia? hypernatremia? What are the symptoms of each?

30. What is hypokalemia? hyperkalemia? What are the symptoms of each?

31. What are the functions of potassium in the body?

32. Under what conditions may serum calcium level change?

33. What are the sources of magnesium for the body? What are the symptoms of a magnesium deficiency? What might cause such a deficiency?

34. What are the functions of phosphate ions in the body? iron ions? copper ions? calcium ions? chloride ions?

35. A lab reports the magnesium ion concentration as 3.0 mmol/L. Express this as milliequivalents per liter.

36. If the colloid osmotic pressure of the plasma drops to 20 mm Hg, all other pressures remaining constant, what will be the effect on fluid flow? What is such a condition called?

37. What clinical tests can distinguish between metabolic acidosis and respiratory acidosis? between metabolic alkalosis and respiratory alkalosis?

38. Name two causes for metabolic acidosis. How does the body compensate for this condition?

39. Why is NaCl used to treat metabolic alkalosis?

40. What causes kwashiorkor? What are its symptoms?

41. Determine whether serum albumin would be elevated or lowered in the following disorders: dehydration, malnutrition, nephrosis, and chronic hepatic insufficiency. Explain your answers.

42. Is the concentration of chloride ions higher in arterial or venous blood? Explain.

43. What is the source for sulfate ions in the blood?

44. The production of ammonia is impaired in a malnourished person. Discuss what effects this would have on the renal regulation of acid-base balance.

Questions 45 to 47 pertain to the following scenario:

A 28-year-old homeless man is rushed to the county hospital medical center. He is comatose and in respiratory depression. The emergency department nurse recognizes this patient as having a previous history of drug use including heroin and barbiturates. The arterial blood gases show a pH of 7.21; total CO_2 of 52 mm Hg; and a HCO_3^- of 28 mEq/L.

45. What is this patient's acid-base status?

46. What is the cause of the acid-base imbalance?

47. Could you expect improvement if ventilatory therapy were initiated?

Questions 48 to 50 pertain to the following scenario:

A 73-year-old man with a history of idiopathic pulmonary fibrosis is admitted to the hospital emergency department with a fever of 38.9° C and severe shortness of breath. His respiratory rate is 60/min. X-ray evaluation shows that pneumonia has developed in his left lower lobe. The patient has been on a regimen that includes corticosteroids, digoxin, diuretics, and low-sodium diet. The result of his blood chemistry studies are shown below. Because of his continued severe respiratory distress, the patient finally agrees to ventilatory therapy; the resulting changes in pH, pCO_2, and HCO_3^- are also shown in tabular form below.

These values lead his physician to change the ventilator settings so that there will be more retention of CO_2 by the patient, who is also being given small doses of morphine and diazepam (Valium). In 6 hours the blood values have changed. With further ventilator adjustments over the next 3 days, the patient's condition improves.

Parameter	On Admission	On Ventilatory Therapy	After 6 Hours	Final
pH	7.43	7.55	7.31	7.42
pCO_2 (mm Hg)	63 (8.53)	40 (5.33)	78 (10.5)	50 (6.66)
HCO_3^- (mEq/L)	41	35	40	30
Creatinine (mg/dL)	0.5			
BUN (mg/dL)	4			
Total CO_2 (mmol/L)	43			
Cl^- (mEq/L)	90			
Na^+ (mEq/L)	145			
K^+ (mEq/L)	3.5			

48. To what would you attribute the patient's laboratory values at admission: creatinine, BUN, Cl^-, Na^+, K^+?

49. What would you have expected if the patient had not undergone ventilatory therapy?

50. Explain the change in the acid-base balance of this patient.

Questions 51 to 53 pertain to the following scenario:

An 18-year-old man with a 10-year history of juvenile-onset diabetes is admitted to the hospital. He has required 90 units of insulin daily to maintain his blood glucose level. On admission his blood pressure is 90/20 mm Hg; deep respirations, 35/min; pulse, 120/min. Laboratory test results are as follows: serum glucose, 65 mmol/L (1200 mg/dl); hematocrit, 45 percent; sodium, 132 mmol/L; potassium, 6.5 mmol/L; BUN, 26 mmol/L (74 mg/dl); pH, 6.75; and pCO_2, 11 mm hg (1.33 kPa). The patient's plasma is positive for ketones.

Ten units of regular insulin are given intravenously and 10 U/hr are given by intravenous infusion pump. His serum glucose concentration has been falling at an approximate rate of 5.6 mmol/L/hr. In 8 hours his respirations and blood pH are normal following an intravenous injection of $NaHCO_3$, with fluid and electrolyte replacement.

51. What is the mechanism of this acid-base imbalance?

52. Explain the reason for the sodium bicarbonate injection and the rationale for fluid replacement.

53. Which electrolytes would you replace intravenously?

Practice Test

1. The chloride shift occurs from the _____.
   a. blood cells to the tissues
   b. tissues to the blood cells
   c. plasma to the blood cells
   d. blood cells to the plasma

2. Albumin is most important for maintaining _____ in the blood.
   a. carbon dioxide transport
   b. acid-base balance
   c. osmotic pressure
   d. oxygen transport

3. The ion involved in the contraction of the heart is _____.
   a. Fe^{2+} b. I^- c. K^+ d. Na^+

4. A patient on an anticoagulant drug should avoid taking vitamin _____.
   a. A   b. B   c. E   d. K

5. The ion necessary for normal clotting of the blood is _____.
   a. Fe^{2+} b. Ca^{2+} c. K^+ d. Na^+

6. The main buffer system in the blood uses _____.
   a. chloride   b. phosphate
   c. bicarbonate   d. protein

7. Both pH and HCO_3^- increase during _____.
   a. respiratory alkalosis
   b. respiratory acidosis
   c. metabolic acidosis
   d. metabolic alkalosis

8. The globulins that include the antibodies are the _____.
   a. alpha   b. beta   c. gamma   d. delta

9. The acid-base imbalance characterized by a decrease in pH, pCO_2, and HCO_3^- is _____.
   a. metabolic alkalosis
   b. metabolic acidosis
   c. respiratory alkalosis
   d. respiratory acidosis

10. The plasma protein involved in blood clotting is _____.
    a. lipase   b. fibrinogen
    c. heparin   d. albumin

Vitamins

- To become familiar with the roles of the fat-soluble and the water-soluble vitamins

- To become familiar with the sources, properties, structures, daily requirements, need for, and effects of a deficiency of the various vitamins

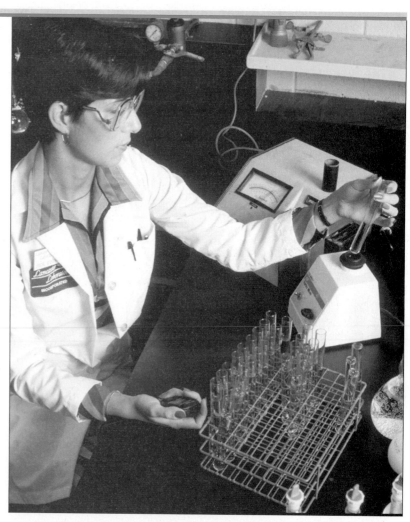

A biochemist analyzes the niacin content of various foods. Vitamins are generally not produced by the human body, yet they play an essential role as enzyme cofactors. Thus it is important to eat a balanced diet to receive adequate amounts of the essential vitamins.

Animals fed on a diet consisting only of purified carbohydrates, fats, proteins, minerals, and water will lose weight and develop certain deficiency diseases. Something else must be administered to sustain normal life. The additional substances required are called vitamins.

The name *vitamin* was originally *vitamine* because the first one found was an amine, hence the name *vital amine*, or *vitamine*. Subsequent studies of other such substances showed that they were not all amines, so the "e" was dropped.

Vitamins are similar to hormones in many ways. Vitamins and hormones are carried by the bloodstream to the various parts of the body where they are needed. Vitamins and hormones are required by the body only in extremely small amounts. Neither vitamins nor hormones furnish energy by themselves, although vitamins function with certain enzymes to control energy changes in the body. One important difference between vitamins and hormones is that most vitamins must be supplied in the diet, whereas hormones are synthesized by the body.

Vitamins are divided into two major groups—fat-soluble vitamins and water-soluble vitamins. The fat-soluble vitamins include vitamins A, D, E, and K and are usually found associated with lipids in natural foods. They are transported in the bloodstream by the lipoproteins. Fat-soluble vitamins are not excreted in the urine but do appear in the feces.

The water-soluble vitamins include vitamin C and the B-complex vitamins. Although the vitamins have a letter designation and sometimes a subscript in addition, such as vitamin B_1, the chemical names are becoming more widely used. For example, the chemical name for vitamin B_1 is thiamine.

More and more people eat a balanced diet, take vitamins, or are under medical care. This has helped to increase life expectancy.

31-1 Fat-Soluble Vitamins

Vitamin A

Source Vitamin A is found in fish liver oils, butter, milk, and to a small extent in kidneys, fat, and muscle meats. The precursor of vitamin A (the substance from which vitamin A can be made) is called provitamin A and is found in yellow fruits and vegetables such as peaches, apricots, sweet potatoes, carrots, and tomatoes and in leafy green vegetables.

Structure Vitamin A is a high molecular mass alcohol known as retinol. Vitamin A has an all-trans structure (see page 270).

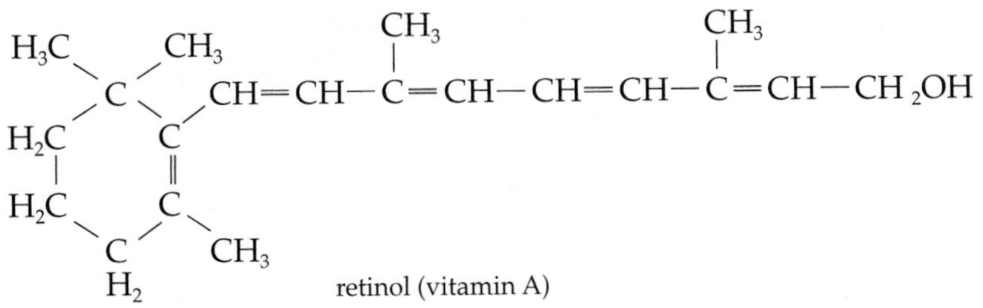

retinol (vitamin A)

Provitamin A is a compound that can be converted into vitamin A. One such provitamin A is β-carotene. The conversion of β-carotene into retinal (vitamin A aldehyde) and then to retinol (vitamin A) is shown in Figure 31-1.

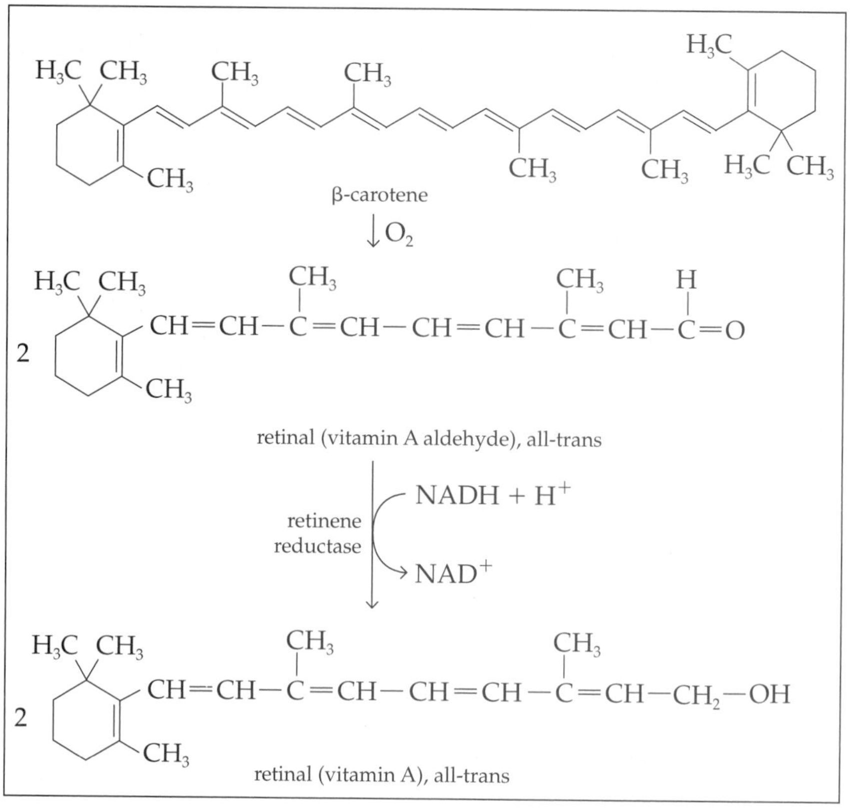

FIGURE 31-1
Conversion of β-carotene to vitamin A.

It should be noted that one molecule of β-carotene produces two molecules of vitamin A. Because β-carotene must be metabolized to produce vitamin A, β-carotene is only one sixth as effective as oral vitamin A itself. Provitamin A is transformed into vitamin A in the intestinal walls of some animals, including rats and pigs, but in the liver in humans.

Properties Vitamin A is soluble in fats but not in water. It is stable to heat, acid, and alkali but is destroyed by oxidation. (Recall that double bonds are usually quite susceptible to oxidation.) Ordinary cooking does not destroy vitamin A. The vitamin A present in butter is destroyed when the butter turns rancid (becomes oxidized).

Daily Requirement The recommended daily dosages of vitamin A for the normal adult male and female are 1000 and 800 retinol equivalents (5000 and 4000 IU), respectively. One retinol equivalent is 1 μg of retinol or 6 μg β-carotene. One international unit (IU) of vitamin A is equivalent to 0.3 μg of retinol or 0.6 μg of β-carotene. The daily requirements of vitamin A are increased to 1000 retinol equivalents (5000 IU) during pregnancy and to 1300 retinol equivalents (6000 IU) during lactation. A child requires 400 to 700 retinol equivalents (2000 to 3300 IU) daily.

Vitamin A is necessary for normal growth and development, reproduction, mucus secretion, and lactation. It is necessary for the synthesis of the membranes around the lysosomes and the mitochondria and acts to regulate membrane permeability. Vitamin A plays an important role in the functioning of the retina.

Vitamin A plays a role in maintaining the integrity of the epithelial tissues, particularly because of its antioxidant properties. Some evidence has shown that there is a reverse relationship between the vitamin A content of the diet and the risk of some types of cancer that occur in the epithelial tissues.

Retinol is bound to cellular retinol binding protein and is transported to other parts of the cell where it can then bind with nuclear proteins. It is believed that this latter binding is involved in the control of the expression of certain genes. The requirement of vitamin A for normal reproduction may be due in part to this function.

Note that some of the retinol in Figure 31-1 is converted into retinoic acid. Retinoic acid acts in glycoprotein synthesis and thereby promotes the growth and differentiation of tissues.

Effect of Deficiency Vitamin A was discovered by the observation that certain animals did not grow on a diet low in some animal fats. However, this effect on growth is characterized by a lack of other vitamins as well. A lack of vitamin A causes a shrinking and hardening of the epithelial tissues of the membranes in the eyes, digestive tract, respiratory tract, and genitourinary tract. Such a hardening is called **keratinization.**

When keratinization occurs in the lining of the respiratory tract, the patient is more likely to suffer from colds, pneumonia, and other respiratory infections because of the drying of the membranes.

When keratinization occurs in the eyes, the tear ducts become keratinized and are no longer able to secrete tears to wash the eyes. When this occurs, bacteria are able to attack the corneal tissues of the eyes, producing an infection called **xerophthalmia.** In this disease, the cornea becomes cloudy and does not allow light to pass through, so sight is lost permanently.

An early symptom of the lack of vitamin A is **nyctalopia,** or night blindness. A person with night blindness cannot see very well in dim light because of lack of visual purple (**rhodopsin**) in the retina of the eyes. This pigment is acted on and changed by light and is then regenerated in the presence of retinal, an oxidized form of vitamin A. If there is a lack or deficiency of vitamin A, the rhodopsin is regenerated very slowly; thus, there is an impairment in vision at night.

The role of vitamin A in the visual process and the part it plays in the regeneration of rhodopsin are illustrated in the following simplified cycle.

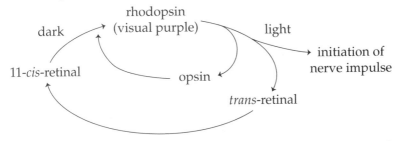

The role of calcium ions in the visual cycle has been elucidated. In the dark, calcium and sodium ions carry a current across the membranes of the retinal rod cells. Ion channels in these membranes are held open by cyclic guanosine monophosphate (cGMP). When a photon of light strikes the eye, cGMP is destroyed, the ion channels close, and the current stops.

Calcium ions control how rapidly the cGMP is replaced by binding to the enzyme that synthesizes cGMP. Therefore, calcium ions control how rapidly the visual system recovers after exposure to light.

A lack of vitamin A can also produce keratinization of the epithelial cells of the genital system and might cause sterility. A lack of vitamin A causes deformities in the teeth of young animals.

All of these symptoms can be cured by the administration of vitamin A.

Effect of Excess Hypervitaminosis A, an intake of vitamin A far in excess of normal daily requirements, has such early symptoms as irritability, loss of appetite, fatigue, and itching. These symptoms usually disappear within a week after the withdrawal of the vitamin.

Acute cases of hypervitaminosis A have been observed following ingestion of polar bear liver, which contains up to 35,000 IU of vitamin A per gram. Symptoms are drowsiness, sluggishness, vomiting, severe headache, and generalized peeling of the skin after 24 hours. Membranes have increased permeability and decreased stability, leading to mitochondrial swelling, lysosomal rupture, and eventually death.

Eating a great many carrots daily may result in carotenosis, a benign condition characterized by a yellowing of the skin.

Storage and Absorption Vitamin A is stored in the liver until it is needed. Mother's milk contains 10 to 100 times as much vitamin A as does ordinary milk. Zinc is necessary for vitamin A to be mobilized from the liver, indicating that vitamin A metabolism may be adversely affected by a zinc deficiency.

Mineral oil should never be taken immediately before or after a meal because it coats the mucosa and interferes with the absorption of carotenes.

Since vitamin A is fat soluble, it cannot be absorbed from the intestinal tract without the presence of bile. Any interference with the flow of bile to the intestines will also cause a deficiency of this fat-soluble vitamin.

Vitamin D

Source Vitamin D is sometimes called "the sunshine vitamin." The richest sources of vitamin D are oils from such fish as cod or halibut and the flesh of such oily fish as sardines, salmon, and mackerel. Milk is not a very good source of vitamin D, although its vitamin D content can be increased by irradiation with ultraviolet light. The vitamin D content of the body can be increased by exposure of the skin to ultraviolet rays from the sun. Care must be taken to avoid overexposure and consequent sunburn.

Structure The D vitamins are a group of sterols with considerable differences in their potency. The two most important vitamins in the D group are vitamin D_2 (ergocalciferol, or activated ergosterol) and vitamin D_3 (activated 7-dehydrocholesterol or cholecalciferol). Vitamin D_1, originally called vitamin D, was found to be a mixture of vitamins D_2 and D_3.

ergocalciferol (vitamin D_2)

Vitamin D_3 is similar to vitamin D_2, the difference being in the structure of the side chain. Vitamin D_3 is made by the irradiation of 7-dehydrocholesterol (equation 31-1), so it is frequently called "activated 7-dehydrocholesterol."

Ergosterol, the precursor of vitamin D_2, occurs in plants, whereas 7-dehydrocholesterol, the precursor of vitamin D_3, occurs in the skin of humans and animals.

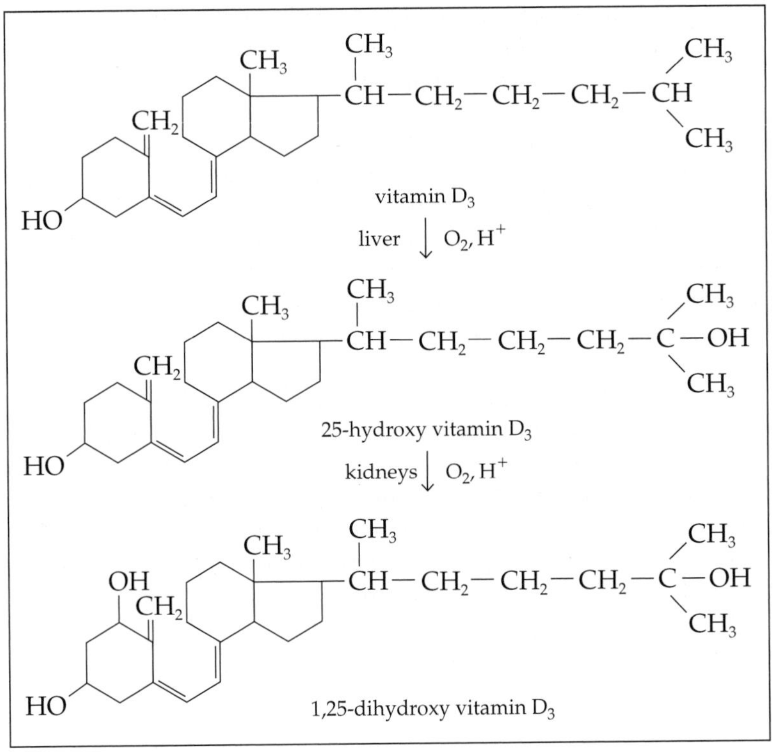

FIGURE 31-2

Vitamin D₃, one of the D vitamins.

7-Dehydrocholesterol in the skin is isomerized to vitamin D_3 under the influence of ultraviolet light.

Vitamin D must acquire two hydroxyl groups to be biologically active. The first hydroxyl group is added in the liver and the second in the kidneys, as indicated in Figure 31-2 for vitamin D_3.

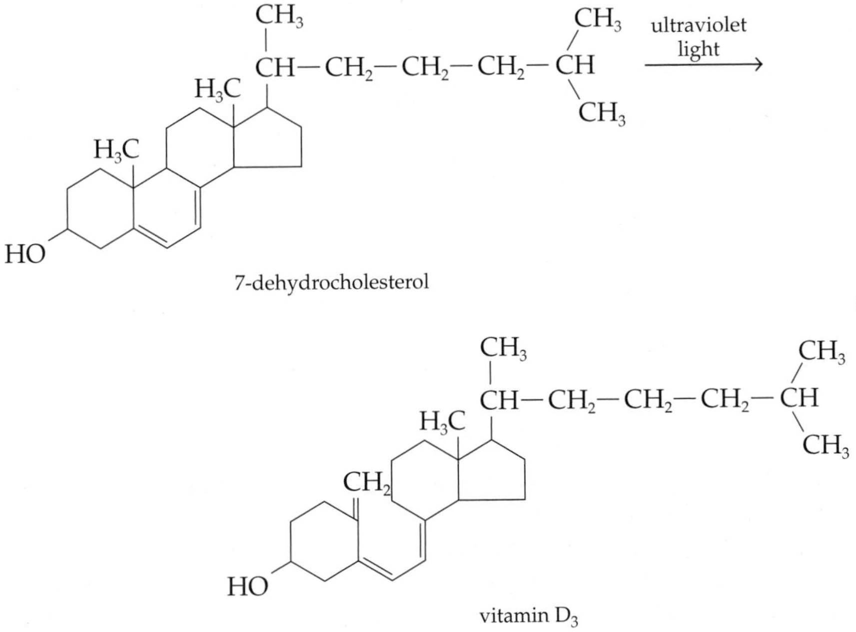

(31-1)

The same reactions occur with vitamin D_2. Patients who cannot convert vitamin D to its active form, because of liver or kidney disease, often develop *osteomalacia, osteoporosis, or other bone diseases. Such patients can now be treated with synthetic active-form vitamin D.

Properties   The D vitamins are soluble in fats and insoluble in water. They are stable to heat, resistant to oxidation, and unaffected by cooking. Vitamin D_2 has a greater potency in humans than vitamin D_3, whereas the reverse is true in chickens.

Daily Requirement   The daily requirement for children is 400 IU. Daily requirements for adults are 200 to 400 IU. One international unit of vitamin D is defined as the biologic activity of 0.025 µg of cholecalciferol.

Physiologic Action   The principal action of vitamin D is to increase the absorption of calcium and phosphorus from the small intestine. It also increases the release of calcium and phosphate in the bones and is necessary for normal growth and development. Vitamin D is required for the proper activity of the parathyroid hormone (see Section 32-8) and so is used therapeutically in the treatment of hypoparathyroidism. A lack of vitamin D may cause hypocalcemia and hypophosphatemia.

Effect of Deficiency   Rickets, a disease primarily of infancy and childhood, was previously thought to be due to a deficiency of vitamin D. This disease is now believed to be due to a lack of sunshine on the skin. Why? Because sunshine is necessary for the synthesis of calciferol in the skin.

Rickets is characterized by an inability to deposit calcium phosphate in the bones. The bones become soft and pliable; they bend and become deformed. The joints become enlarged, and the ribs become beaded. The knobby or beaded appearance of the ribs is called rachitic rosary. A child who has rickets does not grow. He or she develops such symptoms as nervousness, irritability, a bulging abdomen, loss of weight, loss of appetite, anemia, and a delayed development of the teeth. Injection of small amounts of calciferol, or one of its derivatives, or adequate exposure to sunlight can prevent or cure rickets (see Figure 31-3).

A diet low in phosphorus and vitamin D may produce *osteomalacia, or adult rickets. Since adult rickets is a rare condition, this suggests that adults need less vitamin D than children do. When an adult develops rickets, there is no bulging of the joints since the growth of the bones in an adult is already complete. There is, however, some softening of the bones, with accompanying deformities. This disease occurs most often in women after repeated pregnancies during which there was a deficiency of vitamin D.

A lack of calcium and vitamin D in the diet can cause *osteoporosis in the adult. This disease, like osteomalacia, is characterized by decalcification and softening of the bones but to much greater extent.

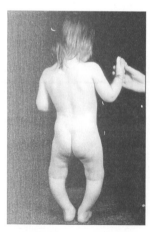

FIGURE 31-3
A child with vitamin D deficiency (rickets).

Effect of Excess   Hypervitaminosis D may show such early symptoms as weakness, lassitude, fatigue, nausea, vomiting, and diarrhea, all of which are associated with hypercalcemia. Later symptoms include calcification of soft tissues, including the kidneys and the lungs. Treatment of hypervitaminosis D consists of the immediate withdrawal of the vitamin, increased fluid intake, a diet low in calcium, and the administration of glucocorticoids.

Vitamin E

Source Vitamin E is found in milk, eggs, fish, muscle meats, cereals, leafy vegetables such as lettuce, spinach, and parsley, and plant oils such as cottonseed oil, corn oil, and peanut oil. Wheat germ oil is particularly rich in vitamin E.

Structure There are several vitamins E. The most important of these is called α-tocopherol; β- and δ-tocopherols are less active, and γ-tocopherol is inactive

As indicated by the ending of the name, *-ol*, vitamin E is an alcohol. The structure of α-tocopherol is

α-tocopherol

Properties α-Tocopherol, the most important and most active of the E vitamins, is a colorless to pale yellow oil. It is soluble in fats and fat solvents but insoluble in water. Vitamin E is stable to heat but is destroyed by ultraviolet light and by oxidizing agents.

The activity of the tocopherols appears to result from their antioxidant properties. They are very effective in preventing the oxidation of vitamin A and unsaturated fatty acids. There is also evidence that vitamin E functions as a cofactor in oxidative phosphorylation reactions. Vitamin E is believed to protect the lung tissues from damage by oxidants present in polluted air.

Vitamin E also plays an important role in selenium metabolism. Selenium is required for pancreatic function. This includes absorption of lipids along with vitamin E. Selenium is present in the enzyme glutathione peroxidase, which helps destroy peroxides and reduces the requirement of vitamin E for maintenance of membranes. Selenium also aids in the retention of vitamin E in plasma lipoprotein.

Conversely, vitamin E reduces selenium requirements in the body by preventing the loss of that element.

The daily requirement for selenium is 70 μg for a male and 55 μg for a female.

Premature infants suffering from hemolytic anemia show very low levels of α-tocopherol in their blood. When given vitamin E supplements, people suffering from hemolytic anemia improve greatly. It is now believed that vitamin E is essential in infant metabolism. The U.S. Food and Drug Administration requires that commercial milk substitutes sold as infant foods contain adequate amounts of vitamin E. Recent evidence has shown that vitamin E is helpful in the diets of pregnant and lactating women. Vitamin E also is useful in treating older persons suffering from circulatory problems.

Vitamin E is required in higher animals such as cattle and poultry for fertility. The name *tocopherol* is derived from the Greek *tokos* meaning "childbirth" and *pherein* meaning "to bear."

Daily Requirement The international unit of vitamin E is defined as the activity of 1 mg of *dl*-α-tocopherol acetate. The recommended daily requirement of vitamin E is 10 IU for adult men and 8 IU for adult women.

Effect of Deficiency Some animals on a vitamin E-deficient diet develop muscular dystrophy, resulting in paralysis. Administration of vitamin E helps these animals overcome such effects. Vitamin E has not been found effective in treating muscular dystrophy in humans.

The daily requirement for vitamin E increases with a greater intake of polyunsaturated fat. In addition, exposure to oxygen, as in an oxygen tent, or intake of mineral oils can lead to a deficiency of vitamin E and eventually to neurologic disorders.

Severely impaired absorption of fat in the intestines can cause vitamin E deficiency in humans. The symptoms of such a disease are muscular weakness, creatinuria, and fragile red blood cells.

There appears to be a relationship between vitamins A and E. Vitamin E may help in the absorption, storage, and utilization of vitamin A. Vitamin E also appears to offer protection against large overdoses of vitamin A. No definite side effects of large doses of vitamin E are known.

Vitamin K

Source Vitamin K is found in the green leafy tissues of such plants as spinach, cabbage, and alfalfa. Vitamin K is also found in putrefied fish meal, liver eggs, and cheese. Fruits contain very little vitamin K.

Structure   There are two naturally occurring K vitamins, K_1 and K_2. Vitamin K_1 is a yellow oil whose structure is

$$\text{vitamin } K_1$$

Vitamin K_2 differs from K_1 only in the nature of the side chain. Vitamin K_1 is produced in plants and K_2 by intestinal bacteria. A synthetic compound, menadione, exhibits vitamin K activity. It is converted in the body to vitamin K.

Compare the structure of the K vitamins with that of coenzyme Q (page 426), which functions in oxidative phosphorylation.

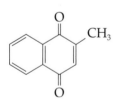

menadione

Properties   The K vitamins are soluble in fats and insoluble in water. They are stable to heat but are destroyed in acidic and alkaline solutions. They are also unstable to light and oxidizing agents.

Daily Requirement   The daily requirement of vitamin K is 70 to 80 µg for a man and 55 to 65 µg for a woman. The average diet supplies sufficient vitamin K; in addition, intestinal bacterial are able to synthesize this vitamin for their host.

Effect of Deficiency   A deficiency of vitamin K will occur only as a result of prolonged use of a broad-spectrum antibiotic coupled with a diet lacking or low in vitamin K. Vitamin K deficiency does occur in breast-fed infants because they do not have intestinal bacteria and because breast milk is a poor source of vitamin K. Vitamin K is usually given as a supplement during the first 5 months after birth. Some infants are born with a deficiency of prothrombin and are subject to bleeding. Other infants have no ability to manufacture vitamin K in the intestine. Without proper care such infants may die of brain hemorrhage. This condition can be alleviated by administering vitamin K to the mother before delivery or to the infant shortly after birth.

A deficiency of vitamin K may also occur when fat absorption is impaired, as in biliary or pancreatic disease or in the atrophy of intestinal mucosa.

Vitamin K is known as the antihemorrhage vitamin. It is necessary for the production of prothrombin in the liver. When there is a deficiency of vitamin K, there is a lack of prothrombin and thus a prolonged clotting time for the blood. Vitamin K is also necessary as a cofactor for oxidative phosphorylation reactions.

Vitamin K is absorbed from the small intestine with the help of bile. In conditions in which bile does not enter the small intestine, such as in obstructive jaundice, vitamin K is not absorbed. This condition leads to a tendency to bleed for a long period of time after an injury or when the individual is undergoing surgery. This effect can be overcome by administering both bile and vitamin K to the patient.

Vitamin K is also used therapeutically as an antidote for anticoagulant drugs such as dicumarol.

Effect of Excess   Large doses of vitamin K can cause hemolysis in infants and aggravate hyperbilirubinemia.

31-2 Water-Soluble Vitamins

The water-soluble vitamins include the B-complex vitamins and vitamin C. The B complex represents a whole series of vitamins, many of which act as cofactors or enzymes in various oxidative reactions. Each B-complex vitamin has a different physiologic activity. The B complex, also called the vitamin B family, comprises the following vitamins:

1. Vitamin B_1 (thiamine)
2. Vitamin B_2 (riboflavin)
3. Niacin
4. Pyridoxine

5. Pantothenic acid
6. Biotin
7. Folic acid
8. Vitamin B_{12} (cobalamin)

Since these vitamins are water-soluble, they can be excreted in the urine and thus they rarely accumulate in toxic amounts. Because they are water-soluble, the B vitamins must be provided continually in the diet. The only exception is vitamin B_{12} (cobalamin) because the human liver can store several years' supply of this vitamin.

Vitamin B_1 (Thiamine)

Source Thiamine occurs in yeast, milk, eggs, meat, nuts, and whole grains. Vegetables and fruits contain very little vitamin B_1. Synthetic vitamin B_1 is now being added to flour and bread to enrich their vitamin content.

Structure   Thiamine has been crystallized as a hydrochloride with the following structure:

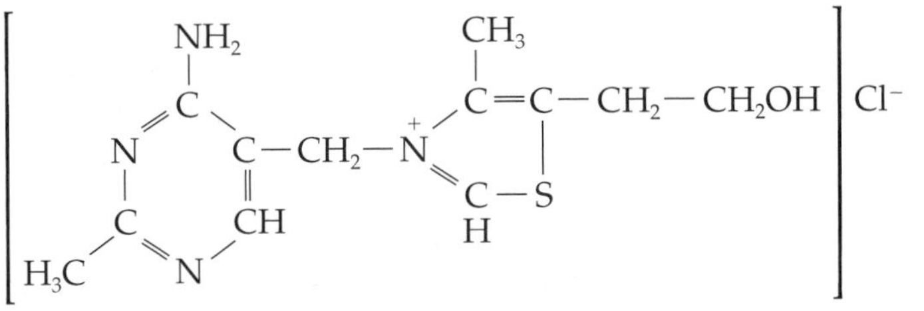

thiamine hydrochloride

Properties   Thiamine is soluble in water and also in alcohol up to 70 percent. It is insoluble in fats and fat solvents. Thiamine is stable in acid solution but is destroyed in alkaline and neutral solutions. Thiamine is quite stable to heat; it can be sterilized for 30 minutes at 120° C without appreciable loss in activity. Thiamine hydrochloride, a salt produced by treating thiamine with hydrochloric acid, is more soluble than thiamine itself and so is generally used whenever this vitamin is required.

Daily Requirement   It is difficult to determine the amount of thiamine required daily. It depends upon several factors. The body's requirement of thiamine increases during a fever, increased muscular activity, hyperthyroidism, pregnancy, and lactation. The thiamine requirement of the body also increases during a diet high in carbohydrates, whereas it decreases with a diet high in fat and protein. Thiamine cannot be stored in the body to any significant degree.

The recommended thiamine daily intake is 0.3 to 0.4 mg for infants, 0.7 to 1.0 mg for children, 1.0 to 1.1 mg for adult females, and 1.2 to 1.5 mg for adult males.

Effect of Deficiency   A deficiency of vitamine B_1 (thiamine) causes a lack of appetite, arrested growth, and loss in weight.

Alcoholism is the most common cause of thiamine deficiency in the United States. This is caused by low vitamin intake plus the high calorie content of alcohol.

As the lack of thiamine continues, a disease called beriberi develops in humans (in animals this disease is called polyneuritis). Beriberi occurs mainly in the Orient, where fish and polished rice (both lacking in vitamin B_1) are the chief diet. In beriberi there is a degeneration of certain nerves leading to the muscles. When pressure is applied along the nerves, severe pain is felt. The muscles served by these nerves become stiff and atrophy from disuse. Cardiovascular symptoms also occur. These are palpitation, tachycardia, an enlarged heart, and an abnormal electrocardiogram. Finally, death may occur because of heart failure. Beriberi can be prevented or treated by a diet containing thiamine.

Another disease caused by a thiamine deficiency is Wernicke's encephalopathy. It occurs frequently in chronic alcoholics who consume little other food.

Thiamine is necessary for the normal metabolism of carbohydrates. The vitamin is changed in the liver to thiamine pyrophosphate (TPP), which acts as a coenzyme (cocarboxylase) for the decarboxylation of pyruvic and α-keto acids and also acts in transketolase reactions. In the Krebs cycle, cocarboxylase is necessary for the conversion of pyruvic acid to acetyl CoA and also for the conversion of α-ketoglutaric acid to succinyl CoA (see Figure 26-10). In thiamine pyrophosphate, the —OH group is replaced by two phosphate groups. During a deficiency of thiamine, pyruvic acid accumulates in the blood and carbohydrates are not properly metabolized.

Thiamine also functions in the utilization of pentoses in the hexose monophosphate shunt and in some amino acid syntheses.

Vitamin B_2 (Riboflavin)

Source   Riboflavin occurs in many of the same sources as thiamine. It is found in yeast, milk, liver, kidney, heart, and leafy vegetables. Cereals contain very little riboflavin unless it is added artificially.

Structure   Riboflavin or vitamin B_2 has been found to consist of a five-carbon sugar alcohol (ribitol) and a pigment (flavin). Its structure is

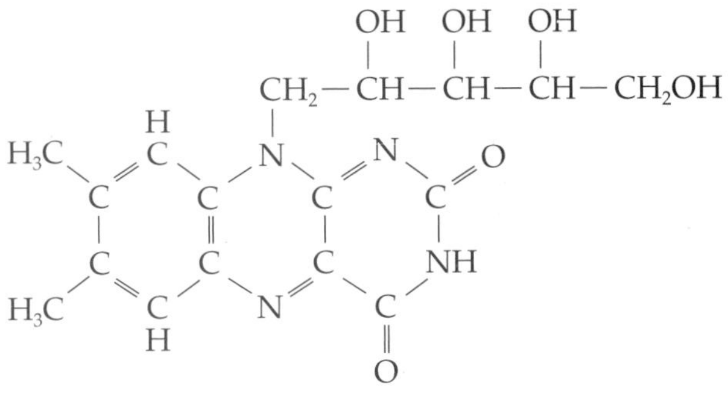

riboflavin

Properties   Riboflavin is an orange-red crystalline solid, slightly soluble in water and alcohol but insoluble in fats and fat solvents. In water solution riboflavin forms a greenish yellow fluorescent liquid. Riboflavin is destroyed by light and alkaline solutions but is fairly stable to heat and so is not destroyed by cooking.

Riboflavin acts as a coenzyme in two different forms: flavin adenine dinucleotide (FAD) and flavin mononucleotide (FMN). These riboflavin coenzymes act as acceptors for the transfer of protons between NAD^+ and $NADP^+$ and the cytochromes, which transport electrons in the mitochondria (see Section 26-10).

Daily Requirement   The daily requirement of riboflavin is 0.8 to 1.2 mg for children, 1.4 to 1.8 mg for adult men, and 1.2 to 1.3 mg for adult women, with increased amount needed during pregnancy and lactation.

Effect of Deficiency   A deficiency of riboflavin in humans produces lesions in the corners of the mouth (cheilitis), inflammation of the tongue (glossitis), and lesions on the lips and around the eyes and nose. There is also an inflammation of the skin (dermatitis) and a clouding of the cornea of the eye.

Because riboflavin is so sensitive to light, newborn infants with hyperbilirubinemia, when treated with phototherapy, show signs of riboflavin deficiency. This conditions persists even when riboflavin supplements are given.

In rats a deficiency of riboflavin produces dermatitis, clouding of the corneas, and loss of hair.

Niacin

Source   Niacin (formerly known as nicotinic acid or vitamin B_5) is widely distributed in plants and animals. It is found in liver, kidney, and heart as well as in yeast, peanuts, and wheat germ. Milk, eggs, and fruit contain some niacin but are generally classified as poor sources of that vitamin.

Structure and Properties   Niacin and niacinamide (nicotinamide), which it readily forms in the body, have the following structures.

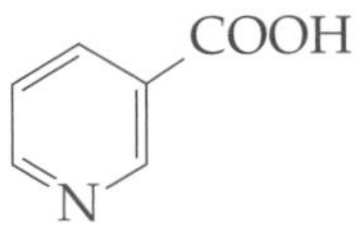

niacin
(nicotinic acid)

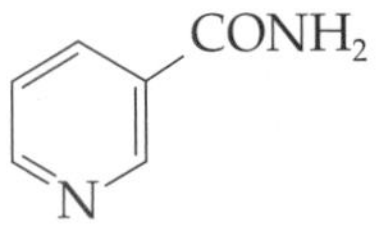

niacinamide
(nicotinamide)

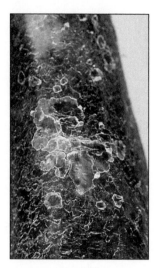

FIGURE 31-4

The roughened skin, dark in patches, that is characteristic of the niacin deficiency disease pellagra.

Niacin is slightly soluble in water but quite soluble in alkali. It is insoluble in fats. Niacin is stable to alkalis and acid and to heat and light, and it is not destroyed by cooking.

Niacinamide, along with thiamine and riboflavin, serves as a coenzyme in tissue oxidation. It functions in the mitochondria in the form of NAD$^+$ (nicotinamide adenine dinucleotide) and NADP$^+$ (nicotinamide adenine dinucleotide phosphate) (see page 426).

Daily Requirement The recommended daily intake of niacin is 15 to 20 mg for males and 13 to 15 mg for females, with a slight increase in requirements for adolescents and during pregnancy and lactation. However, these requirements can be greatly affected by the protein of the diet because the amino acid tryptophan can supply much of the body's needed niacin (60 mg of tryptophan equals 1 mg of niacin). Some niacin may also be synthesized by intestinal bacterial action and thus become available for use in the body.

Effect of Deficiency Niacin was originally called nicotinic acid or the antipellagra factor. The word pellagra comes from the Italian words *pelle agra* meaning rough skin. A deficiency of niacin in humans produces serious consequences, and a deficiency in this vitamin is usually accompanied by a deficiency in other substances also. In pellagra there is a dermatitis (skin rash or lesions) and an inflammation of the mouth and tongue (glossitis). These symptoms are accompanied by diarrhea and then dementia (see Figure 31-4).

A lack of niacin in dogs produces tongue lesions called black tongue. A diet containing niacin is effective in curing pellagra in humans and also niacin deficiency diseases in animals.

Pellagra was once quite common in the southern states where the diet consisted chiefly of corn and fat pork. Corn has a low tryptophan content and can give very little niacin. Fat pork also has very little niacin. Thus, there was a deficiency of this vitamin, leading to pellagra. With an improvement in the diet, especially with the addition of foods containing niacin (or tryptophan), pellagra is not as common in the United States as it was previously.

High doses of niacin (nicotinic acid) but not niacinamide can produce such symptoms as skin flushing, gastrointestinal distress, and pruritus as well as lower serum cholesterol levels.

Case Study 31-1 Niacin Deficiency and Pellagra

A 23-year-old man in Arkansas is seen by his physician with many of the symptoms of pellagra: a swollen tongue, nervous disturbances, and dermatitis. A dietary history reveals that the man eats mostly sweet corn, with only small amounts of protein. His roommate has no similar problems, but his diet, although high in sweet corn, is usually accompanied by beans.

Questions

1. Niacin can be synthesized in humans by tryptophan. Why is it considered a vitamin?
2. What are the tryptophan contents of proteins in corn and beans?
3. Explain the difference in these roommates' health and their diets.

Pyridoxine

Source Pyridoxine is found in yeast, liver, egg yolk, and the germ of various grains and seeds. It is also found to a limited extent in milk and leafy vegetables.

Structure Pyridoxine was originally called vitamin B_6, or the rat antidermatitis factor. Subsequent work showed that vitamin B_6 is a mixture of pyridoxine, pyridoxal, and pyridoxamine. The generally accepted term for these compounds is *pyridoxine* because these compounds are readily interconvertible.

Daily Requirement The recommended daily allowance of pyridoxine for adult men is 2.0 mg and for adult women 1.6 mg per day.

Effect of Deficiency A deficiency of pyridoxine in rats produces dermatitis in the paws, nose, and ears. A deficiency in dogs and pigs produces anemia. If the deficiency of pyridoxine is continued for a long period of time, these animals suffer from epileptiform fits. A deficiency of this vitamin in adults produces such symptoms as dermatitis, sore tongue, irritability, and apathy. A diet containing pyridoxine (or vitamin B_6) will alleviate these symptoms.

The generation of niacin from the amino acid tryptophan requires the presence of the coenzyme pyridoxal phosphate. Therefore, pellagra frequently accompanies pyridoxine deficiency.

A widely used antituberculosis drug, isonicotinic acid hydrazide (isoniazid) can induce isolated pyridoxine deficiency by forming a compound with pyridoxal. This compound is excreted in the urine, thus producing a deficiency of pyridoxal.

The antihypertensive drug hydralazine and the chelating agent penicillinamine will also cause pyridoxine deficiencies. Pyridoxine should not be taken by patients receiving levodopa treatment for Parkinson's disease because pyridoxine enhances the drug's deactivation.

Pyridoxal phosphate and pyridoxamine phosphate serve as coenzymes for the decarboxylation of amino acids, taking part in the reactions occurring primarily in the gray matter of the central nervous system. It is believed that a deficiency of these coenzymes interferes with decarboxylation reactions in the central nervous system and so leads to epileptiform seizures. Pyridoxal phosphate and pyridoxamine phosphate also serve as coenzymes in amino acid metabolism. Furthermore pyridoxine is involved in the absorption of zinc by the intestines.

pyridoxine

pyridoxal

pyridoxamine

Pantothenic Acid

Source Pantothenic acid has a widespread distribution in nature. Its name comes from the Greek word meaning "from everywhere." Good sources of pantothenic acid are egg yolk, yeast, kidney, and lean meats. Other fairly good sources are skimmed milk, broccoli, sweet potatoes, and molasses.

Structure and Properties Pantothenic acid is a viscous yellow oil, soluble in water but insoluble in fat solvents such as chloroform. It is stable in acid and alkaline solution. Pantothenic acid is one of the constituents of coenzyme A (CoA) (see page 425), which is involved in the metabolism of carbohydrates, fats, and proteins and in the synthesis of cholesterol (see page 488).

The structure of pantothenic acid is

pantothenic acid

The daily human requirement of pantothenic acid for adults is 4 to 7 mg and 3 to 7 mg for children, amounts which are easily met with an ordinary diet.

Effect of Deficiency Pantothenic acid was originally known as chick antidermatitis factor because it was a substance that prevented dermatitis in chicks.

Rats and dogs who were given a diet deficient in pantothenic acid showed a loss of pigmentation from their hair. The black hair of such animals turned gray but returned to its original black color upon the addition of pantothenic acid to the diet. There is no evidence that this vitamin is of significant value in restoring hair color in humans. While a deficiency of pantothenic acid is rare, it has been found in prisoners of war as *burning foot syndrome*.

A deficiency of pantothenic acid in animals causes degeneration in the adrenal cortex and a failure in reproduction.

Biotin

Biotin, another member of the vitamin B complex, is widely distributed in nature. Rich sources of this vitamin are liver, egg yolk, kidney, yeast, and milk. Biotin was formerly known as the anti-egg-white injury factor. This name was given to it because rats fed raw egg white failed to grow and also developed dermatitis. Raw egg white contains a protein, avidin, that combines with biotin and renders it unavailable to the animal.

The structure of biotin is

$$
\begin{array}{c}
O \\
\parallel \\
C \\
HN \qquad NH \\
\mid \qquad \mid \\
HC \!-\!\!-\! CH \\
\mid \qquad \mid \\
H_2C \qquad CH\!-\!CH_2\!-\!CH_2\!-\!CH_2\!-\!CH_2\!-\!COOH \\
\diagdown S \diagup
\end{array}
$$

biotin

An artificially produced deficiency of biotin in humans causes scaly dermatitis, nausea, muscle pains, and depression. These symptoms are rapidly relieved by the administration of a diet containing biotin.

Biotin is supplied by the action of intestinal bacteria, in humans as well as animals, so that a deficiency of this vitamin is unlikely in most cases, except on a severely restricted diet. Biotin functions as a coenzyme for carboxylation reactions in the formation of fatty acids (see page 485).

Daily Requirement The recommended daily requirement for biotin is 30 to 100 µg for adults and 20 to 100 µg for children.

Folic Acid

Folic acid (folacin) occurs in green leaves, yeast, liver, kidney, and cauliflower. The structure of folic acid (which contains a pteridine nucleus, *p*-aminobenzoic acid, and glutamic acid) is

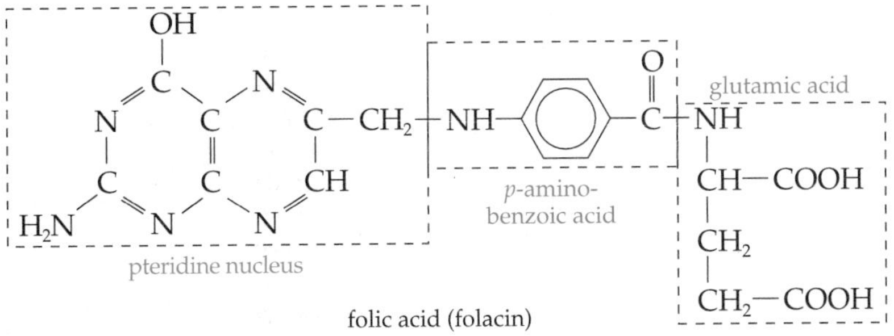

folic acid (folacin)

Actually, this compound, which is also called pteroylglutamic acid (PGA) is only one of several related compounds in the folic acid group. Others are pteroic acid, pteroyltriglutamic acid, and pteroylheptaglutamic acid.

The daily requirement for folic acid is 200 µg for a male and 180 µg for a female.

A deficiency of folic acid in adults causes *megaloblastic anemia and gastrointestinal disturbances.

A reduced intake of folic acid during pregnancy can lead to neural-tube defects, one of the most common major birth defects in the United States. It affects 2 of every 1000 babies. The primary results of such defects can include spina bifida, in which the spine is not closed, and anencephaly, in which most of the brain is missing.

Folic acid in its reduced form, tetrahydrofolic acid, acts as a coenzyme for the transfer of methyl groups in the formation of such compounds as choline and methionine.

Folic acid is primarily included in reactions involving transfer of methyl groups. This type of reaction plays an essential role in the synthesis of hemoglobin, nucleic acids, and methionine.

Treatment with antileukemic drugs such as methotrexate involves inhibiting the conversion of folic acid into its active form tetrahydrofolate. Without this folate, cancer cells (as well as other cells) cannot grow.

Vitamin B_{12} (Cobalamin)

Source   Plants do not contain vitamin B_{12} (cobalamin). Microorganisms are able to synthesize it. The best sources of cobalamin are liver, kidney, fish, eggs, milk, oysters, and clams.

Structure and Properties   Cobalamin is an odorless, tasteless, reddish crystalline compound, soluble in water and alcohol and insoluble in fat solvents such as ether and acetone. One unusual property of this vitamin is that it contains the element cobalt (4.35 percent).

A cyano group (CN^-) is usually attached to the central cobalt atom during isolation of the vitamin. This compound is known as cyanocobalamin. The cyano group must be traded for a methyl group in the body before cobalamin can be activated. The structure of cyanocobalamin is shown on the following page.

Daily Requirement   The recommended daily amount of vitamin B_{12} (cobalamin) is 2.0 µg.

Effect of Deficiency   Vitamin B_{12} is important in the transfer of methyl groups. Thus, many symptoms of vitamin B_{12} and folate deficiency are similar. Vitamin B_{12} is involved in the maintenance of the myelin sheath, in the synthesis of nucleic acids and hemoglobin, and in the metabolism of lipids and carbohydrates. Pernicious anemia is an anemia that resembles megaloblastic anemia. Pernicious anemia has several neurologic complications. It results from a dietary lack of vitamin B_{12} (strict vegetarianism) or of the intrinsic factor.

Vitamin B_{12} (cobalamin) is also called the anti-pernicious anemia factor. It is absorbed from the small intestine in the presence of hydrochloric acid and the intrinsic factor (see page 439). In the absence of the intrinsic factor, vitamin B_{12} cannot be absorbed, and this leads to pernicious anemia. Injection of a small amount of cobalamin will produce remarkable improvement in that disease.

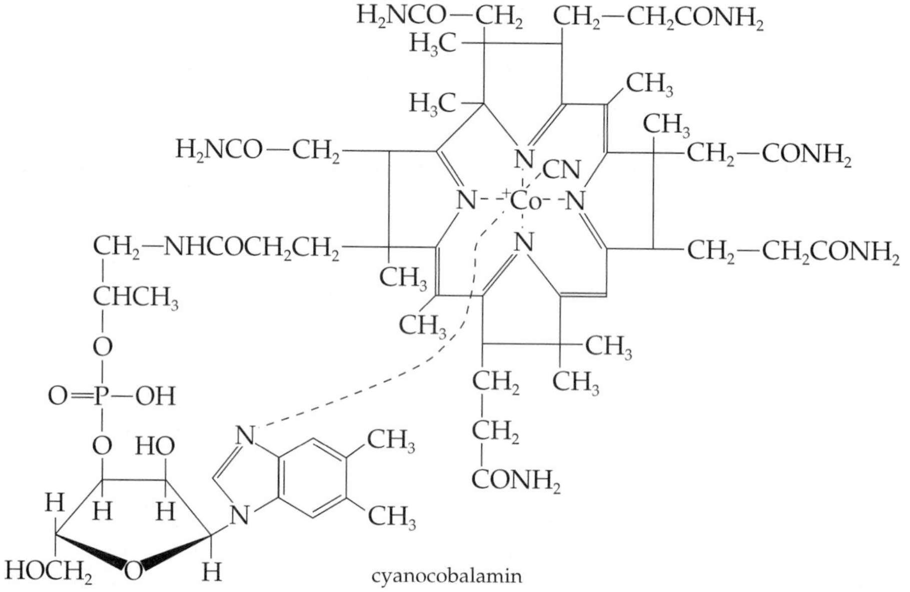

cyanocobalamin

Derivatives   Coenzyme B_{12} is one of several coenzymes derived from vitamin B_{12}. These coenzymes are required for hydrogen transfer and isomerization in the conversion of methyl malonate to succinate, thus involving both carbohydrate and fat metabolism.

Coenzyme B_{12} differs from vitamin B_{12} in that a 5-deoxyadenosine group replaces the CN group attached to the central cobalt.

Related Substances

There are several substances frequently classified along with the B vitamins although they are not truly vitamins. Among these are choline, inositol, *p*-aminobenzoic acid, and lipoic acid.

Choline   Choline is an essential substance as far as the body is concerned, although it is generally not classified as a vitamin because it can be synthesized in the required amounts in the body. It is required in much larger amounts than vitamins.

Choline is a viscous, colorless liquid, soluble in water and alcohol but insoluble in ether. Its structure is

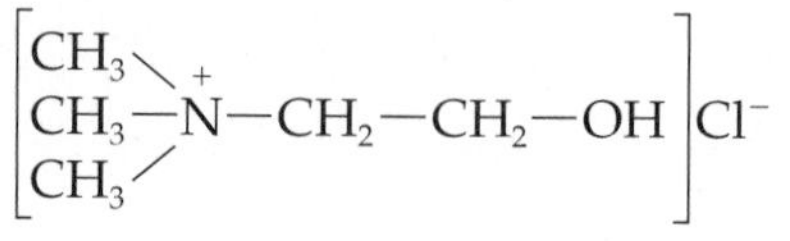

choline

A deficiency in choline leads to such symptoms as fatty liver. Young rats on a diet deficient in choline also had hemorrhagic degeneration of the kidneys. Older rats who survived these symptoms developed cirrhosis. Chicks and young turkeys develop perosis, or slipped tendon disease.

Choline is a constituent of lecithin (phosphatidylcholine) and sphingomyelin and is important in brain and nervous tissue. A deficiency of choline will not develop in a person on a high-protein diet because proteins supply the amino acids from which the body can synthesize this compound.

Choline is a constituent of acetylcholine, which is present in nerve cells and aids in the transmission of nerve impulses by the following means: When a nerve cell is stimulated, it releases acetylcholine. This acetylcholine in turn stimulates the adjacent nerve cell to release acetylcholine. This process continues as a chain reaction until the impulse reaches the brain. Once a nerve cell has passed its impulse on to the next cell, the enzyme acetylcholinesterase hydrolyzes acetylcholine into acetic acid and choline. The nerve cells then use these two compounds to regenerate the acetylcholine for the next impulse.

Patients with Alzheimer's disease are deficient in the enzyme acetylase, which is necessary for the production of acetylcholine in the brain. The only definite proof of Alzheimer's disease has been by autopsy. However, studies have shown that patients with this disease, or those presymptomatic of this disease, have cerebrospinal fluid levels of amyloid β-protein precursor (APP) 3½ times lower than individuals without this disease. It is hoped that testing for APP in the cerebrospinal fluid will help in the diagnosis of individuals predisposed to or having Alzheimer's.

Inositol   It was found that mice on a synthetic diet containing all of the known vitamins still failed to grow. Also there was an effect on their hair and impaired lactation. Pantothenic acid, which affects hair in mice, did not help these animals. However, the symptoms were overcome by the addition of a compound obtained from cereal grain. This compound was inositol. Inositol is required for the growth of yeasts, mice, rats, guinea pigs, chickens, and turkeys.

Inositol is found in liver, milk, vegetables, yeast, whole grains, and fruits. The molecular formula for inositol is $C_6H_{12}O_6$, so it is an isomer of glucose.

Inositol, along with folic acid and p-aminobenzoic acid, is not a true vitamin. However it is included with the B group even though no specific role in human nutrition has been established.

inositol
(hexahydroxycyclohexane)

p-**Aminobenzoic Acid**   *p*-Aminobenzoic acid (PABA) is a growth factor for certain microorganisms. It forms part of the folic acid molecule and is believed to be necessary for the formation of that vitamin. However, humans are incapable of using p-aminobenzoic acid to produce folic acid. p-Aminobenzoic acid is formed when folic acid is hydrolized.

p-Aminobenzoic acid is not regarded as a vitamin for humans, but it is included here because some microorganisms need it for the synthesis of folic acid.

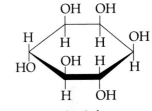

p-aminobenzoic acid
(PABA)

Lipoic Acid   Lipoic acid was first detected in the studies of growth of lactic acid bacteria. It is fat soluble and so was named lipoic acid. Its structure is

$$CH_2-CH_2-CH-(CH_2)_4-COOH$$

lipoic acid

As far as is known, lipoic acid is not required in the diet of higher animals, and no deficiency effects have been noted.

Actually, lipoic acid is not a true vitamin; however, because its coenzyme function in carbohydrate metabolism is closely related to that of thiamine, it is classified along with the B-vitamin group. Lipoic acid functions, along with thiamine, in the initial decarboxylation of α-keto acids to form acetyl CoA for the Krebs cycle.

Vitamin C (Ascorbic Acid)

Source Fresh fruits and vegetables, including oranges, lemons, grapefruit, berries, melons, tomatoes, and raw cabbage, are excellent sources of vitamin C. Dry cereals, legumes, milk, meats, and egg contain very little of this vitamin.

Structure and Properties Vitamin C, ascorbic acid, is a white crystalline substance soluble in water and alcohol but insoluble in most fat solvents. It is a strong reducing agent and is easily oxidized in air, especially in the presence of such metallic ions as Fe^{3+} or Cu^{2+}. Ascorbic acid is rapidly destroyed by heating. For this reason, cooking foods in copper pots should be avoided because this will destroy the vitamin C content of the foods being cooked.

Vitamin C can easily be oxidized to dehydroascorbic acid. Both ascorbic acid and dehydroascorbic acid are biologically active, and both have been synthesized in the laboratory.

ascorbic acid   dehydroascorbic acid

Daily Requirement The recommended daily allowance of ascorbic acid for an adult is 60 mg; the amount is increased to 80 mg during pregnancy and 100 mg during lactation.

Effect of Deficiency Plants and all animals except guinea pigs, humans, and other primates are able to synthesize ascorbic acid and are resistant to diseases caused by a lack of this vitamin. A deficiency of vitamin C produces a disease known as scurvy. The symptoms in humans are swollen, bleeding gums, pain in the joints, decalcification of the bones, loss of weight, and anemia.

A deficiency of vitamin C prevents the body from forming and maintaining the intercellular substance that cements the tissues together. A lack of this intercellular substance in the capillaries leads to rupturing and subsequent hemorrhaging in these vessels and to the formation of weak bones and atrophy of the bone marrow, accompanied by anemia, and also accounts for loosening of the teeth and spongy gums. All of these symptoms are relieved by the addition of ascorbic acid to the diet.

Vitamin C is required for synthesis of collagen, degradation of the amino acid tyrosine, synthesis of epinephrine, proper function of bile acids, absorption of iron, and as an antioxidant.

Case Study 31-2   Ascorbic Acid Deficiency

A 37-year-old housewife with a history of severe depression is seen by her physician with a complaint of pain in her legs. She notes that recently her legs have become discolored. A dietary history reveals that she has not been eating much because of her depression. She is unable to walk or stand unaided. Physical examination shows extensive areas of hemorrhage on her calf muscles. No other hemorrhagic sites are found, and her gums are within normal limits. Laboratory test results show a 12.5 µmol of ascorbic acid/100 mg leukocytes. Her physician prescribes ascorbic acid (3.97 mmol/day). After 16 days her level of urinary excretion of ascorbic acid is 2.26 mmol/24 hours.

Questions

1. What are the biochemical functions of ascorbic acid?
2. What structural similarities are there between ascorbic acid and monosaccharides?
3. How does ascorbic acid work to prevent hemorrhage into the tissues?
4. Why was this patient treated with a high daily intake of ascorbic acid?

Summary

The body requires vitamins in addition to carbohydrates, fats, and proteins but in much smaller amounts. Vitamins must be supplied in the diet, whereas hormones are synthesized by the body.

Vitamins are divided into two types: fat-soluble and water-soluble.

Vitamin A is found in fish liver oils, in butter, and in milk. Vitamin A is an alcohol. The recommended daily adult dose of vitamin A is 1000 retinol equivalents (5000 IU) for the adult male and 800 retinol equivalents (4000 IU) for the adult female. Vitamin A has an all-trans structure.

A lack of vitamin A produces keratinization in the membranes of the eyes, digestive tract, respiratory tract, and genitourinary tract. Nyctalopia is also caused by a deficiency of vitamin A. Vitamin A is stored in the liver.

Vitamin D is found in fish liver oils such as cod or halibut. The vitamin D content of milk is increased by irradiation. The D vitamins are sterols with a structure similar to that of cholesterol. Vitamin D functions to increase the absorption of calcium and phosphorus from the small intestine. Vitamin D also functions in the deposition of calcium phosphate in the teeth and bones. A lack of vitamin D produces rickets.

Vitamin E is found in many foods; wheat germ is particularly rich in this substance. The E vitamins have antioxidant properties and act as a cofactor in oxidative phosphorylation reactions. Vitamin E prevents sterility in animals, but in humans it acts as an antioxidant.

Vitamin K is known as the antihemorrhage vitamin. It is necessary for the production of prothrombin in the liver.

Vitamin B_1 (thiamine) occurs naturally in yeast, milk, and whole grains and also may be made synthetically. A deficiency of this vitamin causes a lack of appetite, arrested growth, and weight loss. A prolonged deficiency of vitamin B_1 leads to the disease known as beriberi (or polyneuritis in animals). Thiamine is also necessary for the normal metabolism of carbohydrates. This vitamin acts as a coenzyme for the decarboxylation of pyruvic acid. The coenzyme, known as cocarboxylase, is necessary in the Krebs cycle for the conversion of α-ketoglutaric acid to succinyl CoA.

Vitamin B_2 (riboflavin) is found in the same sources as thiamine. Riboflavin acts as a coenzyme in two different forms: flavin adenine dinucleotide (FAD) and flavin mononucleotide (FMN). These coenzymes act as acceptors for the transfer of protons between NAD^+ and $NADP^+$ and the cytochromes.

Niacin (or nicotinic acid) is widely distributed in nature. A deficiency of this vitamin produces a condition known as pellagra. Niacin is an important constituent of two coenzymes: nicotinamide adenine dinucleotide (NAD^+) and nicotinamide adenine dinucleotide phosphate ($NADP^+$). These coenzymes are involved in most oxidation-reduction reactions in the mitochondria.

Pyridoxine is involved in the decarboxylation of amino acids and is also required for certain transaminase reactions.

Pantothenic acid is one of the constituents of CoA, which is involved in the metabolism of carbohydrates, fats, and proteins as well as in the synthesis of cholesterol.

Biotin functions in the activation of carbon dioxide for carboxylation reactions in the formation of fatty acids.

Folic acid is concerned with the transfer of methyl groups in the formation of such compounds as choline and methionine.

Vitamin B_{12}, cobalamin, is called the anti-pernicious anemia factor. It is also involved in the synthesis of certain amino acids and of choline and is necessary for the formation of coenzyme B_{12}.

Choline is a constituent of lecithin and so is important in brain and nerve tissue. Choline is also a constituent of acetylcholine, which aids in the transmission of nerve impulses.

p-Aminobenzoic acid (PABA) is formed when folic acid is hydrolyzed. This substance is a growth factor for certain microorganisms.

Inositol is required for the growth of yeasts, mice, and rats.

Lipoic acid is closely related to thiamine in the initial oxidation of α-keto acids.

Vitamin C, ascorbic acid, can be synthesized by plants and most animals but not by humans. A lack of this vitamin produces a disease known as scurvy. A lack of ascorbic acid also causes a lack of the intercellular substance that cements the tissues together.

Questions and Problems

1. Which vitamins are water soluble? fat soluble?
2. Compare vitamins with hormones.
3. What are the sources of vitamin A? What are the effects of a deficiency? of an excess?
4. What is β-carotene? How does the body use it?
5. Where is vitamin A stored in the body? How is it absorbed?
6. Describe the role of vitamin A in the visual cycle.
7. What are the sources of vitamin D? Why is it called the sunshine vitamin? What type of compound is it? Are all D vitamins equally potent?
8. What are the daily adult requirements of vitamin A? vitamin D? vitamin K?
9. What are the functions of vitamin D in the body? What are the effects of a lack of this vitamin? of an excess?
10. What are the sources of vitamin E? What is its name? its structure? its properties? its functions?
11. What are the sources of vitamin B_1? What are its properties? its functions?
12. What are the effects of a lack of thiamine?
13. What are the sources of riboflavin? niacin? biotin?
14. What are the effects of a deficiency of niacin? of pyridoxine? of riboflavin? How does the body use these substances?
15. What are the sources of pantothenic acid? its properties? effects of a deficiency?
16. What are the functions of lipoic acid? folic acid? *p*-aminobenzoic acid? inositol? cobalamin?
17. What are the sources of vitamin C? its properties? its structure? effects of a deficiency?
18. What is carotenosis?
19. How does vitamin E affect selenium metabolism?
20. How would a deficiency of water-soluble vitamins effect the Kreb's cycle?
21. Why are some vitamins toxic if taken in excess?
22. What type of evidence would be necessary to justify a need for a new vitamin?

Questions 23-27 pertain to the following scenario:

A 76-year-old man comes to his physician's office with a complaint of a painful, swollen right leg. The day after the symptoms developed he was unable to walk. The patient is admitted to the hospital for acute thrombophlebitis and treatment with heparin is initiated. Several days later a coumarin-type anticoagulant is prescribed. The patient's prothrombin time was 11 seconds before the coumarin was started; this gradually increases to 37 seconds as the dosage is increased. A daily maintenance dose of coumarin is established to keep the prothrombin time between 30 to 40 seconds. The patient improves and the heparin is discontinued. The patient is discharged with a supply of coumarin and told to have his prothrombin time tested weekly. After 2 weeks the man decides that the weekly trip to the clinic is inconveniencing his family too much and stops coming for follow up. However, he continues taking the coumarin. Five weeks later the man calls his physician and reports that his urine is bright red. The physician admits the patient to the hospital. The patient's prothrombin time had increased to 73 seconds, so a water-soluble vitamin K analog, menadione, is given by injection. The coumarin anticoagulant is discontinued and additional menadione and vitamin K are given. The patient's prothrombin time decreases to 13 seconds and the urine color returns to yellow.

23. How did this thrombus dissolve?

24. How do the coumarin drugs decrease the tendency of blood to coagulate?

25. How does vitamin K overcome the action of the coumarin drugs?

26. Why was menadione given first rather than natural vitamin K?

27. Why should this patient avoid aspirin-type compounds? List three aspirin-type compounds.

Questions 28 to 30 pertain to the following scenario:

A comatose 7-month-old infant is brought to the emergency room. Physical examination shows no response to painful stimuli in this pale and flaccid infant. His length is in the fortieth percentile and his weight (5.5 kg) is below the thirtieth percentile. Developmental history reported by the mother indicates that the infant has poor head control and had not learned to turn over. The mother states that the infant was born at term without complications and that she has breastfed. The mother also reveals that she has been a strict vegetarian for 8 years and does not take vitamins. The infant's EEG is abnormal and his laboratory test results show that the hemoglobin level is 5.5 g/L and

vitamin B_{12} level is 19 pg/mL (normal range is 150 to 1000 pg/mL). The folate, glucose, and electrolyte levels are within normal limits. The mother's laboratory test results show that her breast milk is low in vitamin B_{12} (75 pg/mL; normal range is 1000 to 3000 pg/mL). The infant is given Vitamin B_{12} 1 mg/day IM for 4 days. On the fifth day he is alert and smiling. His repeat EEG is now normal.

28. What are good dietary sources of B_{12}? How much did the vitamin B_{12} injections exceed the RDA?

29. Why is folate deleted from vitamin preparations?

30. Explain how the absorption of vitamin B_{12} differs when given intramuscularly.

Questions 31 to 34 pertain to the following scenario:

A 60-year-old man is admitted to the hospital for treatment of shortness of breath and weakness. He has a 35-year history of alcoholism and was admitted previously for megaloblastic anemia with scurvy. His admission laboratory test results show:

hematocrit	28%
reticulocytes	22%
leukocytes	4000/mm^3
folate	2.2 nmol/L (1 × 10^{-9} g/mL)
B_{12}	0.080 nmol/L (116 × 10^{-12} g/mL)

Three months after recovery the patient enters a metabolic study. He is started on a low-folate diet and on the tenth day his reticulocyte count is down to 11.6%. He is then started on 32 oz of wine per day. On day 15, the reticulocyte count is 0.6%. On day 22, he begins intramuscular injections of folate at 75 μg/day. No change was seen in the reticulocyte, leukocyte, or platelet counts. The folate dose is then increased. Several days later the reticulocyte count increases to 18.5% along with an increase in the platelet and leukocyte counts.

31. What is the minimal daily requirement for folic acid? Is the improvement of the megaloblastic anemia the result of the increased folate in the diet or the removal of the alcohol? How does this contrast when the patient was taking both folate and consuming wine during the study. What conclusions can you draw?

32. What is the relationship of alcohol and folic acid to hematopoiesis?

33. Why might adding folate to wine be dangerous? Consider the methyl-trap hypothesis.

34. Why is megaloblastic anemia less common in beer drinkers?

Practice Test

1. Niacin is important in the prevention of _____.
   a. rickets
   b. scurvy
   c. pellagra
   d. beriberi

2. Pellagra is caused by a lack of _____.
   a. inositol
   b. pyridoxine
   c. riboflavin
   d. niacin

3. A vitamin that is not fat soluble is vitamin _____.
   a. A
   b. K
   c. C
   d. D

4. The vitamin used by the body to make CoA is _____.
   a. pantothenic acid
   b. riboflavin
   c. biotin
   d. pyridoxine

5. Rickets is caused by a lack of vitamin _____.
   a. A
   b. B_1
   c. C
   d. D

6. The only vitamin that contains a metallic element is B_{12}. It contains _____.
   a. Fe
   b. Mg
   c. Co
   d. Zn

7. Which of the following vitamins is toxic if taken in excess?
   a. A
   b. B_2
   c. C
   d. K

8. A lack of which vitamin can lead to hemorrhaging?
   a. B_1
   b. C
   c. E
   d. K

9. A vitamin used in the visual cycle is vitamin _____.
   a. A
   b. B_1
   c. C
   d. E

10. Beriberi is caused by a lack of vitamin _____.
    a. B_1
    b. B_2
    c. B_6
    d. B_{12}

Hormones

Objectives

- To understand the classification of hormones

- To become aware of receptor sites and the effects of abnormalities in those sites

- To become familiar with the cardiac, pineal, and gastrointestinal hormones

- To understand the role of the hormones of the pancreas and the effects of a deficiency or excess

- To understand the role of the hormones of the thyroid and the effects of a deficiency or excess

- To become familiar with the hormones of the parathyroid glands and the adrenal glands and the effects of abnormalities in those glands

- To understand the role of the pituitary hormones of the pituitary gland and the effects of the various releasing and release-inhibiting factors from the hypothalamus

- To understand the function of the sex hormones

- To become aware of the role of neurotransmitters

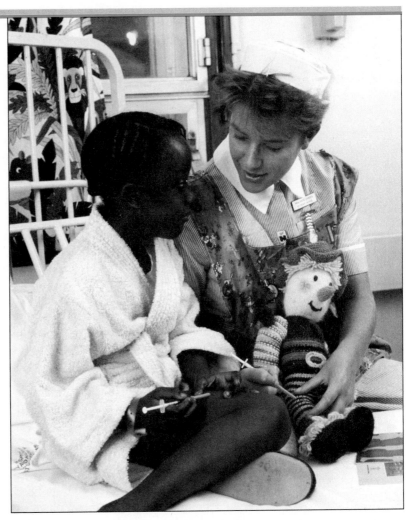

Hormones are an important means of intercellular communication and are involved in eliciting and coordinating a variety of cellular functions, including metabolism. Insulin, for example, is a hormone required by cells for glucose uptake. Without it, cells starve. Diabetes mellitus is a condition in which the body does not produce insulin. The diabetic child pictured is being taught how to inject herself with insulin.

Hormones exert a very important influence on the regulation of body processes. Hormones are produced in the endocrine glands and are secreted directly into the bloodstream. They may be proteins, peptides, amino acid derivatives, or steroids.

There is an internal balance and interaction among the various endocrine glands. Hormones act in several ways. Some stimulate RNA production in target cell nuclei and thus increase the production of enzymes. Some hormones stimulate enzyme synthesis in the ribosomes through the translation of information carried by messenger RNA. Others are involved in the transportation of various substances across membranes.

Both vitamins and hormones are necessary in only very small amounts; however, unlike vitamins, hormones are produced in the body. Hormones are produced in a particular organ and are carried by the bloodstream to some other body part, where they cause a specific physiologic effect. The level of hormones in the bloodstream ranges from 10^{-6} M to 10^{-12} M.

32-1 Classification of Hormones

Hormones may be classified according to the mechanism of their action. Group I hormones bind to intracellular receptors, and Group II hormones bind to cell-surface receptors.

Group I hormones are lipophilic and are usually derived from cholesterol. Such hormones have a long plasma half-life and use blood transport proteins. Examples of Group I hormones are the estrogens, androgens, progestins, glucocorticoids, mineralocorticoids, calcitriol, and the thyroid hormones. Group I hormones diffuse through the plasma membranes of all cells but interact only with the specific receptor sites on their target cells.

Group II hormones are hydrophilic, do not use blood transport proteins, and have a short plasma half-life. Group II hormones, the larger group of hormones, bind to cell-surface receptors, which then use intracellular (second) messengers to relay their message to the target cells.

The type of second messenger involved determines the subclasses of Group II hormones.

Group IIA hormones: The second messenger is cAMP. Cyclic adenosine monophosphate (cAMP) is present in almost every type of body cell. It is produced through the activity of the enzyme adenylate cyclase (found in the membranes of the cell walls) on ATP. cAMP is not found in plants.

When a Group IIA hormone reaches its target cell, it interacts with the cell membrane receptors that are specific for that hormone. The adenylate cyclase in the cell then triggers the production of cAMP from ATP. The cAMP thus released activates the enzyme systems that in turn catalyze protein synthesis by the target cell, which produces the characteristic effects of that cell.

The scheme of the reactions is shown in Figure 32-1.

The concentration of cAMP in the cells is controlled primarily by two methods: (1) by regulating its rate of synthesis and (2) by changing it into an inactive form, AMP, through the action of certain enzymes.

Abnormalities in the metabolism of cAMP may explain the effects of certain diseases. For example, the bacteria that cause cholera produce a toxin that stimulates the intestinal cells to accumulate cAMP. The excess cAMP instructs the cells to secrete a salty fluid. The accumulation and subsequent loss of large amounts of this salty fluid and the resulting dehydration, if unchecked, can cause cholera to be fatal.

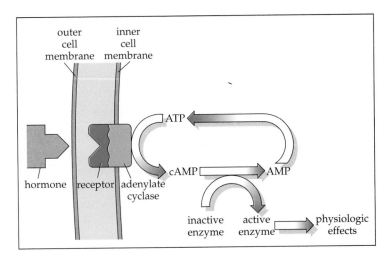

FIGURE 32-1
Mechanism of hormone action.

cAMP activates a protein kinase that catalyzes phosphorylation and thus activates RNA polymerase. RNA polymerase, in turn, stimulates RNA production, which then functions as a messenger for protein synthesis (see Figure 23-4). In this way, cAMP influences the synthesis of protein.

The role of cAMP in the breakdown of liver and muscle glycogen to glucose is indicated on page 458. The structure of cAMP is as follows.

cyclic adenosine monophosphate (cAMP)

Examples of Group IIA hormones are adrenocorticotropic hormones (ACTH), parathyroid hormone, follicle stimulating hormone (FSH), melanocyte stimulating hormone (MSH), luteinizing hormone (LH), thyroid stimulating hormone (TSH), glucagon, and calcitonin.

Group IIB hormones: The second messenger is cGMP. Cyclic guanosine monophosphate, cGMP, is similar to cAMP. It is involved in the actions of the F prostaglandins, whose effects run counter to those of the E prostaglandins. It has been suggested that cAMP and cGMP function reciprocally in regulating cellular activity. An example of a Group IIB hormone is the atrial natriuretic factor (ANF).

Group IIC hormones: The second messenger is calcium or phosphatidylinositides, or both. Examples of this subclass of hormones are vasopressin, oxytocin, gastrin, and cholecystokinin.

Group IID hormones: The second messenger is unknown. Among these hormones are insulin, prolactin, and the growth hormone (GH).

32-2 Abnormalities in Receptor Sites

Abnormalities in hormone receptor sites can cause certain diseases. Such abnormalities can be divided into three categories:

1. *Presence of certain antibodies.* Antibodies directed against a specific hormone receptor site can cause such diseases as asthma, Graves's disease, and myasthenia gravis.
2. *Absence of receptor sites.* The absence or defects present in hormone receptor sites can cause such diseases as congenital nephrogenic diabetes insipidus, testicular feminization syndrome, and vitamin D–resistant rickets type II.
3. *Abnormal receptor regulation.* Obesity can be caused if the binding of insulin to its receptor site is increased, and diabetes mellitus type II can be caused if the binding of insulin to its receptor site is decreased.

32-3 Cardiac Hormones

The atrium of the heart produces atriopeptin, a cardiac hormone involved in fluid, electrolyte, and blood pressure homeostasis. Atriopeptin is stored in the *cardiocytes as the prohormone atriopeptinogen, which consists of 126 amino acids. The hormone atriopeptin, which has a chain of 28 amino acids, is continually released in small amounts and so is always present in the blood.

Elevated vascular volume triggers a series of events. First, more atriopeptin is released, which causes the kidneys to increase the rate of renal blood flow and thereby the rate of glomerular filtration. Thus urine volume is increased, as is sodium excretion. Renin activity (see Section 24-7) is decreased, and so is the production of aldosterone (see page 590); however, the amount of vasopressin is increased (see page 590). The resulting decrease in vascular volume causes a negative feedback that decreases the levels of atriopeptin.

32-4 Hormone of the Pineal Gland

The pineal gland is a cone-shaped body deep in the back of the brain. For years it was thought to be a *vestigial organ with no function. However, in 1958 a hormone called *melatonin* was isolated from this gland. Melatonin is an unusual hormone in that it is more prevalent in the body during the hours of darkness than during daylight hours. That is, melatonin production rises as darkness falls. The brain interprets this increase in melatonin as a signal to go to sleep for the night. The light of dawn shuts down melatonin production.

It has been found that patients suffering from melancholia have unusual melatonin rhythms. In these patients, maximum amounts of the hormone occur at dawn or midnight instead of at about 2 A.M. as in healthy individuals.

Travelers given melatonin before long flights do not suffer from jet lag, indicating that somehow melatonin keeps the mind in synchronization with the outside world.

In children melatonin inhibits sex hormones and so restrains sexuality. As the melatonin levels fall during puberty, the activity of the sex hormones rises.

32-5 Hormones of the Gastrointestinal System

The pyloric mucose produces a hormone called *gastrin*. Gastrin is absorbed into the bloodstream and is carried back to the stomach, where it stimulates the secretion of hydrochloric acid in the gastric juice and also produces greater motility of the stomach. Gastrin also stimulates the secretion of pepsin and the intrinsic factor (see page 439). Gastrin is a polypeptide containing 17 amino acids.

Secretin is formed in the mucosa of the small intestine when the acidic chyme enters the duodenum. Secretin stimulates the pancreas to release the pancreatic juices into the small intestine. Secretin is a polypeptide containing 27 amino acids, 14 of which are identical to those found in glucagon (page 580).

Cholecystokinin (CCK) is secreted when fat enters the duodenum. CCK stimulates the contraction and emptying of the gallbladder into the small intestine. It also stimulates pancreatic enzyme secretion. CCK is a polypeptide consisting of 33 amino acids.

In addition to these gastrointestinal hormones, several others are involved in the gastrointestinal system. Among these are the following:

1. Motilin, which stimulates gastric motility
2. Gastric inhibitory polypeptide (GIP), which inhibits gastric acid secretion and gastric motility and stimulates release of insulin by the pancreas
3. Vasoactive intestinal polypeptide, which stimulates pancreatic bicarbonate secretion

32-6 Hormones of the Pancreas

The pancreas secretes digestive enzymes (see Chapter 25) and four hormones—insulin, glucagon, somatostatin, and pancreatic polypeptide. Insulin and glucagon are involved in carbohydrate metabolism. Somatostatin is involved in local regulation of insulin and glucagon secretion. Pancreatic polypeptide affects gastrointestinal secretion.

Pancreatic polypeptide contains 36 amino acids. It is produced by the F cells of the pancreas. Secretion of pancreatic polypeptide is increased during a high-protein diet, during fasting, exercise, and acute hypoglycemia. Secretion is decreased by somatostatin and intravenous infusion of glucose. The function of pancreatic polypeptide is unknown.

Insulin

Insulin is a protein secreted by the β cells of the islets of Langerhans in the pancreas. Insulin has been isolated and crystallized. Crystalline insulin contains a small amount of zinc, which it obtains from the zinc-rich tissues in the pancreas. Small amounts of chromium are also needed for the synthesis of insulin. The amount of chromium in the body decreases with age. The amino acid sequence in insulin is indicated in Figure 22-2.

Insulin increases the rate of oxidation of glucose, facilitates the conversion of glucose to glycogen in the liver and the muscles, and increases the synthesis of fatty acids, protein, and RNA. Insulin decreases blood sugar to its normal fasting level after it has been increased by digestion of carbohydrates by facilitating the transportation of glucose through the membranes into the cells.

Normally the blood sugar level rises to 120 to 130 mg per 100 mL of blood after a meal containing carbohydrates. This level is returned to its normal fasting value by the action of insulin. If the islets of Langerhans are underactive or degenerated, little or no insulin is produced, so the blood sugar level remains high. This condition is termed hyperglycemia and is associated with diabetes mellitus. In diabetes mellitus, there is an increased blood sugar level; glucose appears in the urine, and there is formation of ketone bodies, accompanied by acidosis. Injection of insulin will produce a rapid recovery from these same symptoms. It will not, however, cure diabetes because these same symptoms will reappear after a short time. Thus insulin must be taken by diabetics for the rest of their lives.

Since insulin is a protein, it cannot be taken orally because it would be digested. Therefore, insulin is given by subcutaneous injection. Injections of insulin must be given two or three times a day to a patient with diabetes mellitus. However, when insulin is combined with protamin (a protein), the product, called protamine zinc insulin, is utilized much more slowly and is effective for more than 24 hr. Thus only one injection is needed daily.

If too large a dose of insulin is administered, the blood sugar level falls far below its normal fasting value. This condition is called hypoglycemia and is characterized by symptoms such as dizziness, nervousness, blurring of vision, and then unconsciousness. Such a state is called insulin shock and may be relieved by the administration of sugar, either orally or by injection. Hypoglycemia also occurs when there is a tumor on the islets of Langerhans in the pancreas.

Insulin is degraded primarily in the liver and kidneys by the enzyme glutathione insulin transhydroxylase.

Insulin is available in sterile solutions in which 1 mL contains 100 units (U-100); 1 unit of insulin is the amount required to reduce the blood sugar level of a normal 2-kg rabbit after a 24-hr fast from 120 to 45 mg per 100 mL.

Although insulin cannot be taken orally, there are certain *hypoglycemic substances that will lower the blood sugar level. These substances are effective in treating adult diabetes but not juvenile diabetes. They function by stimulating the β cells of the islets of Langerhans to produce insulin. Examples of such hypoglycemic agents are glyburide (Micronase), troglitazone (Rezulin), and metformin (Glucophage).

Recent evidence indicates that these so-called hypoglycemic agents have many toxic side effects. Adult patients with a mild type of diabetes frequently can control it just as well by means of dietary management.

Glucagon

When crude insulin was first used to lower blood sugar levels, it was noted that a temporary hyperglycemia occurred first and then, soon afterward, came the hypoglycemic effects expected of insulin. The unknown substance present in the crude insulin that gave the hyperglycemic effects was called the hyperglycemic glycogenolytic factor, or HGF. This factor is now known as *glucagon*. Glucagon has been isolated and crystallized. It is a polypeptide containing 29 amino acids in a single chain.

Glucagon has a different arrangement of amino acids than insulin. Unlike insulin, it contains no disulfide bridges. Glucagon contains methionine and tryptophan, which insulin does not. However, insulin contains cysteine, proline, and isoleucine, whereas glucagon does not.

Glucagon is formed in the α cells of the islets of Langerhans in the pancreas. Significant amounts of glucagon also come from A cells in the stomach and other parts of the gastrointestinal tract.

Glucagon causes an increase in the sugar content of the blood by stimulating phosphorylase activity in the liver. The reactions involved are as follows:

$$\text{glycogen + phosphate} \xrightarrow{\text{phosporylase}} \text{glucose 1-phosphate}$$

$$\text{glucose 1-phosphate} \xrightarrow{\text{phosphoglucomutase}} \text{glucose 6-phosphate}$$

$$\text{glucose 6-phosphate} \xrightarrow{\text{phosphatase}} \text{glucose + phosphate}$$

Glucagon increases the formation of cAMP, which activates phosphorylase, and so increases the glucose content of the blood, causing hyperglycemia.

Glucagon stimulates the formation of glucose from amino acids (gluconeogenesis). It also increases the release of potassium ions from the liver. In the adipose tissue, glucagon increases the breakdown of lipids to fatty acids and glycerol.

Glucagon has been used to treat hypoglycemic effects due to an overdose of insulin and insulin shock induced in the treatment of psychiatric patients.

Glucagon is also used in a diagnostic test for glycogen storage disease.

32-7 Hormones of the Thyroid Gland

The thyroid gland is an H-shaped gland consisting of one lobe on each side of the trachea with a piece of tissue connecting the two lobes. In the adult the thyroid gland weighs approximately 25 to 30 g.

The hormones of the thyroid gland regulate the metabolism of the body. They also affect the growth and development of the body.

The thyroid is filled with many small follicles that contain *colloid*, which contains the thyroid gland's stored hormones. Actually, the thyroid is the only endocrine gland in the body that is capable of storing appreciable amounts of hormone.

The thyroid gland contains the element iodine—one of the elements necessary for the proper functioning of the body. Iodine exists in the body in two different forms: as iodide ions and in the thyroid hormones.

Older, indirect methods for evaluating activity such as basal metabolic rate (BMR), protein-bound iodine (PBI), and butanol-extractable iodine (BEI) have been replaced by radioimmunoassay methods that are direct and more reliable.

The colloid of the thyroid contains the protein *thyroglobin,* a glycoprotein. This protein liberates *triiodothyronine* (T_3) and *thyroxine* (T_4), the principal thyroid hormones. T_3 is the major active form of the thyroid hormone and is three to five times as biologically active as T_4. Some mono- and diiodothyronine are also formed, but these compounds are quickly deiodinated in the bloodstream and the freed iodine is used to form more thyroglobin. The structures of these hormones are

thyroxine (T_4)   triiodothyronine (T_3)

The C cells in the thyroid gland produce the hormone *calcitonin* (thyrocalcitonin), which, along with the parathyroid hormone, regulates the calcium ions in the blood. It has been shown that the parathyroid hormone sustains the blood supply of calcium ions whereas calcitonin prevents the blood calcium ion concentration from rising above the required level.

Pure calcitonin was first isolated in 1968, and its structure was determined shortly thereafter. This hormone is a polypeptide containing a single chain of 32 amino acids. Calcitonin has now been synthesized in the laboratory, and its mode of action in the body has been investigated. It produces its effect by inhibiting the release of calcium ions, from the bone to the blood. The release of calcitonin is stimulated by high levels of Ca^{2+} in the blood.

Medullary thyroid carcinoma, a disorder of the C cells of the thyroid, causes an abnormally high production of calcitonin.

Hypothyroidism

Hypothyroidism is a condition in which the thyroid gland does not manufacture sufficient thyroxine for the body's needs. It is usually caused by a lack of iodine in the diet, particularly in parts of the country where the water and foods contain little iodine. Hypothyroidism may also be due to a disease of the thyroid gland or to its congenital absence.

The symptoms of hypothyroidism are sluggishness, weight gain, slower heartbeat, reduced metabolic rate, and loss of appetite. Hypothyroidism can be remedied by the use of iodized salt as part of the normal diet or by the use of synthetic thyroid hormones.

Cretinism   If the thyroid gland is absent or fails to develop in an infant, the effects produced are called cretinism and the individual is called a cretin. Cretins have greatly retarded growth, both physically and mentally. They are usually abnormal dwarves with coarse hair and thick dry skin, and are obese with protruding abdomens. They are also underdeveloped mentally and sexually.

A cretin may develop normally if given thyroid hormones before he or she reaches adulthood (see Figure 32-2).

If the thyroid gland should atrophy after an individual reaches adulthood, the same symptoms as in cretinism appear, except that the individual remains adult in size. One very noticeable symptom is in the development of thick, coarse, dry skin. Such a condition is known as *myxedema. Persons with myxedema are

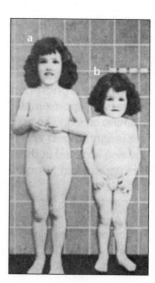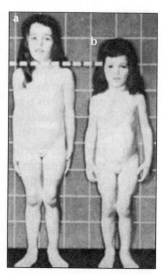

FIGURE 32-2

Cretin B is much shorter than her twin, A. After therapy with thyroid hormones the difference, although present, is greatly reduced.

also sluggish, have a lower pulse and metabolic rate and lower body temperature and are frequently anemic. They are also very sensitive to cold.

Myxedema can usually be cured by the administration of thyroxine.

Simple Goiter

Simple *goiter, also called colloid goiter or endemic goiter, is a condition in which the thyroid gland enlarges, usually because of a lack of iodine in the diet. The decreased production of thyroid hormones causes an increased production of TSH (thyroid stimulating hormone), which in turn overstimulates the thyroid gland. Simple goiter is accompanied by a definite increase in the amount of colloid material in the thyroid gland and also an increase in the size of the neck itself.

Simple goiter occurs in areas where there is a deficiency of iodine in the food and drinking water. The condition can be successfully prevented or cured by the addition of iodine compounds (usually iodized salt) to the diet.

Hashimoto's Disease Hashimoto's disease is a type of hypothyroidism in which all aspects of thyroid function may be impaired. The disease is caused by an attack on the thyroid gland by the body's own immune system.

Hyperthyroidism Hyperthyroidism occurs when the thyroid gland produces excess thyroxine. The symptoms are an increased metabolic rate, bulging of the eyes (exophthalmos), nervousness, loss of weight, a rapid, irregular heartbeat, and an elevated body temperature. Such a condition is also called Graves's disease, Basedow's disease, or exophthalmic goiter. Hyperthyroidism may also be due to a tumor in the thyroid gland (toxic adenoma or Plummer's disease).

Hyperthyroidism can be controlled or cured by surgical removal of part of the thyroid gland, by the oral administration of radioactive iodine, or by the use of antithyroid drugs. Hypertrophy of the endocrine glands can lead to toxic adenomas with a malignant potential if not treated promptly.

Case Study 32-1 Hyperthyroidism

A 44-year-old woman complains to her physician of weight loss, palpitations, and weakness. The physician notices that the patient has a goiter and exophthalmos. The patient admits that she is very irritable and is intolerant of heat. As the physician performs a physical examination of the patient he notices that she has bilateral eyelid lag with an enlarged thyroid gland and an audible bruit over the right lobe. Laboratory test results show a hematocrit level of 37%. The T_3 and T_4 levels are grossly elevated and the ^{131}I uptake by the thyroid gland is very high at 18% in 4 hours. The physician diagnoses the patient's condition as hyperthyroidism.

Questions

1. Explain how T_3 and T_4 serum levels are measured. How are they related to the thyroid?
2. Explain how TRH and TSH are involved in the regulation of thyroid hormone production and secretion.
3. In hyperthyroidism, what is the mechanism of increased ^{131}I uptake?

Radioactive iodine, used in the treatment of hyperthyroidism, is usually administered in the form of sodium iodide, NaI. The body converts the inorganic radioactive iodide into thyroglobin in the thyroid gland, thus subjecting that gland to radiation that will cut down its activity.

32-8 Hormones of the Parathyroid Glands

There are four small parathyroid glands attached to the thyroid gland. In humans these glands are reddish brown and together weigh 0.05 to 0.3 g. In early experimental thyroidectomies (removal of the thyroid glands) in animals, the parathyroid glands were also inadvertently removed. This caused the death of the animals.

The parathyroid glands produce *parathyroid hormone*, a hormone that influences the metabolism of calcium and phosphorus in the body. This hormone is a protein with a molecular mass of approximately 9500 and consists of a single polypeptide chain of 84 amino acids. The parathyroid gland cannot store this hormone, so the hormone is synthesized and secreted continuously. Recall that calcitonin (Section 32-7) is also involved in the regulation of calcium. Administration of vitamin A decreases parathyroid hormone, possibly by increasing calcium uptake into the parathyroid gland.

Surgical removal of the parathyroid glands causes hypoparathyroidism, characterized by such symptoms as muscular weakness, irritability, and tetany, owing to a decrease in the calcium content of the blood plasma. Death occurs because of convulsions caused by the lack of calcium. At the same time as the calcium content of the plasma is decreasing, the calcium content of the urine is also decreasing and the phosphate content of the plasma is increasing.

The symptoms of hypoparathyroidism can be relieved by treatment with vitamin D or calcium salts, or both.

Hyperparathyroidism is an increase in the production of hormones by the parathyroid glands. It is usually caused by a tumor of those glands (parathyroid adenoma) and produces such symptoms as decalcification of the bones, followed

Case Study 32-2 Hyperparathyroidism

A 60-year-old male hemodialysis patient is admitted to the hospital with complaints of abdominal pains, back pain, tiredness, constipation, and myalgia. The patient has been on dialysis for 10 years but has been relatively healthy, with only hypertension. Laboratory test results showed[†]:

Calcium	3.0 mmol/L
Phosphorus	0.96 mmol/L
Alkaline phosphatase	19 units (King-Armstrong)

A urine analysis shows that calcium oxalate crystals are present. On a controlled diet for calcium, his 24-hour urine collection results range between 9.5 to 11.3 mmol. The patient has a parathyroid adenoma, which is then removed. Postoperatively the calcium was 2.5 mmol/L and the phosphorus was 1.2 mmol/L.

Questions

1. How does the parathyroid hormone regulate calcium and phosphorus levels?
2. How does calcitonin and vitamin D affect calcium and phosphate?

[†]*See back of book for normal laboratory values.*

by deformation and fractures of the bones, nausea, and polyuria. Deposits of calcium occur in soft tissues, and renal stones frequently occur.

In hyperparathyroidism, the calcium content of the blood plasma is high and the phosphate content low. The extra calcium in the blood is obtained from the bones, thus causing the bones to become decalcified. Urine calcium, phosphate, and cAMP are increased.

Hyperparathyroidism is usually treated by the surgical removal of the tumor of the parathyroid glands.

32-9 Hormones of the Adrenal Glands

The adrenal (suprarenal) glands are located close to the upper pole of the kidneys and weigh about 3 to 6 g each. The adrenal glands are divided into two distinct portions: the cortex, which is the outer portion, and the medulla, which is the inner portion (see Figure 32-3). Each of these portions is distinct both structurally and physiologically, and each produces its own hormones.

Hormones of the Adrenal Cortex

The hormones of the adrenal cortex are steroidal and fall into three categories:

1. Glucocorticoids, which primarily affect the metabolism of carbohydrates, lipids, and protein. Examples are corticosterone, cortisone (11-dehydroxycorticosterone), and cortisol (hydrocortisone).
2. Mineralocorticoids, which primarily affect the transportation of electrolytes and the distribution of water in the tissues. The most potent of this group is aldosterone.
3. Androgens or estrogens, which primarily affect secondary sex characteristics. The principal androgen is dehydroepiandrosterone.

The steroids of the adrenal cortex function in the cell nucleus for the synthesis of RNA and protein. The steroid nucleus contains four fused carbon rings, numbered as shown at the right.

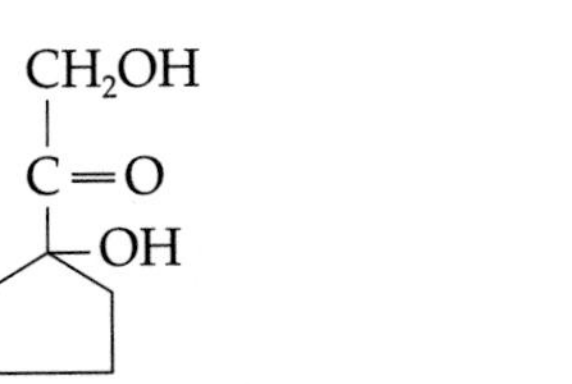

steroid nucleus

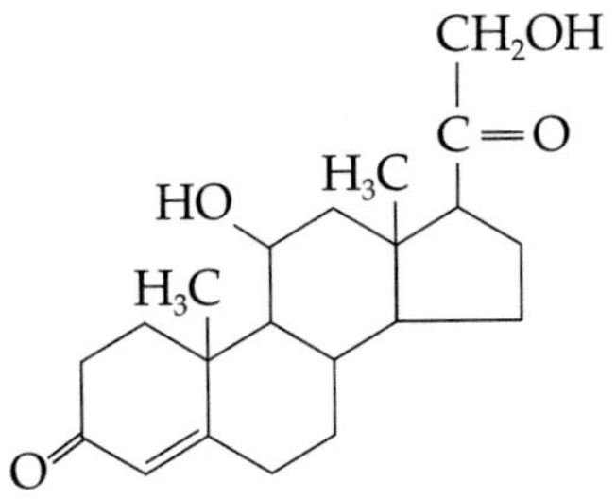

corticosterone

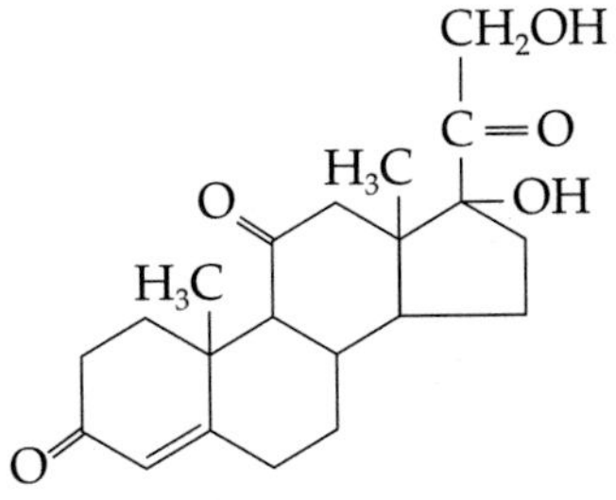

cortisone

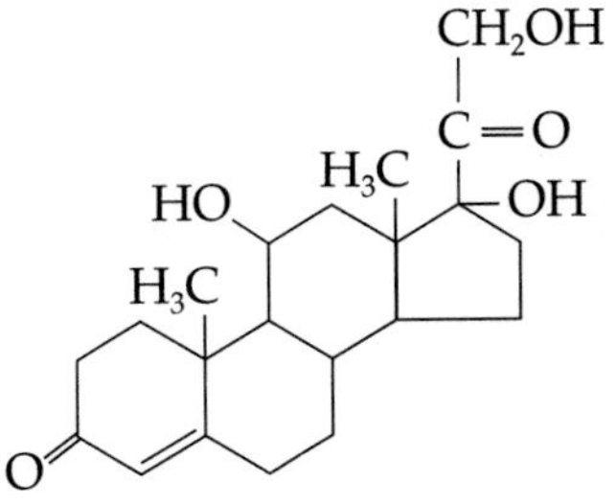

cortisol

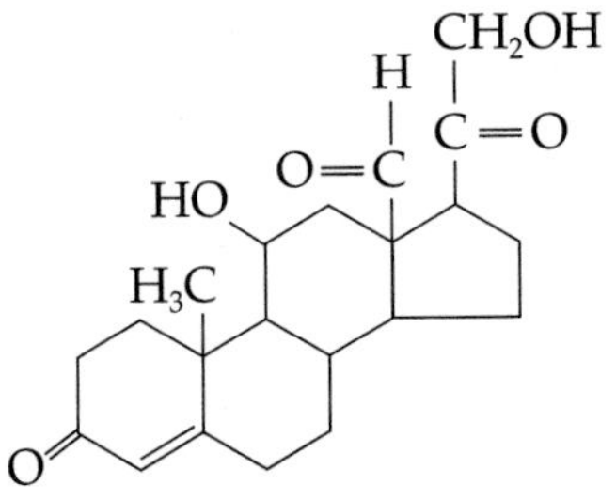

aldosterone

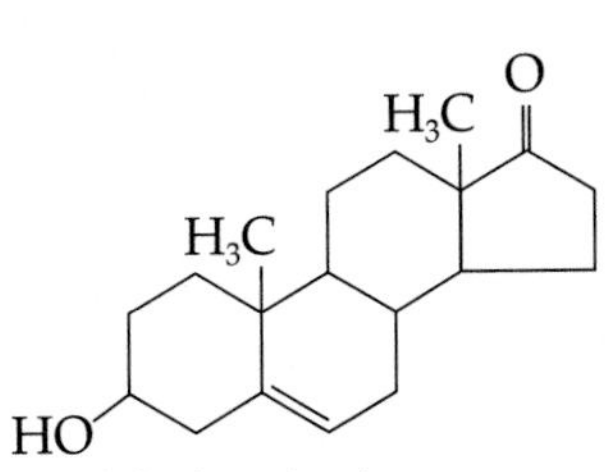

dehydroepiandrosterone

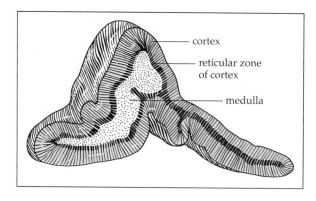

cortex

reticular zone of cortex

medulla

FIGURE 32-3

Drawing fo adrenal gland showing crotex and medulla.

The adrenal steroid hormones are released into the blood plasma as they are produced, so very little, if any, is stored within the adrenal cells.

Effect of a Deficiency of Adrenal Cortex Hormones   If the adrenal glands are removed from an animal, it will soon die because of a lack of the hormones produced by the adrenal cortex. In humans, a hypofunctioning of the adrenal glands, because of a tuberculosis of those glands or in association with pernicious anemia, diabetes, or hypothyroidism, results in Addison's disease. It is characterized by an excessive loss of sodium chloride in the urine, low blood pressure, low body temperature, hypoglycemia, elevation of serum potassium, muscular weakness, a progressive brownish pigmentation of the skin, nausea, and loss of appetite.

The preceding symptoms, except for the pigmentation, are due to a lack of the salt- and water-regulating hormones—the *mineralocorticoids*, primarily aldosterone. These hormones stimulate the kidney tubules to reabsorb sodium ions. If there is a lack of these hormones, sodium ions are not reabsorbed and are eliminated in the urine. Along with the sodium ions, chloride ions are also eliminated so that the urine is abnormally high in sodium chloride. Accompanying the sodium chloride loss by the body is a loss of water through osmosis. This loss in water in turn decreases both blood volume and blood pressure. The blood becomes more concentrated, and excretion of urea, uric acid, and creatinine is decreased, thereby increasing the concentration of these substances in the blood. Dehydration occurs rapidly and death occurs from circulatory collapse.

Another function of the hormones of the adrenal cortex is to stimulate the process of gluconeogenesis, the formation of glucose from amino acids. This process takes place in the liver. In the absence of the adrenocortical hormones, the *glucocorticoids*, the blood sugar falls and the glycogen stores in the liver and the muscles decrease considerably.

All of these symptoms, except the skin discoloration, can be relieved by the use of cortisonelike compounds—either naturally occurring extracts or synthetically prepared ones.

Effects of Hyperactivity of the Adrenal Cortex   Hyperactivity of the adrenal cortex (hyperadrenocorticism) is caused by a tumor on the adrenal cortex, by an overdosage of cortisone or ACTH, or by an increased production of ACTH (see page 589).

In children, hyperactivity of the adrenal cortex is manifested by early sexual development. Hyperactivity of the adrenal cortex in the adult female, which occurs more often than in the adult male, causes a decrease in feminine characteristics. The voice deepens, the breasts decrease in size, the uterus atrophies, and hair appears on the face. In the adult male, hyperactivity of the adrenal cortex is manifested by an increase in male characteristics—an increase in amount of body hair, increased size of sex organs, deeper voice.

Other effects of hyperadrenocorticism are hyperglycemia and glycosuria, retention of sodium ions and water, increased blood volume, edema, depletion of potassium ions (hypokalemia), and excessive gluconeogenesis.

Hormones of the Adrenal Medulla

Even though the medulla of the adrenal glands secretes three hormones—epinephrine (adrenaline), norepinephrine (noradrenaline) and dopamine—it is not essential to human life. That is, it can be removed without causing death.

Dopamine and norepinephrine are precursors of epinephrine and account for 80 percent of the adrenal medulla hormones.

The hormones of the adrenal medulla (particularly epinephrine) are necessary for the body's adaptation to acute and chronic stress.

Under conditions of stress, epinephrine does several things. It rapidly provides fatty acids as the primary fuel for muscle action. It increases glycogenolysis and gluconeogenesis in the liver and decreases glucose uptake in muscle and other organs. It also depresses insulin release to preserve glucose for the central nervous system. Other effects are increased blood flow to the brain, stimulation of heart activity, increased rate of breathing, constriction of arterioles of the skin, and simultaneous dilation of the arterioles of the skeletal muscles.

Epinephrine also has the following effects.

1. It relaxes the smooth muscles of the stomach, intestines, bronchioles, and bladder. This relaxing effect on the muscles of the bronchioles makes epinephrine especially useful in the treatment of asthma and hay fever.

2. Epinephrine is sometimes used during minor surgery when it is administered along with a local anesthetic. The constriction of the arterioles by epinephrine prevents the anesthetic from spreading too rapidly from the site of the injection. Epinephrine can be injected directly into the heart muscle when that organ stops beating or when it does not start to beat in a newborn baby.

3. Epinephrine production is increased during anxiety, fear, or other stress. This extra epinephrine in turn causes a rise in blood sugar, frequently exceeding the renal threshold. In this case, glucose appears in the urine. Such a condition is called emotional glycosuria and disappears as soon as the stress is relieved.

Norepinephrine raises blood pressure by constricting the arterioles. It does not affect the heart itself and does not relax the muscles of the bronchioles as does epinephrine. Norepinephrine is found in the sympathetic nerves where it acts as a neurotransmitter.

The medulla of the adrenal glands rarely becomes diseased. To date, no deficiency effects of its hormones are known. However, certain tumors of the medulla of the adrenal glands stimulate these glands to produce excess hormones. The symptoms caused are intermittent hypertension leading to permanent hypertension and eventually to death from such complications as coronary insufficiency, ventricular fibrillation, and pulmonary edema.

32-10 Hormones of the Pituitary Gland

The pituitary gland (*hypophysis*) is located at the base of the brain. It consists of two parts: the anterior and intermediate lobes (the *adenohypophysis*) and the posterior lobe (the *neurohypophysis*). Each of these parts secretes or releases its own hormones. The pituitary gland has often been called the "master gland" of the body because it seems to exert a direct influence on most of the other endocrine glands. It is now known that almost all of the secretory activity of the pituitary gland is controlled by a small area of the brain known as the *hypothalamus* (see Figure 32-4).

Hormones of the Anterior Lobe

Six hormones have been isolated from the anterior lobe of the pituitary gland.

1. Growth hormone (GH)
2. Thyrotropic hormone (TSH)
3. Adrenocorticotropic hormone (ACTH)
4. Prolactin, or lactogenic hormone (LTH)
5. Luteinizing hormone (LH), also known as the interstitial cell-stimulating hormone (ICSH)
6. Follicle-stimulating hormone (FSH)

Growth Hormone   Growth hormone (GH), also called somatotropin, is a protein with a molecular mass of about 21,500 and contains 191 amino acids. It is present in much greater quantities than other pituitary hormones. Growth hormone stimulates the growth of the long bones at the epiphyses, stimulates the growth of soft

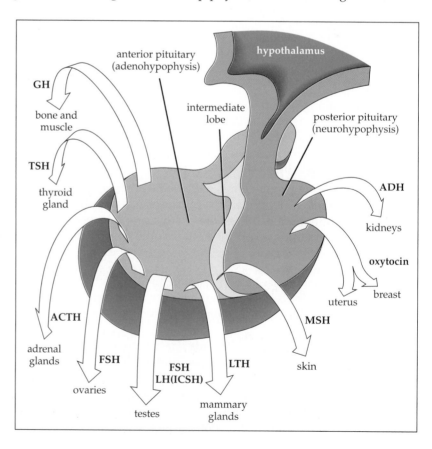

FIGURE 32-4
Hormones of the pituitary.

tissue, and increases the retention of calcium ions. The growth hormone also increases protein synthesis, leading to a positive nitrogen balance. In the muscle, growth hormone increases amino acid transportation across membranes, also leading to increased synthesis of protein, DNA, and RNA. Growth hormone causes the mobilization of fatty acids from fat deposits, providing cellular fuel. In the muscles, growth hormone antagonizes the effects of insulin; it inhibits glucose metabolism by muscle tissue.

Growth hormone also stimulates the mammary glands and increases the production of somatomedins from the liver. Somatomedins, in turn, foster sulfate incorporation into the cartilage.

Underactivity of the anterior lobe of the pituitary gland in children leads to a deficiency of growth hormone, causing dwarfism. These children develop normally but do not grow in size. Unlike cretins, they are not mentally retarded, but they may be sexually underdeveloped.

Overactivity of the anterior lobe of the pituitary gland, possibly caused by a tumor, results in an overproduction of growth hormone. When this occurs in a child, growth is stimulated and gigantism results. If the overactivity of this gland occurs during adulthood, the individual does not grow in size, but the bones in the hands, feet, and face grow, producing a condition known as *acromegaly.

Secretion of growth hormone is regulated by the growth hormone-releasing factor (GH-RF) from the hypothalamus.

Thyrotropic Hormone   The thyrotropic hormone is also called the thyroid stimulating hormone (TSH). This hormone is a glycoprotein with a molecular mass of approximately 30,000. The release of TSH from the pituitary gland is regulated by the thyrotropic-releasing factor (TRF) from the hypothalamus.

A deficiency of TSH causes the thyroid gland to atrophy. As a result, thyroxine production ceases and the metabolic rate drops.

If this hormone is injected into an animal, the symptoms of hyperthyroidism appear—increased metabolic rate, increased heart rate, and *exophthalmos.

Adrenocorticotropic Hormone   The adrenocorticotropic hormone (ACTH) is a polypeptide with a molecular mass of 4500. ACTH contains 39 amino acids, but it has been shown that only the first 23 are required for activity. The remaining 16 vary according to the animal source. The release of ACTH by the pituitary gland appears to be regulated by the corticotropin-releasing factor (CRF) of the hypothalamus in response to various biologic stresses. ACTH has been prepared synthetically and is used medically.

ACTH stimulates the synthesis and release of the hormones by the adrenal cortex. ACTH, like many other hormones, controls its target tissue through cAMP. Administration of ACTH to a normal person causes retention of sodium ions, chloride ions, and water, elevation of blood sugar, and increased excretion of potassium ions, nitrogen, phosphorus, and uric acid.

Prolactin   Prolactin (lactogenic hormone or luteotropin [LTH]), was first identified by its property of stimulating the formation of "crop milk" in the crop glands of pigeons. The arrangement of the amino acids in prolactin has been determined. It is a protein with a molecular mass of about 23,000.

Prolactin initiates lactation. In mammals, a hormone produced by the placenta stimulates the growth of the mammary glands and at the same time inhibits the secretion of prolactin. At parturition the inhibiting effect of the placenta is not present, so prolactin is secreted and thus initiates lactation.

Luteinizing Hormone Luteinizing hormone (LH) is also known as the interstitial cell-stimulating hormone (ICSH). This hormone is a glycoprotein with a molecular mass of about 40,000. It was the first pituitary gonadotropin whose sequence of amino acids was precisely determined.

LH stimulates the development of the testes in males and also causes an increased production of testosterone. In male animals who have had the pituitary gland surgically removed (hypophysectomy), administration of LH increases the weight of the seminal vesicles and also of the ventral lobe of the prostate gland.

In females, LH plays an important role in ovulation. It not only causes the production of the corpus luteum but sustains it and stimulates the production of progesterone. In hypophysectomized females, an injection of LH stimulates repair of interstitial ovarian tissues and also increases ovarian weight.

Follicle-Stimulating Hormone Follicle stimulating hormone (FSH) is a glycoprotein with a molecular mass of 25,000. It stimulates and initiates the development of the follicles of the ovaries and prepares those follicles for the action of LH. It also stimulates the secretion of estrogen. In males it causes the growth of the testes and stimulates the production of spermatozoa.

Gonadotropic Hormone The gonadotropic hormones LH and FSH (and also prolactin) are secreted by the anterior lobe of the pituitary gland. Removal of the anterior lobe in a male causes atrophy of the testes, prostate gland, and seminal vesicles. Removal in a female causes atrophy of the ovaries, uterus, and fallopian tubes.

LH and FSH as well as TSH consist of an α and a β chain of amino acids. The α chain is the same in all three, so biologic specificity must reside in the β chain. Both LH and FSH stimulate cAMP synthesis in appropriate organs.

Hormone of the Intermediate Lobe

The pars intermedia or intermediate lobe of the pituitary gland secretes a hormone called *intermedin* or the *melanocyte-stimulating hormone* (MSH). It is a polypeptide with α and β parts.

MSH increases the deposition of melanin in the human skin, thus producing darker pigmentation. When the adrenal cortex is underactive, as in Addison's disease, more MSH is produced. This leads to an increased synthesis of melanin with the accompanying brown pigmentation of the skin. Epinephrine and, even more strongly, norepinephrine inhibit the action of the melanocyte stimulating hormone.

Hormones of the Posterior Lobe

The posterior lobe of the pituitary gland contains two hormones—*vasopressin* and *oxytocin*—which are produced in the hypothalamus and stored in this lobe of the pituitary gland.

Vasopressin Vasopressin, also called the antidiuretic hormone (ADH), stimulates the kidneys to reabsorb water. When the water content of the body is high, very little of this hormone is secreted and more water is eliminated. Conversely, when the water content of the body is low, more of this hormone is secreted, causing the kidney tubules to reabsorb more water. Thus, this hormone serves to regulate the water balance in the body.

Diabetes insipidus (see page 512) is cased by the absence of the antidiuretic hormone, which results in excessive daily elimination of water (up to 30 L). Diabetes insipidus can be controlled by the administration of ADH.

Vasopressin, or ADH, stimulates the peripheral blood vessels to constrict and cause an increase in blood pressure. Because of this effect, it has been used to overcome low blood pressure caused by shock following surgery.

Vasopressin is a polypeptide whose arrangement of amino acids is indicated at the right.

Cys—Tyr—Phe
|　　　　　|
Cys—Asn—Gln
|　　　　　|
Cys—Arg—Gln

vasopressin

Oxytocin　Oxytocin is also a polypeptide whose structure is similar to that of vasopressin, except that isoleucine occurs in place of phenylalanine and leucine in place of arginine.

Oxytocin contracts the muscles of the uterus and also stimulates the ejection of milk from the mammary glands. Oxytocin is used in obstetrics when uterine contraction is desired.

32-11　Hormones of the Hypothalamus

The hypothalamus secretes certain neurohormones, some of which act as stimulators and others as inhibitors for the secretion of hormones by the anterior pituitary. These hormones are also called *releasing hormones* or *factors*. They have a relatively short life span in the bloodstream. Their half-life is about 2 to 3 minutes compared to 10 to 15 min for the growth hormone. Therefore, levels of releasing factors are 100 to 1000 times lower in the bloodstream than those of growth hormone.

The first of the hypothalamic hormones to be isolated (and later synthesized) was the thyrotropin releasing hormone (TRH), also known as the thyrotropin releasing factor (TRF), which controls the release of the thyroid stimulating hormone (TSH). TRH is a tripeptide containing three amino acids—glutamic acid, histidine, and proline. The structure of TRH is as follows:

TRH, thyrotropin releasing hormone
(pyroglutamylhistidylproline amide)

TRH is highly specific and causes an increase of TSH within 1 minute. It can be administered orally. TRH is used to distinguish between lesions in the pituitary and the hypothalamus. TRH also affects the central nervous system. Synthetic analogs of TRH have been prepared that enhance the effect on the central nervous system while not affecting the thyrotropin releasing influence.

The second hypothalamic releasing factor that was isolated was the luteinizing hormone-releasing hormone (factor), LHRH, which controls the release of the luteininzing hormone from the anterior pituitary. It is a decapeptide (10 amino acid chain), whose amino acid sequence is as follows:

Glu-His-Trp-Ser-Tyr-Gly-Leu-Arg-Pro-Gly

An injection of LH-RH causes an increase in circulating LH in 1 to 2 min. LH-RH also increases the amount of FSH.

Another releasing factor is corticotropin (ACTH) releasing hormone, CRH, which has recently been synthesized. It contains 41 amino acids, as shown in Figure 32-5. Other releasing factors are growth hormone–releasing hormone (factor), GH-RH (see Figure 32-5); prolactin releasing hormone, PRH; melanocyte stimulating hormone releasing hormone (factor), MSHRH.

In addition to the releasing factors, the hypothalamus also contains *release-inhibiting factors*—hormones that inhibit the release of the "releasing factors."

Among the release inhibiting factors are the growth hormone release inhibiting factor, GHRIH or GIF. GIF, also called somatostatin, is a tetradecapeptide (14 amino acid chain; see Figure 32-5). This factor inhibits the release of the growth hormone. GIF also inhibits the release of insulin (see page 579), glucagon, gastrin, TSH, and FSH.

Other release inhibiting factors are prolactin releasing hormone–inhibiting hormone (or factor), PRIH or PIF; and follicle stimulating hormone–release inhibiting factor, FSHRIH or FSHRIF. Additional release inhibiting factors have been postulated but have not been isolated or identified.

The hypothalamus also produces two hormones, which are stored in and secreted by the posterior lobe of the pituitary gland. They are vasopressin, the antidiuretic hormone (ADH), and oxytocin (see page 591).

In addition, the hypothalamus contains polypeptides called *endorphins*. One of these compounds, β-endorphin (see Figure 32-6), exhibits morphinelike activity in the brain and has been used in the treatment of psychiatric patients. In rats α-endorphin produces analgesia, whereas γ-endorphin induces violent behavior.

A subclass of endorphins is the *enkephalins*, which are pentapeptides. However, the enkephalins act separately in the body instead of in conjunction with the endorphins.

Two predominant enkephalins in the brain differ only in the end amino acid:

Tyr-Gly-Gly-Phe-Leu   leucine enkephalin
Tyr-Gly-Gly-Phe-Met   methionine enkaphalin

Morphine and other opiates bind to opiate receptors in the brain. Endorphins and enkaphalins also bind to these same receptors.

The enkaphalins are involved with the sensation of pain. It is believed that enkaphalins bind to the opiate receptors and prevent the transmission of pain impulses. Morphine also binds to opiate receptors and increases the pain-killing effect of enkephalins.

Opiates such as morphine are addictive. Why? In the cells, the synthesis of cAMP (see page 577) is catalyzed by the enzyme adenylate cyclase. Opiates and enkephalins inhibit the production of this enzyme so that the amount of cAMP in

FIGURE 32-5

Structures of some hypothalamic releasing and release-inhibiting hormones.

CRH	Ser-Gln-Glu-Pro-Pro-Ile-Ser-Leu-Asp-Leu-Thr-Phe-His-Leu-Leu-Arg-Glu-Val-Leu-Glu-Met-Thr-Lys-Ala-Asp-Gln-Leu-Ala-Gln-Gln-Ala-His-Ser-Asn-Arg-Lys-Leu-Leu-Asp-Ile-Ala
GRH	Tyr-Ala-Asp-Ala-Ile-Phe-Thr-Asn-Ser-Tyr-Arg-Lys-Val-Leu-Gly-Gln-Leu-Ser-Ala-Arg-Lys-Leu-Leu-Gln-Asp-Ile-Met-Ser-Arg-Gln-Gln-Gly-Glu-Ser-Asn-Gln-Glu-Arg-Gly-Ala-Arg-Ala-Arg-Leu
GHRIH (somatostatin)	⌐————————— S ————————— S ———⌐ Ala-Gly-Cys-Lys-Asn-Phe-Phe-Trp-Lys-Thr-Thr-Phe-Ser-Cys

the cells decreases. Therefore, the cells try to compensate by synthesizing more of the enzyme. When this occurs, more opiate must be used to have the same pain-killing effect. Eventually, the body will adjust to the new amount of opiate and again cAMP will begin to increase. Thus, the addict requires more and more of the drug.

Conversely, if the opiate is withdrawn, the synthesis of adenylate cyclase is no longer inhibited and large amounts of cAMP are produced. It is the high con-centration of cAMP that produces the symptoms of withdrawal.

The increased production of endorphins at about 20 to 30 minutes into a run may be responsible for a "runner's high" and for the addiction to running experi-enced by many runners.

It is believed that acupuncture works by triggering nerve impulses that stim-ulate the release of endorphins and enkephalins.

32-12 The Female Sex Hormones

The ovary secretes two different types of hormones. The follicles of the ovary secrete the follicular or estrogenic hormones. The corpus luteum that forms in the ovary from the ruptured follicle secretes the progestational hormones.

Estrogenic or Follicular Hormones

The maturing follicles of the ovaries produce the estrogenic hormones, which are also called *estrogens*. These hormones are *estradiol, estrone,* and *estriol.* Of these three hormones, estradiol is the parent compound and also the most active. The other two hormones are derived from estradiol. Estriol is the main estrogen found in the urine of pregnant women and also in the placenta. Estrone is in metabolic equilibrium with estradiol. Note the similarities of structures of these three hormones. One international unit (1IU) of estrogen activity is equal to 0.1 mg of estrogen.

Estradiol, estrone, and estriol are concerned with the maturation of the eggs (ova) and the maintenance of the secondary sex characteristics. In lower animals, the estrogens produce estrus, the urge for mating. The estrogens also suppress production of FSH, the follicle stimulating hormone, by the pituitary gland. FSH initially starts the development of the follicle. Once the follicle begins to develop, further production of FSH is not needed and is inhibited by the estrogenic hor-mones. However, the estrogenic hormones stimulates the production of LH (luteinizing hormone).

estradiol   estrone   estriol

Conjugated estrogens for oral administration contain a mixture of estrogens obtained from natural sources. They are naturally occurring sodium salts and so are water soluble. One such estrogen is Premarin. Estrogens can be used to treat women with underdeveloped female characteristics; they are also used to treat some symptoms associated with menopause and osteoporosis.

Progestational Hormones

The corpus luteum is produced in the follicle after the matured ovum is discharged into the uterus. The corpus luteum produces a hormone called *progesterone*. This hormone causes development of the endometrium of the uterus, preparing the uterus to receive and maintain the ovum, and it stimulates the mammary gland. Progesterone inhibits estrus, ovulation, and the production of LH, the hormone that initially stimulated ovulation, which led to the formation and maintenance of the corpus luteum. If the ovum is not fertilized, the corpus luteum breaks down and menstruation follows. If the ovum is fertilized, progesterone from the corpus luteum aids in the development of the placenta.

Progesterone is excreted as *pregnanediol* and is found in the urine.

In addition to progesterone, the corpus luteum also produces another hormone, *relaxin*, which is a polypeptide with a molecular mass of 5521. Relaxin also occurs in the placenta. It causes ligaments of the symphysis pubis to distend, increases dilation of the cervix in pregnant women at parturition, and also helps, along with estrogen and progesterone, to maintain gestation.

progesterone

pregnanediol

32-13   The Male Sex Hormones

Male hormones are produced primarily in the testes, although small amounts are also produced in the adrenal glands.

The male hormones are called *androgens*. They consist of testosterone and dihydrotestosterone, with the former being the major steroid secreted by the adult testes. The principal metabolic product of testosterone is dihydrotestosterone (DHT), the active form of the hormone in many tissues.

testosterone

DHT

The male hormones, mainly testosterone and dihydrotestosterone, are involved in (1) spermatogenesis, (2) development of the male sex organs, (3) development of secondary male characteristics, (4) male-pattern behavior, and (5) gene regulation. Note that the structures of the male sex hormones are similar to those of the female sex hormones. Both are derived from a common substance, cholesterol.

A deficiency in testosterone synthesis is called hypogonadism. If this occurs before puberty, secondary male characteristics fail to develop. If it occurs in the adult, many of the male characteristics regress.

32-14 Hormones of the Kidney

In addition to their excretory function, the kidneys act as endocrine glands. Among the hormones produced by the kidneys are *renin* (Section 24-7), which increases the force of the heartbeat and constricts the arterioles, and *erythropoietin* and *erythrogenin*, which cause the bone marrow to stimulate the production of red blood cells. The kidneys play an important role in the activation of vitamin D (see Figure 31-2). The kidneys also produce prostaglandins PGE_1, PGE_2, PGE_{20}, and possibly others. PGE_1 antagonizes the effects of vasopressin; PGE_2 increases excretion of sodium; and PGE_{20} decreases venous tone.

32-15 Hormones That Regulate Calcium Metabolism

Calcium ions regulate a number of body processes. Among these are coagulation of blood, enzyme reactions, secretory processes, neuromuscular excitability, plasma membrane transport, bone and teeth mineralization, recovery of the visual cycle, and release of hormones and neurotransmitters. To ensure that these processes function normally, calcium ions must be controlled within very narrow limits.

Plasma calcium exists in three forms—complexes with organic acids (6 percent); protein bound, mostly with albumin (47 percent); and ionized calcium ion (47 percent).

Three hormones control the body's calcium. They are the parathyroid hormone (see Section 32-8), calcitriol, and calcitonin (see page 582).

Calcitriol can be produced in the skin from vitamin D. The amount of calcitriol is related directly to the amount of sunlight on the skin and inversely to the amount of skin pigmentation. It is also related to age due to age-related loss of 7-dehydrocholesterol and hence of vitamin D in older persons. Calcitriol is also produced in the liver, the kidneys, and the intestines.

Calcitriol stimulates intestinal absorption of calcium and phosphates. It is made from vitamin D_3 in the liver and kidneys. Thus, a lack of calcitriol is caused by a lack of vitamin D_3 and is another aspect of deficiency of that vitamin (see page 558).

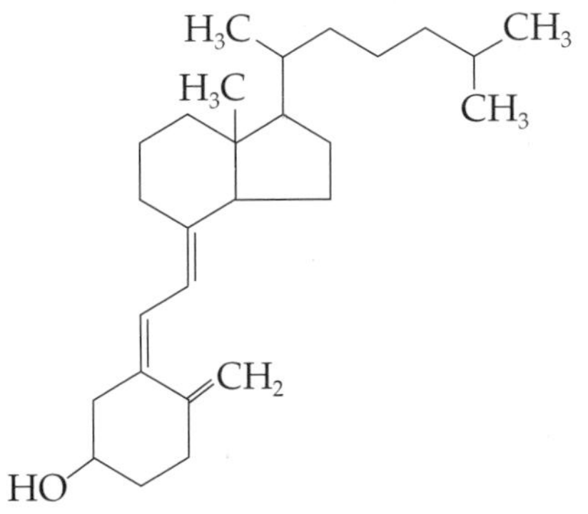

vitamin D_3 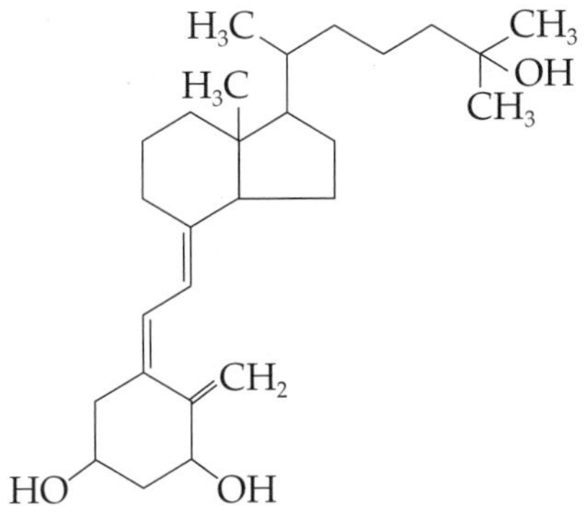   calcitriol

32-16 Neurotransmitters

Hormones carry chemical messages from the endocrine glands to the specific body part that those hormones affect. Neurotransmitters carry chemical messages from one nerve cell to another. Although hormones can travel a great distance in the

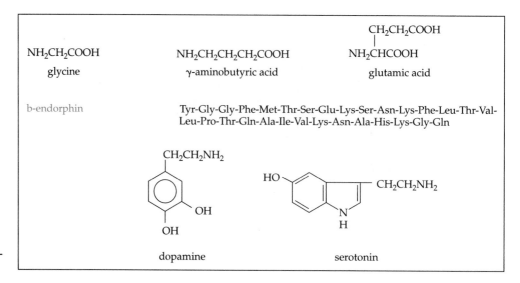

FIGURE 32-6

Structures of some neuro-
transmitters.

body, neurotransmitters only travel very short distances from one neuron to the neighboring one. Whereas hormones are produced in the endocrine glands, neurotransmitters are produced in the neurons.

Some neurotransmitters are simple amino acids, some are polypeptides, and others are in a group called catecholamines (see Figure 32-6). When a nerve impulse travels along a neuron, it comes to a synapse, a space between that neuron and the adjacent one. The impulse cannot cross that gap. A neurotransmitter is needed to carry the impulse across (see Figure 32-7). It should be noted that the parasympathetic nervous system uses acetylcholine as a neurotransmitter, whereas the sympathetic nervous system uses norepinephrine.

When the impulse reaches the presynaptic membrane, that membrane becomes depolarized and allows calcium ions in rapidly. The calcium ions cause the fusion of the vesicle membranes with the presynaptic membrane, thereby releasing the contents of those vesicles, the neurotransmitters. After the release of the neurotransmitters, the vesicles reform rapidly.

The neurotransmitters travel across the synaptic cleft and bind to receptor sites on the postsynaptic membrane, where they alter the permeability of that membrane to sodium and potassium ions and so produce a new impulse that can travel along that neuron.

The neurotransmitter must be removed or deactivated so further transmission of nerve impulses can occur. For example, acetylcholine, a neurotransmitter, is deactivated by the enzyme choline acetyltransferase and forms choline and acetic acid. Monoamine neurotransmitters such as serotonin, epinephrine, norepinephrine, and dopamine are deactivated by the oxidoreductase monoamine oxidase (MAO).

MAO inhibitors have been used in the treatment of hypertension and depression, but great care must be taken with their use.

Serotonin uptake inhibitors (SUIs) such as Prozac or Zoloft are used in the treatment of depression. They achieve their results in 7 to 10 days without the side effects of MAOIs.

Recent studies have shown that nitric oxide, NO, acts as one of the body's messenger molecules. NO is formed in the brain (and in other tissues) from the amino acid arginine under the influence of the enzyme nitric oxide synthase (see page 139). The presence of calcium ions is necessary in this process.

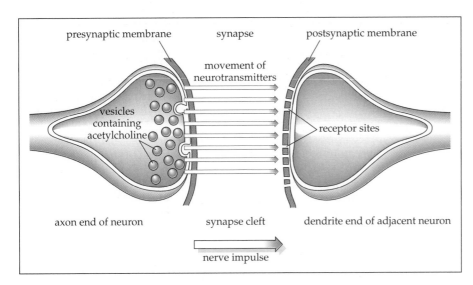

presynaptic membrane   synapse   postsynaptic membrane

movement of neurotransmitters

vesicles containing acetylcholine

receptor sites

axon end of neuron   synapse cleft   dendrite end of adjacent neuron

nerve impulse

FIGURE 32-7

Transmission of nerve impulse between neurons.

NO is released from the neurons and diffuses into adjacent neurons. Its receptor site is the iron at the center of the enzyme that forms cyclic guanosine monophosphate, cGMP. By binding to that iron, NO causes a three-dimensional change in the shape of the enzyme, thereby increasing its activity in the formation of cGMP.

Carbon monoxide, CO, long known as a poisonous gas because of its affinity for hemoblobin (see page 141), can also act as a neurotransmitter in the brain. It has been found that heme oxygenase, a hormone in the brain, liberates CO. The CO in turn binds to the heme in guanyl cyclase, thereby stimulatling the production of cyclic guanosine monophosphate, cGMP. It has been postulated that CO normally maintains the cGMP levels in several regions of the brain.

Summary

Hormones are produced in the body's endocrine glands and are secreted directly into the bloodstream, which carries them to the body parts on which they produce their effects.

Hormones are classified according to the mechanism of their action. Group 1 hormones bind to intracellular receptors; Group II hormones bind to cell-surface receptors.

The pyloric mucosa produces gastrin, which stimulates the secretion of hydrochloric acid and also produces greater motility of the stomach.

The mucosa of the small intestine produces secretin, which stimulates the pancreas to release the pancreatic juices, and cholecystokinin, which stimulates the contraction and emptying of the gallbladder.

The pancreas secretes insulin, which increases the rate of oxidation of glucose and also facilitates the conversion of glucose to glycogen in the liver and the muscles.

The pancreas also secretes glucagon, which causes an increase in blood sugar content by stimulating phosphorylase activity in the liver.

The hormones of the thyroid gland regulate the metabolism of the body and also affect the growth and development of the body. The thyroid hormones contain the element iodine. The hormonal iodine carried by the blood is termed protein-bound iodine, or PBI. Hypothyroidism occurs when the thyroid gland does not manufacture sufficient thyroxine. The symptoms of hypothyroidism are sluggishness, weight gain, slower heartbeat, reduced metabolic rate, and loss of appetite.

If the thyroid gland is absent or fails to develop in an infant, the effect produced is called cretinism. If the thyroid gland in an adult should atrophy, the effect produced is called myxedema. If the thyroid gland enlarges, the condition is called simple goiter.

If the thyroid gland produces excess thyroxine, hyperthyroidism results; the symptoms are increased metabolic rate, bulging of the eyes, nervousness, weight loss, and a rapid, irregular heartbeat. Certain drugs may be given to counteract the effects of an overactive thyroid gland.

Another hormone of the thyroid gland, calcitonin, keeps the calcium concentration of the blood from becoming too high by inhibiting the release of calcium ions from the bone.

The parathyroid glands produce a hormone, parathyroid hormone, which influences the metabolism of calcium and phosphorus in the body.

The adrenal glands are divided into two portions: the cortex and the medulla.

The hormones of the adrenal cortex are divided into three categories: the glucocorticoids, which primarily affect the metabolism of carbohydrates, fat, and protein; the mineralocorticoids, which primarily affect the transportation of electrolytes and the distribution of water in the body; and the androgens or estrogens, which primarily affect secondary sex characteristics. A hypo-functioning of the adrenal glands produces Addison's disease, which is characterized by an excessive loss of NaCl in the urine, low blood pressure, elevation of serum potassium, muscular weakness, brownish pigmentation of the skin, nausea, and loss of appetite.

The medulla of the adrenal glands secretes epinephrine and norephinephrine. Norephinephrine is a precursor of epinephrine. Epinephrine relaxes the smooth muscles, elevates blood pressure, and causes glycogenolysis in the liver.

The pituitary gland has three lobes: anterior, intermediate, and posterior. The flow of hormones from the pituitary gland is under the control of the hypothalamus.

The hormones of the anterior lobe of the pituitary gland are the growth hormone (GH), the thyrotropic hormone (TSH) the adrenocorticotropic hormone (ACTH), lactogenic hormone (LTH), luteinizing hormone (LH), and follicle stimulating hormone (FSH).

The posterior lobe of the pituitary secretes the hormones vasopressen and oxytocin. Vasopressin stimulates the kidneys to reabsorb water. Oxytocin contracts the muscles of the uterus and stimulates the ejection of milk from the mammary glands.

The pars intermedia (intermediate lobe) of the pituitary gland secretes the hormone intermedin, which increases the deposition of melanin in the skin.

Thy hypothalamus secretes neurohormones, some of which stimulate and others of which inhibit the secretion of hormones by the anterior pituitary.

The ovaries secrete two different types of hormones. The follicles secrete the follicular or estrogenic hormones. The corpus luteum that forms in the ovary from the ruptured follicle secretes the progestational hormones.

The male sex hormones are produced in the testes. The principal male hormone is testosterone. Both male and female sex hormones are derived from cholesterol.

The hypothalamus also contains endorphins and enkephalins, which have morphinelike activity.

The kidneys, in addition to their excretory function, also act as endocrine glands.

Calcium ions regulate many body processes. Three hormones that regulate calcium are the parathyroid hormone, calcitriol, and calcitonin.

Neurotransmitters carry chemical messages from one nerve cell to another.

Questions and Problems

1. How are hormones classified? Give an example of a hormone in each classification.

2. Where is gastrin produced? secretin? CCK? What is the function of each?

3. Which hormones are produced by the pancreas?

4. Why can insulin not be taken orally? What metallic element does it contain? What is protamine insulin?

5. What are the functions of insulin in carbohydrate metabolism?

6. Will insulin cure diabetes? Why or why not?

7. What is hypoglycemia? hyperglycemia? What are the symptoms of each? What can be done to overcome this condition?

8. What is the function of glucagon?

9. List the symptoms of diabetes mellitus.

10. The thyroid hormones contain which unusual element?

11. What are the symptoms of hypothyroidism? hyperthyroidism?

12. What is a cretin? What causes myxedema?

13. What are the symptoms of a goiter? How is it treated?

14. What are the functions of the parathyroid hormone? What are the symptoms of hypoparathyroidism? hyperparathyroidism?

15. What three types of hormones are produced by the adrenal cortex? What are their functions? Give an example of each.

16. What causes Addison's disease? What are its symptoms?

17. Indicate the symptoms of a lack of the adrenocorticotropic hormone. What are the symptoms of hyperadrenocorticism?

18. Name the hormones of the adrenal medulla. What are the functions of each?

19. Name the hormones of the anterior lobe of the pituitary gland. What are the functions of each?

20. Compare dwarfism with cretinism.

21. Name the hormones of the posterior lobe of the pituitary gland. What is the function of each?

22. What are the symptoms of an overactive anterior lobe of the pituitary gland?

23. What is MSH? Where is it formed? What is its function?

24. What types of hormones are produced by the hypothalamus? What do they do? What type of structure do they have?

25. Why is the pituitary gland no longer considered to be the "master gland" of the body?

26. Where are the male sex hormones produced? the female sex hormones?

27. Discuss the role of NO and CO as neurotransmitters.

28. Which hormones are produced by the kidneys? What are their functions?

29. How do enkephalins prevent transmission of pain?

30. Why are opiates addictive?

31. Why would a patient with hyperparathyroidism be prone to bone fractures and calcium-containing kidney stones?

32. Why would having only a large quantity of black coffee for breakfast produce an elevated level of free fatty acids in the plasma?

33. Why can hormones such as cortisol be given orally, whereas other hormones such as glucagon and ACTH must be given by injection?

34. Why does pregnancy sometimes cause glucose intolerance and hyperthyroidism?

Questions 35 to 37 pertain to the following scenario:

A 10-year-old girl with respiratory distress is brought to the emergency department. Physical examination shows her to be wheezing and her respiratory rate is 38. The girl had been playing outside for most of this warm, sunny day. She has a history of allergy to dust, mold, and pollen. A diagnosis of bronchial asthma is made, and she is given epinephrine and aminophylline, which reduce her symptoms. She is able to return home in 3 hours.

35. What effect does epinephrine have on plasma glucose and free fatty acids?

36. How are β-adrenergic receptors, cyclic nucleotides, and protein phosphorylation related to the action of epinephrine?

37. After epinephrine administration, which enzymes are involved in the metabolism and inactivation of the epinephrine?

Practice Test

1. The hypothalamus produces _____.
   a. calcitonin
   b. TST
   c. MSH
   d. oxytocin

2. A hormone involved in carbohydrate metabolism is _____.
   a. serotonin
   b. insulin
   c. epinephrine
   d. calcitriol

3. Which of the following is *not* a hormone of the anterior lobe of the pituitary gland?
   a. LTH
   b. TSH
   c. MSH
   d. GH

4. A hormone that contains the element iodine is _____.
   a. T_3
   b. insulin
   c. calcitriol
   d. somatostatin

5. A hormone involved in fluid-electrolyte balance is _____.
   a. thyroxin
   b. oxytocin
   c. insulin
   d. atriopeptin

6. An example of a neurotransmitter is _____.
   a. calcitriol
   b. nitrous oxide
   c. insulin
   d. epinephrine

7. A symptom of hyperthyroidism is _____.
   a. cancer
   b. Hashimoto's disease
   c. myxedema
   d. exophthalamos

8. Early sexual development may be caused by an overproduction of _____.
   a. intermedin
   b. adrenalin
   c. oxytocin
   d. ACTH

9. Male hormones are called _____.
   a. adrenergic
   b. estrogens
   c. androgens
   d. melatonins

10. The pineal gland produces what hormone?
    a. ACTH
    b. melatonin
    c. androgen
    d. somatostatin

Glossary

acetal Compound formed by the reaction of one molecule of an aldehyde with two molecules of alcohol. The general formula is

$$R-\overset{\displaystyle OR'}{\underset{\displaystyle OR''}{C}}-H$$

acidosis Condition in which the pH of the blood drops from 7.35.

acromegaly Pathologic enlargement of the bones of the face, hands, and feet resulting from an overactive anterior lobe of the pituitary gland.

activation energy Minimum amount of energy molecules must possess in order for a reaction to occur.

adhesion The attraction or joining of two dissimilar substances.

aerobic Requiring oxygen.

aerosol Gaseous suspension of fine solid or liquid particles.

AIDS Acquired immune deficiency syndrome is a T cell immunodeficiency in previously healthy adults in association with opportunistic infection of Kaposi's sarcoma.

alkalosis Condition in which the pH of the blood rises from 7.45.

alloy Solid solution of two or more metals.

ALT Alanine transaminase.

alveoli Air sacs in the lungs.

amoebicide Substance that destroys amoebae.

amphipathic Molecule having a hydrophobic and a hydrophilic end.

amphoteric Capable of reacting either as an acid or a base.

anaerobic Not requiring the presence of oxygen.

analgesic A pain killer.

anaphylaxis A reaction of immediate hypersensitivity that results from sensitization of mast cells by antibodies following exposure to antigen.

anion A negatively charged ion.

antibody A protein molecule that is released by a plasma cell and binds to an antigen.

antimetabolite Substance that inhibits utilization of metabolites.

antineoplastic agents Agents that inhibit the growth of malignant cells.

antipyretic Anything that reduces a fever.

antiseptic Substance capable of destroying disease-causing microorganisms.

antispasmodic Capable of preventing spasms or convulsions.

antitoxin An antibody capable of neutralizing a poison of biologic origin.

apoenzyme Protein part of an enzyme.

asphyxiation Unconsciousness or death caused by lack of oxygen.

AST Aspartate transaminase.

astringent Substance that draws together or contracts tissue.

atherosclerosis Thickening of arterial wall, characterized by deposition of fatty substances.

auscultation Act of listening through a stethoscope.

autoradiograph Self-picture taken by a radioactive substance.

bactericidal Capable of killing bacteria.

bacteriostatic Capable of inhibiting growth of bacteria without destroying them.

bradycardia Slower-than-normal heart rate.

bronchioles Thin-walled extensions of the bronchial tubes in the lungs.

bruit Abnormal sound heard on auscultation.

BUN Blood urea nitrogen.

calculi Stones, such as those found in the gallbladder or kidney.

capillary action Rise of fluid through a small opening.

carcinogenic   Cancer-causing.

cardiocyte   A heart cell.

cathartic   Laxative.

catheter   Slender, flexible tube for insertion into a body channel.

cation   Positively charged ion.

chain reaction   Self-sustaining nuclear reaction yielding products that cause further reactions of the same kind.

chelating agent   substance that aids in covalently binding two or more nonmetallic atoms to a central metal ion.

chemotherapeutic agent   Chemical used in treatment of disease, particularly cancer.

cholangiography   Radiography of bilary ducts after administration of contrast medium.

cholecystogram   X ray of gallbladder.

chorea   A nervous disorder marked by uncontrollable and irregular body movements.

chromogen   A substance without color that can be transformed into a colored dye by chemical reaction.

coenzyme   Small molecule that combines with the apoenzyme to form an active enzyme.

cohesion   Property of a substance sticking to itself.

colostrum   The first secretion of the mammary glands at the termination of pregnancy.

crenation   Shrinking of red blood cells in a hypertonic solution.

curie   Unit of radiation; 1 curie (Ci) equals 37 billion nuclear disintegrations per second.

diabetes mellitus   Disorder of carbohydrate metabolism characterized by inadequate secretion or utilization of insulin.

diuretic   Substance that increases output of urine.

eczema   Noncontagious inflammation of the skin, marked by the outbreak of lesions that become encrusted and scaly.

edema   Accumulation of fluid in the tissues.

electroencephalogram   Graphic record of the electrical activity of the brain.

electrophoresis   Separation of charged particles under the influence of an electric field.

emphysema   Pathologic enlargement of the alveoli in the lungs.

emulsification   making a colloidal suspension of one liquid in another liquid.

encephalitis   Inflammation of the substance of the brain.

endothermic   Absorbing energy.

equilibrium   Dynamic state in which the rates of opposing reactions are equal.

erythroblastosis fetalis   A hemolytic disease in newborns that results from the development in the mother of anti-Rh antibody in response to Rh-positive fetal blood.

erythrocyte   Red blood cell.

euphoria   Feeling of well-being.

excoriated   Chafed.

exogenous   Developed from external causes.

exophthalmos   Abnormal protrusion of the eyeball.

exothermic   Releasing energy.

fission   Splitting of the nucleus of an atom.

flaccid   Soft and limp.

flatulence   Excessive gas in the stomach and intestine.

free radical   A particle with an unpaired electron.

fusion   Combining of small nuclei to make a larger one.

genome   an organism's genetic material.

glycosidic linkage   Bond formed between two monosaccharides.

goiter   Enlargement of the thyroid gland visible as a swelling in the front of the neck.

granulocyte   A white blood cell (neutrophil, basophil, or eosinophil) that contains cytoplasmic granules.

halogen   Any one of group VII nonmetals.

hematocrit   Ratio of volume of cells in a given volume of blood that was centrifuged.

hematopoietic system   System responsible for the formation of blood cells.

hemiacetal   compound formed by the reaction of one molecule of an aldehyde with one molecule of an alcohol; the general formula is:

$$R - \overset{\displaystyle OR'}{\underset{\displaystyle OH}{\vert\ \ C\ \vert}} - H$$

hemiketal   Compound formed by the reaction of one molecule of a ketone with one molecule of alcohol; general formula is

$$R - \overset{\displaystyle OR''}{\underset{\displaystyle OH}{\vert\ \ C\ \vert}} - R'$$

hemolysis   Destruction of red blood cells caused by a hypotonic solution.

hemostasis   Stoppage of bleeding or blood flow.

hertz   Unit of frequency equal to 1 cycle per second.

hydrophilic   Having an affinity for water.

hydrophobic   Antagonistic to water.

hydrostatic   Referring to the pressure liquids exert.

hyperbaric   High pressure oxygen.

hypercalcemia   Increased serum calcium concentration.

hypercholesterolemia   Presence of abnormal amounts of cholesterol in the blood.

hyperglycemia   Abnormally high blood sugar level.

hyperlipidemia   An increased level of lipoproteins in the blood.

hypertension   High blood pressure.

hypertonic   Salt concentration higher than that of the blood.

hypnotic   Sleep inducer.

hypoacidity   Lower-than-normal acidity.

hypoglycemia   Abnormally low blood sugar level.

hypotonic   Salt concentration less than that of the blood.

hypoxia   Deficiency in the amount of oxygen reaching body tissues.

interstitial fluid   Fluid between tissues.

idiopathic   Of unknown causes.

isoelectric point   pH at which amino acid is neutral.

isotonic   Having the same salt concentration as the blood.

IUPAC   International Union of Pure and Applied Chemistry.

jaundice   Yellowish pigmentation of the skin caused by deposition of bile pigments.

ketal   Compound formed by the reaction of one molecule of a ketone with two molecules of alcohol; general formula is

$$R - \underset{\underset{OR'''}{|}}{\overset{\overset{OR''}{|}}{C}} - R'$$

ketoacidosis   Acidosis caused by accumulation of ketone bodies.

lacrimator   Substance that causes production of tears.

lacteals   Lymphatic vessels arising from the villi of the small intestine.

laking   Bursting of red blood cells so that hemoglobin is released into the plasma.

Lesch-Nyhan syndrome   A purine metabolic defect that causes severe mental retardation.

lymphatics   Vessels that carry lymph.

lipoprotein   A compound that contains both lipid and protein.

macroglobulinemia   Disease characterized by an increase in blood serum viscosity and presence of highly polymerized globulins.

macromolecule   A very large molecule containing hundreds or thousands of atoms.

mammography   X-ray examination of the breasts.

medulla   Inner part of a biologic structure, such as the adrenal medulla.

megaloblastic   Referring to very large immature cells.

metastases   Spreading of disease from original sites.

mitochondria   Microscopic bodies, found in all cells, that play an important part in metabolic reactions and energy production.

multiple myeloma   Disease of the bone marrow.

mutagenic   Causing biologic mutation.

myalgia   Pain in the muscles.

myocardial infarction   Formation of dead tissue resulting from an obstruction of blood vessels supplying the myocardium.

myxedema   Disease caused by decreased activity of the thyroid gland in adults.

neoplasms   Abnormal new growth of tissues; tumors.

nephritis   Inflammation of the kidneys.

occlusion   Substance containing trapped liquid and gaseous material.

oliguria   Very small amount of urine formation.

osmolarity   Measure of the concentration of particles in solution.

osmosis   Flow of solvent through a semipermeable membrane until concentrations on both sides are equal.

osteolysis   Bond destruction.

osteomalacia   Softening of the bones because of a deficiency of vitamin D, calcium, or phosphorus.

osteomyelitis   Inflammation of the bone marrow.

osteoporosis   Reduction in the quantity of bone.

pancreatitis   Inflammation of the pancreas.

paranoia   Chronic psychosis characterized by delusion of persecution or of grandeur.

parenchymal cells   Functional cells of an organ.

pathogen   Disease-producing organism.

plasmolysis   Shrinking of red blood cells because of addition of a hypertonic solution.

pneumonitis   Inflammation of the lungs.

polyuria   Large amount of urine formation.

positron   Positive electron

proteinuria   Presence of more than 1 g/L of protein in urine.

pruritus   Itching.

psychedelic   Substance that causes hallucinations.

psychoneurosis   Neurosis based on emotional conflict.

rad   Unit of radiation equal to 100 ergs per gram of irradiated tissue.

rem   Unit of radiation equivalent to the absorption of 1 roentgen by a human.

refractory   Something that is not easily treated.

refractory period   A time interval, such as the interval following the excitation of a neuron or the contraction of a muscle, during which repolarization occurs.

resonance   Property exhibited by a compound that is represented by two or more structures differing only in the position of the valence electrons.

roentgen   Unit of radiation involving x rays that produce 1 electrostatic unit of positive or negative charge in 1 cm^3 of air.

semipermeable membrane   A selectively permeable membrane.

side effect   Usually associated with pharmacologic results of therapy unrelated to the objective.

specific heat   Amount of heat required to raise the temperature of 1 g of a substance 1° C.

spectrophotometer   An instrument for measuring the intensity of the various wavelengths of light transmitted by a substance or solution.

substrate   Molecule on which an enzyme acts.

surfactant   Surface-active agent.

synapse   Region between the axon of one neuron and the dendrite of an adjacent neuron.

syndrome   A group of signs and symptoms that collectively indicate a disease.

tachycardia   Excessively rapid heart beat.

tetrahedral   Having four sides with equal triangular faces.

therapeutic   Having healing powers.

thromobophlebitis   Presence of a thrombus in a vein.

thromboses   Blood clots.

tincture   Alcoholic solution.

ultrasonic   Very high frequency sound waves.

Van Gierke's Disease   A glycogen storage disease.

vasoconstrictor   Substance that constricts blood vessels.

villi   Projections arising from a mucous membrane.

virucide   Substance that destroys viruses.

viscosity   A measure of resistance to flow.

volatile   Easily vaporized.

Answers to Odd-Numbered Questions and Problems

1. Kilogram; meter; second; kelvin

3. 10^{-6}; 10^{-2}; 10^3; 10^6; 10^{-3}

5. (a) 3; (b) 3; (c) 5; (d) 3

7. Meter

9. 230° F; 383 K

11. (a) 1580 mL; (b) 0.1226 g; (c) 0.00058 m; (d) 1250 µg; (e) 0.408 L; (f) 12.66 mL; (g) 2040 g; (h) 2,000,000 mg; (i) 0.187 g; (j) 0.0001 km

13. 63.0 in; 154 lb; 101.5° F

15. 157.5 cm; 54.5 kg

17. 10.20 g

19. 5400 cm^3; 5400 mL; 5.4 L; 0.0054 kL

21. 3000 g; 6.6 lb

23. 3.64 mL

25. 10^{-6} sec; 10^3 watts; 10^{-3} amps; 10^6 curies

27. 1200 mg; 400 mg

29. (a) 163 cm; (b) 54.5 kg; (c) 1.57 m^2; (d) 628 µg

31. 0.570 mL

33. $-40°$

Chapter 2   (page 40)

1. Element: iron; calcium
   Compound: water, boric acid, zinc oxide, sugar
   Mixture: air, table salt
   Chemical: flammability, reactivity

3. (a) Gases are compressible; liquids and solids are incompressible
   (b) Gases have low density; liquids and solids may have high or low density
   (c) Gases have no definite volume; liquids and solids have definite volumes
   (d) Gases and liquids have no definite shape; solids have definite shape

   (e) Gas particles have unrestricted motion; liquid particles have somewhat restricted motion; solid particles have highly restricted motion
   (f) Gases expand greatly when heated; liquids and solids expand slightly when heated

5. Elements and compound must be homogeneous; mixtures maybe homogeneous

7. 400 Kcal

9. Law of conservation of energy states that energy is neither created or destroyed during a chemical reaction; law of conservation of matter and energy states that matter and energy cannot be created or destroyed but they can be converted from one to the other

11. (a) Cu, Fe; (b) Ca, P; (c) Cr; (d) I

13. (a) Ca; (b) Co; (c) C,H,N,O

15. Gases expand more because gas particles do not interact with each other

17. Metals conduct heat and electricity, are shiny, ductile, and malleable
    Nonmetals usually do not conduct heat and electricity, are brittle, not ductile or malleable, and are not shiny

19. Melting causes the particles to move faster and farther apart; boiling allows the particles to become independent of each other

21. Calories is larger; 1 cal = 4.18 joules

23. 200 Cal

25. All the energy is being used to change the particles from one state to the other

27. Body heat is used to convert from liquid to gas; alcohol evaporates more easily

29. The food is frozen and placed in a vacuum to remove the water vapor; both freezing and sublimation are involved

31. Melting; condensation; condensation

33. 3018 Kcal

35. Running, 4.7 hrs; walking, 17.8 hrs; swimming, 7.6 hrs; biking, 24 hrs

Chapter 3 (page 59)

1.

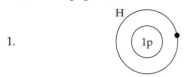

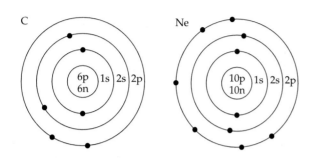

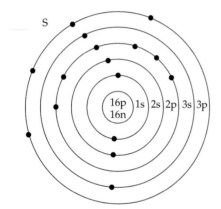

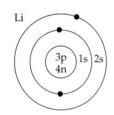

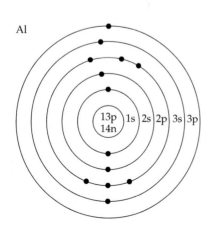

3. (a) 2 electrons; 6 levels   (b) 6 electrons; 4 levels
  (c) 7 electrons; 3 levels   (d) 2 electrons; 4 levels
  (e) 4 electrons; 3 levels   (f) 3 electrons; 2 levels
  (g) 2 electrons; 7 levels   (h) 2 electrons; 4 levels
  (i) 2 electrons; 6 levels   (j) 2 electrons; 4 levels
  (k) 5 electrons; 4 levels   (l) 8 electrons; 4 levels

5. Families; periods

7. Nucleus

9. Identical chemical properties, but different physical properties

11. All transition element are metals because all have metallic properties

13. Metals: from left to lower right; nonmetals: upper right; noble gases: last column on right side

15. Yes; table will be extended to cover all new elements

17. (a) 14 electrons; (b) 29; (c) 14; (d) IVA; (e) 3rd; (f) nonmetal; (g) Si

19. Atomic masses are the average of all naturally occurring isotopes for a particular element

21. The higher the abundance of a particular isotope, the closer the atomic mass will be to the mass of the isotope

23. Radium has a structure similar to that of calcium, which is a component of bone

25. A group number is the number of valence electrons in an atom

27. Mg: 2, 8, 2; As: 2, 8, 18, 5; Cl: 2, 8, 7

29. In general, elements that are solid at room temperature have melting and boiling points that increase across a period, whereas the melting and boiling points for gases decrease across a period

31. Halogens: group VIIA; alkali metals: group IA; alkaline earth metals: group IIA; transition metals: group B

33. 116

Chapter 4 (page 77)

1. Molecule is a combination of 2 or more atoms

3. Number of electrons in the outer shell of an atom

5. A metal ion is smaller than its atom because the excess positive charge pulls in the electrons; nonmetal ion is larger than its atom because the excess electrons repel and expand

7. Anions are negative ions: Cl^-, O^{2-}
  Cations are positive ions: Na^+, Mg^{2+}

9. Charge is determined by the number of electrons lost or gained

11. (a) sodium bromide   (b) magnesium sulfate
  (c) calcium sulfide   (d) potassium nitrate
  (e) potassium sulfide   (f) aluminum iodide

13. An ion with two or more elements: NH_4^+, SO_4^{2-}, NO_3^-

15. (a) dinitrogen pentoxide   (b) iodine bromide
  (c) carbon disulfide   (d) phosphorus tribromide
  (e) sulfur trioxide   (f) diarsenic trioxide

17. O, oxygen atom; O_2, oxygen molecule; O_3, ozone; CO, carbon monoxide; CO_2, carbon dioxide; S_8, sulfur molecule

19. F; best attractor of electrons

21. Magnesium has only two valence electrons

23. No; if same element involved in the covalent bond, then nonpolar; if different element is part of the bond, then polar

25. (a) trigonal
    (b) trigonal bipyramidal
    (c) tetrahedral
    (d) bent
    (e) linear
    (f) trigonal bipyramidal
    (g) trigonal
    (h) pyramidal
    (i) tetrahedral
    (j) octahedral

Chapter 5   (page 96)

1. (a) $2\,S + 3\,O_2 \longrightarrow 2\,SO_3$
   (b) $Mg + 2\,HCl \longrightarrow MgCl_2 + H_2$
   (c) $C + O_2 \longrightarrow CO_2$
   (d) $4\,Al + 3\,O_2 \longrightarrow 2\,Al_2O_3$
   (e) $Zn(OH)_2 + H_2SO_4 \longrightarrow ZnSO_4 + 2\,H_2O$
   (f) $Ca + 2\,AgNO_3 \longrightarrow Ca(NO_3)_2 + 2\,Ag$
   (g) $2\,N_2 + 5\,O_2 \longrightarrow 2\,N_2O_5$
   (h) $K_2SO_4 + BaCl_2 \longrightarrow BaSO_4 + 2\,KCl$

3. SO_3 40.0% S   60.0%   O
   CO_2 27.3% C   72.7%   O
   Al_2O_3 52.9% Al   47.1%   O

5. 5 moles; 3.0×10^{24} molecules

7. 512 g $ZnCl_2$

9. (a) Nature of reacting substances: reactions involving ions usually faster than reactions that require breaking covalent bonds:
      $$H^+ + OH^- \longrightarrow H_2O \qquad \text{fast reaction}$$
   (b) Temperature: increase in temperature increases reaction rate; a fever increases metabolic rate in a patient
   (c) Concentration: increased concentration increases reaction rate; giving pure oxygen to a patient increases oxygen uptake
   (d) Surface area: increased surface area increases reaction rate; flour dust can act as an explosive

11. 32.06 g

13. 1.2×10^{24} molecules; 3.6×10^{24} atoms

15. 70 g

17. Left: add heat or NH_3; remove N_2 or H_2
    Right: add N_2 or H_2; remove heat or NH_3

19. Dynamic: forward and reverse reactions continue but concentration of reactants do not change

21. (a) 122; (b) 162; (c) 152; (d) 142; (e) 294; (f) 342

23. Yes, if molecule formula is the simplest formula:
    $$H_2O;\ CO_2;\ H_2SO_4$$

25. 138 g K_2PtCl_4; 11.3 g NH_3

27. Urea, 46.7%N; $\underline{NH_3, 82.4\%N}$; NH_4NO_3, 35%N; $NaNO_3$, 16.5%

29. 454; 2.64×10^{-4} moles; 1.59×10^{20} molecules

31. Vitamin C: 3.41×10^{-4} moles; 2.05×10^{20} molecules
    Folic acid: 9.07×10^{-7} moles; 5.46×10^{17} molecules
    Vitamin A: 5.24×10^{-6} moles; 3.15×10^{18} molecules

33. (a) 66 Kcal; (b) 1.5 g; (c) 1.1 g

Chapter 6   (page 112)

1. Oxidizing agent   Reducing agent
   (a) O_2 S
   (b) HCl   Mg
   (c) $CuSO_4$ Fe
   (d) Cl_2 NaBr

3. Ethyl alcohol is oxidized by potassium dichromate (orange) to produce chromic sulfate (green); the greater the change from orange to green the more alcohol present

5. (a) $2\,FeCl_3 + SnCl_2 \longrightarrow 2\,FeCl_2 + SnCl_4$
   (b) $3\,P + 5\,HNO_3 + 2\,H_2O \longrightarrow 5\,NO + 3\,H_3PO_4$
   (c) $C_{12}H_{22}O_{11} + 12\,O_2 \longrightarrow 12\,CO_2 + 11\,H_2O$
   (d) $2\,Al + 3\,H_2SO_4 \longrightarrow Al_2(SO_4)_3 + 3\,H_2$

7. *See Table 6-2, page 110

9. Yes; reversible in theory only, not in practical sense

11. Ethyl alcohol is oxidized and acts as the reducing agent; potassium dichromate is reduced and acts as the oxidizing agent

13. $Pb + PbO_2 + 2\,H_2SO_4 \longrightarrow 2\,PbSO_4 + 2\,H_2O$
    Pb oxidized; PbO_2 reduced

Chapter 7   (page 126)

1. (a) Volume occupied by a gas is mostly empty space
   (b) Gas molecules do not attract each other
   (c) Gas molecules are in rapid motion and move in straight lines
   (d) Distance between the molecules is very great
   (e) Molecules of gas collide with no loss of energy
   (f) An increase in temperature will increase the average kinetic energy of the molecules; to maintain a constant pressure, the size of the container must be increased; reverse also true
   (g) Gas molecules are far apart, an increase in pressure will force them closer together; reverse also true

3. 1120 mmHg

5. 3.47 ft^3

7. Diffuses from higher concentration to lower concentration

9. When the pressure inside the respirator is decreased, the air in the lungs expands, forcing the diaphragm down; when the pressure in the respirator is increased, the volume of air in the lungs is decreased, allowing the diaphragm to move upward; this alternate increase and decrease in pressure enables the patient to breathe

11. CO_2, lower molar mass

13. Gas molecules are moving rapidly; when the molecules strike walls of a container they exert a pressure from the force of the strike

15. A barometer consists of a glass tube filled with mercury, the open end of which is inverted into a dish of mercury. The mercury in the tube falls until the pressure of the air just balances the pressure of the column of mercury in the tube. As air pressure increases or decreases, the mercury level in the barometer rises or falls correspondingly.

17. When the bulb is squeezed, the pressure of air inside the bulb is increased. This increased pressure is transmitted to the mercury in the cuff. As the blood exerts a pressure against the cuff, this change in pressure is indicated on the dial attached to the sphygmomanometer.

19. The solubility of a gas in a liquid at a given temperature is proportional to the pressure of that gas. Since cancer cells are more susceptible to chemotherapy at higher oxygen levels, the higher pressure inside the hyperbaric chamber allows for higher oxygen levels in the blood due to increased solubility of oxygen at the higher pressure.

21. 0.2 moles

23. 25 g/mole

25. 12 K

27. 46 balloons

29. 1.1 L; 29 g/L

Chapter 8 (page 144)

1. Oxygen: colorless, odorless, tasteless, slightly heavier than air, slightly soluble in water, moderately reactive at room temperature, very reactive at high temperature; reacts with almost all elements to form oxides

3. Laboratory: heating potassium chlorate and electrolysis of water
   Commercial: distillation of liquid air

5. passing air or oxygen through an electrical discharge; no, too toxic

7. (a) plants convert CO_2 into O_2 during photosynthesis
   (b) $2\,Na_2O_2 + 2\,CO_2 \longrightarrow 2\,Na_2CO_3 + O_2$
   $4\,KO_2 + 2\,CO_2 \longrightarrow 2\,K_2CO_3 + 3\,O_3$

9. Oxygen is used for patients with pulmonary diseases, for patients with hypoxia, for cases of asphyxia, or for those who have nearly drowned. care must be taken not to allow smoking or active electrical equipment in rooms in which oxygen is in use

11. Decomposes in light

13. CO prevents hemoglobin from carrying oxygen; CCl_4 vapors are highly toxic (liver)

15. Colorless, odorless, tasteless gas; heaver than air; slightly soluble in water; supports combustion, so danger of fire must be prevented

17. Carbon dioxide stimulates the respiratory center of the brain, which increases the rate of breathing

19. Metals form oxides; active metals form peroxides; very active metals form superoxides

21. Used to slow down oxygen usage by various organs to prevent organ damage during brain, transplant, and open heart surgery

23. Overall reactions are the reverse of each other; the reaction pathway for each is very different

25. Nasal catheter provides oxygen via tube to the nose; oxygen tent provides an oxygen environment around patient

27. Both gases are highly toxic, but hydrogen sulfide is much easier to detect because of its "rotten egg" odor

29. Giving oxygen under increased pressures; used to treat cancer, gangrene, and carbon monoxide poisoning

31. Blocks incoming ultraviolet radiation via conversion of $O_3 \longrightarrow O_2$

33. Smokers have smaller babies with a higher infant mortality rate; higher rates of lung cancer, emphysema, and cardiovascular disease

35. Acidic rain due to the presence of nitrogen and sulfur oxides; damage to crops, trees, fish, buildings, humans, cars, and sculptures

37. Helium-oxygen mixture is used to reduce the chances of the bends occurring since helium is less soluble in the blood than nitrogen

Chapter 9 (page 160)

1. Amount of heat needed to change the temperature of 1 g of water 1° C; joule

3. Ice is less dense than cold water

5. The presence of two polar oxygen hydrogen bonds and the bent shape of the oxygen molecule combine to make water a polar molecule

7. Hydroxides or bases; acids

9. Bicarbonates of calcium and magnesium, which can be removed by boiling; other soluble compounds of calcium and magnesium that can be removed by sodium carbonate, borax and water softeners

11. Water heated above natural temperatures; it decreases the amount of dissolved oxygen and speeds up metabolic processes in fish and microorganisms; it may be fatal to some forms of marine life

13. A salt combined with a definite amount of water of crystallization; loses the water

15. Water is heated to boiling and changed to steam; the steam passes through a condenser, which converts the steam back into liquid water, which is collected; distillation removes all dissolved and suspended particles

17. The positive sodium ion interacts with the partially negative oxygen in water; the negative chloride ion interacts the the partially positive hydrogen in water

19. Hydrogen bonding makes it much more difficult for water molecules to evaporate; this lowers the vapor pressure of water and raises the boiling point

21. Allows the breakdown of complex food molecules into smaller, less complex molecules

23. The higher the vapor pressure, the lower the boiling point

25. Cl_2 destroys bacteria; lime and aluminum sulfate are used to remove suspended material from water

Chapter 10   (page 183)

1. (a) 10 g of boric acid dissolved in sufficient water to make 250 mL
   (b) 1 g of $KMnO_4$ dissolved in sufficient water to make 1 L
   (c) 224 g of KCl dissolved in sufficient water to make 1.5 L
   (d) 0.9 g of NaCl dissolved in sufficient water to make 100 mL

3. Solvent: material in which something is dissolved
   Dialysis: separation of solute particles from colloidal particles by means of a semipermeable membrane
   Diffusion: process whereby a substance moves from an area of its higher concentration to a region of less concentration
   Osmosis: diffusion of water through a semipermeable membrane from a weaker solution (more water) to a stronger solution (less water)
   Isotonic: solution with the same solute concentration

5. The scattering of light by colloidal particles when a beam of light is passed through a colloidal dispersion; the rapid random motion of colloidal particles caused by molecules striking the molecules of the suspending medium

7. Temperature; surface area; stirring

9. The artificial kidney machine consists of a long cellophane tube wrapped around itself to form a coil; the coil is immersed in a temperature-controlled solution whose chemical composition is carefully regulated to the patient's need

11. Colloidal particles are smaller than the openings in filter paper but larger than the pores in a membrane

13. Gels are colloidal dispersions with a strong attraction among the colloidal particles and the suspending liquid; gelation in water is an example
   Sols are colloidal dispersions with little attraction among the colloidal particles and the suspending liquid; starch in water is an example

15. A device that generates an aerosol mist consisting of large water particles that can penetrate into the trachea and large bronchi; used for inhalation therapy; must avoid bacterial contamination and excess addition of water to the patient's lungs

17. 4.2 g

19. Pressure increases the solubility of gases in a solvent but has little affect on solid or liquid solutes

21. Approximately 830 g

23. Viscosity: a measure of the resistance to flow for a liquid
   Adhesion: attraction of unlike molecules
   Cohesion: attraction of like molecules
   Capillary action: rise of water in a capillary tube due to surface tension and adhesion

25. Dissolve 18 g of $C_6H_{12}O_6$ into enough water to make 200 mL of solution

27. Saltwater causes reverse osmosis from the cells because of the higher concentration of particles in the saltwater

29. Percent solution: grams per 100 mL solution
   Milligrams percent: milligrams per 100 mL solution

31. (a) HCl; $NaNO_3$
   (b) NaCl; KNO_3
   (c) gas solubilities decrease with an increase in temperature
   (d) about 5 grams
   (e) about 490 grams

33. 0.86% NaCl, 0.03% KCl, 0.033% $CaCl_2$
   0.148 M Na^+, 0.004 M K^+, 0.003 M Ca^{2+}, 0.158 M Cl^-
   7.96 atm

35. 34 mL

Chapter 11   (page 191)

1. Strong electrolytes are completely ionized in water: HCl, NaOH

3. Lower the freezing point of a solution because of the increased number of ions present

5. Nonelectrolytes decrease the freezing point and increase the boiling point of a solution, but not as much as an electrolyte

7. Anions are negative ions: Cl^-, SO_4^{2-}, NO_3^-, OH^-
   Cations are positive ions: H^+, Na^+, Ca^{2+}, Fe^{3+}

9. (a) When electrolytes are placed in water, the molecules break up into ions
   (b) Some ions have a positive charge, others a negative charge
   (c) The sum of the positive charges equals the sum of the negative charges
   (d) The conductance of electricity by solutions of electrolytes is due to the presence of ions
   (e) Nonelectrolytes do not conduct electricity because of the absence of ions
   (f) The greater effect of electrolytes on boiling point and on freezing point is due to the increased number particles (ions) present

11. Conduction of electricity requires freely moving ions, which occurs with molten NaCl but not solid NaCl

Chapter 12   (page 210)

1. (a) Acids — Yield H^+ when placed in water; have a sour taste; react with indicators; neutralize bases

   (b) Bases — Yield hydroxide ions in solution; have a bitter taste and a slippery feeling; react with indicators; neutralize acids

3. Neutralization: reaction between acids and hydroxide bases producing water and a salt

   pH: $-\log (H^+)$

   Acid: proton donor

   Base: proton acceptor

   Titration: a procedure for determining the concentration of a solution by allowing a carefully measured volume of the solution to react with a second solution of known concentration

5. Calcium hydroxide, an antacid and antidote used for oxalic acid poisoning; magnesium hydroxide, an antacid and laxative; ammonium hydroxide, a heart and respiratory stimulant

7. Antacids: calcium carbonate, magnesium trisilicate, magnesium hydroxide, aluminum hydroxide, aluminum sodium dihydroxycarbonate

9. Acid spill: wash thoroughly with water and neutralize with sodium bicarbonate.

   Base spill: large amounts of water

11. Reaction of a compound with either the hydrogen ion or hydroxide ion derived from water; a solution that maintains a constant pH upon the addition of either an acid or a base

13. $NaHCO_3 + HCl \longrightarrow NaCl + CO_2 + H_2O$

    $NaHCO_3 + NaOH \longrightarrow Na_2CO_3 + H_2O$

    $Na_2CO_3 + BaCl_2 \longrightarrow BaCO_3\downarrow + 2\,NaCl$

    $Cu(NO_3)_2 + Zn \longrightarrow Zn(NO_3)_2 + 2\,Cu$

15. See Table 12-10

17. $Ca(OH)_2$ produces OH^- ions in solution; $C_2H_4(OH)_2$ does not

19. Now considered a poison

21. Neutralizes the acid and precipitates the oxalate ion as calcium oxalate

23. Strong acids and bases are almost completely ionized in solution; weak acids and bases are only slightly ionized in solution

25. (a) 7; (b) below 7; (c) above 7; (d) 7; (e) above 7

27. They maintain a constant pH for the blood cells so enzymes can function best

29. 6.8

31. Since blood pH is above 7, would expect HCO_3^- because it removes H^+

33. pH = 5; acidic

Chapter 13   (page 236)

1. Alpha particles have very little penetrating power; they are very dangerous if inhaled or ingested but relatively harmless externally

3. Gamma rays are very penetrating and are far more harmful from external sources than internal ones

5. (a) $^{214}_{82}Pb$; (b) ^1_0N; (c) $^{17}_8O$; (d) ^7_3Li

7. Time required for half of the atoms in a sample to decay

9. Short half-life to minimize radiation exposure

11. Relative biologic effectiveness:

    $$RBE = \frac{\text{dose of gamma radiation from cobalt-60}}{\substack{\text{dose of radiation required to produce} \\ \text{same biologic effect}}}$$

13. The patient is given a radioactive element that will concentrate in the area under study or in abnormal tissues; a external count or scan is passed over the area of interest and the amount of radiation is measured; I^{131}, Tc-99m

15. Boron-10 concentrates in tumor cells; neutron-capture causes the boron to emit alpha particles which destroy cancerous cells

17. Roentgen; rad; rem; RBE; curie

19. Use of x rays to produce an image on opaque paper by a photoelectric process; used in mammography and to search for foreign bodies such as plastic, wood, and glass

21. Radioactive element naturally present in the body or ingestion of radioisotopes

23. Shielding; increasing distance from radioactive source; limiting exposure

25. (a) decrease of 2; (b) increase of 1; (c) no change

27. Magnetic resonance imaging; a sample is placed in a strong magnetic field and subjected to radio waves of a frequency appropriate to the nucleus of the element being studied

29. Lengthens the shelf-life and reduces the need for preservatives

31. Fission: splitting large atomic nuclei into smaller nuclei

    Fusion: combining small nuclei into larger nuclei

33. Splits molecules into two charged particles

    $$H_2O \xrightarrow{\text{radiation}} e^- + H_2O^+$$

    $$H_2O \xrightarrow{\text{radiation}} H^+ + OH^-$$

    disrupts chemical process inside the cell

35. A device used to measure and detect radiation; it contains a crystal of sodium iodide mixed with a small amount of thallium iodide; when the crystal is hit by radiation, a flash of light is produced and measured

37. Metastable

39. 17,100 years

Chapter 14 (page 245)

1. Aliphatic, aromatic, heterocyclic

3. See Table 14-1

5. Structural formulas show the actual bonding structure of the molecule; molecular formulas only provide atom ratios in the compound; structural formulas are needed because organic compounds have the possibility of isomers

7. Tetrahedral shape with a bond angle of 109.5 degrees

9. No; hydrogen can only have one bond

11. (a) aliphatic   (b) aliphatic   (c) heterocyclic
     (d) aliphatic   (e) heterocyclic   (f) aromatic
     (g) aromatic   (h) aromatic

Chapter 15 (page 262)

1. Butane; hexane

3. (a) 2-methylbutane   (b) 2,3,4-trimethylhexane
     (c) dimethylpropane   (d) 3-ethylheptane

5. Petroleum and natural gas

7. (a) 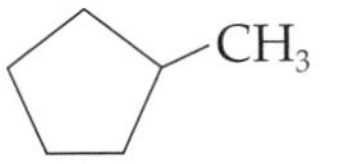   (b)

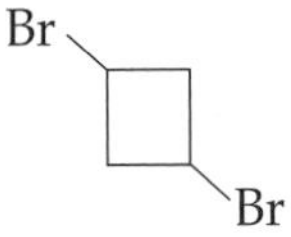

9. 2,2,4-trimethylpentane

$$2\,C_8H_{18} + 25\,O_2 \longrightarrow 16\,CO_2 + 18\,H_2O + energy$$

11. Lead poisoning

13. (a) $2\,C_8H_{18} + 25\,O_2 \longrightarrow 16\,CO_2 + 18\,H_2O$
     (b) $CH_4 + Br_2 \longrightarrow CH_3Br + HBr$

15. (a)

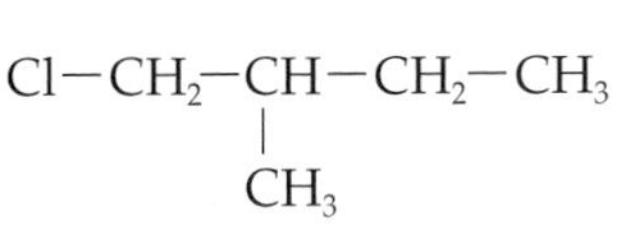

1-chloro-2-methylbutane

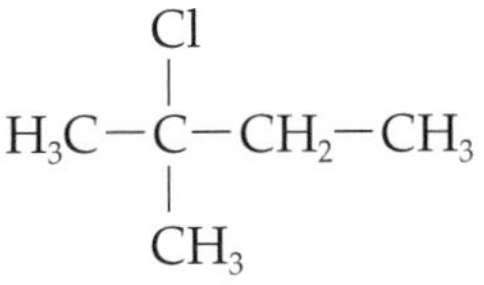

2-chloro-2-methylbutane

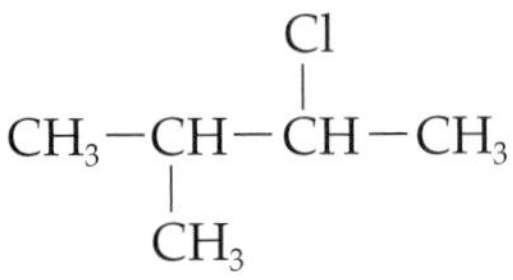

2-chloro-3-methylbutane

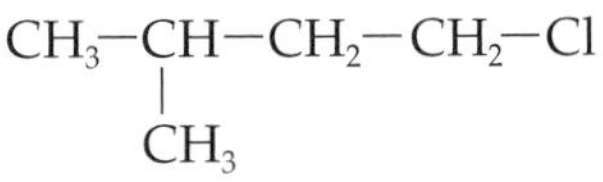

1-chloro-3-methylbutane

(b) 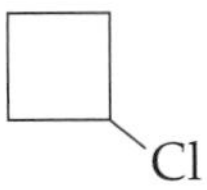
chlorocyclobutane

17. (a)

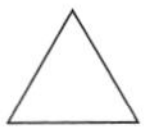

(b)

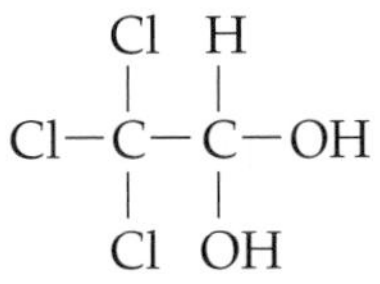

(c)

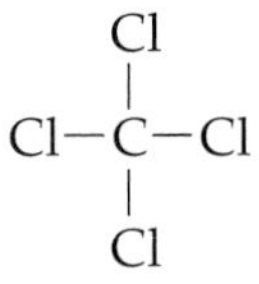

(d)

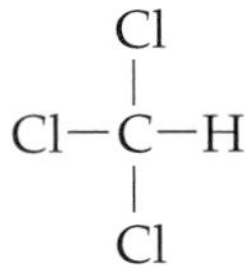

(e)

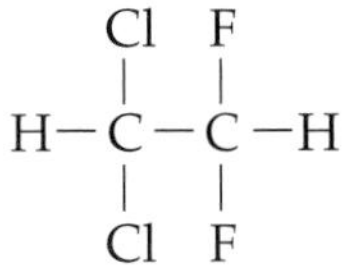

(f)

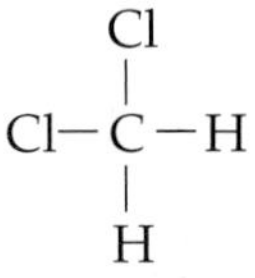

(g)

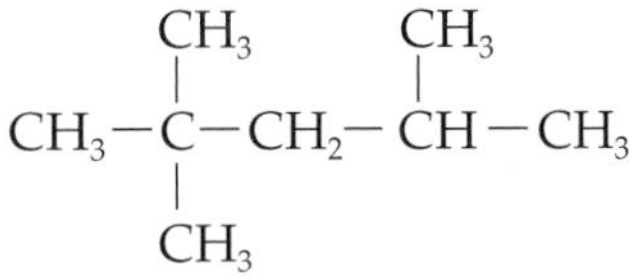

Chapter 16 (page 279)

1. (a) alkene; (b) alkyne; (c) alkene; (d) alkane

3. Single bonds between carbon atoms; one or more multiple bonds between carbon atoms

5. (a) 1-chloro-2-butene
     (b) 4-methyl-2-pentyne
     (c) 1,2-dichloropropene

7. Complete combustion produces carbon dioxide; incomplete combustion produces carbon monoxide

9. Cis: attached to the same side
     Trans: attached to the opposite side

11. (a) 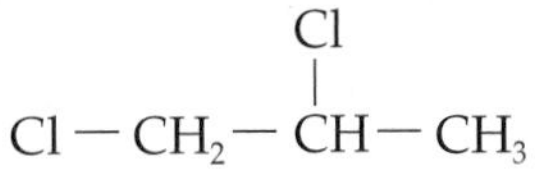   1,2-dichloropropane

(b) 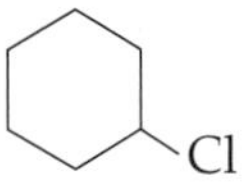   chlorocyclohexane

(c) 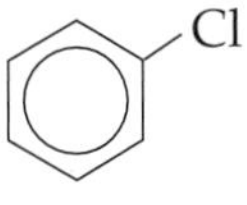   chlorobenzene

(d) 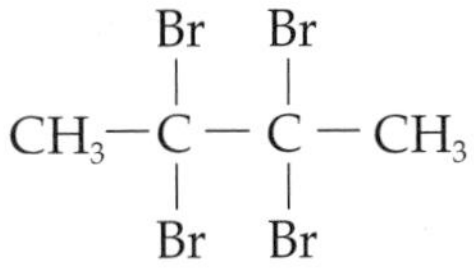   2,2,3,3-tetrabromobutane

13.

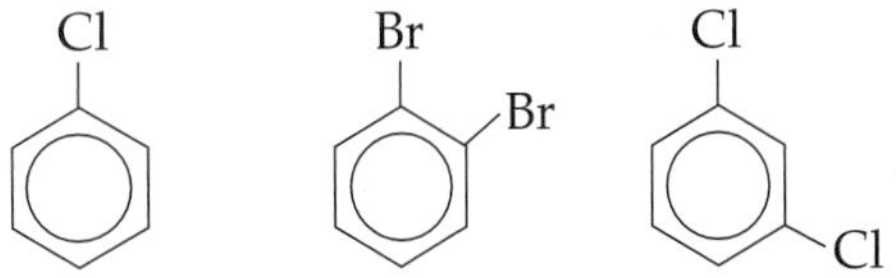

15. All six carbons in benzene are equivalent

17. Chlorocyclohexane; chlorobenzene; 1,2-diodobenzene
19. Positions are either alpha or beta
21. Five rings; produced when coal, wood, paper are burned; carcinogenic
23.

chlorobenzene   2-chlorotoluene

3-chlorotoluene   4-chlorotoluene

25. (a)   (b)

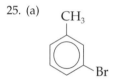

(c)

$$CH_2=\overset{\displaystyle Cl}{\overset{|}{CH}}$$

(d) $HC\equiv CH$

Chapter 17 (page 294)

1. R—OH; polar; nonelectrolyte; variable solubility in water
3. Causes blindness and death
5. Works at a slower rate and penetrates all the way through the cell
7. Fermentation of sugar or addition of water to ethylene; twice the percentage by volume of alcohol in the solution
9. Constituent of fat (triglycerides)
11. Primary, methanol or methyl alcohol; secondary, 2-propanol or isopropyl alcohol; tertiary, 2-methyl-2-propanol.
13. Dehydration of an alcohol
15. Advantages: easy to give; excellent muscle relaxant; very little effect on respiration, blood pressure, pulse rate
   Disadvantages: may produce nausea; irritating to membranes of respiratory tract; flammable
17. Ethyl alcohol, which prevents the liver from oxidizing the ethylene glycol until after the ethylene glycol is eliminated in the urine
19. Both are disinfectants
21. $CH_3CH_2CH_2OCH_2CH_2CH_3$; CH_3—CH=CH_2
23. From carbon monoxide and hydrogen; used as solvents; from the fermentation of blackstrap molasses

25.

$$CH_3-\overset{\displaystyle CH_3}{\overset{|}{\underset{\displaystyle H}{\underset{|}{C}}}}-CH_2-CH_2-\overset{\displaystyle CH_3}{\overset{|}{CH}}-CH_2-OH$$

2,5-dimethyl-1-hexanol

$$CH_3-\overset{\displaystyle CH_3}{\overset{|}{\underset{\displaystyle H}{\underset{|}{C}}}}-CH_2-CH_2-\overset{\displaystyle CH_3}{\overset{|}{\underset{\displaystyle OH}{\underset{|}{C}}}}-CH_3$$

2,5-dimethyl-2-hexanol

$$CH_3-\overset{\displaystyle CH_3}{\overset{|}{CH}}-CH_2-\overset{\displaystyle}{\underset{\displaystyle OH}{\underset{|}{CH}}}-\overset{\displaystyle CH_3}{\overset{|}{CH}}-CH_3$$

2,5-dimethyl-3-hexanol

27. (a)

$$CH_3-(CH_2)_5-\overset{\displaystyle CH_3}{\overset{|}{CH}}-CH_2-OH \qquad 1°$$

(b)

$$CH_3-\overset{\displaystyle CH_3}{\overset{|}{CH}}-\overset{\displaystyle CH_3}{\overset{|}{\underset{\displaystyle OH}{\underset{|}{C}}}}-CH_3 \qquad 3°$$

(c)

2°

Chapter 18 (page 310)

1.

$$\overset{\displaystyle O}{\overset{\|}{R-CH}} \qquad \overset{\displaystyle O}{\overset{\|}{R-COH}} \qquad \overset{\displaystyle O}{\overset{\|}{R-C-R}}$$

3.

$$CH_3-CH_2-CH_2-OH \xrightarrow{[O]} CH_3-CH_2-\overset{\displaystyle O}{\overset{\|}{CH}}$$

5. The reaction of aldehydes with a Cu^{2+} solution will produce a red precipitate of Cu_2O
7. Acetone
9. An acid;

$$CH_3-\overset{\displaystyle O}{\overset{\|}{CH}} \xrightarrow{[O]} CH_3-\overset{\displaystyle O}{\overset{\|}{COH}}$$

11. $CH_3COOH + NaOH \longrightarrow CH_3COONa + H_2O$
   sodium acetate

13. (a) 3,4-dichloropropanal
   (b) propanone
   (c) 2-pentanone
   (d) chloroethanoic acid

15. Ketones cannot be oxidized

17. A compound formed by the reaction of one molecule of a ketone with one molecule of an alcohol; molecule of a ketone with two molecules of an alcohol; important in the structure of saccharides

19. (a)

$$CH_3-CH_2-CH_2-\overset{\displaystyle O}{\overset{\displaystyle \|}{C}}-CH_3$$

2-pentanone

(b)

$$CH_3-CH_2-CH_2-\overset{\displaystyle O}{\overset{\displaystyle \|}{C}}-OH$$

butanoic acid

Chapter 19   (page 326)

1. $R-NH_2$ $R-\overset{\displaystyle O}{\overset{\displaystyle \|}{C}}OR$ $R-\overset{\displaystyle O}{\overset{\displaystyle \|}{C}}NH_2$

3. Primary amines, one R group; secondary amines, two R groups; tertiary amines, three R groups

5. A compound with both an amine and acid functional group: NH_2-CH_2-COOH

$NH_2-CH_2-CH_2-COOH$

proteins

7. $R-\overset{\displaystyle O}{\overset{\displaystyle \|}{N}}H_2$ an amide bond linking two amine acids together

9. Ethanamide

11.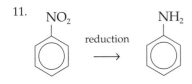

13. See Problem 17 for possible answers

15. (a) aminoethane
    (b) N-methylaminomethane
    (c) N,N-diethylaminoethane
    (d) 2-hydroxyaniline

17. (a) alcohol, amine, amide, ketone
    (b) amide, sulfur bond, acid
    (c) amide, sulfur bond, acid
    (d) alcohol, amine
    (e) amide, amine
    (f) amine
    (g) alcohol, amine
    (h) alcohol, hemiacetal
    (i) alcohol, ether
    (j) alcohol
    (k) alcohol, amide, acid
    (l) amine, sulfur bond

(m) ketone, amine, amide, alcohol, sulfur bond, acid salt
(n) ether, amine, amide
(o) amine, fluorocarbon, sulfur bond, (sulfonamide)
(p) ether, (nitrile), amine
(q) acid salt, ether, ketone, alcohol

19. Ethylethanoate; ethyl acetate

21. (a) angina; (b) counterirritant; (c) local anesthetic

Chapter 20   (page 354)

1. A polyhydroxyl aldehyde or ketone or a substance that yields these compounds on hydrolysis; they were considered to be hydrates of carbon because of the 2 to 1 ratio of hydrogen to oxygen

3. Aldoses are carbohydrates that contain the aldehyde group; ketoses are carbohydrates that contain the ketone group; numbers of carbons are indicated by prefixes such as tri-, tetr-, pent-, hex-

5. Rotates the direction of the polarized light

7. No; chiral carbons have four different groups attached to them

9. The configuration of the alcohol group on the carbon next to the end carbon containing the primary alcohol; D on the right, L on the left; most are D.

11. A polysaccharide that yields hexoses on hydrolysis; one that yields pentose on hydrolysis; carbohydrates that differ at one carbon in configuration

13. Monosaccharides, glucose; disaccharides, sucrose; polysaccharides, starch

15.

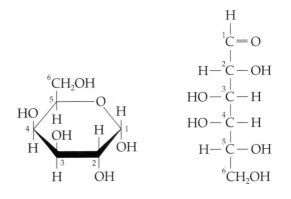

15. cont'd

α-D-glucose†   D-glucose

All have the molecular formula $C_6H_{12}O_6$

17. Fruit juices, particularly grape; fruit juice and honey; glycolipids and glycoproteins; germinating grain; table sugar; milk

19. Sorbitol

21. Reduces the copper to copper (I) oxide; used to test for the the presence of hexoses

23. No; since sucrose does not have a free carbonyl group it cannot be reduced

25. Lack of fermentation will distinguish lactose from maltose and sucrose; ability to reduce an alkaline copper (II) solution will distinguish between maltose and sucrose

27. Lactose and maltose have 1,4 linkages, which leaves an aldehyde free for reaction with copper(II)
Sucrose has a 1,2 linkage and no free aldehyde

29. Starch(blue) $\longrightarrow$ erythrodextrins(red) $\longrightarrow$ maltose (colorless) $\longrightarrow$ glucose(colorless)

31. Three-dimensional representation of structure in which the horizontal lines represent bonds extending forward from the plane and the vertical lines represent bonds extending backward from the plane; a way to show the structure of a sugar with the five- and six-numbered rings being represented as flat

33. Compound formed by the reaction of an aldehyde with an alcohol
Compound formed by the reaction of a ketone with an alcohol

35. The —OH group can be above or below the plane of the ring

37. (a) ketotriose; (b) aldohexose; (c) aldotetrose

Chapter 21   (page 377)

1. Carbohydrates are usually soluble in water, lipids are not; lipids one soluble in nonpolar organic solvents, carbohydrates are not; lipids yield fatty acids upon hydrolysis or combine with fatty acids to form esters; carbohydrates yield monosaccharides upon hydrolysis

3. Straight-chain organic acid; unsaturated fatty acids contain one or more double bonds

5. Even, because of the fatty acid synthesis pathway

7. Amount of unsaturation; oils

9. Three fatty acids and glycerol; salts of three fatty acids and glycerol

11. Hydrogenation; converts oils into solids (vegetable shortenings)

13. Soaps are salts of fatty acids, detergents are salts of long-chain alcohol sulfates; soaps form calcium and magnesium precipitates with hard water, detergents do not

15. A soap that contains a germ-killing compound

17. Reaction of unabsorbed fatty acids with calcium ions to form insoluble calcium compounds

19. Dipalmitoyl lecithin, surface active agent on inner surface of lungs
Phosphatidyl ethanolamine, blood clotting phosphatidylcholine fat metabolism

21. They form a bilayer, which helps control passage into and out of the cell

23. Deposition of excess lipids (particularity cholesterol and triglycerides) from the bloodstream; reduce the plasma lipid concentration by reducing lipid intake or using antihyperlipidemic drugs

25. Aids in the absorption of fatty acids from the small intestine and precursor to other steroid molecules

27. Yes; they form protective coatings for plants and animals

29.

prostacyclin

Prostacyclin is a potent inhibitor of platelet aggregation and is a powerful vasodilator

prostaglandin E_1 (PGE$_1$)

Prostaglandin E_1 (PGE$_1$) is now used to strengthen babies born with cyanotic congenital heart disease ("blue ba-bies") to prepare them for corrective surgery

TXB$_2$

Thromboxanes are potent aggregators of blood platelets and have a profound contractive effect on a variety of smooth muscles

Chapter 22 (page 397)

1. C, H, N, O; also S, P, metals

3. Nitrogen in the air is fixed by bacteria into soil nitrates, which are picked up by plants and changed into plant protein; plant protein becomes animal protein, then waste, which is denitrified and changed back to air nitrogen

5. Compounds containing the acid and amine functional groups; see Table 22-2

7. pH at which there are an equal number of positive and negative ions; minimum solubility

9. Amide bond binding two amino acids together; many amino acids joined together

11. Alanine, arginine, asparagine, cysteine, glutamic acid, glutamine, glycine, histidine, isoleucine, leucine, lysine, phenylalanine, proline, serine, threonine, tyrosine, valine

13. Proteoses, peptones, polypeptides, tripeptides, dipeptides, amino acids

15. See Tables 22-5 and 22-6

17. 100% alcohol only coagulates outer protein layer

19. Energy; hormones, enzymes, oxygen transport, antibodies, nerve transmission, hereditary, muscular activity, blood pressure maintenance, most body tissues

21. Polar amino acids contain R groups such is ⁻OH, ⁻SH, ⁻NH₂, ⁻COOH and are more water soluble

23. Because the amino acids are all of the ʟ configuration

25. Globular proteins are "ball"-shaped and are either soluble or form a colloidal dispersion in water; fibrous protein are coiled or stretched out and are insoluble in water

27. Storage, ferritin; transportation, hemoglobin, serum albumin; enzyme, pepsin; hormone, insulin

29. Nitrogen in proteins

31. Injection to avoid peptide digestion

33. lys.ala.his.thr.pro.leu.ala.tyr

Chapter 23 (page 417)

1.

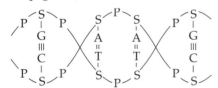

Ordinary chemical bonds and weaker hydrogen bonds

3. Messenger RNA: transmission of genetic information from DNA to site of protein synthesis

 Ribosomal RNA: forms ribosomes

 Transfer RNA: holds specific amino acid for incorpora tion into a protein molecule

5. Chromosomes contain genes and genes are made of DNA, which holds the hereditary information

7. Cystic fibrosis, phenylketonuria, sickle cell anemia, galactosemia

9. A synthetic RNA composed only of uracil was substituted for mRNA in a system; the polypeptide formed contained only phenylalanine. Thus, since the RNA had the code UUU, the corresponding code of AAA on DNA specifies phenylalanine

11. A miscopying of the genetic code or omission or change in sequence of the letters of the code

13. (a) Lack of the enzyme hexosaminidase A; red spots in the retina, muscular weakness, death

 (b) Lack of the enzyme tyrosinase; very white skin and hair and extreme sun sensitivity

 (c) Lack of a blood clotting protein; bleeding

 (d) Lack of the enzyme phenylalanine hydroxylase; severe mental retardation

 (e) Lack of the enzyme uridyl transferase; cataracts, liver failure, mental retardation

 (f) Lack of the protein ceruloplasmin; liver and kidney damage

15. Cytosine contains an amine group

Chapter 24 (page 436)

1. An increase in temperature increases its rate of reaction (if temperature is too high, the enzyme ceases to function); a decrease in temperature decreases the rate; each enzyme has an optimum pH at which it functions best

3. Proteins

5. Type of reaction catalyzed

7. Absolute; stereo; linkage; reaction; groups

9. Enzyme for the hydrolysis of nucleic acids, ribonuclease enzyme for the transfer of a functional group, transaminases enzymes for the interconversion of optical, geometric, or structure isomers, retinal isomerase

11. A coenzyme made up of pantothenic acid, adenine, ribose, phosphoric acid, and mercaptoethanolamine; essential for metabolism of carbohydrates, fats, proteins; a coenzyme found in the mitochondria, it functions in electron transport and oxidative phosphorylation

13. Nonspecific inhibitors affect all enzymes in the same manner; specific inhibitors affect only one enzyme or group of enzymes

 Competitive inhibitors bind to the active site; noncompetitive inhibitors bind to the enzyme at sites other than the active site

15. Use of chemicals to destroy infectious microorganisms and cancerous cells without damaging the host cells; antibiotics such as penicillin and tetracycline and antimetabolites such as sulfanilamide

17. Inactive precursors of enzymes; different forms of the same enzyme; key metabolic enzymes whose activity can be changed by molecules other than the substrate; to ensure that biologic processes remain coordinated at all times

19. In the lock-and-key theory the active site is rigid; in the induced-fit theory the active site is flexible and can change upon binding to a substrate

21. Measure the reaction rate as a function of substrate concentration: if reaction rate increases with substrate concentration, the inhibitor was competitive; if reaction rate remains the same, inhibitor was noncompetitive

Chapter 25 (page 450)

1. Digestion breaks down large molecules into simple molecules; simple foods such as monosaccharides, minerals, and vitamins do not need further breakdown

3. Water, mucin, salivary amylase, lingual lipase, K^+, Ca^{2+}, HCO_3^-, SCN^-; pH range of 5.75 to 7.0; lubricates, buffers, starts breakdown of starches and triglycerides

5. 1 to 2; HCl

7. Gastric lipase is inactive at pH of the stomach

9. Gastric juice: pepsinogen (pepsin), protein hydrolysis and gastric lipase, triglyceride hydrolysis
   Intestinal juice: aminopeptidase and dipeptidase, peptide hydrolysis
   Phosphoglyceridase: hydrolysis of phosphoglycerides; nucleases; hydrolysis of nucleoproteins; phosphatase; hydrolysis of organophosphates

11. Emulsification of fats; increase the effectiveness of pancreatic lipase; aid in the absorption of fatty acids; stimulate intestinal motility

13. Cholesterol precipitate

15. Jaundice is produced, causing yellow pigmentation of the skin

17. Small intestinal enzyme that converts trypsinogen into trypsin

19. Digestion of carbohydrates begins in the mouth with ptyalin, which acts on starch; digestion is completed in the small intestine through the acid of enzymes in the pancreatic and intestinal juices; absorbed through villi in the small intestine
   Digestion of lipids begins in the mouth with gastric lipase, which starts the hydrolysis of triglycerides; digestion is completed in the small intestine where pancreatic lipase, aided by bile, completes the hydrolysis of triglycerides into fatty acids and glycerol; absorbed through the villi in the small intestine
   Digestion of protein begins in the stomach through the action of pepsin, continues in the small intestine through the action of trypsin and chymotrypsin and absorption through the villi of the small intestine

21. Indole and skatole

23. Bile contains bile salts, bile pigments, and cholesterol; bile removes many drugs and poisons from the body and aids in the emulsification of fats, which promotes digestion of fats

25. High dietary cholesterol intake; liver damage; inability for any reason that bile is not able to increase the solubility of cholesterol; infections

27. Little effect since the liver will still secrete bile; a decreased intake of fat may be required

Chapter 26 (page 476)

1. 70 to 100 mg glucose/100 mL of blood; blood glucose level above which sugar appears in the urine is about 180 mg glucose/100 mL of blood

3. Glucose

5. Glucose in the urine; very high blood glucose levels

7. Organic phosphates such as ATP and ADP

9. Glycolysis, oxidative phosphorylation

11. Metabolism of glucose into pyruvate; anaerobic; it supplies most of the body's energy for anaerobic muscular contraction

13. Glycogenesis

15. It provides five-carbon sugars for synthesis of nucleic acids and $NADP^+$

17. Fuel for the cycle

19. See Figure 26-10

21. Transportation of glucose across cell membranes; oxidation of glucose in the cells: transformation of glucose to glycogen (glycogenesis) in the muscle and also in the liver; depresses the production of glucose (glycogenolysis) in the liver; promotes the formation of fat from glucose; produced by the beta cells of the islets of Langerhans in the pancreas

23. See Figure 26-11

25. Promotes glycogenolysis during periods of emotional stress; medulla of the adrenal glands

27. No; as the level during surgery is not the average of the mixture; probably due to decreased glycolysis during surgery

29. Both lower pH and lower levels of 2,3 DPG will produce less O_2 saturation of hemoglobin

Chapter 27 (page 491)

1. Through lacteals of villi of small intestine into lymphatics to thoracic duct to bloodstream to liver

3. Removal of two carbons at a time from the end of the fatty acid chain; acetyl CoA; enters the citric acid cycle

5. Carbohydrate, 4 kcal; fat, 9 kcal

7. Must be reduced by a dehydrogenase

9. See Figure 27-3; acetoacetic acid is broken down into the other two; liver

11. Ketone bodies in the blood; ketone bodies in the urine; ketone bodies in both blood and urine

13. All cells, particularly brain and nerve cells; liver 50%, gut 15%, skin 35%; precursor of severe steroids

15. Conversion of glucose to fat; in the mitochondria and the cytoplasm

17. Form the framework of cell membrane

19. Increase: glucagon, ACTH, epinephrine, growth, cortisol, thyroxine
   Decrease: insulin

Chapter 28 (page 507)

1. Carbohydrate as glycogen; fat as fat; no protein storage
3. To build new tissue; to replace old tissue; to form hemoglobin, enzymes, and hormones; to produce energy
5. 0.8 g/kg of body weight; increased metabolic states such as fevers
7. A reaction whereby an amino group from an amino acid is transferred to a keto acid; to manufacture amino acids the body needs
9. Heart muscle; elevated levels may indicate a myocardial infarction
11. Urine, urobilin; feces, stercobilin
13. Accumulation of bilirubin in the blood; excessive hemolysis, bile duct blockage, liver disease
15. Toxic amine produces by large intestinal bacteria
17. Direct infusion of a concentrated solution containing glucose, amino acids, or a hydrolysate of a nutritionally complete protein, essential minerals (except iron), parenteral form of fat emulsion (if necessary), vitamins; for patient unconscious or unable to take food by mouth
19. Hyperammonemia caused by a lack of the enzyme carbamoylphosphate synthetase; increased ammonia blood levels; citrullinemia caused by decreased level or lack of the enzyme arginine succinate synthetase; large amounts of citralline in urine
21. All can be made from pyruvate
23. Tyrosine
25. Ultracentrifuge is used to separate lipoproteins, based on differences in densities
27. Phenlypyruvate is the deamination product of phenylalanine
29. Oxaloacetate is used to make glucose, which reduces the catabolism of acetyl CoA in the citric acid cycle; acetyl CoA is converted into ketone bodies instead, which produce ketosis and acidosis
31. Yes, indicative of liver disease or damage; yes, for same reason
33. Indirect due to less ability of bilirubin to form the glucuronide
35. Loss of blood; damaged liver reduces blood clotting ability
37. NAD^+; linked enzymes are necessary
39. High familial incidence; either increased formation of uric acid or decreased renal excretion, which results in crystals of urate being deposited
41. No; most are synthesized
43. Blocks conversion of purines into uric acid
45. Any condition in which the plasma triglyceride and cholesterol levels are elevated after a 12-hour fast
47. Much of the fatty acid used by the liver to make triglycerides is synthesized from dietary carbohydrates; fewer carbohydrates lead to reduced triglyceride production

Chapter 29 (page 519)

1. Kidneys; they also regulate water, electrolyte, and acid-base balance in the body
3. Blood pressure
5. Drugs that promote water and salt loss through the urine; ethyl alcohol; caffeine, thiazide
7. Controls renal reabsorption of water; large urine output
9. Urine volume and solute concentration; diet, fever, blood pH; precipitation of calcium phosphate
11. Purine metabolism

13. Deposits of uric acid and urates in joints and tissues
15. From arginine, methionine, and glycine; creatinine is the breakdown product of creatine; creatinine contains a cyclic amide group
17. Abnormal amounts of creatine occur in the urine; causes include muscular dystrophy, myasthenia gravis, poliomyelitis, hyperthyroidism

    Protein in the urine: causes include liver, kidney, heart disease
19. Increases in hyperparathyroidism; decreases in hypoparathyroidism and renal diseases
21. Glucose in the urine; exercise, high carbohydrate meal, diabetes, renal or liver disease
23. Inadequate carbohydrate intake or utilization; starvation, pregnancy, diabetes
25. Obstruction of bile flow to the liver

Chapter 30 (page 549)

1. Blood serum is blood plasma without the fibrinogen for clotting
3. 4.5 to 5.0 million per mm^3; carry oxygen to the cells

   5 to 10 thousand per mm^3; fight infections
5. Neutrophils, basophils, eosinophils, lymphocytes, monocytes
7. 7.35 to 7.45; four and a half times that of water; 1.054 to 1.060 g/mL
9. Conjugated protein; hemorrhaging
11. Failure in red blood cell production; lack of vitamin B_{12} or intrinsic factor
13. Albumin: osmotic pressure of the blood

    Globulin: antibodies, transport of molecules and ions

    Fibrinogin: blood clotting
15. Accumulation of water in tissue; decreased plasma proteins, heart disease, a terminal illness, shock

17. When a blood vessel is cut, the tissues liberate thromboplastin; in the presence of thromboplastin and Ca^{2+}, prothrombin is changed to thrombin, which in turn acts on fibrinogen to change it to fibrin, the clot

19. Reduces production of prothrombin, which is required for clotting

21. See Figure 30-8

23. Ability of blood buffers to neutralize acid; pH of blood

25. Bicarbonate, phosphate, protein buffers; plasma; inside the red blood cells

27. Thirst (ingestion), metabolism; as perspiration, as urine, through lungs, through feces

29. Hyponatremia: vomiting; diarrhea; hormonal disorders; starvation; burns; kidney damage; diuretics; cold, clammy extremities; lowered blood pressure; weak, rapid pulse; oliguria; muscular weakness; cyanosis; fingerprinting over the sternum

    Hypernatremia: deficient water intake; excessive sweating; excessive water output; poor renal excretion; excessive intake of sodium salts; hyperactivity of the adrenal cortex; cerebral diseases; dry, itchy mucous membranes; intense thirst; oliguria or anuria; rough, dry tongue; elevated temperature; tachycardia; edema; cerebral disturbances

31. Maintain osmotic pressure in the cell; electric potential of the cell; cell size; heart contraction, nerve impulse transmission

33. Green plants

    Magnesium deficiency: muscular tremors; convulsions, delirium, delusions, disorientation, hyperirritability, elevated blood pressure

    Magnesium excess: sedation, coma, respiratory paralysis, cardiac arrest

35. 6 mEq/L

37. Blood bicarbonate and carbon dioxide levels for both; see Figure 30-4

39. Increased renal excretion of bicarbonate

41. Elevated with dehydration due to decreased amount of water in plasma

    Lowered with malnutrition and chronic hepatic insufficiency due to less protein synthesis, and lowered with nephrosis due to protein leakage into urine

43. Metabolism of sulfur containing amino acid

45. Respiratory acidosis

47. Yes; decreased CO_2 levels

49. Increased CO_2 levels; decreased respiration

51. Diabetes produces metabolic acidosis due to ketone bodies

53. Potassium levels will drop as pH increases; if drops below 3.5 mEq/L may need to give KCl; other electrolytes have returned to normal

Chapter 31 (page 572)

1. Water-soluble: B-complex, C
   Fat-soluble: A, D, E, K

3. Fish liver oils, buttermilk: precursors found in yellow fruit and vegetables; deficiency causes shrinking and hardening of epithelial tissues in eyes, digestive tract, and genitourinary tract; excess causes irritability, loss of appetite, fatigue, itching; severe overdose can cause death

5. In the liver; absorbed with aid of bile

7. Fish oils, irradiated milk, sunlight; produced by exposure of skin to ultraviolet light from the sun; steroids; no, D_2 and D_3 most potent

9. Absorption of calcium and phosphate from the small intestine, release of calcium and phosphate from the bone, promotion of proper activity of parathyroid hormone; deficiency results in rickets, osteomalacia, osteoporosis; excess causes calcification of soft tissues

11. Yeast, milk, eggs, nuts, whole grains
    Soluble in water and 70% alcohol, only stable in acid solution; heat stable
    Carbohydrate metabolism, amino acid synthesis, and hexose monophosphate shunt

13. Riboflavin: yeast, milk, liver, kidney, heart, leafy vegetables
    Niacin: liver, kidney, heart, yeast, peanuts, wheat germ
    Biotin: liver, egg yolk, kidney, yeast, milk

15. Egg yolks, yeast, kidney, lean meats, skim milk, broccoli, sweet potatoes, molasses, viscous yellow oil
    Water-soluble, stable in acid and alkaline solutions
    Burning foot syndrome

17. Fresh fruits and vegetables, dry cereals, milk, meats, eggs
    Soluble in water and alcohol, insoluble in fat solvents
    Strong reducing agent rapidly destroyed by heat
    Deficiency: scurvy

19. Reduces selenium requirements by preventing selenium loss

21. Can be stored in fat tissues

23. Naturally by the body

25. Coumarin acts as competitive inhibitor of vitamin K; giving vitamin K overcomes this competitive inhibition

27. Aspirin-type compounds interfere with blood clotting, e.g., Excedrin, Bufferin, Bayer aspirin

29. Folate masks the symptoms of pernicious anemia

31. 200 µg; removal of alcohol; as more wine is consumed, more folate is needed

33. Will mask B_{12} deficiency, which can lead to pernicious anemia

Chapter 32 (page 598)

1. Classified according to the mechanism of their action:
   Group I binds to intracellular receptors: estrogens, androgens

   Group II binds to cell-surface receptors: cAMP

3. Insulin, glucagon

5. Increases rate of oxidation of glucose, facilitates glycogenesis, and increases synthesis of fatty acids, protein, and RNA

7. Hypoglycemia: low blood sugar level; faintness, weakness, dizziness, tremors, anxiety, hunger, "cold sweats," mental confusion, motor incoordination; oral or iv glucose (sugar). Hyperglycemia: high blood sugar levels: glucose in the urine, ketone bodies, acidosis; dehydration; insulin

9. Increased blood sugar level, sugar in urine ketone bodies in blood; acidosis

11. Hypothyroidism: sluggishness, weight gain, bradycardia, reduced metabolic rate, loss of appetite
    Hyperthyroidism: increased metabolic rate, weight loss, rapid, irregular heartbeat, elevated temperature, bulging of the eyes, nervousness

13. Increase in size of thyroid and neck; adding iodine to the diet

15. Glucocorticoids affect metabolism of carbohydrates, fats, and protein; corticosterone, cortisone, cortisol. Mineralocorticoids affect transportation of electrolytes and distribution of water; aldosterone. Androgens or estrogens; affect secondary sex characteristics; dehydroepiandrosterone

17. Lack of adrenocorticotropic hormone, see question 16; hyperadrenocorticism: decrease in feminine characteristics, increase in masculine characteristics, body hair, sex organs, voice

19. Growth hormone: growth of long bones and soft tissue, increases calcium retention, protein, DNA, RNA synthesis, fatty acid mobilization, antagonizes effects of insulin. TSH : stimulates thyroid gland to produce thyroxine. ACTH: stimulates synthesis and release of hormones by the adrenal cortex. LTH: initiates lactation. LH: stimulates development of testes and production of testosterone in males; role in ovulation in female. FSH: development of ovarian follicles and estrogen secretion in females; growth of testes and production of spermatoza in males.

21. Vasopressin: stimulates renal reabsorption of water; oxytocin: contracts muscles of uterus

23. Melonocyte-stimulating hormone: intermediate lobe of the pituitary gland; increases the deposition of melanin in the skin

25. Controlled by the hypothalamus

27. Nitric oxide is released by a neuron, diffuses to an adjacent neuron, binds to an enzyme which increases the formation of cGMP. Carbon monoxide is liberated by heme oxygenase and attracts to guanyl cyclase resulting in increased production of cGMP

29. By binding to opiate receptors in the brain

31. Elevated products of parathyroid hormone breaks down bone. results in weak bone structure and kidney stone formation

33. Peptide hormones would be digested if taken orally

35. Increase both glucose and free fatty acids

37. Monoamine oxidase, catechol-O-methyltransferase

Answers to Practice Tests

Chapter	1	2	3	4	5	6	7	8	9	10	11	12	13	14	15	16	17	18	19	20
1	a	c	c	d	d	d	d	c	c	b										
2	c	c	c	a	c	b	c	c	d	b										
3	b	b	c	a	b	a	b	c	c	c										
4	b	b	b	b	b	b	a	d	c	a										
5	c	c	b	a	a	c	b	c	d	b										
6	d	a	a	b	c	b	c	c	c	b										
7	d	b	b	b	a	b	b	a	a	a										
8	c	b	c	c	d	b	d	c	b	a										
9	c	a	b	d	c	d	d	a	c	c										
10	c	b	b	a	b	c	c	c	b	a										
11	d	b	d	c	b	b	b	d	b	a										
12	c	a	c	a	b	d	c	a	c	a	c	b	d	d	b	a	d	c	a	b
13	b	c	c	d	d	d	a	d	c	d										
14	b	a	d	a	c	b	b	d	c	a										
15	c	b	b	b	d	a														
16	a	b	b	a	c	d	c	b	c											
17	c	b	c	b	d	b	c	a	a	d	c									
18	d	a	b	a	d	a	c	b	a	d										
19	a	c	b	a	c	b	d	d	b	b										
20	d	c	d	b	c	d	d	b	a	d										
21	d	d	c	c	d	a	c	c	c	d										
22	c	b	a	c	c	c	d	a	b	d										
23	a	d	d	d	c	c	d	b	b	b										
24	a	c	c	c	b	c	a	b	d	b										
25	d	d	d	a	b	c	a	b	d	b										
26	b	c	a	b	b	c	b	d	b	c										
27	d	d	d	b	b	a	c	d	a	b										
28	b	a	d	a	d	b	c	d	d	a										
29	b	c	a	a	c	b	b	a	a	b										
30	d	c	c	d	b	c	d	c	b	b										
31	c	d	c	a	d	c	a	d	a	a										
32	d	b	c	a	d	d	d	d	c	b										

Credits for Photographs

Fig./Page	Photographer/Source
Fig. 13-10	P. Robert/SYGMA
Chapter 14	**Introduction to Organic Chemistry**
p. 238	Account Phototake/Phototake NYC
p. 245, middle	Scott Camazine/Photo Researchers, Inc.
Chapter 15	**Saturated Hydrocarbons: The Alkanes and Their Halogen Derivatives**
p. 247	Scott Camazine/Photo Researchers, Inc.
Fig. 15-3	Tom McCarthy
Chapter 16	**Unsaturated Hydrocarbons and Their Halogen Derivatives: Alkenes, Alkynes, and Aromatic Compounds**
p. 264	Paul Silverman/Fundamental Photographs
Fig. 16-1	Jim Olive/Peter Arnold, Inc.
Chapter 17	**Alcohols, Thiols, Phenols, and Ethers**
p. 281	Dan McCoy/Rainbow
Chapter 18	**Aldehydes, Ketones, and Carboxylic Acids**
p. 296	Richard Megna/Fundamental Photographs
Chapter 19	**Esters, Amines, and Amides**
p. 312	Van Bucher/Photo Researchers, Inc.
Chapter 20	**Carbohydrates**
p. 329	Robert Mathena/Fundamental Photographs
Fig. 20-1 A B	Richard T. Hutchings
Fig. 20-5	Courtesy Jean Clough, St. Francis Hospital School of Nursing, Evanston, Ill.
Chapter 21	**Lipids**
p. 356	Karen Leeds/The Stock Market
Fig. 21-3 A	© Reproduced with permission. Photo courtesy of the American Heart Association, 1996. Copyright American Heart Association.
Fig. 21-3 B	© Reproduced with permission. Photo courtesy of the American Heart Association, 1996. Copyright American Heart Association.
Chapter 22	**Proteins**
p. 379	CNRI/Science Photo/Photo Researchers, Inc.
Chapter 23	**Nucleic Acids**
p. 399	Joel Gordon Photography
Fig. 23-1	UPI/Corbis-Bettmann
Fig. 23-9 A	Stuart Kenter
Fig. 23-9 B	Courtesy Richard F. Baker, University of Southern California Medical School, Los Angeles
Chapter 24	**Enzymes**
p. 419	CNRI/Science Photo Library/Photo Researchers, Inc.
Chapter 25	**Digestion**
p. 438	Joe Baraban/The Stock Market
Fig. 25-2	Department of ClinicalRadiology, Salisbury District Hospital/SPL/Photo Researchers, Inc.
Chapter 26	**Metabolism of Carbohydrates**
p. 452	Blair Seitz/Photo Researchers, Inc.
Chapter 27	**Metabolism of Fats**
p. 478	Jean-Marc Barey/Agence Vandystadt/Photo Researchers, Inc.
Fig. 27-7 A	© Reproduced with permission. Photo courtesy of the American Heart Association, 1996. Copyright American Heart Association.
Fig. 27-7 B	© Reproduced with permission. Photo courtesy of the American Heart Association, 1996. Copyright American Heart Association.

Fig./Page	Photographer/Source
Fig. 27-7 C	© Reproduced with permission. Photo courtesy of the American Heart Association, 1996. Copyright American Heart Association.
Chapter 28	**Metabolism of Proteins**
p. 492	Fundamental Photographs
Chapter 29	**Body Fluids: Urine**
p. 509	Peter Cull/Science Photo Library/Photo Researchers, Inc.
p. 514	Reprinted from the Clinical Slide Collection on the Rheumatic Diseases, © 1991, 1995. Used by permission of the American College of Rheumatology.
Fig. 29-2	American College of Rheumatology
Fig. 29-3	Courtesy of Leonard Berlin, MD, Rush North Shore Medical Center Radiology Department, Skokie, Ill.
Chapter 30	**Body Fluids: Blood**
p. 521	Jon Feingersh/The Stock Market
Fig. 30-1	Dr. Tony Brain/Science Photo Library/Photo Researchers, Inc.
Fig. 30-4	Carl Purcell/Photo Researchers, Inc.
Chapter 31	**Vitamins**
p. 552	Blair Seitz/Photo Researchers, Inc.
Fig. 31-3	Biophoto Associates/Photo Researchers, Inc.
Fig. 31-4	Custom Medical Stock Photo
Chapter 32	**Hormones**
p. 575	St. Bartholomew's Hospital/Science Photo Library/Photo Researchers, Inc.
Fig. 32-2 A B	Courtesy Warner-Chilcott Lab, Morris Plains, NJ
Color Insert	
Plate 1	Larry Mulvehill/Photo Researchres, Inc.
Plate 2	Larry Mulvehill/Photo Researchres, Inc.
Plate 3	SKA, Inc.
Plate 4	Geoff Tompkinson/Science Photo Library/Photo Researchrs, Inc.
Plate 5	Chris Jones/The Stock Market
Plate 6	Will McIntyre/Science Source/Photo Researchers, Inc.
Plate 7	Phillipe Plailly/Eurelios/Science Photo Library/Photo Researchers, Inc.
Plate 8	N. Durrell McKenna/Photo Researchers, Inc.
Plate 9	Jim Olive/Peter Arnold, Inc.
Plate 10	Zeve Oelbaum/Peter Arnold, Inc.
Plate 11	Brookhaven National Laboratory
Plate 12	Chris Priest/Science Photo Library/Photo Researchers, Inc.
Plate 13	Larry Mulvehill/Photo Researchers, Inc.
Plate 14	Ulrike Welsch Photography
Plate 15	Chris Priest/Science Photo Library/Photo Researchers, Inc.
Plate 16	Brownie Harris, Photographer
Plate 17	Robert Frerck/Woodfin Camp & Associates
Plate 18	Alexander Tsiaras/Science Source/Photo Researchers, Inc.
Plate 19	Ulrike Welsch Photography
Plate 20	Picker International
Plate 21	Howard Sochurck/Woodfin Camp & Associates
Plate 22	Wellcome Dept. of Cognitive Neurology/Science Photo Library/Photo Researchers, Inc.
Plate 23	Hank Morgan/Rainbow
Plate 24	Custom Medical Stock Photo
Plate 25	Hank Morgan/Science Source/Photo Researchers, Inc.
Plate 26	Custom Medical Stock Photo
Plate 27	Hank Morgan/Photo Researchers, Inc.
Plate 29	Dan McCoy/Rainbow

Functional Groups

Functional Groups

Class of Compound	Functional Group	Example		
		Formula	IUPAC Name	Common Name
Alkane	$-\overset{\mid}{\underset{\mid}{C}}-\overset{\mid}{\underset{\mid}{C}}-$	$CH_3CH_2CH_3$	Propane	Propane
Alkene	$-\overset{\mid}{C}=\overset{\mid}{C}-$	$CH_3CH{=}CH_2$	Propene	Propylene
Alkyne	$-C{\equiv}C-$	$HC{\equiv}CH$	Ethyne	Acetylene
Alcohol	$-OH$	CH_3OH	Methanol	Methyl alcohol
Ether	$-O-$	$CH_3CH_2OCH_2CH_3$	Ethoxyethane	Diethyl ether
Thiol	$-SH$	CH_3SH	Methanethiol	Methyl mercaptan
Aldehyde	$-C{\overset{\displaystyle O}{\underset{\displaystyle H}{{<}}}}$	CH_3CHO	Ethanal	Acetaldehyde
Ketone	${>}C{=}O$	$CH_3\overset{O}{\overset{\|}{C}}CH_3$	Propanone	Acetone
Acid	$-\overset{O}{\overset{\|}{C}}-OH$	$CH_3\overset{O}{\overset{\|}{C}}-OH$	Ethanoic acid	Acetic acid
Amine	$-NH_2$	CH_3NH_2	Methanamine	Methylamine
Amide	$-\overset{O}{\overset{\|}{C}}-NH_2$	$CH_3\overset{O}{\overset{\|}{C}}-NH_2$	Ethanamide	Acetamide
Ester	$-\overset{O}{\overset{\|}{C}}-O-$	$CH_3\overset{O}{\overset{\|}{C}}-OCH_2CH_3$	Ethylethanoate	Ethyl acetate

Normal Laboratory Values

Normal Laboratory Values

Hematology

Hematocrit	Men: 40%-50%
	Women: 37%-47%
Hemoglobin (b)	Men: 14-18 g/dL (2.09-2.79 mmol/L)
	Women: 12-16 g/dL (1.86-2.48 mmol/L)
Platelets	150,000-400,000/L
Prothrombin time	11-14.5 seconds
Red blood cells	Men: 4.5-6.2 million/µL
	Women: 4-5.5 million/µL
Reticulocytes	0.2%-2% of red blood cells
Sedimentation rate	Less than 20 nm/h
White blood cells	5000-10,000/µL

Blood

Usual reference range

Alkaline phosphatase (s)[†]	5-13 U (King-Armstrong)
ALT[‡]	0-45 IU/L
Ammonia (p)	5-50 µmol/L (10-80 µg/dL)
AST[‡]	0-41 IU/L
Bicarbonate (s)	24-28 mmol/L (24-28 mEq/L)
Bilirubin (s)	2-20 µmol/L (0.2-1.2 mg/dL)
BUN[†] (s,p)	2.9-7.1 mmol (8-20 mg/dL)
Calcium (s)	2.1-2.6 mmol/L (8.5-10.3 mg/dL)
Chloride (s,p)	96-106 mmol/L (96-106 mEq/L)
Cholesterol (s,p)	3.9-5.72 mmol/L (150-220 mg/dL)
CO_2 content (s,p)	24-29 mmol/L (24-29 mEq/L)
Copper (s,p)	16-31 µmol/L (100-200 µg/dL)
Creatine (s)	Men: 15-40 µmol/L (0.2-0.6 µg/dL)
	Women: 40-75 µmol/L (0.6-1.0 µg/dL)

[†] b = whole blood; p = blood plasma; s = blood serum.
[‡] *ALT* = alanine aminotransferase; *AST* = aspartate aminotransferase; *BUN* = blood urea nitrogen; *NPN* = nonprotein nitrogen.

Continued.

Normal Laboratory Values—cont'd

Blood	Usual reference range
Creatine kinase isoenzyme: MB	7 IU/L
Creatine kinase isoenzyme: BB	0
Creatine kinase isoenzyme: MM	5-70 IU/L
Creatinine (s,p)	50-100 µmol/L (0.6-1.2 mg/dL)
Fatty acids, free (s)	0.3-0.8 mmol/L (0.3-2.0 mg/dL)
Galactose-1-P-uridyl transferase (in RBC)	22-31 units/g hemoglobin
Glucose (s,p)	3.6-6.1 mmol/L (65-110 mg/dL)
Iron (s)	9-31 µmol/L (50-175 µg/dL)
Ketone bodies, total (s)	0.3-2 mg/dL
Lactate (b)	0.44-1.8 mmol/L (4-16 mg/dL)
NPN (b,s)	15-25 mg/dL
Osmolality	280-296 mosmol/kg water for adults
	260-285 mosmol/kg water for children
pCO_2	35-45 mm Hg
pO_2	80-100 mm Hg
pH	7.35-7.45
Phosphate, inorganic (s)	1-1.5 mmol/L (3-4.5 mg/dL)
Phospholipid (s)	145-200 mg/dL (1.45-2 g/L)
Potassium (s,p)	3.5-5 mmol/L (3.5-5 mEq/L)
Protein, total (s)	6-8 µg/dL
Pyruvate (b)	70-114 µmol/L (0.6-1 mg/dL)
Sodium (s,p)	136-145 mmol/L (136-145 mEq/L)
Specific gravity (b)	1.056
Specific gravity (s)	1.0254-1.0288
Specific gravity (p)	1.059
Triglycerides (s)	1.9 mmol/L (<165 mg/dL)
Uric acid (s,p)	Men: 0.18-0.54 mmol/L (3-9 mg/dL)
	Women: 0.15-0.46 mmol/L (2.5-7.5 mg/dL)
	Children: 1.5 g/L
Vitamin A (s)	0.53-2.1 µmol/L (15-60 µg/dL)
Vitamin B_{12} (s)	>148 pmol/L (>200 pg/mL)
Vitamin C (p)	23-85 µmol/L (0.4-1.5 mg/dL)
Vitamin D (s) as:	
25-Hydroxycholecalciferol	19.4-137 nmol/L (8-55 ng/mL)
1,25-Dihydroxycholecalciferol	62-166 pmol/L (26-65 pg/mL)
24,25-Dihydroxycholecalciferol	2.4-12 nmol/L (1-5 ng/mL)
Zinc (s)	7.65-22.95 µmol/L (50-150 µg/L)

Index